H51 MOD

THE CAMBRIDGE HISTORY OF SCIENCE

VOLUME 7

The Modern Social Sciences

Volume 7 of *The Cambridge History of Science* provides a history of the concepts, practices, institutions, and ideologies of the social sciences (including behavioral and economic sciences) since the eighteenth century. The authors offer original, synthetic accounts of the historical development of social knowledge, including its philosophical assumptions, its social and intellectual organization, and its relations to science, medicine, politics, bureaucracy, religion, and the professions. The 43 chapters include inquiries into the genres and traditions that formed social science, the careers of the main social disciplines (psychology, economics, sociology, anthropology, political science, geography, history, and statistics), and international essays on social science in Eastern Europe, Asia, Africa, and Latin America. The volume also features essays examining the involvement of the social sciences in government, business, education, culture, and social policy. This is a broad cultural history of social science that analyzes the participation of the social disciplines in the making of the modern world. The contributors, world leaders in their respective specialities, engage with current historiographical and methodological controversies and stake out positions of their own.

Theodore M. Porter is Professor of the History of Science in the Department of History at the University of California, Los Angeles. He is the author of *The Rise of Statistical Thinking, 1820–1900* (1986) and *Trust in Numbers: The Pursuit of Objectivity in Science and Public Life* (1995) and coauthor of *The Empire of Chance: How Probability Changed Science and Everyday Life* (1989).

Dorothy Ross is the Arthur O. Lovejoy Professor of History at Johns Hopkins University. She is the author of *G. Stanley Hall: The Psychologist as Prophet* (1972) and *The Origins of American Social Science* (1991) and editor of *Modernist Impulses in the Human Sciences, 1870–1930* (1994).

THE CAMBRIDGE HISTORY OF SCIENCE

General editors
David C. Lindberg and Ronald L. Numbers

VOLUME 1: *Ancient Science*
Edited by Alexander Jones

VOLUME 2: *Medieval Science*
Edited by David C. Lindberg and Michael H. Shank

VOLUME 3: *Early Modern Science*
Edited by Lorraine J. Daston and Katharine Park

VOLUME 4: *Eighteenth-Century Science*
Edited by Roy Porter

VOLUME 5: *The Modern Physical and Mathematical Sciences*
Edited by Mary Jo Nye

VOLUME 6: *The Modern Biological and Earth Sciences*
Edited by Peter Bowler and John Pickstone

VOLUME 7: *The Modern Social Sciences*
Edited by Theodore M. Porter and Dorothy Ross

VOLUME 8: *Modern Science in National and International Context*
Edited by David N. Livingstone and Ronald L. Numbers

David C. Lindberg is Hilldale Professor Emeritus of the History of Science at the University of Wisconsin–Madison. He has written or edited a dozen books on topics in the history of medieval and early modern science, including *The Beginnings of Western Science* (1992). He and Ronald L. Numbers have previously coedited *God and Nature: Historical Essays on the Encounter between Christianity and Science* (1986) and *When Science and Christianity Meet* (2003). A Fellow of the American Academy of Arts and Sciences, he has been a recipient of the Sarton Medal of the History of Science Society, of which he is also past-president (1994–5).

Ronald L. Numbers is Hilldale and William Coleman Professor of the History of Science and Medicine at the University of Wisconsin–Madison, where he has taught since 1974. A specialist in the history of science and medicine in America, he has written or edited more than two dozen books, including *The Creationists* (1992) and *Darwinism Comes to America* (1998). A Fellow of the American Academy of Arts and Sciences and a former editor of *Isis*, the flagship journal of the history of science, he has served as the president of both the American Society of Church History (1999–2000) and the History of Science Society (2000–1).

THE CAMBRIDGE HISTORY OF SCIENCE

VOLUME 7

The Modern Social Sciences

Edited by

THEODORE M. PORTER

DOROTHY ROSS

CAMBRIDGE
UNIVERSITY PRESS

PUBLISHED BY THE PRESS SYNDICATE OF THE UNIVERSITY OF CAMBRIDGE
The Pitt Building, Trumpington Street, Cambridge, United Kingdom

CAMBRIDGE UNIVERSITY PRESS
The Edinburgh Building, Cambridge CB2 2RU, UK
40 West 20th Street, New York, NY 10011-4211, USA
477 Williamstown Road, Port Melbourne, VIC 3207, Australia
Ruiz de Alarcón 13, 28014 Madrid, Spain
Dock House, The Waterfront, Cape Town 8001, South Africa

http://www.cambridge.org

First published 2003

Printed in the United States of America

Typeface Adobe Garamond 10.75/12.5 pt.   *System* LATEX 2_ε [TB]

A catalog record for this book is available from the British Library.

Library of Congress Cataloging in Publication Data
(Revised for volume 7)
The Cambridge history of science
p.   cm.
Includes bibliographical references and indexes.
Contents: – v. 4. Eighteenth-century science / edited by Roy Porter
v. 5. The modern physical and mathematical sciences / edited by Mary Jo Nye
v. 7. The modern social sciences / edited by Theodore M. Porter and Dorothy Ross
ISBN 0-521-57243-6 (v. 4)
ISBN 0-521-57199-5 (v. 5)
ISBN 0-521-59442-1 (v. 7)
1. Science – History.   I. Lindberg, David C.   II. Numbers, Ronald L.
Q125 C32   2001
509 – dc21
2001025311

ISBN 0 521 59442 1 hardback

CONTENTS

Contents
xiii

NOTES ON CONTRIBUTORS

MITCHELL G. ASH is Professor of Modern History at the University of Vienna, Austria. He was a Fellow at the Institute for Advanced Study, Berlin, and is a Full Member of the Berlin-Brandenburg Academy of Sciences and Humanities. His publications on the history of modern psychology and modern science in Germany and the United States include *Gestalt Psychology in German Culture, 1890–1967: Holism and the Quest for Objectivity* (1995).

JORGE BALAN is currently Program Officer with The Ford Foundation in New York. His contribution to this volume was written when he was a Senior Researcher at Centro de Estudios de Estado y Sociedad (CEDES) and Professor at the University of Buenos Aires, both in Argentina. His most recent book is *Politicas de reforma de la education superior y la universidad latinoamericana* (2000).

ROBERT C. BANNISTER is Scheuer Professor of History (emeritus) at Swarthmore College, Pennsylvania. His publications include *Ray Stannard Baker: The Mind and Thought of a Progressive* (1965), *Social Darwinism: Science and Myth* (1979), *Sociology and Scientism: The American Search for Objectivity, 1880–1940* (1987), and *Jessie Bernard: The Making of a Feminist* (1991).

ELAZAR BARKAN is Chair of the Cultural Studies Department and Professor of History and Cultural Studies at Claremont Graduate University. He is the author of six books, including *The Guilt of Nations: Restitution and Negotiating Historical Injustices* (2000), *Modernism and Primitivism* (in Hebrew, 2001), and *The Retreat of Scientific Racism* (1993).

ANDREW E. BARSHAY is Professor of History and Chair of the Center for Japanese Studies at the University of California, Berkeley. His publications include *State and Intellectual in Imperial Japan* (1988, Japanese trans. 1996) and "Postwar Social and Political Thought, 1945–1990," in *Modern Japanese Thought* (ed. Bob Tadashi Wakabayashi, 1998).

JOHN CARSON is Assistant Professor of History at the University of Michigan. His publications include "Minding Matter/Mattering Mind: Knowledge and the Subject in Nineteenth-Century Psychology," *Studies in the History and Philosophy of the Biological and Biomedical Sciences*, 30 (1999), 345–76, and "Army Alpha, Army Brass, and the Search for Army Intelligence," *Isis*, 84 (1993), 278–309. He is currently working on a book entitled *Making Intelligence Matter: Cultural Constructions of Human Difference, 1750– 1940*.

TERRELL CARVER is Professor of Political Economy at the University of Bristol, England. His recent publications include *Engels after Marx* (with Manfred Steger, 1999) and *The Postmoderm Marx* (1998). He is currently working on a book on men in political theory.

PARTHA CHATTERJEE is Director of the Centre for Studies in Social Sciences, Calcutta, and Visiting Professor of Anthropology at Columbia University, New York. His books include *Nationalist Thought and the Colonial World* (1986) and *The Nation and Its Fragments* (1993). He is a founding member of the journal *Subaltern Studies*.

ALAIN DESROSIÈRES is a statistician in the Institut National de la Statistique et des Études Économiques (INSEE), the French statistical office. His research is about the history and sociology of the production and the uses of statistics, both official and scientific. His book, *The Politics of Large Numbers: A History of Statistical Reasoning*, appeared in English translation in 1998.

JAMES FARR is Professor of Political Science at the University of Minnesota. He is editor of *Political Science in History* (1995) and *Discipline and History* (1993), as well as author of several studies in the philosophy of social science and the history of political thought.

ELLEN FITZPATRICK is Professor of History at the University of New Hampshire. She is the author of *History's Memory: Writing America's Past, 1880–1980* (2002), *Endless Crusade: Women Social Scientists and Progressive Reform* (1990), and *America in Modern Times* (with Alan Brinkley, 1997) and has edited several volumes and essays.

JAN GOLDSTEIN is Professor of Modern European History at the University of Chicago, where she is also a member of the Committee on Conceptual and Historical Studies of Science. Her books include *Console and Classify: The French Psychiatric Profession in the Nineteenth Century* (1987), *Foucault and the Writing of History* (1994), and *The Post-Revolutionary Self: Competing Psychologies in France, 1750–1850* (forthcoming).

BETTINA GRANSOW is Assistant Professor of Chinese Studies at the Institute for East Asian Studies, Free University of Berlin. Her book, *Geschichte der chinesischen Soziologie*, appeared in 1992. Her current research concerns internal migration in China, and methodologies of social assessment of Chinese development projects.

JOHAN HEILBRON is a sociologist at the Centre de Sociologie Européenne in Paris and an associate professor at Utrecht University in the Netherlands. His publications include *The Rise of Social Theory* (1995) and *The Rise of the Social Sciences and the Formation of Modernity* (coedited with Lars Magnusson and Björn Wittrock, 1990).

SUSAN HERBST is Professor and Chair of the Department of Political Science at Northwestern University. She is author, most recently, of *Reading Public Opinion: How Political Actors View the Democratic Process* (1998) and coauthor of *Public Opinion* (1999), an interdisciplinary textbook. She is writing a book on representations of American public opinion in popular culture from 1920 to 1960.

ELLEN HERMAN is Associate Professor of History at the University of Oregon and the author of *The Romance of American Psychology: Political Culture in the Age of Experts* (1995). She is currently working on a book about child adoption and the modern human sciences.

DAVID A. HOLLINGER is Preston Hotchkis Professor of History at the University of California at Berkeley. His books include *Science, Jews and Secular Culture: Studies in Twentieth-Century American Intellectual History* (1996).

JAROMÍR JANOUŠEK is Professor of Psychology at Charles University in Prague. He was a Fellow at the Center for Advanced Study in Behavioral Sciences from 1990 to 1991 and is the author of *Practice and Knowledge* (1963), *Social Communication* (1968), and *Joint Activity and Communication* (1984) and coauthor of *Methods of Social Psychology* (1986) and *Psychological Atlas* (1993).

ADAM KUPER is Professor of Social Anthropology at Brunel University in London. His books include *The Invention of Primitive Society* (1988), *Anthropologists and Anthropology: The Modern British School* (3rd ed., 1996), *Among the Anthropologists* (1999), and *Culture: The Anthropologists' Account* (1999).

MICHAEL E. LATHAM teaches history at Fordham University. He is the author of *Modernization as Ideology: American Social Science and "Nation Building" in the Kennedy Era* (2000). His research explores the relationship between American intellectual and cultural history and American foreign relations.

HARRY LIEBERSOHN is the author of *Fate and Utopia in German Sociology, 1870–1923* (1988) and *Aristocratic Encounters: European Travelers and North American Indians* (1998). He is Professor of History at the University of Illinois, Urbana-Champaign.

ELIZABETH LUNBECK is Professor of History at Princeton University. She is the author of *The Psychiatric Persuasion: Knowledge, Gender, and Power in Modern America* (1994) and the editor of several collections of essays and an edition of an early psychoanalytic case history. She is currently working on a history of psychoanalytic practice in the United States before 1920.

PETER MILLER is Professor of Management Accounting at the London School of Economics and Political Science. His publications in the fields of accounting, management, and sociology include *Domination and Power* (1987) and several coedited volumes: *The Power of Psychiatry* (1986), *The Foucault Effect: Studies in Governmentality* (1991), and *Accounting as Social and Institutional Practice* (1994).

MARY S. MORGAN is Professor of History of Economics at the London School of Economics and also holds a chair in the History and Philosophy of Economics at the University of Amsterdam. She is the author of, among other works, *The History of Econometric Ideas* (1990) and is currently writing a book on the twentieth-century development of economics as a modeling science.

ANTOINE PICON is Professor of the History of Architecture and Technology at the Harvard University Graduate School of Design. He is the author of *French Architects and Engineers in the Age of Enlightenment* (1988, English trans. 1992) and *L'Invention de l'Ingenieur Moderne: L'Ecole des Ponts et Chaussees, 1747–1851* (1992). He is currently writing on the history of the Saint-Simonian movement and on the relations between technology and utopia.

THEODORE M. PORTER is Professor of the History of Science in the Department of History at the University of California, Los Angeles. His books include *The Rise of Statistical Thinking, 1820–1900* (1986) and *Trust in Numbers: The Pursuit of Objectivity in Science and Public Life* (1995). He is currently writing a book on the early career of Karl Pearson.

JULIE A. REUBEN is a professor at the Harvard Graduate School of Education. She is the author of *The Making of the Modern University: Intellectual Transformation and the Marginalization of Morality* (1996) and is currently working on a book entitled *Campus Revolts: Politics and the American University in the 1960s*.

JACQUES REVEL is Professor of History at the Ecole des Hautes Etudes en Sciences Sociales, Paris. His fields are European cultural history from the sixteenth to the nineteenth century and historiography. Among his books are *The Vanishing Children in Paris* (with A. Farge, 1988), *Histoire de la France* (with A. Burguière, 4 vols., 1989–93), *Jeux d'échelles* (1996), and *Histoires: French Construction of the Past* (with Lynn Hunt, 1996).

MARIE-CLAIRE ROBIC, a geographer, is Research Director of the National Center of Scientific Research (CNRS) and is attached to the laboratory

Géographie-cités (Paris). She is coeditor of *Géographes face au monde: L'Union géographique internationale et les congrès internationaux de géographie* (1996), *Le Tableau de la géographie de la France de Paul Vidal de la Blache: Dans le labyrinthe des formes* (2000), and *Géographes en pratiques (1870–1945)*: *Le terrain, le livre, la Cité* (2001).

ROSALIND ROSENBERG is the Ann Whitney Olin Professor of History at Barnard College. She is the author of *Beyond Separate Spheres: Intellectual Roots of Modern Feminism* (1982) and *Divided Lives: American Women in the Twentieth Century* (1992), as well as articles on gender, law, and comparative feminism. She is currently at work on a book entitled *Changing the Subject: Women at Columbia and the Invention of Gender*.

DOROTHY ROSS is Arthur O. Lovejoy Professor of History at Johns Hopkins University. She is the author of *G. Stanley Hall: The Psychologist as Prophet* (1972) and *The Origins of American Social Science* (1991) and editor of *Modernist Impulses in the Human Sciences, 1870–1930* (1994).

ALAIN ROUSSILLON is a researcher in politics at the Centre National de la Recherche Scientifique in Paris. He has spent several years in Egypt, as Vice-Director of the Centre d'Etudes et de Documentation Economiques, Juridiques et Sociales, and in Morroco as Director of the Centre Jacques Berque. His main focus is on social reform and related issues and on Arabic travel writings.

MARGARET SCHABAS is Professor of Philosophy at the University of British Columbia. She is the author of *A World Ruled by Number: Jevons and the Rise of Mathematical Economics* (1990) and articles in *Isis, History of Political Economy, Dialogue, Studies in History and Philosophy of Science*, and *Public Affairs Quarterly*. Her forthcoming books are *Nature in Classical Economics: Oeconomies in the Age of Newton* (coedited with Neil De Marchi) and *Hume's Political Economy*.

OWEN SICHONE is Senior Lecturer in Social Anthropology at the University of Cape Town in South Africa. His work on southern African political culture includes two edited books, *Democracy in Zambia: Challenges for the Third Republic* (1996) and *State and Constitutionalism in Southern Africa* (1998). His current research interests are migration, globalization, and xenophobia in South Africa.

IRINA SIROTKINA is Senior Researcher at the Institute for the History of Science and Technology, Russian Academy of Sciences. She is the author of *Diagnosing Literary Genius: A Cultural History of Psychiatry in Russia, 1880–1930* (2002).

KEITH TRIBE taught sociology and economics at Keele University from 1976 to 2000 and was Alexander von Humboldt Fellow at the University of

Heidelberg and at the Max Planck Institut für Geschichte, Göttingen, from 1979 to 1985. He is the author of *Land, Labor and Economic Discourse* (1978); two books on German economic discourse, *Governing Economy* (1988), and *Strategies of Economic Order* (1955); the editor of *Economic Careers: Economics and Economists in Britain, 1930–1970* (1998), and translator of Wilhelm Hennis, *Max Weber's Science of Man* (2000).

STEPHEN TURNER is Graduate Research Professor and Chair of the Department of Philosophy at the University of South Florida, Tampa. He is the author of *The Search for a Methodology of Social Science: Durkheim, Weber, and the Nineteenth-Century Problem of Cause, Probability, and Action* (1986) and coauthor of *Max Weber: The Lawyer as Social Thinker* (1994); *Max Weber and the Dispute Over Reason and Value: A Study in Philosophy, Ethics, and Politics* (1984); and *The Impossible Science: An Institutional Analysis of American Sociology* (1990). He recently edited the *Cambridge Companion to Weber* (2000).

PETER WAGNER is Professor of Social and Political Theory at the European University Institute in Florence, Italy, and Professor of Sociology at the University of Warwick. His recent books include *A History and Theory of the Social Sciences: Not All That Is Solid Melts into Air* (2001), *Theorizing Modernity: Inescapability and Attainability in Social Theory* (2001), and *Le travail et la nation: Histoire croisée de la France et de l'Allemagne* (coeditor, 1999).

JOHNSON KENT WRIGHT is Associate Professor and Director of Graduate Studies in the Interdisciplinary Humanities Program at Arizona State University. He is the author of *A Classical Republican in Eighteenth-Century France: The Political Thought of Mably* (1997), as well as essays on early modern and modern historiography.

EILEEN JANES YEO is Professor of Social and Cultural History at the University of Strathclyde in Glasgow. Her publications include "Henry Mayhew as a Social Investigator," in *The Unknown Mayhew* (coedited with E. P. Thompson, 1971), and *The Contest for Social Science: Relations and Representations of Gender and Class* (1996).

GENERAL EDITORS' PREFACE

In 1993, Alex Holzman, former editor for the history of science at Cambridge University Press, invited us to submit a proposal for a history of science that would join the distinguished series of Cambridge histories launched nearly a century ago with the publication of Lord Acton's fourteen-volume *Cambridge Modern History* (1902–12). Convinced of the need for a comprehensive history of science and believing that the time was auspicious, we accepted the invitation.

Although reflections on the development of what we call "science" date back to antiquity, the history of science did not emerge as a distinctive field of scholarship until well into the twentieth century. In 1912 the Belgian scientist-historian George Sarton (1884–1956), who contributed more than any other single person to the institutionalization of the history of science, began publishing *Isis*, an international review devoted to the history of science and its cultural influences. Twelve years later he helped to create the History of Science Society, which by the end of the century had attracted some 4,000 individual and institutional members. In 1941 the University of Wisconsin established a department of the history of science, the first of dozens of such programs to appear worldwide.

Since the days of Sarton historians of science have produced a small library of monographs and essays, but they have generally shied away from writing and editing broad surveys. Sarton himself, inspired in part by the Cambridge histories, planned to produce an eight-volume *History of Science*, but he completed only the first two installments (1952, 1959), which ended with the birth of Christianity. His mammoth three-volume *Introduction to the History of Science* (1927–48), a reference work more than a narrative history, never got beyond the Middle Ages. The closest predecessor to *The Cambridge History of Science* is the three-volume (four-book) *Histoire Générale des Sciences* (1957–64), edited by René Taton, which appeared in an English translation under the title *General History of the Sciences* (1963–4). Edited just before the late-twentieth-century boom in the history of science, the Taton set quickly became dated. During the 1990s Roy Porter began editing the very useful Fontana History of Science (published in the United States as the Norton History of Science), with volumes devoted to a single discipline and written by a single author.

The Cambridge History of Science comprises eight volumes, the first four arranged chronologically from antiquity through the eighteenth century, the latter four organized thematically and covering the nineteenth and twentieth centuries. Eminent scholars from Europe and North America, who together form the editorial board for the series, edit the respective volumes:

Volume 1: *Ancient Science*, edited by Alexander Jones, University of Toronto

Volume 2: *Medieval Science*, edited by David C. Lindberg and Michael H. Shank, University of Wisconsin–Madison

Volume 3: *Early Modern Science*, edited by Lorraine J. Daston, Max Planck Institute for the History of Science, Berlin, and Katherine Park, Harvard University

Volume 4: *Eighteenth-Century Science*, edited by Roy Porter, late of Wellcome Trust Centre for the History of Medicine at University College London

Volume 5: *The Modern Physical and Mathematical Sciences*, edited by Mary Jo Nye, Oregon State University

Volume 6: *The Modern Biological and Earth Sciences*, edited by Peter Bowler, Queen's University of Belfast, and John Pickstone, University of Manchester

Volume 7: *The Modern Social Sciences*, edited by Theodore M. Porter, University of California, Los Angeles, and Dorothy Ross, Johns Hopkins University

Volume 8: *Modern Science in National and International Context*, edited by David N. Livingstone, Queen's University of Belfast, and Ronald L. Numbers, University of Wisconsin–Madison

Our collective goal is to provide an authoritative, up-to-date account of science – from the earliest literate societies in Mesopotamia and Egypt to the beginning of the twenty-first century – that even nonspecialist readers will find engaging. Written by leading experts from every inhabited conti-nent, the essays in *The Cambridge History of Science* explore the systematic investigation of nature, whatever it was called. (The term "science" did not acquire its present meaning until early in the nineteenth century.) Reflecting the ever-expanding range of approaches and topics in the history of science, the contributing authors explore non-Western as well as Western science, applied as well as pure science, popular as well as elite science, scientific practice as well as scientific theory, cultural context as well as intellectual content, and the dissemination and reception as well as the production of scientific knowledge. George Sarton would scarcely recognize this collaborative effort as the history of science, but we hope we have realized his vision.

David C. Lindberg
Ronald L. Numbers

ACKNOWLEDGMENTS

The chapters in this volume were first presented in two working conferences organized by the editors, one at the Clark Library of UCLA and one at the Woodrow Wilson International Center for Scholars in Washington, D.C., which aided immensely in the coordination of the project and the development of common themes. We thank the Clark Library and the Wilson Center for their logistical assistance. For the financial support that made these events possible we thank the Spencer Foundation, the UCLA Center for Seventeenth- and Eighteenth-Century Studies, and the National Science Foundation (grant SBR-9703894). Our editorial advisors for the volume were Charles Camic, Mary Furner, Henrika Kuklick, Richard Olson, Peter Reill, Roger Smith, and Donald Winch. We thank them for their help with refereeing the papers and their valuable advice on revisions.

Theodore M. Porter
Dorothy Ross

I

INTRODUCTION
Writing the History of Social Science

Theodore M. Porter and Dorothy Ross

How do we write the history of social science? There are problems even with
the name. In English alone, "sciences of man," "moral sciences," "moral and
political sciences," "behavioral sciences," and "human sciences" have been
among its many predecessors and competitors. Their proliferation reflects
the unsettled nature of this broad subject matter. All are capable of giving offense, both by exclusion and by inclusion. Many have long and contradictory
histories.

Consider the career of the "moral sciences." The phrase "*sciences morales
et politiques*" was introduced in France about 1770. In 1795 it was enshrined
as the official label for the "second class" of the Institut de France (the former
Académie des Sciences was the first class), until this nest of critics was reorganized out of existence by Napoleon in 1803. Restored in 1832, the official
institution of the moral and political sciences was now suitably conservative, emphasizing philosophy and individual morality. John Stuart Mill, an
admirer of Auguste Comte's "sociology," included in his enduringly influential 1843 treatise on logic a section aiming to "remedy" the "backward state
of the moral sciences" by "applying to them the methods of physical science, duly extended and generalized." A German translation of Mill's work
rendered "moral sciences" as *Geisteswissenschaften* – not the first use of that
German term, but an influential one. It referred to the sciences of *Geist*, which
could be translated back into English as "spirit" or "mind." In German, this
remained a standard label until well into the twentieth century. It was understood to indicate that such studies had a moral and spiritual character, quite
unlike the sciences of nature.

In French and English, there has been more emphasis on the continuity
of scientific knowledge. David Hume, among others, argued in the eighteenth century that politics could be a science. "Political economy," especially in Enlightenment Scotland, was part of a broad effort to comprehend the moral and historical dimensions of human society. It had gained
wide acceptance by the early nineteenth century and was appreciated for

its contribution to the art of governing. The usual German term, "national economy," evoked this political dimension still more clearly, while the French campaign to replace it with "social economy" implied a certain discontent with mere politics. Such also was the tendency of "social science," a term that first gained currency in French, having been introduced just prior to the French Revolution. It expressed an increasingly widespread view that politics was conditioned by something deeper. Social science aimed to comprehend the forces of progress and their instabilities in a way that reduced neither to an individualistic, psychological dimension nor to the domain of state and government. In this respect, it provided an enduring model for "scientific" investigation of the human domain.

In English, the "social sciences," now plural, emerged in the late nineteenth century, above all in the United States, and that umbrella term remains in common use. But any word or phrase presuming to name so disparate an endeavor was bound to create controversy. For a time, it seemed possible that social knowledge would not require such synthetic labels, because it would be united in a single field. This was Comte's vision for "sociology," and in the later nineteenth century some envisioned "anthropology" in the same way. More recently, the challenge to "social sciences" has come overwhelmingly from those who would secede from them. Psychologists have been the least happy with that phrase, pressing often to be grouped with the biologists, or, if they had to keep the company of sociologists and anthropologists, insisting at least on a rival adjective. The term "behavioral sciences" gained wide currency in the mid twentieth century in North America, but not in Europe. Indeed, the object of behaviorism can scarcely be called social, and its late-twentieth-century decline in favor of "cognitive" and physiological orientations only accentuated the differences. Neither can economics be described straightforwardly as a social science, and economists often claim a higher standing for their field. "Social, behavioral, and economic sciences" has begun to emerge as a bureaucratic designation. We have only to add "political," "cultural," "demographic," and "historical" to embrace all of those university disciplines lying outside the professional schools that are neither humanities nor sciences of nature nor mathematics. But this is taxonomic splitting run amok.

The French language offers an appealing alternative, the *sciences humaines*, or human sciences. The term dates back at least to the seventeenth century. During the Enlightenment it was more or less synonymous with *sciences de l'homme* (sciences of man), then a very common designation and one that remains acceptable in French, though it has become officially sexist in English. *Sciences humaines* regained its currency in the 1950s, and was particularly favored by Georges Canguilhem and Georges Gusdorf. They used it to refer to a broadly philosophical tradition of inquiry, embodying a humanistic vision that provided an alternative to the work of technocratic specialists who divide up the human domain – indeed, who carve up *l'homme* himself, the better to manage him.[1] Michel Foucault adopted the name, but associated it more darkly with professional and administrative forms of knowledge. The phrase "human science" has spread to English mainly because of Foucault's extraordinary impact on the academic humanities. Roger Smith used it as the title of a synthetic historical work emphasizing the history of psychology in relation to a wide domain of social thought and investigation.[2] In English, at least, "human science" remains a category of the scholarly observer, mostly unknown to "human scientists," if such there be. Its provenance is ill defined. Psychology and psychiatry are central to it, along with ethnography. Studies of language, literature, art, and music are often included, and the vast domain of medicine occupies the borderlands. The more mathematical fields, notably economics, are sometimes excluded, ostensibly as inhuman sciences.

Although the term "human science" has its attractions, we have not chosen it for this volume. We have also resisted the temptation to multiply terms. While we recognize, and indeed emphasize, the diversity of the social sciences, we are impressed also by their family resemblances, at least from a cultural and intellectual standpoint. One of the crucial ambitions of this volume is to show what is gained by bringing their histories together, if not in a single narrative, then at least in a group of intersecting essays. So it is not just in order to save ink that our title names its topic with only one adjective. We have chosen "social."

There is also some question about "science," which has long been understood to imply a certain standard of experimental or conceptual rigor and of methodological clarity. In English, especially in the twentieth century, the claim to scientific status has meant the assertion of some fundamental resemblance to natural science, usually regarded even by social scientists as the core of "real" science – as temporally prior and logically exemplary. Historically, however, this appears to be something of a misapprehension. Although science has long referred to natural or human knowledge as opposed to revelation, theology had a better claim to the status of science during the Middle Ages than did the study of living things, or even the study of matter in motion. During the seventeenth and eighteenth centuries, an assortment of names was used for various branches or aspects of natural knowledge, including "natural philosophy," "natural history," "experimental physics," and "mixed mathematics." "Science" was too nebulous to be useful, especially in English, until about 1800, when it emerged as the standard name for the organized

---

[1] Claude Blanckaert, "L'Histoire des sciences de l'homme. Principes et périodisation," and Fernando Vidal, "La 'science de l'homme': Désirs d'unité et juxtapositions encyclopédiques," in *L'Histoire des sciences de l'homme: Trajectoire, enjeux et questions vives*, ed. Claude Blanckaert, Loïc Blondiaux, Laurent Loty, Marc Renneville, and Nathalie Richard (Paris: L'Harmattan, 1999), pp. 23–60, 61–78.

[2] Roger Smith, *The Fontana History of the Human Sciences* (London: Fontana Press, 1997). (In the United States, *The Norton History of the Human Sciences*.)

pursuit of knowledge. Early-nineteenth-century social science was bound up with this same endeavor. Few in 1830 doubted that political economy was a science; even its critics attacked it on other grounds. Politics had reasonable claims to be a science, as did theology; so it was not immoderate for inchoate fields like sociology, anthropology, or statistics to march under the same banner. In German, *Wissenschaft* imposed more strenuous requirements, but somewhat different ones. There, the model science was philology, a linguistic and literary study, whose dignity derived from its relation to an important subject area and its use of rigorous, scholarly methods. The modern practice of attacking fields of inquiry by denying their scientific credentials was un-common until late in the nineteenth century, and it remains more plausible in English than in most other languages.

The possibility of a more restricted meaning of "science" emerged in the same period, and debates about the status of social knowledge were centrally involved in defining it. Consider the role of social science in the origins of modern philosophy of science. In the 1820s, Comte initiated a massive effort to define the methods and historical progression of the sciences. His main purpose was to announce the discovery, and define the standing, of sociology. He rejected decisively the idea that social science should adopt the same methods as astronomy, physics, or physiology. Yet at the same time he defined a hierarchy of knowledge, with social science dependent for its formulation on all the sciences that had gone before. And despite his claims for the inclusion of social knowledge, he made of "science" something special and exclusive. There had been, he argued, no science of physics before the seventeenth century, no true chemistry before Lavoisier. The origins of physiology were still more recent, and the founder of scientific sociology was, to cast aside false modesty, himself. Theology and metaphysics were not part of positive science, but its predecessors and its antithesis. Law, literature, and rhetoric could never occupy this hallowed ground. Thus, while Comte formulated his philosophy in order to vindicate sociology and to define its place within science, he insisted also on a highly restrictive sense of "science," a standard the social sciences could not easily meet.

In practice, the natural sciences don't conform well to philosophical pre-scriptions either. But Comte's language, echoed and elaborated by Mill, en-couraged the idea that science stands for a methodological ideal, which social science has but imperfectly realized. In scholarly and popular discussions of science, including discussions of the history of science, social science has often been regarded as an ambiguous case, and partly for that reason as a marginal one. We might put this differently. Social science is, in a way, a doppelgänger of science. The "doubles" of science – among them engineering and medicine as well as social science – represent the practicality of science, and so have embodied much of its significance for the larger culture. They have often been less abstract and more engaged, thereby testing the boundaries of sci-ence. These applications and extensions have sometimes been embraced and sometimes shunned by those who speak for science. In part because of its very marginality, social science has taken the scientific ideal very seriously, and if that ideal fails as description, it retains a certain normative potency. The "scientific method," for example, has been of particular interest to social scientists questing for the mastery or certainty of "true" science. Talk of method in natural science has been shaped in part by these social discussions, though scientists often invoke method to explain why social disciplines are not scientific. Historians and philosophers of science often argue, and rightly, that nothing like a rigorous or unitary method is to be found in the actual practice of science, but that does not make such talk inconsequential. It supports the prestige of science, helps to shape its identity, and sometimes forms its conscience.

In historical writing, the disposition to exclude has traditionally been a powerful one. Histories of science written by natural scientists often omit the social disciplines entirely. Philosophical histories of science have often undertaken first to study the most successful fields, which could then serve as models for the rest. The new professional historians of science had begun by the 1960s to reshape the field in ways that would seem to favor a greater inclusiveness. They refused to take for granted the narrative of ceaseless progress that had guided most of their predecessors. They wanted to treat their topic naturalistically, to avoid enshrining it as a privileged category. This has come to mean viewing science through the lens of historicism, as a social formation, to be studied as one would study other social formations. Especially since the 1970s, historians have often taken a more critical view of science than is customary among scientists themselves. Many have wanted to understand the validity of science in relation to the shared assumptions and material and social practices of particular communities, not as timeless and transcendental truth. They have been especially critical of what George W. Stocking, Jr., the historian of anthropology, first referred to as the Whig interpretation of science.[3] The name derives, by analogy, from a complacent view of British political history, characterized in a well-known study by Herbert Butterfield. The Whig view of science regards discoveries that comport with our current knowledge as natural and laudable, and condemns the prejudices and misconceptions that could have led scientists to believe what we now take to be false. Since the 1960s, the conventional practice has been to avoid this teleological view of scientific progress, insisting instead on what is called "symmetry" of explanation.

Historical writing on science has, nevertheless, continued to recognize in practice, if not always in theory, a conventional hierarchy of the sciences. Before 1960, historians of science worked mainly on medieval or early modern astronomy, mechanics, and optics, generally understood as the points of

[3] George W. Stocking, Jr., "On the Limits of 'Presentism' and 'Historicism' in the Historiography of the Behavioral Sciences" (1965), in his *Race, Culture, and Evolution: Essays in the History of Anthropology* (New York: Free Press, 1968), pp. 1–12.

origin for modern science. Modern physical science rose to prominence in
the historical writing of the 1960s and 1970s, and the history of biology has
flourished since 1970. The social sciences, like the applied and engineering
sciences, have been accepted into the history of science more slowly, and
have participated only partially in its dynamic. The subordinate status of
social science is replicated in its historiography, which is often regarded as
less advanced than that of science proper.

Thomas S. Kuhn's *Structure of Scientific Revolutions* (1962) helped to support and yet also to erase that separation. Kuhn noted that it was in part
the absence of agreement in the social sciences that had alerted him to the
existence of paradigms in the natural sciences – agreed-upon frameworks
of theory and practice that enabled and constrained the normal practice of
science. Yet he later blurred the bright line he had previously drawn, and his
signal demonstration of the historical construction of science has stimulated
inquiry into the social sciences, as well.[4]

The debate between "internalist" and "externalist" analysis in sociology
and the history of science has had important implications for the standing of
social science. During the 1970s, "externalism" generally meant an emphasis
on the development of scientific institutions, as an alternative to a focus on scientific ideas. Paradoxically, the institutions in question in these "externalist"
accounts were scientific ones, and were often treated as autonomous. In a
way, this implied a narrower understanding of science than that reflected
in some of the older intellectual histories that linked scientific conceptions
to broadly philosophical ideas – and also one that tended to exclude social science. Kuhn's name – increasingly against his own inclinations – was
usually invoked by the externalists in this notoriously slippery debate, and
their narrow focus drew some support from his work, which concentrated
on the character of scientific communities and left unspecified their relationship to wider intellectual and political currents. By 1980, "externalism" was
more likely to refer to attempts to use social factors to explain the acceptance
of new scientific truth claims. But most advocates of this "new" sociology of
science sought something more impressive than the "social construction" of
social science, which was often criticized in related terms. And their program
has tended increasingly to a micro-view of laboratories as sites of a distinctive set of discourses and of their own special material cultures. It may be
questioned who is really the "internalist."

The history of the social sciences, now formalized by a Forum on the
History of the Human Sciences within the History of Science Society,
is distinguished by its close attention to methods and ideas, its careful

[4] Thomas Kuhn, *The Structure of Scientific Revolutions*, 2nd ed. enlarged (Chicago: University of
Chicago Press, 1970), p. viii, Postscript; Gary Gutting, ed., *Paradigms and Revolutions: Appraisals and
Applications of Thomas Kuhn's Philosophy of Science* (Notre Dame, Ind.: University of Notre Dame
Press, 1980).

contextualization, and its success in showing how the social sciences have mattered, avoiding the severe limits of purely local studies by bringing to bear on larger historical issues a tightly focused historical analysis. Its insights are not limited to social science. Much of the most exciting work on what we might call the culture and the sensibility of science has involved the history of the social sciences. The common context or shared cultures of natural and social investigation has been explored in historical studies of Malthus, Darwin, and social Darwinism; the sciences of energy and economics; statistical thinking and the development of quantitative methods; laboratory instrumentation and ideals of precision; and positivism and objectivity, to give only a few notable examples.[5]

Historians of science are not the only people to write the history of the social sciences. Practitioners of the social sciences were the first historians of their disciplines, although historical purpose was subordinated to social scientific aims. Writing history was generally an exercise in disciplinary self-definition, linking the modern discipline to selected forebears and legitimating a certain kind of disciplinary practice. A number of such texts achieved considerable historical distinction and have remained useful works, such as Edwin G. Boring's *History of Experimental Psychology* (1929, 1957), Joseph Schumpeter's *History of Economic Analysis* (1954), and Joseph Dorfman's five-volume *The Economic Mind in American Civilization* (1946–59). Still, these works suffered from Whiggish assumptions, and only Dorfman, an institutionalist, linked economic doctrine to a deep political and cultural context. They hardly made a dent in social scientists' ignorance of their own histories that had been one of the consequences of the dehistoricization of the social sciences, especially in the United States.

A new wave of historical interest that emerged in the 1960s, led by social scientists outside the mainstreams of their disciplines, saw the establishment of journals and university centers in the history of psychology and economics. Clinical psychologists formed the core of historical interest in psychology, with Robert I. Watson founding the *Journal of the History of the Behavioral Sciences* (1965), a separate division of the American Psychological Association (1966), and a program at the University of New Hampshire (1967).[6] Economists at Duke University, long a center of historical

---

[5] Robert Young, *Darwin's Metaphor: Nature's Place in Victorian Culture* (Cambridge: Cambridge University Press, 1970); M. Norton Wise, "Work and Waste: Political Economy and Natural Philosophy in Nineteenth-Century Britain," *History of Science*, 27 (1989), 263–301, 391–449; and 28 (1990), 221–61; Theodore M. Porter, *The Rise of Statistical Thinking, 1820–1900* (Princeton, N.J.: Princeton University Press, 1986); Jill Morawski, ed., *The Rise of Experimentation in American Psychology* (New Haven, Conn.: Yale University Press, 1988); Ruth Benschop and Douwe Draaisma, "In Pursuit of Precision: The Calibration of Minds and Machines in Late Nineteenth-Century Psychology," *Annals of Science*, 57 (2000), 1–25.

[6] Mitchell G. Ash, "The Self-Presentation of a Discipline: History of Psychology in the United States between Pedagogy and Scholarship," in *Functions and Uses of Disciplinary Histories*, ed. Loren Graham, Wolf Lepenies, and Peter Weingart (Boston: D. Reidel, 1983), pp. 143–89.

economics, and a group of English historians who had just begun a newsletter in the history of economic thought collaborated to found the journal *History of Political Economy* (1969). Deliberately choosing the term "political economy" to counter the narrowed scientific focus of postwar economics, they urged the value of history in an ahistorical and uncritically technocratic age.[7] The historical character of this work, and of subsequent initiatives in sociology,[8] varied widely, from the ahistorical search for elements useful to current theory and practice, to sophisticated research agendas informed by intellectual history and by the history and sociology of science.

These social science disciplinary milieux were soon invaded and augmented by a new generation of professional historians. George Stocking was a pioneer figure, a young historian studying ideas of race in the United States who was drawn deeply into the history of anthropology. Psychology also attracted considerable historical talent, and the interchange of historical sophistication and specialized social science knowledge raised the standards of scholarship. An historian like Stocking and a psychologist like the Canadian Kurt Danziger became, so to speak, fully bilingual.[9]

Most professional historians who became interested in the social sciences were less committed to the dialogue of a particular social science discipline than to the discourses of the historical profession and the public sphere. The social sciences emerged as an historical topic largely because of their influence on postwar society, governance, and culture, particularly in the United States.[10] With their technocratic expertise and scientific claims, the social sciences were also a ready target for the "unmasking" mood that followed the radicalism of the 1960s. Historians found in the social science project professional self-interest, elitist desires to exercise "social control," and structural class and institutional constraints on knowledge.[11] By the 1980s, Foucault's work had drawn attention to the coercion exercised by the very processes

[7] Crawfurd D. W. Goodwin, Joseph J. Spengler, and Robert S. Smith, "Avant-Propos"; "Robert Sidney Smith, 1904–1969"; and A. W. Coats, "Research Priorities in the History of Economics," all in *History of Political Economy*, 1 (Spring 1969), 1–18.

[8] The *Journal of the History of Sociology* appeared intermittently from 1978 to 1987. *Cheiron* and the JHBS welcomed all of the social sciences, but only sociology and anthropology maintained a presence alongside psychology. A Research Committee in the History of Sociology and its newsletter, part of the International Sociological Association, also attracted American and European scholars.

[9] See particularly Stocking, *Race, Culture, and Evolution*, Prefaces and chap. 1; and Danziger, *Constructing the Subject* (Cambridge: Cambridge University Press, 1990), Preface, Introduction.

[10] Early and characteristic works are Mark H. Haller, *Eugenics: Hereditarian Attitudes in American Thought* (New Brunswick, N.J.: Rutgers University Press, 1963); the essays of John C. Burnham, since collected in *Paths into American Culture* (Philadelphia: Temple University Press, 1988); Nathan G. Hale, Jr., *Freud and the Americans: The Beginnings of Psychoanalysis in the United States, 1876–1917* (New York: Oxford University Press, 1971).

[11] A sophisticated pioneering work in this vein is Mary O. Furner, *Advocacy and Objectivity: A Crisis in the Professionalization of American Social Science, 1865–1905* (Lexington: University Press of Kentucky, 1975).

of rationality deployed by the human sciences.[12] Although a critical stance persisted, as these views were absorbed into historical discourse a wider variety of historians, with a wider spectrum of interpretive purposes, brought the history of the social sciences into their work.

Professional historians were not alone in bringing a new dimension of critique to the history of the social sciences. All participants in this diverse field were affected by the self-examination that gripped the humanities and social sciences during these decades, as knowledge claims in all the disciplines were thrown into doubt.[13] The reflexive interest of social scientists in their history was in part a facet of this larger movement of self-examination, which encouraged the effort of social scientists to come to grips with the historical character of their own domain. The historical discipline, always adjacent to and sometimes allied with the social sciences, scrutinized its own quest for objectivity and narrative strategies. Historicism was often figured as the philosophical ground of the new intellectual movement, but it did not valorize the professional historian's construction of experience.[14] Indeed, historians often used concepts and analyses borrowed from the social sciences, and narratives of modernity developed by the social sciences structured their stories. In the largest sense, the history of the social sciences invites reflection on the ways in which historians and social scientists are mutually implicated in each others' work.

We thus enter into the task of this volume with considerable pride in the intellectual tools at our command and a heightened awareness of their complexity and provisionality. As the work in this volume shows, there are now rich and powerful models for historical work in the social sciences. Authors in this field, however, have not always been aware of one another, and some perhaps have discovered only recently that all along they have been writing this species of prose. We believe that the history of social science is not merely a residual category, that its object has a cultural coherence, and that its pursuit is important for history. We have assembled authors from a variety of backgrounds and encouraged them to take seriously the methods and the intellectual content of social science, while considering at the same time the ways in which it has shaped and been shaped by a larger culture. The essays display differing balances among these objectives, as indeed they must.

We have planned this volume with an eye to the balance and range of the whole, and not just to the quality and comprehensiveness of the parts. It is, of course, impossible to be comprehensive. The four parts of this book

---

[12] See, for example, Nikolas Rose, *The Psychological Complex: Psychology, Politics and Society in England, 1869–1939* (London: Routledge, 1985).

[13] See Quentin Skinner, ed., *The Return of Grand Theory in the Human Sciences* (Cambridge: Cambridge University Press, 1985).

[14] Hayden White, *The Content of the Form: Narrative Discourse and Historical Representation* (Baltimore: Johns Hopkins University Press, 1987); Peter Novick, *That Noble Dream: The "Objectivity Question" and the American Historical Profession* (Cambridge: Cambridge University Press, 1988).

concentrate on different regions and periods. Part I, on the origins of social science, is concerned mostly with Europe, while Part II, on the modern disciplines, and Part IV, a collection of case studies illustrating the larger societal importance of social science, are somewhat biased toward the United States. Because it was impossible in these parts to do justice to much of the rest of the world, we have included a separate section on the internationalization of the social sciences, with essays on eastern Europe, Asia, Africa, and Latin America. Our authors themselves come from many disciplines, though most work in history and the history of science. Some topics, such as the development of the modern disciplines, draw heavily from historical writing in the United States, while others, especially those concerning the period before 1870, reflect British, French, and other European traditions of scholarship. The internationalization of social science, fittingly, engages historical understandings from around the world. Increasingly, the entire field of history of social science does so.

This volume in the Cambridge History of Science does not and could not present a collection of introductory articles representing the state of a well-demarcated field. We are aware of no work, whether singly or collectively authored, that has aspired to present such a wide historical view of the social sciences. The essays included here examine the history of the social sciences over some three centuries and many countries, attending to their knowledge and methods, the contexts of their origin and development, and the practices through which they have acted on the world. Our aim throughout has been to present the social disciplines not as a natural, inevitable solution to the organization of knowledge or the administration of modernity, but as problems – historically contingent, locally variable, always in flux, often contested, and yet as real sites of power in the world. We conceive of this book, too, not as reflecting the settled state of a field, but as something provisional, the product of a rich dialogue that, we hope, will be further advanced by its appearance.

Part I

SCIENCES OF THE SOCIAL TO THE
LATE NINETEENTH CENTURY

2

GENRES AND OBJECTS OF SOCIAL INQUIRY, FROM THE ENLIGHTENMENT TO 1890

Theodore M. Porter

"Social science" entered the vocabulary of the West near the end of the eighteenth century, first of all in the United States and France. Many of its early enthusiasts, well into the nineteenth century, aspired to a single, unified science of the social, in stark contrast to the multiple disciplines that were taking shape by 1900. We might be tempted to frame the history of social science as a relentless process of advancing specialization, just as the history of natural science has often been conceived as a sequence of disciplinary separations from a once-unified philosophy. But such an understanding is no more satisfactory for social than for natural knowledge. Not least among its shortcomings is its privileging of the pure life of the intellect, the *vita contemplativa,* over the interventions and engagements of scientific life in practice. Social science has from its earliest beginnings aimed to administer and to change the world as well as to understand it. It did not spring forth from the head of humanity only, but from the body as well – from law, medicine, politics, administration, and religion, as well as from philosophy. Both intellectually and institutionally, it has always been diverse.

Seeing social science as part of philosophy has, nevertheless, some decided advantages over the most influential opposing view, disciplinary Whiggism, which regards each of the modern fields of knowledge as if they have always been coherent specialties. Strict disciplinary history encourages – if it does not require – a narrowness of perspective that leaves few openings for an inclusive cultural understanding. It can lead also to the rather absurd view that makes Aristotle the first psychologist, the first anthropologist, and one of the first sociologists, economists, and political scientists. Could a single Aristotle have so many essences? Yet, though no political scientist, he certainly had a politics; and if his philosophy ranged over much of the human (as well as the natural) terrain, he did not put everything into a single comprehensive work. We need to find a balance between intellectual unity and disciplinary fragmentation as ways of thinking about social knowledge in the centuries before the emergence of modern specialties.

Part I of this volume concerns the period up to the late nineteenth century when social science, if not amorphous, lacked well-defined institutional strructures. This chapter introduces the social sciences in Europe and North America from about 1700 up to the beginning of this disciplinary project. It aims first of all to provide a loose periodization of the early history of social science, and of the broader historical changes that made it seem both possible and necessary. The chapter starts during the period of the Enlightenment, when discourses of nature and reason began to be applied more systematically to "man" and society, often in the spirit of criticism or reform. The French Revolution of 1789 marked an important shift, in which social progress came to seem both more powerful and more threatening, opening up a new problematic in thinking about modern societies. A second transition, of particular consequence for the practices of social science, took place roughly during the decade of the 1830s, as the economic and social changes of industrialization became visible to everyone, and social science emerged as a tool for managing as well as for understanding the problems of this new era. The chapter then proceeds to investigate the ways in which social science was defined in relation to contemporaneous understandings of natural science, which was important both as a positive and as a negative model. It concludes by considering briefly how the meanings of "discipline" and "profession" in social science were changing during the 1870s and 1880s.

THE "SCIENCES OF MAN" IN THE EARLY MODERN PERIOD

Although there were no social science disciplines before the nineteenth century, there were recognized European traditions of thought and practice concerned with politics, wealth, the senses, distant peoples, and so on. Since we are interested here in practical and political life as well as in academic learning, it is perhaps best to speak of genres or discourses, with the understanding that deeds as well as words are at issue. The genres corresponding to our social sciences were disparate. Early modern treatises on the human capacity to acquire knowledge, or on the ideal polity, were largely distinct from writings on coinage, political arithmetic, or the physical features and customs of faraway peoples. Much of what we call "anthropology" was to be found in travel narratives and medical treatises. Thinking and understanding were largely philosophical topics, until late Enlightenment medical authors introduced a rival discourse of the brain. Political writings could be philosophical as well as legal or historical, but they rarely were dissolved into general philosophy, even when they involved explicit metaphysical and epistemological assumptions.

If we are not too shy about anachronism, the following might be identified as the defining objects of some important discourses concerning what early modern writers called "man": populations, economies, states, bodies, minds, and customs. Each was closely related to one or several topics of natural philosophy, and none was sharply marked off from politics, from religion, or from moral reasoning. In the eighteenth and nineteenth centuries, even within the European world, the genres of social inquiry were highly variable and interrelated in complex ways. On the one hand, they were often tightly imbricated. One could scarcely write on population in the eighteenth or even the nineteenth century without considering economies, governments, and customs. Assumptions and beliefs about thinking and human behavior were used to support political systems and to explain the functioning of an economy. Almost any interpretation of man, at least before the eighteenth century, presumed an understanding of the biblical story of creation, and of doctrines of sin and salvation.

Also, the subject matter of social science was not neatly divided up. Even within Europe, the genres were often defined as much by a field of debate as by agreement on key methods and doctrines; they varied from place to place, and sometimes came into competition. Among the economic studies, British "political economy" was not quite the same as French "Physiocracy," and was quite different from German "cameralism." "Psychology," a term used mainly in German lands, was no more in accord with English writings on sensation and reflection than was Leibniz's philosophy with that of Locke or Newton. The German and Italian science of statistics, the empirical study of the state, was largely distinct from the study of politics, a more philosophical discourse about how states should be governed. By 1800, statistics had begun to be overrun by population numbers, until then the business of "political arithmetic," which had exalted them as an index of the quality of government, and often interpreted them theologically. Writing about the customs of diverse peoples was closely tied to an understanding of their climates, and often also of their bodies, which comprised a principal topic of anthropology.

"Social science," as we argue in the Introduction, is even now an unsettled category, and a contested one. Three centuries ago it was less contested, in part because it was still more unsettled: There was no rubric like "social science" under which these discourses could be arrayed, and toward which they could direct their grand methodological ambitions. The forms of knowledge that we call "social" were not then rivals, because their objects as well as their methods were largely distinct. This did not prevent encyclopedic intellects from working seriously in two or several of these genres, though usually in separate publications, and interactions among them were as rich and interesting then as they are now. Still, only in the eighteenth century did an idea of "science of man," "moral science," or "*sciences morales et politiques*" begin to reconfigure these diverse inquiries – to unite then into a family, which could then squabble. This also is when "philosophical history" as a comprehensive outline or natural history of the progress of "civilization" was initiated, especially in France and Scotland. Out of this tradition developed the idea of society as a proper object of science. Our history begins with this period.

ENLIGHTENMENT SCIENCES OF ECONOMY, POPULATION, AND STATE

"The Enlightenment" is usually taken to refer to a movement of criticism and reform, against the authority of church and aristocracy, in favor of "nature" and "reason." Following the logic of this characterization, the rise of social science during the same period has been represented as almost inevitable.[1] But this depends on some subtle questions of definition. By the standard even of the nineteenth century, most Enlightenment social writings appeared light and popular rather than profound and scientific. It is not simply that there were there no professional structures – no university degree programs to offer formal training and credentials in the moral sciences. These were still rare, during the eighteenth century, even in "natural philosophy." Natural science, however, had at least its academies and societies, its gatherings of experts, and its journals, which had no equivalents in social science before the French Revolution in 1789. The Enlightenment sciences of man were mostly public or bureaucratic discourses rather than specialized ones.

It would be overly fussy as well as anachronistic to define social science as necessarily a specialized, technical discourse. The birth of social science has much to do with liberalizing political moves and the growth of a public sphere. The Enlightenment, as an intellectual and social movement, depended on increasingly free public discussion, and on mechanisms for the circulation of ideas. To be sure, most men and women of the eighteenth century remained illiterate, and only a few had access to the ideas of Enlightenment. Yet by the late eighteenth century an informed public had emerged in the leading countries of Europe. The French *philosophe* and mathematician Condorcet (1743–1794) presented Gutenberg's invention of the printing press as a signal event in the history of progress, since it allowed knowledge to advance without ever being lost. Never had the presses been so busy, and never had they reached so wide an audience, as in his own time. The growth of newspapers was particularly significant in opening up a public space. New institutions such as coffeehouses, salons, and Masonic lodges also provided opportunities for relatively free discussion of issues and events. The nascent moral sciences were a part of this same world.

They were not, however, wholly at ease with it. Cultural historians have taken a keen interest in the circulation of books and journals during the late eighteenth century, especially in France, in quest of that historical grail, a

[1] Peter Gay, *The Enlightenment: An Interpretation*, vol. 2: *The Science of Freedom* (New York: Knopf, 1969).

convincing account of the ties between the Enlightenment and the French Revolution. They have found that the enduring works of political philosophy and social science by the likes of Charles-Louis Montesquieu (1689–1755), Jean-Jacques Rousseau (1712–1778), and even Voltaire (1694–1778) sold relatively few copies, while early blockbuster novels and various productions of the gutter press reached wide audiences.[2] The search for a science of politics and society meant an effort to rise above appeals to passion and ignorance. The *philosophes* saw it as spreading light (*lumières*), and much of their writing was for a select audience – not of specialists, but of the enlightened. In some ways, the move to a language of social science was designed to undercut the authority of mere political will, and to replace it with something more detached and objective: simultaneously to vindicate human freedom and to subject it to standards of reason. Characteristic, if a bit extreme, is Condorcet's mathematics of elections and judicial decisions, which acknowledged the claims of public opinion while devising mechanisms to assure that it would lead to rational decisions rather than to dogmatic or arbitrary ones.[3]

It would be a mistake to suppose that the credibility of Enlightenment social theory rested only or even mainly on its similarities to mathematics and the sciences of nature. The assertion of natural rights in the political writings of authors such as John Locke (1632–1704) and Rousseau, and in crucial documents of the American and French Revolutions, owed more to moral doctrines of "natural law," which concerned the just political order, than to Cartesian or Newtonian laws of nature. Montesquieu, often portrayed as the founder of sociology or at least of social theory, was very much interested in natural science, especially physiology, but his problematic came chiefly from a different set of sources. He had been trained in the law, and made his profession as a jurist. Donald R. Kelley writes that the pioneers of social science were "not the cosmologists who belatedly shifted their gaze from the heavens to the human community but rather . . . the law-makers who were confronted by the predicaments of human society."[4]

These lawmakers were not, however, deprived of theoretical resources. Natural law meant more than law as handed down by tradition in a particular place – "positive law"; it stood for an immutable ideal, a system of obligations and of rights deriving from human nature. It had been cultivated most notably in early modern Europe by the Dutch statesman Hugo Grotius, by Samuel

[2] Roger Chartier, *The Cultural Origins of the French Revolution* (Durham, N.C.: Duke University Press, 1991).

[3] Keith Michael Baker, *Inventing the French Revolution: Essays on French Political Culture in the Eighteenth Century* (Cambridge: Cambridge University Press, 1990), pp. 98, 153–66, 189; Peter Wagner, "Certainty and Order, Liberty and Contingency: The Birth of Social Science as Empirical Political Philosophy," in *The Rise of the Social Sciences and the Formation of Modernity: Conceptual Change in Context, 1750–1850, Sociology of the Sciences Yearbook, 1996*, ed. Johan Heilbron, Lars Magnusson, and Bjorn Wittrock (Dordrecht: Kluwer, 1998), pp. 241–63.

[4] Donald R. Kelley, *The Human Measure: Social Thought in the Western Legal Tradition* (Cambridge, Mass.: Harvard University Press, 1990).

Pufendorf, advisor to German and Swedish rulers, and by Locke, who in the 1680s worked out a philosophical rationale for overthrowing an unjust monarch. These writers were impressed by the analogies between the natural and social orders, and sought to understand human nature as something universal. In this way, they hoped to provide a general framework for political society during the turmoil of the seventeenth century. Their work became known in France in Montesquieu's time, and his *Spirit of the Laws* (1748) undertook to explain the relation of natural law – presumed to be universal – not simply to positive law, which varies greatly from place to place, but to its "spirit."[5] Thus, despite or even because of his moral universalism, Montesquieu was led to examine and explain the customs and practices of particular places in a way that has been called sociological.

While natural law, with its moral orientation, was distinct from belief in laws of nature, understood as independent of human purposes, these often intersected. Grotius took the geometry of his contemporary Galileo as a model for moral reasoning, and when Anne-Robert-Jacques Turgot (1727–1781) urged on King Louis XVI of France the wisdom of governing, like God, by general laws, he evidently drew on both traditions. Political economy, too, involved natural justice as well as naturalism. Adam Smith (1723–1790) argued influentially that regulation was not required to coordinate an economy or to assure a standard of quality of manufactures. In a commercial system, individuals served the public interest even as they worked to advance their own. This formulation, which derived from French arguments for a free economy (*laissez-faire*), involved a move away from theological explanations, which saw labor as necessarily sinful, the outcome of Adam's fall. The guild insistence on systems of apprenticeship and detailed regulation of artisanal trades was thus gradually supplanted by a focus on the order produced by self-interested behavior and social customs.[6]

Smith's work also undercut arguments for a "moral economy" that would, for example, limit prices in times of scarcity and guarantee the subsistence of workers who had no property. Smith and David Hume (1711–1776), political economists of the Scottish Enlightenment, provided a defense of commercial society against the Christian utopia of a community of goods, and also against the idealization of the virtuous ancient republic, with its citizenry of free, independent farmers. While they played down the need for political intervention in economic matters, their writings were profoundly engaged with moral questions. Against the republican tradition, which looked to the simple societies of the remote past as exemplars of virtue, they held that ancient tribes were rude and barbarous. Social progress took place in stages

[5] Johan Heilbron, *The Rise of Social Theory*, trans. Sheila Gogol (Minneapolis: University of Minnesota Press, 1995), pp. 96–9; see also Richard Olson, *Science Deified and Science Defied: The Historical Significance of Science in Western Culture* (Berkeley: University of California Press, 1990), chap. 5.
[6] William Sewell, *Work and Revolution in France* (Cambridge: Cambridge University Press, 1980).

associated with a sequence of economic systems; Smith identified these as hunting, shepherding, agriculture, and finally commerce. The advantages of a modern commercial economy, they argued, accrued to the poor as well as to the rich. The advance of wealth did not undermine, but rather enhanced, moral virtues such as prudence and honesty.[7]

There were, naturally, others who opposed these arguments about the benefits of commerce, and we should be wary of identifying "economics" or "social science" exclusively with those who looked to the self-regulating mechanisms of a market economy. Many doubted the adequacy of a "natural order" and aimed instead to develop tools to guide and stimulate production. Among the many alternative eighteenth-century modes of addressing economic matters, cameralism deserves particularly to be noticed here. Cameralism was first of all a German science, concerned specifically with measures for increasing the flow of revenue into state treasuries. Just as ethnography was often practiced as part of a project that would survey biological species and peoples simultaneously – in one notable case, to advance the Russian settlement of Siberia[8] – so cameralism joined economics to the realm of science and industry. Mining and agriculture as well as property and exchange were included within its ambit. Since the cameralists believed, along with most economic writers, that only nature was really productive, the generation of wealth through commerce had for them an alchemical aspect.[9] Their business was to supply knowledge and advice about how to increase prosperity and how to tax it. While they regularly published textbooks and tracts, they wrote more for an audience of bureaucrats and rulers than for lay readers.[10]

Cameralism was characteristic in many ways of the practical, utilitarian turn of Enlightenment universities in northern Europe. This administrative impulse was no mere application of social science, but a powerful force in shaping it. It provided rich opportunities for the growth of technical methods and formalized expertise. Early modern quantitative studies of populations, sometimes under the label "political arithmetic," had grown highly mathematical by the end of the eighteenth century. To be sure, the advice proffered

---

[7] Istvan Hont and Michael Ignatieff, eds., *Wealth and Virtue: The Shaping of Political Economy in the Scottish Enlightenment* (Cambridge: Cambridge University Press, 1983), esp. chapters by Hont and Ignatieff, John Dunn, John Robertson, and J. G. A. Pocock; Albert O. Hirschman, *The Passions and the Interests: Political Arguments for Capitalism before Its Triumph* (Princeton, N.J.: Princeton University Press, 1977).

[8] Han F. Vermeulen, "Origins and Institutionalization of Ethnography and Ethnology in Europe and the USA, 1771–1845," in Han F. Vermeulen and Arturo Alvarez Roldán, *Fieldwork and Footnotes: Studies in the History of European Anthropology* (London: Routledge, 1995), pp. 39–59.

[9] Pamela Smith, *The Business of Alchemy: Science and Culture in the Holy Roman Empire* (Princeton, N.J.: Princeton University Press, 1994).

[10] Keith Tribe, *Governing Economy: The Reformation of German Economic Discourse, 1750–1840* (Cambridge: Cambridge University Press, 1988); David Lindenfeld, *The Practical Imagination: The German Sciences of State in the Nineteenth Century* (Chicago: University of Chicago Press, 1997); Franco Venturi, *Settecento riformatore*, 7 vols. (Torino: G. Einaudi, 1969–90), English trans. of vol. 1 by R. Burr Litchfield as *The First Crisis* (Princeton, N.J.: Princeton University Press, 1989).

by mathematicians about state-run lotteries and pension schemes was often ignored. Studies of mortality, and of the potential advantages of inoculation against smallpox, were more influential, perhaps in part because these results were addressed to a wider public, not only to monarchs, and could be acted on in a decentralized way. In the last decades before the French Revolution, an alliance of mathematicians and administrators in France undertook to use the most advanced tools of mathematical probability to estimate the population of France. Of course functionaries and bureaucrats – then as now – were less interested in theoretical ruminations than in numerical data. It suited them, however, to work in a relatively closed social space with privileged experts. Such probabilistic population estimates, developed in a context of administrative secrecy, were largely abandoned in the nineteenth century, when most official numbers had to be published.[11]

ENLIGHTENMENT SCIENCES OF MINDS, BODIES, AND CULTURES

The "sciences of man" in the eighteenth century were associated above all with questions that are now called psychological, questions about what was then called "human nature." Roger Smith, in his comprehensive history of the "human sciences," writes: "To quote references to human nature in the eighteenth century is a bit like quoting references to God in the Bible: it is the subject around which everything else revolves."[12] The subject was closely linked to natural philosophy, especially because one of its central ambitions was to understand the human ability to acquire and use empirical knowledge. Voltaire, in his *Letters on England* (1733), read Newton's achievement as the vindication of Baconian method – of science founded on experience, not on mathematical deduction. Voltaire also included in his *Letters* a chapter on Locke's *Essay Concerning Human Understanding* (1690). Locke, though not a religious skeptic, sought a naturalistic account of human nature. So, significantly, he spoke of mind rather than of soul, and he described the mind as essentially plastic, forming its ideas from sensations and reflection. At birth, it was like a blank slate. Hence men were made good or evil by their education, and were not captive to original sin. The *Essay* thus supplied a

[11] Lorraine Daston, *Classical Probability in the Enlightenment* (Princeton, N.J.: Princeton University Press, 1988); Andrea Rusnock, "Biopolitics: Political Arithmetic in the Enlightenment," in *The Sciences in Enlightened Europe*, ed. William Clark, Jan Golinski, and Simon Schaffer (Chicago: University of Chicago Press, 1999), pp. 49–68; Eric Brian, *La Mesure de l'État: Géomètres et administrateurs au XVIIIe siècle* (Paris: Albin Michel, 1994); Eric Brian, "Mathematics, Administrative Reform and Social Sciences in France at the End of the Eighteenth Century," in *Rise of the Social Sciences*, ed. Heilbron, Magnusson, and Wittrock, pp. 207–24; Alain Desrosières, *The Politics of Large Numbers: A History of Statistical Reasoning* (Cambridge, Mass.: Harvard University Press, 1998).
[12] Roger Smith, *The Fontana History of the Human Sciences* (London: Fontana Press, 1997), p. 216. (In the United States, *The Norton History of the Human Sciences*.)

weapon in Enlightenment struggles against the moral and institutional power of the Church, as well as a rationale for systematic schooling.

As Jan Goldstein argues, Lockean psychology was widely accepted among Enlightenment *philosophes* in France as well as Britain, and it formed the basis for further inquiry. Among the most intriguing challenges was the quest to gain access somehow to original human nature, before it was shaped and corrupted by society. This project was allied to an influential form of political theory, also deriving from Locke (and from Thomas Hobbes before him), which posited a "state of nature" prior to the "social contract" that had established society. The political philosophers disputed as to whether this involved an enviable state of freedom (Rousseau) or a nasty struggle of all against all (Hobbes). One could ask, too, about the development of the faculty of perception. Denis Diderot (1713–1784) wondered in his *Letter on the Blind* (1749) whether a blind man suddenly given sight would be able to distinguish visually between a cube and a sphere. It would be yet more interesting to know how people would act and think if raised wholly outside of society. A number of so-called wild boys or wild men, discovered in forests and wastelands during the eighteenth century, were examined to shed light on this crucial question. The increasing, though still very limited, exposure of Europeans to anthropoid apes provided occasion to ponder whether these animals had the human capacity to learn language and to reason. Explorers were fascinated by the customs, and especially what they took to be the exotic sexual practices, of distant peoples. At the very end of the century, the French Société des Observateurs de l'Homme (Society of Observers of Man) undertook expeditions to study human nature, still thought of as uniform, under the most primitive conditions.[13]

Doctrines of race had little standing in the moral sciences of the eighteenth century. The French Enlightenment acquired a reputation among its nineteenth-century critics as materialistic, a charge that was mostly false and wholly imprecise. Beginning with some of Locke's more radical followers, materialistic psychologies were sometimes advanced as political or religious critiques, but this was a fringe position. Medical explanations of drunkenness and sexual excess, and medical or penological treatments of crime and madness, were displaced in the late eighteenth century from the body to the mind or spirit – an antimaterialistic move.[14] In France, Etienne Bonnot de Condillac's analysis of human sensory capacities was taken up at the end of

---

[13] Ibid., chaps. 7–8; George W. Stocking, Jr., *Race, Culture and Evolution: Essays in the History of Anthropology* (New York: Free Press, 1968), chap. 2; Sergio Moravia, *La Scienza dell'uomo nel Settecento* (Bari: Guis Laterza & Figli, 1970).

[14] Roy Porter, "Medical Science and Human Science in the Enlightenment," in *Inventing Human Science: Eighteenth Century Domains*, ed. Christopher Fox, Roy Porter, and Robert Wokler (Berkeley: University of California Press, 1995), pp. 53–87, esp. pp. 73–4; also, in same volume, Gary Hatfield, "Remaking the Science of Mind: Psychology as Natural Science," pp. 184–231; Graham Richards, *Mental Machinery: The Origins and Consequences of Psychological Ideas, 1600–1850* (Baltimore: Johns Hopkins University Press, 1992), p. 135.

the century by medical authors, who were particularly interested in the materiality of mental faculties – that is, in the brain. This movement culminated during the Napoleonic period, from about 1800 to 1815, in an alliance of physiological and cultural investigations of mind and morals. But this was more nearly vitalism than materialism, for living matter was not inert but self-organizing, infused with life and spirit.[15] Medicine was at the center of a late Enlightenment revolt against mathematics and in support of the sciences of life, initiated at mid century by the comte de Buffon and by Diderot. Still, the union of medical and moral inquiry was perhaps the most aggressive version of the science of man in 1789, when its questions were made much more pressing by the outbreak of the French Revolution. That alliance was prominently defended and significantly transformed within social science during the ensuing decades.

SOCIAL SCIENCE IN AN AGE OF REVOLUTION, 1789–1830

Few of the leading French *philosophes* survived to 1789. The mathematician Condorcet, permanent secretary of the Academy of Science, might be called the voice of the Enlightenment during the Revolution. It was a precarious role. Voltaire, Rousseau, Condillac, Turgot, d'Alembert, and Diderot all died between 1778 and 1784. Historians have often regarded the Enlightenment as waning or even as having finished some years before the outbreak of the Revolution. In the politically polarized climate after 1789, a career like that of Voltaire or Diderot, based on appeals to universal reason, was scarcely possible. There were, however, younger intellectual figures working in the 1780s who might have been remembered as spokesmen for enlightenment under other circumstances. Recently, some historians have emphasized the intellectual continuities across the divide of 1789.[16] These are important and real. For the history of the institutions and practices of social science, the decade of the 1830s marks a still more decisive transition. The ideological significance of the French Revolution for social science was, however, without parallel. Unruly passions, threatening political stability, inspired a pervasive sense of danger. Social science became more urgent, and often more ideological, looking to the past, or to science, in order to comprehend what seemed the precarious circumstances of modernity.

While Enlightenment *philosophes* disapproved of arbitrary acts of power, many maintained a favorable opinion of absolute monarchy, since it held out the prospect of immediate reform, if only the king could be brought around. Condorcet viewed the Revolution in just these terms, indeed as an

---

[15] Moravia, *Scienza dell'uomo*; Martin S. Staum, *Cabanis: Enlightenment and Medical Philosophy in the French Revolution* (Princeton, N.J.: Princeton University Press, 1980).
[16] Heilbron, Magnusson, and Wittrock, eds., *Rise of the Social Sciences*.

unparalleled opportunity, and he engaged actively in revolutionary politics. He opposed the radical Jacobins on such questions as whether to execute Louis XVI, and he worried about democratic excesses. He sought a political system based on relatively wide participation that would yet place men of cultivation and of science, men like himself, in positions of responsibility. He envisioned a state founded on natural and social science. Science would form the core of a system of universal education, with the elite Academy of Science at its summit. It would also form the basis of administration. He drew up plans for a vast statistical apparatus, one suited to a countable society of free and independent citizens. The state, henceforth acting on the basis of full information and rational methods, would naturally advance the public good.[17]

By 1800, such aspirations had come to seem wildly utopian to many. Indeed, as early as 1790 Edmund Burke dismissed them in such terms in his *Reflections on the Revolution in France.* Burke interpreted the Revolution as a consequence of irresponsible men, shallow ideologues, provoking abrupt changes in a social organism – the state – whose natural development is slow and gradual. Half a century later, Alexis de Tocqueville (1805–1859) attributed the excesses of the Revolution to the influence of detached intellectuals, men without actual experience of government. This was, in a way, an indictment of social science, at least in its utopian form. And Condorcet, for example, had indulged in a good bit of utopianism, having written, while in hiding from the Terror of Robespierre and the Committee on Public Safety, his famous *Sketch of a Historical Picture of the Progress of the Human Spirit.* This was a story of unilinear progress, driven by the advance of knowledge, through ten stages, of which the last and most glorious was signaled by the very Revolution that was hunting him down.

Condorcet's historical account was, as Antoine Picon's chapter shows, very much an Enlightenment document, part of the intellectual shift that had displaced utopia ("no place") from somewhere in space (far away) into time, the near or distant future. The influence of this genre, introduced around 1750 by Condorcet's mentor, Turgot, testifies to a new sense of historical dynamism, one that survived and flourished in the nineteenth century, but often in a nonlinear form that we might call dialectical. The key figure here was Claude Henri de Saint-Simon (1760–1825), an aristocratic opportunist who had fought with George Washington in America, and later supported the French Revolution. (He in fact used the occasion of the Revolution to enrich himself by dealing in Church lands.) In 1793 he took the peasant name of Bonhomme, which however did not keep him from being imprisoned, though he escaped with his life. Subsequently he set up house in Paris near the new Ecole Polytechnique, established to educate an elite corps of technical experts. There, his personality and his patronage attracted a circle of brilliant

---

[17] Keith Baker, *Condorcet: From Natural Philosophy to Social Mathematics* (Chicago: University of Chicago Press, 1975), pp. 57, 200, 207, 262, 272, 303.

young students of mathematics, lost souls seeking stable meanings in an age of turmoil.

Saint-Simon did not endorse the Old Regime, or aspire to return to it. He instead conceived of modern history in terms of what he called "organic" and "critical" periods. The medieval Church, he held, with its communitarian ethos and its union of spiritual and scientific knowledge, had supported a social order that was admirable, but unsustainable. In the fifteenth century, Europe had entered an age of fragmentation and individualism. The rise of Protestantism and then of secular criticism marked the demise of the old order. It was finally extinguished by the French Revolution, leaving a condition of spiritual emptiness. A new organic order must arise, in which the primacy of a social ethic over an individualistic one would be restored. Saint-Simon announced the inevitability of this new order, and at the same time worked to create it. He identified first science, then industrial organization, and finally a "new Christianity" as its basis. The Saint-Simonians repudiated the terrible anarchy of the critical period, which could be justified only as a necessary destructive phase to clear the way for a better future.[18]

There were others, of course, who doubted that revolution could be a harbinger of anything good. Those who envisioned the ideal future as a return to the wholeness of the past rarely marched under the banner of science, but others sought a science of society in order to understand and control the unruly impulses of the modern age. This was, in some ways, the Saint-Simonian ideal too, even if the expression was utopian. Auguste Comte (1798–1857), another recruit from Polytechnique, and Saint-Simon's most famous and most rebellious disciple, was writing already in the 1820s of the indispensable role of religion in the new scientific order. Man (and especially woman) is not at bottom coldly rational, but rather spiritual and emotional. Comte eventually gave him, and her, an object of reverence and a calendar of festivals and commemorations in his "religion of humanity." He explicitly discarded personal freedom as a burden on the individual and a chaotic force in society. As Peter Wagner has remarked, social science during this period did not so much express the liberty and contingency of the modern period as seek to rein it in.[19] Even in the United States, where 1776 was celebrated as a triumph, political economists viewed the European experience anxiously, hoping that the American republic could avoid the endemic social strife of the Old World. Freedom, while a blessing, had to be held within bounds.

The French tradition of administration by engineers defined the locus of a powerful tradition of social and economic science in the nineteenth

---

[18] Frank E. Manuel, *The Prophets of Paris* (Cambridge, Mass.: Harvard University Press, 1962); Frank E. Manuel and Fritzie P. Manuel, *Utopian Thought in the Western World* (Cambridge, Mass.: Harvard University Press, 1979).

[19] Peter Wagner, "Certainty and Order, Liberty and Contingency: The Birth of Social Science as Empirical Political Philosophy," in *Rise of the Social Sciences*, ed. Heilbron, Magnusson, and Wittrock, pp. 241–63.

century. Tocqueville interpreted the Revolution as an acceleration of central-izing tendencies that already were pronounced under the Old Regime; and the analytical style of savants and engineers, treating social questions as prob-lems to be solved, exemplifies this continuity in one of its forms.[20] Planning and economic analysis after the Revolution fell increasingly to Polytechnique engineers. Around midcentury, Frédéric Le Play of the highly elite Corps des Mines initiated his method based on detailed monographs to comprehend the domestic economies of miners, artisans, and laborers. Such information could be used by employers and local notables as a guide to charity and organization. This was social science as a set of pragmatic tools rather than as utopian vision. Champions of rational administration under the rubric of science had exploited what opportunities they could during the revolutionary period, though in the end their successes were modest. In the latter years of the Napoleonic wars, and especially after 1815, when the French monarchy was restored and a new conservative order was imposed on Europe, the in-fluence of this ideal was much diminished. It began slowly to revive in the 1820s, especially in France and Britain, in the more sober guise of statistics.

The transformation of grand ideals into prosaic bureaucracy is exempli-fied in the career of radical utilitarianism. Jeremy Bentham's most important programmatic tract was published precisely in 1789. Drawing from a com-monplace of Enlightenment psychology, that it is human nature to pursue pleasure and avoid pain, and from a utilitarian ethic that pursued "the greatest happiness for the greatest number," Bentham proposed an uncompromising program of rationalistic reform. It was his mission to abolish custom and tradition and to cut through the obfuscation that surrounded them in favor of whatever would advance the general welfare. Bentham's grand schemes at-tracted a band of followers, the philosophic radicals, whose designation gives some idea of the extent of their ambitions. Among the most influential was James Mill, who converted Bentham to democracy with the argument that governing classes would never enact a program to benefit the whole popula-tion until the masses had the vote. This was no otherworldly movement, but a pragmatic campaign with an effective slogan, and sometimes self-consciously anti-utopian. Many of the radicals, including Mill and his more famous son, John Stuart Mill (1806–1873), subscribed to the arguments of Thomas Robert Malthus, who taught the relentless, geometric increase of population, and the need somehow to check its growth before the pressure of want and misery became too severe. The overarching ambition of the philosophic radicals was to enact laws backed up by inducements sufficient that each individual would be led by self-interest to act in a manner that would advance the collective happiness. Bentham's program included an elaborate calculus of the punish-ments necessary to outweigh the attraction of every particular crime, taking

---

[20] Brian, *Mesure de l'Etat*; Antoine Picon, *L'Invention de l'ingénieur moderne: L'Ecole des Ponts et Chaussées, 1747–1851* (Paris: Presses de l'Ecole Nationale des Ponts et Chaussées, 1992).

into account the probabilities of arrest and of conviction. As an effective program of reform, driven by a new breed of bureaucratic experts such as Edwin Chadwick, philosophic radicalism came into its own in the 1830s as one important constituent in the growth of the British state.

THE MANAGEMENT OF SOCIAL AND ECONOMIC CHANGE, 1830–1880

Between 1830 and 1848, the leading nations of Europe faced urgent new social problems. While contemporary economic historians commonly trace the beginning of the Industrial Revolution in England back to the eighteenth century, usually to around 1760, that dating requires the wisdom of hindsight. If some of the more acute British economic observers during the era of Adam Smith recognized their time as one of advancing prosperity, none identified any startling changes, much less revolutionary ones. By early in the nineteenth century, improvements in machinery had become sufficiently manifest to draw the wrath of the Luddites, who saw industrial power as robbing laborers of their work. Political economists in the era of Malthus and of David Ricardo (1772–1823) remained pessimistic, but steam engines and other mechanical substitutes for labor gave them some hope. The changes that we call the industrialization of England began to appear remarkable only after 1815. As the Napoleonic wars drew to a close, and as European trade opened up once more, the power of English industry began to reshape and disrupt Continental economies, especially in the textile trades. The onset of industrialization in France, Germany, and the Low Countries is usually attributed to the beginning of railroad construction, no earlier than 1830.

In Britain, the 1830s was the decade of "the social question." By 1840, it had become pressing also on the Continent. Economic change brought economic dislocation. It involved a massive flow of people from farms to cities, sometimes with a crucial ethnic dimension, as in the Irish migrations to England. Changing patterns of work altered family arrangements, drawing women and children into factories and mines. A cholera epidemic swept through Europe in 1832. Urban squalor, crime, and disease seemed to threaten good order, especially in this unsettled political situation. The British moved away from repression in favor of reform during the 1830s. The Revolution of 1830 in France replaced the descendants of the Old Regime ruling family, the Bourbons, with a constitutional monarch, and the Belgian revolution of the same year brought independence from the Netherlands. The post-1830 franchise in France was, however, very limited, and the "great reform" of 1832 expanded it only modestly in Britain. Strikes and mass movements expressed the dissatisfaction of many working people with their new circumstances. Britain faced the possibility of revolution through the late 1830s and early 1840s, while most Continental nations experienced real revolutions in 1848.

The utilitarian spirit of social reform in Britain during the early nineteenth century, and especially from about 1830 to 1850, marked a turn away from the historicizing perspectives that had been developed, especially in Scotland, during the latter part of the eighteenth century.[21] That turn was much less pronounced in France; and in Germany, as Kent Wright's chapter shows, historicism survived through the revolutionary period and then flourished as never before, particularly as a discipline of history grew up in the new German research universities. The sciences of mind, however, offered a largely ahistorical frame for the moral and political debates of the early nineteenth century. As Jan Goldstein's chapter underscores, they were partly continuous with the Enlightenment. The associationist perspective of Locke and Condillac defined the psychology of the English utilitarians, of Bentham and the Mills and, in alliance with brain physiology, of the French current of moral science called *idéologie* that extended the central commitments of the late Enlightenment into the Napoleonic period.

Phrenology, which developed into a considerable movement during the early nineteenth century, involved at first a more radical materializing move. Goldstein discusses its changing political resonances, from a subversive science of brain that cast doubt on the Christian doctrine of the soul in the 1820s, to a more popular but less threatening language of self-improvement in the 1840s. In some ways, though its theoretical basis was entirely different, its career paralleled that of mesmerism, which had become popular during the 1780s. The two were sometimes joined by midcentury, peddled by traveling lecturers who gave demonstrations and told fortunes. Both phrenology and mesmerism engaged with central questions of religion, politics, and professional authority. Both grew up on the margins of the authorized science of British gentlemen, French academicians, and German professors, and both lost credibility among elite "men of science" as their appeal moved down the social hierarchy. Their popularity came to be seen as a defect in the public understanding of science, and sometimes as a social danger, to be controlled by means of better science.

Mental science, like other forms of social knowledge, could be deployed as a tool of regulation and administration. This aspect of social science, emphasized most tellingly by the French philosopher Michel Foucault, is central for understanding its history during this period.[22] As Elaine Yeo argues in her chapter, social science was not automatically controlled by elite scientists, reformers, and officials, but was contested, often very effectively. Working-class alternatives to gentlemanly sciences, both social and natural, flourished in the decades following 1830. Here was another dimension to

[21] J. W. Burrow, *Evolution and Society: A Study in Victorian Social Theory* (Cambridge: Cambridge University Press, 1970).

[22] Michel Foucault, *Discipline and Punish: The Birth of the Prison*, trans. Alan Sheridan (New York: Vintage Books, 1979).

the problem of managing populations – the struggle for social science was also a struggle for social power. It was, on the whole, an unequal struggle, with working people at a considerable disadvantage. But the threat to social order that they seemed to present gave the intellectual problem of social science an important ideological aspect: Working people should perhaps be taught a catechism of political economy as well as of Christianity. Such anxieties animated the drive for valid social knowledge among gentlemen and professionals, who regarded the working classes less as credible makers of social science than as its proper objects.

Social science, then, developed during the middle third of the nineteenth century above all as a liberal, reformist answer to the upheavals of the era. It was less autonomous vis-à-vis government and urban life than it would become in twentieth-century universities. Some influential works of political economy were, as Margaret Schabas argues, relatively detached and analytic, and small but increasing numbers of political economists in Britain and elsewhere were hired to teach in universities. "National economy" and statistics were overlapping specialties in German universities, linked to but rarely subsumed under the sciences of state, or *Staatswissenschaften*. In Britain, statistics – and later political economy – was allocated a section at the British Association for the Advancement of Science within two years of its founding in 1831. But its meetings were too political and contentious for the natural scientists, and "statistical and economic science" was always regarded as peripheral to the mission of the larger body. Whether or not at universities, in whatever country, those who claimed the mantle of political economy, social science, statistics, or sciences of state and politics were almost invariably engaged in the practical work of reform, admin- istration, and political action. Social science was not itself a calling, but a charitable activity or a manner of exercising some other profession or office.

Statistics was in many ways the characteristic social science of the mid nine- teenth century. Its theoretical ambitions were less grand than those of political economy or of Comte's "sociology," but this actually placed it in closer ac- cord with the prevailing view of science, especially in Britain. Statistics was resolutely empirical. Between about 1830 and 1850, it came to be defined in terms of its use of numbers, as the quantitative science of society. Only occasionally, as in the program of the Belgian astronomer and statistician Adolphe Quetelet (1796–1874), was it linked to mathematical probability. The decisive preference of statisticians for empirical data over theoretical or mathematical formulations was ideally calculated for bureaucratic users and for a politically engaged middle-class audience.

Much statistical number gathering was performed officially. Several European nations introduced rudimentary censuses around 1800, following Sweden (1749) and the United States (1790). During the 1830s, many of the leading nations of Europe (but not the United States) created permanent census offices. Statistical bureaus concerned with trade, industry, health, military recruitment, and crime were set up at about the same time. These efforts are very much a part of the history of social science, not only because they provided indispensable sources of data, but also because their leaders often took an active role in interpreting the figures – which often meant propagandizing for public education, for example, or for improved sanitation.

There also were private statistical organizations, including a flurry of municipal statistical societies in Britain during the 1830s. Most failed within a few years, but those in London and Manchester survived and eventually prospered, as did an American society, founded in Boston in 1839. Their entry on the world stage, as perhaps the first enduring associations devoted to social science, was somewhat hesitant. In London, the statists were so worried about becoming politicized that they passed a notorious self-denying ordinance: Statistics must be a science of facts; "its first and most essential rule" is "the exclusion of opinion."[23] This was not the brave claim of incipient technocracy, but a gesture of humility before an unreachable ideal of objectivity. Neither their sense of science, nor the demands of the official positions held by some, kept the "statists" from issuing vigorous appeals for certain reforms. The Manchester society was involved above all with municipal improvement in that first industrial city, where it acted with considerable assurance.

Both of the British societies were relatively secure in relation to their objects of study. They or their representatives traveled from door to door, inquiring into the lives of working people, paupers, criminals, and immigrants. As Eileen Yeo shows, the surveys were designed to observe such people from above, to produce records of their behavior, and to find ways to make them behave more responsibly. This was perhaps the most vital mission of social science for the rest of the nineteenth century and beyond, not only in Britain but also in much of Europe and North America.

The institutional forms, to be sure, varied enormously, with volunteerism most prominent in Britain and the United States, and professorial activism especially strong in Germany. Empirical study – featuring, but not limited to, statistics – was central also to the (British) National Association for the Promotion of Social Science (NAPSS), which attracted an august membership including titled nobles and government ministers; it was active for about three decades beginning in 1857. An interlocking cluster of French institutions, most of them associated with high officials of illustrious *corps* such as Mines and Ponts et Chaussées (the state civil engineering corps), flourished during the later nineteenth century. The Statistical Society of Paris (founded in 1860) was among the first. The leadership of these organizations was probably no less influential, and indeed more effective, than that of the NAPSS.

---

[23] Michael Cullen, *The Statistical Movement in Early Victorian Britain: The Foundations of Empirical Social Research* (New York: Barnes and Noble, 1975); Theodore M. Porter, *The Rise of Statistical Thinking, 1820–1900* (Princeton, N.J.: Princeton University Press, 1986).

The Ponts engineer and social reformer Emile Cheysson, a follower of Le Play, had his hand in almost all of them. An American version of the NAPSS, the American Social Science Association, was established in 1865, but it never really succeeded. About two decades later, social science became a very important affair in the United States, with university professors and municipal reform organizations assuming leading roles.[24]

Colonial social science, too, became an important endeavor during the nineteenth century. Colonial administrators developed, through publications as well as letters, a kind of anthropology, which they were able sometimes to put into practice.[25] In India, for example, the British state tried out interventions that would never have been tolerated at home; and colonial populations, like working-class ones, were in a poor position to resist the inquiries of statisticians. Among the notable, if ironical, achievements of colonial social science was the reduction of the variegated Indian caste system to a uniform, official set of categories.[26]

Intellectually, statistics was most closely affiliated with political economy. In a broad sense, the statisticians of early-nineteenth-century Britain and France took for granted the legitimacy of markets and of free enterprise in their efforts to understand and remedy the ills of an industrializing society. Yet abstract economics, in the tradition defined by Jean-Baptiste Say and, above all, by Ricardo, had little use for empirical numbers. By no means did political economists discard the moral and political dimensions of their subject. But political economy came to be associated, in France as well as in England, with liberalizing moves to discard state restrictions on production, labor, and trade and to discredit institutions, such as poor relief, designed to soften the effects of social inequality. Malthus argued that liberal poor relief was counterproductive, that it increased the level of misery by encouraging marriage and reproduction where the means of support were lacking. Some Evangelicals, while doubting this, yet preferred the severity of laissez-faire as a form of divine penance.[27] In any case, critics such as Thomas Carlyle denounced

---

[24] Lawrence Goldman, "A Peculiarity of the English: The Social Science Association and the Absence of Sociology in Nineteenth-Century Britain," *Past and Present*, 114 (1987), 133–71; Sanford Elwitt, *The Third Republic Defended: Bourgeois Reform in France, 1880–1914* (Baton Rouge: Louisiana State University Press, 1986); Thomas Haskell, *The Birth of Professional Social Science: The American Social Science Association and the Nineteenth-Century Crisis of Authority* (Urbana: University of Illinois Press, 1977); Dorothy Ross, *The Origins of American Social Science* (Cambridge: Cambridge University Press, 1991).

[25] David Hoyt, "The Surfacing of the Primitive: Social Welfare, Colonial Management, and Ethnographic Discourse, 1870–1914" (PhD dissertation, University of California at Los Angeles, 1999).

[26] See Partha Chatterjee's chapter in this volume; see also Bernard S. Cohn, "The Census, Social Structure, and Objectification in South Asia," in his *An Anthropologist among the Historians and Other Essays* (Delhi: Oxford University Press, 1987), pp. 224–54.

[27] Donald Winch, *Riches and Poverty: An Intellectual History of Political Economy in Britain, 1750–1834* (Cambridge: Cambridge University Press, 1996); Boyd Hilton, *The Age of Atonement: The Influence of Evangelicalism on Social and Economic Thought, 1785–1865* (Oxford: Clarendon Press, 1988).

classical political economy as a "dismal science." Its severity, exemplified by the new poor law of 1834, was often attributed to its strict reliance on deduction, unmoderated by attention to empirical facts and human lives.

Beginning in the 1870s, when the theory of political economy began to be rewritten in mathematical form, authors such as the statistician and economist William Stanley Jevons (1835–1882) would argue that economics should be mathematical, because its data were quantitative. But Jevons's mathematical economics provided no more opportunity for the insertion of data than had Ricardo's. Some critics of capitalism in the 1830s and 1840s appreciated the Ricardian style for its uncompromising logic, which seemed to reveal the fundamental injustice of the capitalist system. Karl Marx was, in this sense, also a Ricardian, although he, like the popular economists who wrote for a working-class audience, deployed Ricardian assumptions in order to display the immorality of capitalism and to vindicate a radically different economic system.[28] Middle-class social reformers, by contrast, were critical of classical political economy on methodological grounds. They were joined by some natural scientists, for whom dedication to an ethic of empirical and experimental precision counted for more than logical or mathematical rigor. Often, scientific objections supported political ones. Among the founding members of the London Statistical Society was William Whewell, an influential naturalist and philosopher of science, who energetically backed the efforts of his friend Richard Jones to establish political economy on an empirical basis. Ambivalence, or worse, about classical political economy was one of the reasons for founding a statistical society in the first place.[29]

Economy and statistics came together in Germany, especially after its unification in 1870. This alliance presupposed a radical rejection of Ricardian classical theory, what the Germans called *Manchestertum*, in favor of a more historical approach that supported a statistical economics. The common locus of this activity was the Verein für Sozialpolitik or Social Policy Association, led by Gustav Schmoller. Some of its members held positions in official statistical agencies, but the most prominent among them were professors. For them, academic positions were perfectly compatible with passionate advocacy, at least until the early twentieth century, when Max Weber criticized their reformism in the name of objectivity. Their nineteenth-century opponents called them *Kathedersozialisten*, professorial socialists. In some ways they perpetuated the German Enlightenment ideal, discussed in Keith Tribe's chapter, of economic understanding as a practical tool of state. But these professors

---

[28] Noel W. Thompson, *The People's Science: The Popular Political Economy of Exploitation and Crisis* (Cambridge: Cambridge University Press, 1984).

[29] Theodore Porter, "Rigor and Practicality: Rival Ideals of Quantification in Nineteenth-Century Economics," in *Natural Images in Economic Thought: Markets Read in Tooth and Claw*, ed. Philip Mirowski (Cambridge: Cambridge University Press, 1994), pp. 128–70; Lawrence Goldman, "The Origins of British 'Social Science': Political Economy, Natural Science and Statistics, 1830–1835," *Historical Journal*, 26 (1983), 587–616.

had more independence than their eighteenth-century predecessors, and their writing was more frankly political. The great problem of their time was the social dislocation brought about by German industrialism, which advanced at an unprecedented pace in the later nineteenth century. With it came mass working-class parties, made possible by Bismarck's moves toward universal male suffrage, but then officially prohibited for more than a decade.

German social democracy became, as Terrell Carver's chapter shows, not only a Marxist party, but also the nucleus of Marxism as a social science. Marx himself had developed a range of approaches to the social changes of his time – philosophical, historical, and economic. After the failed revolutions of 1848, he turned increasingly to economics, and by the time the first volume of *Das Kapital* appeared, in 1867, he had developed several related arguments for the inevitable collapse of capitalism from within. At the same time, he devoted immense labors to the analysis of official British statistical reports, whose integrity he took almost for granted, and he was actively involved in the organization of an international labor movement. Intellectuals allied to labor parties, especially in Germany, developed Marxism into a social science tradition in its own right. While most historical economists opposed the radical solutions proffered by the representatives of social democracy, they were not unaffected by its critiques. They were also impressed by the socialist threat to political order, which they hoped to disarm through sensible measures of social amelioration. Thus they favored state activity on behalf of farmers and workers, including social insurance and the right to organize. They were, in some ways, the intellectual founders of the welfare state or "social state." Against what seemed the mirrored ideological extremisms of classical economics and radical socialism, they argued for statistical and historical study, for letting experience decide which interventions were effective.

The historical economists developed a language and a set of concepts intended to contrast their form of science (*Wissenschaft*) with natural science (*Naturwissenschaft*). This latter stood for determinism – an absence of personal or political agency – and for timeless uniformity rather than historical change. They relentlessly denounced as misguided Quetelet's ambition to turn statistics into a "social physics," with its inflexible "statistical laws."[30] They thus understood social science to be in opposition to natural science, following a tradition that, as Johan Heilbron discusses, goes back to the eighteenth century. It was, however, mechanics that they rejected, not biology. The sciences of life seemed less strictly deterministic, more compatible with expert guidance and reform, than mathematical physics. The identification of scientific models for social science was, here and in general, a political and ideological as well as an intellectual decision.

---

[30] Porter, *Rise of Statistical Thinking*, pp. 162–92, 240–55.

NATURALISM AND ANTI-NATURALISM
IN SOCIAL SCIENCE

"Society" acquired a meaning in the nineteenth century that it had lacked in the eighteenth. Social thinkers came to understand society as a dynamic, progressive, possibly unstable entity that was, in a way, more fundamental than the state. Quetelet's German critics argued that he had made a fundamental mistake by seeking to understand society in terms of the characteristics of the "average man," as if the properties of individuals translated directly into the characteristics of a nation. They rejected also his "mechanical" conception of the natural laws of society, which left no room for effective social reform. These objections were, in a way, misguided, since Quetelet unmistakably saw his science as the instrument of rational improvement. But they were provoked by his mechanical analogies. To confuse society with physics became, in the course of the nineteenth century, increasingly unacceptable, especially in Germany.

By contrast, the ascent of neoclassical economics, especially in Britain and America, signified a rejection of historicism and an endorsement of mechanics as a scientific model. Because of the simplicity of its basic doctrines, classical political economy had always appeared amenable to a mathematical formulation. When, in the early 1870s, Jevons in Manchester and Léon Walras in Lausanne achieved this through their theories of marginal utility, they relied on physical understandings and analogies. Walras, in particular, in deriving the mathematics of general equilibrium in economics, followed very closely the mechanics he had learned decades earlier as a student.[31]

Psychology emerged from a more complex field of natural and philosophical models. In the eighteenth century, David Hume and David Hartley both discussed the mind in Newtonian terms. In the nineteenth century, the "psychophysics" announced by Gustav Theodor Fechner (1801–1887) formed the basis of a new laboratory discipline. His aims were metaphysical – to break down the dualism of mind and matter by establishing the laws linking human sensations to physical stimuli. The mechanical world was to be reconfigured not as something external, but as an element of human experience, inextricably joined to mind. As an experimental program, psychophysics came to be supported by a new kind of laboratory, developed primarily by Wilhelm Wundt (1832–1920). His was full of electrical and physiological instruments used to test reaction times and the duration of elemental thought processes, as well as the ability of his subjects to distinguish different weights, colors, and degrees of brightness. Wundt has often been called the founder of scientific psychology, and he defined the field for a generation of American students. While many were won over to an

---

[31] Philip Mirowski, *More Heat than Light: Economics as Social Physics, Physics as Nature's Economics* (Cambridge: Cambridge University Press, 1989).

ethic of measurement, few, if any, looked to physics to provide a basis for psychological theories.[32]

Biology, not physics, was the crucial point of reference for the nascent social sciences in the nineteenth century. Biology, to be sure, was not a single thing, but many. It also remained somewhat inchoate during this period – as with social science, the biological disciplines were institutionalized mainly in late-nineteenth-century American universities. Moreover, the traffic in methods and analogies went both ways. In no way can the shape of social science be explained as a mere consequence of biological advances. It was rather a case of mutual adaptation and differentiation, occurring in diverse ways and at various levels.

Anthropology, as one of the premier "sciences of man" in the decades around 1800, especially in France, readily combined the biological and the moral. Studies of bodies, minds, and customs were thought to lead to complementary conclusions. Medicine was the core of the project, and in the early nineteenth century doctors advanced a variety of medical programs for a science of society. One, drawing from an ancient trope that likened the political order to the human body, involved a social physiology. Another, allied with the public health efforts of the 1830s and 1840s, pursued what the French called "public hygiene." A successful journal was published under this name, the *Annales d'hygiène publique*, whose avowed mission was not only to improve public health, narrowly conceived, but also to combat such social maladies as crime and scarcity. This effort, unlike the physiological one, was linked to the contemporaneous statistical movement.

Auguste Comte, who was perhaps uniquely well informed about the sciences of his day, warmly applauded the new physiology of the early nineteenth century, and especially the work of Xavier Bichat (1771–1802). Johan Heilbron's chapter explores this connection. Bichat and his contemporary Georges Cabanis claimed to demonstrate experimentally that the vital capacities of the living body must elude the calculations of mechanical science. They argued aggressively that physiology was concerned with a distinctive class of phenomena, and that it should be autonomous from the physical sciences. At the same time, Bichat presented this medical understanding as the proper basis for a science of society, a social physiology. The argument was not simply reductive; his intention was to demonstrate that society, too, stood above the mathematical sciences of dead nature. Comte took the new physiology as a model for his own project. The science for which he invented the name "sociology" should also be independent of those less difficult sciences that had gone before it. He argued for a hierarchy of autonomous sciences, ordered in time in a progression from lesser to greater complexity: mathematics, astronomy, physics, chemistry, physiology, sociology. According to Comte,

---

[32] Smith, *Fontana History*, chap. 14.

Bichat's mistake was his failure to extend to sociology the same consideration that he had demanded for physiology.

Comte was, as Stephen Turner's essay shows, an important and influential philosopher of science, and especially of social science. His understanding of society was a vitalistic one, at least in the sense of being antimechanical, and thus preserved an analogy to the biological understanding of the organism. One expression of this outlook was Comte's skepticism about the value of mathematics for social science, and his consequent criticism of Condorcet. Mathematics, he argued, was appropriate for less complex sciences, such as physics, but was unsuited to physiology and almost useless in sociology. He opposed also the empirical use of numbers in statistics: Numbers might conceivably be useful after sociological theory had clarified its fundamental concepts, but could never provide the means for sociology to become scientific in the first place. What his contemporary and rival Quetelet called "social physics" he dismissed as "mere statistics." For all that, he could not block the assimilation of statistics into the science he claimed to have discovered, sociology.[33] Most notably, in the 1890s Émile Durkheim used statistics as the basis for his sociological study of suicide. Durkheim preserved, however, the vitalistic temper of Comte's view of society, and he insisted, as would Comte, on a proper sociological classification before the numbers could be interpreted. Durkheim also deployed Comte's fundamental distinction, borrowed from the medicine of the early nineteenth century, between the normal and the pathological, as a basis for assessing the health of a whole society.[34]

Comte's dim view of psychology is analyzed in Jan Goldstein's chapter. His preference for phrenology, that is, for a physiological approach to the mind, appears in a certain sense reductionistic, but behind it lay his devout anti-individualism, reflecting his understanding that society is like an organism. He also disliked political economy, for several reasons. Among them was his belief, which might be called holistic, that social science should not be broken into parts. Mill, his admirer in many respects, disagreed on this point, endorsing political economy as a special science concerned with just one important aspect of human behavior, the pursuit of personal gain. Comte inspired a historical school of political economy in Britain that was more or less contemporaneous with the German one. Its members demanded that the economic domain be understood as part of a larger science of society, not reduced to abstract propositions about production, consumption, and trade.

Theories of evolution defined another important field of interaction between biology and social science, one with rather different political resonances. Throughout the nineteenth century, from Jean-Baptiste Lamarck to Ernst Haeckel and beyond, theories of biological evolution were less

---

[33] Ian Hacking, *The Taming of Chance* (Cambridge: Cambridge University Press, 1990).
[34] Georges Canguilhem, *On the Normal and the Pathological* (1943), trans. C. R. Fawcett (Dordrecht: Kluwer, 1978).

mechanical than purposeful, involving a teleological progression of species toward greater perfection. Among social evolutionists of the late nineteenth century, this understanding remained more influential than Charles Darwin's theory of natural selection, even if *Origin of Species* (1859) had made evolution scientifically respectable. The paradigmatic social evolutionist of the late nineteenth century was Herbert Spencer, who regarded biological and social progress as parallel instances of a more general law, a tendency for homogeneous matter to become increasingly complex and differentiated. Darwin himself put particular emphasis on his discovery of a mechanism of natural selection, one that required neither design nor purpose in nature. A classic body of historical scholarship links the discovery of natural selection to the harshness of capitalistic society and of Malthusian social doctrine in England during the period when Darwin came to maturity and formulated his theory, the late 1820s and 1830s.[35] Darwin certainly did learn something crucial from Malthus's theory of population. But the doctrine of natural selection had only a modest role in nineteenth-century social theories, and indeed was not widely supported even in biology.

The larger significance of biological evolution for the social and human sciences involved, rather, the credibility that it gave to biological interpretations of human culture. It was not the driving force in this story, but it did provide a framework that many found satisfying for interpreting the diversity of human peoples. Among the most crucial doctrines with which it was linked was that of race. As Elazar Barkan argues in his chapter, the language of race goes back to the Enlightenment, but it was then a comparatively soft concept, not sharply distinguished from the effects of experience and culture. A variety of factors in the early nineteenth century conspired to sharpen racial doctrines, not least the effort to defend slavery in the American South against increasingly forceful moral objections. In opposition to the Bible, which taught the common descent of all humans (monogenism), there grew up polygenic theories of human origins. The statistical impulse to weigh and measure was mobilized in anthropometric studies, often of human skulls, in order to support the doctrine of racial distinctiveness – or sometimes, as with Quetelet, in order to challenge it. In the 1830s Darwin, by family background a monogenist, took an interest in the races of man as an example of biological differentiation within a species over time. Later, he believed his theory of evolution to have settled the issue in favor of common descent, without however excluding the possibility of significant biological differences among the races.

One of the principal sources of ethnography, as Harry Liebersohn's chapter shows, was travel descriptions of distant peoples. Its scientific claims were enhanced by the establishment of "ethnological" societies in France,

---

[35] Robert Young, *Darwin's Metaphor: Nature's Place in Victorian Culture* (Cambridge: Cambridge University Press, 1985).

the United States, and England, all within a few years of 1840. There was also a strong tradition of ethnographic writing in Germany. Ethnologists developed a hierarchical division of labor, which endured until about 1900, in which mere reports from observers in the field were regarded as something like raw material. Elite ethnologists were not prohibited from traveling, but they earned their standing through the more bookish activity of gathering up publications and correspondence and synthesizing them into accounts of the modes of living, legends, religious beliefs, and marriage patterns of primitive peoples. This contrasted with the premier sciences of nature, notably geology, for which mere observing and collecting were also of low status, but which required their elite members to go into the field regularly to work out the stratigraphy of a significant site. The spirit of ethnographic writing was generally condescending but sympathetic. As with the isolated rural populations of Europe, whose stories were collected by folklorists in order to preserve some traces of this vanishing way of life, ethnography was associated with an effort to preserve native peoples, or at least their traces, at a time when European exploration, trade, conquest, and settlement threatened them with biological or cultural extinction.

Early ethnologists were interested also in the physical characteristics of the peoples they studied. But the move to assign primacy to the biological came later. It was expressed most sharply in the almost simultaneous formation of societies called "anthropological" in Paris and London around 1860. In these years the name "anthropology" came to signify a specifically biological approach to "man." The correspondence in time of the anthropological societies with *Origin of Species* was mostly coincidental, for most "anthropologists" opposed Darwin, and some of his prominent supporters, including T. H. Huxley, were active in the British ethnological society. A broadly cultural ethnology and this stridently racialist anthropology remained in competition for about two decades. When they came together in Britain, it was largely on the terms of the ethnologists, though under the name of anthropology.[36] Still, the late nineteenth century marked the rise of strict doctrines of racial separateness and hierarchy. This racism is an important part of what "social Darwinism" has come to mean; and while there was little specifically Darwinian about it, it was expressed biologically, and sometimes in the language of evolutionary progress through competitive struggle.

Methodological writings such as Durkheim's on sociology have encouraged modern readers to suppose that the beginning of professional social science meant the creation of an autonomous social domain, or indeed of separate domains of society, economy, culture, and mind. Indeed, an organized revolt against biological reductionism was important in some cases. But biology had immense prestige and influence in the social sciences of

---

[36] George W. Stocking, Jr., *Victorian Anthropology* (New York: Free Press, 1987); Laurent Mucchielli, *La Découverte du social: naissance de la sociologie en France* (Paris: Editions la Découverte, 1998).

the late nineteenth century. Its significance was not merely a matter of the "influence" of evolutionary biology or physiology. More important, it manifested itself in the form of hybrids of biological and social theories and practices, such as Herbert Spencer's evolutionary sociology, Francis Galton's eugenic campaign to improve mankind by selective breeding, the racialism against which Franz Boas fought for anthropology, and the Lamarckian elements of Sigmund Freud's psychoanalysis. The new theories were caught up in policy battles as well. In the late nineteenth century, biology was most often deployed in support of conservative and elitist understandings, rather than on the side of social mutability and reform.

DISCIPLINED INTERVENTIONS: PROFESSIONALS AND REFORMERS

This introduction, like the chapters in Part I, emphasizes the interrelations among the various social science traditions, the importance of their links with the sciences of nature, and above all their practical role in reform, administration, and ideology. A longer essay might have given more attention to social science in relation to law, religion, and philosophy. Part of the purpose is to subvert the anachronistic practice of writing this history as if societies, minds, cultures, and economies had always been studied in distinct traditions of thought and practice that developed into the familiar fields of the present day. Dorothy Ross's introduction to Part II shows that the formation of modern disciplines was gradual, and sometimes discontinuous, even in the twentieth century. But the themes presented here cannot be summed up as a battle against the disciplines, as if social knowledge were, through the late nineteenth century, loose and unstructured. I do call for an historicist approach, one that recognizes the changing structures, boundaries, aims, and practices of social knowledge. But this historicism applies equally to our study of more contemporary knowledge, which, however disciplined, should also be understood as part of a larger cultural, intellectual, political, and administrative history. The chapters in Part I, and throughout the book, examine social science through a broad lens, in order to relate inside and outside, knowledge and society, and in the end to blur the boundaries between them. From this standpoint, there is a kind of unity to social science even when its disciplinary divisions seem almost impermeable: The social sciences, collectively, participate in something much broader.

Social science disciplines were not invented in fin-de-siècle America. One may doubt whether any field was ever so worried about its independence as was the German science of statistics. Even before C. G. A. Knies published, in 1850, his programmatic volume on "statistics as an autonomous science [*selbständige Wissenschaft*]," and still more afterward, a flood of publications asked how this science could be defined and practiced so as to deserve a separate existence. To be sure, all of this reflective writing owed much to the incompleteness and instability of this disciplinary formation, and to pervasive disagreement about its definition. Yet statistics had to be a specialized science, with its own object and methods; the structure of German university chairs in the nineteenth century almost required this.[37] The formation of a French school of geography under the Third Republic, and of British political economy after 1870, also serve to reveal that what Dorothy Ross calls here the "disciplinary project" was not unique to America, and was not invented during the 1880s. Yet each of these sciences – however much their practitioners aspired to create distinct fields of teaching and research – was designed to play an important role in the life of the nation. Looking inward was not inconsistent with looking outward. Disciplines helped to give credibility to social knowledge, and to nourish technical methods that could be crucially important for economics and politics.

There was, in the end, something distinctive, perhaps even epochal, about developments in the United States at the end of the nineteenth century. The unprecedented scale of the new American research university, and the relative weakness of its traditional elites, permitted social science to assume there a unique structure and role. Yet the significance of social science was also growing in Europe at this time, and for broadly similar reasons, even if the Europeans did not follow, and sometimes actively disapproved of, the American form of social science institutionalization. Effective sciences of society seemed indispensable to deal with the immense economic, social, and political changes of the "second industrial revolution" of the late nineteenth century, which were particularly decisive in Germany and the United States. Social science, both within and outside the universities, was very much involved with issues of migration, urban poverty, industrial labor, popular radicalism, and economic fluctuations. As Alain Desrosières argues, the welfare state evolved in conjunction with new kinds of data and new forms of social science. The connection between social science and Western modernity was perhaps recognized most acutely outside the West – in Japan and China, for example – but the point is a general one. Historians need to recognize the evolving methods and intellectual content of social science, and its changing institutional forms, not merely or mainly as a set of internal intellectual developments, but in relation to a much larger set of changes that have affected the entire world.

---

[37] Theodore Porter, "Lawless Society: Social Science and the Reinterpretation of Statistics in Germany," in, *The Probabilistic Revolution*, vol. 1: *Ideas in History*, ed. Lorenz Krüger, Lorraine Daston, and Michael Heidelberger (Cambridge, Mass.: MIT Press, 1987), pp. 351–75.

3

SOCIAL THOUGHT AND NATURAL SCIENCE

Johan Heilbron

Among the intellectual traditions that have helped to form modern social science, natural philosophy and natural science stand out. The emerging social sciences have also drawn in important ways from humanist philosophy, juridical scholarship, political tracts and treatises, Christian theology, travel accounts, and literary and moral essays. But the natural sciences have provided an enduring set of models for modern social science, models that go well beyond suggestive analogies and illustrative metaphors. Their formative influence was particularly salient during the period addressed here, from the Enlightenment to the last third of the nineteenth century.[1]

NATURALISM AND MORAL PHILOSOPHY

In the eighteenth century, the new natural philosophy came to be seen in Europe as the most reliable and authoritative system of knowledge. Inescapably, it was considered relevant to political thought and moral philosophy as well. In its most basic form, natural philosophy meant the search for natural principles and laws, in place of supernatural agencies. Applied to the domain of moral philosophy, the naturalistic outlook generally fulfilled a similar function: It allowed for a shift away from Christian doctrines toward secular models, yet offered reliable knowledge by which one could evade the relativistic consequences of the "skeptical crisis" of the sixteenth and seventeenth centuries.[2]

---

[1] For general overviews, see Christopher Fox, Roy Porter, and Robert Wokler, eds., *Inventing Human Science* (Berkeley: University of California Press, 1995); Roger Smith, *The Fontana History of the Human Sciences* (London: Fontana Press, 1997) (in the United States, *The Norton History of the Human Sciences*); Johan Heilbron, Lars Magnusson, and Björn Wittrock, eds., *The Rise of the Social Sciences and the Formation of Modernity* (Dordrecht: Kluwer, 1998).

[2] Richard Popkin, *The History of Scepticism From Erasmus to Descartes*, rev. ed. (Berkeley: University of California Press, 1979).

Among the traditions that grew out of this naturalistic quest for knowledge of human nature and human society was modern natural law, initiated by Hugo Grotius (1583–1645). It provided the predominant general framework for questions of state and society during the seventeenth and much of the eighteenth centuries. Natural law theorists like Thomas Hobbes (1588–1679) and Samuel Pufendorf (1632–1694) developed elaborate systems of moral duty and political obligation based upon what they took to be permanent features of human nature, such as the concern for self-preservation. Sometimes connected to natural jurisprudence were the various forms of state science that emerged in the process of early modern state building: political economy, political arithmetic, and the cameral sciences. Moral essays, concerned predominantly with private issues such as morality and manners rather than with government or legislation, represented yet another intellectual genre. Theories of human nature typically provided a conceptual foundation for elaborating moral and political norms. Until the Enlightenment, however, this rarely entailed any extensive study of social and political realities.

While references to natural philosophy were frequent, they were neither uniform nor uncontested. Invoking natural science often involved the use of mechanical metaphors and an image of the world as a well-ordered machine, but it did not exclude organic analogies. Some proponents of natural philosophy insisted on the primacy of observation and experience, but others preferred rational deduction. Measurement and quantification were indispensable to the scientific method for some, but were ignored by many others. So, even when these early modern discourses remained within a shared naturalistic framework, there was uniformity neither of method nor of content.

If the Enlightenment was a formative period for the social sciences, this was fundamentally because a secular intelligentsia now explicitly claimed, and effectively exercised, the right to analyze any subject matter, however controversial, independent of established authorities and official doctrines. The flourishing discourses on political, moral, and economic issues displayed their reliance on factual evidence and detail in a way that had been alien to natural law systems. One symptom of renewal was the introduction of new terms for what had previously been known as moral philosophy or natural jurisprudence. The expression "moral and political sciences" first appeared in France in the circle of the Physiocrats during the 1760s. "Social" and "society" gained currency during the same period, both in France and in Scotland. The expression "social science" was coined during the revolutionary period in the writings of Sieyès, Condorcet, and other members of the Société de 1789. It generally referred to a broadly conceived science of government and legislation. Only after three decades was the expression properly translated into English as "social science" (in place of "moral science"). Its introduction into German-speaking countries came a bit later still.[3]

NATURAL SCIENCE AND SOCIAL THOUGHT

The significance of natural science for the social sciences can be characterized in terms of three distinct trends. Each was marked by a particular intellectual strategy, drawing from a characteristic stance in regard to the natural sciences.

The first of these involved the attempt to construct a social science immediately derived from, or directly based on, the natural sciences. The aim was to apply natural science methods and its modes of conceptualization consistently to the domain of the social sciences. Some of these efforts were derived from mathematical and mechanical disciplines, others from the life sciences. The distinction between the two became pertinent during the last decades of the eighteenth century and generated a major intellectual dispute. John Stuart Mill noted in the 1860s that all speculations concerning government and society bear the impress of two conflicting theories. In the mechanical conception, human institutions are seen in the same light as a steam plough or a threshing machine. This mode of thinking is atomistic and by the analogy to mechanical contrivances is informed by schemes for rational design. The rationalist-mechanical conception was opposed to theories expressed in terms of organic growth. In the latter, institutions appear as spontaneous products of growth, and social science is seen as a branch of natural history rather than of social engineering.[4]

The second trend grew from the differentiation of natural science and its epistemological consequences. The rise of vitalist currents in the life sciences during the late Enlightenment had a critical impact on science as a whole by contributing to the demise of a unitary conception of natural philosophy. In its place arose a fundamental split between animate and inanimate bodies, and later a more differentiated view, which reflected the emerging structure of scientific disciplines. Once biology had been conceived as a general science of life, distinct from physics, the underlying argument could be transferred to the field of social science. Thus did Auguste Comte distinguish social science from biology, as biology had been separated from chemistry and physics. Social science, for Comte, was a relatively autonomous endeavor, with a subject matter of its own and a specific method of study. Disciplinary

[3] Keith Michael Baker, "Enlightenment and the Institution of Society: Notes for a Conceptual History," in *Main Trends in Cultural History*, ed. Willem Melching and Wyger Velema (Amsterdam: Rodopi, 1994), pp. 95–120; Brian W. Head, "The Origins of 'La science sociale' in France, 1770–1800," *Australian Journal of French Studies*, 19 (1982), 115–32; Johan Heilbron, *The Rise of Social Theory* (Cambridge: Polity Press, 1995).

[4] John Stuart Mill, *Considerations on Representative Government* (1861), in *Utilitarianism, Liberty, Representative Government* (London: Everyman's Library, 1964); Werner Stark, *The Fundamental Forms of Social Thought* (London: Routledge, 1962).

differentiation in the natural sciences thus provided the social sciences with a second option for scientization, one that rejected emulation of the established sciences in favor of a search for specific principles proper to each particular science. This strategy was explicitly antireductionist, though it remained within a naturalistic framework.

The third trend is represented by opposition to the prevailing forms of naturalism in the human sciences. The elaboration of these humanistic or cultural alternatives made natural science, with its insistence on mechanical laws and causal models, an object of criticism.

Although these three trends overlapped in time, they were, by and large, successive phases. The scientific model of moral and political discourse preceded the trend toward disciplinary differentiation, which in turn came before the elaboration of a full-blown countermodel of cultural science or *Geisteswissenschaft*. Newer trends, however, did not simply replace the older ones, but rather served to broaden the scope of epistemological possibilities.

The scientific conception of moral philosophy was strongest in England, Scotland, and France, although it was obviously not restricted to these countries. Its apogee was in France from about 1770 to 1830, when Paris was the scientific capital of Europe. The most scientistic designations for the social sciences were coined in French during these years: "social mathematics," "social mechanics," "social physics," and "social physiology." The second trend of differentiation was rooted specifically in vitalist currents in the life sciences, manifested in various countries but elaborated most systematically in France, where vitalism had a particular impact, both in biology and sociology. The culturalist countermovement sprang up in several nations but was particularly strong in Germany. Whereas English and French critics of natural science models were often literary figures outside of the academic system, German opponents of scientific naturalism developed an alternative within the walls of academia. Against what they saw as the antihistorical reductionism of natural science, they advanced an interpretative or hermeneutic methodology as the proper basis of a cultural science.

THE SCIENTIFIC MODEL OF MORAL
AND POLITICAL THEORY

Of these three trends, scientization was the oldest and indeed the primary one.[5] Early examples go back at least to the beginnings of the Scientific

[5] Theodore M. Porter, "Natural Science and Social Theory," in *Companion to the History of Modern Science*, ed. Robert C. Olby, Geoffrey N. Cantor, John R. R. Christie, and M. J. S. Hodge (London: Routledge, 1990), pp. 1024–43; Richard Olson, *The Emergence of the Social Sciences, 1642–1792* (New York: Twayne, 1993); I. Bernard Cohen ed., *The Natural and the Social Sciences* (Dordrecht: Kluwer, 1994); Sabine Maasen, Everett Mendelsohn, and Peter Weingart, eds., *Biology as Society, Society as Biology: Metaphors* (Dordrecht: Kluwer, 1995).

Revolution. Grotius admired Galileo and tried to follow a mathematical ideal of demonstration in his system of natural law. Thomas Hobbes applied a geometrical style of reasoning to a mechanical definition of interacting individuals, all moved by the same concern for their own interest. Newtonianism gave a new impetus to the drive for a natural science of the moral world. Newton was a recurrent reference in eighteenth-century moral and political discourse, the renewal of which was led by the Scottish moral philosophers and French *philosophes*. For the Scots, moral philosophy was to be transformed into an uncompromising empirical science. That, in any case, was David Hume's (1711–1776) message when he presented his *Treatise on Human Nature* (1739–40) as an "attempt to introduce the experimental method of reasoning into moral subjects." The ambition was not entirely novel, and Hume was not the only candidate to be the Newton of the moral sciences, but he played an exemplary role for many of his compatriots. In a country depoliticized by the union with England of 1707, there was great appeal in approaches that transcended the boundaries of classical political theory.

The Scottish philosophers analyzed politics and legislation as fundamentally dependent on economic structures and corresponding forms of morality and manners. They viewed the interconnections within a historical model of four stages, progressing from hunting to shepherding to agriculture and then to commerce. This developmental pattern, from rudeness to refinement, forms the common background for Adam Smith's theory of commercial society, Adam Ferguson's *An Essay on the History of Civil Society* (1767), and John Millar's *The Origin of the Distinction of Ranks* (1771).[6]

For Hume and Adam Smith (1723–1790), such a historical scheme of the development of civil society was the very consequence of their scientific stance. Both rejected arguments from an assumed "state of nature" that implied contractual agreements as the basis of human institutions. Hume saw no ground for belief in the existence of a state of nature prior to society. As a merely hypothetical construct, it was incompatible with the precepts of experimental science. Contracts and other legal rules, in his view, must be conventional rather than natural.

If the science of man is to be truly experimental, Hume argued, we cannot go beyond experience. "We must therefore glean up our experiments in this science from a cautious observation of human life, and take them as they appear in the common course of the world." Where experiments of this kind are "judiciously collected and compared, we may hope to establish on them a science, which will not be inferior in certainty, and will be much superior in utility to any other of human comprehension."[7] For Hume, human history

[6] Gladys Bryson, *Man and Society* (Princeton, N.J.: Princeton University Press, 1945); Ronald Meek, *Social Science and the Ignoble Savage* (Cambridge: Cambridge University Press, 1976); Istvan Hont and Michael Ignatieff, eds., *Wealth and Virtue* (Cambridge: Cambridge University Press, 1983).
[7] David Hume, "Introduction," in *A Treatise of Human Nature*, ed. Lewis A. Selby-Bigge (Oxford: Clarendon Press, 1978).

was to moral philosophy what experiments were to natural philosophy. The argument led away from speculations about the state of nature and natural principles toward a historical science of human society. One of its central concerns was the possibility of progress and the explanation of the relative advance or stagnation of nations. This was a central question for Adam Smith in his *Wealth of Nations* (1776), and it stimulated Adam Ferguson (1723–1816) and others to produce histories of civil society, understood as natural histories of man in his social state. Their work helped to develop a new understanding of history as a cumulative, progressive movement through time.

From the Scottish point of view, many *philosophes* fell short of the proper standards of social philosophy. The main exception was Charles de Secondat, baron de Montesquieu (1689–1755). Montesquieu's pioneering *De l'esprit des lois* (1748) was widely admired for having demonstrated, as Hume put it, that "the laws have, or ought to have, a constant reference to the constitution of governments, the climate, the religion, the commerce, the situation of each society."[8] In place of deductions from an original principle, he had carefully uncovered the connections between government and the "general spirit" of the nation, a spirit that was shown to have a variety of causes, both physical and moral. Because Montesquieu's investigations were so thorough, John Millar (1735–1801) called him the Lord Bacon of the moral sciences – considering that, after all, Adam Smith was its Newton. Montesquieu, Hume, Smith, and a number of others had broken away from central features of the natural law tradition in favor of what they defended as a more empirical and scientific approach. By focusing on the interdependencies of climate, commerce, morality, and government, Enlightenment theorists challenged the conventional centrality of politics and religion. The notion of "society" and the adjective "social" came into use precisely to designate the broadening scope of moral and political discourse.

PHYSICAL AND PHYSIOLOGICAL MODELS

Other Enlightenment efforts drew on the natural sciences in a more specific way, conceptualizing the social world in a language derived from either the physical or the life sciences. These strategies became particularly salient in France during the latter decades of the Old Regime, and continued to prevail during the revolutionary period and its immediate aftermath. A crucial impetus had come from the reform policies initiated by Anne-Robert-Jacques Turgot (1727–81) when he served as minister from 1774 to 1775. The *philosophe* and mathematician M. J. A. Nicolas de Caritat, marquis

---

[8] David Hume, *Enquiries Concerning Human Understanding and Concerning the Principles of Morals*, ed. Lewis A. Selby-Bigge (Oxford: Clarendon Press, 1902), p. 197.

de Condorcet (1743–1794) was his chief scientific advisor, and many natural philosophers became involved in administrative reform and plans for modernizing the French state. Condorcet stressed the urgency of adapting scientific methods to the analysis of state matters. The moral sciences, he announced, must "follow the same method" as the natural sciences; they "ought to acquire a language as exact and precise, and should reach the same level of certitude."[9]

For Condorcet, the probability calculus provided the means to achieve this end, and he pioneered the use of mathematical techniques to analyze voting procedures and judicial decisions. In one of his last programmatic essays on the application of calculation to the moral and political sciences, in 1793, he called for a new branch of science, "social mathematics." Pierre-Simon Laplace (1749–1827), Condorcet's close colleague and rival in the Academy of Science, resumed the work in his classic *Traité analytique des probabilités* (1812). Some members of the Laplace school continued the project, but their way of working soon fell into disrepute. There was, however, one direct heir to the project of a social mathematics who was quite successful: Adophe Quetelet (1796–1874).

The Belgian astronomer and statistical entrepreneur met the Laplacians during his Parisian stay in 1823. Back in Brussels, Quetelet set up an observatory similar to the one he had studied in Paris, and increasingly turned his attention to statistics, drawing from the proliferating numbers collected by state bureaus. No longer restricted to revenue and population, as they were in the tradition of political arithmetic, the numbers came to include moral and social matters as well. What has been called the statistical enthusiasm of roughly 1820 to 1850 generated a new faith in the regularities of these numbers. Beneath the apparent diversity of specific events and individual acts, it seemed, were to be found patterns with astounding stability.[10]

Quetelet waxed eloquent on these points. From undeliverable letters in the Paris post office to the most impulsive and unruly acts of individuals, everywhere he found astounding regularities at the level of aggregate rates. The statistics of homicide and suicide were paradigmatic, and their lawlike collective behavior suggested that they were subject to immutable laws. For Quetelet, statistics allowed a science of society in the form of a genuine "social mechanics" or "social physics," based on the stability of averages. Variation, being trivial, was arrayed according to the astronomer's law of errors. By reducing the science of man to the science of the average man, *l'homme moyen*, he found statistical laws to compare with those of celestial mechanics.

[9] Condorcet, "Discours de réception" (1782), in *Oeuvres de Condorcet* (Paris: Firmin Didot Frères, 1847–9), vol. 1; Keith M. Baker, *Condorcet: From Natural Philosophy to Social Mathematics* (Chicago: University of Chicago Press, 1975); Éric Brian, *La mesure de l'État* (Paris: Albin Michel, 1994).

[10] Theodore M. Porter, *The Rise of Statistical Thinking, 1820–1900* (Princeton, N.J.: Princeton University Press, 1986), and *Trust in Numbers* (Princeton, N.J.: Princeton University Press, 1995); Alain Desrosières, *La politique des grands nombres* (Paris: La Découverte, 1993).

When the historian Henry Thomas Buckle (1821–1862), in his widely read *History of Civilization in England* (1857), invoked "the undeviating regularity" of the moral world, he had Quetelet's statistical determinism in mind.

Independent of the social mathematicians, utilitarian philosophers reasoned in a style that was equally modeled on the physical sciences. They started from a simple, unambiguous principle, a kind of axiomatic truth, from which they deduced both theoretical and political consequences. For Claude-Adrien Helvétius (1715–1771), whose *De l'esprit* (1758) was a critical response to the abundant complexities of Montesquieu's work, self-interest was the characteristic of all human conduct and the proper equivalent of gravity in the moral world. Jeremy Bentham (1748–1832) and James Mill (1773–1836) were similarly convinced of the need for such a plain starting point to ground social and political theories. Mill regarded complexity in matters of government as an "infallible sign" of imperfection.

The founding principle of the Utilitarians was interest or utility: Human beings seek pleasure and avoid pain, and human conduct is therefore universally guided by ideas and feelings associated with these stimuli. Bentham announced in his *Introduction to the Principles of Morals and Legislation* (1789) that nature had placed mankind under "the governance of two sovereign masters: pain and pleasure." The proposition was not merely descriptive. As Elie Halévy put it, the morality of the Utilitarians was little more than their economic psychology put into the imperative. Their behavioral model, furthermore, was equally valid for individuals and for the polity as a whole. All should promote the increase of happiness and the reduction of pain. Moral arithmetic, based on the principle of the "greatest amount of happiness of the greatest number," thus provided the means for assessing public institutions.

Various writers of the eighteenth century had made suggestions as to how this mode of thinking might be developed. In the early nineteenth century, Bentham and Mill made Utilitarianism into an intellectual movement for reform, a "philosophic radicalism." They proposed various reform projects, such as Bentham's notorious model prison, the Panopticon. As proponents of the calculus of pleasures and pains, the Utilitarians were critics of church establishments and traditional authority, generally opposing the subjection of the many to the few.[11] While their work cut across various fields, including ethics, associationist psychology, law, and philosophy, their preference for deductive reasoning and physical analogies survived primarily in political economy – the field described by William Stanley Jevons as "the mechanics of utility and self-interest."

The other way of emulating natural science was to draw on the life sciences. This orientation gradually became more prominent, overshadowing

---

[11] Elie Halévy, *The Growth of Philosophic Radicalism* (London: Faber and Faber, 1928); Stefan Collini, Donald Winch, and John Burrow, *That Noble Science of Politics* (Cambridge: Cambridge University Press, 1983).

the exemplary role of classical mechanics. The life sciences offered two separate traditions of thought: medicine and natural history. The medical program for the science of man had been proclaimed in a most uncompromising way by Julien Offray de La Mettrie (1709–1751). His notorious *L'homme-machine* (1747) was one of the first sustained attempts to overcome the dualism of body and soul. Human consciousness and conduct had to be explained by bodily arrangements and physical needs, and no longer in terms of immaterial substances. This line of thought was reformulated by various authors during the last decades of the eighteenth century. Many were suspected of medical materialism, since they seemed to deny the existence of a soul, but their ideas received considerable attention from the reading public. The doctrine of phrenology, fashionable all over Europe during the first decades of the nineteenth century, attests to the popularity of medical models of the mind.

A particularly influential and long-lived medical tradition was initiated by the Montpellier physician Paul-Joseph Barthez with his *Nouveaux éléments de la science de l'homme* (1778). Barthez (1734–1806) broke from the mechanical conceptions of Hermann Boerhaave and La Mettrie by advocating vitalism as the basis of the science of man. His ambitions were taken up systematically by Pierre-Jean-Georges Cabanis (1757–1808), a leading French physician of the revolutionary period. Although physicians had traditionally been concerned with health and illness, Cabanis saw medicine as providing a scientific basis for the entire domain of the human sciences. Mathematics was of no use here, since the variability of thoughts, feelings, and passions did not allow quantification. Cabanis examined the biomedical basis of mental phenomena in a series of well-known lectures, published as *Rapports du physique et du moral de l'homme* (1802).[12]

Cabanis set out from the principle that humans are sensory beings, open to internal and external impressions. External impressions were processed into ideas, while internal ones formed instincts. Feelings generally resulted from a combination of the two. None of this was mechanical. It depended on the organization of the body, on how the organs operated and interacted with each other. Cabanis differentiated his model according to age, sex, temperament, habits, and climate. This emphasis on habits and climate, including occupational peculiarities, supported a sustained attention in the Montpellier tradition to the circumstances of human life. The possible effects of changes in these circumstances were of special interest during the revolutionary years.

Cabanis's psychophysiological research program was one of the cornerstones of the work of the *idéologues*, a group of moderate revolutionary

[12] Elizabeth A. Williams, *The Physical and the Moral* (Cambridge: Cambridge University Press, 1994); Stephen Jay Gould, *The Mismeasure of Man* (New York: Norton, 1981).

intellectuals.[13] The philosopher Antoine-Louis-Claude Destutt de Tracy (1754–1836) proposed to transform philosophy into the science of ideas – *idéologie*, as he called it. The old metaphysics was to be replaced by a rigorously scientific program for which Cabanis's biomedical theories provided the basis. The *idéologues* also considered Cabanis's work to be of vital significance for the effective reform of education and health care. Closely affiliated with the *idéologues* was the Société des Observateurs de l'Homme (1799–1805), a learned society whose aim was to "observe the physical, intellectual and moral aspects of mankind." Its members were predominantly medical doctors, naturalists, and explorers (among them Lamarck, Cuvier, Cabanis, Pinel, and Bougainville). Notions of human anatomy and physiology were an integral part of their ethnographical work. The comprehensive *science de l'homme* of Barthez and Cabanis was continued during the Restoration by Broussais, and was defended against spiritualist philosophers by Auguste Comte. Finally, it was eclipsed after the mid nineteenth century, when a range of specialties, including psychiatry, public hygiene, physical anthropology, and ethnography, took its place.

In a more metaphorical sense, the notion of "organization" had further implications. Regarding organisms as organized bodies, as distinct from the brute matter of mechanics, implied that studying their organization was the essential method of analysis. This idea of naturalists and physicians was appropriated by Claude-Henri de Saint-Simon (1760–1825), who proclaimed that human societies were also organized bodies. The science of society should be transformed into a "social physiology," defined as the "science of social organization." Within this physiological framework, Saint-Simon distinguished critical from organic periods of history. Organization was characteristic of organic periods and the desideratum of critical ones.

The basic feature of this stance was contained in the image of society as a physiological process. This implied a natural and spontaneous order, with a minimal role for government apart from a kind of medical supervision; legislation was comparable to public hygiene. This apolitical tendency was linked to the political isolation of the *idéologues* during the Napoleonic years, when many were dismissed from their official functions. Their journal, the *Décade philosophique*, finally ceased to exist in 1807. The physiological imagery undoubtedly had a special appeal to these men, now removed from the political center, and no longer inclined to conceive of their work as the science of the legislator.

---

[13] Sergio Moravia, *Il pensiero degli idéologues* (Firenze: Nuovo Italia, 1974); Marc Regaldo, *Un milieu intellectuel: la Décade philosophique (1794–1807)* (Paris: Champion, 1976); Jean Copans and Jean Jamin, eds., *Aux origines de l'anthropologie française* (Paris: Le Sycomore, 1978); Robert Wokler, "Saint-Simon and the Passage from Political to Social Science," in *The Languages of Political Theory in Early Modern Europe*, ed. Anthony Pagden (Cambridge: Cambridge University Press, 1987).

EVOLUTIONARY THOUGHT

Evolutionary social theories are often understood to be derived from bio-logical evolution, but this is seriously misleading, particularly for the period prior to Darwin's *Origin of Species* (1859). From the late Enlightenment until the last decades of the nineteenth century, biological and social theory largely evolved in a common context.[14] Evolutionary thinking in the life sciences owed as much to the human sciences as it did to biology.

Understandings of progressive change over longer periods of time were rooted specifically in what is usually called "philosophical" history.[15] The core concepts of this tradition were progress and perfectibility. The notion of progress was defined by the late-seventeenth-century battle between what were called the Ancients and the Moderns. The Moderns argued that the new natural philosophy attested to the progress of the human mind. Whereas it might not be possible to observe progress in literature or art, they suggested, advances in science and technology were unmistakable. This was the view of Francis Bacon and Bernard de Fontenelle. It was broadened during the Enlightenment by Turgot and further elaborated by Condorcet in his posthumous *Esquisse d'un tableau historique des progrès de l'esprit humain* (1795), a tribute to human perfectibility through the advancement of knowledge. Widely read as a heroic testament of the Enlightenment, Condorcet's work was the basic reference for Saint-Simon's and Comte's doctrines of social progress. It also helped to provoke Thomas Robert Malthus's (1766–1834) strongly anti-utopian *Essay on the Principles of Population* (1798).

Attacking Condorcet's optimistic vision of an indefinite perfectibility, Malthus argued that the operation of natural laws could well produce misery and starvation, not progress. Due to the sexual appetite of man, populations tend to grow at a geometrical rate, while food supplies can increase only arithmetically. The structural imbalance made poverty, and sometimes starvation, natural aspects of the human condition. The Malthusian law of population was a recurrent issue in many nineteenth-century debates; it provided Darwin with the clue for his theory of natural selection.

Social writings of the late Enlightenment also advanced the historicization of natural history. In defiance of the Linnaean program of collecting and classifying, Buffon wrote extensively on geology and cosmology, and had a more historical understanding of life. In *Les époques de la nature* (1778), he envisaged the historical development of the Earth and its inhabitants as

[14] Robert Young, *Darwin's Metaphor* (Cambridge: Cambridge University Press, 1985); Peter J. Bowler, *Evolution: The History of an Idea* (Berkeley: University of California Press, 1984); Robert J. Richards, *Darwin and the Emergence of Evolutionary Theories of Mind and Behavior* (Chicago: University of Chicago Press, 1987); George W. Stocking, Jr., *Victorian Anthropology* (New York: Free Press, 1987).

[15] Richard F. Jones, *Ancients and Moderns* (St. Louis, Mo.: Washington University Press, 1936); Jean Dagen, *L'histoire de l'esprit humain dans la pensée française de Fontenelle à Condorcet* (Paris: Klincksieck, 1977).

the unifying principle of natural history. Natural historians, Buffon argued, should in this respect follow "civil historians." His work had an enormous impact, particularly on Lamarck's studies of the transformation of species. Chronologies were introduced, and historical sequences became a guiding principle for organizing data of the natural world.

The new conceptions of natural history further reinforced the historicization of social science. Developmental or evolutionary theories in the broad sense became the prevailing form of the science of society in the nineteenth century. After the American and French Revolutions, and in response to ongoing industrialization and urban growth, social theories came to be fundamentally concerned with the causes and consequences of these deap-seated transformations. Alexis de Tocqueville and Auguste Comte, Karl Marx and Herbert Spencer, Henry Thomas Buckle and Henry Sumner Maine, all grappled with the historical characteristics of modern society – with its principles of change, and with its future direction. In that sense all of them were evolutionary thinkers, although few of them were evolutionists proper.

The best-known representative of evolutionism and one of the most widely read intellectuals of the nineteenth century was Herbert Spencer (1820–1903).[16] An evolutionist before Darwin's *Origin*, he did much to popularize the term "evolution" and to make progressive change the common denominator of all natural processes. From the maturation of an embryo to the development of human society and the evolution of the solar system, all things evolve from the simple to the complex through successive differentiation. Evolution, in other words, is the natural and necessary process of change from incoherent homogeneity to coherent heterogeneity. Because differentiation leads to higher levels of integration and coordination, evolution is practically synonymous with progress. This optimistic vision of progress as a "beneficent necessity" did not come from a single source. The idea that development means progress through differentiation combined Adam Smith's harmonious view of the division of labor with the embryology of Karl Ernst von Baer (who had used the terminology of homogeneity and heterogeneity).

Spencer's view of evolution was thus much broader than either Comte's sociological or Darwin's biological theory. It had the status of a cosmic law and formed the core of his all-embracing system of synthetic philosophy. The outline of this universal philosophy of evolution was presented in the essay "Progress: Its Law and Cause" (1857) and systematically developed in his *First Principles* (1862). There followed a series of multivolume works in which he applied the model successively to various domains – biology, psychology, sociology, and ethics.

---

[16] John D. Y. Peel, *Herbert Spencer: The Evolution of a Sociologist* (London: Heineman, 1971); Mike Hawkins, *Social Darwinism in European and American Thought, 1860–1945* (Cambridge: Cambridge University Press, 1997).

Spencer's social thought, as presented in his sociological studies, was entirely cast in the organic idiom. The features of social organization result neither from divine Providence nor from great law givers; they are the consequences of the ever-growing social organism. Spencer and many organicists after him took the analogy literally and worked out detailed correspondences between human society and other organisms.

Social change, according to Spencer, was linked especially to the transition from military to industrial society. In the first type of society, integration derives from a controlling center; in the latter, it is the spontaneous effect of individuals cooperating on the basis of a division of labor. For Spencer, the market was the primary model of the advanced type of integration. Since social evolution was natural and progressive, he strongly favored laissez-faire politics. Although this liberal stance is identified with what has come to be known – rather imprecisely – as "social Darwinism," it was not based on the mechanism of natural selection and its assumed beneficial effects. Spencer placed his political faith in natural growth and evolutionary progress, not so much in selection or the elimination of the unfit.

A DIFFERENTIAL EPISTEMOLOGY

The development of the life sciences and the fundamental criticism of mechanical models eventually gave rise to another mode of scientization of the social sciences. The unitary view of nature, expressed in mechanical metaphors and in the idea of a great chain of being, tended to give way to a dichotomy between inanimate and animate bodies, between matter and life. The common properties of living organisms were subsequently defined as the object of "biology," a term coined in the 1790s. As the general science of life, biology served to unify previously distinct domains, such as botany, zoology, and medicine. These were now more clearly separated from "physics," a term that also received a new, narrower meaning.

This process of differentiation contributed to the decline of a unitary conception of natural philosophy. The vitalists, in particular, had fought for their independence against mechanical and reductionist programs, of which the Laplace school was the prime example in physics and chemistry. Around the 1800s, then, a shift was perceivable from a relatively unified natural philosophy with various branches toward a division into scientific disciplines: mathematics, physics, chemistry, and biology. Encompassing terms such as "nature" and "reason" lost some of their appeal. Philosophy itself tended to become a discipline – a superior one to be sure, but a discipline nonetheless. Having previously stood for a general notion of systematic knowledge, philosophy was now redefined as a specialty for the purpose of transcendental analysis (Kant), or for analyzing ideas (as in Destutt de Tracy's *idéologie*), or simply as the "specialty of generalities," in Auguste Comte's phrase.

This process of differentiation and disciplinary division transformed the Enlightenment legacy and raised the problem of unity and difference in science in an entirely new manner. This was the central question of Comte's *Cours de philosophie positive* (1830–42). Auguste Comte (1798–1857) is best known for his idea that human knowledge develops through three states or stages: the theological, the metaphysical, and the scientific. In the positive or scientific stage, knowledge is concerned merely with laws or lawlike regularities. Since these laws are "relations of similarity and succession," there can be no positive knowledge, either of the intimate nature of things (essences, substances) or of first and final causes. The search for laws is the common characteristic of positive science, and Comte is commonly remembered for his obsession with invariable regularities and for his unfailing belief in having discovered the law of human society.

This reputation, however, is too restrictive and in an important sense misleading. What Comte's *Cours* actually contains is less a unified than a differential theory of science.[17] This differential theory was a favorable response to newly emerging scientific fields such as biology and social science, as well as to the recent developments in the physical sciences of heat, light, and electricity, which had diverged from the Laplacian program. Himself trained in the mathematical sciences at the École Polytechnique, Comte obtained a thorough knowledge of the life sciences as well. He cherished the ambition of developing an encompassing theory of science in an age of differentiation. This theory would provide a proper foundation for social science and, as such, a sound basis for political and moral reform.

The message of the *Cours*, in brief, was that the sciences shared the ambition of uncovering laws, but that they did so in various ways, following different methods. Considering the positive sciences in their actual diversity, there was no way they could be reduced to one basic type – neither to mechanics, as the Laplacians had claimed, nor to some form of general physiology, as some biologists had supposed. Rather than following a uniform model and a single method, each fundamental science had its own methods and research procedures – and necessarily so, for the complexity of their subject matters varied greatly. Astronomy was concerned with the geometry and mechanics of celestial bodies. Physics was already a more complex and less unified science: It could not be reduced to mechanics, although physical phenomena (light, heat, electricity, magnetism) were simple enough for mathematical description. Chemistry studied matter at the level of molecular composition and decomposition; in addition to the laws of mechanics and physics, these processes were subject to "chemical affinities." Biologists studied organisms whose conduct could not be explained by physical forces or chemical affinities, since it depended primarily on their complex

---

[17] Heilbron, *Rise of Social Theory*; Mary Pickering, *Auguste Comte: An Intellectual Biography*, vol. 1 (Cambridge: Cambridge University Press, 1993).

structural organization. Human societies, finally, were the most complex of all.

The sciences, then, composed a series of increasing complexity and decreasing generality. The main question of the entire *Cours de philosophie positive* was how recent developments in the sciences could be interpreted in view of this scheme. Contrary to current views, the central issue of the *Cours* was neither how science could be demarcated from metaphysics, nor how a logical or methodological foundation might be constructed for the unity of science. Comte's analysis had a different purpose. It explained in great detail how and why different methods prevailed in the various sciences: the experimental method in physics, the comparative method in biology, the historical method in sociology.

As a consequence, Comte forcefully rejected the use of mathematics in biology and sociology. Whereas in chemistry, mathematics was still of limited use, in biology the "enormous numerical variations" of the phenomena and the "irregular variability of effects" made mathematical techniques useless. This argument, borrowed mainly from vitalists such as Xavier Bichat (1771–1802), applied even more decisively to the social sciences. Comte accordingly rejected Condorcet's social mathematics, and he ridiculed Quetelet's social physics as "mere statistics."

Emphasizing what we would now call the relative autonomy of the sciences, Comte elaborated an ingenious and indeed pioneering differential theory of science. He did so mainly in opposition to reductionism. The consequences of this view for the social sciences were already formulated in his early notes. Instead of founding the social sciences on one of the natural sciences, it was more fruitful to follow indirectly the example of biology. Biology was a distinct science of life; its distinctiveness suggested both a differentiated comprehension of natural science and a program for reconceptualizing the aims and claims of social science. As vitalists had done for biology, Comte founded his sociology on the specific and irreducible properties of its subject matter. Because human beings have the capacity to learn, the progress of knowledge is the basis for the development of human society, and the law of the three stages is the core of sociology. Every historical stage has its own problems and possibilities; political and educational reform must be based on the requirements of each particular stage.

Independent of Comte's other contributions, whether philosophical or political, his differential theory of science had a formative impact on biology and sociology in France.[18] The program of the Société de Biologie (1848) was drawn up by a pupil of Comte, Charles Robin, and was directly inspired by his interpretation of the life sciences. The sociology of Émile Durkheim

---

[18] Georges Canguilhem, *Études d'histoire et de philosophie des sciences* (Paris: Vrin, 1983), pp. 61–98; W. M. Simon, *European Positivism in the Nineteenth Century* (Ithaca, N.Y.: Cornell University Press, 1963).

(1858–1917) likewise followed Comtean principles. Durkheim's formula that social facts must be explained by other social facts (and not by biological or psychological facts) was a more empirically minded translation of Comte's differential epistemology.

CULTURALISM AND SOCIAL SCIENCE

The promise and prestige of the natural sciences did not remain uncontested. Countermovements to the naturalistic understanding of human society became an intellectual force in the course of the nineteenth century. In retrospect, Vico and Herder can be seen as the pioneers of this approach.[19] In his *La scienza nuova* (1725), Giambattista Vico (1668–1744) proposed a science of human history that diverged in a fundamental sense from the predominant models. Drawing on Renaissance scholarship and natural jurisprudence, Vico set out to create an historical science of the "world of nations" in which cultural forms have a primary significance. For Vico, these cultural forms – poetry, myth, language, law – are not simply given, but are created by men. Precisely because they are man-made, our knowledge of them is, in a sense, deeper and more truthful than our knowledge of nature. Along these lines, Vico proposed an understanding of the main epochs in human history and advocated a new science to account for it. By implication, he suggested a genuine reversal of the intellectual hierarchy: The human sciences would henceforth crown the edifice of knowledge.

Whereas Vico's work was long neglected, Johann Gottfried Herder (1744–1803) became an influential figure in the historical and philological sciences in Germany. His four-volume *Ideen zur Philosophie der Geschichte der Menschheit* (1784–91) was read as the leading contribution to a new understanding of human societies. Each society, each people, is marked by a peculiar cultural spirit, a *Volksgeist*, expressed in its customs, myths, and folktales. The task of the human sciences is to uncover the peculiarities of this spirit, especially in its linguistic expressions.

Although Herder himself did not conceive it in this way, his work contributed to an emerging culturalist understanding of human societies, a tendency that was strongly reinforced by the Romantic reaction. Chateaubriand's glorification of poetic and religious sensibility was a violent revolt against the newly won authority of science and against what he saw as the tyrannical rule of scientists. Similar suspicions were voiced by Coleridge and Wordsworth, and in the satirical mode by Thomas Carlyle in *Sartor Resartus* (1831). Conservative theorists like Bonald, who mocked the redefinition of moral science

---

[19] Isaiah Berlin, *Vico and Herder* (London: Hogarth Press, 1976) and *The Crooked Timber of Humanity* (London: John Murray, 1990).

as a "branch" of anatomy and physiology, considered that a "war" had broken out between literature and science.[20]

But the criticism of naturalistic models was not always directed against the sciences. Herder's work, and more generally the German movement of *Naturphilosophie*, vividly opposed mechanistic and empiricist positions, but not naturalism per se. It was only in the mid nineteenth century, when *Naturphilosophie* as a rival version of naturalism had disintegrated, that a consistent alternative to the naturalistic program emerged. One of the founding fathers was the historian Johann Gustav Droysen (1808–1884), who synthesized the tradition of historical scholarship and the hermeneutic methods of text interpretation. This synthesis, in explicit opposition to Anglo-French views that the science of history required lawlike regularities, became the starting point for a new conception of the human sciences.[21] Wilhelm Dilthey (1833–1911) provided the classic formulation in his *Einleitung in die Geisteswissenschaften* (1883), which was further developed by Wilhelm Windelband (1848–1915) and Heinrich Rickert (1863–1936). Their work constructed the encompassing dichotomy of *Geistes-* and *Naturwissenschaften*, opposing interpretation and explanation as the fundamentally different methods of, respectively, the idiographic and the nomothetic sciences. By challenging a natural-science ideal that itself remained powerful, representatives of the cultural or hermeneutic sciences produced a series of new questions for the social sciences. As Max Weber (1864–1920) and Georg Simmel (1858–1918) recognized, these were questions not of naturalism but of culturalism.

---

[20] Wolf Lepenies, *Between Literature and Science*, trans. R. J. Hollingdale (Cambridge: Cambridge University Press, 1988).

[21] Manfred Riedel, *Erklären oder Verstehen?* (Stuttgart: Klett-Cotta, 1978).

4

CAUSE, TELEOLOGY, AND METHOD

Stephen Turner

The model of social science established in methodological writings of the 1830s and 1840s formed an ideal that has endured to the beginning of the twenty-first century. Subsequent authors have been obliged to excuse the social sciences for their failure to achieve this ideal model of science, to reinterpret the successes of social science in terms of it, or to construct alternative conceptions of social science in contrast to it. The ideal was worked out in two closely related texts, Auguste Comte's (1798–1857) *Cours de Philosophie Positive*[1] and John Stuart Mill's (1806–1873) *A System of Logic*.[2] The positive achievement of these texts was to clarify the application of the notion of "law" to the subject matter of social science. Their negative achievement was to eliminate, as much as possible, the role of teleological thinking (explanation appealing to purposes or "final causes") from the study of the social realm.

The subject of this chapter will be the reformulation of the ideas of cause and teleology before and during the period of Mill and Comte, and its aftermath up to the early twentieth century in the thinking of several founding figures of disciplinary social science. The discussion to be examined here focused on the problem of the sufficiency of causal explanations, and particularly on the question of whether some particular fact could be explained without appeal to purpose. In response to such questions, the defenders of the new conception attempted to replace older terms with new ones, replacing "purpose" with "function," for example. While they did not always achieve the clarity for which they aimed, they did establish the terms of the modern discussion of method in the social sciences.

---

[1] Auguste Comte, *Philosophie Première: Cours de philosophie positive*, ed. Michel Serres, François Dagonet, and Allal Sinaceur (Paris: Hermann, 1975); *The Positive Philosophy of Auguste Comte*, trans. Harriet Martineau (New York: Calvin Blanchard, 1855).

[2] John Stuart Mill, *A System of Logic Ratiocinative and Inductive* (1843), in *Collected Works*, ed. J. M. Robson (Toronto: University of Toronto Press, 1974).

TWO MODELS OF LAW

Methodological writings on social science set out from the traditional tenets of natural law theory, a teleological or purposive mode of theorizing about the social world. The key idea of the older conception is captured in the writings of the Ecclesiastical political thinker Richard Hooker (1553–1600): By "the law of nature . . . we sometimes mean that manner of working which God has set for each created thing to keep." Every person and thing was supposed to have an essence reflecting divine or natural purposes. The term "destiny" was used for the process by which the end was contained in the nature of the person or thing. "Every thing both in small or in great fulfilleth the task which destiny hath set down," as Hooker quoted Hippocrates. "Natural agents" do this "unwittingly"; for voluntary agents, the law is "a solemn injunction" to fulfill the tasks for which they have been created.[3] This distinction marked the divide between the human and the physical.

The metaphysics of natural law theory held that the world consisted of a variety of beings and objects whose essence disposed them toward the fulfillment of higher purposes. The larger hierarchy of purpose answered the question, "Why does thing x exist?" The manifest "natures" of things were evidence that creation is purposeful. The model could be applied to both the physical and the human worlds, taking account of the difference in the essential characters of humans and things, and the difference in how they are governed by natural law.

This style of explanation was eventually undermined by two logical difficulties. The first was its circularity. The explanations operated by treating a particular state – health, harmony, rest, stability, perfection, full development or growth – as an inherent goal, that is, as a part of the nature of the person or thing whose behavior was to be explained. The task or purpose was inherent in the essential nature; the essential nature explained what the person or thing did to fulfill this purpose or task. But matters were not quite so simple. All acorns do not grow into oaks; they do so only if a great many conditions are met. The "true end" is thus a *potential* effect or a tendency, which is distinguished from other potential effects by the fact that it requires no external cause.

One can often appeal to many possible explanations for the failure of a cause to produce an effect. In practice, the "nature" of something, and hence also its true purpose, could be established only theoretically, that is to say, only by using unobservable facts. Much of the discussion of "final causes" in the period following the Scientific Revolution, accordingly, focused on the question of whether one could identify essential natures or

---

[3] Richard Hooker, *Laws*, in *The Works of Mr. Richard Hooker with an Account of His Life and Death by Isaac Walton*, vol. 1, 7th ed., arranged by Rev. John Keble, revised by Rev. R. W. Church and Rev. F. Paget (Oxford: Clarendon Press, 1888), Book I, chap. 3, sect. 2, pp. 206–8.

purposes. Typically, a distinction was made between manifest purposes or natures, which were visible, and hidden purposes, which could be known only theoretically.[4] Revealing hidden purposes amounted to revealing the purposive order imposed by God on the universe. René Descartes (1596–1650) commented that "there is considerable rashness in thinking myself capable of investigating the impenetrable purposes of God."[5] The sentiment was echoed by theological Augustinians.[6] But if the larger purposive order of the universe were knowable, even hidden purposes could be understood and identified.

The second difficulty involved the relation between final causes and other kinds of causes, and especially the relation between final causes and what Aristotle called "efficient causes." "Final causes," purposes, or tasks, in this model, were not competitors of "efficient causes" but operated *through* efficient causes, as Aristotle himself had pointed out.[7] One of David Hume's (1711–1776) examples of causal knowledge – that I know, on the basis of experience, that bread is nourishing – exemplifies the point.[8] If bread did not nourish, that is, if it did not have the "efficient" causal effect, it could not serve the purpose of nourishing. The dependence of final on efficient causes was not quite reciprocal, since there was no problem of circularity for efficient causes. Final causes were commonly regarded as necessary to complete our understanding of the processes advanced by efficient causes, but this "completion" could also be seen as superfluous. That is, the asymmetry between the two forms of causation allowed for the elimination of final causes, but not of efficient causes.

Final causes were only gradually removed from the standard scientific picture of the physical world in the period following the Scientific Revolution. The first step was to argue that final causes serve no explanatory purpose, because they add nothing to efficient causes or laws. Newton's maxim that no more causes are to be admitted than those that are both true and sufficient to explain the appearances, which was enthusiastically propounded by such eighteenth-century figures as Thomas Reid (1710–1796),[9] makes the burden clear.[10] But physicists were circumspect about arguing directly for the complete elimination of final causes from the natural universe. One

---

[4] Cf. Pierre Gassendi, in *The Philosophical Writings of Descartes, Vol. II,* trans. John Cottingham, Robert Stoothoff, and Dugald Musdoch (Cambridge: Cambridge University Press, 1984), p. 215.

[5] Ibid., p. 39.

[6] Leszek Kolakowski, *God Owes Us Nothing: A Brief Remark on Pascal's Religion and on the Spirit of Jansenism* (Chicago: University of Chicago Press, 1998).

[7] Aristotle, *The Works of Aristotle,* trans. and ed. David Ross (Oxford: Clarendon Press, 1930), *Physica, Vol. II,* 195a, and *De Partibus Animalium, Vol. V,* 642a.

[8] David Hume, *Enquiries Concerning the Human Understanding and Concerning the Principles of Morals,* 2nd ed., ed. L. A. Selby-Bigge (Oxford: Clarendon Press, 1902), sect. IV, pt. II, p. 37.

[9] Larry Laudan, "British Methodological Thought," in his *Science and Hypothesis: Historical Essays on Scientific Methodology* (Dordrecht: D. Reidel, 1981), p. 92.

[10] Reid, *Essays on the Intellectual Powers of Man,* in *Works of Thomas Reid, D. D.,* 6th ed., ed. William Hamilton (Edinburgh: Maclachan and Stewart, 1785), vol. 1, p. 235.

exception was Descartes, who described final causes as "totally useless in physics."[11]

TELEOLOGY DURING THE ENLIGHTENMENT

Teleological explanation and the teleological worldview came increasingly under pressure during the eighteenth century, a development that owed much to the proliferation and "abuse" of final causes. In Germany, especially, as theology became possible outside the control of the Church, teleological thinking was carried to conclusions that were logical, but ludicrous. The philosopher Christian Wolff (1679–1754), for example, argued at some length that the sun shone so that people could more easily go about their work in the streets and fields.[12] Voltaire (1694–1778) mocked an unnamed contemporary work that held that "the tides are given to the ocean so that vessels may enter port more easily."[13]

Enlightenment thinkers were drawn in several directions in the face of these problematic arguments. They generally agreed that teleology had been abused in the past. But they were impressed with the idea that organisms are understandable *only* teleologically, only in terms of some internal principle or nature that cannot be reduced to mechanism; and they relied freely on the idea of human nature, characterized by inherent purposes, in their political reasoning. Even the most naturalistic *philosophes* wrote routinely and unselfconsciously in teleological ways about the natural course of history. They spoke of "forces" that assured its inevitability, and insisted on a fundamental similarity between the laws of social science and the laws of physics and biology.[14]

The philosopher who finally grasped the nettle was Immanuel Kant (1724–1804), who began his career as an enthusiastic proponent of a teleological physical universe, but who eventually rejected it. His position on "universal history" was more cautious; he refused to commit to the reality of teleological forces, but urged nevertheless that history had to be understood as a teleological process. How could Kant have it both ways? He articulated in his mature writings an argument that teleological explanations are always circular and, in consequence, cognitively different from mechanical laws. In his *Critique of Judgement*, he posed the question of whether an organism as a whole can be explained in an entirely causal way, like a mechanical system. He argued that it cannot. This "insufficiency" argument was then, and continued to be,

---

[11] *Philosophical Writings of Descartes, Vol. II*, p. 39, cf. also p. 258.
[12] Christian Wolff, "Deutsche Theology" (1725), in *Gesammelte Werke* (New York: Hildesheim, 1962), vol. 1, pp. 74–5.
[13] Voltaire, *Philosophical Dictionary* (1764), trans. H. I. Woolf (New York: Knopf, 1924), pp. 133–5.
[14] Frank E. Manuel and Fritzie P. Manuel, *Utopian Thought in the Western World* (Cambridge, Mass.: Harvard University Press, 1979).

the basic argument in favor of teleological accounts. But Kant then argued that the notion of purpose can, properly speaking, be applied only to the free actions of intelligent beings: When we apply it to organisms, we do so only in a metaphorical or analogical sense, that is to say, *as if* they had purposes. He introduced the notion that "an organized natural product is one in which every part is reciprocally both means and ends."[15] But "means" and "ends" serve only as analogical terms here. So Kant's solution to the conflict between cause (in the sense of mechanical causality) and teleology is to assign them to different categories of thought. To identify purposes in nature requires us to go beyond the sensible world, the world that we can subject to observation or experiment. Purposes are matters of *our* concern, as intelligent beings, rather than something in the physical world itself.[16] Comte radicalized this insight by historicizing it: He relegated teleological thinking to a stage in the historical development of thought, rendering it unnecessary and even retrograde.

THE REPLACEMENT OF TELEOLOGY

Comte was a self-conscious revolutionary. He saw himself as completing the project of expelling final causes from science by extending it to social science. "The Positive philosophy is distinguished from the ancient . . . by nothing so much as its rejection of all inquiring into causes, first and final; and its confining research to the invariable relations which constitute natural laws."[17] For Comte, this meant the thoroughgoing elimination from all of science of theologico-metaphysical notions – notably, the notion of a purposive universe – in all of their forms, manifest and hidden. He distinguished himself as a thinker by ferreting out hidden teleological usages and systematically replacing them with positive laws. His project was unprecedented in scope, and relentlessly pursued.

Comte's core sociological idea, his law of the three stages, itself contained the idea of the elimination of final causes. Like much else in Comte's work, the thought behind the law was not original. The basic idea had been present in Anne Robert Jacques Turgot's (1727–1781) account of the development of physics:

> Before knowing the connection of physical facts with one another, noth-ing was more natural than to suppose that they were produced by beings intelligent, invisible, and like to ourselves . . . when philosophers perceived the absurdity of these fables, . . . they fancifully accounted for phenomena

---

[15] Immanuel Kant, *The Critique of Judgement*, trans. James Creed Meredith (Oxford: Clarendon Press, 1928), div. I, sect. 66, pp. 24–6.

[16] Kant, *Critique of Judgement*, div. I, sect. 68, pp. 26–7.

[17] Harriet Martineau, *The Positive Philosophy of Auguste Comte* (New York: Calvin Blanchard, 1858), p. 799.

by abstract expressions, by essences and faculties, which indeed explained nothing, but were reasoned from as if they were real existences. It was only very late that, from observing the mechanical action of bodies on one another, other hypotheses were inferred, which mathematics could develop and experience verify.[18]

Comte refined and greatly extended this reasoning by classifying the sciences and arguing that each scientific area progressed successively through three stages. The first was one of superstition and animism, a stage that he called theological, marked by the appeal to "fictitious entities." There followed an intermediate stage, which he called metaphysical, in which explanations appealed to abstract entities or forces, such as "momentum" (and "cause" itself, in any sense other than the strict sense of invariable relations). Finally, in the positive stage, these ideas were eliminated, and purely predictive laws constituted the whole of what was taken to be scientific in that domain.

Physics had, for the most part, arrived at the positive stage: One no longer asked what "caused" gravitation, for example, precisely because one recognized that the answer to such a question was inevitably either theological or metaphysical. Biology had not quite reached this stage. Final causes and other pseudo-explanations abounded, often in concealed forms. Social science was even further from liberation from pseudo-explanation. Comte took this liberation as his task.

The notion of the positive stage was a powerful critical tool. It led to questions about scientific concepts in the sciences that had not yet reached this stage. Were "life" and "organism" metaphysical notions? Could such notions be replaced, or rather, could they be freed of their metaphysical connotations? These were problems that concerned Comte greatly in his accounts of the development of these fields, accounts that occupy much of the *Cours*. Hypotheses and fictions especially interested him, in part because of the contemporaneous controversy over the wave theory of light, in which he was an active disputant. He argued that the use of hypotheses, and even of fictions, is often necessary in science at certain stages of inquiry, but he insisted that in the end hypotheses had to be supported by sensory evidence.

Comte thus envisioned science as consisting of complex theoretical arguments that could be verified. In sociology, he believed, theoretical arguments and ancillary hypotheses had a large role to play. There were no readily accessible and unproblematic laws in social science. But Comte proposed a new way of establishing them. One first constructed generalizations from selected cases and examples. The generalizations based on these few cases were then combined with more general ideas to produce a more complex analysis than could be produced by simple induction or deduction alone. This was a strategy that could deal with exceptions: The general idea formed

---

[18] Quoted in Manuel and Manuel, *Utopian Thought*, pp. 848–9, n. 23.

the basic law; then a secondary law could be constructed that explained the exceptions or conditions under which the primary law applied. He contrasted this approach with that of his Enlightenment predecessors, who argued for the inevitability of progress on the ground that the forces favoring progress outweighed the forces opposing it, and would thus prevail in the long run. Comte, instead, theorized about the *conditions* for progress.

Mill grasped immediately the significance of Comte's general strategy, which he christened "the inverse deductive method." Mill described the method as being

> chiefly applicable to the complicated subjects of history and statistics: a process differing from the more common form of the Deductive Method in this – that instead of arriving at its conclusions by general reasoning, and verifying them by specific experience (as is the natural order in the deductive branches of physical science), it obtains its generalizations by a collation of specific experience, and verifies them by ascertaining whether they are such as would follow from known general principles.[19]

The phrase "history and statistics" is critical in the quoted passage, for the terms represent, for Mill, the almost intractably complex factual material of the social sciences. The basic strategy of the "inverse deductive method" in the face of complexity is one of simplification and selection, and Mill saw that both were characteristic of social science.

Mill's approach to these issues strained to avoid a conclusion that seems to follow naturally from one of his own arguments. The reasons for the relative wealth of nations, he argued, could not be determined causally – not because the differences were not governed by causal laws, but because of their complexity. One major source of complexity was this: In the case of differences of this sort, many causes have small effects, which contribute to the whole but which cannot in any practical way be aggregated:

> [T]he effects of the separate causes . . . are intermingled with, and disguised by, the homogenous and closely allied effects of other causes . . . some of which cancel one another, while others do not appear distinguishably, but merge in one sum . . . [so that] there is often an insurmountable difficulty in tracing by observation any fixed relation whatever.[20]

There is no guarantee that the inverse deductive method will produce results in such cases; and if the causes always appear in complex intermixtures, there is no way to identify the laws that govern the causal relationships in the first place. Mill also recognized that causal relationships might themselves be irreducibly probabilistic in character.

---

[19] John Stuart Mill, *Autobiography and Literary Essays, Vol. I*, in *Collected Works*, ed. J. M. Robson (Toronto: University of Toronto Press, 1974), p. 219.
[20] Mill, *System of Logic*, Book III, chap. 10, p. 443.

Mill nevertheless believed that in *some* cases we can isolate the causes and determine the form of the relations and the mode of combination of the effects. There was thus hope for the problem of complexity produced by statistics: the hope that in *many* cases we might be able to identify major causal relationships, produce "approximate generalizations" governing them, and then explain exceptions in terms of interfering causes. Social science, for Mill, thus resembled the science of the tides, which can never be reduced to a general theory. Although the main effects are understood, and predictions from these main effects are both possible and valuable, they are nevertheless subject to local causes of diverse kinds.

Economics, though deductive in form, could be seen as empirical because its laws, despite their failure to predict satifactorily, were firmly based on introspective psychology and supported by such natural experiments as the economic policies of governments provide. But economic phenomena are influenced by many noneconomic causes, so economics and the rest of the social sciences could be only inexact sciences.

TELEOLOGY IN ITS MANY FORMS

Resistance to the causal picture of the social world was intense but divided, and was associated with a variety of philosophical currents, including the movement of German Idealism, which opposed the determinism implied by a causal conception. Methodological writing more narrowly construed was frequently linked to broader cultural issues and, especially in Germany, to nationalism. German writers regularly denounced French positivism and, in economics, English "*Manchestertum.*" Yet antinaturalism, antiempiricism, and antipositivism did not mean opposition to social inquiry in any systematic or rigorous sense. Even overt forms of teleological thinking were not always opposed to social science. Empirical social inquiry could be, and sometimes was, understood as pointing to the hidden teleological order of God's Creation. Christian Wolff, whom we have already encountered as one of the more extreme "abusers" of teleology, wrote a Preface to Johann Peter Süssmilch's important compilation of statistics, which promised to reveal the divine order through statistics of birth and death.[21] A century later, the economics of the German historical school was equally teleological and, in the case of Wilhelm Roscher, even theistic, yet also determinedly "scientific" and engaged with the problem of the nature of historical and economic knowledge. Why did teleological thinking, contrary to the expectations of Comte and Mill, not only survive but continue as a vital part of the social sciences?

---

[21] *Die göttliche Ordnung in den Veranderungen des menschlischen Geschlechts aus der Geburt, dem Tode und der Fortpflanzung desselben erwiesen* (Berlin, 1741). Cf. Jacob Viner, *The Role of Providence in the Social Order* (Princeton, N.J.: Princeton University Press, 1972).

Teleology survived the Enlightenment in three main forms: the retention of purposive language as applied to the actions of individuals, the organic analogy, and historical teleology. This latter referred sometimes to the belief that particular nations had particular developmental paths, sometimes to the idea that history had a discernable direction and end. Historical relativism arose from the idea that these differences included the realm of intellect, so that there was no single path of intellectual progress. Instead, people of different historical periods and national traditions had fundamentally different world outlooks.

The idea that each nation or culture had its own intrinsic nature, and that consequently each had a distinctive intellectual destiny or path of development, had emerged already in the contemporary response to the Enlightenment in the writings of Johann Gottfried Herder (1744–1803) and Johann Georg Hamann (1730–1788). The case for fundamental cultural differences could be separated from the teleological idea of destiny. The powerful movement of neo-Kantianism, which dominated philosophy in the German-speaking world from 1860 to 1920, understood such distinctions as differences in fundamental presuppositions. Because such presuppositions are unprovable, this made a case for relativism. Relativism was in turn brought to bear on methodological issues, especially in the writings of Max Weber.

THE ORGANIC ANALOGY

The organic analogy produced the greatest confusion, because the language it employed could be interpreted either causally or teleologically. The asymmetry between cause and teleology discussed earlier, together with the general methodological consideration that nothing unnecessary should be included in an explanation, meant that a successful causal interpretation made teleological explanation superfluous. Comte's struggle against teleology included many attempts to absorb and explain, in nonteleological terms, the phenomena that the defenders of teleology held to be proof positive of the ineliminability of purposes. He and Mill attempted to show how such notions as "consensus" could be understood causally, and to substitute notions such as "harmony," a physical term, for teleological conceptions.[22] One effect of these efforts was to turn organic analogies and talk of "function" into the common property of both sides. Some important thinkers of the next period, such as Herbert Spencer (1820–1903) and Durkheim, are in the end difficult to classify. Both vigorously rejected teleology, but both employed many terms used by teleologists and suggested that they could be understood causally. It was thus possible for them to use the organic analogy in order to evade

[22] Stephen Turner, *The Search for a Methodology of Social Science: Durkheim, Weber, and the Nineteenth-Century Problem of Cause, Probability, and Action* (Dordrecht: D. Reidel, 1986), pp. 22–7, 53.

the question of whether organic explanations were neccessarily teleological. Whether they slipped into teleological reasoning unwittingly is a matter of legitimate dispute. Spencer, however, almost certainly did. He remarked that in his own *Social Statics* "there is everywhere manifested a dominant belief in the evolution of man and society. There is also manifested the belief that this evolution is . . . determined by the incidence of conditions, the actions of circumstances. And there is further . . . a recognition of the fact that organic and social evolutions, conform to the same law."[23] But his discussions of the law have little to do with the incidence of conditions, and much to do with "general laws of force."[24] These undergird the general principle that progress is "the evolution of the simple into the complex, through successive differentiations."[25]

"Evolution" is a highly ambiguous term in this context: Is it teleological or causal? There is good reason to be confused. As his expositors have said of Spencer's *Social Statics*, he "almost seems to see the social state as a fulfillment of a preexisting disposition, and he continually asserts an identity between processes in which the outcome is predetermined (like an embryo's maturation) and those in which it is not (like socialization or social evolution)."[26] Spencer freely employed the language of "essences" and "natures" (though apparently without regarding such usages as anything more than commonsensical). He appears even to fall into the teleologists' problem of circularity, as when he treats empirical exceptions to his generalizations as "incidental" facts, which do not relate to the "nature" of society.[27] His confusion was not resolved by other writers who employed the analogy.

French discussion of science in the mid nineteenth century was dominated by the issue of "vitalism," the doctrine that life was purposive and could not be reduced to mechanical explanation. Even the influential physiologist Claude Bernard wrote in his notebooks that "one must be a materialist in form and a vitalist at heart."[28] In France, the issue of organicism could not easily be evaded. The founding figure of French sociology, Émile Durkheim, was a careful reader of Comte and Spencer, as well as of German psychological and legal theorists who were concerned with issues of cause and teleology. He was philosophically tutored by a thinker, Emile Boutroux, who had sought to preserve a teleological understanding of the physical universe.[29] Not surprisingly, Durkheim was sensitive to the implications of teleological usages, and

[23] Herbert Spencer, *Essays Scientific, Political and Speculative*, vol. 2 (New York: Appleton, 1901), p. 137.
[24] Ibid., p. 138.
[25] Herbert Spencer, *Selected Writings*, ed. J. D. Y. Peel (Chicago: University of Chicago Press, 1972), p. 40.
[26] J. D. Y. Peel, "Introduction," in Spencer, *Selected Writings*, pp. xxxviii.
[27] Ibid., pp. xxxviii–xxxix.
[28] Francisco Grande and Maurice B. Visscher, *Claude Bernard and Experimental Medicine* (Cambridge, Mass.: Schenkman, 1967), p. 119.
[29] Émile Boutroux, *The Contingency of the Laws of Nature*, trans. Fred Rothwell (Chicago: Open Court, 1920), pp. 193–4.

especially to the issue of the reducibility of apparently purposive holistic phenomena to mechanistic explanation. His commitment to the idea of cause was clear. But he also attempted to account causally for collective phenomena, and intermittently employed an analogy between society and organisms.

Durkheim's meaning should be clear from a comment he made in accounting for the "maintenance" of social institutions. Employing a notion that we may recognize from Kant, who spoke of the reciprocity of means and ends, he suggested that "if more profoundly analyzed, [the] reciprocity of cause and effect might furnish a means of reconciliation which the existence, and especially the persistence, of life implies."[30] Thus Durkheim promoted a causal interpretation of the social organism. He also made a considerable effort to redefine such concepts as "normal" and "pathological" in nonteleological ways, as well as to use words such as "function" rather than "purpose" and to construe these words causally.

Durkheim's novel contribution to the methodological discussion arose from his twist on the issue of irreducibility, which had a long history in the French context, stemming from Comte's emphasis on the irreducibility of one discipline to others. He conceded that "social facts" were both irreducible to individual facts – sui generis – and also irreducibly mental. Typically such arguments, in the hands of such influential contemporaries as the German Ferdinand Tönnies (1855–1936), had led directly to the claim that society was a purposive being. Durkheim concluded, rather, that both the "collective consciousness" and the individual consciousnesses were governed by laws that were reducible neither one to the other nor to the laws of some other science, such as biology.

DECISION AND INTENTIONALITY: WEBER AND THE MARGINALISTS

The idea of human purpose had a different course, one that turned the defenders of intentional language and of the irreducibility of intentions to causes toward an alternative methodological tradition. Historically, the problem of determinism and free will is at its root. The most prominent methodology grounded in human freedom is hermeneutics, the idea that the understanding of action is methodologically analogous to the interpretation of texts, as intention is to meaning. The intellectual background of these ideas is exceptionally rich, including Kantian ideas of the freedom of the will, the "science" of Biblical interpretation, the irrationalism of Hamann, legal notions of action rooted in Roman law, and even a tension in Mill's own account of social science.

Mill supposed that reasons were causes, and that reasons were accessible to introspection. It is one of the oddities of the history discussed here that this

[30] Durkheim, *Rules of Sociological Method*, p. 144.

now little-regarded idea was the basis of his model of the relations of the social sciences, which, in contrast to Comte, made psychology a basic science.[31] Yet the fuller development of the notion of psychological causation led away from the notion of reasons as causes. The problem arose directly, in a special form, within economics itself, but the issue became apparent only with the marginal revolution in economics. Classical economics was largely unconcerned with choice and decision, or for that matter with "rationality." The focus was on "factors" of production and commodities, and on the constraints imposed by Malthusian forces governing demand for food and the physical difficulties of production.[32] These are readily construed as causes. The effect of the marginal revolution was to shift attention to individual choices. Contemporary critics, such as Thorstein Veblen, who had written his dissertation on Kant's *Critique of Judgement*, recognized that this amounted to a reversion to teleological thinking, ignoring the general tide against teleology in science.[33]

There were, however, two very different methodological directions in which such an emphasis on choice, free will, and intentionality could lead. One was toward the construction of abstract models of the economic agent. The marginalists posited individual rational agents, pursuing self-selected purposes, whose separate decisions led to aggregate patterns of equilibrium. Thus they assumed a particular abstracted teleology at the individual level to explain the teleological properties of the market. The strategy raised the question of the application of the model to the reality it purported to explain, as well as the question of circularity that was characteristic of teleological theorizing. Perhaps economic choices depended on culture. In that case, historical understanding would require intuitive insight into the mental worlds of the persons who were the subject of historical inquiry, an idea associated with hostility to abstraction, but that also came to be associated with historical relativism.

Max Weber, whose significance in German thought was comparable to Durkheim's in France, provided a comprehensive critique and synthesis of these ideas in his methodological writings. Even if one could have "a sort of 'chemistry' if not mechanics of the psychic foundations of social life," he wondered, would it have consequences "for our knowledge of the historically given culture or any phase thereof, such as capitalism, in its development and cultural significance?"[34] His answer was that it would not, because terms like "capitalism" are cultural in character.

Weber understood "culture" as "a finite segment of the infinity of the world process, a segment on which human beings confer meaning and significance."

---

[31] Robert C. Scharf, *Comte after Positivism* (Cambridge: Cambridge University Press, 1995).

[32] John Stuart Mill, *Principles of Political Economy*, ed. W. J. Ashley (London: Longmans, Green, 1929).

[33] Thorstein Veblen, "Why Is Economics Not an Evolutionary Science?" (1898), in *The Portable Veblen*, ed. Max Lerner (New York: Viking Press, 1948), pp. 215–41.

[34] Max Weber, *The Methodology of the Social Sciences*, trans. and ed. Edward A. Shils and Henry A. Finch (New York: Free Press, 1949), p. 75.

Different cultures or epochs confer different "meaning and significance" on different finite segments. The questions of social science are themselves questions that begin with what is meaningful and significant for us, and from our point of view. So the "knowledge of cultural reality" that the social sciences seek "is always knowledge from particular points of view."[35] But Weber also argued that the social sciences were causal and necessarily employed abstraction, and this led him to a complex position. He rejected teleological thinking and spared no effort in rooting it out, violently attacking the teleological formulations of the German historical school in economics as well as the teleology implied by collective concepts of the state and law. But at the same time he defended explanation of what he called meaningful social action in terms of human intentions.

Trained as a lawyer, Weber pointed out that legal reasoning about responsibility was causal, and argued that this kind of reasoning, properly understood, was relevant to and sufficient for the kinds of factual historical questions that arise within cultural points of view. The causal character of these questions should be understood in this way: Determinations of causality or responsibility do not require scientific laws; they require only a judgment that in a class of similar cases, subtracting a given condition would have lowered the probability of the outcome. This kind of reasoning could be applied to such historical questions as the question of the contribution of Protestantism to the rise of capitalism, in which case of course it would necessarily be hypothetical. But the model also allows explanations of ordinary intentional action as simultaneously intentional and causal. Intentions are attributed by showing that the sequence of events of which the act is a part is intelligible or meaningful as an action of a particular kind. Causal responsibility is shown by establishing that it would have some probability of producing the outcome.[36]

Causal and "meaningful" or intentional considerations are coequal in Weber's model of social science explanation, at least in principle, with interpretation being tested by probability. In practice, interpretation, and especially the task of testing interpretations against the course of events, predominated. Most meaningful interpretations of action correspond to some degree of predictive probability. But, in historical analysis as in a courtroom, many hypotheses about motives do not fit the facts. So Weber's accommodation of intentional explanation to causal analysis had the effect of raising the status of interpretation.

THE PERSISTENCE OF TELEOLOGY

The struggle against teleological explanation had profound consequences for social science, but they were not the consequences that Comte had

---

[35] Ibid., p. 81.
[36] Ibid., pp. 167–75.

anticipated. The project of stripping science of its teleological elements was difficult, perhaps impossible to carry through consistently. So it is not surprising that the problem of cause and teleology persisted in the social sciences. But it did so in many forms, such as the continuing critique of "positivism" and scientism in the methodological literature of the social sciences, and the conflicts between interpretive and quantitative approaches, each rooted in earlier reactions to a causal law model. At least one major current in one of the social sciences, Straussianism, has involved the self-conscious restatement and updating of arguments made in Descartes' time on behalf of teleology.

Even if such disputes no longer employ the language of the earlier struggle against teleology, they are often not far removed from it. The most technical domain of social science methodology today, the application of artificial intelligence to the problem of determining when statistical relationships are "structural," is the site of a dispute over whether wholly mathematical criteria can ever distinguish cause from correlation – an argument that Comte would eagerly have joined. Even the complexities that arise in Spencer's thought have present-day analogues. Rational choice theory in the social sciences, for example, is explicitly teleological, but seeks a nonteleological grounding in evolutionary biology, which is perhaps itself teleological. The question of whether one's teleology is legitimate or merely circular is now commonly stated in terms of the existence of "feedback mechanisms." Ironically, Voltaire would have recognized this argument, and might well have rejected it for begging the question of the origins of such mechanisms.

5

UTOPIAN SOCIALISM AND SOCIAL SCIENCE

Antoine Picon

During the nineteenth century, utopian socialism was most often interpreted as an essentially political phenomenon. Few commentators took seriously its ambition to create a new science of man and society. Yet the invention of such a science was one of the fundamental claims of Saint-Simon, Fourier, Owen, and their disciples who saw a scientific understanding of society as a prerequisite for its reconstruction.

At the turn of the century, Émile Durkheim was among the first to stress the role of utopian socialism in the emergence of the social sciences.[1] He considered Saint-Simon, the mentor of Auguste Comte, to be the true founder of sociology. Since the time of Durkheim, the importance of utopian socialism in the birth of the social sciences has been widely recognized.[2] This role is, however, difficult to assess accurately. Utopian socialism was, after all, the inheritor of eighteenth-century reflections regarding man and society. These reflections were in turn indebted to a long tradition of utopian writings dealing with social organization, beginning with Thomas More's *Utopia*, published in 1516.[3] To what extent did Saint-Simon, Fourier, and Owen break with the Enlightenment and its utopian component to mark a new era in social thought?

Another justification for a more thorough inquiry lies in the definition of the social sciences given by the utopian socialists. Although meant to be a departure from the philosophical tradition, their idea of science was still imbued with philosophical and even metaphysical conceptions. Extending far beyond the limits of our contemporary social sciences, Saint-Simon's, Fourier's, and Owen's doctrines appear in retrospect as a disconcerting combination of

---

[1] Émile Durkheim, *Le Socialisme; sa définition, ses débuts; la doctrine saint-simonienne* (Paris: F. Alcan, 1928).

[2] Barbara Goodwin, *Social Science and Utopia: Nineteenth-Century Models of Social Harmony* (Sussex: Harvester, 1978).

[3] Frank E. Manuel and Fritzie P. Manuel, *Utopian Thought in the Western World* (Cambridge, Mass.: Harvard University Press, 1979).

brilliant intuition and oversimplification, of original thought and prejudice. Given the ambiguities of these doctrines, as well as the wide range of issues addressed by them, it would be simplistic to reduce their contribution to the emergence of disciplines such as sociology and anthropology or to their influence on such figures as Auguste Comte and John Stuart Mill. Neither can Saint-Simon, Fourier, and Owen appear as mere forerunners of "scientific socialism," as Marx and Engels used the term in their *Manifesto of the Communist Party*. The relation between utopian socialism and social science must, rather, be placed within the broad framework of nineteenth-century cultural history.

THE ENLIGHTENMENT LEGACY

Whereas Owen readily acknowledged the influence of eighteenth-century philosophy on his thought, Saint-Simon and Fourier often presented their doctrines as reactions against the shortcomings of the Enlightenment. However, Saint-Simon's preoccupation with a new encyclopedia and Fourier's fascination with the Newtonian law of mutual attraction revealed their debts to the eighteenth century, as did Owen's faith in individual perfectibility, a belief inspired by his reading of Helvétius. Above all, the utopian socialists inherited the ambition of constructing a science of man and society. Expressed by philosophers like Turgot and Condorcet, and later continued by the main upholders of their thought, the Idéologues, this ambition was one of the chief legacies of the Enlightenment.

The notion of progress, the collective advancement of humanity, was another key piece of the heritage. It implied the redefinition of history as an itinerary leading from the primitive origins of civilization to its present complexity. The present appeared, in turn, as the anteroom to a still more brilliant future. Turgot had already conceptualized history as progress in his *Tableau philosophique des progrès successifs de l'esprit humain* of 1750 and in his *Discours sur l'histoire universelle et sur les progrès de l'esprit humain* of 1751. During the French Revolution, Condorcet extended and systematized it in his *Esquisse d'un tableau des progrès de l'esprit humain*.[4] Published shortly after Condorcet's death in 1794, the *Esquisse*, with its evocation of the future wisdom and happiness of mankind, created an agenda for Saint-Simon, who at the beginning of his intellectual career intended to complete Condorcet's broad historical picture.

More complex was the filiation between the eighteenth-century vision of society as the result of a voluntary contract between men, and the utopian socialists' organic conception of the social bond. Although seemingly

---

[4] Keith M. Baker, *Condorcet: From Natural Philosophy to Social Mathematics* (Chicago: University of Chicago Press, 1975).

contradictory, the two visions assumed that social organization was highly malleable. The arbitrariness of legal agreements and the adaptability of life both reflected this flexibility. The conviction of Saint-Simon, Fourier, and Owen that society could be shaped according to different patterns was also a tribute to the Enlightenment. Such social experiments as Fourier's Phalansteries and Owen's Harmonies presupposed the extreme diversity of human institutions, laws, and customs, a recognition sustained by travelers' accounts and theorized by philosophers like Diderot in his *Supplément au voyage de Bougainville* of 1772. That book, however ironic its tone, has a marked utopian dimension, especially in its preoccupation with the sexual freedom of the Polynesians.

The utopian form flourished in the later eighteenth century, and during that period it displayed some novel features. One of these was a deep commitment to universality. Most previous utopian writings had stressed the singularity of the ideal society rather than its generic character. Thomas More, the creator of the genre, had named his utopia from the Greek *ou* and *topos*, meaning "negation" and "place," respectively. Utopia was literally to be found nowhere. More's utopia was intended not as a positive example, but as a critique of the existing social order. Only such a purpose could explain why a fervent Catholic such as More would assign so many pagan habits to the citizens of his Utopia. Through their search for universality, eighteenth-century utopias began to acquire a new meaning. They came to represent models to be imitated all over the world. The broadly egalitarian perspective of Enlightenment anthropology regarding physical and moral dispositions played a role in this shift. Utopia could be truly universal, since the fundamental needs and capacities of men were the same everywhere.[5]

An important consequence of this shift from singularity to universality, from nowhere to everywhere, was a gradual displacement of utopia into history.[6] Whereas utopias had previously been described as contemporary kingdoms, they were now often relocated into the future, as the final stage of human progress. Published in 1770, Sébastien Mercier's *L'An 2440* displays this tendency in its evocation of a futuristic Paris. Two decades later, Restif de la Bretonne followed Mercier's example with *L'An 2000*. The trend toward the future culminated with Condorcet's Atlantide. Named to recall Bacon's *New Atlantis*, the Atlantide utopia represented the final stage reached by humanity in the philosopher's broad historical trajectory.

From the desire to build a science of man and society to the redefinition of utopia as universal model, the influence of the Enlightenment on the utopian socialists should not be underestimated. Were Saint-Simon, Owen,

[5] Michèle Duchet, *Anthropologie et histoire au siècle des Lumières: Buffon, Voltaire, Rousseau, Helvétius, Diderot* (Paris: Robert Laffont, 1971).
[6] Cf. Jean Marie Goulemot, *Le Règne de l'histoire: Discours historiques et révolutions XVIIᵉ –XVIIIᵉ siècle* (Paris, 1975; new edition Paris: Albin Michel, 1996), esp. pp. 263–94.

and Fourier original? Their originality was a matter not only of ideas and opinions, but also of moral sensibilities. The fathers of utopian socialism showed a common tendency to adopt a prophetic tone.

THE PROPHETS OF A NEW GOLDEN AGE

The three founding figures of utopian socialism, Claude Henri Saint-Simon (1760–1825), Robert Owen (1771–1858), and Charles Fourier (1772–1837), were very different one from another.[7] The first came from an aristocratic background, whereas the two others belonged to the common people. Saint-Simon had begun as an army officer in the American Revolution before turning to real estate speculation. Ruined by the end of the French Revolution, he survived by working as a clerk. The only Englishman of the trio, Owen had been a successful manufacturer at the head of the New Lanark factory before entering the ranks of social reform in England and America. For most of his life, Fourier remained an obscure shop assistant.

Above all, the conceptions of the ultimate social organization developed by the three men diverged. Saint-Simon's concern with a large single industrial society ruled as a peaceful army of workers was incompatible with Owen's and Fourier's proposals for strictly limited agrarian communities. Inhabitants of Owen's Harmonies were supposed to lead a rather austere life, whereas Fourier's Phalansteries would allow all sorts of pleasures.

Saint-Simon, Owen, and Fourier nevertheless adopted a common prophetic tone when contrasting the present forlorn state of humanity with its future happiness, with the new and definitive Golden Age to be ushered in by their principles. Like the Romantic philosophers and writers, their contemporaries, the founding fathers of utopian socialism were able to discern a gleaming future through the mists and shadows of the present.[8] But their prophetic inspiration was also the consequence of their tragic vision of early-nineteenth-century European society. Contrary to the Enlightenment philosophers, whose speculations remained generally somewhat abstract, Saint-Simon, Owen, and Fourier were acutely aware of the distress of their time. The political and social changes brought about by the French Revolution and the English Industrial Revolution figured prominently in this pessimistic assessment of the present. In the utopian socialists' eyes, the science of man and society was not just an intellectual challenge, but an urgent effort to ward off social chaos.

---

[7] Frank E. Manuel, *The New World of Henri Saint-Simon* (Cambridge, Mass.: Harvard University Press, 1956); Frank Podmore, *Robert Owen: A Biography* (London: Allen and Unwin, 1906); Serge Dupuis, *Robert Owen: Socialiste utopique 1771–1858* (Paris: CNRS, 1991); Simone Debout, *L'Utopie de Charles Fourier* (Paris: Payot, 1979); Jonathan Beecher, *Charles Fourier: The Visionary and His World* (Berkeley: University of California Press, 1976).

[8] Paul Bénichou, *Le Temps des prophètes: Doctrines de l'âge romantique* (Paris: Gallimard, 1977).

CLASSES, HISTORY, AND SOCIAL SCIENCE

The very different pictures of the Golden Age given by Saint-Simon, Owen, and Fourier were rooted in contrasting visions of man. With the exception of Fourier's extravagant and precise study of human passions, these visions remained somewhat unformed. Although he had written a *Memoire sur la science de l'homme* at the beginning of his intellectual career, indicating that such a science should be based on the contemporaneous medical studies of Vicq d'Azyr, Cabanis, and Bichat, Saint-Simon never proposed a specific conception of man. To judge from the various hints provided in his writings, he seemed to interpret man as an essentially active creature, the nature and degree of this activity varying strongly from one individual to another. Saint-Simon's anthropology was anything but egalitarian. Theoretical equality between individuals was, by contrast, a fundamental principle for Owen, even if his Harmonies were to be severely hierarchical. This led him to emphasize man's capacity to improve himself through proper education, though that proposal was never worked out in detail.

Improving man was not on Fourier's agenda. He boasted of taking man as he was instead of trying to change him. For Fourier, this meant studying the various passions that drove humanity. With its fascination for numbers, its sophisticated catalogue of human inclinations, and its often provocative character, Fourier's "mechanics of passions" was an ambitious attempt to deal with man from an entirely new scientific perspective.

Despite their contradictory visions of man, the utopian socialists agreed on the organic character of the social bond. This implied a vision of society other than the eighteenth-century definition of a mere association of individuals. In France, the political instability created by the Revolution seemed indeed to demonstrate that a permanent social order could not be founded on individualism. The growing social tensions experienced in Britain because of the Industrial Revolution suggested the same conclusion. Thus, the restoration of an organic social order was among the priorities of Saint-Simon, Owen, and Fourier.

The utopian socialists were not alone in this critical assessment of the shortcomings of individualism. Conservative thinkers such as Joseph de Maistre and Louis de Bonald shared this perspective. But while the latter turned to transcendent religious and anthropological principles, to Providence and the family, Saint-Simon, Owen, and Fourier focused on social classes. The notion of class was not entirely new. In his *Esquisse*, Condorcet had applied it to the priests, for example. But the notion, formerly marginal in the philosopher's perspective, now acquired a fundamental importance.

Although Saint-Simon's characterization of the various social classes remained imprecise, it is in his work that they played the most decisive role. The consideration of social classes – such as the industrialists, a class which he defined as the "mass and union of men devoted to useful works" – freed him from the eighteenth-century belief in the constant interaction between psychological and social considerations. A new social science based solely on the study of collective functions and behavior was thus made possible, a science that his former disciple Auguste Comte would later call sociology. The accent placed on social function and class was accompanied by a renewed interest in history. In contrast to the faculties of the individual, on which eighteenth-century authors such as Condorcet had focused, the features of social class were historically determined. The new social science was to be founded on historical knowledge. Its ambition was to decipher the laws of evolution at work in the history of mankind, laws that implied the advent of a new Golden Age.

Whereas Condorcet was mainly concerned with the continuous progress of science and technology, the utopian socialists' vision of history was based on the identification of a series of organic stages, such as pre-Christian antiquity and the Middle Ages. According to Saint-Simon, those stages were separated by periods of cultural and social uncertainty and unrest. The Reformation was, for him, such a period, one that had led to eighteenth-century critical philosophy, to the ruin of Christianity, and eventually to the French Revolution. The Golden Age that he announced was to bring cultural and social unrest to a definitive end, replacing it with a new organic order. In many respects, Comte's Positive Age was to play a similar role.

The utopian socialists' emphasis on social class was of course among the reasons that Marx could see them as forerunners of "scientific socialism." Marx shared their dynamic vision of society based on class struggle. Saint-Simon, Owen, and Fourier were acutely aware of the conflicts developing in the early industrialized societies. They saw class struggle not as a temporary characteristic of a period of incertitude and unrest, but as a dynamic principle of historical evolution. The prophetic tone they chose to adopt was partly a consequence of this conviction. Marx followed them also in stressing the intimate relation between economic and social organization. Like the triumph of the Marxist proletariat, the utopian socialists' Golden Age was to be based on the radical reform of production. Contrary to the Marxist doctrine, however, this reform was not to be initiated by the proletariat. The first truly communist nineteenth-century utopia was to be developed later by Etienne Cabet (1788–1856).[9] By contrast, Saint-Simon, Owen, and Fourier remained deeply committed to a conception of social change based on the leading role of an elite. Severely criticized by Marx and Engels in their *Manifesto*, this elitist attitude was later denounced by twentieth-century liberals because of its technocratic implications.

---

[9] Jules Prudhommeaux, *Icarie et son fondateur Etienne Cabet: Contribution à l'étude du socialisme expérimental* (Paris: Édouard Cornély, 1907); Christopher H. Johnson, *Utopian Communism in France: Cabet and the Icarians, 1839–1851* (Ithaca, N.Y.: Cornell University Press, 1974).

TOWARD A RELIGION OF HUMANITY

The part played by the utopian socialists in the emergence of notions and themes that were to become fundamental for the social sciences must not lead us to disregard the more extravagant features of their doctrines, such as their intention to replace Christianity with a new religion. Though religion was not prominent in Saint-Simon's early writings, it emerged as essential in his *Nouveau Christianisme*, published in the year of his death.[10] As for Owen, he turned to spiritualism rather late in life. The religious dimension was nevertheless an important aspect of early-nineteenth-century utopian socialism. Saint-Simon's, Owen's, and Fourier's disciples, with their various attempts at the creation of new cults, were in that respect even more radical than their masters. Although often inspired by the Catholic hierarchical order and by its appealing ritual, the new religions differed from Christianity in avoiding worship of a remote God. Humanity and its achievements – or, in the Saint-Simonian case, a pantheistic association between humanity and the rest of the universe – were to replace the former Christian deity.

The project to create a religion of humanity was to a large extent a consequence of the ambition to establish a new organic order, to restore a true community transcending individual differences and interests. Such a goal could not be achieved merely by appealing to the intellect, since most men are ruled not by their minds but by their hearts. This view had already been articulated at the very beginning of the nineteenth century by Chateaubriand in his *Génie du Christianisme*. In the utopian socialists' perspective, only religion could fill the gap between the general and abstract understanding of the elite and the more intuitive and emotional capacities of the people. Efficiently spreading a new social credo as a means to insure its observance was not, however, the only issue at stake. At a more profound level, it was also a matter of reconciling man's intellectual and emotional natures. Neglected at first by Auguste Comte, such an objective was to play a greater role after Comte's encounter with Clotilde de Vaux around 1842. Like Saint-Simon, Owen, Fourier, and their disciples, the creator of positivism would then start the transformation of his philosophy into a religion.[11]

The unity of culture was also at stake, a unity that was jeopardized by the growing gap between the exact sciences and other types of cultural production. In his *Esquisse*, Condorcet had insisted on the link between religious beliefs, the state of scientific knowledge, and the various cultural achievements of a given society. By the end of the Revolution, the same line of thought could be found in Charles Dupuis's *De l'Origine de tous les cultes*.

---

[10] Henri Desroche, "Genèse et structure du *Nouveau Christianisme* saint-simonien," Introduction to Henri De Saint-Simon, *Le Nouveau Christianisme et les écrits sur la religion* (Paris: Le Seuil, 1969), pp. 5–44.

[11] Mary Pickering, *Auguste Comte: An Intellectual Biography*, vol. 1 (Cambridge: Cambridge University Press, 1993).

The utopian socialists' religious preoccupation was an expression of their ambition to restore the fundamental unity of culture that had characterized organic periods such as the Middle Ages. In this respect, Comte would prove more realistic than his utopian forerunners. Positivism would never attempt to merge the various types of knowledge into a single body of scientific knowledge.[12]

Although the attempt to found new religions was abandoned by the social sciences of the late nineteenth and early twentieth centuries, the utopian socialists had again anticipated some of the most fundamental concerns of social scientists. From Ferdinand Tönnies to Émile Durkheim, the replacement of tightly bound communities by looser systems of social relations during the passage from traditional to industrial societies became a major concern of sociology. Like Saint-Simon's, Owen's, and Fourier's writings, the rapidly developing sociological literature was permeated by a dull nostalgia for what had been lost in this passage.[13] Moreover, the relationship between religion, culture, and social organization was becoming a major sociological subject. If Max Weber's *Protestant Ethic and the Spirit of Capitalism* was fundamentally indebted to the tradition of German historical economics,[14] Émile Durkheim's *Les Formes élémentaires de la vie religieuse* had more to do with the utopian socialist heritage, through the mediation of Auguste Comte.

In an industrialized world in which the exact sciences and their technological applications were gradually replacing religion as the ultimate source of spiritual legitimacy, though lacking its emotional appeal, one may even wonder whether the social sciences were not attempting to occupy an intermediary position between pure scientific reason and emotion. The utopian socialists had tried to fill precisely that position at the beginning of the nineteenth century. The desire to combine scientific rigor and emotional fulfillment remained a concern of the social sciences.[15]

RESHAPING EDUCATION, FAMILY, AND SEXUALITY

For the utopian socialists, science and action were intimately linked. This link was especially strong with regard to such subjects as education, the family, and sexuality. Regarding education and the family, Owen and Fourier were more radical than Saint-Simon, as they proposed a collective upbringing of children that would weaken the traditional family structure. In Fourier's

---

[12] Annie Petit, "Heurs et malheurs du positivisme: Philosophie des sciences et politique scientifique chez Auguste Comte et ses premiers disciples (1820–1900)" (PhD dissertation, Université de Paris I-Sorbonne, 1993).

[13] Robert A. Nisbet, *The Sociological Tradition* (New York: Basic Books, 1966).

[14] Wilhelm Hennis, *La Problématique de Max Weber* (Tübingen, 1987; French translation Paris: PUF, 1996).

[15] Wolf Lepenies, *Between Literature and Science: The Rise of Sociology* (Cambridge: Cambridge University Press, 1999).

doctrine, that structure was further threatened by a sexual life allowing for the expression of all types of human passion. Curiously enough, the views expressed in Fourier's *Nouveau monde amoureux* were generally rejected by his disciples, but exerted a profound influence on the Saint-Simonians in the early 1830s.

The emancipation of women was a major concern for the second generation of utopian thinkers that claimed to follow Saint-Simon, Owen, and Fourier. Most of the women attracted to utopian socialism were soon disappointed, however, by the superficiality of their male comrades' commitment to their cause. Former Saint-Simonian, Owenite, and Fourierist women nevertheless played a fundamental role in the emergence of feminism as a political and social movement.[16]

Collective education and women's emancipation were part of a larger agenda aiming at a drastic reshaping of social relations. Consistent with the utopian socialists' condemnation of individualism, such a reshaping was meant to suppress or at least to weaken attachments that could impede the formation of a true collective spirit, from social prejudices instilled by parents to exclusive love. Was this agenda totalitarian? That has been argued by many authors, including Friedrich von Hayek and Hannah Arendt, who often liken the utopian socialists' ideas with the program of twentieth-century communism.[17] It is difficult to draw conclusions, to compare doctrines that were never applied on a large scale to actual regimes that lasted for decades in many countries. One cannot but be struck, however, by the contrast between the libertarian tone used by Saint-Simon, Owen, Fourier, and their disciples, and the severe discipline of mature Eastern European and Asiatic communism.

This libertarian tone is all the more surprising because liberty was not invoked as a fundamental value by the founding fathers of utopian socialism or by their direct descendants. They held that a proper social organization would make individual initiative unnecessary. Determinist in essence, their social science would supplant politics and its half-measures, as well as economic liberalism, its egoistic inspiration and its trail of miseries. In this respect, their science was far from Condorcet's conception, which allowed for human free

---

[16] Maria Teresa Bulciolu, *L'Ecole saint-simonienne et la femme: Notes et documents pour une histoire du rôle de la femme dans la société saint-simonienne 1828–1833* (Pise: Goliardica, 1980); Carol A. Kolmerten, *Women in Utopia: The Ideology of Gender in the American Owenite Communities* (Bloomington: Indiana University Press, 1990); Bernadette Louis, ed., *Une Correspondance saint-simonienne: Angélique Arnaud et Caroline Simon (1833–1838)* (Paris: Côté-femmes éditions, 1990); Benoîte Groult, *Pauline Roland ou comment la liberté vint aux femmes* (Paris: Robert Laffont, 1991); Michèle Riot-Sarcey, *De la Liberté des femmes: Lettres de dames au Globe (1831–1832)* (Paris: Côté-femmes, 1992); Michèle Riot-Sarcey, *La Démocratie à l'épreuve des femmes: Trois figures critiques du pouvoir 1830–1848* (Paris: Albin Michel, 1994).

[17] Hannah Arendt, *Le Système totalitaire* (1951) (French translation Paris: Le Seuil, 1972), p. 72; Friedrich A. Von Hayek, *The Counter-Revolution of Science* (1952) (new edition New York: Free Press of Glencoe; London: Collier-Macmillan, 1955); George Iggers, *The Cult of Authority: The Political Philosophy of the Saint-Simonians* (The Hague: M. Nijhoff, 1958; reprinted, 1970).

will and action. Both in Europe, where it had first appeared, and in America, where Owen and Fourier found numerous disciples, the history of utopian socialism was marked by a recurring tension between a determinist vision of history and a more positive assessment of human agency.

SOCIAL EXPERIMENTS AND FAILURES

The wide influence of Saint-Simon's, Owen's, and Fourier's doctrines seems quite comprehensible in retrospect, given the tensions of early-nineteenth-century European and American society. Nevertheless, the extent of this success surprised many of their contemporaries. On his deathbed, Saint-Simon was surrounded by a few friends only. By the early 1830s, under the guidance of Saint-Amand Bazard and Prosper Enfantin, Saint-Simonianism had attracted hundreds of engineers, lawyers, and physicians, not to speak of the thousands of workers who followed the Saint-Simonian preaching in Paris, Lyon, Metz, and Toulouse.[18] Luckier than his older counterpart, Owen was able to observe the diffusion of his ideas in England and America. The rise of Fourierism was even more spectacular. By the 1840s, it had become influential in France, and the history of American Fourierism was about to begin with the conversion of the Brook Farm community to Phalansterian ideals. Dozens of Phalansteries would be founded in the following years throughout the United States.[19]

Following their initiators' preoccupation with social experiments, Saint-Simonians, Owenites, and Fourierists tried to create new conditions of life and work. Most of these attempts were, however, short-lived. Beyond the mere impracticability of general schemes such as organizing the working class as a peaceful army ruled by a new type of theocracy, as with the Saint-Simonians, or building harmonious and self-sufficient agrarian communities, as with the Owenites and Fourierists, other factors accounted for this series of failures. In the Saint-Simonian case, the fundamental ambiguity of the movement played a role. Because of their proposals regarding the modernization of the French banking system and the construction of railways, the disciples of Saint-Simon had attracted not only utopians dreaming of a new and better world, but also practical minds such as the bankers Emile and Isaac

---

[18] Sébastien Charléty, *Histoire du saint-simonisme (1825–1864)* (Paris: P. Hartmann, 1931); Henri René D'Allemagne, *Les Saint-simoniens 1827–1837* (Paris: Gründ, 1930); see also the five issues of the journal *Economies et sociétés* published under the title "Saint-simonisme et pari pour l'industrie," vol. 4, nos. 4, 6, 10; vol. 5, no. 7; vol. 7, no. 1 (1970–3); Jean Walch, *Bibliographie du saint-simonisme* (Paris: Vrin, 1967); Philippe Régnier, "De l'Etat présent des études saint-simoniennes," in *Regards sur le saint-simonisme et les saint-simoniens*, ed. Jean René Derré (Lyon: Presses universitaires de Lyon, 1986), pp. 161–206.

[19] Carl J Guarneri, *The Utopian Alternative: Fourierism in Nineteenth-Century America* (Ithaca, N.Y.: Cornell University Press, 1991).

Pereire and the engineer and entrepreneur Paulin Talabot.[20] Thus, Saint-Simonianism adumbrated socialism and the type of authoritarian capitalism that would develop during the Second Empire. The tension generated by the movement's dual nature was not easy to overcome.

On a more general level, utopian socialism was appealing insofar as its promises were in profound accordance with the aspirations of its time, particularly with the desire to make the new economic and social competition compatible with the restoration of collective and altruistic values. But once it became clear that these aspirations could as well be pursued using more traditional means, such as political action, the decline of the utopian movements was rapid. In France, for instance, the Republican party was able to attract many former utopians during the late 1840s. A similar process occurred in the United States, where Fourierism gradually lost its relevance as a viable alternative to political activism.

At their apex, utopian socialist movements emphasized practical issues, thus neglecting the scientific ambitions of their founding fathers. This neglect was especially pronounced in America, where the creation of communities absorbed the greater part of the available energies. The construction of a new science of man and society nevertheless remained an official goal. After the collapse of the utopian socialist movements, some of their old members became involved in scientific societies created for the same purpose. In France, for example, a former Saint-Simonian, Gustave d'Eichtal, became an active member of the Société Ethnologique, which was created in 1839.[21] Former American Fourierists played a similar role in the American Social Science Association, which was founded in 1865 by the Massachusetts humanitarian reformer Frank Sanborn.[22] Generally speaking, their contributions to this type of enterprise remained modest. Utopian socialism perhaps played a greater role as a counterexample than as a direct source of inspiration. Its failures seemed to demonstrate in particular the need to separate reflection and action. After Durkheim and Weber, the split between academic disciplines such as sociology and reformist activism was to serve as a guide for the further development of the social sciences.[23]

Were the utopian socialists the true founding fathers of nineteenth-century social science? The answer remains ambiguous. On the one hand, Saint-Simon, Owen, and their followers paved the way for Auguste Comte and his positive sociology by focusing on such problems as the collective

---

[20] Bertrand Gille, *La Banque en France au XIXᵉ siècle* (Genève: Droz, 1970); R. B. Carlisle, "Les Saint-simoniens, les Rothschild, et les chemins de fer," *Economies et sociétés*, 5 (1971), 1185–1214; Jean Walch, "Les Saint-Simoniens et les voies de communication," *Culture technique*, no. 19 (1989), 285–94.

[21] W. H. Chaloner and B. M. Ratcliffe, *A French Sociologist Looks at Britain: Gustave d'Eichtal and British Society in 1828* (Manchester: Manchester University Press, 1977), p. 148.

[22] Guarneri, *Utopian Alternative*, p. 400.

[23] Antoine Savoye and Bernard Kalaora, *Les Inventeurs oubliés: Le Play et ses continuateurs aux origines des sciences sociales* (Seyssel: Champ Vallon, 1989).

history of humanity and the study of society as a system of functions and of classes fulfilling those functions. Their focus on class struggle served to inspire Marx. On the other hand, their practical contribution to the emerging social sciences remained limited. Their characterization of society was based on general assumptions rather than on more specific material, such as case studies and surveys. As a whole, one might be tempted to interpret utopian socialism as a kind of prehistory of our contemporary social sciences rather than as an early stage of their history in the strict sense. In raising issues such as the weakening of the social bond and the social importance of religion, Saint-Simon, Owen, and Fourier were probably creating an agenda for sociology rather than answering its questions.

As a more positive way to assess the role played by utopian socialism, one can observe that many of the issues it raised exceeded the scope of the emerging social sciences. The disciples of Saint-Simon, for instance, paid attention to the emerging notion of networks. Extending their reflections far beyond the transportation networks that were developing at the time, they tended to interpret society itself as a series of interconnected networks.[24] The Saint-Simonians were interested in global issues, such as relations between the Occident and the Orient, and they did not take for granted the superiority of Europe over the rest of the world.[25] Fourier's interest in sexual liberation would become a major theme for later social scientists. The rediscovery of Saint-Simon's and Fourier's works in the 1960s was to a large extent a consequence of this evolution.

Finally, the most unruly features of the utopian socialists' doctrines, such as Saint-Simon's and Fourier's cosmologies,[26] may also be integrated into this positive assessment. Saint-Simonians and Fourierists were included in the notes left by Walter Benjamin for a book he never completed on nineteenth-century Paris.[27] The book was intended as a demonstration that capitalism and the rationalization process it implied had a mythical, almost dreamlike dimension. On the eve of the industrial revolution, utopian socialism was perhaps one of the best expressions of this mythical dimension, which was also to permeate the emerging social sciences. If not the transmigration of souls, then the cult of progress and the belief in absolutely positive social facts, as well as in permanent historical laws that could illuminate the future of mankind, were perhaps among those founding myths.

---

[24] Cf. Pierre Musso, *Télécommunications et philosophie des réseaux: La Postérité paradoxale de Saint-Simon* (Paris: PUF, 1997).

[25] Magali Morsy, ed., *Les Saint-simoniens et l'Orient: Vers la Modernité* (Aix-en-Provence: Edisud, 1989); Philippe Régnier, *Les Saint-simoniens en Egypte (1833–1851)* (Cairo: Amin F. Abdelnour, 1989); Ghislaine Alleaume, "L'Ecole polytechnique du Caire et ses élèves: La Formation d'une élite technique dans l'Egypte du XIXe siècle" (PhD dissertation, Université de Lyon II, 1993).

[26] Michel Nathan, *Le Ciel des fouriéristes: Habitants des étoiles et réincarnations de l'âme* (Lyon: Presses Universitaires de Lyon, 1981).

[27] Walter Benjamin, *Paris capitale du XIXe siècle: Le Livre des passages* (Frankfort, 1982; French translation Paris: Cerf, 1989).

6

SOCIAL SURVEYS IN THE EIGHTEENTH AND NINETEENTH CENTURIES

Eileen Janes Yeo

Jesus Christ was born while Mary and Joseph were on their way to be counted in an imperial census, in order to be taxed.[1] From antiquity onward, the state has played an active part in social survey work. By the sixteenth century, according to the *Oxford English Dictionary*, the word "survey" meant a state-conducted inventory of property, provisions, or people in order to raise revenue or a military force. However, starting in the seventeenth century, and well entrenched by the nineteenth, a different set of purposes for studying populations had also evolved, and the process of taking surveys began to pass into the hands of other social groups as well. Now voluntary enthusiasts as well as state bureaucrats were becoming concerned with statistics, in the sense not only of facts useful to the state but also of tabulated facts that would depict "the present state of a country," often "with a view to its future improvement."[2]

This chapter will explore some key developments and discontinuities in the history of large-scale quantitative social surveys, mainly in Britain and France. Others have told this story in terms of conceptual and methodological discoveries leading toward truly scientific modern surveys. I will instead examine the historical practices of social inquiry considered scientific in their own times, and argue that these investigations were also shaped by social imperatives, even in ostensibly neutral areas like statistical method.[3] The chapter begins with the introduction of the census around the time of the

---

[1] John Rickman, "Thoughts on the Utility and Facility of Ascertaining the Population of England" (1796) in David V. Glass, *Numbering the People: The Eighteenth-Century Population Controversy and the Development of Census and Vital Statistics in Britain* (Farnborough: Saxon House, 1973), p. iii.

[2] Sir John Sinclair's popular definition in *A Code of Political Economy, Founded on the Basis of Statistical Inquiries* (Edinburgh, 1821), p. xii; Alain Desrosières, *La Politique des Grands Nombres: Histoire de la Raison Statistique* (Paris: Éditions La Découverte, 1993), pp. 28–9, 35–6.

[3] For contested historiography, see Martin Bulmer, Kevin Bales, and Kathryn Kish Sklar, eds., *The Social Survey in Historical Perspective, 1880–1940* (Cambridge: Cambridge University Press, 1991), chap. 1, pp. 62–3.

French Revolution, and ends with the move to professionalization around the
time of the First World War. It considers the investigative focus on groups such
as the working classes and the poor, who were seen as important indicators
of national well-being and who can sometimes be glimpsed responding from
their own point of view.

Vision is integral to the "survey." An early synonym for survey was
"surview" (*surveu*), which involved a location in a visual field and in a power
relationship. The observers were positioned at a height and at a distance,
where they would obtain an overview of the whole, indeed a commanding
view, which became a qualification for the exercise of command. But surveys
are not like original sin, forever tainted by their historical origin. Indeed,
one of the important aspects of the social survey story is the active contestation that surrounded inquiries of all kinds. Social surveys were an important
part of social science in its nineteenth-century meaning as an empirical,
action-oriented science of happiness or improvement. As such, surveys were
contestable activity.[4]

POPULATION SURVEYS, ANCIENT AND MODERN

The need to conduct the earliest type of survey, the population census, became increasingly urgent from the eighteenth century onward, ultimately
for opposite reasons in Britain and France. As Michel Foucault has observed, modern states rest their legitimacy on their power to guarantee
life rather than to inflict death by means of execution or war.[5] This concern with the vitality of populations developed in two phases. Before the
French Revolution, a convergence of assumptions from religion and political economy highlighted population size. Theology, whether Catholic or
Protestant, took literally the injunction in Genesis to "be fruitful and multiply," a view exemplified in Rev. Johann Peter Süssmilch's *Divine Order*
(*Die Göttliche Ordnung*, 1741). Both mercantilists, stressing the importance
of trade, and Physiocrats, emphasizing wealth in land, thought of a large
population as crucial. The need to count the population and assess the pattern of its growth became urgent, but the task was beset with considerable
difficulty.

Old Regime surveys often met resistance from people opposed to higher
taxes and, sometimes, to "impious enumerations" that "outraged the Creator."[6] Also, the findings of such inquiries were considered state secrets and

---

[4] See Eileen Yeo, *The Contest for Social Science: Relations and Representations of Gender and Class*
(London: Rivers Oram, 1996), pp. x–xi.

[5] Michel Foucault, *The History of Sexuality*, vol. 1: *An Introduction* (1976), trans. R. Hurley
(Harmondsworth: Penguin, 1981), p. 136.

[6] Fernand Faure, "France," in *The History of Statistics: Their Development and Progress in Many Countries*,
ed. John Koren (New York: Macmillan, 1918), pp. 258–9.

rarely divulged. A survey conducted in 1697 by the Duc de Beauvillier was leaked, then summarized by Sébastian de Vauban in 1709, and used virtually unaltered for more than fifty years to argue that the size of the French population was static or declining. The myth of stagnation or underpopulation took deep hold in the French psyche at that time and has remained there ever since.[7] In Britain, there was controversy over whether the population had increased or decreased after the great fire in London (1666) and the Glorious Revolution (1688). This prompted thinkers like Sir William Petty (1623–1687) to calculate population growth (sometimes starting with Noah and the Flood) in a new inquiry that Petty called political arithmetic, a forerunner of demography.[8]

Enthusiasts and officials eager to number the population had to rely on their own ingenuity in place of comprehensive information. During this period there was no reluctance to utilize samples and multipliers of various kinds to reach conclusions about the national picture. In France, parish curates reported vital statistics to local officials, who conducted head counts in selected parishes and calculated a ratio between the mean number of births over the preceding six years and the total population in those parishes. They then determined the national population by multiplying the total number of births in France by the ratio.[9] In Britain, calculations were based on lists of taxpayers or bills of mortality. The defects of British records were so well recognized that parliamentary bills were introduced, in 1753 and 1758, to authorize an annual population census and the national collation of vital statistics. Both met defeat. The opposition attacked these attempts to "molest and perplex every single family in the Kingdom"; Sir William Thornton lambasted the bill as "totally subversive of the last remains of English liberty" and warned that he would subject any inquisitive enumerator to "the discipline of the horse pond."[10]

Nevertheless, in both countries, the force of events was moving to overcome such resistance by the end of the eighteenth century. In France, Enlightenment *philosophes* – and their opponents – insisted that the government under which, in Rousseau's words, "the citizens do most increase and multiply, is infallibly the best." But they argued that the population had fallen dramatically because of the degeneracy of the ancien régime. Jean-Jacques Rousseau (1712–1778) lambasted modern morals, targeting women who "turn to the prejudice of the species the attraction given for the sake of multiplying it. This practice, added to the other causes of depopulation, presages the impending

[7] Albert Soboul, *La Civilisation et la Revolution Française* (Paris: Arthaud, 1970), vol. 1, chap. 6; Faure, "France," pp. 250–5; Jacques Dupaquier, *Histoire de la Population Française* (Paris: PUF, 1988), vol. 2, pp. 30–43.

[8] William Petty, *Several Essays in Political Arithmetick* (1755) (London: Routledge, 1992).

[9] Eric Brian, *La Mesure de l'Etat: Aministrateurs et Géomètres aux XVIIIe Siècle* (Paris: Albin Michel, 1994).

[10] Quoted in Glass, *Numbering the People*, p. 20.

fate of Europe." He urged "experts in calculation" to count down the Old Regime.[11]

After the French Revolution, most states abandoned statisical secrecy. This was a critical turning point; nations resting their authority on a "rational" rather than "traditional" basis began to depend on what has recently been called a "knowledge base." They collected empirical information in order to formulate policy and monitor performance, and invited wide publicity and public discussion of surveys as evidence of their open style of government, their commitment to the public good, and, in democratic states, their representativeness and accountability to the people. The United States Constitution required a decennial census from 1790 onward precisely in order to ensure the equal apportionment of congressional seats. In Italy, statistics even gave some reality to a theoretical entity that was still to be created by a process of unification. Prussia's already elaborate machinery, created by Dr. Ernst Engel, was further replicated in the cities and states that were unified as the Kaiserreich in 1871.[12]

A torrent of statistics poured out in France, regardless of the pendulum swings between republic and monarchy, as each government tried to secure itself and expose the deficiencies of the previous regime. In 1801, a Service de la Statistique Générale was created, and Minister of the Interior J. A. Chaptal initiated a general enumeration of population and resources to be carried out by the new departmental *préfets*, who would be trained in statistical investigation while they familiarized themselves with the people they were to govern. The need of the Napoleonic state to gain credibility affected the choice of metholodogy. Suggestions for a more mathematically driven practice that reasoned from sampling were rejected as involving only a small knot of professional Paris *calculateurs*, which might smack of ongoing secret and abusive central power. Moreover, it was considered important not only to monitor the impact of reforms throughout the nation but also to enlist local élites into the nation-building project. However, local capitalists, landowners, and professional men were unwilling to have their own "private" spheres interrogated. In the end, the common people (*"ce qu'on appelle ici le peuple"*) became the acceptable object of scrutiny.[13]

[11] Jean Jacques Rousseau, *Emile; or, On Education* (1762), trans. A. Bloom (Harmondsworth: Penguin, 1991), p. 14, and "The Social Contract," in *Social Contract*, ed. Ernest Barker (Oxford: Oxford University Press, 1947), p. 280.

[12] Gianfranco Poggi, "The Modern State and the Idea of Progress," in *Progress and Its Discontents*, ed. Gabriel A. Almond, Martin Chodorow, and Roy Harvey Pearce (Berkeley: University of California Press, 1982), pp. 346–7; Michael Lacey and Mary Furner, "Social Investigation, Social Knowledge and the State," in their *The State and Social Investigation in Britain and the United States* (Cambridge: Cambridge University Press, 1993), pp. 5–7; Silvana Patriarca, *Numbers and Nationhood: Writing Statistics in Nineteenth-Century Italy* (Cambridge: Cambridge University Press, 1996), pp. 6–7; Ian Hacking *The Taming of Chance* (Cambridge: Cambridge University Press, 1990), pp. 18, 20.

[13] Marie-Noëlle Bourguet, "Décrire, Compter, Calculer: The Debate over Statistics during the Napoleonic Period," in *The Probabilistic Revolution*, ed. Lorenz Krüger, Lorraine Dalston, and Michael Heidelberger (Cambridge, Mass.: MIT Press, 1987), vol. 1, pp. 309–11.

In Britain, the impact of the French Revolution also focused the investigatory gaze on the laboring poor, but with dire foreboding about their fertility. During the decade of the Napoleonic wars, ruling-class alarm escalated, triggered by widespread food riots, intense radical activity, rebellion in Ireland, and mutiny in the fleet. The gentry and middle class, who had sometimes been at political loggerheads, now closed ranks to stabilize the nation. Urgent attempts were made to get an analytical as well as a political grip on the situation. *An Essay on the Principles of Population* was published in 1798 by the pioneer political economist and demographer Rev. Thomas Malthus (1766–1834), who challenged prevailing theological wisdom about population numbers as well as optimistic Enlightenment beliefs, such as Condorcet's belief in progress. Malthus argued that the laws of nature, that is, the general laws through which God acted, caused population to increase faster than the food supply in order to stimulate man, who is innately sluggish, to activity. For Malthus, unimpeded population growth would lead to national disaster. The remedy for the imbalance between population and subsistence lay in the capacity of the poor to exert moral restraint on their fertility. In 1803, Malthus brutally declared that if a man "cannot support his children, they must starve."[14] These shocking ideas were not readily accepted, but anxiety about the laboring poor in a context of political disorder prompted renewed demands for a national population census, which was actually carried out in 1801.

SOCIAL STATISTICS AND THOROUGHGOING ENTHUSIASM, 1830–1850

A proliferation of government and voluntary survey work on an unprecedented scale characterized the age of statistical enthusiasm. In 1833, the Statistique Générale de la France was revived, and from 1836 conducted quinquennial censuses that focused on families and households, no longer using samples or multipliers. The Académie des Sciences Morales et Politiques, with a section of Economie Politique et Statistique, was also revived. In Britain, new state agencies came into being, including the statistical section at the Board of Trade (1833) and the registrar general's office (1837), and learned bodies such as the London (later Royal) Statistical Society (1834), were founded. Londoners exulted at the "tendency to confront the figures of speech with the figures of arithmetic."[15] This was far more often a matter of comprehensive investigation than of probabilistic estimates. The most influential champion of a mathematical statistics, the Belgian savant and

---

[14] Thomas Robert Malthus, *An Essay on the Principle of Population* (1797), *Second Essay* (1803), ed. Patricia James (Cambridge: Cambridge University Press, 1989), vol. 1, p. 205, vol. 2, p. 105.

[15] *Journal of the Statistical Society of London*, 1 (1839), 8; Bertrand Gille, *Les Sources Statistiques de l'Histoire de France des Enquêtes du XVIIe siècle à 1870* (Geneva: Librairie Droz, 1964).

government statistician Adolphe Quetelet (1796–1874), did not ultimately practice what he preached. He assumed the lawfulness of the social world and urged the creation of a social physics (*physique sociale*) that would utilize quantitative methods to discover and express those laws. Deeply fearful of social disturbance, which he had directly experienced when troops had invaded his observatory during the upheavals of 1830, he sought regularities, constant forces of nature, that could withstand the perturbational forces unleashed by revolution.[16]

The consistency of the French crime statistics (published from 1827 onward) convinced him that large-scale regularity prevailed in every social domain, and that statistical laws were true when applied to groups even if false in relation to a specific individual: "The greater the number of individuals, the more the individual will is submerged beneath the series of general facts which depend on the general causes according to which society exists and is conserved."[17] He gave body to the mean in the form of his most famous construct, "*l'homme moyen*," the average man. This abstract being was the average of all human attributes in a given country, an epitome of the national character analogous to the center of gravity in physics. As deviations from the average necessarily cancelled themselves out whenever a great number of instances was considered, the mean was the significant type and had physical characteristics (easily measurable) and moral characteristics (more problematic) that developed over a lifetime. *L'homme moyen morale* could be calculated most easily, Quetelet suggested, from the crime statistics, divided by population numbers. Yet, despite his manifestoes, Quetelet almost never used mathematics in his statistical work but instead translated his quest for social order into the more mundane business of collecting, classifying and correlating facts.[18]

The dedication to thoroughness also characterized the upsurge of voluntary survey work, which focused not so much on the search for national character as on the pressing agenda of social pathology and class conflict. In France, Britain, and the United States, a striking feature of the nongovernmental investigations of the period was the focus on disorder in large cities. Particularly between 1830 and 1848, French survey work spotlighted what is now called the underclass and was then named *Les Classes Dangereuses de la Population dans les Grandes Villes,* the title of Dr. H. A. Frégier's classic study (1840). Doctors grouped around the *Annales d'Hygiene Publique* (1829–53), like their statistician counterparts in Britain, were mobilized by the cholera epidemic of 1832. Envisaging society as an organism and utilizing a medical language of health and disease, they considered cholera to be yet another

[16] Theodore M. Porter, *The Rise of Statistical Thinking* (Princeton, N.J.: Princeton University Press, 1986).
[17] Quetelet (1832) quoted ibid., p. 52.
[18] Desrosières, *Politique*, p. 206 and chap. 3.

symptom, along with political disruption and moral decay, of disorder in the social body. Surveys to diagnose the manifold symptoms of social illness extended also to Alexandre Parent-Duchâtelet's fifteen-year-long inquiry into prostitution in Paris (1836) and his crusade to rid the social body of blockages, such as dead flesh in sewers, and contaminants, including putrid discharges from the bodies of prostitutes. Dr. Louis Villermé produced a two-volume *Tableau de l'État Physique et Moral des Ouvriers Employé dans les Manufactures de Coton, de Laine et de Soie* (1840), which depicted the poorest factory workers living in the cellars of Lille as subhuman and indiscriminately "stacked" into "impure beds." His outlook resembled that of cholera doctor James Kay-Shuttleworth, whose 1832 study of Manchester cotton workers, along with his activity in the local statistical society, helped shift the focus away from the industrial scene.[19]

In Britain, the urban statistical societies, which appeared from 1833 onward (and which had American cousins in Boston and New York), were largely composed of the rising local bourgeoisie; only the London society had a predominantly professional membership on the French or American pattern. The British statisticians laid claim to local political authority on the grounds of their science and their service among the local working population, evinced not least by their social surveys. With part of their time, the local societies acted as embryonic town councils collecting civic statistics. But they spent most of their time making large-scale residential surveys of the local working classes, with a view toward improving their condition.[20] The Manchester society even apologized for not visiting *all* members of the working class, although the 4,102 families "below the rank of shopkeepers" included every such household in Dukenfield, Staleybridge, and Ashton-under-Lyne. There was no question of sampling. Completeness was mandatory, not only to ensure reliability, but also as a measure of social service and evidence that an overview (for governance) had been achieved.

The surveys focused upon "moral and intellectual statistics," not poverty. They emphasized facts about housing that they believed had implications for moral order, like the number of rooms, number of beds, and number of people in them. British investigators, like the French, thought that overcrowding and confusion, particularly in sleeping arrangements, which "indiscriminately" mixed sex, age, and family groups, were a potent index of disorder. Despite claiming that they asked no questions about wages or working conditions, because they had detected a "disposition to mislead or to resent inquiry" on these subjects, they persisted with questions about the ratio of people to

---

[19] James Kay-Shuttleworth, *The Moral and Physical Condition of the Working Classes Employed in the Cotton Manufacture in Manchester* (1832) (London: Cass, 1970); Villermé, *Tableau*, vol. 1, p. 83; Alexandre Parent-Duchâtelet, *La Prostitution à Paris au XIXe Siècle* (1836), ed. Alain Corbin (Paris: Le Seuil, 1981), pp. 12–14; for the inquiry movement, see Gérard LeClerc, *L'Observation de l'Homme: un Histoire des Enquêtes Sociales* (Paris: Le Seuil, 1979); Yeo, *Contest*, p. 63.

[20] See Yeo, *Contest*, pp. 64–76, for their surveys.

beds, which provoked equal resistance.[21] Their selective questions reflected their commitment to laissez-faire economics, which prohibited interference in the industrial system, and their hostility to working-class combinations, especially trade unions, that broke economic laws. However, they were deeply concerned with social discipline, which they undertook to influence by providing churches and schools and pressing for state aid to education.

They also supported scientific philanthropy. From the period of the French Revolution onward, more systematic monitoring of working-class households became common, both in villages and towns, on the model pioneered by Evangelicals like Hannah More (1745–1833). This trend was reinforced by the Malthusian Rev. Thomas Chalmers (1780–1847), a Christian political economist influential in British and American philanthropy for over a century. In 1820, Chalmers began a famous experiment in his Glasgow parish, relying on deacons who regularly visited the homes of the poor and exercised "the privilege of a strict search and entry upon the question of every man's state, who should claim relief."[22] His work was patterned on the German Elberfeld system, where men had been the visitors. Women took an important role in the Anglo-American "science of the poor," not least in the Charity Organisation Society, founded in 1869, which perfected the investigative method of casework and later helped to establish professional training in social work. Surveillance as well as survey, the close-up picture as well as the panorama, was the continuing outcome of impulses from many quarters to restructure the lives of the poor.

SOME EPISODES OF CONTESTATION

Not surprisingly, some of the objects of scrutiny openly contested such survey practices. The early socialist movement refused to prioritize urban residential conditions and sexual behavior as the pressing issues and instead pushed for the collection of "really useful knowledge." Their "social science" involved a critical analysis of the capitalist system and a blueprint for an alternative "New Moral World," which would restructure social as well as economic institutions in order to promote happiness for the majority. Socialists attacked the statisticians for wasting time on "laborious exhibitions of truths, tabulated and figured, which in the gross, are generally known and felt." The socialists, trade unions, friendly societies (insurance collectives), and Chartists (who agitated for universal suffrage) all collected statistics for their own purposes.

[21] Manchester Statistical Society, *Report . . . on the Condition of the Working Classes in an Extensive Manufacturing District in 1834, 1835, and 1836* (London: James Ridgway, 1838), p. 14; James Kay to Thomas Chalmers, Manchester Statistical Society Appendix, Manchester Central Library, item 4; Bristol Statistical Society, *Proceedings of the Second Annual Meeting* (Bristol, 1838), p. 10.

[22] Thomas Chalmers, *On the Sufficiency of the Parochial System, without a Poor Rate, for a Right Management of the Poor* (Glasgow: William Collins, 1824), p. 110; Yeo, *Contest*, pp. 8–9, 66–7, discusses scientific philanthropy.

The Chartist census of the regions in 1839 asked questions about the family's combined wages and the cost of living, topics that they considered vital to well-being, but which were being ignored by the statisticians.[23]

Because of such new working-class perspectives, correlations of a new kind became possible and for a short time were carried into the middle-class statistical world, not least by Henry Mayhew (1812–1887). He conducted perhaps the first poverty survey in 1849 and 1850, suggesting a causal relation between the industrial system and poverty. Starting with a hypothesis that low wages were a key cause of poverty, he devised a method of interviewing a representative cross-section of workers in a trade and developed a complex way of calculating wages that took factors like unemployment into account. He took seriously the point of view of his respondents, while recognizing that workers and employers had different biases: "Workpeople are naturally disposed to imagine that they get less than they really do, even as the employer is inclined to fancy his workmen make more than their real gains."[24]

The "true" working class briefly attracted the attention of French investigators. There were increasing complaints during the 1840s that neither voluntary investigators nor the state were seeking really useful facts about labor conditions. For example, the *Enquête Industrielle*, conducted fitfully by the minister of commerce between 1830 and 1847, tried to track economic prosperity by soliciting information only from industrialists. In the charged political atmosphere of 1848, socialists pressured the Constituent Assembly to order an *Enquête* focused on the working and living conditions of Parisian laborers. In response, the Paris Chamber of Commerce undertook a rival inquiry, published as *Statistique de l'Industrie à Paris, 1847–8*. This elite group of businessmen, manufacturers, and economists were intent on providing an alternative analysis of the impact of industrial capitalism. Rather than depicting workers as oppressed by capitalists, the *Statistique* saw small family enterprises both as the units of production and as matrices of moral development, where women functioned not only as a disciplining force but also as a symbol of class order (when they stayed at home). This polemical "reply to the socialists, in the guise of a scientific report" was the only survey to be published under the rigid censorship of the Second Empire.[25]

State survey work in Britain increasingly presented itself as comprehensive and objective. Professionalizing civil servants such as Edwin Chadwick

[23] David Rowe, "The Chartist Convention and the Regions," *Economic History Review*, 2nd ser., 22 (1969), 58–9, 71–2; Statistical Committee of the Town Council, "Report upon the Condition of the Town of Leeds and of Its Inhabitants," *Journal of the Statistical Society of London*, 2 (1839); *Northern Star*, 6 (13 Feb. 1841).

[24] See Eileen Yeo, "Mayhew as a Social Investigator," in *The Unknown Mayhew*, ed. E. P. Thompson and Eileen Yeo (London: Merlin, 1971), pp. 153, 54–64.

[25] Joan Scott, "Statistical Representations of Work: The Politics of the Chamber of Commerce's Statistique de L'Industrie à Paris, 1847–8," in *Work in France: Representations, Meaning, Organization and Practice*, ed. Steven Kaplan and Cynthia Koepp (Ithaca, N.Y.: Cornell University Press, 1986), pp. 354–63; Hilde Rigaudias-Weiss, *Les Enquêtes Ouvrières en France entre 1830 et 1848* (Paris: PUF, 1936).

(1800–1890) argued that only disinterested state officials could harmonize private and public interests. Chadwick wanted impartial investigatory bodies to collect authoritative facts as the basis for legislation. Then state inspectors would enforce the law while collecting yet more facts.[26] Under Chadwick's watchful eye, between 1832 and 1846 over 100 royal commissions inquired into such key issues as the condition of women and children in various industries and the health of towns. Inspectors "spread like contagion." In Britain and France, domestic census taking became routinized, especially after 1851, while metropolitan countries also took stock of their growing empires abroad. The most ambitious inquiries were the decennial censuses of India, starting in 1871 and undertaken in the name of efficiency and welfare reform.[27] All this state apparatus gave authority and the appearance of neutrality to what was often contestable knowledge. For example, the British census regarded home-based women as productive workers at midcentury, but by 1881 had started to move them into an "unoccupied class" of unproductive dependents, a designation that feminists all over the Western world were disputing at the turn of the twentieth century.[28]

Perhaps the most dramatic responses were directed at the imperial surveys. The early Indian censuses aroused not only the familiar fears of higher taxes and military conscription, but also suspicions that their real aim was to find wives for British soldiers.[29] As a result, in some places a spurt of marriages took place before census night; in others, young girls were returned as older women, or else not declared at all. Equally vexed was the issue of caste. The census authorities asked for caste affiliation, despite the difficulties of standardizing a classification across the country, and ranked the castes in order of "social precedence." Nationalists complained that this actually intensified the rivalry of castes and constituted a clear attempt to divide and rule. The British relished the princely power to decide this ranking of castes but found in time that others could play the game for their own advantage, as Indian groups began to lobby for better positions that would deliver immediate benefits in terms of jobs.[30] Another significant reaction in India and also in the Philippines was the use of indigenous dramatic forms to respond to the census. In Lahore, a comedy entitled *Census* played to packed houses. It caricatured the enumerator for taking the job without pay, made fun of rumors that the sexes were to be equalized by killing spare men, and mocked

[26] Yeo, *Contest*, pp. 76–8.
[27] Beverly, *Report on the Census of Bengal, 1872*, pt. 1, p. 1. See Hacking, *Taming of Chance*, p. 17, for France, the United States, and other imperial powers.
[28] Desley Deacon, "Political Arithmetic: The Nineteenth-Century Australian Census and the Construction of the Dependent Woman," *Signs*, 2:1 (1985), 29–32.
[29] Dandapani Natarajan, *Indian Census through a Hundred Years* (New Delhi: Registrar General, 1971), pp. 285–6, 283, 294.
[30] Ibid., pp. 287, 305–6; Bernard S. Cohn, "The Census, Social Structure and Objectification in South Asia," in his *An Anthropologist among the Historians and Other Essays* (Delhi: Oxford University Press, 1987), pp. 242–50.

behavior like that of the zealous man who numbered the flies among the living things in his household. In the Philippines, the census was directly challenged not only by guerillas, but also by a genre of nationalist melodrama that pictured the woman-nation and her patriot protector being threatened by an outside interloper male, the United States of America.[31]

MIDCENTURY EXPERTISE AND THE WORKING CLASSES

During the mid nineteenth century, the state monopolized large-scale social inquiry. Voluntary effort was dominated by experts, now including women, who defined the branches of meliorist social science. In Britain, public health physicians, reforming lawyers, slum clergy, and women philanthropists presented themseves as indispensable diagnosticians of social ills in the areas of sanitary, reformatory, and moral science, including education. Together with social economy, which addressed industrial and labor questions, these fields structured the departmental divisions in the National Association for the Promotion of Social Science in Britain (1857), the Brussels-based Association International pour le Progrès des Sciences Sociales (1862), and the American Social Science Association (1865), and helped to shape the concerns of the eight International Statistical Congresses held between 1855 and 1881. Such initiatives led to more internationally collaborative and standardized activity. These associations usually did not undertake surveys themselves; they received information about social problems and remedial "experiments" from experts, that is, from people in positions of administrative responsibility in state and voluntary organizations, including labor movements.

Within these bodies, a divided view of the working class usually prevailed. On the one hand, there were the "perishing and dangerous" classes, who were also denigrated as "immoral sewerage" or the "residuum," using public health or biological imagery. Usually urban and sometimes homeless, these poor people were the particular focus of the new remedial sciences. On the other hand, there were the "true" working classes, who were characterized in part by their membership in labor organizations, which were now regarded with tolerance. A vision of a well-functioning social system in which trade unions and capitalists could achieve negotiated agreements, with the help of arbitration services when necessary, prompted demands for information that could facilitate the process. The British Social Science Association carried out only one survey, into trades societies and strikes (1860), and strongly lobbied for an industrial inquiry as part of the 1871 census. Under Carroll D. Wright (1840–1909), who was prominent in the American Statistical Society and the

---

[31] Vincent Raphael, "White Love: Surveillance and Nationalist Resistance in the U.S. Colonization of the Philippines," in *Cultures of United States Imperialism*, ed. Amy Kaplan and Donald Pease (Durham, N.C.: Duke University Press, 1993), pp. 204–14; Natarajan, *Indian Census*, p. 294.

Social Science Association, the Massachusetts Bureau of Statistics of Labor (1869) began collecting information about wages and budgets.

Despite a new era of class cooperation in some social science bodies, the intensifying demands for working-class citizenship and increasing labor militancy provoked deep anxiety elsewhere. Both in France and in Britain there were accompanying developments in the focus of survey work. In opposition to Quetelet's preoccupation with averages, there arose a new interest not only in variation and variety of types, but also in minorities of excellence. Intellectual currents in biology helped shape this agenda, especially following the publication of Charles Darwin's *Origin of Species* in 1859. But the Anglo-French political context also helped to shift the focus.

In Britain, the anxious debate surrounding the 1867 Reform Bill, which gave the vote to a minority of working men, awakened real fears about "leaps into the dark" and about preserving social elites from extinction. Sir Francis Galton's major eugenic work, *Hereditary Genius*, appeared in 1869, the same year as Matthew Arnold's *Culture and Anarchy*. Galton (1822–1911) identified the educated professional classes as biologically superior stock, the key to national greatness. Utilizing the bell-shaped "normal" curve, as it would soon be called, he put the spotlight on the nature and effects of variation, and especially on the extremes of genius and worthlessness. The bulging hump of the curve, however, earned Galton's disdain: "Some thorough-going democrats may look with complacency on a mob of mediocrities, but to most other persons they are the reverse of attractive."[32]

In France, after the short-lived workers' commune of 1871, which was to haunt the imagination even of progressives like Émile Durkheim, those with political and cultural power felt the overriding need to put society again under the control of responsible élites. In 1876, Adolphe Bertillon launched a sharp attack on Quetelet in an influential essay on "La Théorie des Moyennes en Statistiques." He chipped away at the usefulness of *l'homme moyen* in social analysis, arguing that the mathematically derived traits of the average man were rarely to be found in actual individuals. He also made Galton-like noises against the belief that the average man could represent any ideal of moral or intellectual perfection; rather, such a man would be *"le type de la vulgarité."*[33] This article put the final nail in the coffin of Quetelet's reputation.

Yet the French wanted more working-class babies, however vulgar or mediocre. The chronic lament over depopulation became noisier after military defeat in 1871 and the loss of Alsace to Germany; it reached a crescendo in 1896, when census figures revealed that deaths had outstripped

[32] Francis Galton, "President's Address," *Journal of the Anthropological Institute*, 18 (1889), 407; see also his "The Possible Improvement of the Human Breed" (1901 Huxley Lecture), in his *Essays in Eugenics* (London: Eugenics Education Society, 1909), pp. 8–11, 19–20; and his *Hereditary Genius: An Inquiry into Its Laws and Consequences* (1869), 2nd ed. (London: Macmillan, 1892).

[33] Bernard-Pierre Lécuyer, "Probability in Vital and Social Statistics: Quetelet, Farr, and the Bertillons," in *Probabilistic Revolution*, ed. Krüger, Daston, and Heidelberger, vol. 1, pp. 330–1.

births. Bertillon's son Jacques, a physician and statistician, helped to found the National Alliance for French Population Growth in 1896 and developed a new science called demography.[34] In Britain, the more selective breeding strategies urged by eugenists proved to be too extreme for most social analysts. They supported instead a new science of social hygiene, which emphasized the importance of environment as well as heredity in developing the vigor of a nation.

INTERNATIONAL COMPETITION / INTERNATIONAL COMPARISON, 1880–1915

The physical efficiency of working people, both inside and outside the country, became an obsession in the late nineteenth century as economic and imperialist rivalry between Western nations reached a climax. To the more familiar fears, about labor militancy and the "residuum" of slum dwellers or immigrants, were added social guilt about poverty and eugenic panic over the possible degeneration of the national race. Since the vitality of the population was judged crucial for national competitiveness, there was now real impetus to compare the condition of the working classes in the various competing countries. This driving concern led eventually to methodological breakthroughs in social survey technique.

When Charles Booth started his massive survey of the *Life and Labour of the People of London* in 1886, he refused "the representative method," as sampling was then called, and chose comprehensiveness. The owner of a Liverpool shipping company, Booth had an intellectual and an actual cousinship to some of the businessmen who had originated statistical societies and created the door-to-door survey.[35] By the time chocolate manufacturer Seebohm Rowntree tested Booth's findings in provincial York in 1899, the country was becoming obsessed with physical efficiency, following revelations about the unfitness of many army recruits during the Boer War. Using the new science of nutrition, Rowntree set his "poverty line" at the budget necessary to keep a family in "a state of merely physical efficiency." Rowntree wanted to use York as a pointer to the national picture, but his assertion that "25 to 30 percent of the town populations of the United Kingdom are living in poverty" failed to convince the influential Interdepartmental Committee on Physical Deterioration (reporting in 1904).[36]

Instead, the baton was seized by professionals – government and university statisticians – who developed sampling methods to sharpen national pictures

---

[34] Karen Offen, "Depopulation, Nationalism, and Feminism in Fin-de-Siècle France," *American Historical Review*, 89 (1984), 658–9.

[35] Bulmer, *Social Survey*, chapters by Kevin Bales and E. P. Hennock.

[36] E. P. Hennock, "The Measurement of Urban Poverty: From the Metropolis to the Nation, 1880–1920," *Economic History Review*, 2nd ser., 40 (1987), 215–16; Seebohm Rowntree, *Poverty: A Study of Town Life* (London: Macmillan, 1900); Yeo, "Mayhew," pp. 88–95.

and to facilitate international comparisons. Issues about the representative method were thrashed out from 1895 onward in the International Statistical Institute (founded in 1883) by well-known figures like A. N. Kiaer, the head of the Norwegian Statistical Service.[37] Statistical innovation was most rapid in relation to the labor and poverty "problems," with the U.S. Bureau of Labor Statistics (founded in 1885) setting the international pattern. Under the direction of Carroll Wright, the Bureau produced a continuous, if misleading, series of average wages and retail prices covering the period 1860–91.[38] The French Office du Travail, created in 1891 within the Ministry of Commerce, undertook a range of inquiries into wages, unemployment, strikes, living conditions, and, in the tradition of Frédéric Le Play, family budgets. In Britain, the Labour Department gained a new sophistication with the arrival of Hubert Llewellyn Smith (1864–1945), who had studied mathematics at Oxford before moving to the Toynbee Hall social settlement and joining Charles Booth's survey team. Smith hired trained statisticians to devise an index of some 100 British towns, which made possible comparisons between British real wages and their European and American counterparts. They developed index numbers to express the range of variation among cities and to represent changes over time.

The key British figure to apply the representative method to social statistics was the mathematician A. L. Bowley (1869–1957). He developed techniques of random sampling and used the mathematics of probability and standard deviation tests to calculate the margin of error. This permitted quick and relatively cheap comparative local studies, which could then be matched against government indices to find their place in a national picture. His survey of five percent of working-class households in Reading, Northampton, Warrington, and Stanley produced a pathbreaking national analysis in *Livelihood and Poverty* (1915).[39] Bowley broke new ground in yet another way. His academic status, as a teacher of statistics at the London School of Economics and at Reading University, enabled him to create courses of training in the discipline and to promote the professionalization of social statistics.

WOMEN AND SOCIAL SURVEYS

The focus on poverty, the concern over the quality of the race, and the trend toward professionalization all proved helpful to women investigators,

[37] Alain Desrosières, "The Part in Relation to the Whole: How to Generalise? The Prehistory of Representative Sampling," in *Social Survey*, ed. Bulmer, Bales, and Sklar, p. 232; Roger Davidson, *Whitehall and the Labour Problem in Late-Victorian and Edwardian Britain: A Study in Official Statistics and Social Control* (London: Croom Helm, 1985), chap. 4.

[38] Mary Furner, "Knowing Capitalism: Public Investigation and the Labor Question in the Long Progressive Era," in *The State and Economic Knowledge: The American and British Experiences*, ed. Mary Furner and Barry Supple (Cambridge: Cambridge University Press, 1990), pp. 253–4.

[39] Arthur L. Bowley, *Livelihood and Poverty* (London: Bell, 1915); Hennock, "Measurement," pp. 220–3.

especially in Britain and America. During midcentury they had argued that woman's special qualities – her intuitive understanding, her affinity with the moral aspects of life, her caring commitment to individuals and practical action – ought to be added, on the communion of labor principle, to men's abstract intelligence and capacity to plan and command large-scale institutions and reforms.[40] Bringing the law of love into science, they insisted, would create a "stereoscopic view" and generate true social progress. They sometimes constructed themselves as social mothers, making family issues their special concern and providing reparenting where necessary through their active social work. As mothers and children moved higher on the national agenda at the turn of the twentieth century, women investigators could engage in survey work that was now deemed to be of national importance. They could also argue in favor of training that would make such social inquiry and social work more effective; thus they opened new professionalized career paths for women.

The investigative spotlight fell both on women workers and on mothers. The Women's Industrial Council in Britain, arising out of the women's trade union movement, conducted a series of surveys of industrial conditions. The most famous, *Married Women's Work* (1915), came to the unorthodox conclusion that working women could offer more to their children than dependent married mothers in very poor homes.[41] The Fabian Women's Group conducted a five-year-long investigation of the weekly budgets of some forty Bermondsey housewives with an income of *Round About a Pound a Week* (1912–13), concluding that it was impossible to "maintain a working man in physical efficiency and rear healthy children on the amount of money which is all these same mothers have to deal with."[42] In the United States, social settlements in urban neighborhoods were investigative powerhouses. In Chicago, the *Hull House Maps and Papers* (1895) contained the results of surveys that mapped the ethnic, racial, social, and economic dimensions of the local ward (see Chapter 35). By contrast, British women social workers resisted survey activity in favor of the close-up picture available through casework, which seemed a more direct expression of personal service.[43]

University-educated American women such as Edith Abbott, a pioneer figure in social work training, mobilized historical and economic analysis to illuminate working women's oppression. In a comparable way, the British Fabian Women's Group was convinced that women had not studied pressing

---

[40] Yeo, *Contest*, chaps. 5, 9; William Leach, *True Love and Perfect Union: The Feminist Reform of Sex and Society* (London: Routledge, 1981); Kathryn Kish Sklar, *Florence Kelley and the Nation's Work: The Rise of Women's Political Culture, 1830–1900* (New Haven, Conn.: Yale University Press, 1995), pp. 69–70.

[41] Clementina Black, ed., *Married Women's Work: Being the Report of an Enquiry Undertaken by the Women's Industrial Council* (London: Bell, 1915), p. 7; Ellen Mappen, *Helping Women at Work: The Women's Industrial Council, 1889–1914* (London: Hutchinson, 1985).

[42] Maud Pember Reeves, *Round About a Pound a Week* (1913) (London: Virago, 1979), p. 145.

[43] Dorothy Keeling, *The Crowded Stairs: Recollections of Social Work in Liverpool* (London: National Council of Social Service, 1961), p. 114.

issues "scientifically in their own interests. The available material is represented by the male investigator with his own unavoidable sex bias." Finding a home in the borderline discipline of economic history, Fabian academics such as Mabel Atkinson committed themselves to breaking new intellectual ground: "The economic history of this country from the point of view of the workers, to say nothing of the women workers, has yet to be written."[44]

While most women investigators, like their male counterparts, did little to enlist the perspectives of their subjects into the work, there were exceptions. Jane Addams (1860–1935) of Hull House believed that women's true contribution to social investigation would be to function as participant interpreters. They could explain the culture of social groups to one another, especially the views of parties involved in the familiar dyads of power relations: workers and capitalists, for example, or city authorities and ethnic communities.[45] In Britain, the Women's Co-operative Guild's general secretary, Margaret Llewelyn Davies (1861–1944), developed an investigative practice of self-representation. Continually asked to give the views of her organization, she tried to elicit the ideas of the members instead, using extensive questionnaire work that also asked informants to provide their own explanations and points of view. Her book *Maternity: Letters from Working Women* (1915) is perhaps the best-known example of this practice.[46]

PROFESSIONALIZATION VERSUS COMMUNITY SELF-STUDY

Thus, while the pendulum was swinging toward professionalization, populist ambition was also strong during the pre-war period. The civic survey movement in Britain, which had a more expert-led analogue in the American social survey movement, aimed to engage local citizens in the study and planning of their own cities. Patrick Geddes (1854–1932), the key ideologue of the British movement, established his headquarters in the aptly named Outlook Tower, positioned high above Edinburgh with commanding views. Geddes rejected the idea that any power was attached to such "supervision"; he said he was simply adopting Aristotle's ideal of a city that could be seen in its entirety all at once. Local people were to study their communities in terms of history and ecology, and to participate in a "Social Survey proper" of the people, "their occupation and real wages, their family budget and culture-level." The research would culminate in local exhibitions, using visual aids including

---

[44] Fabian Women's Group, *Three Years Work, 1908–1911* (London: Fabian Society, [1911]), p. 12.

[45] Dorothy Ross, "Gendered Social Knowledge: Domestic Discourse, Jane Addams, and the Possibilities of Social Science," in *Gender and American Social Science*, ed. Helene Silverberg (Princeton, N.J.: Princeton University Press, 1998).

[46] Women's Co-operative Guild, *Maternity: Letters from Working Women* (1915), ed. M. Llewelyn Davies (London: Virago, 1978); see also Yeo, *Contest*, pp. 266–7.

pictures and maps, to display the findings and the options for appropriate future development.

Not only was Geddes keen on a "synoptic vision," he wanted to recruit everyone to this way of seeing. He felt that those previously excluded from public life, (e.g., workers, women, schoolchildren) had a special contribution to make. "The essential matter for all of us," wrote Geddes, "is to become more and more of surveyors ourselves."[47] However, this vision was undermined by the very professionalization that it ostensibly challenged. In Britain, Geddes's most responsive "community" consisted of local authorities, teachers, and professional town planners. In the United States, where the survey relied more on expert direction, the voluntary helpers also tended to be other professionals rather than "average citizens."[48] The slow and by no means one-way process whereby both social statistics and social surveys became professionalized activities, undertaken by trained experts working in government, market research, and university posts, belongs to the twentieth century. The nineteenth century, as we have seen, was characterized by the involvement of a wider range of social groups and institutional settings, which made social surveys a more visible part of a contested politics of knowledge.

[47] Patrick Geddes, "A Suggested Plan for a Civic Museum (or Civic Exhibition) and Its Associated Studies," *Sociological Papers* (1906), 203; see also his *Cities in Evolution* (1915), new and rev. ed. (London: Williams and Norgate, 1949), pp. 157, 86, 122; Martin Bulmer, "The Decline of the Social Survey Movement and the Rise of American Empirical Sociology," in *Social Survey*, ed. Bulmer, Bales, and Sklar, pp. 295–7.

[48] Stephen Turner, "The Pittsburgh Survey and the Survey Movement: An Episode in the History of Expertise," in *Pittsburgh Surveyed: Social Science and Social Reform in the Early Twentieth Century*, ed. Maurine Greenwald and Margo Anderson (Pittsburgh: University of Pittsburgh Press, 1996), pp. 37–9.

7

SCIENTIFIC ETHNOGRAPHY AND TRAVEL, 1750–1850

Harry Liebersohn

The period from the late eighteenth to the mid nineteenth century forms a distinctive era in the history of scientific ethnographic writing. A double framework of technological and political change demarcates its beginnings. On the technological side, advances in mathematics and scientific instrument making facilitated accurate navigation over the thousands of miles of a world sea voyage.[1] On the political side, the era opens with the British victory over the French in the Seven Years' War (in its North American theater, the French and Indian War), which was ratified by the Treaty of Paris, signed in 1763. This conclusion to one contest set off a new round of competition between the two great powers, who now played out their rivalry in the vast, hitherto imperfectly charted expanse of the Pacific.

State-sponsored French and British voyages soon set out to scour the far side of the globe for layover stations on the journey to Asia. Louis Antoine de Bougainville (1729–1811), a mathematical prodigy who had served Montcalm's expedition during the disastrous concluding phase of the struggle for North American hegemony, led a world voyage from 1766 to 1769. On the British side, James Cook (1728–1779), who had distinguished himself as a surveyor-hydographer in Newfoundland, led three scientific voyages around the world from 1768 to 1771, from 1772 to 1775, and from 1776 until his death in Hawaii. These and other "scientific voyages" of the late eighteenth century served imperial aims by providing accurate charting of island locations and coastlines, one of the most remarkable achievements of the officers and scientists who risked their lives on wind-driven odysseys to the ends of the earth.[2] During

---

[1] Marie-Noëlle Bourguet and Christian Licoppe, "Voyages, mesures et instruments: une nouvelle expérience du monde au siècle des lumières," *Annales*, 52 (1997), 1115–51.

[2] Bernard Smith, *European Vision and the South Pacific*, 2nd ed. (New Haven, Conn.: Yale University Press, 1985); Lynn Withey, *Voyages of Discovery: Captain Cook and the Exploration of the Pacific* (Berkeley: University of California Press, 1989); more generally, David P. Miller and Peter H. Reill, eds., *Visions of Empire: Voyages, Botany, and Representations of Nature* (Cambridge: Cambridge University Press, 1996); Nicholas Jardine, James A. Secord, and Emma C. Spary, eds., *Cultures of Natural History* (Cambridge: Cambridge University Press, 1996).

the French Revolutionary and Napoleonic eras, there was a slowdown in the quest for scientific and political mastery of the Pacific, despite notable exceptions such as the Baudin world voyage of 1800–03. With the restoration of peace on the European continent after 1815, a new phase of scientific voyaging began. Russian and American long-distance voyagers now competed with the French and British for trade and colonies. Russia, its empire already extended to Kamchatka, also sent out scientific expeditions, beginning with the world voyage of the *Nadeshda* and *Newa*, commanded by Adam Krusenstern (1770–1846), from 1803 to 1806. While the United States did not outfit a comparable naval voyage, the Lewis and Clark expedition (1804–6) had the same ambition to use the latest scientific knowledge to map out a large unknown territory and open it up to colonization; like the Bougainville, Cook, and Krusenstern voyages, it was the first of a succession of officially sponsored scientific expeditions.[3] These were the equivalent of today's voyages to outer space, a race for prestige as well as for material gain. Private travelers, too, made scientific journeys. The most important of these was Alexander von Humboldt (1769–1859), whose careful preparation and precise methods of observation served as a model for other nineteenth-century travelers.[4]

To define a period roughly from 1750 to 1850 as a unified era in the history of scientific ethnography and travel cuts across conventional categories of cultural history by splicing together the Enlightenment and Romantic eras. While these categories remain useful, they also overlap in travel accounts: A proto-Romantic sympathy with native peoples and longing for exotic places colors reports of the late eighteenth century, while an enlightened interest in scientific methodology and empirical accuracy carries over into the first half of the nineteenth century. Characteristic of the entire era is a tension between Enlightenment sense and Romantic sensibility.

NETWORKS OF KNOWLEDGE

We simplify the process of acquiring ethnographic knowledge beyond recognition if we imagine that well-trained experts simply gathered information and took it back home. It would be more accurate to view them as operating within makeshift *networks of knowledge* spun around the world, with many nodules supporting their published reports.[5] In "new" areas like the Pacific, a

---

[3] William H. Goetzmann, *Exploration and Empire: The Explorer and the Scientist in the Winning of the American West* (New York: Knopf, 1966).

[4] Michael Dettelbach, "Humboldtian Science," in Jardine, Secord, and Spary, eds., *Cultures of Natural History*, pp. 287–304.

[5] See Bruno Latour, *Science in Action: How to Follow Scientists and Engineers through Society*, chap. 6 (Cambridge, Mass.: Harvard University Press, 1988); see also Mary Louise Pratt, *Imperial Eyes: Travel Writing and Transculturation* (London: Routledge, 1992).

moment of first contact quickly gave way to a systematic mapping of islands and the creation of a standardized body of lore about the peoples of Oceania.

The specialized case of ethnography gives a radical twist to the significance of the networks of knowledge; for they not only transmitted, but were also inseparable from the production of, knowledge about "savages." European conceptions of these peoples – and what we can know today about their histories – was inseparable from the networks pulsating from metropolis to periphery and back. What the travelers knew depended on the natives they talked to; what they could report was conditioned by their patrons and audiences at home. They were not independent agents, but mediators.[6]

At home, learned societies like England's Royal Society (founded in 1660) and the American Philosophical Society (founded in 1768) helped to organize voyages. Entrepreneurs of science – such as Joseph Banks (1743–1820) (after his return from Cook's first voyage), Alexander von Humboldt (after his return from his own voyage to South America), and the Göttingen professor of anatomy Johann Friedrich Blumenbach (1752–1840) – recommended the scientists qualified to serve on them.[7] Monarchs and governments provided ships and permission papers for travel, and granted the privilege of publishing an account after one's return from an official voyage. Missionary societies, such as the Nonconformist Evangelicals, imbued their emissaries with a powerful ideological zeal that combined belief in the Kingdom of God and in the superiority of British civilization; missionaries' reports from South Africa, for example, blended with secular accounts to provide the European public with its images of Africa.[8] Quite apart from the need to appeal to a reading public, scientists had to trim their accounts of voyages to provide suitable propaganda for their patrons. One obstreperous scientist, Johann Reinhold Forster (1729–1798), lost his permission to write a narrative account of the second Cook voyage. While rebuke was rarely so severe, these scientists did not write in an atmosphere free of political and economic constraints.[9]

Once they embarked on their voyages, travelers encountered new forms of dependency. The pitiless discipline and rough society of the ship was often an ethnographic adventure in its own right. Sailors could ruin specimens

[6] Richard H. Grove, *Green Imperialism: Colonial Expansion, Tropical Island Edens, and the Origins of Environmentalism, 1600–1860* (Cambridge: Cambridge University Press, 1995).

[7] John Gascoigne, *Joseph Banks and the English Enlightenment: Useful Knowledge and Political Culture* (Cambridge: Cambridge University Press, 1994) and *Science in the Service of Empire: Joseph Banks, the British State and the Uses of Science in the Age of Revolution* (Cambridge: Cambridge University Press, 1998); Hans Plischke, *Johann Friedrich Blumenbachs Einfluss auf die Entdeckungsreisenden seiner Zeit* (Abhandlungen der Gesellschaft der Wissenschaften zu Göttingen, Philologisch-Historische Klasse, Dritte Folge, Nr. 20) (Göttingen: Vandenhoeck & Ruprecht, 1937).

[8] Jean and John Comaroff, *Of Revelation and Revolution: Christianity, Colonialism, and Consciousness in South Africa*, vol. 1 (Chicago: University of Chicago Press, 1991).

[9] James E. McClellan III, *Science Reorganized: Scientific Societies in the Eighteenth Century* (New York: Columbia University Press, 1985); Michael E. Hoare, *The Tactless Philosopher: Johann Reinhold Forster (1729–98)* (Melbourne: Hawthorn Press, 1976).

and instruments; captains could suspect the scientists, who did not fit neatly into the ship's hierarchy, of not respecting their authority. By the time they reached the peoples of the Pacific, scientists could consider who was better company – their fellow Europeans on the ship or the natives they encountered in island societies. The naturalist and man of letters Adelbert von Chamisso (1781–1838), disgusted by the crudeness of the Russian sailors and needled by his captain on the *Rurik* scientific voyage of 1815–18, preferred the *ali'i* or aristocrats of Hawaii and the atoll dwellers of Micronesia.[10]

Their dependence did not lessen, but only changed, when they were in the "field," for scientists needed to work with informants and guides. In North America, the economy of the fur trade had engendered a whole *métis* world of hunters and traders who moved back and forth between Native Americans and Euro-Americans. Sacagawea (ca. 1786–1812), guide to Lewis and Clark, has been sentimentalized (and her role sometimes exaggerated), but she exemplifies an entire class of local guides who made it possible for Europeans to cross unknown lands and seas.[11] Travelers also had to take into account the demands of local leaders; for example, the painter and writer George Catlin (1796–1872), who made a visual survey of native peoples of North America, once angered Mandan Indian warriors in the upper Missouri valley when he painted a village good-for-nothing before turning his attention to them.[12] Polynesian political elites influenced scientists through their skill as hosts: Quick to flatter foreign visitors and to satisfy their demands for women, water, and food, they were coauthors of the European myth of Pacific paradises. Europeans believed the myth at their peril, as the mutineers on the *Bounty* discovered when, without cannon to back them up, they ventured onto Tahiti and found themselves caught in a crossfire of local rivalries.[13] For Europeans to suppose that natives would submit without challenge to their imperial will was to underestimate the resilience of native politics.

The ethnographies of the late eighteenth and early nineteenth centuries, then, do not simply transcribe travelers' impressions of the things they have witnessed. Rather, they capture a many-sided drama involving actors across the world, all of them contending to dominate the "truth" about encounters among strange peoples.

---

[10] Adelbert von Chamisso, *A Voyage around the World with the Romanzov Exploring Expedition in the Years 1815–1818 in the Brig Rurik* (1836), trans. and ed. Henry Kratz (Honolulu: University of Hawaii Press, 1986).

[11] James P. Ronda, *Lewis and Clark among the Indians* (Lincoln: University of Nebraska Press, 1984); Richard White, *The Middle Ground: Indians, Empires, and Republics in the Great Lakes Region, 1650–1815* (Cambridge: Cambridge University Press, 1991).

[12] George Catlin, *North American Indians* (1841), ed. Peter Matthiessen (New York: Penguin, 1989), p. 112. Originally published as *Letters and Notes on the Manners, Customs, and Conditions of the North American Indians Written during Eight Years' Travel (1832–1839) amongst the Wildest Tribes of Indians of North America*.

[13] Greg Dening, *Mr. Bligh's Bad Language: Passion, Power and Theatre on the Bounty* (Cambridge: Cambridge University Press, 1992).

NARRATIVES OF KNOWLEDGE

Scientists were not just mediators in the extended geographic networks of periphery and metropolis; they were also independent-minded intellectuals who formed their own views of the things that they saw and, indeed, sometimes developed a belief that they were bearing witness to world-historical events for a European public. The pattern of commentary that most directly linked travelers to the politics of their age was the discourse of emancipation. European society before the French Revolution was organized into birth-ordered ranks of varying degrees of fluidity, from the flexible distinctions of English society to the legally enforced hierarchies of the Continent and the near-slave conditions of peasant servitude in Russia. During the late eighteenth century, thinkers across Europe debated the rights and wrongs of social hierarchy and speculated about whether man in a state of nature had lived in conditions of equality, conditions that if restored would add dignity to the lives of the many and bring an end to the corrupt rule of an empowered few.

Philosophes of the French Enlightenment drew on travel writing to validate their criticisms of politics at home and of colonial administration overseas. The most famous such work was Jean-Jacques Rousseau's (1712–1778) *Discourse on Inequality* (1755). Often simplified in the course of the polemics it has engendered since its publication, the *Discourse* both affirmed the natural equality of man and accepted the inevitability of inequality in a technologically advanced, state-governed society. Rousseau turned to the reports of travelers for empirical evidence about social organization during earlier stages of social evolution.[14] Hardly less brilliant was Denis Diderot's (1713–1784) *Supplement to Bougainville's Voyage* (1796).[15] Bougainville's news of a "New Cythera" called Tahiti was a sensational topic of discussion for the republic of letters; Diderot took it as the starting point for a utilitarian critique of Europe's sexual mores, in which Christian prohibitions on sexuality would give way to a eugenics program for breeding a healthy and intelligent population. More broadly, the *philosophes* made use of travel literature to point out the political virtues of man in his natural state and to criticize the abuses resulting from privilege. Their aims were usually reformist. Voltaire's *L'Ingénu* (1767), for example, used outside perspectives as a corrective to the defects of a European society whose fundamental superiority was never in doubt.[16]

The first scientific travelers to the Pacific were aware that they had an opportunity to test the Rousseauist notion that indigenous societies were

---

[14] Jean-Jacques Rousseau, *A Discourse on Inequality*, trans. Maurice Cranston (New York: Penguin, 1984).

[15] Denis Diderot, *Supplément au Voyage de Bougainville*, ed. Herbert Dieckmann (Geneva: Droz, 1955); see esp. Dieckmann's Introduction, p. xxxix. Peter Jimack, *Diderot: Supplément au Voyage de Bougainville* (London: Grant and Cutler, 1988).

[16] Voltaire, *Zadig/L'Ingénu*, trans. John Butt (New York: Penguin, 1964); Michèle Duchet, *Anthropologie et histoire au siècle des lumières: Buffon, Voltaire, Rousseau, Helvétius, Diderot* (Paris: Maspero, 1971).

naturally egalitarian. They returned skeptical, with cautionary tales for their contemporaries. Bougainville records how he was entranced by the seeming equality of the Tahitians, only to become aware that theirs was actually a highly stratified society.[17] A more searching examination of the issue was George Forster's (1754–1794) account of the second Cook voyage around the world (on which he served as his father's assistant). As the ship wandered from island to island, Forster treated the Pacific as a kind of social scientific laboratory, comparing the effects of wealth and climate on various forms of human government. His *Voyage Round the World* (1777) observed a cycle that moved from republican purity to aristocratic corruption. Forster admired the egalitarian simplicity of the Marquesans and regretted the Tahitians' decline into a hierarchical system of plantation exploitation by an aristocratic warrior elite.[18] He returned to Europe with a profound knowledge of the diversity of human societies, but with his republican convictions unshaken.

From one end of the globe to the other, travelers commented on the institution of slavery. Stories about its horrors found a wide international readership. One of the best-sellers of the day was John Gabriel Stedman's (1744–1797) *Narrative of a Five Years Expedition against the Revolted Negroes of Surinam* (1796). The book wove together Stedman's account of his reluctant participation in Surinam's civil war (between former slaves and masters), his love affair with a slave woman, and scientific observations on the nature and economy of the region.[19] Another best-selling commentator on slavery was the autobiography of Olaudah Equiano (ca. 1750–1797). A native of Guinea, Equiano was captured and sold into slavery but eventually gained his freedom and became a well-known man of letters in England. He wrote his life story as a conversion narrative, a rhetoric well calculated to ingratiate him with, and to humanize him for, his English reading public.[20] Travelers' reports on slavery and other abuses of European colonialism provided the empirical evidence for the Abbé Raynal's (1713–1796) *Philosophical History of the Two Indies* (1770). This work was not written by Raynal alone but was a collaborative effort of the late Enlightenment, with Diderot among its authors; a vast critique of European colonialism, it was unremitting in its condemnation of slavery in the name of universal human rights.[21] The

---

[17] Louis-Antoine de Bougainville, *A Voyage Round the World*... (1771), trans. J. R. Forster (Amsterdam: Da Capo, 1967).

[18] Georg Forster, *Werke*, vol. 1: *A Voyage Round the World* (Berlin: Akademie-Verlag, 1968); Claus-Volker Klenke with Jörn Garber and Dieter Heintze, *Georg Forster in interdisziplinärer Perspektive* (Berlin: Akademie-Verlag, 1994).

[19] Editors' Introduction to John Gabriel Stedman, *Stedman's Surinam: Life in an Eighteenth-Century Slave Society*, ed. Richard Price and Sally Price (Baltimore: Johns Hopkins University Press, 1992). Originally published as *Narrative of a Five Years Expedition against the Revolted Negroes of Surinam*.

[20] Olaudah Equiano, *The Interesting Narrative of the Life of Olaudah Equiano, Written by Himself*, ed. Robert J. Allison (1789) (Boston: Bedford Books/St. Martin's Press, 1995).

[21] Guillaume-Thomas Raynal, *Histoire philosophique et politique des établissemens et du commerce des européens dans les deux Indes*, 10 vols. (Geneva: Pellet, 1780); Michèle Duchet, *Diderot et L'histoire des Deux Indes, ou l'Écriture Fragmentaire* (Paris: Nizet, 1978).

debate over slavery did not end in 1789; travelers fervently reported both for and against it until its abolition in the British colonies, the French colonies, and the United States. The travel accounts of the Russian expeditions of the day, too, brooded on the evils of human bondage. Again and again, captains and scientists exposed the abusive practices of the Russian-American Company, a government-sponsored fur agency that kidnapped Aleutian and other native men in order to exploit their hunting skills. Krusenstern (already mentioned as commander of the first Russian world voyage), Georg Heinrich von Langsdorff (1774–1852), the ship's doctor, and other learned visitors were appalled by the barbarism of the Russian overseers.[22] These observers were conservative reformers convinced that it was in the Russian empire's self-interest to correct the evils they observed.

A significant number of women went abroad during the late eighteenth and early nineteenth centuries. Their writings were not necessarily enlightened or sympathetic to the native peoples they observed. Janet Shaw, a well-to-do Scottish woman traveling in the British West Indies in 1774 and 1775, integrated plantation slavery into her overall aestheticization of the tropical islands.[23] Yet there were also women travelers who used their experiences abroad as an opportunity to reflect on conditions at home. For example, women travelers to the Middle East were permitted, unlike their male counterparts, to enter the women's quarters of Muslim households. Despite male travelers' portrayal of paternal tyranny, they came away favorably impressed by the autonomy of Muslim women and with a heightened awareness of their own restrictions at home.[24] To appreciate the significance of independent women travelers in an age of emancipation, we should recall that travel was a social practice as well as a preparation for writing. For women, as for men, to leave behind fathers and spouses and to make their own way in a foreign country offered a refreshing pause from paternalistic authority.[25]

After 1789 there was a shift in the political center of gravity of travel writings. *Equality* now became a subject of intensified critique. With the actual institution of republican regimes in Europe itself, a society based on legal equality was no longer a matter of speculation. Salon guests who had once recounted fables about the virtues of natural man now had to contend

[22] A. J. von Krusenstern, *Reise um die Welt in den Jahren 1803, 1804, 1805 und 1806 auf Befehl seiner Kaiserlichen Majestät Alexander des Ersten auf den Schiffen Nadeshda und Newa . . .* (St. Petersburg: Schnoorschen Buchdruckerey, 1811), vol. 2, pp. 113–21; G. H. von Langsdorff, *Bermerkungen auf einer Reise um die Welt in den Jahren 1803 bis 1807* (Frankfurt am Main: Friedrich Wilmans, 1812), vol. 2, pp. 11, 31, 55, 63–4; V. M. Golovnin, *Around the World on the Kamchatka, 1817–1819* (1822), trans. Ella L. Wiswell, Foreword by John J. Stephan (Honolulu: University of Hawaii Press, 1979), p. xxix.

[23] Elizabeth A. Bohls, *Women Travel Writers and the Language of Aesthetics, 1716–1818* (Cambridge: Cambridge University Press, 1995), pp. 46–65.

[24] Billie Melman, *Women's Orients: English Women and the Middle East, 1718–1918 – Sexuality, Religion, and Work* (Ann Arbor: University of Michigan Press, 1992), pt. 2.

[25] Cf. Dennis Porter, *Haunted Journeys: Desire and Transgression in European Travel Writing* (Princeton, N.J.: Princeton University Press, 1991).

in their own lives with the consequences, good and bad, of a democratic revolution. Several refugees from the French Revolution recorded their disappointment with North America after experiencing it firsthand. François-René de Chateaubriand (1768–1848), who had toyed with radical late Enlightenment ideas as a young lieutenant stationed in Paris, records in his memoirs how he lost his illusions of an early Roman republic reborn soon after he actually arrived in the United States in 1791.[26] Constantin-François de Volney (1757–1820), a supporter of the Girondins, had time to rethink his revolutionary convictions while imprisoned by the Jacobins for thirteen months; after his period of exile in the United States, he wrote a debunking account of American Indians in order to deflate the *philosophes'* idealization of them as natural republicans. Whether the supposed egalitarianism of "primitive" peoples was a good or a bad thing was to remain a subject of controversy for decades to come, but in general, chastened liberals as well as conservative travelers campaigned against the eighteenth-century's idealizations of "natural man."[27] The impact of equality on human institutions later provoked comment from Charles Darwin in his 1839 account of the *Beagle* voyage: Darwin attributed the wildness and poverty of the inhabitants of Tierra del Fuego to their insistence on an equal sharing of property and power, which checked, he thought, any formation of a higher culture.[28]

While the supposed egalitarianism of indigenous peoples became more problematic, travel writers and theorists after 1789 developed a new appreciation of *liberty*. Indigenous hunter-warriors were supposed to have preserved the ancient virtues of the archaic Greeks, the early Romans, and the Teutonic forest. This was an old interpretation of indigenous societies; Michel de Montaigne (1533–1592) had proposed it in his essay "On Cannibals."[29] In the eighteenth century, Adam Ferguson (1723–1816) took up the same theme; pointing in *An Essay on the History of Civil Society* (1767) to the martial virtues of the North Americans, Ferguson feared that commercial societies, in the course of their material progress, had lost the hardiness of their ancestors.[30] After 1800, however, there was a renewed emphasis on this theme. The motives were mixed; they included resentment toward the democratic leveling of ancient privileges as well as fear of the effects of commerce and industry. Alexis de Tocqueville (1805–1859) made ingenious use of this concept of freedom in the first volume of *Democracy in America* (1835). Drawing on both

[26] François-René de Chateaubriand, *Mémoires d'outre tombe*, ed. Maurice Levaillant and Georges Mouliner (1849) (Paris: Gallimard, 1951), vol. 1, p. 220.

[27] Constant-François de Volney, *Oeuvres*, vol. 4: *Tableau du climat et du sol des États-Unis d'Amérique*..., 4th ed. (1803) (Paris: Parmantier, 1825), pp. 371–463.

[28] Charles Darwin, *Voyage of the Beagle* (1839), ed. Janet Browne and Michael Neve (New York: Penguin, 1989), p. 184.

[29] Michel de Montaigne, *Essays* (1595), trans. J. M. Cohen (New York: Penguin, 1958), pp. 105–19.

[30] Adam Ferguson, *An Essay on the History of Civil Society* (Edinburgh: Millar and Caddel; London: Kincaid and Bell, 1767), p. 143; Fania Oz-Salzberger, *Translating the Enlightenment: Scottish Civic Discourse in Eighteenth-Century Germany* (Oxford: Clarendon Press, 1995).

his personal observations and his reading, from Tacitus to James Fenimore Cooper, he portrayed American Indians as the epitome of the hunter dedicated to personal freedom. Tocqueville employed the theme as both an ideal and a warning: He admired the Indians' warrior virtues (and was disgusted by the settlers' and the government's treatment of them), but he also thought that their supposed inability to adapt to the tedium of bourgeois routine should serve as a warning to another warrior elite, his fellow aristocrats.[31]

COMPARATIVE METHODS

One general characteristic of the period 1750–1850 is the search for valid *comparative methods* of inquiry. Europeans gathered an ever-growing number of "scientific" reports about indigenous peoples. Evaluating the fantastic clutter of skulls, costumes, vocabularies, adventure stories, economic reports, and other souvenirs was not easy. Scientists and travelers tried out competing schemes for creating comparisons across space that would bring order to the newly discovered diversity of human societies. As Foucault has observed, the search for structural principles was a widespread feature of the sciences after 1800.[32] One may also note a certain proto-professionalization after the turn of the century, with the formation of societies in which the work of collecting and comparison began to take place. The newly founded geographic societies had a special interest in native societies: The first was founded in Paris in 1821, followed by a society in London in 1830 and one in Berlin in 1832. We lack detailed studies of these organizations, but even a cursory look at their early membership lists and publications reveals that they were organizations for the notables of their time, bringing together aristocrats, powerful ministers of state, generals, and leading scientists. Their members' curiosity ranged widely, from surveys of natural resources to cultural and linguistic inquiries.[33] By the time the Société Ethnologique de Paris and the Ethnological Society of London were founded (in 1839 and 1843, respectively), a crystallization of interest in the comparative study of native societies was already under way.[34]

[31] Alexis de Tocqueville, *Democracy in America* (1835–40) (New York: Harper and Row, 1988), pp. 316–39, and "A Fortnight in the Wilds," in his *Journey to America*, trans. George Lawrence, ed. J. P. Mayer, revised in collaboration with A. P. Kerr (Garden City, N.Y.: Doubleday, 1971), pp. 350–403; Harry Liebersohn, *Aristocratic Encounters: European Travelers and North American Indians* (Cambridge: Cambridge University Press, 1998).
[32] Michel Foucault, *The Order of Things* (1966) (New York: Vintage, 1973), pp. 221, 263–94.
[33] For the Société de Géographie, see the *Bulletin de la Société de Géographie* (1822–); Dominique Lejeune, *Les sociétés de géographie en France et l'expansion coloniale au 19. siècle* (Paris: Albin Michel, 1993), pp. 9–45. For the Royal Geographical Society, see *The Journal of the Royal Geographical Society of London* (1832–); Ian Cameron, *To the Farthest Ends of the Earth: 150 Years of World Exploration by the Royal Geographical Society* (New York: Dutton, 1980). For the Gesellschaft für Erdkunde, see *Jährliche Übersicht der Thäatigkeit der Gesellschaft für Erdkunde in Berlin* (1834–).
[34] George W. Stocking, Jr., *Victorian Anthropology* (New York: Free Press, 1987), pp. 243–5.

From a variety of different perspectives, both travelers and stay-at-home men of learning attempted to study natives by developing a racial science. Johann Caspar Lavater (1741–1801) sketched a science of "physiognomy" that would deduce psychological from facial and other physiological character-istics. While never terribly respectable, physiognomy gave expression to a widespread, often implicit travelers' belief that one could judge the character and culture of foreign peoples by their external appearance – whether they were neatly dressed, clean, had a proud bearing, and conformed to European conceptions of physical beauty.[35] A more direct influence on scientific travel was Johann Friedrich Blumenbach's (1752–1840) gathering of a collection of skulls from around the world and his attempt to develop from it a classifica-tion of human types. His dissertation *On the Natural Variety of Man* (1775, 2nd ed. 1781) distinguished five such types: a normative "Caucasian" type flanked by the Mongolian, the Ethiopian, the American, and – added in the second edition to accommodate new travelers' gifts from the Pacific – the Malaysian. Blumenbach did not think of these as fixed races, but as heuristic distinctions to account for differences caused by climate and other environ-mental factors.[36] His classification of physical types was a contribution – on the side of human unity – to the contemporaneous debate over whether human beings derived from a common ancestry ("monogenism") or had di-verse origins ("polygenism"). The Bristol physician James Cowles Prichard (1786–1848), raised in a Quaker family and deeply imbued with a religious belief in the unity of mankind, defended monogenist explanations of human diversity in his *Researches into the Physical History of Man* (1813). Prichard enriched his *Researches* with added ethnographic material over the following decades, defending his views against an increasingly self-confident polygenist counter-current.[37]

Environmentalist thinkers more directly analyzed the external factors conditioning differences of phenotype and social structure. Drawing on the Forsters and other eighteenth-century predecessors, Alexander von Humboldt provided a widely influential model for the study of the total life conditions of a place, beginning with climatological zones and extend-ing through natural resources and political conditions to an all-encompassing understanding of the character of a particular people.[38] Humboldt's influence

[35] Johann Caspar Lavater, *Physiognomik. Zur Beförderung der Menschenkenntniss und Menschen-liebe* (1783), 4 vols. (Vienna: Sollinger, 1829); Mary Cowling, *The Artist as Anthropologist: The Representation of Type and Character in Victorian Art* (Cambridge: Cambridge University Press, 1989).

[36] Gascoigne, *Joseph Banks and the English Enlightenment*, pp. 149–58; Johann Friedrich Blumenbach, *On the Natural Varieties of Mankind (De Generis Humani Varietate Nativa)* . . . (1836), trans. and ed. Thomas Bendyshe (New York: Bergman, 1969).

[37] James Cowles Prichard, *Researches into the Physical History of Man* (1813), ed. George W. Stocking, Jr. (Chicago: University of Chicago Press, 1973); Stocking, *Victorian Anthropology*, pp. 48–53.

[38] Janet Browne, *The Secular Ark: Studies in the History of Biogeography* (New Haven, Conn.: Yale University Press, 1983).

diminished after midcentury, but in the 1880s Franz Boas (1858–1942), one of the founders of modern anthropology, affirmed his intellectual affinity with him.[39] Subsistence theories offered another approach to the conditioning of native societies and cultures. Thomas Malthus (1766–1834) gave extended attention to Tahitians and other indigenous peoples from around the world in the fourth edition of his famous analysis of the subsistence constraints on population growth, *An Essay on the Principle of Population* (1st ed. 1798).[40] Scottish philosophers such as Ferguson and Lord Kames (1696–1782), along with Anne-Robert-Jacques Turgot (1727–1781) and Rousseau in France, assigned "savages" to the category of hunter and explained other aspects of "savage" life, notably its political institutions, on the basis of this economic activity. They often had North American Indians in mind (and among Indians, the Iroquois) as people either to be admired for their Spartan virtues or to be condemned for their supposed cruelty and indolence. With their attention to processes of production and property relations, and their incorporation of "savages" into evolutionary theories of society, their subsistence theories anticipated Marxian analyses of indigenous societies.[41]

One of the most striking achievements of the period after 1800 was the development of comparative linguistics. From India, reports made their way back to Europe that Latin and Greek had structural affinities with Sanskrit. Following on the earlier work of William Jones (1746–1794), Friedrich Schlegel's (1772–1829) *On the Language and Wisdom of the Hindus* (1808) announced the program of a comparative linguistics that could unearth the prehistoric relationships among different peoples.[42] Decades later, one of the culminating works of this Romantic school was Wilhelm von Humboldt's (1767–1835) essay *On Language* (1836),[43] a work he could not have written without the help of travelers, including his brother Alexander, who helped him with South American Indian languages, and Adelbert von Chamisso, who helped him with Hawaiian. Humboldt's essay documents many of the unresolved tensions of the period. On the one hand, he sought a normative language for mankind, which he found in the Indo-European family; on the other hand, he had a Romantic appreciation of the diversity of languages as enriching the expressive possibilities of humanity. This was the side of his speculations that was later taken up by Edward Sapir (1884–1939)

---

[39] Smith, *European Vision*; Franz Boas, "The Study of Geography" (1887), in his *Race, Language, and Culture* (Chicago: University of Chicago Press, 1982), pp. 639–47.

[40] Thomas Malthus, *An Essay on the Principle of Populations . . .*, 2 vols., 4th ed. (London: J. Johnson, 1807).

[41] Ronald Meek, *Social Science and the Ignoble Savage* (Cambridge: Cambridge University Press, 1976).

[42] Friedrich Schlegel, *Über die Sprache und die Weisheit der Indier. Ein Beitrag zur Begründung der Altertumskunde*, ed. E. F. K. Koerner, Introduction by Sebastiano Timpanaro (Amsterdam: Benjamins, 1977).

[43] Wilhelm von Humboldt, *On Language: The Diversity of Human Language-Structure and Its Influence on the Mental Development of Mankind* (1836), trans. Peter Heath, Introduction by Hans Aarsleff (Cambridge: Cambridge University Press, 1988).

and other linguistic relativists, who pointed out that American Indian and other native languages might have expressive and cognitive capacities different from, and in some ways superior to, those of Western languages.[44]

By the mid nineteenth century, the conditions of European scientific travel had been dramatically transformed. As late as the late 1820s and 1830s, a voyage around the world was still a rare and dangerous adventure. Thomas Cook started organizing trips to Scotland in the 1840s, and by the 1870s his firm was leading trips around the world for the British middle classes, who could count on comfortable, predictable vacations and could enjoy the wonders once seen and described only by the intrepid few.[45] This transformation of ocean travel was part of the larger industrialization of British and other European economies on display at the famous Crystal Palace exhibition of 1851. The technological gap between Europeans and native peoples had turned into an enormous gulf, with great consequences for the perceptions of observers. Their technological superiority fed the illusion that Europeans belonged to a different species and that the gap between natives and themselves was unbridgeable.[46] Politics as well as technology served to change the assumptions of ethnographers. For continental Europeans, the Revolutions of 1848 changed the atmosphere of European politics and with it, their perceptions of non-Europeans. Gone was the programmatic yearning for personal freedom and the naive generosity toward peoples around the world of the Romantic era; instead, Europe had entered an age of industrialization and *Realpolitik*. After midcentury, belief in worldwide progress faded, displaced by a vision of a racial hierarchy descending from northern Europeans to the various colonized peoples.

These changes also, however, gave new significance to another preoccupation going back to the Romantic era. Discussions of indigenous peoples centered on equality in the late eighteenth century, on liberty in the early nineteenth century; but what of the third term of the revolutionary triad, fraternity? The Romantics took up this topic, drawing attention to the way in which human beings drew together as cultural communities. Wilhelm von Humboldt's meditations on language served as the paradigm for thinkers such as Heymann Steinthal (1823–1899) and Moritz Lazarus (1824–1903), who looked to indigenous societies as well as to classical antiquity in order to understand how language formed the social psychology of nations. Anxieties over the breakdown of the traditional European social order and the growth of class divisions led intellectuals to give increasing weight to the theme of community, culminating in the publication of Ferdinand Toennies'

[44] John A. Lucy, *Language Diversity and Thought: A Reformulation of the Linguistic Relativity Hypothesis* (Cambridge: Cambridge University Press, 1992); Roger Langham Brown, *Wilhelm von Humboldt's Conception of Linguistic Relativity* (The Hague: Mouton, 1967).

[45] James Buzard, *The Beaten Track: European Tourism, Literature, and the Ways to Culture, 1800–1918* (Oxford: Clarendon Press, 1993), pp. 48–64.

[46] Stocking, *Victorian Anthropology*, pp. 1–6.

(1855–1936) *Community and Society* (1887) and Émile Durkheim's (1858–1917) *The Elementary Forms of the Religious Life* (1912).[47] The Romantics bequeathed the problem of community to later generations of social scientists, which would continue to look to distant times and places for the fulfillment of European ideals.

---

[47] Ferdinand Toennies, *Community and Society*, trans. Charles P. Loomis (New York: Harper and Row, 1963); Émile Durkheim, *The Elementary Forms of Religious Life*, trans. and introd. Karen E. Fields (New York: Free Press, 1995).

8

HISTORY AND HISTORICISM

Johnson Kent Wright

History occupies a singular position among the modern social sciences. It was the first to assume a durable professional shape. The basic canons for modern academic historiography were introduced in Germany early in the nineteenth century. By that century's end, the model of Barthold-Georg Niebuhr and Leopold von Ranke had been widely imitated across western Europe and the United States, establishing the permanent institutional mold of the discipline. The special place of history among the social sciences involves more than mere precedence, however. For historiography was accompanied in its passage toward science by an enabling philosophy of history – or a set of such philosophies – that claimed a unique *privilege* for historical explanation and understanding, with consequences for the entire range of the social sciences.

It was only early in the twentieth century that these philosophies or ideologies of history were first gathered together, retrospectively, under a single rubric, that of "historicism." Although the term was a century old, its release into wider circulation really began with Ernst Troeltsch, who used it, in the years following the First World War, to describe what he saw as the dominant outlook of the preceding century, which had emphasized the decisive place of *change* and *development* in the human realm. Contrasting it with *Naturalismus*, the outlook of the natural sciences, Troeltsch declared *Historismus* to be in "crisis," having issued into antiscientific skepticism and relativism.[1] A decade later, Friedrich Meinecke gave the term a slightly different inflection. Tracing its origins to Johann Gottfried von Herder and Johann Wolfgang von Goethe, Meinecke saw their stress on the *concrete*, the *unique*, and the *individual* as the core of historicism. If his judgment of the tradition was more positive than Troeltsche's – "the rise of historism was one of the greatest intellectual revolutions that has ever taken place in Western thought"[2] – Meinecke

---

[1] Ernst Troeltsch, *Historismus und seine Probleme* (Tübingen: J.C.B. Mohr, 1922) and "Die Krisis des Historismus," *Die Neue Rundschau*, 33 (1922), 572–90.
[2] Friedrich Meinecke, *Historism: The Rise of a New Historical Outlook*, trans. J. E. Anderson and H. D. Schmidt (London: Routledge, 1972), p. liv.

also sharply distinguished historicism from natural-scientific modes of understanding. Indeed, as the term gained widespread currency after midcentury, critics began to charge that historicism was incompatible with any type of genuine science. This view reached its polemical extreme in Karl Popper's *The Poverty of Historicism* (1957), which attacked both "anti-naturalistic" and "pro-naturalistic" historicism, charging the former with "teleology" and "holism," the latter with advancing notions of historical prediction based on faith in illusory "laws" of development.

The combination of anachronism, indefinition, and polemical fervor in these usages has led some to suggest that the concept of "historicism" should simply be given a decent burial. In fact, the term is indispensable to any attempt to account for the passage of historiography toward the status of social science in the eighteenth and nineteenth centuries. This chapter will advance two arguments, in particular, about history, historicism, and social science during the period it surveys. One is to dissent from the common view that historicism was a distinctively nineteenth-century phenomenon, born of a Romantic reaction to an ahistorical Enlightenment. In accord with recent scholarship, historicism will here be assigned a more extensive genealogy, one more directly connected to the Enlightenment. Second, this chapter will stress the close relations between historicism and conceptions of social science throughout the eighteenth and nineteenth centuries. Such tensions as there were between the two tended to be productive ones. The sense of an outright rupture – the "crisis of historicism" of the turn of the century – marked the end of an era.

THE EIGHTEENTH CENTURY: PRECONDITIONS

The emergence of both modern historiography and historicist doctrine was made possible, in the first instance, by the clearing away of older approaches to historical understanding descended from the classical and Christian traditions. The extension of temporal and spatial horizons brought about by the Scientific Revolution left specifically Christian conceptions of "universal history" in ruins. The last major work in that tradition, Bossuet's *Discourse on Universal History*, was published in 1681. The settling of accounts with the legacy of Greco-Roman historiography, on the other hand, was a far more complicated and extended process. No outright rejection of the heritage of Herodotus and Thucydides, Polybius and Tacitus, ever occurred, not least because of the emergence of a rich tradition of *neo*classical historiography in early modern Europe. The works of Machiavelli and his successors down to the eighteenth century faithfully reproduced the chief structural characteristics of ancient historiography: cyclical theories of large-scale change, focused on the alternation of political regimes; a methodological reliance on eyewitness evidence; and a philosophical belief in an invariant "human nature" as a key explanatory principle. The example of James Harrington's *Common-wealth of Oceana* (1656), with its emphasis on economic determinants in history and appeals to Baconian standards of empirical evidence, shows how far the classical republican tradition could advance in the direction of modern social science, even within these constraints. Nevertheless, the emergence of a genuinely *historicist* approach to the past required a breach with the norms of classical historiography.

Let us begin with the question of large-scale historical change. What prompted a move away from classical theories of a cyclical rotation of political regimes? A long tradition holds that the conceptions of directional change and progress in history that first appeared in the eighteenth century should be seen as "secularizations" of Christian notions of salvation and redemption. However persuasive the "secularization thesis" – for which precise mechanisms and vehicles are rarely specified – it happens that seventeenth-century Europe *did* see the arrival of a wholly novel language for interpreting long-term historical development. Troeltsch and Meinecke were later to declare that historicism was born of a revolt against the Western tradition of natural law, which in each of its incarnations – Aristotelian, Stoic, Thomist, early modern – proposed a set of timeless norms based on belief in an unchanging set of human dispositions and traits. In a nice irony, however, recent scholarship has almost entirely reversed this relationship. For it now appears that, far from being a foil for historicism, the tradition of natural law was actually one of its seedbeds. The pivotal figure here was Samuel Pufendorf (1632–1694). He represented his predecessor Hugo Grotius as the founder of a "modern" school of natural jurisprudence, aimed at combating the moral and epistemological skepticism of Montaigne and Charron. Tempering Grotius's optimism with a realism inspired by Hobbes, Pufendorf historicized the natural "sociability" of mankind in relation to the successive stages of property regime.[3] The conceptual vocabulary pioneered by Pufendorf at the end of the seventeenth century was then widely diffused during the eighteenth, most notably through the translations of Jean Barbeyrac (1674–1744), who integrated Locke's more radical English natural law into the tradition as well. In this form, the "modern" theory of natural law supplied something like the deep structure of Enlightenment social thought, forming the foundation for the major stadial theories of historical development of the latter half of the eighteenth century.

By far the most important of these was the "four-stages" theory, which, once it had emerged from the cocoon of natural jurisprudence, found mature expression in the hands of a remarkable gallery of French and Scottish

---

[3] On the "modern" tradition of natural jurisprudence, see Richard Tuck, "The 'Modern' Theory of Natural Law," in *The Languages of Political Theory in Early Modern Europe*, ed. Anthony Pagden (Cambridge: Cambridge University Press, 1987), pp. 99–119; on Pufendorf, see Istvan Hont, "The Language of Sociability and Commerce: Samuel Pufendorf and the Theoretical Foundations of the 'Four Stages' Theory," ibid., pp. 253–76.

thinkers.[4] Its earliest statements appeared in the 1750s – in France, in the writing of Turgot, Quesnay, Helvétius, and Gouget, and in Scotland, in Dalrymple and Kames. The major presentations of the theory then came in the great masterpieces of the Scottish Enlightenment – Adam Ferguson's *An Essay on the History of Civil Society* (1767), John Millar's *The Origin of the Distinction of Ranks* (1771), and Adam Smith's *Wealth of Nations* (1776). For all their differences, these works expressed a common conviction that economic "modes of subsistence" were the determining instance in social life, and that there was a general tendency for these modes to evolve through specific, progressive stages – in one of Smith's versions, "first, the Age of Hunters; secondly, the Age of Shepherds; thirdly, the Age of Agriculture; and fourthly, the Age of Commerce." The explanation for this procession was typically sought at two levels. Four-stages theorists generally started from the *intentional* explanation of individual actions, grounded in rationalist or utilitarian conceptions of human nature. They then proposed essentially *causal* explanations at the collective level, where the aggregation of these actions produced consequences unintended by any individual or group, especially in the transition from one mode to the next. The result, in Ferguson's famous formula, was "establishments, which are indeed the result of human action, but not the execution of any human design."[5] The four-stages theory shed light in every direction, pointing forward to what would become the separate sciences of economics, sociology, and anthropology. Indeed, in Millar's hands, the theory yielded a precocious sociology of gender. But it did not exhaust the field. The second half of the eighteenth century saw any number of spectacular examples of conjectural history of this kind, from the savage indictment of civilization in Rousseau's *Discourse on Inequality* of the 1750s, to its passionate defense in Kant's *Idea for a Universal History from a Cosmopolitan Perspective* and Condorcet's *Sketch for a Historical Tableau of the Progress of the Human Mind* of the 1790s.

These theories of stadial development were in many ways the major achievement of Enlightenment social thought, its lasting contribution to the modern social sciences. If historicism is to be understood – in a definition made famous by Maurice Mandelbaum – as "the belief that an adequate understanding of the nature of any phenomenon and an adequate assessment of its value are to be gained through considering it in terms of the place which it occupied and the role which it played within a process of development,"[6] then these were among its founding documents. But this

---

[4] Ronald L. Meek, *Social Science and the Ignoble Savage* (Cambridge: Cambridge University Press, 1976); Peter Stein, *Legal Evolution: The Story of an Idea* (Cambridge: Cambridge University Press, 1980).

[5] Adam Ferguson, *An Essay on the History of Civil Society*, ed. Fania Oz-Salzberger (Cambridge: Cambridge University Press, 1995), p. 119.

[6] Maurice Mandelbaum, *History, Man, and Reason: A Study in Nineteenth-Century Thought* (Baltimore: Johns Hopkins University Press, 1971), p. 42.

accounts for only one element in mature historicism. What of the other side – of what Georg Iggers has called the "core of the historicist outlook," the assumption that "there is a fundamental difference between the phenomena of nature and those of history, which requires an approach in the social and cultural sciences different from those of the natural sciences"?[7] For the great stadial theorists of the French and Scottish Enlightenments drew no sharp distinction between nature and history, such that explanations of change and development in the human realm would demand a different methodology altogether. On the contrary, the typical move of the stadial theorists was to extend a basically Newtonian model of explanation, moving from general "laws" and "principles" to the identification of specific causal mechanisms, from the natural to the human world. The four-stages theory, in particular, amounted to a discovery of the basic "laws of motion" of the social world, and it was typically presented as such. Where, then, should we find the source for the other side of historicism, the emphasis on the *distinctive* character of historical explanation?

Natural jurisprudence, as it happens, also provided a context in which this theme could develop. In this case, however, the truly creative achievements lay in idiosyncratic performances on the critical margins of natural law theory. Pride of place here, at least retrospectively, belongs to the Neapolitan jurist Giambattista Vico (1668–1744), who devoted a lifetime, working in obscurity on the periphery of European intellectual life, to developing a "new science" of the "nature of nations." At first glance, the successive editions of the *New Science* (1725, 1730, 1744) appear to mark a step backward. Proceeding from a theological critique of modern natural jurisprudence, Vico presented a model of "ideal eternal history" – a stylized recapitulation of the history of ancient Rome – whose upshot was a cyclical theory of historical change. At the core of this apparently retrograde program, however, was a revolutionary methodological principle. Having begun as a Cartesian, committed to the unity of the sciences, Vico soon made an about-face, arguing not just for the autonomy of historical from natural-scientific understanding, but for its *superiority* as well, in terms of the certainty of its knowledge. The key was Vico's famous assertion that "the true and the made are convertible" – in effect, that human affairs are open to a distinct mode of comprehension "from the inside," as it were, that lies beyond the reach of the sciences of nature. This pragmatics of historical explanation was based on the assumption of a universal human nature. Yet its practical effect was a novel accent on the *plasticity* of the latter. The real emphasis of the *New Science* was on the sheer variety of forms of social life, as societies pass through each of Vico's stages.

The impact of the *New Science* was delayed until a century later, when it was rediscovered and celebrated with remarkable intensity during the

[7] Georg G. Iggers, *The German Conception of History: The National Tradition of Historical Thought from Herder to the Present*, rev. ed. (Middletown, Conn.: Wesleyan University Press, 1983), pp. 4–5.

age of classical historicism. Its themes were not entirely lost on the eighteenth century, however. Four years after the last edition of the *New Science*, Montesquieu (1689–1755) published *On the Spirit of the Laws*, which was immediately recognized as the greatest work of political and social thought of the age. Defining all law – divine, natural, and positive or social – as "the necessary relations deriving from the nature of things," Montesquieu was formally committed to a Cartesianism that he almost immediately rescinded in practice. For "the intelligent world is far from being as well governed as the physical world. . . . The reason for this is that particular intelligent beings are limited by their nature and are consequently subject to error; furthermore, it is in their nature to act by themselves."[8] This was a rationale for a specifically *human* science of agency and irrationality, aimed at explaining the "spirit" of human laws. On this basis, Montesquieu set forth a universal taxonomy of three "forms of government" – republican, monarchical, and despotic – each governed by a single subjective "principle" – virtue, honor, and fear, respectively. In the foreground, the taxonomy featured a global theory of geographical determinism that consigned "despotism" to the East. In its background was the silhouette of an historical account of the transition from the classical world of virtuous republics to the commercial monarchies of modern Europe, grounded in an early version of stadial theory. But the overall effect of *On the Spirit of the Laws* was not far from that of Vico's masterpiece – a model for a new science of society, one capable of a sympathetic understanding of the whole range of human variety and difference.

Rousseau joined Montesquieu in defining an enlarged sense of the variability of human nature, and in proposing a new method for grasping its differences. But the full harvest of these ideas, joining them to a novel conception of historical development, was to emerge from the German *Aufklärung* rather than from the French Enlightenment. Here the major achievement was the speculative philosophy of history of Johann Gottfried Herder (1744–1803). Like Vico, Herder posited a universal human nature but stressed its radical plasticity in different geographical, political, and cultural settings. The starting point for his own philosophical anthropology was an intense critique of the partitive "faculty psychology" of the French Enlightenment. For Herder, human individuals and the collectivities they formed were unique totalities, each qualitatively distinct from the rest. At the same time, the essential identity of human nature guaranteed that historical change was directional for Herder, no less than for the stadial theorists, whose works he knew well. But there was no real precedent for Herder's focus on the political *nation* and ethnic *Volk* as the central subjects of development, rather than on economic modes or structures. He advanced also a novel conception of development, combining an organicist model of change drawn from contemporary vitalist

---

[8] Montesquieu, *The Spirit of the Laws*, trans. Anne M. Cohler, Basia Carolyn Miller, and Harold Samuel Stone (Cambridge: Cambridge University Press, 1989), p. 4.

biology with the claim that human collectivities typically progress toward ever greater *self*-determination over time. The methodological result of this conception of human agency, finally, was a hermeneutic program calling on historians to recapture the uniqueness and diversity of their historical objects by creatively "entering into" their subjective experiences and motivations.

The upshot of Herder's philosophy of history was a theoretical charter for historicist practice, just slightly before the fact. Indeed, all of the elements that went into the classical historicisms of the nineteenth century – large-scale theories of change and development across time, methodologies of hermeneutic understanding unique to the "human" sciences – had made their first appearance by the end of the eighteenth century. At the same time, a glance at the actual historiography of the epoch, even the most advanced, shows that a genuine synthesis of these elements, one capable of effecting a fundamental alteration in the practice of history itself, had not yet emerged. The second half of the eighteenth century did indeed see a remarkable flowering of narrative history. The masterpieces of which the age could boast included Voltaire's *The Century of Louis XIV* (1751) and *Essay on Customs* (1754), David Hume's *History of England* (1754–62), William Robertson's *History of the Reign of the Emperor Charles V* (1769) and *History of America* (1777), and Edward Gibbon's *The Decline and Fall of the Roman Empire* (1776–88). In two respects, these works can be called at least proto-historicist. What made the writing of Voltaire, Hume, Robertson, and Gibbon "philosophical" in the eyes of contemporaries was, on the one hand, their innovative use of theories of development – drawn variously from Montesquieu, from the classical republican tradition, and from the stadial theorists described earlier – and, on the other, their unprecedented exploitation of source material, including the extra-European sources now increasingly available. The shape of a new historiography, aimed at large-scale explanation and confident in its use of the widest and remotest sources, had come into view.

Yet these historians were far from believing that historical interpretation represented a privileged or even a unique mode of understanding of the social world, distinct from the natural sciences. One reason is that they held to traditional conceptions of "human nature." This point is often exaggerated, as if they regarded "human nature" as fixed and unvarying. Still, the idea was far more likely to serve as *explanans* in their work than as *explanandum*, as it tended to do in the more historicist approaches of Vico and Herder. Not surprisingly, it was in late-eighteenth-century Germany, where the influence of Leibniz and Herder could be felt directly, that the initial steps toward a genuine professionalization of historiography were taken. In particular, the new University of Göttingen, founded in 1737, played host to a distinguished group of historians, including Johann Christoph Gatterer and August Ludwig Schlözer. They brought into precarious balance in historiography itself, for perhaps the first time, what were to become the two sides of mature historicism: a decisive emphasis on the mastery of the original records of the

past, buttressed by a philosophical insistence on the uniqueness and individuality of historical phenomena, and a constant concern to integrate these
sources into causal explanations of long-term development. They invoked
mechanisms ranging from the geographical and structural determinisms of
Montesquieu and the stadial theorists to the more "spiritual" forms of agency
featured in Herder's writing.[9]

THE RANKEAN REVOLUTION: CLASSICAL HISTORICISM

A different historical context, however, was required to convert this practice
into a durable and reproducible model. The full "modernization" of historiography and of historicist doctrine came as a direct result of the political and
ideological turmoil that overtook Europe in the wake of the French Revolution. The scene was Prussia, whose defeat at the hands of Napoleon had
introduced the "reform era" – a remarkable attempt to modernize the Prussian
polity and society "from above." The renovation of the educational system,
from elementary *Volksschule* to university, was undertaken by Wilhelm von
Humboldt (1767–1835), whose key achievement was to preside over the creation of the new University of Berlin in 1810. Humboldt's contribution to
history went beyond the provision of institutional shelter, however. For the
central role of Berlin in the historiographic revolution of the first half of
the century could be traced to the particular model of science theorized
by Humboldt and articulated directly into the structure of the university.
"Science" here was *Wissenschaft*, which referred both to the collective enterprise of scholarship and learning and, in the plural, to the specific disciplines that contributed to it. There was no hierarchy distinguishing these
Wissenschaften in terms of the certainty or value of the knowledge they generated, or of the dignity attaching to their pursuit. But Humboldt did establish
certain methodological distinctions between the *Naturwissenschaften* and the
"historical" or "human" sciences. The latter were no less dependent on empirical evidence than the former. The human sciences were set apart, however, by
the specific character of their historical sources – records of the lives of unique
totalities, individual and collective – and by the key role of irrationality in
human affairs. Their proper method was to proceed from objective historical
facts to a grasp of their interconnection, necessity, and meaning, by means
of a specific kind of intuitive "understanding" or *Verstehen*. "The truth of all
that happens requires the addition of that above mentioned invisible element
of every fact and this the writer of history must add."[10]

---

[9] Peter Hanns Reill, *The German Enlightenment and the Rise of Historicism* (Berkeley: University of
     California Press, 1975).
[10] Wilhelm von Humboldt, "On the Tasks of the Writer of History," in his *Gesammelte Schriften*
     (Berlin: B. Behr, 1903–36), vol. 4, pp. 35–6, cited in Iggers, *The German Conception of History*, p. 60.

Under the banner of this methodological prescription, which assigned to history the role of interpretation and synthesis, a galaxy of remarkable scholars assembled at Berlin. Among them, of course, was G. W. F. Hegel, whose career was devoted to an extravagantly ambitious attempt to unify all the "sciences" according to the dictates of one of them, philosophy. This enterprise can properly be characterized as a philosophical or "intellectual" historicism.[11] The revolution in the writing of history that was launched at Berlin around the same time, however, was distinct from the Hegelian project. Indeed, the primary goal of its chief architect, Leopold von Ranke (1795–1886), was precisely to establish the *autonomy* of historiography as a scholarly enterprise, to render it irreducible to any other discipline, especially philosophy. Ranke arrived at Berlin only in 1825, after the publication of his first major book, *Histories of the Latin and Teutonic Nations from 1494 to 1535*. The methodological revolution announced in its pages, that of a new source criticism or *Quellenkritik*, was of course not uniquely that of Ranke. Its basic model was drawn directly from the adjacent disciplines of classical philology, biblical criticism, and legal history. Ranke himself paid particular tribute to Georg Barthold Niebuhr, who had lectured at Berlin in the early years of the university, before moving on to the University of the Rhineland at Bonn. Niebuhr's *Roman History* (1811–12), which founded the modern study of ancient Rome, has some claim to being the first work of modern positive historiography.

Ranke's distinction, in the first instance, was simply to extend the methods of *Quellenkritik*, pioneered in classical and legal studies, to the field of modern European history. He owed his initial fame, and his accession to his chair at Berlin, above all to the second volume of the *Histories of the Latin and Germanic Nations*, a methodological appendix to the narrative of the Italian wars recounted in the first. In it, he critically reviewed the work of previous historians – staging a famous confrontation between Guicciardini and Machiavelli, to the advantage of the latter – in order to make a case for a historiography based solely on the immediate evidence of the past, whether archival, epigraphic, or archeological. Ranke's subsequent canonization as "founder" of modern scientific history depended, in no small measure, on the conformity of his own practice as an historian to this norm. His career as a researcher amounted to a decades-long voyage of discovery through the archives of the major states of western Europe – Italy, Austria, Germany, France, and Britain – that left him, at its end, with an unparalleled knowledge of the sources of early modern history. At the same time, Ranke also developed the research seminar, whose purpose was to train students in the critical evaluation and use of historical evidence. Gatterer had experimented with an earlier version at Göttingen, but it was Ranke's reintroduction and

[11] Michael Forster, *Hegel's Idea of a Phenomenology of Spirit* (Chicago: University of Chicago Press, 1998), pt. 3.

systematization of the procedure at Berlin after 1833 that made it an indispensable component in the formation of professional historians.

At the same time, this enterprise was sustained by a theoretical and metaphysical vision that was far from being positivist. Ranke's historicism had two sides. On the one hand, the autonomy and distinction of historiography as a science did indeed rest on its grasp of objective fact. His famous aspiration in the Preface to the *Histories of the Latin and Teutonic Nations* to "show how things really happened" was to become a mantra for historians of every stripe. He returned to the point even more forcefully in his demolition of Guicciardini: "We on our side have a different concept of history: naked truth, without embellishment, through an investigation of the individual fact, the rest left to God, but no poeticizing, no fantasizing."[12] The polemical targets of these remarks are often overlooked. One was the traditional conception of history as a *magistra vitae*, pressing historiography into devotional or didactic service. But the real menace was the totalizing systems of Fichte and Hegel, which threatened to absorb historiography into a vast philosophic design. History remained a distinct enterprise: "There are only two ways of acquiring knowledge about human affairs – through the perception of the particular, or through abstraction; the latter is the method of philosophy, the former of history."[13] As the ambiguity of the adverb in "*wie es eigentlich gewesen*" suggests, however (the phrase may be translated, equally accurately, as "as it *actually* happened" or "as it *essentially* happened") Ranke's conception of "particularity" lay squarely in the historicist tradition descending from Herder and the Göttingen historians through Humboldt. In Ranke's case, this belief in the sanctity of the unique and the individual ultimately rested on theological grounds. "Every epoch is immediate to God," he wrote, in one of a hundred variations on the same theme. "In this way the contemplation of history, that is to say of *individual life in history* acquires its own particular attraction, since now every epoch must be seen as something valid in itself and appears highly worthy of consideration."[14]

This was only one side of Ranke's historicism. The other was a vision of historical development, concentrated resolutely on the political histories of the great nation-states of western Europe, from their first appearances in the Dark Ages down to the present. The consistency of this focus over his career is deeply impressive. Ranke's early major works – *Histories of the Latin and Teutonic Nations*, his study of Ottoman and Spanish relations in the sixteenth century, and the *History of the Popes* (1834, 1836) – surveyed the history of the

[12] Leopold von Ranke, *Histories of the Latin and Germanic Nations from 1494 to 1535*, 2 vols. (Leipzig, 1824), vol. 2, p. 18, as cited and translated by Felix Gilbert in *History: Politics or Culture? Reflections on Ranke and Burckhardt* (Princeton, N.J.: Princeton Univerity Press, 1990), pp. 19–20.
[13] Leopold von Ranke, "A Fragment from the 1830s," in *The Varieties of History: From Voltaire to the Present*, ed. Fritz Stern (New York: Meridian, 1956), pp. 58–9.
[14] Ranke, *The Theory and Practice of History*, ed. Georg G. Iggers and Konrad von Moltke (Indianapolis: Bobbs-Merrill, 1973), p. 58.

entire western European set of nations, at the moment of their transition from feudal to absolute monarchy. In his maturity, Ranke turned to the individual fates of these nations, writing separate histories of Germany, France, and England. He concluded with a *Universal History*, which attempted, prematurely, to extend this vision around the globe. His concentration on the *state* as an object of study – its political development, and its diplomatic and martial contention with other members of a set of nations – was never exclusive. But by comparison to the capacious range of eighteenth-century conjectural and narrative history, whose embrace included the cultural evolutionism of Voltaire and the economic determinism of Smith and Ferguson, Ranke's focus on political history represented a definite narrowing. At the same time, the actual shape of his politics – a Restoration conservatism, which retreated toward theological reaction over time – has not served Ranke well. Still, the constriction of vision was inseparable from his overall achievement in providing a model for cumulative professional historiography. For it was precisely in political history – the level at which determining structure and subjective agency meet – that the bulk of the sources of the European past lay most readily to hand. This is what permitted the exemplary fusion of explanation and evidence in Ranke's work that has formed the basis for professional historiography ever since.

Only in the second half of the nineteenth century did the deliberate imitation of the Rankean model of "scientific" historiography get under way. Eventually it gained canonical status, not only in Germany but also in France, the United States, and Britain. In the meantime, the more traditional forms of historical practice also evolved in an historicist direction, independent of the German model. In France and England, the still preprofessional and "prescientific" character of the major historiography permitted something closer to an eighteenth-century latitude with regard to theories of development. The work of François Guizot (1787–1874) is an outstanding case in point. The causal pluralism of his *History of Civilization in Europe* (1828), mingling a sociology of economic conflict and a hermeneutic of values and principles, made him an heir both to the four-stages theorists and to Montesquieu and Herder. His use of a comparative method, mediating between abstract models and particular instances, was to influence such disparate successors as Tocqueville and Marx. But the bulk of historiographical energy during this period was devoted to narrating the *nation*, though embodying political values very different from those of Ranke and the "Prussian" school. In France, Chateaubriand's historicism was answered by a remarkable set of liberal historians, including Guizot, Mignet, and Thiers, who charted the advance of the principle of liberty through French history. Their work was succeeded by the populist historicism of Jules Michelet (1798–1875), the chief rediscoverer of Vico during this period, whose quasi-mystical sense of the evolving identity of the French "people" can be set beside Ranke's conception of the Prussian state. In England, the work of Henry Hallam

(1777–1859) and Thomas Babington Macaulay (1880–1859) introduced an alternate tradition of liberal historicism, later to be called the "Whig interpretation" of English political history. The analagous founding figure in the United States was George Bancroft (1800–1891), who had taken his doctorate at Göttingen. Although the bulk of this nationalist historiography remained preprofessional – some of it, as with Guizot, was even written by political leaders – it was frequently accompanied by new collective enterprises for the gathering of historical evidence. This activity was increasingly sustained by state sponsorship, from the French Ecole des Chartres, founded in 1821, to the great German and English collections of medieval sources of the 1830s and 1840s. By midcentury, the evidentiary foundations for modern European history had been durably established.

THE LATER NINETEENTH CENTURY: DIFFUSION AND DEVELOPMENT

There were two major developments in this field in the second half of the nineteenth century. One was the completion of the professionalization of historiography in western Europe and the United States: the establishment of academic chairs, the creation of degree-granting programs, the founding of disciplinary associations, the launching of specialist journals. This process was everywhere seen as a matter of raising history for the first time to the dignity of a "science." In nearly every case, this involved the deliberate imitation of the Rankean model of historiography, though with significant variations in the understanding of the kind of "science" it embodied. At the same time, Rankean historiography never entirely monopolized the field. The second half of the century also saw creative work, by less conventional historians, in a recognizably historicist mode. Beyond historiography proper, something like a second great era of grand theories of large-scale historical development arrived, a period to rival the Enlightenment itself.

In Germany, Ranke himself enjoyed an impressive longevity, retiring from Berlin only in 1871. Well before this, however, his retrograde politics, rooted in the Restoration, had left him increasingly isolated. The upheavals of midcentury inspired the emergence of an emphatically *liberal* historiography, represented above all by the career of Georg Gervinus. From here the torch passed, in the 1850s, to a distinctively "Prussian school" of historical writing, which balanced liberalism with a decisively *nationalist* accent and whose leading lights were Friedrich Dahlmann, Johann Gustav Droysen, and Heinrich von Sybel. But all their work was conducted in the manner established by Ranke, combining a commitment to rigorously "objective" primary research, a passionate belief in the centrality of the state in modern history, and a growing sense of professional solidarity. It was Sybel who founded the main German professional organ, the *Historische Zeitschrift*, in 1859. Perhaps the consummation of Rankean historiography in practice came not in German history but in that of ancient Rome, in the spectacular career of Theodor Mommsen. But Johann Gustav Droysen's (1808–1884) *Outline of the Principles of History* (1857–83) is today regarded as its supreme theoretical expression – the philosophical defense of the autonomy of historical science that Ranke never wrote. Indeed, Droysen sharply criticized Ranke and his immediate followers for bending the stick too far in an "objectivist" direction in their cult of primary sources.

Genuine historical understanding, for Droysen, certainly began with the objective facts disclosed in the sources, which were then to be placed in their proper material and political contexts. From there, however, he called on the historian to proceed to a psychological reconstruction of the intentions and purposes of the historical actors involved, and finally to a totalization of these in terms of the collective "ethical forces" that gave them meaning over time. The "communities of spirit" to which these "forces" gave rise – ideas for which Droysen was equally indebted to Humboldt and Hegel – ranged from the "natural" (family and *Volk*), to the "ideal" (language, art, science, and religion), to the "practical" (economy and state). History was the science of the growth and development of such communities. Its *differentia specifica* was a form of *understanding* well beyond the grasp of philosophy, which aimed at "abstract cognition" outside of time. It diverged also from natural science, which approached the temporal in terms of lawlike repetition rather than of "ceaseless progress" (*ratlose Steigerung*), which was the stuff of historical change. Droysen launched a famous attack on Henry Thomas Buckle's *History of Civilization in England*, not for its evolutionism but for its *naturalism* – its attempt to eliminate intentionality and purpose from the explanation of large-scale historical development. Droysen's *Outline*, by contrast, can be seen as the theoretical climax of nineteenth-century historicism – as Hayden White has recently suggested, "the most sustained and systematic defense of the autonomy of historical thought ever set forth."[15]

Between the time of the first and last versions of Droysen's *Outline*, the professionalization of historiography outside of Germany had gotten under way, nearly everywhere under the inspiration of the German example. The decisive step was the full entry of history into university systems. In France, the threshold was marked by the creation of the Ecole Pratique des Hautes Etudes in 1868. There, Gabriel Monod and other German-trained scholars promoted the notion of history as a fully scientific academic discipline, shifting the center of gravity in training from the lecture to the Ranke-style seminar. In American universities, chairs in history were first established in the 1850s. In the 1870s, Herbert Baxter Adams, who had studied at Heidelberg,

---

[15] Hayden White, "Droysen's *Historik*: Historical Writing as Bourgeois Science," in his *The Content of the Form: Narrative Discourse and Historical Representation* (Baltimore: Johns Hopkins University Press, 1987), p. 99.

presided over the creation of the PhD program at Johns Hopkins University, a widely imitated model. William Stubbs and John Robert Seeley, who became Regius Professors of History at Oxford and Cambridge in 1866 and 1869, respectively, promoted the idea of German-style "scientific" history in Britain. Academic emplacement was punctually followed by the creation of the chief national journals for the propagation of the new scholarship, on the model of the *Historische Zeitschrift*: the *Revue historique* was launched in 1876, the *English Historical Review* in 1886, and the *American Historical Review* in 1895. These moves were sealed, finally, by the appearance of major theoretical statements, in the form of manuals and manifestos. Bernheim's *Handbook of Historical Method* (1889) and Langlois and Seignebos's *Introduction to Historical Studies* (1898), widely diffused in English translation, are chief examples of the former. The most famous instance of the latter, marking perhaps the climax of the whole process of the professionalization of history, was J. B. Bury's inaugural lecture at Cambridge in 1902, "The Science of History."

What sort of "science" was thus theorized? There is no doubt that Anglo-French conceptions of science were distinct from those that informed the German notion of *Wissenschaft*. Alternate philosophic cultures, empiricist or rationalist, ensured that "scientific" historiography in France, England, and the United States would assume a less idealist cast than it had in Germany. Still, they were not utterly disparate. The leading figures in the professionalization of history in these countries all acknowledged the inspiration of the German model, and most paid particular tribute to Ranke. The suggestion that their stance involved an "almost total misunderstanding" of the philosophical outlook of the latter presumes a stark contrast between German historicism and Anglo-French "positivism" for which there is little warrant.[16] There were, indeed, historians in France, England, and the United States whose commitments to a positivist unity of the sciences put them beyond the pale of any kind of historicism. The earliest and most notorious example was Buckle, whose *History of Civilization in England* (1857, 1861) proposed a model for scientific historiography, identifying general "laws" of change and development, that in effect canceled the autonomy of *historical* explanation altogether. In France, Hippolyte Taine played a not dissimilar role, promoting an alternate version of historiographic positivism. In 1891, Karl Lamprecht unleashed the German equivalent of the Buckle controversy with his *German History*, a frontal assault on the Rankean establishment. But these figures were distinguished precisely by their isolation from the established historiography of the epoch, whose main currents flowed in the direction marked by Rankean historicism.

This is especially clear in Bury's inaugural address, coming as it did from the native land of empiricism. Looking back at the process by which history

---

[16] Phrase used by Peter Novick, *That Noble Dream: The "Objectivity Question" and the American Historical Profession* (Cambridge: Cambridge University Press, 1988), p. 26.

had been "enthroned and sphered among the sciences" in the nineteenth century, Bury traced its beginnings to Niebuhr and Ranke, whose achievements owed less to their promotion of "objective" documentation than to their discovery of "the idea of human development." Calling this notion "the great transforming conception, which enables history to define her scope," Bury concluded in the voice of authentic historicism: "The world is not yet alive to the full importance of the transformation of history (as part of a wider transformation) which is being brought about by the doctrine of development . . . but we need not hesitate to say that the last century is not only as important an era as the fifth century B.C. in the annals of historical study, but marks, like it, a stage in the growth of man's self-consciousness."[17]

The "doctrine of development" was never the sole possession of historians, however. The overwhelming bulk of the new professional historiography of the second half of the nineteenth century was devoted to a single object, the emergence and evolution of the modern nation-state. But the same period also saw a great flowering of theories of large-scale historical development, extending well beyond the narrowly political focus of Rankean historiography, that were to have a lasting impact on the shape of the modern social sciences. The most sweeping and extravagant of these new stadial theories was also the earliest – the vision set forth by Auguste Comte (1798–1857) in his *Course in Positive Philosophy* (1830–42). Heir to Condorcet and Saint-Simon, Comte divided human history into three progressive stages, "theological" (extending roughly to the Reformation), "metaphysical" (ending with the French Revolution), and "positive" (projected from the present into the future), each subject to a distinct kind of social causation. An isolated figure during the first half of the century, Comte was joined in the second half by any number of competing theorists of development. The successor to Savigny in the history of law was Henry Maine (1822–1888), whose *Ancient Law* (1861) drew a distinction between "stationary" and "progressive" societies, tracing an evolution in the latter from "status" to "contract" as the central social institution. Beginning with his essay "Progress: Its Law and Cause" (1857), Herbert Spencer (1820–1903) sketched a comprehensive theory of social development, describing a movement from incoherent homogeneity to coherent heterogeneity through three social stages, together with a general evolution from military to industrial society. Edward Burnett Tylor's *Primitive Culture* (1871) also posited a development through three successive technological stages – savagery, barbarism, and civilization – as did Lewis Henry Morgan's still more elaborate *Ancient Society* (1877), which ended with pioneering treatments of the evolution of the state, the family, and property.

These names only scratch the surface: A dozen others could be mentioned. Were these grand theorists of social change all "historicists"? There is no doubt that their work conformed to the definition set forth by Mandelbaum, cited

[17] John Bagnell Bury, "The Science of History," in *The Varieties of History*, ed. Stern, pp. 214–15.

earlier, and – in some cases – to the more polemical usage popularized by Karl Popper. While each of these theorists of social evolution or development appealed to empirical evidence, they typically operated at a considerable distance from contemporary historiography. If none of these thinkers eschewed intentional explanation altogether, none relied on a methodology unique to *historical* understanding, of the hermeneutic kind central to the professionalization of historiography. Indeed, Comte – who launched the term "positivism," after all – and Spencer, among others, were willing to express an outright hostility toward historiography proper, condemning it for its methodological emphasis on the unique and the individual.

For this reason, the influence on historiography of figures such as Comte and Spencer, Morgan and Maine, was less important in the long run than that of theorists and historians who occupied a fertile middle ground between overarching theory and conventional academic historiography. Two pioneers in the history of culture, in fact, occupied opposite ends of the historicist spectrum. In France, Numa Fustel de Coulanges (1830–1889) promoted a distinctive brand of "scientific" historiography, exemplified in his study of *The Ancient City* (1864) and his later contributions to French history. In the German cultural zone, Jacob Burckhardt (1818–1897), whose career was formed under Ranke, produced three masterpieces of what can best be described as "aesthetic," as opposed to scientific, historicism in his *The Age of Constantine the Great* (1852), *The Culture of the Renaissance in Italy* (1860), and *Greek Cultural History* (1898).

But by far the most influential theorists of this era, at least from a late-twentieth-century standpoint, were two figures who stood at a much further remove from professional historiography. One was Alexis de Tocqueville (1805–1859), whose career, oscillating between political activism and scholarly withdrawal, conformed to an earlier pattern. Tocqueville's two masterpieces, *Democracy in America* (1835–40) and *The Old Regime and the Revolution* (1856) are not always read as documents of historicism, but they more than meet the definition. Both works were sustained by a sweeping vision of social development – the inexorable, wrenching transition from "aristocratic" to "democratic" society that defined modernity, in Tocqueville's eyes. Both works brought this vision to earth in an extraordinary combination of intentional and causal explanation, in effect founding modern "political psychology" – based, it should be added, on an impressive command of primary source material, contemporary and archival.[18] The other major figure was, of course, Karl Marx (1818–1883), whose contributions to modern social science, together with those of Friedrich Engels and of later Marxist thinkers, are treated at length elsewhere in this volume. Here it is enough to note that the conceptual centerpiece of Marx's historical materialism, the notion of a "mode of production," was itself a historicist device par

[18] Jon Elster, *Political Psychology* (Cambridge: Cambridge University Press, 1993), chaps. 3–4.

excellence. It was designed simultaneously to chart the uniqueness and diversity of forms of social life, and to grasp their place within a process of progressive development encompassing the better part of human history.

The impact of Tocqueville and Marx on the actual writing of history lay far in the future, well after the "crisis of historicism" announced by Troeltsch. For him, the "crisis" was one of relativism: The doctrine of development at the core of the historicist outlook risked depriving historical understanding itself of any objective basis. This anxiety, focused primarily on contemporary German philosophers of history, was in a sense a local instance of a wider critical debate. It included the methodological battles swirling around the works of Durkheim, Weber, and other founders of modern sociology and, at the upper reaches of philosophy, the "crisis of the European sciences" famously identified by Edmund Husserl. At a further remove, such doubts are often traced back to Friedrich Nietzsche (1844–1900), intellectual comrade to Burckhardt, whose essay *On the Advantage and Disadvantage of History for Life* (1874) launched a critical assault on contemporary historicist culture at the moment of its triumph. Cataloguing the variety of approaches to historical understanding of his day – "monumental," "antiquarian," "critical" – Nietzsche measured each in terms of its contribution to "life" and found all of them wanting. The antidote to an oppressive obsession with the past, he suggested, was the cultivation of other attitudes – the "unhistorical" and the "super-historical."

Among other things, Nietzsche's tract is a reminder that it is more accurate to speak of a variety of historicisms, with loosely overlapping congruent themes, than of a unitary intellectual tradition. At the core of the historicist outlook have always been two distinct notions: a conception of large-scale historical *development* as a central explanatory device, and a claim that the particular nature of historical phenomena, described variously as "unique" and "individual," "intentional" and "purposive," requires a method of hermeneutic *understanding* different from the causal explanations typical of the natural sciences. As we have seen, conceptions of historical development and of hermeneutic methodology emerged separately in the epoch of the Enlightenment, finding expression in such disparate traditions as the four-stages theory of the French and Scottish Enlightenments and the philosophies of history of Vico and Herder. A unique synthesis of historicist doctrine then enabled Ranke and his followers to create a model for "scientific" historiography in the first half of the century. In the second half, the model was widely imitated, promoting history, at least for the moment, to the front rank of the social sciences.

Beyond professional historiography there flourished a wide variety of other historicisms, whose impact, in some cases, was deferred until the twentieth century. As Peter Reill has suggested, the core elements of historicism always stood in tension with one another. In isolation, a strong conception of historical development is difficult to reconcile with a stress on the uniqueness and individuality of historical phenomena.[19] The variants of historicism surveyed in this chapter maintained a precarious balance between these two elements. Indeed, the "crisis" that overtook these various traditions at the end of the nineteenth century involved a separation and isolation of the themes of development and individuality that cast doubt on the scientific status of both. If, a century later, this "crisis" sometimes seems well-nigh permanent, it is worth stressing that the achievements of what might be described as historicism's heroic age are also still with us.

[19] Reill, *German Enlightenment*, p. 214.

9

BRINGING THE PSYCHE INTO
SCIENTIFIC FOCUS

Jan Goldstein

Human beings have probably always cultivated knowledge about their own cognitive and affective processes, knowledge that might be called, in the broadest sense of the term, "psychological." Over the *longue durée*, such knowledge has been stored, accumulated, and reworked within a variety of discursive pigeonholes, among them philosophy, religion, and literature. But only with the Scientific Revolution and the Enlightenment of the seventeenth and eighteenth centuries did Western Europeans begin to specify the foundations of their hitherto multiform knowledge of the psyche and to codify it with the special kind of rigor called science.[1] Only later still would they attempt to create for it a new, exclusive pigeonhole bearing the name "psychology." This chapter treats the early phase of the endeavor to bring cognitive and affective processes into scientific focus; it leaves off around 1850, before the advent of concerted efforts to create and institutionalize the unitary academic discipline of "psychology."[2]

The history narrated here is necessarily a heterogeneous one, a kind of patchwork. This is true not only because of the predisciplinary and hence somewhat inchoate condition of the particular bodies of knowledge that constitute its subject matter, but also because of the approach that the chapter takes to the category of science. A positivist approach would assume that the criteria of scientific knowledge are clear and universal and hence that the history of psychology can and should be narrated as a teleological progress leading from faulty, methodologically unsound propositions to verifiable scientific ones. Such a history would, in other words, possess a distinctive and forceful plot line. This chapter, by contrast, treats "science" more capaciously as an historical category, a native category of the country of the past, and

---

[1] Gary Hatfield, "Remaking the Science of Mind: Psychology as Natural Science," in *Inventing Human Science: Eighteenth-Century Domains*, ed. Christopher Fox, Roy Porter, and Robert Wokler (Berkeley: University of California Press, 1995), pp. 184–231.

[2] Roger Smith, "Does the History of Psychology Have a Subject?," *History of the Human Sciences*, 1 (1988), 147–77, esp. 156.

places on equal footing the diverse bodies of psychological knowledge that Western Europeans regarded as scientific during the eighteenth and early nineteenth centuries. It will, accordingly, emphasize the contestation among these psychologies, their competitive bids for recognition and legitimation, instead of constructing a narrative around the inevitable victory of the "really scientific" one. The intertwining of psychology and politics will thus be one of its central themes – both in the general sense in which politics refers to the allocation of power, and in the more specific sense in which national political communities and regimes choose to institutionalize one or another form of knowledge.

In keeping with this political theme, the chapter will also pay special attention to the operationalization of these early psychological sciences: their application to concrete social practices, their conversion into social technologies, their invocation to validate practices of otherwise dubious origin. Knowledge is, to be sure, embedded from the outset in its sociopolitical context, and it develops in complex ways in the course of practice. But once a particular theory has been codified, disseminated, and even reduced to a set of convenient formulas, the relatively straightforward process of its operational *reinsertion* into its context is a common pattern. From the vantage point of this chapter, such operationalization is significant in two ways. First, by its deliberate and inherently risky nature, it provides further evidence of contemporaries' convictions about the scientific reliability of the psychological theories in question. Second, since operationalization is a direct measure of the authority that a given theory commands, it underscores the political dimension of psychology.

Newly emergent scientific psychologies also had to negotiate their relations with older forms of psychological knowledge, especially those propounded under the rubrics of philosophy and religion. Sometimes they declared themselves compatible, sometimes they set themselves in opposition. In either case, those intrinsically political negotiations are germane to this chapter.

The highly influential positivist concept of science, which has not presided over the writing of this chapter, belongs to the chapter in another way. It was first advanced during the period under consideration here, having been proposed in a series of public lectures in Paris during the 1830s by Auguste Comte (1798–1857), then a decidedly marginal figure. Drawing on a current of philosophical speculation that reached far back into the previous century,[3] Comte defined science both as the consummate method of inquiry and as the stage in human history during which that method of inquiry was ascendant. The method in question renounced all a priori knowledge and postulation of ultimate causes, confining itself to the sensory observation of phenomena and to the discernment of the lawful regularities among them. Popularizing

---

[3] Keith Michael Baker, *Condorcet: From Natural Philosophy to Social Mathematics* (Chicago: University of Chicago Press, 1975), chaps. 2–3.

the adjective "positive" as a badge of scientific honor, Comte drew up a list of sciences in the order in which they had achieved (or would subsequently achieve) "positive" status. Starting with mathematics and ending with sociology, he pointedly omitted psychology altogether. Efforts to obtain systematic knowledge about the mind had, he said, been so long cultivated under the rubric of metaphysics that they were, from a positive standpoint, ineradicably tainted. Their lack of an observable object of investigation, their airy and insubstantial character, destroyed for these would-be psychologies any claim to an autonomous scientific existence. They would, however, acquire scientific status through the vehicle of another science, that of the physiology of the brain and nervous system, to which all of the evanescent phenomena discussed by the metaphysicians would one day be securely reduced.[4]

Comte's belief that promoting positive science required an understanding of the *history* of science gave him an overview of the field of competing psychologies that was probably unique in his era and that offers a valuable primary source to the early-twenty-first-century historian. While rejecting Comte's definition of science and his triumphalist narrative of the history of science, this chapter will make use of Comte's insights about the politics of psychology in the early nineteenth century.

THE PREEMINENCE OF SENSATIONALIST PSYCHOLOGY

The labor of bringing the psyche into scientific focus was first achieved in the modern period by a philosophical theory now called sensationalism or empiricism. Identified most strongly with the work of John Locke (1632–1704) in Britain and of the abbé de Condillac (1715–1780) in France, sensationalism understood the human mind at birth to be a tabula rasa or blank slate. Its contents derived from the impingement of external reality on the sense receptors, producing mental impressions, subsequently fashioned into ideas, that were in turn amenable to an infinity of combinations. The theory differed in its details among the various thinkers who pursued it. Locke, for example, invoked the dual principles of sensation and reflection to account for the contents of a fully developed mind, while Condillac aimed at greater parsimony. By attributing the invention of linguistic signs to the elementary mental faculties, he was able to make do with sensation alone.[5]

But variations apart, the systematic nature of sensationalism, its motif of beginning at the beginning and building up a complex mental picture

---

[4] Auguste Comte, *Cours de philosophie positive*, 6 vols., 2nd ed. (Paris: J.-B. Baillière, 1864), vol. 3, lesson 45. The text was originally presented orally in December 1837. An abridged version in English translation can be found in Gertrud Lenzer, ed., *Auguste Comte and Positivism: The Essential Writings* (New York: Harper Torchbooks, 1975), pp. 182–94.

[5] Georges Le Roy, Introduction to *Oeuvres philosophiques de Condillac*, 3 vols. (Paris: PUF, 1947), vol. 1, p. xv.

from irreducibly simple elements, argued forcefully for the scientific status of the theory to many contemporaries.[6] Also eminently scientific, in an era in which Newtonian physics had become the regnant model of science, was the epistemological modesty of sensationalism. Its proponents, including Locke and Condillac, who made memorable pronouncements to this effect, refused to appeal to ultimate causes or metaphysical principles, declaring themselves necessarily satisfied with a humbler, localized, and empirically grounded explanation of the human capacity for reasoning.[7] Those eighteenth-century Europeans persuaded of the scientific authority of sensationalism demonstrated their serious intellectual allegiance to it by making it a theory particularly fertile in operationalizations.

Thus, even before Locke published his magisterial *Essay Concerning Human Understanding* (1690), Thomas Hobbes (1588–1679) had employed a rudimentary theory of sensationalism as the grounding of *Leviathan* (1651), his attempt at a scientific theory of politics. Self-consciously seeking in this pre-Newtonian era a model from the physical sciences, Hobbes pressed into service the "resolutive-compositive" method of Galileo. He offered it as a justification for his strategy of dissolving political society into its component atomistic individuals and, in turn, dissolving those individuals into the forces that putatively propelled them into action. Here he enlisted a version of sensationalism, postulating that the motions of individual human beings were, as one commentator has put it, "the effects of a mechanical apparatus consisting of sense organs, nerves, muscles, imagination, memory and reason, which apparatus moved in response to the impact (or imagined impact) of external bodies on it."[8] Subsequent theorists of sensationalist psychology would refine Hobbes's theory. But the basic sensationalist credo was certainly articulated by Hobbes, who announced on the very first page of the first chapter of *Leviathan* that "there is no conception in a mans mind, which hath not at first, totally, or by parts, been begotten upon the organs of Sense."

What I have called Hobbes's operationalization of sensationalism was, to be sure, the employment of one theory to give birth to another theory – in this case, Hobbes's famous brief in favor of constitutional absolutism. But other operationalizations of sensationalist psychology achieved the full-scale transition into the realm of practice, especially pedagogical and psychiatric practice.

Thus, under the aegis of sensationalism, the eighteenth century saw – at least in England, its American colonies, and France – something of a pedagogical mania. Locke provided the specific as well as the general impetus: His book *Some Thoughts Concerning Education* (1693), already reprinted twenty

---

[6] Isabel F. Knight, *The Geometric Spirit: The abbé de Condillac and the French Enlightenment* (New Haven, Conn.: Yale University Press, 1968), pp. 27–8.

[7] Baker, *Condorcet*, chaps. 2–3.

[8] C. B. Macpherson, Introduction to Thomas Hobbes, *Leviathan* (New York: Penguin, 1968), pp. 25–8, at p. 28.

times by 1764,[9] spelled out the implications for child rearing of the psychological and epistemological model he had proposed a few years earlier in his *Essay*. "I imagine the Minds of Children as easily turned this way or that, as Water," he argued, so that the "little, and almost insensible Impressions on our tender Infancies have very important and lasting Consequences."[10] Given children's extreme plasticity, parents should take firm control of their upbringing, keeping them as much as possible in the parents' own company and deliberately minimizing the imponderable influence of servants.[11] (As this reference to servants indicates, Locke addressed his practical pedagogical advice to an elite stratum of society, despite the abstract, universalistic claims of his psychology.) And given the moral neutrality of the tabula rasa, as opposed to the wicked will that Puritans ascribed to infants, Locke criticized the use of physical restraints on young children. So influential was his opinion that he has been credited with the widespread abandonment of swaddling in eighteenth-century England, a change of mores in which that country led the rest of Europe. Locke's pedagogical dictum may have had even wider ramifications: By making babies accessible to adult cuddling and caresses, the demise of swaddling encouraged in eighteenth-century England the precocious development of a new kind of the nuclear family dedicated to the cultivation of affectionate ties.[12]

Locke was equally concerned that children be accorded moral freedom. According to his psychology, children could not be taught proper conduct by rules, "which will be always slipping out of their Memories," but only by means of practical repetition leading to habituation and especially by means of parental example. With respect to the latter, and in the masculinist spirit of his era, he focused attention on the dynamics of the father–son relationship. He instructed the father to "do nothing before [the son] which you would not have him imitate." If you punish your son for behavior that he sees you yourself practice, Locke contended, the youngster will develop an embittered attitude toward authority; he will believe that your severity toward him stems not from a kindly paternal concern to correct his faults but from the "Arbitrary Imperiousness of a Father, who, without any Ground for it, would deny his Son the Liberty and Pleasures he takes for himself." Such paternal "Imperiousness" will cause a son, once he is grown and guided by his own reason, secretly to wish for the death of his father.[13]

The ideal father that Locke derived from his psychology, one who was obliged to practice what he preached and who exacted obedience only on

---

[9] James L. Axtell, ed., *The Educational Writings of John Locke* (Cambridge: Cambridge University Press, 1968), "Checklist of Printings," pp. 98–9.

[10] John Locke, *Some Thoughts Concerning Education*, in *Educational Writings*, ed. Axtell, pp. 114–15.

[11] Ibid., p. 164.

[12] Lawrence Stone, *The Family, Sex and Marriage in England, 1500–1800* (New York: Harper and Row, 1977), pp. 424–6; see also p. 264.

[13] Locke, *Thoughts Concerning Education*, pp. 145, 158, 171–2.

rational grounds, had clear political implications. As one scholar has noted, Locke had no wish to "circumscribe paternal authority." His aim was rather to "render it more effective by making it noncoercive." To be genuinely normative, paternal authority had to be based on freely given filial esteem; forced obedience to irrational and overly rigorous demands would only undermine it. Thus, in a political universe consonant with Lockean psychology, there could logically exist no absolute sovereigns, only constitutional ones whose rights were balanced by their duties. Widely and enthusiastically adopted in the households of the American colonies, Locke's psychological and pedagogical teachings may actually have served to foster robust, principled opposition to George III, thus providing one of the conditions of possibility of the American Revolution of 1776.[14]

On the other side of the Channel, Condillac's oeuvre progressed in a manner similar to Locke's, moving from psychology to pedagogy. Decades after writing his psychological treatises, the *Essai sur l'origine des connaissances humaines* (1746) and the *Traité des sensations* (1754), Condillac published the course of study that he had personally devised and implemented while serving as tutor to the prince of Parma. The instructional plan closely followed the pattern of unfolding of the human mental faculties, as Condillac had earlier postulated it. Thus the boy was never presented with pat, abstract generalizations but was rather taught to arrive at such generalizations himself by reasoning from empirical particulars of which he had had direct experience. Condillac summarily dismissed the old saw that children were simply ineducable until some mysterious infusion of rationality occurred at the "age of reason," substituting for it a developmental schema in which learning was possible at all ages. He predicated his instruction on the ability of the teacher to empathize with the pupil, using as the rational basis for such empathy the theory of the sequential generation of the faculties from the first sensations. The theory would enable the teacher to gauge and then to identify imaginatively with the pupil's particular stage of cognitive development. "In order to execute my plan, I must draw closer to my pupil, put myself entirely in his place; I must be a child, rather than a preceptor." The pupil, too, was supposed to acquire psychological sophistication and self-reflexivity, and thus not merely to be educated at the hands of the tutor but to grasp the mechanics of that process of education. "Why then could one not make [the child] notice what is happening within him when he judges or reasons, when he desires or forms habits?"[15]

[14] Jay Fliegelman, *Prodigals and Pilgrims: The American Revolution against Patriarchal Authority, 1750–1800* (Cambridge: Cambridge University Press, 1982), Introduction and chap. 1, esp. pp. 1–2, 13.
[15] Etienne Bonnot de Condillac, "Discours préliminaire" and "Motif des leçons préliminaires," *Cours d'études pour l'instruction du prince de Parme* (1775), in *Oeuvres philosophiques de Condillac*, ed. Le Roy, vol. 1, pp. 397–8, 408.

More widely known in France than Condillac's pedagogical treatise was *Emile, or On Education* (1762), by Jean-Jacques Rousseau (1712–1778). This book so impressed its enthusiastic bourgeois readers with the gravity of their duties as parents ("How I wish I knew more, so that I might give my own children lessons!" one such reader exclaimed) that it even inspired a vogue of maternal breast-feeding among a population accustomed to farming its infants out to wet nurses.[16] Rousseau, who for a time had been a close intellectual companion of Condillac's, was thoroughly familiar with the theories of his fellow *philosophe* and cited them in his own works.[17] But while *Emile* was less devoutly sensationalist than Condillac's *Cours d'études*, it too assumed a child with little in the way of innate intellect, one whose mind was putty in the hands of his energetic and psychologically astute tutor. And, while Rousseau believed in certain natural propensities of the child that an artificially contrived education could preserve, he certainly struck the requisite sensationalist note in the opening chapter of *Emile*: "We are born with the use of our senses, and from our birth we are affected in various ways by the objects surrounding us. As soon as we have, so to speak, consciousness of our sensations, we are disposed to seek or avoid the objects which produce them. . . ."[18]

Just as pedagogy received an impetus from the theoretical vistas opened by sensationalism, so also did the nascent medical specialty of psychiatry. Its founding therapeutic paradigm, the so-called moral treatment, had a curious provenance, for in both England and France it was initially employed not by certified physicians, but by lay healers and barely literate madhouse guards. The technique, never precisely defined, consisted in acting upon the insane by psychological means – sometimes gentle and cajoling, sometimes strict and authoritarian – instead of subjecting them to the battery of physical means (bleedings, purgings, pharmacological preparations) long favored for this purpose by trained physicians.

When in the closing decades of the eighteenth century the French physician and founding father of psychiatry Philippe Pinel (1745–1826) integrated the moral treatment into orthodox medical practice, one of his strategies was to supply the jerry-built therapy with a suitably scientific rationale. To that end, he showed it to be a rehabilitative pedagogy grounded in sensationalist psychology. In the Introduction to his seminal *Traité médico-philosophique sur l'aliénation mentale* (1801), Pinel indicated his philosophical debt to *Idéologie*, as the followers of Condillac called their project. The case

---

[16] Robert Darnton, "Readers Respond to Rousseau: The Fabrication of Romantic Sensitivity," in his *The Great Cat Massacre and Other Episodes in French Cultural History* (New York: Basic Books, 1984), pp. 215–56, esp. pp. 217–22, 235–42, and p. 239 for quotation; George D. Sussman, "The Wet-Nursing Business in Nineteenth-Century France," *French Historical Studies*, 9 (1975), 304–28, esp. 306–7.

[17] Knight, *Geometric Spirit*, chap. 1.

[18] Jean-Jacques Rousseau, *Emile; or, On Education*, trans. Allan Bloom (New York: Basic Books, 1979), p. 39.

histories that larded the body of the text demonstrated his operationalization of Condillac's Lockean principle that madness was primarily a disorder of the imagination. Hence, Pinel employed a variety of quasi-theatrical devices, including deliberately staged scenes and jokes with surprising punchlines, to "strike the imagination strongly," to shake it up and thereby dislodge the erroneous, pathological idea that had taken hold there, to rupture the "vicious chain of ideas." The spectacular aspect of the moral treatment fit well with Condillac's contention that the imagination was a preverbal operation of mind. Hence, though oblivious to logic, it could be influenced by images and display.[19]

A second therapeutic paradigm of the nascent psychiatric specialty was also justified in terms of sensationalist psychology. "Isolation" meant the removal of an insane person from his or her habitual surroundings to the artificial environment of an institution for a stay of some duration. As articulated by Pinel's most important student, J.-E.-D. Esquirol (1772–1840), the technique was held to work in a manner analogous to the theatrics of the moral treatment. The sudden change in environment would "shock" the patient and, by withdrawing the sensory underpinnings of the pathological configuration of ideas then in place, would recreate a mental blank slate on which the institutional personnel could deliberately impress new, salutary ideas. So implicated in the practical life of its era did the theory of isolation become that peers and deputies even cited it on the floor of the French legislature to argue for the passage of the Law of 30 June 1838, which mandated the creation of a national network of asylums for the incarceration and medical care of the insane.[20]

The true heyday of the operationalization of sensationalist psychology in France came, however, not in 1838 but some four decades earlier, during the era of the French Revolution. In 1795 the revolutionaries established a system of national secondary schools, the *écoles centrales*, in which Condillac's psychology functioned as the foundation of the whole curriculum. It also supplied the content of the master propaedeutic course[21] – a practice entirely in keeping with Condillac's belief that, even as they were learning, pupils must actively understand the mental processes that enabled them to learn. Nor was the influence of sensationalist psychology confined to the classroom. The revolutionaries deliberately altered certain practices of everyday life in the hope of creating, by means of sensationalist psychological techniques, a truly regenerated citizen body no longer attached to the crown and Church of the Old Regime and maximally fit for participation in the new nation. Thus, Paris streets were renamed so that the city dweller might encounter at

---

[19] Jan Goldstein, *Console and Classify: The French Psychiatric Profession in the Nineteenth Century* (Cambridge: Cambridge University Press, 1987), chap. 3, esp. pp. 77, 84, 90–3.
[20] Ibid., chap. 8, esp. pp. 285–92.
[21] Robert R. Palmer, *The Improvement of Humanity: Education and the French Revolution* (Princeton, N.J.: Princeton University Press, 1985), chap. 6.

every turn street signs that gave rise to patriotic thoughts and sentiments.[22]
An annual cycle of revolutionary festivals was instituted, as the organizers
explained, so that the sensory bombardments of spectacle and music would
"imprint" the "soft wax" of the participants' minds with a lasting connection
between the idea of the Republic and that of a superabundant richness.
The consequence would be unwavering political devotion and invincible
heroism.[23] The revolutionary calendar was similarly justified by means of
sensationalist psychology. "We conceive of nothing except through images,"
said one of its supporters in the legislature. In this case, the new names of
months, conjuring up the processes and bounties of nature, would purge
time itself of priestly references and support a secular worldview.[24]

THE MESMERIC COUNTERPOINT

Overlapping chronologically with the late-eighteenth-century sensationalist vogue was the fashion for another brand of psychology, associated with
the Viennese physician Franz Anton Mesmer (1734–1815). Strictly speaking,
Mesmer's theory of animal magnetism, which he first expounded in 1774
and presented as a science in the Newtonian manner, was not a psychology, although subsequent elaborations by others would qualify it as such. It
was a holistic theory of bodily health, predicated on the assumption that an
invisible magnetic fluid filled the entire universe and formed the medium
connecting human beings, the Earth, and the celestial bodies. The amount
and distribution of this universal fluid within the individual organism were,
Mesmer held, responsible for its health or sickness. By provoking "crises" in
his patients, who were for this purpose seated around large tubs from which
protruded fluid-dispensing iron rods, he redistributed their portions of fluid
in order to cure whatever ailed them. Hence his aphorism: "There is only
one illness and one healing."[25]

Rebuffed by the Viennese medical establishment, Mesmer moved to Paris
in 1778. He quickly assembled a group of French disciples, and people of all
orders of society were soon flocking to the *baquets*, as the mesmeric tubs were
called. Among these devotees, the ability to succumb to a crisis, or convulsive
seizure, at the *baquet* was prized as a mark of *sensibilité*[26] – the capacity for

---

[22] Abbé Henri Grégoire, *Système de dénominations topographiques pour les places, rues, quais, etc. de toutes les communes de la République* (Paris: Imprimerie nationale, 1794), p. 10.

[23] Mona Ozouf, *Festivals and the French Revolution*, trans. Alan Sheridan (Cambridge, Mass.: Harvard University Press, 1991), chap. 8, esp. p. 203.

[24] See the report of Philippe Fabre d'Eglantine, 24 October 1793, in *Procès-verbaux du Comité d'instruction publique de la Convention nationale*, 7 vols., ed. James Guillaume (Paris: Imprimerie nationale, 1891–1959), vol. 2, pp. 697–706.

[25] Henri E. Ellenberger, *The Discovery of the Unconscious: The History and Evolution of Dynamic Psy-chiatry* (New York: Basic Books, 1978), pp. 55–74.

[26] See Antoine François Jenin de Montègre, *Du magnétisme animal et de ses partisans* (Paris: D. Colas, 1812), p. 4.

intense feeling, thought by the Montpellier vitalists to be rooted in nervous physiology, that formed the core value of late-eighteenth-century European pre-Romanticism.[27]

The frequency of unruly mesmeric gatherings in the French capital alarmed the government of Louis XVI. That the convulsions spread from one person to another as if by epidemic contagion suggested that this mode of healing might, if widely employed, undermine social and political order.[28] Also feeding royal anxiety was the national scope of the movement, the result of an aggressive expansion that drew on the organizational model and some of the membership of the Masonic lodges.[29] In 1784, the crown prudently appointed a commission of scientists and physicians to subject the phenomenon of animal magnetism to thorough investigation.

During the previous decade, Mesmer had actively solicited hearings from the Academy of Sciences, the Royal Society of Medicine, and the Paris Faculty of Medicine. All three bodies had summarily rejected his theories. The royal commission, too, returned a negative verdict, but this one was more carefully reasoned and more widely publicized than its predecessors. The commissioners would lend no credence to Mesmer's universal fluid; and while they accepted the authenticity of the convulsions produced by the treatment, they ascribed them to the overstimulated imaginations of the clientele seated around the *baquet*.[30] Oddly enough, the emphasis placed by the hostile, debunking commissioners on the fundamentally psychological nature of mesmerism was consonant with later characterizations by its most influential advocates.

Pivotal in shifting the definition of mesmerism was the work of Mesmer's disciple A.-M.-J. Chastenet, marquis de Puységur (1751–1825). Although initially accepting the universal fluid theory and, in a pastoral variant on the *baquet*, ministering to his peasants by connecting them with ropes to an old elm near his chateau, Puységur gradually abandoned the founder's interpretation of mesmeric phenomena. He hypothesized that Mesmer's canonical procedures induced a "magnetic sleep" or somnambulism – an altered, trancelike state of consciousness in which the subject became markedly more susceptible to influence. Consequently, he identified the curative agent as the magnetizer's will and the power it exercised over the mesmerized patient.[31] Puységur thus reclassified mesmerism from a cosmological to a psychological theory, a move that led to the abandonment of the *baquet* as an extraneous

[27] Janet Todd, *Sensibility: An Introduction* (London: Methuen, 1986).
[28] Jan Goldstein, "Moral Contagion: A Professional Ideology of Medicine and Psychiatry in Eighteenth- and Nineteenth-Century France," in *Professions and the French State, 1700–1900*, ed. Gerald L. Geison (Philadelphia: University of Pennsylvania Press, 1984), pp. 181–222.
[29] Charles Coulston Gillispie, *Science and Polity in France at the End of the Old Regime* (Princeton, N.J.: Princeton University Press, 1980), pp. 278–9.
[30] *Rapport des commissaires chargés par le Roi de l'examen du magnétisme animal* (Paris: Imprimerie royale, 1784).
[31] Ellenberger, *Discovery of the Unconscious*, pp. 70–2.

piece of equipment. He thereby established a basic parity between mesmerism and sensationalist psychology, turning them into competing discourses.

Hence, once the elderly Dr. Minoret, a character in Balzac's novel *Ursule Mirouët* (1841), becomes convinced of the verifiability of the mesmeric trance, his Enlightenment credo is decisively shaken: "Founded on the theories of Locke's and Condillac's followers, the whole of his scientific system was now in ruins." Minoret's newfound belief in mesmerism even leads the way to his religious conversion.[32] Though fictional, Minoret is historically representative. Earlier proponents of mesmerism in France, analogizing the universal fluid to gravity, had tended to emphasize the strict rationalism of Mesmer's doctrine and to portray it as carrying out a Newtonian-style conquest of a new domain for science. Postrevolutionary proponents like the fictional Minoret tended to give a different inflection to these scientific claims. Stressing the *subtilité* of the fluid and its participation in another, imperceptible world, they surrounded mesmeric science with a religious aura.[33]

Affinity with religion was also the hallmark of mesmerism's early-nineteenth-century career in the United States, where its introduction in the 1830s by the French lecturer Charles Poyen coincided with the peak of the Second Great Awakening. Like Minoret, Poyen's New England listeners were often "converted from materialism to Christianity by the facts in Animal Magnetism," as one of them testified in a letter to a Boston newspaper. According to a recent historical account, American mesmerism was by the 1850s straddling a fine line between sacrament and scientific psychology and partaking of the dynamics of each.[34]

As the July Monarchy setting of Balzac's novel would suggest, the 1784 royal commissioners' report had hardly dealt a death blow to animal magnetism in France. Nor would a second official condemnation, that of the Royal Academy of Medicine in 1840, achieve that end. Rather, rebaptized "hypnotism" by the Scottish surgeon James Braid in his book *Neurypnology; or, the Rationale for Nervous Sleep* (1843),[35] mesmeric phenomena were destined for a long and prominent career – one in which the claims of "elite" and "popular" science constantly interacted and blended – in France, elsewhere in Europe, and in the United States, for the rest of the nineteenth century.[36]

The French persecutory style with respect to animal magnetism was not everywhere the rule. Indeed, the reception of animal magnetism in Germany demonstrates a significant difference between the early-nineteenth-century

---

[32] Honoré de Balzac, *Ursule Mirouët*, trans. Donald Adamson (New York: Penguin, 1976), pp. 101–3, at p. 101.

[33] Robert Darnton, *Mesmerism and the End of the Enlightenment in France* (Cambridge, Mass.: Harvard University Press, 1968), chap. 5.

[34] Robert C. Fuller, *Mesmerism and the American Cure of Souls* (Philadelphia: University of Pennsylvania Press, 1982), pp. 22, 68, 75.

[35] Alan Gauld, *A History of Hypnotism* (Cambridge: Cambridge University Press, 1990), p. 281.

[36] Alison Winter, "Mesmerism and Popular Culture in Early Victorian England," *History of Science*, 32 (1994), 317–43.

German and French scientific cultures. Regarding French Enlightenment-style rationality as arid and mechanical, German physicians and scientists gravitated toward the spiritual, organic, and Romantic conceptions of *Naturphilosophie*, an intellectual trend probably reinforced for nationalist reasons after Napoleon's 1806 defeat of Prussia. Hence, official German scientific circles tended to accord a more sympathetic welcome to animal magnetism than had their French counterparts. Rejection by the 1784 royal commission encouraged the embrace of mesmerism by the radical political fringe in pre-Revolutionary France, especially by embittered hack pamphleteers who saw Mesmer's rejection as emblematic of the closed corporate structure of the Old Regime, which could find no place for meritorious outsiders.[37] By contrast, a Prussian police ordinance of 1812 legitimized mesmeric practice (though by certified physicians only), and a Prussian government commission of inquiry into the claims of animal magnetism arrived at favorable conclusions in 1816. Academic recognition soon followed. Dr. K. C. Wolfart, a leading German proponent of mesmerism, treated patients in a tastefully furnished salon (containing two large *baquets* of his own design) that became a meeting place of the Berlin intelligentsia. In 1817, he was made a full professor at the University of Berlin, and a state-subsidized clinic for the magnetic treatment of the poor was set up under his direction.[38] No wonder that French magnetists described the situation of their science in Germany as "entirely established," enviously noting that "*there* its existence can no longer be called into question."[39]

THE PSYCHOLOGICAL PLAYING FIELD ACCORDING TO AUGUSTE COMTE

When in the late 1830s Comte issued his provocative pronouncement that the mind belonged to biology, he also surveyed the contemporary playing field of competing psychological discourses. There he found three main contenders. The old stalwart, sensationalism, which had consistently prided itself on its antimetaphysical stance, appeared to the founder of positivism to be an outmoded "metaphysical" system. It earned that pejorative label by dint of its reliance on introspection or, in Comte's phrase, "interior observation," which he ridiculed as "absurd." "In order to observe, your intellect must pause from activity; yet it is this very activity that you want to observe." He rejected also its lopsided preoccupation with the intellectual faculties to the exclusion of

---

[37] See Darnton, *Mesmerism*, chap. 3.
[38] Gauld, *History of Hypnotism*, chap. 4, esp. pp. 88–9; Ellenberger, *Discovery of the Unconscious*, p. 77; Annelise Ego, *"Animalischer Magnetismus" oder "Aufklärung": Eine mentalitätsgeschichtliche Studie zum Konflikt um ein Heilkonzept im 18 Jahrhundert* (Würzburg: Königshausen & Neumann, 1991), pt. 3.
[39] Introduction to *Le Propagateur du magnétisme animal, par une société de médecins*, 1 (1827), v–xvi, at p. vi, my emphasis.

the affective ones.[40] A new but equally "metaphysical" theory of psychology had also come on the scene, a certain "deplorable psychological mania that a famous sophist [has] . . . succeeded in inspiring in French youth."[41] This was the philosophical psychology of Victor Cousin (1792–1867). The third contender was phrenology, which Comte associated with the names of two Germans, Franz Joseph Gall (1758–1828) and Johann Gaspar Spurzheim (1776–1832). Comte was convinced that the future belonged to phrenology, the theory through which psychology would finally shed its metaphysical baggage and attain positivity.

Although Comte's evaluation of the competitors circa 1830 was highly polemical, his identification of them was apt. Indeed, a thinker as antipathetic to Comte as Hegel had drawn up much the same list in the late 1820s, when presenting his own philosophy of mind in lectures to his Berlin students.[42] Comte's rendition of the psychological playing field will be used to organize the rest of this chapter.

COUSINIAN PSYCHOLOGY IN EUROPEAN CONTEXT

Contrary to Comte's predictions, Cousin's philosophical "eclecticism" became the dominant psychology in France through most of the nineteenth century. Yet, ironically, at the point of its introduction it aroused intense suspicion, in part because it bore so strongly the imprint of non-French influences.

Cousin conceived of eclecticism as a reaction against sensationalism, which he derisively called "sensualism." In his view, the widely accepted psychology of the Enlightenment bore responsibility for the anarchical excesses of the French Revolution and the political instability that had subsequently plagued France. According to this argument, the main fault of sensationalism lay in its inability to ground a strong and cohesive self, or *moi*. Building up consciousness through the successive accumulation of atomistic sensations, it had access to no overarching principle of unity. Cousin regularly cited, with an incredulity verging on outrage, Condillac's definition of the self as "a collection of sensations." Furthermore, the consciousness thus posited was essentially passive, coming into existence only as a response to sensory prodding. Finally, sensationalism denied the independent spiritual principle that anchored an immortal soul: If the psychology of Locke and Condillac was too metaphysical for Comte, it was entirely too materialist for Cousin. In sum, sensationalism vitiated human moral responsibility. It had consequently

---

[40] Lenzer, ed., *Comte*, pp. 184–5; see also p. 80 for Comte's 1830 introductory lecture to the *Cours de philosophie positive*.

[41] Comte, *Cours de philosophie positive*, lesson 45.

[42] G. W. F. Hegel, *Philosophy of Mind, being Part Three of the Encyclopaedia of the Philosophical Sciences* (1830) (Oxford: Clarendon Press, 1971), esp. pp. 147, 183.

nurtured in late-eighteenth-century France a generation of revolutionaries –
people incapable of setting limits, propelled into action by their fantasies,
unrestrained by the fear of divine retribution in the afterlife.

The philosophical choices made by Cousin in order to rectify the historical
situation thus diagnosed were shaped by his particular political commitments.
He was affiliated with a group called the Doctrinaires, who, upon coming
to power during the July Monarchy (1830–48), stood for the cautious and
conservative liberalism of the *juste-milieu*. They supported a constitutional
monarchy with property qualifications for voting that enfranchised no one
beneath the upper reaches of the bourgeoisie. Their motto was, "Establish
authority first, then create liberties as counterweights."[43] Hence, Cousin
sought a psychology that would be consonant with a stable and, above all,
moderate French polity, one whose deliberate middlingness would make it
proof against renewed revolution.

In his effort to repair a dangerously defective sensationalism, Cousin
looked abroad for inspiration. His teacher Pierre-Paul Royer-Collard, also
a Doctrinaire, had earlier introduced the writings of the Scottish commonsense school, especially those of Thomas Reid (1710–1796), into his Sorbonne
philosophy course. Cousin would second that reliance on the Scots, who had
from his perspective the compelling advantage of imputing activity to consciousness. Reid's *Inquiry into the Human Mind on the Principles of Common
Sense* (1764) had strategically revised sensationalist psychology in ways that
anticipated Kant, while scrupulously avoiding the residual skepticism that
would be the hallmark of Kant's critical philosophy. According to Reid, the
component units of psychological life were not atomistic sensations producing atomistic ideas but instead the relational principles that Reid called
judgments (e.g., causality, induction). These, he argued, must already be
present before the senses could operate. Such judgments were self-evident,
prior to experience; they came from "our own nature" and hence were termed
"common sense." Thus modifying the sensationalist account of mental life,
the psychology of the eighteenth-century Edinburgh school went some distance toward repairing the fatal political flaws that Cousin had discerned
in the doctrines of Locke and Condillac. But by steadfastly refusing to re-
instate metaphysics, Reid failed, in the view of the Cousinians, to go far
enough.[44]

More innovative – and far more controversial – were Cousin's borrowings from German Idealism, which enabled eclecticism to go the whole
distance to an embrace of metaphysics. The mature Cousin was fully aware
of the allegation made against him: that his philosophy was a wholesale

[43] Pierre-Paul Royer-Collard as quoted in Dominique Bagge, *La conflit des idées politiques en France
sous la Restauration* (Paris: PUF, 1952), p. 100.
[44] "Reid (Thomas)," in *Dictionnaire des sciences philosophiques*, 2nd ed., ed. Adolphe Franck (Paris:
Hachette, 1875), pp. 1468–72; *Victor Cousin, les Idéologues, et les Ecossais: Colloque international de
février 1982* (Paris: Presses de l'Ecole Normale Supérieure, 1985).

importation from Germany and hence an offense to patriotism.[45] Certain modern commentators have construed his relationship to Germany more sympathetically, arguing that the unknown terrain of German intellectual culture served Cousin as a kind of mirror facilitating the invention of a new French philosophical identity.[46] In any case, Cousin's first trip to Germany in 1817 was a philosophical grand tour during which he met, among others, Hegel, Schelling, and Friedrich Schlegel. From Hegel he would derive an insistence that the history of philosophy must be the foundation of philosophy, as well as the Hegelian dialectic, with its *Aufhebung* of conflicting opposites, which Cousin converted into a far less rigorous syncretism. From Fichte, he derived the emphasis on the ego and its titanic metaphysical powers, the vocabulary of *moi* and *non-moi* that would mark his psychology.

Derivative as a thinker, Cousin possessed real genius as an academic entrepreneur and institution builder. He succeeded in training and placing in teaching posts a "regiment" (as contemporaries called it) of loyal disciples, who fanned out from Paris to form a network covering the entire country. He entrenched his version of psychology not only in the universities but also, by a national decree of 1832, in every *lycée* in France, making it the first substantive part of the philosophy curriculum – a position it essentially maintained throughout the nineteenth century.

Psychology instruction à la Cousin had two major components. First, the student had to learn about the a priori existence of the *moi* and to gain mastery of those introspective techniques that would enable him to apprehend his own *moi* directly. Introspection would reveal the *moi* to be a spontaneously active entity, a pure volition. This important knowledge would serve both to empower and to inculcate moral responsibility in the bourgeois male adolescents who exclusively formed the student body of the *lycées* for most of the century. (Tellingly, although no internal logic dictated that Cousinian-style selfhood be confined either to a social elite or to a single gender, Cousinian educational practice identified possession of a *moi* as an upper-class male prerogative.)[47] To the great relief of those who worried about the future of France, this new generation would be inoculated against the passive, flimsy, and random aggregation of sensations that passed for a self in sensationalist psychology! Second, the student would learn that psychology was, in Cousin's phrase, the "vestibule" to ontology. The inward, introspective turn would be followed by an outward turn to the structure of the universe at large, in which

[45] Victor Cousin, *Fragmens philosophiques*, 2d ed. (Paris: Ladrange, 1833), "Préface de la deuxième édition," p. xxx.

[46] Michel Espagne and Michael Werner, Introduction to *Lettres d'Allemagne: Victor Cousin et les hégéliens* (Tusson: Du Lérot, 1990).

[47] Jan Goldstein, "Saying 'I': Victor Cousin, Caroline Angebert, and the Politics of Selfhood in Nineteenth-Century France," in *Changing History: Politics, Culture and the Psyche*, ed. Michael S. Roth (Stanford, Calif.: Stanford University Press, 1994), pp. 321–35, and her "Eclectic Subjectivity and the Impossibility of Female Beauty," in *Picturing Science, Producing Art*, ed. Peter Galison and Caroline Jones (New York: Routledge, 1998), pp. 360–78.

the student would grasp eternal verities about the True, the Beautiful, and the Good (as Cousin's official *lycée* textbook was called). The willful, active self would thus be hedged round with venerable norms putatively expressive of its own nature, ensuring that its activity would be devoted to maintaining rather than offending the status quo.

For our purposes here, it is important to underscore Cousin's contention that, despite its overt political resonances, his psychology was fully scientific – not in the speculative sense of Hegelian *Wissenschaft* but in the Baconian, empirical and inductive sense. Repeating the formula of the Scottish commonsense school, he was fond of saying that his psychology had the epistemological status of physics. It differed only as a function of the different phenomena to be observed, which were external to man in the latter case and carried within him in the former, necessitating that they be illuminated by the interior light of consciousness.[48] It is certainly a testament to the authority of science in the nineteenth century that Cousin clung to this scientific self-representation, that he wished to be included on Comte's playing field even while he violently disagreed with Comte about the nature of psychology.

THE PERSISTENCE OF SENSATIONALISM:
ASSOCIATIONIST PSYCHOLOGY IN
NINETEENTH-CENTURY BRITAIN

If the dominant school of early-nineteenth-century French psychology agreed with Comte about the dangerously retrograde nature of sensationalism, its counterpart across the Channel unapologetically continued the Lockean legacy. In his *Essay*, Locke had briefly discussed the principle he called the "association of ideas," relegating it to those customary, chance, or otherwise aberrant instances of ideational linkage that eluded rational explanation.[49] It fell to David Hartley, more than a half-century later, to rescue that same principle from marginality and to construct in his *Observations on Man* (1749) a science of psychology on the Newtonian model, in which the role played by gravity was assigned to the association of ideas. It was from Hartley that Jeremy Bentham (1748–1832), the founder of the Utilitarian psychology that would have such a brilliant career in nineteenth-century Britain, learned one of his own central principles: that happiness could be treated as the sum of simple pleasures united by association.[50]

---

[48] Cousin, "Préface de la deuxième édition," p. viii. See also the text of Cousin's 1816 course, reprinted in his *Premières essais de philosophie*, 3rd ed. (Paris: Librairie nouvelle, 1855), p. 134; and "Ecossaise (Ecole)," in *Dictionnaire des sciences philosophiques*, ed. Franck, pp. 425–8.

[49] John Locke, *An Essay Concerning Human Understanding*, ed. P. Nidditch (Oxford: Oxford University Press, 1975), Book II, chap. 33 (pp. 394–401).

[50] Elie Halévy, *The Growth of Philosophic Radicalism*, trans. Mary Morris (Boston: Beacon, 1955), pp. 7–8.

The principle of the association of ideas has, according to one historian of science, two components: (1) complex mental phenomena are formed from simple elements derived ultimately from sensations, and (2) the mechanism of their formation depends on the similarity and/or repeated juxtaposition of the simple elements in space and time.[51] Following the lead of the French Enlightenment *philosophe* Helvétius, Bentham highlighted the stable psychological association of certain experiences with pleasure or pain. He used the associationist principle both as a foundational axiom about human behavior – that human beings unfailingly act to maximize their pleasure and minimize their pain – and, by extension, as an art-and-science[52] of morals and legislation. Boldly identifying the "is" with the "ought," Bentham insisted that the sole test of a good moral precept or a good law was that it conduced to the greatest happiness of the greatest number, as measured by the relative amounts of pleasure and pain that it brought in its wake. As a psychologist aspiring to scientific status, Bentham meant this "felicific calculus" quite literally, going so far as to list seven axes along which pleasure and pain could be quantified: intensity, duration, certainty, propinquity, fecundity, purity, and extent. His stark reduction of human mental life to the quest for pleasure and the avoidance of pain entailed a revision of the lexicon of psychology. Bentham deleted such apparently key psychological terms as motive, interest, and desire, declaring them "fictions" – that is, pleasures and pains masquerading under other, fancier names.[53]

Curiously enough, despite the radicalism of his position on morals and legislation, Bentham began his political life as a conservative. During his Tory phase, he believed that once properly enlightened, an unreformed Parliament dominated by landed aristocrats would hasten to apply his Utilitarian principles to the business of lawmaking. Experience disabused him of this view, and a fateful meeting with James Mill (1773–1836) in 1808 converted him from Toryism to a belief in democracy. Closed corporations such as political aristocracies, he now held, were by definition hostile to the principle of general utility. At the same time, he persuaded Mill that a lucid theory of representative government had to be couched in Utilitarian terms. An even exchange resulted: "Bentham gave Mill a doctrine, and Mill gave Bentham a school."[54]

Under the aegis of that school, Bentham finally exerted on practical affairs the impact he had long sought. In fact, no less dominant a trend than the growth of political liberalism in early-nineteenth-century Britain was strongly inflected by Benthamism and its associationist psychology. Even the Millite

---

[51] Robert M. Young, "Association of Ideas," in *Dictionary of the History of Ideas*, ed. Philip P. Wiener, 5 vols. (New York: Scribner's, 1973–4), vol. I, pp. 111–18, esp. p. 111.

[52] The term is that of M. P. Mack, *Jeremy Bentham: An Odyssey of Ideas* (New York: Columbia University Press, 1963), chap. 6.

[53] Ibid., chap. 5, esp. pp. 229, 247.

[54] Halévy, *Growth of Philosophic Radicalism*, pp. 251–64, at p. 251.

tactics used to lobby for the Great Reform Bill of 1832, which expanded parliamentary suffrage to take urban growth into account, were shaped by Benthamite psychological reasoning. Like all people, reasoned Mill, rulers acted to maximize and thus perpetuate their power; hence, concessions could be wrested nonviolently from an oligarchy only if it could be persuaded that those concessions were in its own self-interest. On such grounds, the Millites made intimidation by the threat of revolution their standard (and eminently successful) tactic for convincing sitting MPs to extend the vote to previously disenfranchised Britons.[55]

As Benthamite psychology increasingly pervaded early-nineteenth-century British culture, it increasingly became the target of cultural criticism. One particularly privileged observer was John Stuart Mill (1806–1873), whose father, James, had subjected him since early childhood to a thoroughgoing Benthamite pedagogical regimen. ("My course of study," he recalled, "had led me to believe that all moral feelings and qualities . . . were the results of association; that we love one thing, and hate another, . . . through the clinging of pleasurable or painful ideas to those things. . . .")[56] In his *Autobiography*, J. S. Mill described as "the crisis of my mental history" his paralyzing realization at the age of twenty that "the whole course of my intellectual cultivation had made precocious and premature analysis the inveterate habit of my mind," leaving him devoid of spontaneous "feelings in sufficient strength to resist the dissolving influence of analysis."[57] A similar critique of the Benthamite habit of mind was lodged, though in the name of the working classes rather than the intelligentsia, by Charles Dickens in his novel *Hard Times* (1854). Dickens scathingly depicted the schoolmaster Thomas Gradgrind, charged with inculcating Utilitarian principles in the children of the working classes, as having "a rule and a pair of scales and a multiplication table always in his pocket, ready to weigh and measure any parcel of human nature, and tell you exactly what it comes to." Gradgrind dedicated himself to banishing wonder from the psychological repertories of his pupils, and therein "lay the spring of the mechanical art and mystery of educating the reason without stooping to the cultivation of the sentiments and affections."[58] To counter the emotionally desiccating effects of Benthamite psychology, young Mill placed himself on a supplementary diet of Wordsworthian poetry and German thought propounding holistic conceptions of the personality.[59] His private experience might be read as a vindication of Comte's dictum: As influential as Benthamite psychology was in shaping early-nineteenth-century

---

[55] Joseph Hamburger, *James Mill and the Art of Revolution* (New Haven, Conn.: Yale University Press, 1963), chap. 2, esp. p. 23; chap. 3, pp. 50–73.

[56] John Stuart Mill, *Autobiography* (1873) (Indianapolis: Library of the Liberal Arts, 1957), p. 88.

[57] Ibid., p. 90.

[58] Charles Dickens, *Hard Times* (1854) (New York: New American Library, 1961), Book I, chap. 2, p. 12; chap. 8, p. 56.

[59] Mill, *Autobiography*, pp. 95–7, 105, 112–13.

British society, this prolongation of the vogue of sensationalism was, from at least one perspective, a cul-de-sac.

PHRENOLOGY: A PSYCHE FOR THE MASSES

Phrenology, whose imminent triumph Comte predicted, was a theory well qualified to seduce a positivist. It held that mind and brain were equivalent; that the brain was not a unitary organ but was comprised of some thirty different organs, each controlling a single intellectual or affective trait; that the size of each brain organ reflected the strength of that trait in the individual's personality; and that brain organ magnitude not only could be revealed by postmortem autopsy but also was externally visible in the cranial protuberances, or bumps, of the living human being.[60] If, then, one embraced phrenological principles, the elusiveness and the interiority of mental life, so frustrating to a science of observation, acquired a pleasing solidity and externality. Moreover, the phrenological complement of brain organs recognized emotional as well as rational attributes, thus obviating the problem that Comte had discerned in sensationalism and that John Stuart Mill's painful personal history exemplified.

The trajectory of phrenology was in many ways parallel to that of mesmerism, which it postdated by about three decades. Phrenology, too, was Viennese in origin. Its Baden-born founder, Gall, received his medical degree in Vienna and began lecturing on phrenology there. Gall also migrated to Paris, hoping, as Mesmer had hoped, to get a more sympathetic hearing in the French capital. Like mesmerism, but with greater success, phrenology made incursions into official science. It confirmed, even if it had not devised, the existence of a disease entity called monomania, which French alienists used in the 1820s and 1830s to argue for a psychiatric presence in criminal court.[61] It made its way, very briefly, into the inner sanctum when a course on the subject was taught at the Paris Faculty of Medicine in 1836 by F.-J.-V. Broussais (1772–1838), the controversial founder of "physiological medicine." It was taken seriously enough by reputable people that in 1842 the eminent physiologist Pierre Flourens (1794–1867), a member of both the Académie des Sciences and the Académie Française, devoted a highly influential pamphlet to attacking it. Strikingly, Flourens addressed not technical scientific issues but general moral and religious ones: Phrenology was wrong because its monism was incompatible with the freedom of the will and the immortality of the soul.[62]

---

[60] Georges Lanteri-Laura, *Histoire de la phrénologie: L'homme et son cerveau selon F.-J. Gall* (Paris: PUF, 1970); Angus McLaren, "A Prehistory of the Social Sciences: Phrenology in France," *Comparative Studies of Society and History*, 23 (1981), 3–22.

[61] Goldstein, *Console and Classify*, chaps. 5, 7.

[62] Pierre Flourens, *Examen de la phrénologie* (Paris, 1842).

Flourens's intervention is usually taken as marking the disappearance of phrenology from the scene of French establishment science. In the end it, like mesmerism, was cultivated most intensively under popular auspices. The task of disseminating it fell, in both France and Britain, not to prestigious educational institutions, but to itinerant popular lecturers and makeshift adult education courses. So deep was the sociological affinity between phrenology and mesmerism that the two theories could even be combined. At demonstrations of so-called phreno-mesmerism in British mechanics' institutes during the 1840s, the phrenological bumps of a mesmerized subject were touched, causing that subject to perform behaviors associated with the brain organs in question. Such demonstrations apparently had great persuasive power, converting large numbers of spectators to the truth of phrenology.[63]

The distinctive visual modality of phrenology, epitomized by its easy-to-read maps of the brain and, especially in Britain, by the ubiquitous white plaster cast of a head inscribed with a complement of brain organs, has caught the attention of historians, inspiring them to advance richly suggestive and entirely plausible hypotheses about the social meaning of phrenology. One line of interpretation stresses the perceived utility of Gall's science for survival in the new, anonymous world of the nineteenth-century metropolis. If one must have everyday encounters and business dealings with persons whose identities are not vouchsafed in advance by a close-knit community, what better self-protection than "reading" these strangers' skulls for information about any vicious propensities?[64] Thus Dr. Broussais himself acknowledged that the "art of dissimulating has been carried so far in our present state of civilization" that the experience through which we gradually learn the true character of another person almost always comes too late. To the rescue comes phrenology, quickly alerting us to and rendering decipherable the "external, positive signs" of our fellows' intellectual and affective makeup. [65]

Another line of historical interpretation emphasizes the literal, almost aggressive superficiality of phrenology – that is, its relocation of the psychological domain to the visible surface of things.[66] Coupled with the phrenologists' claim that their science was within everyone's grasp and with their proselytization of the masses, this superficiality functioned in both France and Britain as a mocking challenge to the introspective philosophical psychology championed by the academic elite. Phrenology, after all, required no long period of study in a selective educational institution; it stood, furthermore, for the removal of depth from mental life. In these ways it engaged, through its very

[63] Roger Cooter, *The Cultural Meaning of Popular Science: Phrenology and the Organization of Consent in Nineteenth-Century Britain* (Cambridge: Cambridge University Press, 1984), p. 150.

[64] Judith Wechsler, *A Human Comedy: Physiognomy and Caricature in 19th-Century Paris* (Chicago: University of Chicago Press, 1982).

[65] François-Joseph-Victor Broussais, *Cours de phrénologie* (Paris: Baillière, 1836), Preface.

[66] Steven Shapin, "Phrenological Knowledge and the Social Structure of Early Nineteenth-Century Edinburgh," *Annals of Science*, 32 (1975), 219–43, esp. 239.

structure as a theory, in a form of social combat, defying elite pretensions, making a case for democratization.

Although Gall remained in Paris from his arrival in 1807 until his death in 1825, thus apparently naturalizing phrenology on French soil, his science never flourished as a popular movement in France to the extent that it did in Britain. Gall's erstwhile collaborator, Spurzheim, personally introduced phrenology into Britain in 1815 and continued to lecture there until shortly before his death in 1832. Also, the British phrenological movement rapidly acquired skillful indigenous leadership in the person of a young Edinburgh lawyer, George Combe (1788–1858). But the striking receptivity of Victorian Britons to phrenology undoubtedly had less to do with the talents of those who propounded it than with the close fit that was contrived between the theory and British social and political attitudes.

In a word, phrenology meshed perfectly with the belief in self-help that guided the behavior of both the middle and working classes of nineteenth-century Britain in the context of a laissez-faire economy and a nonrevolutionary political tradition.[67] At first glance, phrenology's central postulate about the innateness of brain organs might seem to render it a deterministic, fatalistic doctrine. But in fact it combined physiological innateness with a strong emphasis on environmental plasticity. An individual's initial organological configuration was, to be sure, a given, but education could and should be applied to increase the size of organs governing positive traits. Prominent organs for negative traits were more problematic; they might, with special training, be prevented from growing, but, for safety's sake, surveillance of their owners was advised. Indeed, phrenology made education a less haphazard business than formerly, an investment more likely to yield a return, because children's innate talents could be infallibly pinpointed for cultivation from an early age. The self-help aspect of phrenology was already present in Gall and Spurzheim's founding texts, but it was accentuated for the British public by Spurzheim and Combe, who added to the roster of brain organs new ones that spoke to the value of work discipline: "conscientiousness," "time," "order," "concentrativeness." Spurzheim also streamlined the classification of brain organs, arranging them into genera and species and into a hierarchy that assigned low standing to the sex drive, politely renamed "amativeness," thus reinforcing the primacy of work and the necessity for delayed gratification.[68]

The dissemination of phrenology in Britain was a two-stage process, affecting first the middle classes, and especially the physicians among them, who starting in the 1820s used it as an instrument of self-assertion against the gentlemen then dominating the professions. By the 1840s its main locus

---

[67] Cooter, *Cultural Meaning of Popular Science*; Terry Parssinen, "Popular Science and Society: The Phrenology Movement in Early Victorian Britain," *Journal of Social History*, 7 (Fall 1974), 1–20.
[68] Cooter, *Cultural Meaning of Popular Science*, pp. 78–9, 116–17.

of propagation had shifted from middle-class phrenological societies to mechanics' institutes; indeed, the former sometimes even donated their used phrenological paraphernalia to the latter. Such material aid was, moreover, hardly disinterested since the middle classes had a stake in the ideological transfer it facilitated. Working-class adoption of phrenology meant a channeling of popular energies away from movements that disputed bourgeois hegemony and toward the internalization of individualistic bourgeois values. According to the most thorough examination of the subject, phrenology succeeded brilliantly in this capacity and can be regarded as an agent of working-class consent to the nineteenth-century British bourgeois order.[69]

The far less impressive popular gains of French phrenologists can perhaps be traced to their particular construction of Gall's theory. If British phrenologists emphasized the congruence of their doctrine with mainstream liberal culture, French phrenologists during the period before the Revolution of 1848 typically stressed its oppositional potential. They read the phrenological map of the brain as an argument for socialism, or for some other form of social organization that placed the collectivity above the individual.[70] This was certainly the view of Comte, who saw the multiple organs of the brain as objective proof that Cousin's unitary self was nothing but a "fiction." He dismissed it as an anachronism in the mid nineteenth century, when sociology was about to achieve positivity, bringing the anti-individualist perspective to the fore.[71] The multiple organs of the brain would, Comte believed, support a social reordering in which the individual's capacities, instead of being fused into the unitary, "metaphysical" subject of rights so dear to classical liberalism, would be dispersed through a network of duties to society.

Phrenology arrived in the United States in the 1830s, just before mesmerism, and in 1840 a disapproving John Quincy Adams listed both psychologies among the "plausible rascalities" contributing to the "bubbling cauldron" of Jacksonian culture.[72] The figure most responsible for the exportation was Spurzheim, whose Boston lectures of 1832 generated wild enthusiasm (but unfortunately coincided with a cholera epidemic, to which the lecturer succumbed). Spurzheim's brief visit was followed six years later by Combe's long, triumphal tour along the Eastern seaboard. The American version of phrenology derived from British rather than French sources and bore the stamp of the individualist ideology prevalent in both Anglophone cultures. "Self-made or never made" was the motto of the influential New York publishing firm of the Fowler brothers that was devoted to

---

[69] Ibid., passim.

[70] Typical is A.-Pierre Béraud, *De la phrénologie humaine appliquée à la philosophie, aux moeurs et au socialisme* (Paris: Durand, 1848).

[71] Comte, *Cours de philosophie positive*, lesson 45.

[72] Quoted in Charles Colbert, *A Measure of Perfection: Phrenology and the Fine Arts in America* (Chapel Hill: University of North Carolina Press, 1997), p. 1.

the propagation of phrenology.[73] As in Britain and France, phrenology in America was a popular movement, promoted by a fleet of self-appointed itinerant lecturers.

Surveying the principal varieties of psychological knowledge that claimed scientific status in western Europe and North America during the period 1700–1850, this chapter has stressed the contingent nature of the competition among them, what I have called the "political" aspect of their project to bring the psyche into scientific focus. The axes of that competition were multiple: introspective versus biological theories; theories friendly to religion versus anticlerical ones; theories conducive to social change versus socially conservative ones; and perhaps most salient, academic versus popular theories.

That scientific psychology ran on a double track, both academic and popular, from the very beginning is, on reflection, not very surprising. Among the human sciences, psychology stands out as having the greatest immediate relevance to the individual. Other human sciences, such as sociology, political economy, and anthropology, lend themselves readily to the formation of state and social policy but contain little to beguile individuals hoping to improve their own lot. Psychology is more ecumenical in its appeal. It contains resources not only for policy makers, such as French revolutionaries and British philosophic radicals, and for expert practitioners, such as the new psychiatric professionals, but also for ordinary people bent on self-understanding or prone to informed tinkering with their own heads. Thus, alongside the psychologies that during this period received official valorization and institutionalization flourished their counterparts: bodies of knowledge scorned by the academic establishment but eagerly embraced by the laity. The lay fascination with psychology, and the production of psychological sciences intended especially for lay consumption, continue in our own day. From the mesmeric tub and the itinerant lecturer on cranial bumps of the eighteenth and nineteenth centuries to the pop psychology and psychobabble of the early twenty-first century would seem to be a direct line.

---

[73] Madeleine B. Stern, *Heads and Headlines: The Phrenological Fowlers* (Norman: University of Oklahoma Press, 1971), pp. 39, 54.

10

CONTINENTAL POLITICAL ECONOMY FROM THE PHYSIOCRATS TO THE MARGINAL REVOLUTION

Keith Tribe

Political economy was a creation of the European Enlightenment – more specifically, at first, of the French and Scottish Enlightenments. By the early nineteenth century, Adam Smith's *Wealth of Nations* (1776) had been widely acknowledged as the founding text of a new classical economics that treated labor as the source of value and the accumulation of the products of human industry as the path to national wealth. It regarded commercial liberty and civic liberty as joint conditions for progress along this path. "Political economy" was understood in the early nineteenth century as a body of doctrine that identified the principles governing the good order of the body politic, or of wise legislation. "Economics," the modern term that displaced this usage in the late nineteenth century, systematically elaborated these basic principles; they became more arcane and academic, no longer part of the general knowledge of those active in public life and the world of commerce. The economic agents of the classical world – laborer, capitalist, and landlord – contributed in their different ways to the production of commodities, and received revenues – wages, profits, and rents – according to their contributions. The "new economics" of the later nineteenth century replaced these social groups, each with its particular income, with agents linked only by the mutuality of supply and demand, the allocation of resources becoming purely a question of price formation. Each agent sought to maximize its own welfare through a calculus of choice; economics became a logic of optimizing decisions capable of mathematical representation.

The principle turning point in this development is the so-called marginal revolution of the 1870s, during which William Stanley Jevons, Léon Walras, and Carl Menger published books that were based upon a rejection of the classical paradigm, and that shared a common understanding of price formation as the outcome of choices based upon an evaluation of the marginal utility of economic goods. Jevons, Walras, and Menger were university professors, not private scholars like Adam Smith, John Stuart Mill, and David Ricardo; their work was intended primarily for an academic readership. Writing in the three principal world languages of the day – English, French and German – they at first encountered indifference, even hostility, but within twenty years Walras could claim that the new economics had won adherents and had been accepted in all major countries where economic theory was taught.[1]

By the 1920s, the main principles of neoclassical economic theory had become established in a form readily recognizable today. Smith, Ricardo, and Mill had become part of the history of economics, representatives of a coherent classical tradition that had been eclipsed by the new marginalism. As with all revolutionary change, that which precedes the revolutionary divide becomes the "oldthink," the ancien régime of economic discourse, an economic language that we now imperfectly understand and are inclined to make intelligible by translating it into the terms of our "newthink." It is even more difficult to imagine how the new thinking grew out of the old thinking – which, of course, it did, but along a path that has been obscured by our own progress. Between the mid eighteenth and late nineteenth centuries, there occurred major changes in economic language, appearing as a series of cumulative shifts and revisions in the literature that eventually resulted in a major transformation of our conceptions of economic action and market rationality. This chapter offers a story about this transformation that turns upon the contrast between two distinct conceptions of economic order. The first is the natural order of the Physiocrats, in which the regularities of agricultural production are coupled with circulatory imagery drawn from human physiology. The second is the marginal framework opened up by Menger and Walras in the later nineteenth century, a theory of choice and allocation that presupposes a denaturalized world of abstract consumers confronting abstract producers. Economics, then, is about how their interactions are reconciled through a coordinated system of prices in a disembodied world of goods.

ÉCONOMIE POLITIQUE AS THE NATURAL LAW OF CONDUCT

The economic language of early modern Europe was a language of counsel and persuasion directed to rulers, or to those with influence at court. Many of these texts follow the style established in the literature on the art of prudent government, presuming to identify a unique path to wealth and fortune for a ruler, to be realized through the good government of his territories. The means to wealth and power were at the same time the currency of wealth and power: a large, flourishing population. Such a population was capable of paying the taxes and levies that supported the court, the nobility, and the Church; it also furnished manpower for the armies and navies that fought the wars of

---

[1] Léon Walras, *Éléments d'économie politique pure ou Théorie de la richesse sociale*, 2nd ed. (1889) (Auguste et Léon Walras Oeuvres Économiques Complètes VIII) (Paris: Economica, 1988), Preface, p. 16.

succession and accession that were characteristic of the eighteenth century before the American and French Revolutions. The principal exception to this axiom, the economic success of the United Provinces, serves as an additional proof. The commercial success of the Dutch Republic during the seventeenth century provoked wide discussion concerning how a small country could also be a rich country; the city-states of medieval Italy might have provided one developmental model, but the sustained commercial success of a country with a small population was a different matter. Early modern economic discourse was therefore principally oriented toward one or more specific sectors of economic activity: overseas trade, commerce, money and finance, labor organization, agricultural production, taxation and economic regulation, manufacturing and luxury goods. A populous state was either wealthy or potentially so; on this there was little dispute. There were, however, a great many ways in which an individual state might become more prosperous – in part, of course, related to its particular situation and climate, factors to which Montesquieu (1689–1755) drew attention in his *Spirit of the Laws*.[2]

The variety of sectoral factors that played a part in the creation of wealth was in turn modulated by the institutional framework of the early modern territorial state. In northern and central Europe, a German-language literature developed that was directed to the improvement of the economic administration of the ruler's domains, the sole feasible way of increasing a ruler's income given the friction that existed between rulers and the *Stände* over the right of taxation. The work of economic administration was primarily in the hands of officials with legal training, which of course meant university training. Arguments concerning proper administration soon extended to the question of the appropriate training for administrators, which implied an economic, rather than a legal, education. This led to a number of new university chairs established to teach the "oeconomic sciences," although none of the many plans to make such training compulsory for those entering state service was ultimately successful.[3] Nonetheless, many of the new university posts so created survived into the nineteenth century and provided the institutional basis for the development of economic thought in Germany and Austria. The existence of university reading in the "oeconomic sciences" also meant that the German-language literature of economics in the eighteenth century was preeminently a textbook literature, a didactic and restricted discourse to be read in conjunction with lectures delivered by a professor who either had written the textbook himself, or would write one as an aid to his own teaching. Despite the political fragmentation of the German states, German-language economic literature was relatively uniform.

---

[2] Montesquieu, *The Spirit of the Laws* (1748), Introduction by Franz Neumann (New York: Hafner Press, 1949), Book 14, "Of Laws in Relation to the Nature of the Climate"; Book 18, "Of Laws in the Relation they bear to the Nature of the Soil."

[3] See my *Governing Economy: German Economic Discourse, 1750–1830* (Cambridge: Cambridge University Press, 1988).

France, of course, was politically unified, but its power was in rapid decline, as measured by the new economic standard. French economic literature of the eighteenth century was preoccupied with the causes of this decline and its possible remedies. The deep and persistent political and economic crisis of eighteenth-century France under the ancien régime coincided, however, with the Age of Enlightenment, a period of intellectual innovation in the arts and sciences that posited a new, integrated conception of human action and social progress. At midcentury this found expression in Diderot's *Encyclopédie* project, where the leading ideas of the *philosophes* were expounded systematically and at length.

The *Encyclopédie* entry *"Économie (Morale et Politique)"* was written by Jean-Jacques Rousseau (1712–1778) and published in November 1755.[4] At first glance this essay seems something of an oddity, even by the standards of the time. In comparison to Boisguilbert's analysis of the equilibrium between markets and classes some fifty years earlier,[5] or to Cantillon's *Essai sur la nature du commerce en général,*[6] which had had a significant impact upon the group of writers around Vincent de Gournay (1712–1759),[7] Rousseau's presentation appears to be only marginally related to what we would today understand as political economy. The initial definition of his subject in terms of household management leads into an extended discussion of forms of family authority, as compared to those prevailing in the state. But this analogy, drawing on a tradition of argument going back to Aristotle's *Politics,* is introduced only to be rejected as unsuitable, save for one common factor: an obligation upon both heads of household and heads of state to care for the well-being of their respective charges. The political domain is introduced via another analogy, this one borrowed from Hobbes, which conceives of the sovereign as the head of a "body politic," with law and customs forming the brain. Rousseau then proceeds to commerce, industry, and agriculture, "the mouth and stomach which prepare the common substance; public finances are the blood which a wise *economy,* performing the functions of the heart, sends out to distribute nourishment and life throughout the entire body."[8]

This body is also a moral being endowed with a general will, which strives to preserve the health of the body politic both in whole and in part; and this general will directs what Rousseau refers to as "public economy," by which he means administration and police. Wise administration, what was known in Germany as *gute Polizei,* rests on the prudent management of

---

[4] Jean-Jacques Rousseau, "Discourse on Political Economy," in *The Social Contact and Other Later Political Writings,* ed. V. Gourevitch (Cambridge: Cambridge University Press, 1997), pp. 3–38.

[5] Gilbert Faccarello, *Aux origines de l'économie politique libérale: Pierre de Boisguilbert* (Paris: Éditions Anthropos, 1986), chaps. 5–8.

[6] Published posthumously in 1755; see Antoin Murphy, *Richard Cantillon, Entrepreneur and Economist* (Oxford: Oxford University Press, 1986).

[7] Antoin Murphy, "Le Groupe de Vincent de Gournay," in *Nouvelle histoire de la pensée économique,* ed. Alain Béraud and Gilbert Faccarello (Paris: Éditions la Découverte, 1992), vol. 1, pp. 199–203.

[8] Rousseau, "Discourse on Political Economy," p. 6.

"what one has rather than as the means of acquiring what one has not." "The true secret of finances and the source of their increase is to distribute food, money, and commodities in just proportions, according to times and places...."[9] Rousseau did not directly discuss what was later seen as the routine fare of political economy - prices, markets, costs, and profits. He did use public finance as a way of linking state and economy, the patterns of consumption of rich and poor, and the consequent means of assuring equity in taxation. His essay was in many respects typical of an older economic literature whose chief concern was not the augmentation of material wealth, but rather its conservation and proper distribution – underpinned by conceptions of order and equilibrium that were articulated in a material allocation of resources existing independent of monetized exchanges.

The physiological imagery associated with this essay was echoed three years later with the first publication of François Quesnay's (1694–1774) *Tableau Éccnomique*. This was a visual representation of the flows of payments moving from agricultural producers to landed proprietors and thence to producers of manufactured goods, who then in turn recycled their incomes into purchases of manufactures and agricultural products – a flow of the net product from the "productive" to the "sterile" class and back again. This annual movement unites through their incomes and expenditures the three classes of society. Agriculture is considered the unique source of value, which in the course of its circulation between the classes is gradually dissipated.[10] Quesnay's work exemplifies Rousseau's idea of money as the lifeblood of the economic system, but extends it in several important respects.

The origin of circulation is identified in the *Tableau* as agriculture, which is treated as the sole productive sector in the economy. In principle, following each annual harvest, agricultural products would be exchanged for manufactured goods; the "sterile class" of artisans and manufacturers would gain food and wine, while the "productive class" would receive in return manufactured goods. Quesnay interposed a third class, that of landed proprietors, who received the entire net product of the land in the form of rent, returning half of this revenue to the agricultural sector as purchases of food and wine, and transferring the other half as "sterile expenditures" to the manufacturers of luxury goods, who then in turn spent half of their income on food and wine and the other half on manufactured goods - and so on through the year, until the entire system ran down to nothing and the next harvest started the sequence once more. The "sterile expenditures" of the landed proprietors represented withdrawals from the system that should have been invested in agriculture. The *Tableau* was therefore an abstract representation

---

[9] Ibid., p. 27.
[10] The *Tableau* was first published in 1758 and 1759, then appeared in a condensed form in Mirabeau's *Philosophie rurale* (1763) before being recast in mathematical form in 1766.

of the circular flow of economic exchanges, and was also an implicit critique of the economic policy of France under the ancien régime.

Quesnay's first economic publications were two articles in the *Encyclopédie*.[11] "*Fermiers*" outlined the advantages of improved farming for agricultural productivity, while "Grains" condemned the neglect of agriculture and the preoccupation of manufacturing activity with luxury goods. Appended to the latter article was an initial version of Quesnay's fourteen "maxims of economic government," which emphasized the agrarian origin of wealth and the advantages of free trade in the products of agriculture and industry. More important than these early writings themselves was the fact that in July 1757 Quesnay had made a convert of the Marquis de Mirabeau (1715–1789), whose earlier treatise on population had been a great success.[12] Mirabeau had there adopted the then-conventional arguments that the wealth of a nation lay in the size of its population, that luxury consumption diminished wealth, and that agriculture was the most profitable mode of employment. Quesnay persuaded Mirabeau that a large population was not the cause, but the effect, of wealth, and that the proper object of analysis was therefore not population, but wealth. Mirabeau then published in the late 1750s continuations of *L'Ami des hommes* that advocated this new viewpoint, devoting Part 6 to an exposition of the *Tableau*,[13] and later expounding Quesnay's principles at length in his *Philosophie Rurale*.[14] In the course of the 1760s, Quesnay gained a number of other adherents, among them Du Pont de Nemours (1739–1817), who edited a compendium of writings under the title *Physiocratie* and in 1768 published an exposition of this "new science."[15]

The new Physiocratic political economy emphasized the natural foundations of economic activity, identifying agriculture as the source of wealth. Improvements in agriculture were therefore of critical importance to the enhancement of wealth, but improvement required investment, which could take place only if the net produce was not in the course of its circulation diverted into unproductive ends, such as the production of luxury goods for landed proprietors. Circulation within the kingdom should be free of the impediments imposed by special taxes and duties; since all taxes were ultimately funded by the net product, all revenue deemed necessary should be drawn directly from the product as a single tax. Economic government should observe these natural laws; not only should it permit the free circulation of goods within the kingdom, it should also allow free trade in

---

[11] "*Fermiers (Econ. polit.)*," *L'Encyclopédie*, vol. 6 (1756); "Grains (*Economie polit.*)," *L'Encyclopédie*, vol. 7 (1757).

[12] Mirabeau, *L'Ami des hommes ou Traité de la population*, 3 parts (Avignon, 1756).

[13] Mirabeau, *L'Ami des hommes ou Traité de la population*, part 6 (Avignon, 1760), pp. 132ff.

[14] Mirabeau, *Philosophie Rurale, ou Économie générale et politique de l'agriculture*, 3 vols. (Amsterdam, 1763).

[15] Pierre-Samuel Du Pont de Nemours, *De l'origine et des progrès d'une science nouvelle* (Paris, 1768); Pierre-Samuel Du Pont de Nemours. ed., *Physiocratie*, 6 vols. (Yverdon, 1768–9).

raw produce in exchange for the luxury goods of other nations. Rather than being protected by the imposition of duties and prohibitions, domestic agriculture should be subjected to the stimulation that foreign trade can bring. The additional advantage here was that luxury goods could therefore be obtained from overseas producers in exchange for the net product of French agriculture, preserving the equilibrium of the domestic economy while at the same time increasing the disposable net product. This was combined with a radically new conception of economic order:

> The government of the Prince is not, as is commonly thought, the art of leading men; it is the art of providing for their security and for their subsistence through observance of the natural order and physical laws constituting the natural law and economic order, and by means of which existence and subsistence might be assured to Nations and to every man in particular; this object fulfilled, the conducting of men is fixed, and each man leads himself.[16]

As Mirabeau went on to state, "All the magic of a well-ordered society consists in the fact that each works for others while believing that he works only for himself."[17]

The identification of agriculture as the source of wealth is reinforced by the conception of a natural course of circulation among free, self-guiding agents overseen by an administration enjoined not to govern "too much." Economic agents are defined by their relation to this product, being categorized simply as "productive" or "sterile." It is not the *consumption* of luxury goods that is deemed to be economically harmful, a constant refrain encountered in writings of the early eighteenth century; it is the consequences for agricultural advancement of domestic *production* of luxury goods that is the subject of criticism. Inputs to agricultural production are the product either of agriculture itself, or of manufactures. Equilibrium in the system is maintained through the exchange of quantities of goods between sectors; although these exchanges are represented by sums of money, the quantity of goods is assumed to be constant; for whatever the price, the same quantity of grain is consumed.[18] Prices therefore function as a means of representation of the (anterior) material equilibrium of the system, not as a means of coordination in themselves. The quantity of goods in the system, whether grain, manufactured goods, or luxuries, can be increased permanently only by expanding agricultural productivity.

Adam Smith subjected this doctrine to detailed criticism in Book IV, Chap. 9 of his *Wealth of Nations*. Smith rejected the agricultural bias of the doctrine, but recognized the force of many of its arguments in comparison

---

[16] Mirabeau, *Philosophie Rurale*, vol. i, pp. xlij–xliij.
[17] Ibid., p. 138.
[18] Philippe Steiner, *La "science nouvelle" de l'économie politique* (Paris: PUF, 1998), pp. 52–6.

to the ideas of what he called the "mercantile" school, which treated gold or silver as wealth, whose accumulation was to be promoted through artificial restraints upon trade. The Physiocratic system was, he stated, imperfect, but "the nearest approximation to the truth that has yet been published upon the subject of political oeconomy."[19] He distinguished between "productive" and "unproductive" labor, generalizing the original materialist Physiocratic conception to the production of manufactured goods. He consigned all activity not embodied in a physical good to the "unproductive" category. This latter category should not be elided with "luxury goods," for so long as such goods are physical products, work expended upon their creation is "productive labor" under Smith's definition. "Unproductive labor" did not contribute to the formation of capital, it was insubstantial, it represented a service performed of which no physical trace remained. Smith did not deny that all kinds of useful services were performed in society, from the attentions of menial servants to the activities of the king, his officers, and the entire army and navy. But "their service, how honourable, how useful, or how necessary soever, produces nothing for which an equal quantity of service can afterwards be procured."[20] He redefined the origin of value: Instead of being labor performed in a specific sector, it became labor embodied in any material object. "Value" was a sum of money; but unfortunately this did not necessarily coincide with market price. Smith failed to resolve this problem, although the absence of a conception of equilibrium in his system meant that this never threatened the general coherence of his argument.

The quest for a uniform and objective labor standard of value that could be systematically linked to market prices, and therefore could map the social relations of production onto an emergent market equilibrium, was to become a leading preoccupation of classical economists, from Malthus through Ricardo and Mill to Marx. By contrast, the Continental reception of *Wealth of Nations* paid scant regard to this problem; instead, French and German writers placed at the center of their analyses the question of human needs and their satisfaction. Needs might be hierarchized as "necessities," "wants," and "luxuries," but their multiplication with the advance of commercial society raised significant problems of choice. In this way, French and German writers shifted their attention to consumers, away from the producers whose labor was embodied in economic goods, and thus lent them value. This proved to be a decisive shift, which led Continental political economy down a path divergent from that of Anglophone classical economics. By the later 1870s this latter path was looking increasingly like a dead end, as Jevons recognized.

---

[19] Adam Smith, *An Inquiry into the Nature and Causes of the Wealth of Nations* (Oxford: Oxford University Press, 1976), p. 678.
[20] Smith, *Wealth of Nations*, p. 331.

JEAN-BAPTISTE SAY: ECONOMY AND GOVERNMENT

Physiocratic doctrine was a form of social criticism intended to influence those charged with the reform of ancien régime France. Policies adopted should be fitting to the natural order – the natural law of conduct derived from the study of successful societies. This natural order was independent of the form of government; it related to the essential needs which all men shared in common. A political economy was thereby built upon a foundation of human nature, rather than of political order. *The Wealth of Nations* was likewise a treatise aimed first and foremost at the governing classes.[21] The creation of classical economics in the first half of the nineteenth century was everywhere built upon foundations provided by Smith's *Wealth of Nations*. Assimilation of Smith's teachings into established national discourses resulted, however, in divergent readings, the differences between Britain and Continental Europe being especially marked. In Britain, political economy was primarily a matter for self-education, absorbed through leisure-time reading of treatises, primers, and the "improving stories" of Mrs. Marcet. Political economy was not thought to be an especially esoteric form of knowledge. In Continental Europe, by contrast, political economy entered university education as one of the compulsory lecture subjects for law students, many of whom would later enter public administration; its principles became a part of the intellectual armory of the liberal intelligentsia. Smith's work was if anything more widely diffused on the continent; but since facility in the English language was uncommon, *The Wealth of Nations* generally became known through the work of translators and commentators. Here the work of Jean-Baptiste Say played a critical role throughout Europe, for Say wrote in a language that educated Europeans could read, and wrote in a style that was considered more accessible than that of Smith. His *Traité d'Économie Politique*, first published in 1803, was followed by a number of new editions, a "catechism" in 1815, and later by a six-volume "complete course of political economy" (1828–9). A total of fifty-three translations of his works appeared between 1807 and 1836, the first being the translation of the *Traité* published in Germany.[22]

Say was regarded by his British contemporaries as a popularizer, not as an original thinker. It is true that he did not adhere to the cost-of-production model of classical economics, but his emphasis on the fact that production and consumption involved the creation and destruction of *utilities*, not quantities of matter, opened up a perspective in which the prices of goods and services could fluctuate independent of their cost of production. His emphasis on effective demand shifted attention to the role of consumption,

[21] Donald Winch, *Riches and Poverty: An Intellectual History of Political Economy in Britain, 1750–1834* (Cambridge: Cambridge University Press, 1996), pp. 125–36.
[22] Philippe Steiner, "L'économie politique comme science de la modernité," in Jean-Baptiste Say, *Cours d'économie politique et autres essais* (Paris: Flammarion, 1996), pp. 16–17.

rather than production, in determining prices and their fluctuation. These two ideas proved decisive for the later developments of the 1870s.

These important variations on the Smithian canon might have been merely academic, were it not for the manner in which Say conceived knowledge of political economy to be part of the civilizing process, linked to the education of republican citizens. Say was not an "economic liberal," an adherent of the new Smithian political economy; he pursued the distinctly political project, continuous with that of the French Revolution, of inculcating republican manners. The Physiocratic emphasis upon the natural laws of human behavior had played an important part in the debates of the 1790s, as it was quickly perceived that the success of the Revolution depended not only upon the proclamation of rights, but also upon the eradication of the manners of the ancien régime. The Terror was one manifestation of this belief; another was the scheme of free universal education advanced by Sieyes in early 1793. Say worked as a newspaper editor during this period, and in 1798 submitted an entry to a competition offering a prize for the best essay on the question, "What institutions provide a suitable basis for the morals of a people?" Say's essay, later published under the title *Olbie*, argued that education in political economy would enlighten the citizenry as to their "real" self-interest, which in turn would be communicated to the people's legislators. The principles of political economy that Say espoused were indeed formally similar to those of Smith; the difference was in the manner in which these principles were intended to enter the public domain. Smith directly addressed legislators who ruled over subjects; Say addressed badly educated citizens, whose legitimate interests had to be properly articulated if republican government was to succeed. Political economy was the key to the proper articulation of these interests.

The 1803 *Traité* opens with the assertion that political economy should not be confused with politics, and that wealth is largely independent of the prevailing form of political organization – views that gave rise to much adverse comment at the time from Say's republican colleagues. The statement is, however, congruent with the understanding of political economy as a form of enlightenment. The principles of good government are simply distinct from those that determine the formation, distribution, and consumption of wealth; the conduct of government might impede or promote this process, but is not its cause.[23] Throughout the book attention is directed squarely to economic principles, without the kind of historical deviations and policy discussions that had led many to complain about Smith's "lack of system" and long-windedness. Although the tripartite division of the subject into production, distribution, and consumption, one of Say's innovations, was

---

[23] Jean-Baptiste Say, "Discours préliminaire," in his *Traité d'économie politique, ou simple exposition de la manière dont se forment, se distribuent et se consomment les richesses*, 2 vols. (Paris: Deterville, 1803), vol. 1, p. i.

not explicitly imposed upon the book until the second edition in 1814, the definition of production that Say offers was never strictly a Smithian one. As always with elementary treatises and textbooks, the manner in which the material is ordered is as important as the substance of the text itself. The first chapter of the *Traité* argues that nature's gifts to man have no value in themselves until human action endows them with value, the products of agricultural activity being especially important. Following this opening are chapters on manufacture, the nature of capital, and landed property, leading to a general definition of production that was to have far-reaching consequences: "Production is not *creation*; it is the production of utility."[24] Likewise, consumption is the destruction of the utility of what had been produced, not of the object itself.[25]

A modern reader would immediately pause here and ask: What is meant by utility? A discussion of different types of human industry obscured this point in the first edition, but this was rectified in the second edition, so that the treatise now begins with a clarification of "That which is meant by PRODUCTION" in which Say states: "That faculty which certain objects have of satisfying the diverse needs of man one will allow me to call *utility*."[26] Utility was therefore an expression of demand, and the satisfaction of need was limited only by the cost of obtaining goods. The cost of production formed the lower limit of the price of goods, and the upper limit was represented by the available means for purchase – the principle of effective demand. This was in turn linked to a rejection of Smith's distinction between productive and unproductive labor. Say argued that while the work of a doctor, musician, or actor might not be material, it was not unproductive: ". . . and that is a consequence of the meaning he attaches to the word *wealth*; instead of giving this name to all those things which had exchange value, he only gave it to those things with an exchange value *which can be conserved*."[27] As he emphasized, wealth did not consist of products themselves, but of value, an object with exchange value representing utility, the capacity for satisfying a material or immaterial human need.

FROM HUMAN NEEDS TO THE FORMATION OF PRICES

The appearance of Say's *Traité* coincided with a transformation in German economic discourse, which had been dominated throughout the eighteenth century by a natural law tradition that assumed human nature to be inherently disorderly in the absence of proper government. Cameralism treated social order as something that had to be deliberately constructed; only wise and prudent government could bring about the common good that men,

---

[24] Say, *Traité*, vol. 1, p. 24.
[25] Say, *Traité*, vol. 2, p. 338.
[26] Say, *Traité*, 2nd ed. (Paris: A.-A.Renouard, 1814), vol. 1, p. 3.
[27] Say, *Traité* (1803), vol. 1, p. 361.

unaided, were incapable of realizing. This laid a moral imperative upon the state and its officials; an enlightened, rational state must direct its subjects to the realization of their happiness, for by themselves they lacked insight into their own best interests and the means to their realization.[28] Wolffian natural law provided an intellectual legitimation for far-reaching administrative action, in which moral perfection was first defined and then created by the state.

This problematic was demolished by critical philosophy, which reworked natural law into a system based on the presumption that humans possessed the capacity to identify their needs and to purposively conduct themselves so as to meet them. This undermined cameralistic doctrine, although it continued to be taught as a university subject during a sometimes confused transitional period. Slowly the content of the teaching adapted to the new principle of self-guiding human action as the foundation of social order, and a new economic doctrine of human need found its way into the textbooks. During this phase, the German edition of Say's *Traité* appeared.[29] The translator, L. H. Jakob, had written a number of natural law texts during the 1790s, and then in 1805 had published a précis of Say's ideas.[30] Jakob gave heavy emphasis to the natural law roots of Say's argument – for example, by translating Say's "*nature des choses*" as "*Naturgesetze*," although Say's terminology of general and particular facts was adhered to as "*allgemeine und specielle Thatsachen.*"[31] These ideas were imported directly into the German literature by Jakob, both in his translation and in two further editions of his *Grundsätze*. The feature that distinguished this new *Nationalökonomie* from the older cameralism was pithily summarized by Jakob in his textbook: "Alle Einwohner des Staats sind *Consumenten* [All residents of the state are consumers]."[32]

Jakob, like many of his contemporaries, gives due regard to Smith's conception of the labor required to produce or acquire an object as the standard measure of value, but in his textbook he passes straight from this to an exposition of the working of supply and demand upon price that comes straight from Say, including a statement to the effect that it is not the total need, but the "*wirkliche Nachfrage* [real demand]" that has an impact upon the price of a good.[33] From this it follows that the greater the number of buyers, the higher the price, and the fewer the buyers, the lower the price; whereas the fewer the sellers, the higher the price, and the more numerous the sellers, the lower the price.

---

[28] Eckhart Hellmuth, *Naturrechtsphilosophie und bürokratischer Werthorizont* (Göttingen: Vandenhoeck & Ruprecht, 1985), p. 175.

[29] Jean-Baptiste Say, *Abhandlung über die Nationalökonomie, oder einfache Darstellung der Art und Weise, wie die Reichthümer entstehen, vertheilt und verzehrt werden*, 2 Bde., trans. L. H. Jakob (Halle: Ruffsche Verlagshandlung, 1807).

[30] Ludwig Heinrich von Jakob, *Grundsätze der National-Oekonomie oder National-Wirthschaftslehre* (Halle: Ruffsche Verlagshandlung, 1805).

[31] Say, *Traité* (1814), "Discourse Préliminaire," p. xvii; Say, *Abhandlung* , p. ix.

[32] Jakob, *Grundsätze*, §880, p. 480.

[33] Ibid., §200, p. 99.

In 1825 a revised third edition of Jakob's textbook appeared, with a new and important pithy definition:

Anything suitable for the satisfaction of human need is called *a good.*[34]

This thought was taken up by Friedrich Benedikt Hermann (1795–1868) in his treatment of the basic principles of economics, introducing the refinement that an economic good was one that required for its acquisition a money payment, or the sacrifice of labor.[35] Accordingly, a rich nation was not one with a great accumulation of property, but instead one in which all needs are satisfied.[36] The initial discussion of need and its satisfaction is developed here in the context of a discussion of James Steuart and of Say's *Cours* of 1828, arguing that use value is the main feature of a good because of its capacity to satisfy needs. This did not, however, prevent Hermann from developing an analysis of price formation in which the price level for a particular good is made dependent upon the relation of demand and supply, or what is much the same thing, the relation between the number of sellers and the number of buyers, which echoes Jakob's account of prices and effective demand, with the addition of the term "equilibrium" to describe the point where

goods are demanded and supplied in the same quantities.[37]

Given a basic cost that includes the usual rate of interest and entrepreneurial profit, Hermann suggests that if the price falls below the cost then capital and talent will move elsewhere; conversely, when the price prevails above the cost, new entrepreneurs will be attracted, in turn leading to a steady reduction in the price until once more prices and costs are equalized.[38]

By the time Wilhelm Roscher (1817–1894) had published his general textbook at midcentury, the definition of a good and its relation to price had become conventionalized – goods are those things capable of satisfying human needs; economic goods are goods that can be exchanged; the degree of utility (*Brauchbarkeit*) confers value on a good; production is the creation of value and consumption its destruction[39] – basic conceptions that run directly back to Say, rather than to Smith. These ideas are repeated by Hans von Mangoldt (1824–1868) in his own textbook, with the addition of graphically presented demand and supply curves that tend to equilibrium through the workings of a mechanism similar to that outlined by Hermann.[40] As Jakob stated, the

[34] Ludwig Heinrich von Jakob, *Grundsätze der National-Oekonomie, oder Theorie des National-Reichthums*, 3rd rev. ed. (Halle: im Kommission bei Friedrich Ruff, 1825), §880, p. 480.

[35] F. B. Hermann, *Staatswirthschaftliche Untersuchungen* (Munich: Anton Weber, 1832), p. 1.

[36] Ibid., p. 12.

[37] Ibid., p. 67.

[38] Ibid., pp. 4–5, 67–81.

[39] Wilhelm Roscher, *System der Volkswirthschaft Bd. I: Die Grundlagen der Nationalökonomie* (Stuttgart: J. G. Cotta, 1854), pp. 1–5.

[40] Hans von Mangoldt, *Grundriß der Volkswirthschaftslehre* (Stuttgart: J. Engelhorn, 1863), pp. 46ff.–70.

focus on human need throughout the German economic literature of the nineteenth century implied a focus upon the consumer, whose expanding needs were the motor of national wealth. From here it was no great step to the ranking of utilities according to the expression of subjective needs. Thus, as von Mangoldt suggested, value did not inhere in a good, but expressed the relation between a good and a subject.[41] This states clearly for the first time the conception of value that was to become a founding principle of Menger's new theory of "marginal value."

The German texts cited here were products of the university system: Lectures in economics formed part of the compulsory curriculum of law students. The German university routine, wherein each professor was required to teach from a comprehensive textbook, preferably his own, ensured the regular appearance of such texts, which reviewed the existing body of literature and updated the field of study. Textbooks thus served to mark the path along which the subject moved; all of the earlier citations from German economic literature are taken from works used by their authors in conjunction with their lectures. It was different in France, where political economy was not formally incorporated into legal training until the later 1870s. Although there was some teaching of the subject in provincial cities, there were in Paris only two chairs before 1864. All the notable French developments during this period are therefore nonacademic – the regular coverage of economic affairs in the monthly *Journal des Economistes*, which first appeared in 1841 and carried articles, reports on legislation and meetings of economists, reviews, letters, and an economic chronicle; the development of a popular economic literature; and the private writings of teachers and administrators who sought to elaborate new principles of economic science. Léon Walras personifies this French context: Persuaded by his father, a private student of economics, to devote himself to study of the subject in 1858, he for a time worked on the *Journal des Economistes*, then edited and published *Le Travail* with Léon Say, later worked in a bank while pursuing his studies in economics, and in 1870 was finally appointed to a chair in economics at the Academy at Lausanne on the recommendation of a Swiss politician, who had been impressed by his contribution to a convention on taxation held at Lausanne in 1860. This appointment gave him the opportunity to complete and publish, in two parts, his *Éléments d'économie politique pure*.

FROM CLASSICISM TO NEOCLASSICISM

Walras dismissed the definition of value employed by both Smith and Say, adopting instead his father's concept of scarcity – social wealth being defined as consisting of material or immaterial objects that have utility and exist in

---

[41] Ibid., p. 2.

limited quantities, objects available for exchange and capable of multiplication through the application of human industry.[42] Hence, the extension of social wealth is linked to the application of human industry and is facilitated by the division of labor. Although an abundant supply of goods is secured in this way, there remains the possibility that goods might be produced in inappropriate quantities, there being too much production of some scarce goods and not enough of others. Resolution of this problem was a matter of equity in distribution, the appropriation of social wealth being a human fact originating not in the individual will, but in the collective activity of society: "The fact of appropriation is therefore essentially a moral fact, the theory of property is therefore essentially a moral science. *Jus est suum cuique tribuere*, justice consists in rendering to each that which he is due..."[43] Thus Walras's system of production and distribution is not built upon an economic individualism that places social factors to one side; his theory of price formation is conditional upon the existence of specific institutions.

This is clear enough from the manner in which he introduces his conception of market relations. The model that he takes is that of a bourse populated by individuals seeking to buy and sell stock. This is a regulated market: It has a definite location, transactions are made on the basis of complete knowledge of the goods involved and the conditions attached to them, announced by the shouting out of prices and terms by individual buyers and sellers. There are other markets, Walras goes on, that are less well regulated, but which work well enough, such as fruit and vegetable markets, and streets lined with shops, which are, however, rather less effective from the standpoint of competition.

> And so the world can be considered as a vast general market composed of a variety of specialized markets where social wealth is bought and sold, and we are concerned with the laws according to which these sales and purchases tend themselves to be made. To do that, we always suppose a perfectly organized market in respect of competition, just as in pure mechanics one at first supposes frictionless machines.[44]

This understanding informs the various market models that Walras goes on to outline, and its importance should not be underestimated: Price formation can be treated as a function of the interaction of the demand for and the supply of stock. Since individual agents decide on the prices at which they buy and sell, prices emerge out of the interactions of utility maximizing agents, and questions of cost and its components do not arise. This does not, however, mean that the ex ante and ex post distribution of goods is a matter of indifference – we are dealing, argues Walras, with social wealth,

---

[42] Walras, *Éléments*, pp. 46–8.
[43] Ibid., pp. 62, 64.
[44] Ibid., p. 71.

with property, and for such exchanges to function effectively the trading prices must be just prices.

Menger and Walras shared a basic conception of value and price formation, consistent with the line of development sketched here, although of course the forms in which they presented their central ideas were very different. In the Preface to his *Grundsätze der Volkswirthschaftslehre* of 1871, Carl Menger (1840–1921) marks out his basic concepts as follows:

> Whether and under what conditions a thing is *useful* to me, whether and under what conditions it is a *good*, whether and under what conditions it is an *economic good*, whether and under what conditions is has the same *value* for me, and how great the *measure* of this value for me is, whether and under what conditions an *economic exchange* of goods can take place between two economically-active subjects, and the limits set to *price formation* here, all of these things are as independent of my will, as a law of chemistry is from the will of a practical chemist.[45]

Stated in this way, Menger's continuity with his predecessors in the understanding of economic concepts is much more apparent than is the case with Walras. He sought, however, a comprehensive account of human satisfaction. It had become increasingly commonplace in the literature of the nineteenth century to order goods in terms of their degree of importance, and hence of their capacity to satisfy human needs. Menger adopts a different approach: He declares his intent to arrive at an understanding of how humans achieve the most complete possible satisfaction of their needs.[46] Instead of the potential of goods to satisfy human needs, Menger considers the manner in which a human subject's needs are met in such a way as to maximize satisfaction, presenting this in the form of a ranking of needs from the most to the least urgent.[47] Walras had used this principle in the construction of market models in which the expressed needs of individuals were coordinated, developing a complex mathematical system that Menger eschewed. The similarity in their approaches is, however, evident.

Menger defined prices as a means of equalizing exchanges between human subjects. A theory of price should not seek to explain the equality of value between two goods, but must instead explain how economic actors, seeking to maximize their satisfaction, are induced to exchange specific quantities of goods.[48] Prices, therefore, do not reflect the inherent quality of certain goods, but are linked to the subjective estimation of need of the economizing actor; they provide a way to equalize these subjective estimations in a general system of exchange.

[45] Carl Menger, *Grundsätze der Volkswirthschaftslehre* (1871), in *Gesammelte Werke*, vol. I (Tübingen: J. C. B. Mohr [Paul Siebeck], 1968), p. ix.
[46] Ibid., p. 51.
[47] Ibid., pp. 90ff.
[48] Ibid., p. 175.

Menger did not explicitly introduce the term "marginal utility" into the *Grundsätze*, nor did he adopt the mathematical language of his contemporaries Jevons and Walras. There is, however, a clear affinity of purpose in their writings. The "one essential truth,"[49] as F. Y. Edgeworth put it, was that exchange value is equal to the utility of the least useful portion of the commodities exchanged; or, in other words, that price depends upon the relationship between degree of need and degree of utility in satisfying that need. Walras took this idea further, placing this conception of price as an expression of marginal utility in the context of a market in which all prices are in equilibrium, where the demand for and supply of each commodity is therefore a function of the prices of all commodities.

The transition from classical to neoclassical economics turned therefore on a redefinition of the object of economic analysis, both in terms of degree of abstraction (as in Menger) and of the conception of "price" and "market" (as in Walras). The market became the delimited space within which exchanges took place and equilibria were formed; prices were henceforth representations of utilities, quantities of goods and their capacity to satisfy needs that were the means by which markets arrived at a unique equilibrium price that optimized the relationship between suppliers and consumers. Consumer and supplier were linked together in a series of exchanges: The individual human subject, for example, consumed food from the farmer and supplied labor to the manufacturer; the farmer consumed machinery from the manufacturer and supplied food to the market. Prices coordinated these actions, and it was the utility embodied in a material good or a service that was destroyed in the activity of consumption. The principle was equally applicable to production and consumption: Capital was consumed in the production of goods and services, which were then in turn consumed by their purchasers. Agents determined for themselves the nature and degree of effort expended, and were paid according to the marginal product of their labors; the revenue secured in this way was expended on goods and services whose prices were set according to the utility schedules of consumers. The new science of economics propounded the principles, which regulated the process by which the activities of producing and consuming individuals were coordinated through a price mechanism, yielding an optimization of welfare.

[49] Francis Y. Edgeworth, "The Mathematical Theory of Political Economy," *Nature*, 40 (1889), 435.

II

BRITISH ECONOMIC THEORY FROM LOCKE TO MARSHALL

Margaret Schabas

The mercantilist pamphlets of the 1600s are commonly viewed as the first systematic writings on political economy, at least in the English language. While many of these works were unabashed promotions of merchant rights, historians have come to appreciate their rich array of insights on the topics of money, market forces, and the global economy. Two other important traditions of economic inquiry had emerged by the late seventeenth century, fostered by the rise of political freedom and the growth of a scientific culture. The first stems from John Locke's *Two Treatises of Government* (1689–90), which addressed the problems of economic justice and distribution via the fundamental concepts of rights and property. Locke also privileged the economic contract in his state of nature and adumbrated a labor theory of value. The second tradition, exemplified by William Petty's *Political Arithmetic* (1690), devised quantitative measures of economic phenomena, such as the national product of Ireland, the velocity and quantity of money, and the population of London. While Petty's measures were bold and imprecise, they helped draw attention to aggregate phenomena and thus to new empirical relationships.

All three lines of thought spoke to the new capitalist system, which had transformed early modern Europe. As Joseph Schumpeter has rightly observed: "By the end of the fifteenth century most of the phenomena that we are in the habit of associating with that vague word Capitalism had put in their appearance, . . . [and] even then these phenomena were not all of them new."[1] He had in mind the prices of commodities and factors of production, such as the interest rate. I would take this claim one step further and submit that since about 1700 there have been few genuinely new phenomena in economic discourse. I here use the term "phenomenon" as defined by Ian Hacking, something that is "noteworthy, . . . discernible, . . . an event or

---

[1] Joseph A. Schumpeter, *A History of Economic Analysis* (New York: Oxford University Press, 1954), p. 78.

process of a certain type that occurs regularly under definite circumstances."[2] Unlike physicists, for example, who have discovered radically new phenomena, such as electromagnetic induction and x rays, that have prompted new research traditions, economists have essentially been rearranging the same constituents. There have, of course, been many new data in economic inquiry; indeed, every price is a new datum. But the phenomenon of price has been central to economic thinking since antiquity. To be sure, some phenomena, such as value and unemployment, have been given new definitions, which in turn have had significant implications for economic theory. But the key properties of money (quantity, price level and velocity, interest rate), of production and distribution (factor and commodity prices, market forces), and of the national economy (national income, population, employment, balance of trade, exchange rates) were all articulated during the early modern period.

This does not mean that the science of political economy ground to a halt, or that the economy itself ceased to evolve. Theorists have offered new causal accounts of these phenomena, and they have been given greater mathematical refinement. But in contrast to the natural sciences, there have been very few clear and distinct empirical discoveries in economics since the time of Petty and Locke, at least of the sort that have prompted radically new lines of research. The theoretical development of political economy has been much more bound to the armchair, to the working out of the internal logic of previous texts, than to the absorption of new economic events.[3] This peculiarity can be traced in part to the lack of an experimental tradition and in part to the high level of abstraction economists have sought since the discourse emerged. Even the radical theoretical departures associated with the work of William Stanley Jevons (1835–1882) and John Maynard Keynes (1883–1946) were more the product of insights gleaned from logic, psychology, and the philosophy of science than a response to contemporary economic events.[4] One might suppose that practical problems have been the main source of stimuli for economic inquiry, but the record suggests otherwise. Most of the major economists of the past two centuries derived their originality from philosophical sources, although in some cases what helped to give a specific theory currency was its resonance with contemporary economic concerns.

THE EIGHTEENTH CENTURY

Most of the Enlightenment contributors to economic theorizing in the English language were from countries other than England, a fact that might

[2] Ian Hacking, *Representing and Intervening* (Cambridge: Cambridge University Press, 1983), p. 221.
[3] Margaret Schabas, "Parmenides and the Cliometricians," in *On the Reliability of Economic Models: Essays in the Philosophy of Economics*, ed. Daniel Little (Boston: Kluwer, 1995).
[4] Margaret Schabas, *A World Ruled by Number: William Stanley Jevons and the Rise of Mathematical Economics* (Princeton, N.J.: Princeton University Press, 1990); Roderick M. O'Donnell, *Keynes: Philosophy, Economics and Politics: The Philosophical Foundations of Keynes's Thought and Their Influence on His Economics and Politics* (London: Macmillan, 1989). See also Chapter 2 in this volume (pp. 26–32), on the relations of political economy to social and economic issues.

also support the thesis that economic conditions underdetermine the content of economic theory. Both George Berkeley and Richard Cantillon hailed from Ireland, and Bernard Mandeville was from the Netherlands. Scotland, despite its relatively backward economy at the time, produced the majority of the eminent economists of the eighteenth century: John Law in the early part of the century, David Hume in the middle, and James Steuart and Adam Smith in the latter half.

Of those before Adam Smith, David Hume (1711–1776) was the most influential. His many essays on political economy contain insights on the global allocation of money and on the stimulatory consequences of unanticipated inflation. Hume also analyzed trade, population, capital, and interest rates. His *Treatise of Human Nature* (1739) promoted the idea that a science of man was possible insofar as human nature was constant and regular. It also continued Locke's inquiry into the subject of economic justice and distribution. Most notably, Hume explored the question of trust in the formation of commercial contracts, and appreciated the significance of a fully monetarized world. Hume also instantiates the ideological movement identified by Albert O. Hirschman that perceived commerce as the great civilizing force and thus as the best safeguard against political absolutism.[5]

Adam Smith (1723–1790) was greatly indebted to all of these political economists, as well as to Frances Hutcheson and François Quesnay. Although Smith is most celebrated for his *Inquiry into the Nature and Causes of the Wealth of Nations* (1776), his *Theory of Moral Sentiments* (1759) forms an essential part of his system. His wisdom about the workings of the world, like that of Hume, was derived mostly from philosophical reflections on human nature. Smith paid tribute to the Stoics for their notion of a deity removed from the everyday operations of nature, and for treating self-command as the highest of the virtues. For decades, scholars have grappled with the apparent inconsistency between Smith's models of human nature in these two works, whereby individuals are first portrayed as motivated by sympathy for others, and then by self-interest. But there is general agreement now that "Das Adam Smith Problem" has been resolved. Smith appreciated the complexity of human nature and saw different motivating forces operating in different spheres of activity. Furthermore, both sympathy and self-interest are derivative of our more fundamental desire for the approval of others, which we develop through the cultivation of friendship and civil society as well as through the accumulation of wealth and knowledge.[6]

---

[5] Albert O. Hirschman, *The Passions and the Interests* (Princeton, N.J.: Princeton University Press, 1977); Duncan Forbes, *Hume's Philosophical Politics* (Cambridge: Cambridge University Press, 1975); Margaret Schabas, "Market Contracts in the Age of Hume," in *Higgling: Transactors and Their Markets in the History of Economics*, ed. Neil de Marchi and Mary S. Morgan (Durham, N.C.: Duke University Press, 1994).

[6] Laurence Dickey, "Historicizing the 'Adam Smith Problem': Conceptual, Historiographical, and Textual Issues," *Journal of Modern History*, 58 (1986), 579–609; Richard Teichgraeber, *Free Trade and Moral Philosophy: Rethinking the Sources of Adam Smith's Wealth of Nations* (Durham, N.C.: Duke

Among Smith's extant essays is one offering a detailed account of the history of astronomy and paying particular tribute to Newton. But while Smith wished to emulate Newton in the moral realm, his economic theory was mostly discursive and rarely offers an exacting piece of deduction. Even his efforts to support his general principles with empirical evidence were often unsystematic, with material drawn primarily from other books rather than from firsthand observation. Indeed, he seemed to be relatively unaware of the Industrial Revolution, assuming that it was indeed under way by the 1770s. Although he appreciated the value of inventions, he made almost no mention of the recent improvements in textile machinery and steam engines, which were so critical in unleashing the process of industrialization. Moreover, he viewed the agrarian sector as the primary one for capital accumulation.[7]

Smith defined political economy as the "science of the legislator," and thus subordinated his analysis of economic exchange and distribution to the broader questions of political stability and national well-being.[8] His greatness lay less in his specific insights into the theory of prices and distribution than in his overall comprehension of the subject. Within the *Wealth of Nations* one can find discussions of virtually every branch of political economy as it has evolved up to the present, including public finance and economic history (although Smith himself did not recognize all of these branches).

Smith is celebrated for noting the importance of the division of labor in producing economic efficiency, but appeals to the division of labor can be traced back to Plato's *Republic*. His more original insight is his claim that the size of the market limits the division of labor, in the sense of the number of specific trades. The more specialized producers become – the partition of the production of beef, say, into breeders, grazers, drovers, stockmen, and butchers – the larger the scope for trade and for middlemen. Against the popular sentiment of his time, Smith praised those who profited from such transactions.

Smith argued for a labor theory of value, but he acknowledged that the costs of land and capital, along with wages, enter into the formation of prices. He analyzed the factors responsible for the spectrum of wages – training, risk, unpleasantness of the job, and so forth – and distinguished between productive labor, such as farming, and unproductive labor, such as acting, which "perishes in the very instant of its production."[9] But the main thrust

University Press, 1986); Vivienne Brown, *Adam Smith's Discourse: Canonicity, Commerce and Conscience* (London: Routledge, 1994).

7 Charles P. Kindleberger, "The Historical Background: Adam Smith and the Industrial Revolution," in *The Market and the State: Essays in Honour of Adam Smith*, ed. Thomas Wilson and Andrew S. Skinner (Oxford: Clarendon Press, 1976). For the opposite view, see Samuel Hollander, *The Economics of Adam Smith* (Toronto: University of Toronto Press, 1973).

8 Donald Winch, *Adam Smith's Politics* (Cambridge: Cambridge University Press, 1978); Knud Haakonssen, *The Science of a Legislator: The Natural Jurisprudence of David Hume and Adam Smith* (Cambridge: Cambridge University Press, 1981).

9 Adam Smith, *An Inquiry into the Nature and Causes of the Wealth of Nations*, 2 vols., ed. R. H. Campbell and A. S. Skinner (Oxford: Oxford University Press, 1976), vol. 1, p. 331.

of the distinction was to address the problem of capital accumulation and its implications for economic growth.

Smith shared with other Scottish Enlightenment thinkers a preoccupation with the so-called rich country–poor country debate.[10] Why did different nations have different rates of economic development – some progressing, some static, and some even declining? The puzzle was all the more acute insofar as Smith had granted to everyone the same propensity to achieve economic betterment, and had implied that the laws of the marketplace were universal in scope. As a partial answer, he stressed the level of capital investment in the agrarian sector, as well as prevailing institutional and political conditions. China, for example, was relatively stagnant because of its entrenched civil servant bureaucracy. Smith proposed a "natural progress of opulence," whereby a region starts with agriculture, then cultivates manufacturing, and finally engages in overseas trade. But he admitted that, because of the interventions of governments, most regions had inverted the natural order. For this reason Smith was often deemed an advocate of laissez-faire policies, although the last book of his treatise outlines numerous cases for the state provision of public goods, such as education and military protection.

Smith's attention to economic liberty built on the works of Locke, Quesnay, and Jean-Jacques Rousseau. His concept of the "invisible hand" bears a superficial resemblance to Mandeville's insight that private vices unintentionally yield public virtues, but may have owed more to his belief in a Providential order. While only mentioned once in the *Wealth of Nations*, the metaphor of the invisible hand later became a mantra for those who defended the superiority of the competitive capitalist order. Smith argued, more from reason than from evidence, that people would bring about more public benefit unwittingly, while pursuing their own economic ends, than if they set out to do so directly. He justified this argument in part by his belief that individuals knew their own interests better than anyone else, and in part by a deep-rooted faith in market forces.

POPULATION AND ECONOMIC SCARCITY

Eighteenth-century economists looked favorably on population growth as a key indicator of national prosperity. They were also cognizant of economic growth. As Smith observed, even the ordinary English cottager enjoyed more goods than a ruler in Africa. In short, Enlightenment economists painted an optimistic picture of the European states and North American colonies as regions of relative prosperity. A distinct turnabout came with the work

---

[10] Istvan Hont, "The 'Rich Country–Poor Country' Debate in Scottish Classical Political Economy," in *Wealth and Virtue: The Shaping of Political Economy in the Scottish Enlightenment*, ed. Istvan Hont and Michael Ignatieff (Cambridge: Cambridge University Press, 1983); Donald Winch, *Riches and Poverty* (Cambridge: Cambridge University Press, 1996).

of Thomas Robert Malthus (1766–1834), whose *Essay on the Principle of Population* (1798) sent a veritable shock wave across the learned community of western Europe. Working from two simple postulates, the need for food and the passion between the sexes, Malthus argued that unfettered population growth necessarily outstrips food production, and that even England could easily become a region of widespread starvation. His argument was more theoretical than empirical. He posited a tendency, as yet never fully manifest, for human population to grow at a geometric rate, while agrarian output could at best grow only at an arithmetic rate. He also needed the principle of diminishing returns to drive his argument (an insight that he reached explicitly only in 1815), for without it, more persons meant more labor to work the land and hence did not necessarily imply scarcity.

Whatever the merits of Malthus's analysis, it served to alarm his contemporaries about the question of scarcity. There were also numerous policy implications: the abolition of poor relief, a further entrenchment of the Corn Laws, the expansion of religious instruction. Malthus always claimed to be a "friend of humanity," however apparently harsh his insistence that the poor should fend for themselves and learn to exercise "moral restraint." But many of his contemporaries painted him in a less favorable light, and his influence at Westminster remained indirect at best.

Malthus subsequently amassed considerable evidential support for his argument, and issued these findings along with each of the six editions of the *Essay* (the last in 1826). Although these efforts have an ad hoc ring to them, economic historians have since looked kindly on his observational skills, noting that the English population was indeed growing rapidly and that the potential for increased output in the agrarian sector was very modest.[11] But in his *Principles of Political Economy* (1820), Malthus was clearly closer to Smith than to his own contemporaries, featuring agriculture as the key sector rather than manufacturing. John Maynard Keynes would later pay tribute to Malthus for his recognition of the possibility of gluts in the capital sector and for stressing the role of aggregate demand, but these themes were not absorbed into the mainstream at the time.

CLASSICAL POLITICAL ECONOMY

Hume, Smith, and Malthus laid the foundations for the classical theory of political economy, which reigned supreme until the 1870s. Its leading English exponents were David Ricardo (1772–1823) and John Stuart Mill

---

[11] Edward A. Wrigley, "Malthus's Model of a Pre-industrial Economy," in *Malthus and His Time*, ed. Michael Turner (New York: St. Martin's, 1986), p. 16; Anthony M. C. Waterman, *Revolution, Economics and Religion: Christian Political Economy, 1798–1833* (Cambridge: Cambridge University Press, 1991); Samuel Hollander, *The Economics of Thomas Robert Malthus* (Toronto: University of Toronto Press, 1997).

(1806–1873). The classical theory was predicated on a cost-of-production theory of value, with most emphasis being placed on labor costs. Although one finds an increased appreciation for industry and for the spread of machinery, there was still a tendency to view the annual harvest as the time when the leading parameters of the economy, such as commodity and factor prices, were cemented. Moreover, despite the recognition of mechanisms by which prices might adjust fairly quickly, the overall emphasis was on longer periods of reallocation and analysis at the aggregate level. A primary question for debate was the legitimacy of the "natural" distribution among the three groups: landowners, capitalists, and laborers. Smith had ascribed different and conflicting motives to these groups and thus injected an element of disgruntlement into his picture of the world. Such tensions were only intensified in the accounts of nineteenth-century economists.[12]

Among the classical economists, the most revered and controversial book after Smith was Ricardo's *Principles of Political Economy and Taxation* (1817), a brilliant and elegant piece of economic theorizing that spurned Smith's episodic style and loose derivations for an axiomatic-deductive mode of reasoning. Ricardo thus exposed some of Smith's ambiguous claims, particularly those concerning value and wages. Relative values were proportionate to the quantity of labor required in the production of respective goods, not to the labor the good's producer could command in the marketplace. And wage increases, Ricardo demonstrated, are noninflationary, or rather, they do not alter relative prices. True, rising wages would unleash a long chain of adjustment in terms of the allocation of capital and labor, depending upon whether the goods in question were produced with above- or below-average capital-labor ratios of the economy. But under competitive conditions, implying a tendency toward a uniform rate of profit throughout the economy, the original price spectrum would be sustained.[13]

Ricardo's efforts to sort out the theory of prices, given his commitment to a labor theory of value (capital was nothing but crystallized labor, and could thus be included in the calculation of labor costs), led him to posit the hypothetical existence of a measure of value. In principle, such a yardstick would be immune to market conditions and would reflect the average proportion of capital and labor for the entire economy. Ricardo deemed gold to be the best candidate for this measure; in a world with a gold standard, the money price itself would effectively be the correct measure of value. This was an ingenious, if impractical, solution to the long-standing problem of price indexing.

---

[12] Maurice Dobb, *Theories of Value and Distribution since Adam Smith* (Cambridge: Cambridge University Press, 1973); Maxine Berg, *The Machinery Question and the Making of Political Economy, 1815–1848* (Cambridge: Cambridge University Press, 1980).

[13] Neil de Marchi, "The Empirical Content and Longevity of Ricardian Economics," *Economica*, 37 (1970), 257–76; Samuel Hollander, *The Economics of David Ricardo* (Toronto: University of Toronto Press, 1979); Terry Peach, *Interpreting Ricardo* (Cambridge: Cambridge University Press, 1993).

Ricardo also devised a new theory of rent, which effectively reduced it to a transfer payment. Whereas for Smith rent could serve as one component in the formation of prices, for Ricardo it did not. The price of goods was always determined at the margin of production, where no rent was paid, and thus included only the costs of labor and capital. This in turn implied that landlords did not make any legitimate contribution to the national wealth, hence the appeal of Ricardo to subsequent socialist thinkers. Ricardo also identified a tendency for the rate of profit to decline over time, due to population growth and diminishing returns. This meant that more of the net product would go into the hands of the landowners unless measures could be taken to shift resources away from the agrarian sector.

The classical economists downplayed the importance of money in regulating the economy, although some ink was spilled over such questions as the issuing of paper notes and the efficacy of a bimetallic standard. Policy measures focused chiefly on fiscal reform. Ricardo, for example, devoted almost half of his famous text to the subject of taxation, and Mill avidly promoted a flat-rate tax, except for those who become rich in their sleep, through inheritance or the rent of land. Along with the attention to taxation came measures for legal and constitutional reforms. Scores of political economists of the Georgian and Victorian periods, including Ricardo and Mill, served as members of Parliament or were consulted for parliamentary commissions. Nassau Senior was one of the most influential, both on the reform of the Poor Laws (1834) and, via his *Letters on the Factory Act* (1837), on the length of the working day. But the subject on which classical economic theory had the most to say was the Corn Laws, which Ricardo and his followers saw as the greatest impediment to England's prosperity.[14]

The classical economists also took measures to establish their subject in universities and scientific societies. Malthus was the first professor of political economy, with an appointment at the East India College (subsequently Haileybury College) in 1805. Oxford established the Drummond Chair in 1819 (held by Senior). Both Cambridge and University College, London, created teaching posts in political economy in 1828, followed by King's College in 1831. By the latter part of the century, there were dozens of lecturers on the subject at British universities, including nine at Oxbridge alone. The reputation of political economy was also enhanced by the formation of Section F for Statistics (and subsequently Political Economy) at the British Association for the Advancement of Science in 1833, and by the establishment of the Tripos in the Moral Sciences at Cambridge in 1838. Other forums for informed debate were the Political Economy Club (founded in 1821) and the

[14] Raymond Cowherd, *Political Economists and the English Poor Laws: An Historical Study of the Influence of Classical Economics on the Formation of Social Welfare Policy* (Athens: Ohio University Press, 1977); Boyd Hilton, *Corn, Cash, Commerce: The Economic Politics of the Tory Governments, 1815–1830* (Oxford: Oxford University Press, 1977).

London Statistical Society (1834). The contents of the leading periodicals of the day suggest that Britons were captivated by the debates on trade, currency, and labor reforms.[15] Many eminent scientists of the period, notably John Herschel, William Whewell, and Charles Babbage, explicitly recognized and approved the new subject, though not without some qualifications.

Political economy was extremely popular at the time, and found its way – although not always mentioned favorably – into the works of prominent poets and novelists.[16] Jane Marcet's *Conversations on Political Economy* (1816) and Harriet Martineau's *Illustrations of Political Economy* (9 vols., 1834) were the two best-known popular accounts of the subject. Richard Whately's *Easy Lessons on Money Matters* (1833), while intended for children, reached an estimated two million readers. Political economy also spread by means of the mechanics' institutes and other venues for working-class education. Ricardo's implicit criticisms of landowners helped spawn socialist theories – for example, the works of Robert Owen, Thomas Hodgskin, and John Gray. Nevertheless, the prosperity that followed in the wake of the repeal of the Corn Laws (1846) and the slow but gradual improvement in the conditions of factory work were often attributed to the science of political economy. Walter Bagehot, renowned editor of the *Economist*, declared in his tribute on the centenary of Adam Smith's *Wealth of Nations* that "the life of almost every one in England – perhaps every one – is different and better in consequence of it."[17]

JOHN STUART MILL

John Stuart Mill's *Principles of Political Economy* (1848) served as the authoritative text of the mid-Victorian period. While his intention was primarily to settle many of the disputes that Ricardo had sparked rather than to break new ground, his book did much to imbed economic discourse in a broader social philosophy. Inspired by Harriet Taylor (1808–1858), whom he eventually married, Mill also moved toward socialism as he reached middle age. He envisioned a time when the hustle and bustle of economic gain would subside, when humankind might begin to embrace its nonmaterial potential. His *Subjection of Women* (1869) spoke to the amelioration of the political and economic conditions of women.[18]

Mill was the first to write at length on the ontological and epistemological dimensions of political economy, first in his essay of 1836, "On the

---

[15] George J. Stigler, "Statistical Studies in the History of Economic Thought," in his *Essays in the History of Economics* (Chicago: University of Chicago Press, 1965), p. 41.

[16] Gary F. Langer, *The Coming of Age of Political Economy, 1815–1825* (New York: Greenwood Press, 1987).

[17] Walter Bagehot, "The Postulates of Political Economy," in *Economic Studies*, ed. R. H. Hutton (London: Longmans, Green, 1911), p. 1.

[18] Alan Ryan, *J. S. Mill* (London: Routledge, 1974); Stefan Collini, Donald Winch, and John Burrow, *That Noble Science of Politics* (Cambridge: Cambridge University Press, 1983).

Definition of Political Economy and on the Method of Investigation Proper to It," and then in his *System of Logic* (1843). The phenomena of wealth, he argued, are both material and mental and thus draw upon both the laws of physical science and the laws of the mind. This dualism manifested itself in his economic text, where he draws a sharp distinction between the laws of production (which are grounded in the physical) and the laws of distribution (grounded in the mental). Later, under the influence of Alexander Bain (1818–1903), Mill ascribed a much greater role to psychology in the theory of political economy, a step that in certain respects paved the way for the neoclassical doctrine.[19]

Mill argued that the method most appropriate to political economy was Newton's. One began with plausible axioms and hypotheses, undertook derivations, and then sought their verification in the actual world. He acknowledged, however, that political economy was an inexact science, partly because it rested on an unrealistic picture of human nature (the unfettered pursuit of wealth), and partly because of the relative lack of data required for its verification.[20]

Mill's methodological approach was widely influential, even for the early neoclassical economists, but it did not go unchallenged. Thomas Tooke's six volumes on the *History of Prices* (1838–75) helped to launch a statistical approach to the subject.[21] During the same period, Richard Jones and William Whewell called for a more realistic approach to political economy, with emphasis on inductive and historical reasoning. Their sentiments were revived in the 1860s and 70s, notably by Arnold Toynbee, who initiated a school of economic historians at Oxford. But economic theory retained its strong deductive bent and became even more ahistorical with the marginal revolution of the 1870s.[22]

THE MARGINAL REVOLUTION

Jevons's *Theory of Political Economy* (1871) called for a radical displacement of the Ricardo–Mill doctrine, an announcement that has come to be seen as the start of the marginal revolution. Like Léon Walras in Switzerland and Carl Menger in Austria, Jevons sought to replace the classical cost-of-production

---

[19] Fred Wilson, *Psychological Analysis and the Philosophy of John Stuart Mill* (Toronto: University of Toronto Press, 1990); Margaret Schabas, "Victorian Economics and the Science of the Mind," in *Victorian Science in Context*, ed. Bernard Lightman (Chicago: University of Chicago Press, 1997), pp. 72–93.

[20] Daniel Hausman, *The Inexact and Separate Science of Economics* (Cambridge: Cambridge University Press, 1992).

[21] Theodore M. Porter, *The Rise of Statistical Thinking, 1820–1900* (Princeton, N.J.: Princeton University Press, 1986).

[22] Alon Kadish, *The Oxford Economists in the Late Nineteenth Century* (Oxford: Clarendon Press, 1982); John Maloney, *Marshall, Orthodoxy and the Professionalisation of Economics* (Cambridge: Cambridge University Press, 1985).

or labor theory of value with the utility theory of value. Economic classes were abandoned in favor of individual rational agents who might be both workers and owners of stock. Jevons was also keen to promote the use of mathematics in economic theory, especially the calculus. Although earlier economists had turned to mathematics, it was really the efforts of Jevons and his immediate successors – Francis Ysidro Edgeworth, Alfred Marshall, and Walras – that transformed mainstream economics into a mathematical science. Much of the inspiration for this dramatic shift came from new currents in logic and physics rather than from problems internal to the discipline or from specific economic events.[23]

Despite their ostensible similarity, the utility theory of value as cultivated by economists was quite distinct from the moral theory of Utilitarianism as promulgated by Jeremy Bentham. While both theories were predicated on the claim that man is subject to two sovereign masters, pleasure and pain, this insight had implications for the analysis of market phenomena that differed from those pertaining to moral judgments and political reforms. J. S. Mill, though fully at home as a Utilitarian moral philosopher, did not embrace the utility theory of value in his economic writings. Those, such as Senior, who first promoted the utility theory of value in economics often made little effort to cultivate moral theory. Jevons drew a line between higher and lower wants, and proposed that utility in the the economic sense treat only those of the lowest order. Although he also wrote an essay on Utilitarianism, he did not see a need to link it to his economic theory. This was also true of Henry Sidgwick, whose *Methods of Ethics* (1874) was the most prominent treatise on Utilitarianism in late Victorian England, and whose *Principles of Political Economy* (1883) firmly endorsed the Jevonian movement, but who nevertheless did little to wed the two fields of inquiry. Arguably, such a merger came to pass only in the early twentieth century, with A.C. Pigou's analysis of social welfare.

Nevertheless, in a more general sense, utilitarian thinking infused classical political economy from its very inception. Smith frequently referred to the happiness of the "lower orders" of society in his efforts to sort out economic relationships. And Ricardo and the two Mills pursued much the same secular and reformist goals as those articulated by Bentham. Economic theory has never been fully divorced from moral philosophy, any more than theoretical physics has ever severed its ties to natural philosophy and metaphysics.

For a century beginning with the late Enlightenment, the science of political economy was most concentrated and developed in Great Britain. As E. J. Hobsbawm has remarked, the "age of capital" unequivocally belonged to the

---

[23] Philip Mirowski, *More Heat than Light: Economics as Social Physics, Physics as Nature's Economics* (Cambridge: Cambridge University Press, 1989); Bruna Ingrao and Georgio Israel, *The Invisible Hand: Economic Equilibrium in the History of Science* (Cambridge, Mass.: MIT Press, 1990).

English economists, both those at the top level, such as Ricardo and Mill, and also the substantial list of second tier writers, such as Senior.[24] There is no simple explanation of this fact. Insofar as Britain entered a more liberal and progressive era as early as 1688, middle-class and dissenting young men such as Hume, Bentham, and Mill sought fulfillment in political and economic inquiry. But political economy was not just a haven for restless souls. For every dissenter one can cite an economist who defended the status quo, most notably Smith, Malthus, and Marshall.

Perhaps a more significant factor was Great Britain's concomitant rise to global economic power. It is plausible that British intellectuals, as inhabitants of the workshop of the world, would expend time and energy on economic questions and vigorously defend the scientific status of political economy, particularly insofar as that science promoted industry over agriculture. This thesis has an intuitive appeal; and yet, the evidence is ambiguous. For every apologist, there was a more prominent visionary who saw beyond national interests. Moreover, French and German economists were just as able to observe the advent of capitalism and the industrial era, to read the works of Smith and Ricardo, and thus to develop the subject. Indeed, the very fact that their economic development fell behind that of Britain might be viewed as an obvious stimulus to economic inquiry. But while French and German scientists were prodigiously productive during this period, in economics there were far fewer writers of influence, at least when compared to the British.

Furthermore, the British economists sought to lay down the fundamental principles of the science for all time and places; theoretical claims were phrased in such a way as to be detached from the specific controversies and conditions of the time. Smith's magnum opus tells us almost as much about the economic conditions of ancient Rome as it does about contemporary Britain. True, it would be difficult to imagine a work such as Ricardo's without the advent of capitalism, but beyond that there were few historical features that limited the scope and applicability of his analysis. Political economy was much more a literary pursuit than one might suppose, much more bound to an internal reading of texts. While economists were happy to take credit for healthy economies, their aspirations were often channeled toward pure theory. The flourishing of political economy on British soil must be understood, at bottom, as part of an intellectual tradition reaching back to Petty and Locke. Taking root in a stable political system, their formulations of political arithmetic and liberal ideology, respectively, grew into the Victorian science of political economy.

---

[24] E. J. Hobsbawm, *The Age of Capital, 1848–1875* (London: Abacus, 1975), p. 316; Schumpeter, *History of Economic Analysis*, pp. 382–3, 757.

12

MARX AND MARXISM

Terrell Carver

Karl Marx (1818–1883) absorbed and modified, but never rejected, a German intellectual tradition concerning knowledge and science. This tradition, of science as *Wissenschaft*, derives from idealist assumptions about language and truth that contrast with the empiricism of common English usage and of Anglo-American philosophies of science. Moreover, Marx's concept of social science was explicitly political, as was his activity as a social scientist, in contrast to views that social science can be "above politics" or "balanced," that the social scientist can be apolitical or at least neutral between competing political positions. Because of these differences, Marx and Marxism are frequently located as a "Marxist" section or alternative within the various disciplines that have come to constitute the social sciences since his time, although in specific national contexts the social sciences have sometimes been constituted largely within a Marxist frame of reference (e.g., in France) or against a notion of what is Marxist (e.g., in the United States). Yet it is also undeniable that Marxist social science, both substantively and methodologically, has had such a considerable influence on social science generally, and on philosophies of science overall, that the saying "we are all Marxists now" is almost a truism.

WISSENSCHAFT

In the German tradition, *Wissenschaft* refers to knowledge in the broadest sense, provided that it is conceptualized in a systematic way. Thus, the natural or physical sciences (*Naturwissenschaften*) and the social or human sciences (*Geisteswissenschaften*) do not necessarily form separate domains of knowledge derived through distinct methodologies, nor is philosophy strictly distinguished from science in terms of method or content. The most ambitious work within this tradition was undoubtedly that of G. W. F. Hegel (1770–1831). In individual works, and in an encyclopedic resumé, he attempted to present all knowledge systematically, covering human civilization and history, social relations and the state, nature and natural science, logic and method, and human consciousness itself. Besides broadening philosophical inquiry to include any subject of study whatsoever, particularly the politically contentious areas of history and "the state," Hegel also gave the philosopher-scientist an explicitly evaluative task, that of discovering meaning in creation and reconciling consciousness to itself. In his hands, this was a process of finding the positive in the negative, or transcending contradictions, by tracing conceptual relations "dialectically," based upon the claim that they develop toward realization in practice and toward absolute mind in knowledge. In that way, Hegel rejected empiricism, the view that knowledge is derived from sensory experience registered in thought, and he advanced an idealism more ambitious in scope than that of his predecessor Immanuel Kant (1724–1804). Moving beyond the view that concepts and conceptual relations form the matrix through which knowledge must necessarily be apprehended, Hegel at times seems to argue that some kind of universal mind has given existence to, or at least has created meaning within, the development of everything that exists. Less ambitiously and less theologically, he can be read as placing the acquisition of knowledge within a conceptual framework that is social in character and historically developmental.[1]

Completing Hegel's system after his death using manuscripts and lecture notes, as well as determining what evaluations he intended his thoughts to convey and what exactly his method of exposition comprised, was a task that fell to his disciples and commentators in Germany. Did his philosophical method merely reconcile what ought to be with what is already the case, or did it allow critique to reveal what ought to be and so create programs for action? Hegel's own prose was highly ambiguous, and indeed consistently and deliberately so. A conservative reading of his works was used during the 1830s to justify and support the monarchical and other traditionalist regimes in Germany that were hostile to constitutionalism. Democratization was portrayed both as foreign, because it had come with the invading French in the revolutionary wars, and as disruptive, because it promoted popular participation in politics through elected representative institutions and legal limitations on arbitrary power.

Thus, as Marx was growing up, Hegel was a center of controversy in German intellectual and political life. Indeed, the two spheres were largely coincident, as participation in politics was confined officially to a very narrow elite, barely tolerated in the universities, and heavily discouraged elsewhere. Hence, politics was notably intellectualized and often proceeded by using a kind of code. The foundations of knowledge, that is, the character of *Wissenschaft* and the position of the philosopher-scientist, were of greatest

---

[1] The Hegel–Marx philosophical interrelationship is surveyed in David-Hillel Ruben, *Marxism and Materialism: A Study in Marxist Theory of Knowledge*, 2nd ed. (Brighton: Harvester, 1979).

relevance to an overriding issue of the time, namely the truth of Christianity and the nature of Christian belief. This was because the conservative rulers of the German states claimed their political authority to be based on one form or another, of Christianity; all regarded any questioning of their rule as an attack on religious faith, and any criticisms of their authority, such as those made by constitutionalists, as atheistic sedition.

For some conservatives, Hegel's philosophized Christianity marked a dangerous departure from literal orthodoxy, though for some progressives it allowed for spirituality in the world in an up-to-date way. Marx came from a Jewish family that had converted to Lutheranism (in a Catholic region of the Rhineland) for political reasons, but any faith in Christianity, and in traditional authority relations, seems to have been so weak in the young man that it easily evaporated during his university years at Bonn and Berlin (1835–41). By then, Marx had rejected not just conventional Christianity but all religion and religiousness, and had embraced radical political doctrines of popular sovereignty and democratic politics. In terms of a critique of both religion and politics, and of any presumed connection, between the two, he was well to the left of other Hegelians, such as D. F. Strauss (1808–1874), author of a skeptical but pantheistic *Life of Jesus* (1836), and the so-called Young Hegelians, such as Ludwig Feuerbach (1804–1872), author of the atheistic but humanistic *Essence of Christianity* (1841, 2nd ed. 1843).[2]

SYNTHESIS

There is considerable truth in the adage that Marx combined German philosophy, English political economy, and French socialism and revolutionary doctrines in general, though it should be borne in mind that any attempt to disentangle those elements from the compound he created necessarily destroys what is most original in his contributions to social science.[3] It should be evident from the foregoing that his conception of science was formed in a peculiarly Germanic tradition. The other two elements – socialism and political economy – arrived in 1841, when Marx was in contact with the communist Moses Hess (1812–1875), whose book *The European Triarchy* (1840) prefigured the very synthesis later claimed for Marx. While Hegel's *Philosophy of Right* (1821) attempted to deal with economic aspects of society, with social class as a political problem, and with integrative "corporations" as a solution, Hegel's apprehension of the relatively new science of political economy was partial and sketchy, not least because it presupposed an empirical framework of facts, description, causation, and individualism inimical to idealist

[2] See David McLellan, *The Young Hegelians and Karl Marx* (London: Macmillan, 1969).
[3] V. I. Lenin, *Karl Marx: A Brief Biographical Sketch with an Exposition of Marxism* (1918), in *Collected Works*, 4th ed., vol. 21 (Moscow: Foreign Languages Publishing House, 1964), p. 50.

philosophizing. Hegel and his disciples had made some limited headway with Adam Smith (1723–1790), Adam Ferguson (1723–1816), and Sir James Steuart (1712–1780), among others in this largely Scottish school. Hess's prescient insight became Marx's long-term project: Modern industry had created new extremes of wealth and poverty, a new class of impoverished wage laborers, and a new potential for democratic revolution.

Hess's communism was also Marx's goal: an egalitarian society founded on principles of common ownership that remedied the inequalities of income and wealth arising from a system of private property. Again, while Hegel and the Young Hegelians had taken note of social inequality, their proposals to make society orderly and peaceable were either quasi-medieval "estates of the realm" or nostrums for poor relief. The French socialists surveyed by Hess and studied by Marx – Henri Saint-Simon (1760–1825), Charles Fourier (1772–1837), and Étienne Cabet (1788–1856), among others – presented full-fledged utopian schemes for communist societies, even though they differed greatly on the principles and practicalities involved. These ranged from elite managerialism, to work-as-play, to the allegorical-fantastical. Marx swiftly and decisively rejected anything small-scale, colonylike, or religious. Any communism he would support had to be coincident with the mass politics of democracy and the class politics of the industrial age that was soon to sweep across Europe from England. In keeping with this outlook, his social science did not countenance recipes "for the cook-shops of the future,"[4] though he occasionally allowed himself some programmatic thoughts about the short- and long-term goals that communists could properly envisage, extrapolating from present trends and tendencies in ways that were themselves *wissenschaftlich*.

It is possible now to see that Marx's social science was in place – as a projected synthesis – as early as 1842, though this perspective is available to us only because his early articles, manuscripts, and correspondence have now become available. In his own time, Marx's thoughts reached the public only through the vagaries of polemic and journalism, and were therefore filtered by state censorship, editorial demands, publishing economics, and political considerations. It seems that what Marx had in mind was neither a Hegelian philosophical system nor a potboiler in the style of Hess. Rather, he proposed a unified science that was social not just in its subject matter but in its very presuppositions. Natural science, for Marx, was not knowledge of inanimate objects as such, discovered by individuals doing "pure" research, but rather an activity within society itself, producing knowledge that would profoundly influence all humanity through technological applications in industry. Social science would be historical and political in its very foundations, seeing every human phenomenon as developing, rather than static; and it would

---

[4] Karl Marx, *Capital*, vol. 1, trans. Ben Fowkes (Harmondsworth: Penguin Books/New Left Review, 1976), p. 99.

be knowledge for a purpose, namely to promote the emancipation of humankind from class conflict and the transformation of society into a realm of freedom.[5]

Marx was not one for self-characterizing labels. He denied that he was "a Marxist," hardly ever identified himself as a materialist, and was not particularly concerned to distinguish between being a socialist and a communist.[6] On only a few occasions did he characterize his own outlook as one that emphasized the centrality of production in human social life and its progressive development in different modes (such as ancient, asiatic, feudal, and modern bourgeois or capitalist). Arising from the "economic structure," according to Marx, there is a "legal and political superstructure" and corresponding forms of "social consciousness." These have developed through various stages into the modern class struggle and the democratic politics of constitutional revolution. Marx aimed to make the two coincide. From this it followed that a major study of modern industrial production would be central to any convincing social science, and that it would be a critical analysis written to promote the political interests of the working class in a democratizing social revolution.[7]

The political economists whose works Marx read, preeminently David Ricardo (1772–1823), were generally of the view that industrial capitalism was socially progressive, at least in the longer term, and that in order to get to the longer term, it would be necessary, albeit regrettable, to tolerate the poverty and misery from which new wealth and new commodities were generated. By contrast, Marx suspected that capitalism would be subject to economic crises and normative absurdity as the gap between rich and poor widened, and as the gulf between potential productivity and actual production grew more visible. This view was adumbrated independently by the youthful Friedrich Engels (1820–1895) in his "Outlines of a Critique of Political Economy," which Marx published in an edited collection of 1844.[8] It was Marx who took on the task of demonstrating scientifically the correctness of this analysis of capitalism.[9]

CRITIQUE

That work began in earnest in 1844, when Marx began reading the classics of political economy in French or in French translation, as German

[5] See Karl Marx, *Economic and Philosophical Manuscripts of 1844*, in *Collected Works*, vol. 3 (London: Lawrence and Wishart, 1975), pp. 302–4.
[6] Engels to Eduard Bernstein, 2–3 November 1882, in *Collected Works*, vol. 46 (London: Lawrence and Wishart, 1992), p. 356; Engels to Conrad Schmidt, 5 August 1890, in Karl Marx and Frederick Engels, *Selected Correspondence*, trans. I. Lasker, 2nd ed. (Moscow: Progress Publishers, 1965), p. 415.
[7] Karl Marx, Preface to *A Contribution to the Critique of Political Economy* (1859), in *Collected Works*, vol. 29 (London: Lawrence and Wishart, 1987), pp. 261–5.
[8] Marx, *Collected Works*, vol. 3, pp. 418–43.
[9] Terrell Carver, *Friedrich Engels: His Life and Thought* (Basingstoke: Macmillan, 1989), pp. 1–132.

contributions to this science were notably lacking. Promising a thorough critique, along with critiques of "law, morals, politics, etc.," Marx also envisaged a critique of any Hegelian *Wissenschaft* that claimed to show how these subjects were connected. Numerous works of a more directly political character, and pressing domestic considerations, intervened continually in Marx's life, forcing frequent revisions in his plans. At its most extensive, his plan was to write a critique of political economy in six books (covering capital, landed property, wage labor, the state, international trade, and the world market), a critical history of political economy and socialist systems, and a brief historical sketch of the way economic relations had actually developed.[10] What eventually emerged in his lifetime was *Capital*, volume 1 (1867, 2nd ed. 1872, Russian trans. 1872, French trans. 1872–5, 3rd ed. 1883) and a very large number of preparatory and succeeding manuscripts (notably the *Grundrisse*, mostly written in 1857 and 1858), which have been appearing under various editorial regimes since the posthumous publication of *Capital*, volume 2, in 1885 and of volume 3 in 1894, both substantially edited by Engels. The publication history and textual basis of Marx's economic materials is a highly complex and still evolving study, but it will suffice here to say that Marx's intention to produce a "critique of the economic categories" was fulfilled to such a degree that his work counts as a very substantial contribution to social science in two ways.

First, the centrality of productive activities to ordinary life in class-divided societies, and hence to a democratic politics of social change (whether revolutionary or reforming), was evident in what Marx had produced. During the 1840s and 1850s, however, the works that had any circulation at all were generally programmatic announcements, such as the *Manifesto of the Communist Party* (1848) and the "Preface" to *A Contribution to the Critique of Political Economy* (1859), the latter only a slim volume foretelling the larger study. Nonetheless, Marx's perspective on the historical and contemporary importance of social production contrasted with conventional approaches to understanding society and to promoting political change. Broadly speaking, the conventional view was that intellectual schemes – whether traditional, religious, moralistic, liberal, or utopian – were the only way to effect reform, thereby improving society "from above." After Marx, there was necessarily a debate as to whether revolution could progress "from below," arising out of the thoughts and activities of ordinary people in newly industrializing societies.

It is primarily through this debate that Marx's importance in social science has been realized, even though the debaters themselves, from Engels onward, have defined the terms and issues in crucially different ways. At one extreme was "technological" or "economic" determinism, a view that social revolution takes place only in response to an almost autonomous capacity

---

[10] Karl Marx, *Texts on Method*, ed. and trans. Terrell Carver (Oxford: Basil Blackwell, 1975), pp. 29–31.

for change within and between modes of production.[11] This was classically summarized in Karl Kautsky's (1854–1938) *Materialist Conception of History* (1927), which made the international communist revolution dependent on the inevitable collapse of capitalism in advanced countries. On this view, political action should not outpace economic conditions. At the other extreme were "voluntarist" or "workerist" views, which held that class struggle is the means for shifting production from one mode to another. In *What Is to Be Done?* (1902) and *Two Tactics of Social Democracy in the Democratic Revolution* (1905), V. I. Lenin (1870–1924) argued that even a small working class in a backward society, if led by professional revolutionaries, could achieve a national dictatorship of the proletariat and peasantry. This would trigger an international communist revolution, and so destroy the capitalist mode of production. Eduard Bernstein's (1850–1932) "revisionist" *Evolutionary Socialism* (1899) marked out an alternative to both views, arguing that political tasks change as economic development progresses. In Bernstein's view, Marx's concepts of class struggle and proletarian revolution might well be supplanted by a peaceful transformation of state power and economic structures, prioritizing democracy as a means over socialism as an end.

Second, Marx's project was received from the 1870s onward as both a critique of conventional thinking *about* social production (whether classical political economy or the newly developing "marginalist" schools) and a critique of conventional practices *in* social production (such as the production of commodities for profit in a money system of exchange). Marx's social science presumed that conventional economic categories, such as value, money, commodity, and capital, together constitute an intellectual system. Further, it presumed that mundane versions of these categories also instantiate the social practices of ordinary life in commercial societies. Marx's "new materialism" made the relationship between the definitions and models of economic analysis, on the one hand, and the political evaluation of the activities it mirrors and explains, on the other, a foundational issue in social science. However, Marx's outlook was never fully theorized in this regard, and so has been read in widely differing and even contradictory ways.

Marx's linking together of capitalist economic practices and social scientific language has gone largely unappreciated in Marxist economics, which has generally adopted a conventional empiricism. On this view, the concepts of social science are mere constructs reflecting or modeling, preferably in mathematical or at least in formal terms, the structures and processes in society (typically, monetary ones) that count as economic. Eugen von Böhm-Bawerk's (1851–1914) *Karl Marx and the Close of His System* (1898) criticized Marx on this basis and set the stage for an alternative to "bourgeois" economics, though one that also presumed a separation between social scientific

---

[11] G. A. Cohen, *Karl Marx's Theory of History: A Defence* (Oxford: Clarendon Press, 1978).

analysis and political value judgment. Subsequent Marxist economists altered the terms of the analysis in *Capital* to fit the assumptions and methods of economics as it developed after Marx's death, and for many years remained preoccupied with the "transformation problem." This so-called problem involved a formal proof that market prices could be derived from labor inputs, thus demonstrating the truth of Marx's claim that the exchange-value of commodities on the market was in some abstract and general sense a representation of socially necessary labor power expended in production throughout the system. Whether Marx's work actually required this proof, whether it had indeed posed the question at all, or in that way, whether the assumptions required to complete the demonstration were themselves consistent with other aspects of *Capital* in its various volumes, and whether such a proof would have any important political consequences, were all questions raised in a social scientific context. All through the debate, Marx's substantive claim that labor power is a unique commodity in the sense that it can produce more value than is required for its own reproduction, and that therefore surplus value (and ultimately profit) derives from human labor alone, is simply assumed, though there are labor and welfare economists, influenced by Marx, who would set aside that claim as well.

In more deeply hermeneutic approaches to social science, by contrast, Marx is read as discovering a logic that inheres in the concepts, preeminently economic ones, from which individual life-worlds are constructed and through which increasingly frustrating social structures of individual alienation and collective absurdity are generated. This reading has its roots in György Lukács's (1885–1971) *History and Class Consciousness* (1923), which maintained the link between working-class political activity and the communist historical transformation, but made the whole project seem much more problematic. The broadest category of social science in which Marx's categories are appreciated as both analytical and constitutive is sociology. Two early systemizations represent a recurring controversy: N. I. Bukharin's (1888–1938) *Historical Materialism: A System of Sociology* (1921), which held to positivist notions of fact and causality current in the later nineteenth century, and Max Adler's (1873–1937) *Sociology of Marxism* (2 vols., 1930, 1932), which treated Marxian concepts as necessary conditions for any knowledge of society.[12]

PRACTICE

Thus far, Marx's social science has been presented as written in a highly intellectual framework, albeit one that was politicized during Marx's own time,

---

[12] Anthony Giddens, *A Contemporary Critique of Historical Materialism*, vol. 1: *Power, Property and the State*, 2nd ed. (Basingstoke: Macmillan, 1995); vol. 2: *The Nation-State and Violence* (Cambridge: Polity Press, 1985); Jürgen Habermas, *Theory and Practice*, trans. John Viertel (London: Heinemann, 1974).

and then received subsequently by intellectuals, who generally function in an academic environment. That environment, of course, is not apolitical or depoliticized, but rather one focus among others for political activity. In societies where democratic constitutionalism now prevails, as it did not prevail in Marx's Germany of the 1840s, politics and participation have spread outward, most notably into formal structures of partisan elections, governmental policy making, and public accountability. Since Marx's time, there has been worldwide variability in these forms and in their efficacy, as well as periods of regression to authoritarian absolutism; and such transitions continue today. For a brief period Marx was himself an active communist, fighting within popular alliances in western Europe for democratic rights and freedoms, some of which were hard won and quickly lost in the revolutions of 1848 and 1849. Beginning in the 1850s, during which Marx largely eschewed partisan involvement in national politics (because he was an exile in England, and for family reasons), constitutionalism and participatory politics began to make headway, as the struggle for political freedoms became a more popular and less exclusively intellectual preoccupation. Marx's ideas, though derived from political assumptions of the 1840s, became part of these struggles, and were notably conceptualized by Engels as "scientific socialism," a phrase Marx himself never employed.[13]

In an unusual way, then, Marxist social science was not only politicized in its founding principles but also developed by a partisan movement. Ideas from non-Marxist social science, however, are also the very stuff of political programs; in practical terms they have been the essence of numerous policies affecting all areas of social life. Marxist social science is different in two respects: It is overwhelmingly identified with a canon of writings by Marx and Engels, and it was self-consciously adopted as a national "ideology" by certain regimes. Some of these were notable for their huge geographical areas, large populations, imperial proclivities, and strategic significance (e.g., Russia and China). Additionally, and somewhat surprisingly (given Marx's focus on class struggle in Europe), Marxism was also adopted by a number of national liberation movements in countries where capitalism was arriving in the form of Western imperialist penetration, but where local production was still largely that of peasant agriculture (including Cuba, Vietnam, and other colonies or former colonies where Marxist parties have not prevailed, for internal or external reasons). Perhaps this distinctiveness is summarized in the comment that while a sociology of rationality and bureaucracy, like that of Max Weber (1864–1920), may have had more influence on more ordinary lives than the historical and dialectical materialism constitutive of orthodox Marxism, there has never been a political party or organised movement of Weberians.

---

[13] Paul Thomas, "Critical Reception: Marx Then and Now," in *Marx*, ed. Terrell Carver (Cambridge: Cambridge University Press, 1991), pp. 23–54.

Indeed, the socialist movement generally, and class consciousness in the broad sense of democratic struggle, put political questions to Marxists "from below," and this in turn affected the construction of their social science. The "woman question" arose for Marxists in this way, as neither Marx nor Engels was ever explicitly associated with any women's movement nor much involved in controversies about women that were current at the time. Both were aware of contemporary feminisms, but both were essentially reactive to ideas and events; and both rather suspected the various rights- and issue-based women's struggles of being predominantly middle-class and out of touch with members of the exploited working class, both male and female. Splitting the movement along gender lines was not something that Marx or Engels could contemplate in any sense, nor could either admit the proposition that under communism workers (generally conceptualized as male wage-earners) might have significant difficulty instituting emancipated sexual and family relationships for both men and women. Engels's *Origin of the Family, Private Property and the State* (1883) and August Bebel's (1840–1913) *Woman under Socialism* (1883) located a register of "woman questions" (e.g., power and authority in various family forms, sexual relations and reproduction, child-care and domestic labor, female labor in the public sphere) in a framework that owed as much to the theories and assumptions of the historical anthropology of their day as it did to anything conceptual or substantial in Marx's social science.

From a perspective that aims for political neutrality and value-free objectivity, neither Marx's thought nor Marxist doctrine could ever qualify as scientific. However, as recounted here, Marx himself still persuades many readers that human knowledge can never be apolitical, and hence that facts can never be separated from values, as a matter either of individual reflection or of collective practice. If it can be granted that Marx's conception of social science is valid in this respect, then it cannot be discredited simply because it is overtly and fundamentally political. On the other hand, it would be quite a different argument to suggest that it is political import alone that qualifies a proposition as scientific, even if the political import is of a "communist" or "proletarian" character. Nonetheless, a good deal of what claimed to be Marxist social science certainly fell into that trap, particularly works following pronouncements attributed to Joseph Stalin (1879–1953) and to Mao Zedong (1893–1976), both of whom claimed to offer a methodology of dialectic and contradiction that supposedly validated their political programs as scientific and authoritative, whatever the twists and turns of their party lines.[14]

If it is not political import alone that enables Marxists to validate their reasoning as scientific, what then are the protocols, methodological or otherwise, that Marx used in his work that would also be available to others? In mid-career, Marx himself broached the idea of a methodological treatise, or rather

[14] Karl Popper, *The Open Society and Its Enemies*, vol. 2: *The High Tide of Prophecy: Hegel, Marx and the Aftermath* (London: Routledge, 1966).

one dedicated to declaring what he had found useful in Hegel's philosophy.[15] So far as we know, however, he never wrote this down, though there are a number of methodological reflections scattered throughout his works, and an enormous amount of material from which commentators can reconstruct a presumptive methodology. This process began with Engels's two-part review (1859) of Marx's *A Contribution to the Critique of Political Economy*, and it was famously encapsulated in Lukács's dictum of 1923 that "orthodoxy" in Marxism refers exclusively to "method."[16] To understand the development and significance, then, of mainstream Marxist social science, we must return to Marx's writings and to Engels's popularizations, noting that commentators are now inclined to draw a line between the two.

METHOD

From 1859 onward Engels took on the role of reviewing and popularizing Marx's works (though Marx himself had a hand as well), and the two worked together to gain political credence and influence for his ideas through national party organizations (both legal and clandestine), particularly in Germany. They also worked within the inchoate international communist movement of the 1840s, and later through the International Working Men's Association (the so-called First International), which fostered information exchange and transnational cooperation during the 1860s and 1870s. It was Engels's specific achievement to present Marx publicly as both scientist and philosopher, and to support this with a biographical narrative linking Marx's intellectual ambitions to a socialist politics, both national and international. Engels not only summarized what he took to be the essence of Marx's work, but also, more crucially, chose and defined the terms within which most subsequent summaries of Marx have been constructed. In setting the scene for German readers, Engels presented Marx as Germany's premier social scientist precisely because he was expert in French and English political economy, and because his new economics was linked to the nascent proletarian cause. Thus it was Engels who first linked Marx to an innovative scientific method, and who made this an important political issue in socialist politics.

In explicating what he termed Marx's "materialist conception of history," Engels argued for the centrality of "dialectical method." Contrasting the Hegelian idealist dialectic with a materialism of "fixed categories" (developed in the eighteenth century and presupposed by both contemporary natural science and "bourgeois common sense"), Engels announced that Marx had

---

[15] Marx to Engels, 16 January 1858, in *Collected Works*, vol. 40 (London: Lawrence and Wishart, 1983), p. 249.

[16] Frederick Engels, "Karl Marx: *A Contribution to the Critique of Political Economy*," in *Collected Works*, vol. 16 (London: Lawrence and Wishart, 1980), pp. 465–77; Georg Lukács, *History and Class Consciousness*, trans. Rodney Livingstone (London: Merlin Press, 1971), p. 1.

inverted philosophical idealism in order to establish a "new materialist outlook." He had then extracted the "kernel" of Hegel's discoveries in logic, and
so formulated a "new dialectical method." This enabled him to construct
a scientific account of economic developments in history and a scientific
analysis of the contemporary capitalist economy, which was inevitably advancing in Europe and elsewhere. Later in the century, at Marx's graveside in
1883, Engels again eulogized Marx by linking him to a famous intellectual,
Charles Darwin (1809–1882), this time with rather less justification. As in the
comparison to Hegel, Engels implied that Marx's intelligence was superior
to Darwin's because his system was more comprehensive, and because it was
politically supportive of the working class. Marx was credited with discovering the "law of development of human history" (the materialist conception
of history) and the "special law of motion" governing capitalist society (the
theory of surplus value and the falling rate of profit). As expounded in numerous later reviews, prefaces, introductions, and correspondence relating
to Marx's work, Engels's popularizations initially relied on concepts of materialism, idealism, metaphysics, dialectic, interaction, contradiction, and
reflection, defined somewhat idiosyncratically, and later employed concepts
of selection, evolution, and survival adapted from the Darwinians.

    While Marx did sometimes use these terms, suspicions concerning the
accuracy of Engels's summaries surfaced within the international socialist
movement around the turn of the century. Such criticism began with comments made by Bernstein and by Antonio Labriola (1843–1904), and more
influentially in the 1920s by Lukács and by Karl Korsch (1886–1961). However, until recently such suspicions have been treated as little more than
minor doubts and amendments, given that Engels also established a view of
himself as "junior partner" to Marx, and that Engels was also Marx's posthumous editor, literary executor, and political survivor (for twelve years, until
1895).[17] The doubts were in principle important ones, however, as the issues
concerned just what constituted the scientific character of Marx's thought,
and what could be transmitted to a wider social science. The main bone
of contention was the extent to which Marx's thought, and therefore good
social science, was teleological, incorporating a view that the historical process
was in some sense a subject transcending individuals' decisions and leading
humanity involuntarily to an emancipated and therefore classless society.

    These methodological questions were at the root of political disputes aris-
ing within the Marxist wing of the socialist movement. Was proletarian politics a matter of waiting for social conditions to "ripen" in accordance with
"iron laws" of social development that could not, and should not, be defied by
precipitate action? Or was proletarian revolution a process requiring active,

---

[17] Terrell Carver, *Marx and Engels: The Intellectual Relationship* (Brighton: Wheatsheaf, 1983); S. H.
Rigby, *Engels and the Formation of Marxism: History, Dialectics and Revolution* (Manchester:
Manchester University Press, 1992).

even conspiratorial, interventions into politics in order to direct and speed up social change in the requisite way? Was Marxist social science a "reflection" of economic development, which itself proceeds dialectically and inevitably? Or was it, alternatively, a "guiding thread" that assists fallible humans to "make history" amid contingency? Neither Marxists nor commentators have managed to settle such questions. Marx was notably portrayed in the former way by Engels (who was also unwittingly rather inconsistent), and in the latter way by those who found his methods more various, and his outlook less teleological, than Engels had most famously implied.

Another methodological question was the extent to which Marx's method literally incorporated, or inevitably produced, the theory of the capitalist economy expounded in the three volumes of *Capital*. If Marx's deductive argument concerning the labor theory of value does not lead to his conclusions concerning the falling rate of profit, and therefore to his vision of worsening crises and capitalist collapse, what then becomes of the chances for proletarian revolution? If the labor theory of value is false, will capitalist development still prepare the conditions for proletarian uprising? Or will a new agent of social change have to be found in order that capitalism may be overthrown and freedom realized? Alternatively, if the labor theory of value is true, why has capitalism not yet collapsed? Why, when capitalism has come close to collapse in countries with highly productive modern industries, has proletarian revolution, and international workers' solidarity, not been more successful in "winning the battle of democracy," as Marx so confidently predicted that it would? Engels offered little in the way of guidance or explanation on these issues; and it is only very recently that materials have become available that allow his work as editor of *Capital*, volumes 2 and 3, to be judged against Marx's draft manuscripts. The overwhelming tenor of Marx's argument concerning the labor theory of value and the necessary collapse of capitalism is unmistakable. However, if the "critique of the economic categories" is seriously mistaken, how scientific is the method from which it derives? If method is not central to understanding and evaluating Marx, then what is? And if there is anything else relevant to evaluating Marx, what bearing does it have on social science?

SCIENCE

The defence of dialectical materialism, as adumbrated by Engels in his widely circulated tract *Anti-Dühring* (1878) and its abridgement as *Socialism, Utopian and Scientific* (1880), has been largely a theoretical enterprise in the philosophy of science. Though Soviet natural science and social science claimed to apply this method, it is unlikely that significant results emerged as a direct consequence. The classical restatement of Engels's position is in G. V. Plekhanov's *Fundamental Problems of Marxism* (1908). Engels's position itself was explicated further when his manuscript *Dialectics of Nature* (dating from the late 1870s) was posthumously published in 1927. This work was extensively cited in the Soviet Union during the 1930s, and then subsequently up to the 1990s, as the canonical text for Marxist–Leninist science. Dialectical materialism was directly derived from Engels's claim that Marx's materialist dialectic comprised three laws common to "nature, history, and thought": (1) the transformation of quantity into quality, (2) the unity of opposites, and (3) the negation of the negation. The textual and argumentative basis for these claims has been much disputed with respect to Marx's work; and irrespective of that dispute, a claim that all phenomena can in some sense be effectively reduced to, or intelligently explained by, such formulaic generalities has been actively questioned since the 1920s.

Engels's claims do not, however, demand such a stringent reading. For many Marxists, "materialism" slid easily into a view that "economics" – that is, human activity in production, consumption, distribution, and exchange – was the determining factor in a dialectic of social change. While this did not solve the problem of "voluntarism" or agency in political action, it created a framework that linked social science to a positivism of facts and laws, and also to a presumption that, as Marx and Engels commented, "history is the history of class struggles."[18]

Marxist history as a social science extends economic and class analysis to studies of the earliest societies in human civilization; to early modern history, with works such as Engels's *Peasant War in Germany* (1850); and to the circumstances and politics of exploitation and conflict in more recent social structures.[19] In an Introduction (1895) to Marx's republished *Class Struggles in France 1848–1850*, Engels wrote that the historian's job was to demonstrate that political events – struggles between classes and class fractions – were effects of economic causes. These could be derived only by collecting and sifting data some time after the event, something that Marx could not do when writing works of "current history," such as his work on contemporary French politics in 1850.[20] Though the *Manifesto* sketched a world history in outline, Marx himself did not bring his manuscripts quite to the point of considering the "world market." His successors in this tradition responded to the problems of colonialism and imperialism, particularly as they affected the national working classes, by moving historical and political analysis onto the global stage. Among the classics of this genre were Lenin's early *Development of Capitalism in Russia* (1899), Rudolf Hilferding's (1877–1941) *Finance Capital*, Rosa Luxemburg's (1871–1919) *Accumulation of Capital* (1913), and Bukharin's *Imperialism and World Economy* (1915).

[18] Karl Marx and Frederick Engels, *Manifesto of the Communist Party*, in *Collected Works*, vol. 6 (London: Lawrence and Wishart, 1976), p. 482.
[19] Marxist history has a distinguished archival tradition, beginning with Henryk Grossman's journal *Archiv für die Geschichte des Sozialismus und der Arbeiterbewegung* (1911–30), and a notable narrative tradition in works by E. J. Hobsbawm, Christopher Hill, and Perry Anderson.
[20] Carver, *Marx and Engels*, pp. 148–50.

While Engels's overarching dialectical scheme was founded on the view that Marx, the materialist, had "inverted" Hegel, the idealist, Darwin also played a considerable role in Engels's later works of social science, and in what developed as the Marxist branch of social Darwinism. While it is true that Marx praised the *Origin of Species* (1859) for its quality as a work of natural science, he admired it specifically for Darwin's ability to demonstrate a pattern among independent events without resorting to teleology. This makes it unlikely that he saw Darwin's work, or indeed his own, in terms of the laws of development that Engels mentioned in his "graveside speech," though this point is still textually disputed. Engels's manuscript investigation, "The Part Played by Labor in the Transition from Ape to Man" (1876), was published later (1895–6), when social Darwinism had become an important intellectual and political force. Engels attempted to merge Marx's imputed anthropology, which attributed a special role to productive activities in constituting and revolutionizing the different "epochs" of human history, with a Darwinian account of the physiological evolution of primates, including the development of the capacity to use tools and language. Marx's "Anthropological Notebooks" (1880–1) were cited and adapted by Engels in producing his own *Origin of the Family*, though textual investigation has revealed that Marx was largely concerned with excerpting factual materials and Engels with constructing an overall theory and historical schema. Engels's argument attempted to merge a Darwinian theory of sexual selection (as reflected in a supposed history of "marriage forms" in primitive societies) with a class-struggle explanation for both the oppression of women by men and the exploitation of workers by those controlling the means of production.

The Darwinian framework was taken further by Engels when he attempted to argue, in personal correspondence shortly before his death, that Marx's view of history as class struggle was coincident with the Darwinian view that natural selection guarantees "survival of the fittest." Besides raising the familiar problem of a "natural inevitability" in social processes, this argument was a difficult one to reconcile with the obvious facts of working-class deprivation, as documented in Engels's early and reprinted study *Condition of the Working Class in England* (1844), or with the founding theory of proletarian immiseration announced in the *Manifesto* (reprinted in 1872 and in numerous later editions) and further documented in *Capital*, volume 1. In what scientific sense were proletarians the "fittest," and in what political sense was their "survival" a victory?

THEORY

Engelsian orthodoxy raised serious difficulties concerning positivist assumptions about science, particularly when applied to social science, and about fundamental philosophical issues, particularly those relating to the relationship between matter and consciousness, and between material evolution and human progress. While Engels had made Marx resolutely Hegelian, the materialist dialectic he recounted was a peculiarly positivist version of Hegel, encyclopedic in scope and teleological in import. Departing from this framework meant a break with Engels, and by the 1920s a re-Hegelianizing of Marx and Marxism was under way, beginning with Korsch's *Marxism and Philosophy* (1923). This occurred against a background of political frustration and despair, given the failure of proletarian revolution in western Europe, following World War I, and it incorporated a tension between "voluntarist" views and hermeneutic methods, on the one hand, and a Soviet-enforced positivist orthodoxy, on the other. Those who broke with orthodoxy, at least to some degree, turned their attention to the "legal, political, religious, aesthetic or philosophic – in short ideological forms" through which people "become conscious" of class conflict and "fight it out." In effect, this approach tended to presume the influence of the economic "base" in order to concentrate on the "superstructure."[21] If nothing else, this explained away revolutionary failure by suggesting that proletarian consciousness had not yet developed in correspondence with the "economic structure," or that within the superstructure itself "bourgeois" consciousness had temporarily won over sufficient numbers of proletarians so as to block revolt, a situation that Engels (though not Marx) termed "false consciousness."[22] Marx's own views, in contemporary political writings such as the *Class Struggles in France* (1850) and the *Eighteenth Brumaire of Louis Bonaparte* (1852), were difficult to reconcile with any form of technological or economic determinism (even for Engels), precisely because they allowed so much scope for the superstructural influence of ideas and traditions, and even of whim and chance, as class struggle proceeded.

A view that political consciousness, legal and political relations, and forms of property "correspond" to an economic structure in the relations of production is just as much a theory of the former realm (ideas and values) as of the latter ("material" processes of production). Indeed, what was "material" about the realm of economic activity, as conceptualized by Marx, emerged as problematic, in the sense that material objects are commodities only by virtue of their incorporation into a conceptual system instantiated in human practice (e.g., buying and selling "free labor" in order to produce goods for sale at a profit). Without quite putting the supposed materiality of the economy to the test, Marxists in the Frankfurt School launched into investigations of

[21] These formulations come from Marx, Preface to *A Contribution to the Critique of Political Economy*, in *Collected Works*, vol. 29, pp. 263–4; for an alternative translation that departs significantly with respect to the verb "determines" (*bestimmen*), see Karl Marx, *Later Political Writings*, ed. and trans. Terrell Carver (Cambridge: Cambridge University Press, 1996), pp. 158–62.
[22] John Torrance, *Karl Marx's Theory of Ideas* (Paris: Maison des Sciences de l'Homme; Cambridge: Cambridge University Press, 1995).

an historical, philosophical, and psychological character, in order to detail precisely how the political project of proletarian revolution had been so significantly constrained within a complex and important realm of persuasive ideas.

These ideas could be critiqued as ideologies – systems of thought and belief that Marx characterized as partial, misleading, incomplete, and linked to power relations of exploitation and to oppression in other forms. While by no means a unified school, as the name erroneously implies, those working under the aegis of the Frankfurt School also looked outward to Weber's sociology, Sigmund Freud's (1856–1939) psychoanalysis, and even more widely to cultural criticism and aesthetics. Herbert Marcuse's (1898–1979) *Reason and Revolution* (1941) and *One-Dimensional Man* (1964), Max Horkheimer's (1895–1973) essays from the 1930s and 1940s collected as *Critical Theory* (1968), and Theodor Adorno's (1903–1969) *Authoritarian Personality* (1950) and *Negative Dialectics* (1966) stand as landmarks in a revitalization of social science via the idealist tradition, linking action to ideas and consciousness and using interpretative methodologies to understand and explain them.

These developments were to some extent paralleled by Antonio Gramsci's (1891–1937) *Prison Notebooks*, written from 1929 to 1935 but finally published and widely debated beginning in the 1950s. The politics of mass democracy was for Gramsci a serious problem that Marxists had not yet successfully confronted. Preexisting anticommunist cultural influences had already established the "hegemony" of a dominating "bloc" of classes, thus securing widespread consent to the social order. Replacing this bloc with a proletarian bloc, which would rely for its power on the genuine consent of the masses, was essential to his view of social revolution, both in outcome and in process. By dint of the author's circumstances, these ideas and tactics had virtually no influence at the time. But in terms of post-war developments in Marxist social science, the influence of Gramsci on social theory has been considerable, to the point where the emphasis on superstructural struggle, the dominance of capitalist ideologies, and the politicization of forms of social oppression other than class (gender, race, ethnicity, sexuality, etc.) have upset the primacy of class itself as a construct, both in analysis and in politics. For some "post-Marxists," class has as much and as little objective presence and political centrality as the other forms of inequality in society against which reformers and revolutionaries continually struggle.[23]

While at one extreme Marxist social science has now almost dissolved into a politicized realm of cultural studies and a politics of "new social movements," at the other extreme it has come close to painting itself into a corner of structuralist rigidity. Though a positivism based on the presumed materialism of

---

[23] Ernesto Laclau and Chantal Mouffe, *Hegemony and Socialist Strategy: Towards a Radical Democratic Politics*, trans. Winston Moore and Paul Cammack (London: Verso, 1985).

natural objects – and on the allegedly lawlike methods of the natural sciences –
was no longer sacrosanct in Marxist social science (except in regimes where
an Engelsian orthodoxy was enforced), something like positivism staged a
return during the 1960s in the work of Louis Althusser (1918–1990). Arguing
that an "epistemological break" should be recognized in Marx's work, so that
the early influence of Hegelian philosophy can be precisely delimited and a
contrasting "scientific" Marx constructed in chronological and bibliographical terms, Althusser attempted to show that Marx's program was properly
one of conceptualizing social and political life on the basis of an economic
determinism. For Althusser, the early Marx was "humanistic" and therefore
"ideological" or unscientific, concerned with a Hegelian historical narrative
of alienation and emancipation. The later Marx, by contrast, was said to be
a historical materialist and therefore scientific, in the sense that he identified
a hierarchy of practices or structures in society, among which the economic
was causally primary. This causal primacy, though, was only "*ultimately* determining" (as Engels had famously remarked),[24] and the economic structure
was not necessarily dominating or effective at any given time.

While at first glance this view of Marx might seem merely to regenerate
Engels's "materialist conception of history," it did manage to jettison the
materialist metaphysics and reflectionist epistemology that had caused some
embarrassment to orthodox Marxists over the years. Rejecting the empiricism on which Engels had apparently relied, Althusser substituted a view
that knowledge was constructed entirely in thought, and indeed through
theory itself as the manipulation of abstractions, a "theoretical practice." In
addition, Althusser jettisoned the embarrassing problem of human agency
in history by treating individuals as supports or effects of the social formations to which they belong, which are themselves the locus of the causality
that drives change and development. These ideas were expounded in *For
Marx* (1965) and *Reading "Capital"* (1970), which are now chiefly interesting
because of their links to, and differences from, postmodern philosophies of
deconstruction. Deconstructionists are just as suspicious as Althusser was
about overarching narratives of progress and emancipation, and they share
his suspicion of individualisms that privilege a presocial human subject. But
rather than turning to science and theory, as Althusser defined them, they
take a "linguistic turn," following performative and contextual theories of
meaning, and recalling Marx's praise of idealism for developing the "*active*
side."[25]

---

[24] Engels to Joseph Bloch, 21–22 September 1890, in *Selected Correspondence*, pp. 417–19; see also Engels to Schmidt, 27 October 1890, in *Selected Correspondence*, pp. 421–4; and Engels to Heinz Starkenburg, 25 January 1894, in *Selected Correspondence*, pp. 466–8.

[25] Richard Rorty, *Contingency, Irony, and Solidarity* (Cambridge: Cambridge University Press, 1989), pp. 1–22; Jacques Derrida, *Specters of Marx: The State of the Debt, the Work of Mourning, and the New International*, trans. Peggy Kamuf (London: Routledge, 1994); Terrell Carver, *The Postmodern Marx* (Manchester: Manchester University Press, 1998).

RENEWAL

Marx was not the first to make inequalities of income and wealth in society, and the exploitation of modern industrial workers, the objects of social scientific study. Neither was he the first to bring the method of critique to bear on these issues, though he was by far the most influential. Exactly what this method is, and how it relates to alternative methodologies, are questions that he posed rather than answered. Moreover, he raised issues concerning the connection between social scientific reflection and political change that will never be laid to rest, and this in turn brings up fundamental problems regarding the nature of any science as a social activity. Beginning with Engels, Marx has inspired a vast literature of commentary and a wide variety of scientific and political practices. By the 1880s, these had been conceptualized as a Marxist tradition, but by the 1990s this had broken down almost completely, both theoretically and politically. Such fragmentation is not necessarily a sign of weakness in Marx's ideas or in socialist politics, but rather testimony to their continuing relevance and intellectual strength, apart from, and indeed in spite of, efforts to codify his thought as doctrine. His complete works are still in the course of definitive publication, and people of varying philosophical and political views will continue to find him inspiring and enlightening.

Part II

THE DISCIPLINES IN WESTERN EUROPE AND NORTH AMERICA SINCE ABOUT 1880

13

CHANGING CONTOURS OF THE SOCIAL SCIENCE DISCIPLINES

Dorothy Ross

The disciplines recognized in the twentieth century as the social sciences emerged from older branches of knowledge by a process of separation and negotiation between related and overlapping areas of interest. As Theodore Porter points out, some of these lines of inquiry had been relatively continuous genres of writing for centuries, but they were often strands in broader traditions of knowledge and practice – chiefly philosophy, history, and affairs of state – and they were part of the intellectual equipment of liberally educated people, rather than occupations for specialists. Beginning in some cases earlier, but more conspicuously in the late nineteenth and early twentieth centuries, they formed into fields to which specialists devoted their principal efforts and sites for research, reflection, and training. This modern idea of disciplines itself emerged over the course of the nineteenth century, a product of increasing specialization in science, scholarship, and technical expertise; the research ideal pioneered in German universities; and the reconstruction of higher educational systems and administrative institutions in Europe and the United States. University training and credentialing was especially important in solidifying the existence of continuing communities of specialized scholars.[1]

We should not overemphasize the rapidity or pervasiveness of this transformation. In Europe, disciplinary organization was never as firmly established nor as important to the production of social knowledge as it was in the United States; and even there, the course of development was uneven. Still,

---

[1] Charles E. McClelland, *State, Society, and University in Germany, 1700–1914* (Cambridge: Cambridge University Press, 1980); Alexandra Oleson and John Voss, eds., *The Organization of Knowledge in Modern America, 1860–1920* (Baltimore: Johns Hopkins University Press, 1979); Lawrence Stone, ed., *The University in Society*, 2 vols. (Princeton, N.J.: Princeton University Press, 1974); George Weisz, *The Emergence of Modern Universities in France, 1863–1914* (Princeton, N.J.: Princeton University Press, 1983); Peter Wagner and Bjorn Wittrock, "States, Institutions, and Discourses: A Comparative Perspective on the Structuration of the Social Sciences," in *Discourses on Society: The Shaping of the Social Science Disciplines*, ed. Peter Wagner, Björn Wittrock, and Richard Whitley (Dordrecht: Kluwer, 1991), pp. 341–9.

specialized disciplines became a basic feature of the human sciences in the twentieth century and, particularly after the Second World War, an international pattern of intellectual organization.

The term that best captures how most historians – including the authors in this volume – understand the process of disciplinary formation is *project*. To call the formation of social science disciplines a project is to locate it within the contingencies of history. Disciplines were not a product of the automatic progress of science, nor were they "natural" categories. They had to establish themselves as authoritative purveyors of descriptions of the world. The chapters in this section show that process to have been fraught with uncertainties and conflicts. Acting within an already-existing structure of intellectual domains, with its own patterns of authority, social scientists had to "compete for the right to define what shall count as intellectually established and culturally legitimate," not only between disciplinary areas and within them, but also in the public arena.[2] The disciplinary project was also linked to a "professional" one, especially in the United States, where university appointment did not carry with it a traditional role or one that carried civic status, so that professional career lines and expertise were important concerns. Both intellectual and professional considerations interacted in contests for legitimacy, resources, and practical expertise.[3] As Mitchell Ash notes, the history of social science disciplines is a "continuous struggle by multiple participants to occupy and define a sharply contested, but never clearly bounded, discursive and practical field" (p. 252, this volume). The contents and borders of the disciplines that resulted, as Robert Bannister shows for sociology and as all of the authors in this section demonstrate, were the product as much of national cultures, local circumstances, and accidental opportunities as of intellectual logic.

The term "project" also expresses, indeed accentuates, a tension at the heart of all historical explanation. A project is, on the one hand, a shared idea, aspiration, plan, or blueprint. As Bernard Yack explains, it was used in this sense, appropriately for our purposes, by Francis Bacon in the early seventeenth century to denote "a design for improvement through scientific research." The formation of social science disciplines was certainly a project in this

---

[2] See Kurt Danziger, *Constructing the Subject: Historical Origins of Psychological Research* (Cambridge: Cambridge University Press, 1990), chap. 1 and pp. 39–42, for an illuminating discussion of the disciplinary project. Fritz Ringer, *Fields of Knowledge: French Academic Culture in Comparative Perspective, 1890–1920* (Cambridge: Cambridge University Press, 1992), casts this historical process in the terms of Pierre Bourdieu's social cultural theory of the "intellectual field." The quotation is from Ringer, p. 5. Cf. Peter Wagner and Björn Wittrock, "Analyzing Social Science: On the Possibility of a Sociology of the Social Sciences," in *Discourses on Society*, ed. Wagner, Wittrock, and Whitely.

[3] On social science disciplinary formation as professionalization, see Oleson and Voss, eds., *Organization of Knowledge*; Mary Furner, *Advocacy and Objectivity: A Crisis in the Professionalization of American Social Science, 1865–1905* (Lexington: University Press of Kentucky, 1975); Thomas L. Haskell, *The Emergence of Professional Social Science* (Urbana: University of Illinois Press, 1977). Cf. Danziger, *Constructing the Subject*, p. 119, for a "professional project" parallel to the disciplinary one; see also Chapter 30 in this volume for a comparative view.

sense, a shared program, conceived within the modern tradition of science and enlightenment, aimed at social improvement. But, as Yack continues, in a usage taken from the philosopher Martin Heidegger, understanding "always involves the projection of a world of possibilities within which things gain their meaning." A project in this sense is not a plan, but rather a framework, a mode of understanding within which certain kinds of ideas and practices necessarily come into being. In this sense of "project," the program of science and enlightenment aimed at social improvement sets the terms and boundaries within which the social science disciplines form; instead of being a matter of human intentions, it is rather a system of constraints within which intentions and actions arise, for our understanding constructs the world in which we live.[4]

In the spirit of this second meaning of "project," historians of the social sciences have recently emphasized the importance of language as the medium in which meanings are produced, especially those integrated bodies of language called discourses. Discourses of science and enlightenment were the principal medium in which social scientists formed their purposes, and in which they represented those purposes as true and legitimate. Likewise, recent historians have emphasized the importance of practices, what social scientists do and the kinds of tools they use – from the economist's mathematical modeling to the school IQ test, the social survey, and the anthropologist's encounter – as a medium, related to discourse, by means of which these disciplines organize the world.[5] It is in this sense that the French philosopher Michel Foucault argued that the project of the human sciences orders and manages the diverse populations of modern society through its scientific practices and the knowledge they generate. The human sciences become "disciplines" in a double sense, specialized branches of knowledge and agencies of regulatory control.[6]

These new insights can be carried to deterministic extremes. If the social scientific project is presented as a seamless whole, often merely as a thread in a seamless project of modernity, then its trajectory and outcome are predetermined and there is no escape from its constraints. This is a matter of some moment, for those who emphasize constraint also stress its oppressive consequences, citing the desiccating compulsions and totalitarian reach of scientific control. But the social scientific project can be presented instead as one in which discourses and practices are multiple, at least partially divergent

---

[4] Bernard Yack, *The Fetishism of Modernities: Epochal Self-Consciousness in Contemporary Social and Political Thought* (Notre Dame, Ind.: University of Notre Dame Press, 1997), chap. 5, at pp. 116, 117.

[5] See, e.g., Nikolas Rose, *The Psychological Complex: Psychology, Politics and Society in England, 1869–1939* (London: Routledge, 1985); Bruno Latour, *Science in Action* (Cambridge, Mass.: Harvard University Press, 1987); Andrew Pickering, ed., *Science as Practice and Culture* (Chicago: University of Chicago Press, 1992).

[6] Michel Foucault, *Discipline and Punish* (New York: Pantheon, 1978); Jan Goldstein, "Foucault among the Sociologists: The 'Disciplines' and the History of the Professions," *History and Theory*, 23 (1984), 170–92.

in values and effects, and open to historical contingency. As Yack concludes, quoting the Viennese modernist writer Robert Musil, "When we realize that we are not dangling 'from the puppet strings of some hobgoblin of fate, but on the contrary, that we are draped with a multitude of small haphazardly linked weights,' then we regain considerable room for maneuver."[7] Historians typically take Yack's middle course, emphasizing both freedom and determinism, intentions and constraints. Just what the weights are, how large, and how haphazardly they are linked are questions to be put to history.

What sort of project, then, *was* the project of disciplinary formation in the social sciences? We will start with the first period of disciplinary formation, locating the disciplinary program in the search of liberal elites for an authoritative source of reason amid the historical crises of the years 1870 to 1914. The continued search, amid continuing historical crises, accounts for much of the subsequent history of the social science disciplines. Because the disciplinary project was both national and international in scope, because disciplinary borders remained porous and unsettled – and because the chapters in Part II deal with single disciplines – we will pay particular attention to the ways in which ideas, methods, and researchers themselves crossed national and disciplinary lines. During the period between the two world wars, the social science orientations developed in the United States and Europe diverged and then partially rejoined, setting the stage for a period of revitalization, and of boldly reformulated disciplinary ambitions, during the postwar decades. After 1970, these ambitions were severely challenged and the social science disciplinary project itself called into question.

DISCIPLINARY FORMATION, 1870–1914

Liberal elites first formulated social sciences in the late eighteenth century and, although the ideological spectrum widened, they played a central role in sustaining these studies through the nineteenth century and establishing them as disciplines.[8] Members of educated social strata, they embraced the Enlightenment ideal of modernity as a progressive and culminating stage in human history, grounded in individual liberty and guided by scientific social

---

[7] Yack, *Fetishism of Modernities*, p. 40.

[8] For basic information on the topics of this section, see the chapters on the disciplines in this volume; Roger Smith, *The Norton History of the Human Sciences* (New York: Norton, 1997) (in England, *The Fontana History of the Human Sciences*), for psychology, and secondarily for sociology, in all of Europe and in the United States; Dorothy Ross, *The Origins of American Social Science* (Cambridge: Cambridge University Press, 1991), for economics, sociology, and political science in the United States and its European background; Robert C. Bannister, *Sociology and Scientism: The American Quest for Objectivity, 1880–1940* (Chapel Hill: University of North Carolina Press, 1987); John Maloney, *Marshall, Orthodoxy and the Professionalisation of Economics* (Cambridge: Cambridge University Press, 1985); George W. Stocking, Jr., *After Tylor: British Social Anthropology, 1888–1951* (Madison: University of Wisconsin Press, 1995) and his *Race, Culture, and Evolution: Essays in the History of Anthropology* (1968) (Chicago: University of Chicago Press, 1982); David F. Lindenfeld, *The Practical Imagination: The German Sciences of State in the Nineteenth Century* (Chicago: University of Chicago Press, 1997).

knowledge. Allied with capitalist development and representative democracy, liberals had moved by the end of the nineteenth century from the radical edge to the embattled center of political power, where they confronted new issues that forced them to reexamine inherited principles: the sources of moral and social order in secular, urban society; the reconstitution of the nation in an age of democracy and imperialism; the role of the state in managing the new industrial economy; and remedies for "the social question" – the complex of poverty, class conflict, and racial and ethnic diversity created by industrialization and its dislocations. Responding to Marxist critiques of capitalism and conservative critiques of democratic society, social scientists approached their disciplinary tasks through the liberal problematics of individual freedom and social order and through such nationalist problematics as American exceptionalism, French republicanism, and the historical fragmentation of the German state. In the process, they adopted a more interventionist stance toward modern society, setting the stage for, and often participating in, the construction of a widened range of liberal and statist policies.[9] (See Chapter 34 in this volume.)

As the discussion of disciplines has already indicated, these liberal goals were to be achieved through the intellectual authority of specialized scientific communities, organized increasingly into disciplines. Amid the clashing interests and growing irrationalism of industrial society, the social sciences would become authoritative sources of reason. We should note at once the paradoxical logic of this project. Disciplinary specialization promised to enhance the social scientists' combined scientific and political authority by narrowing their focus and distancing them from the pressures of politics. But that strategy cut the taproot to the moral and political world that nourished their project, without insulating them from it, for disciplines are only relatively autonomous from the surrounding world and participate in the national cultures, political conflicts, and social divisions of their milieux.[10] Institutionalized disciplines attenuated class connections, but created their own special class interests. Encouraging intellectual rigor and a measure of detachment, they also discouraged critical reflection on the moral and ideological dimensions that inevitably attached to their work. Although the social sciences were hardly alone in experiencing these tensions related to disciplinary specialization, their project exposed them more relentlessly to such hazards.

---

[9] For the class-based liberal program of these elites, see Ringer, *Fields of Knowledge*, chaps. 1–2; Henrika Kuklick, *The Savage Within: The Social History of British Anthropology, 1885–1945* (Cambridge: Cambridge University Press, 1991), chap. 1; Reba N. Soffer, *Ethics and Society in England: The Revolution in the Social Sciences, 1870–1914* (Berkeley: University of California Press, 1978), Introduction; Ross, *Origins of American Social Science*, pts. 1–2; Peter T. Manicas, *A History and Philosophy of the Social Sciences* (Oxford: Basil Blackwell, 1987), chap. 10; and, for an especially nuanced view, Peter Wagner, "Science of Society Lost: On the Failure to Establish Sociology in Europe during the 'Classical' Period," in *Discourses on Society*, ed. Wagner, Wittrock, and Whitley, chap. 9.

[10] On the erosion of normative perspectives in sociology, see Donald N. Levine, *Visions of the Sociological Tradition* (Chicago: University of Chicago Press, 1995).

Specialization required distancing not only from politics, but also from popular knowledge. Sociology contended with literature, where the realism and social criticism of writers and journalists claimed jurisdiction – most successfully, in England – over the understanding of modern society, and with reform movements that claimed jurisdiction over social intervention.[11] Political science asserted its claims in a field dominated by the actual experience of politicians and citizens; psychologists faced clerical, spiritualist, and commonsense experts. Yet specialists inevitably remained open to popular understanding, especially when they tried to influence it. As the essays in Part IV demonstrate, disciplinary formation did not halt the traffic between formally constructed and popular social knowledge.

The new disciplines borrowed scientific authority from adjacent scholarly domains. One major source of scientific authority was evolutionary biology. Charles Darwin's *Origin of Species* (1859), entering onto a cultural scene already rich in evolutionary ideas, stimulated work in several directions.[12] Darwin's theory made plausible the view of human beings as animals adapting to the environment and the view of society as a kind of organism with mutually adaptive structures and functional needs, so that adaptational, functional, organismic, and evolutionary models attained new legitimacy in psychology, sociology, and anthropology. Anthropologists' "comparative method" placed peoples and races, customs and myths on a vast evolutionary grid and assigned them to stages in a single evolutionary process. The evolutionary perspective provided a technology for evaluating the world that was being created by industrialism and imperialism, with its racial, class, and gender inequalities, using both Eurocentric hierarchical standards and their Romantic subversion. Geography, in turn, drew on cartography and the Earth sciences to construct a global spacial grid for Western hegemony.

Philosophy and history were in many instances as fundamental as the natural sciences in legitimating the human sciences. Moral philosophy, Hegelian philosophy, the neo-Kantian revival, and moves toward pragmatism and philosophies of experience all turned the attention of late-nineteenth-century students of philosophy to the psychological and social domains. In the German neo-Kantian context, for example, psychology was understood as a "philosophical project to provide foundations for rational knowledge."[13] Likewise, the historicism of nineteenth-century culture – developing early in Europe and later in the United States – created the social sciences' chosen problem of modernity, and conclusions from philology, linguistics, and

[11] Wolf Lepenies, *Between Literature and Science: The Rise of Sociology* (Cambridge: Cambridge University Press, 1988).

[12] Smith, *Norton History*, pp. 453–6, makes the strong case for Darwin's catalyzing influence on the social sciences; Peter J. Bowler, in *The Non-Darwinian Revolution* (Baltimore: Johns Hopkins University Press, 1988), and Robert Bannister, in *Social Darwinism* (Philadelphia: Temple University Press, 1979), argue against the direct influence of Darwin on the social sciences during this period.

[13] Smith, *Norton History*, p. 494.

historical jurisprudence undergirded the new evolutionary perspectives and comparative method.

Adjacent disciplines, however, served as both matrix and whetting stone for the social sciences. Younger German and American psychologists moved to make their field an autonomous laboratory science. Sociologists worked to detach their organismic metaphors from identification with biological processes. Political science, sociology, and economics variously drew upon and contested the domain and methods of history. Despite shared intellectual space, disciplinary formation required disengagement from these authorizing fields.

Although the social sciences in time achieved some degree of autonomy, both biology and history remained alternative bases upon which to organize the social science disciplines and exerted considerable, if intermittent, influence. Social scientists read Mendelian genetics, with its sharp distinction between biological inheritance and social usages, as a charter of liberty for their disciplines; but general evolutionary assumptions, racial theories, eugenics, primate studies, and theories of homeostasis and biological systems maintained a biological presence in the social sciences.[14] Historicism remained embedded in the problematics of modernity and the ineradicably concrete and temporal forms upon which social scientists worked. As we shall see, both biology and historicism reemerged at the end of the twentieth century as claimants to the social science project.

So too did economics. Political economy developed early in the nineteenth century, and was reconstructed, beginning in the 1870s, when marginalist economists drew on analogies from physical mechanics to construct a vision of the economy as a self-equilibrating system of market choices. Marginalist analysis formed a seemingly scientific core that allowed the discipline to escape from several decades of criticism for its narrow psychology of self-interest, its inadequate attention to the history and functioning of economic institutions, and its failure to cure the disorders of capitalism. Marginalism helped to reshape classical political economy into the specialized discipline of economics, particularly in Anglo-America, although the emerging neoclassicism remained within a disciplinary matrix of competing historical and institutional schools and continuing political and empirical concerns.

Sociology developed in part from the same sociohistorical context. Beginning in the 1890s, sociologists, often trained as economists, moved beyond economics to examine the social bases of cohesion and progress in modern society. In Germany, the development of economics as an historical discipline and the debate over methods, or *Methodenstreit*, between the partisans of Austrian marginalist economics and German historical economics formed

---

[14] See, e.g., Donna Haraway, *Primate Visions: Gender, Race, and Nature in the World of Modern Science* (New York: Routledge, 1989).

the seedbed for an extraordinarily creative group of sociologists of capitalist society, including Max Weber and Georg Simmel.

Intellectual currents moved across national borders as well as intellectual fields. For the United States, a provincial outpost of European science and scholarship, this movement was crucial. Nine thousand Americans studied at German universities during the nineteenth century, most of them between 1870 and 1900, including key figures in the development of all of the social sciences. In France, the advanced German work caught the attention of Émile Durkheim, while J.-M. Charcot's Paris clinic drew visitors interested in psychopathology from across Europe. In an age of considerable international communication via journals and conferences, the work of scholars in any national context could be important elsewhere.

Institutional support for these disciplinary agendas came chiefly from the expanding university systems that provided centers of graduate education and research, but also from bases outside the universities – banks, trade unions, schools, prisons, hospitals, reform organizations, state bureaus, welfare agencies, museums, and colonial governments – that offered social scientists employment, markets for their expert services, and venues for research. This triangular base – the academy, the market, and political and social institutions – produced severe tensions between, in Mary Furner's phrase, "advocacy and objectivity."[15] Yet these diverse locations also provided the disciplines with a degree of constrained autonomy within which to pursue scientific and practical activities.

Notable among the constraints were narrowed scientific and political vision and masculinization of the disciplines. Universities – whether state institutions, aristocratic corporations as in England, or private institutions funded by capitalists as in the United States – discouraged political, gender, and racial heterodoxy. While many of the popular sites that pursued social science knowledge and practical investigation were well within the liberal fold, others, like the Fabian Society, were more radical, and many of them were also staffed by women. Hull House, a Chicago settlement house staffed by talented women social investigators, was central to the development of the social sciences at the University of Chicago, for example, though never fully utilized or acknowledged. Neither the politically committed and participatory styles of work, nor the gender conventions that identified women with feeling, piety, and the arts, promised scientific authority. The disciplinary project often involved colonizing these sites for academic social science. Practitioners were increasingly required to hold academic degrees, either in the social science disciplines or in practical social science fields (social work, home economics, clinical psychology, counseling) that themselves sought academic legitimacy as sciences.

---

[15] Furner, *Advocacy and Objectivity*.

Political caution narrowed the ideological range of disciplinary discourses; masculine gender norms toughened their scientific style and opened them to conventional representations of women. At the same time, women and radicals were excluded or subordinated by means of selective university appointment and hierarchical systems of credentialing. In the United States, such barriers, coupled with limited access to higher education, also excluded or subordinated racial minorities. Still, local differences, multiple institutional locations, and limited meritocratic standards allowed for important exceptions. The radical economist Thorstein Veblen, W. E. B. DuBois and a later cadre of African-American sociologists, and several Progressive cohorts of women social scientists produced pathbreaking work by reformulating established political, racial, and gender codes or by turning them to their own advantage.[16]

National differences in the structures of university systems had a major impact on disciplinary formation. Beginning in the 1870s and 1880s, Americans rapidly built a decentralized system of private and state colleges and universities, largely committed to a modernized curriculum. New fields bearing the imprimatur of science or of German systematic scholarship, *Wissenschaft*, easily gained recognition, and by 1903 economics, psychology, anthropology, sociology, and political science had founded separate national professional associations. With no traditional faculty bodies standing between professors and the university president, the new universities moved to departmental organization, solidifying disciplinary distinctions. Various combinations continued for a time as joint departments, but they became exceptions to the rule, although imperial efforts by one social science to subordinate others have never ceased.

European university faculties and facilities expanded more slowly, in consultation with traditional faculty bodies and under the tighter rein of central government agencies or, as in England, of conservative private corporations. Hence the social sciences gained fewer appointments in Europe, and newer fields struggled – often in vain – to achieve recognition. In different countries different institutional opportunities and cultural traditions produced different disciplinary outcomes. The field of geography, with roots in national educational systems and imperial ambitions, developed a stronger disciplinary identity in France, Germany, and England than in the United States. Sociology secured a notable, if temporary, academic presence in France and

[16] Eileen Janes Yeo, *The Contest for Social Science: Relations and Representations of Gender and Class* (London: Rivers Oram Press, 1996); Rosalind Rosenberg, *Beyond Separate Spheres: Intellectual Roots of Modern Feminism* (New Haven, Conn.: Yale University Press, 1982); Ellen Fitzpatrick, *Endless Crusade: Women Social Scientists and Progressive Reform* (Oxford: Oxford University Press, 1990); Helene Silverberg, ed., *Gender and American Social Science: The Formative Years* (Princeton, N.J.: Princeton University Press, 1998); Theresa Wobbe, "On the Horizons of a New Discipline: Early Women Sociologists in Germany," *Journal of the Anthropological Society of Oxford*, 25 (Michaelmas 1995), 283–97; Martin Bulmer, Kevin Bales, and Kathryn Kish Sklar, eds., *The Social Survey in Historical Perspective* (Cambridge: Cambridge University Press, 1991).

Germany during these decades, but in England remained centered in social
investigation and was tied to the generalist role of civil servant.[17] Economics
flourished in England, building on an independent and influential tradition
of political economy, but remained subordinated to law and civil service on
the Continent.

The consequences of this differentiated development included not only
a stronger institutional base for the social sciences in the United States,
but also stronger disciplinary borders. What the Europeans lost in disciplinary stability, however, they sometimes gained in richer intellectual milieux. In Europe, philosophy and history were not only competitors for
place but also avenues of advancement; that, and the greater strength of
European traditions of learning and moral reflection, kept philosophy and
history a central part of the education of social scientists even after disciplinary formation, contributing to the intellectual depth and longevity
of the models of social science achieved in Europe during these formative
decades.

BETWEEN SCIENCE AND THE HUMANITIES

Because the means to all the social scientists' disciplinary ambitions – liberal
influence, intellectual authority, and academic place – was science, the effort to construct the social sciences both drew on and sharpened the divide
between scientific and humanistic domains of knowledge.[18] Indeed, social scientists themselves had helped to create this divide, as when Auguste Comte
and John Stuart Mill formalized, in the name of positivism, an invidious
distinction between natural science and other forms of learning. Resting on
the epistemological claim that knowledge comes only from sense experience and logical mental operations, positivism held that only science could
provide valid knowledge; if the social disciplines were to be sciences, they
must develop methods similar to those of the natural sciences. The educated liberal strata who took a leading role in founding the social sciences
believed that science provided the rational tools and, within its own terms
of disinterested objectivity, the moral credo that could underwrite modern
culture.[19]

---

[17] Besides the sources cited in note 1, see Peter Wagner, Carol H. Weiss, Björn Wittrock, and
Hellmut Wollmann, eds., *Social Sciences and Modern States* (Cambridge: Cambridge University Press,
1991).

[18] For general references to this section, see the works cited in note 8.

[19] On the division of knowledge, see the Introduction to this volume; see also David Hollinger,
"The Knower and the Artificer, with Postscript 1993," in *Modernist Impulses in the Human Sciences,
1870–1930*, ed. Dorothy Ross (Baltimore: Johns Hopkins University Press, 1994), pp. 26–53. On
the scientific ethos, see David Hollinger, "James, Clifford, and the Scientific Conscience," in *The
Cambridge Companion to William James*, ed. Ruth Anna Putnam (Cambridge: Cambridge University
Press, 1997), pp. 69–83.

To a considerable degree, positive science was the rallying point for the social science projects of the 1860s, 1870s, and 1880s in England, France, and the United States.[20] In those countries, the new psychologists' turn to laboratory methods, the marginalists' appropriation of physical mechanics, the sociologists' elaboration of evolutionary laws, and the anthropologists' comparative method were believed to put their disciplines on the same footing as mathematics and the natural sciences. Even in political science, natural science lent rhetorical flourishes, and the study of law borrowed its evolutionary organizing principles. But scientism – the demand that the social sciences model themselves on the natural sciences – did not entirely command agreement, nor was it clear what counted as science or how the social scientists' policy and ethical concerns related to it. Particularly in England and the United States, the social sciences attracted men and women steeped in Evangelical piety and religious moral idealism who made science the agent of earthly salvation. In many cases, the positivistic social scientific program absorbed rather than replaced religious goals. In other cases, as in economics and sociology, "ethical" schools formed to oppose hard-line positivist programs.

Beginning in the 1880s and 1890s and continuing for several decades thereafter, these initial disciplinary paradigms were attacked from within, as methods and assumptions faced the critical scrutiny of specialists with divergent intellectual and political commitments. One major casualty was the idea that mankind had undergone a single process of evolutionary advance through uniform stages, thus sending into decline the comparative method in anthropology and evolutionary theories of social development. In economics and psychology, running battles ensued between "schools" claiming different versions of scientific rectitude.

The debate within disciplines was stimulated by and contributed to a broader critique of the possibilities of knowledge. By the end of the century, the grand systematic claims of earlier positivists were being discredited, and the understanding of science itself was being revised. Ernst Mach and Karl Pearson influentially argued that the sense experience on which positivist science relied does not provide a mirror image of reality; science provides descriptions of appearances, formulations useful for orienting ourselves in the world, not access to independently existing objects. In the neo-positivism of these critics, science remained the only genuine knowledge, and its method of abstraction and generalization remained available to the social as well as the natural sciences, but only by paring away extraneous metaphysical assumptions. Under the aegis of neo-positivism, and particularly in the United States, social scientists had concluded by the 1920s that their task was to engage in rigorously empirical investigation, to quantify wherever possible,

[20] See particularly Ringer, *Fields of Knowledge*; Ross, *Origins of American Social Science*; Stocking, *After Tylor*.

usually in the form of statistics, and to build toward a disciplinary framework of universal laws.[21]

A more radical critique of knowledge for the social sciences came from Germany, where positivism was deeply suspect. German social scientific elites were part of an educated middle class that had staked its drive for cultural power on devotion to a humanistic ideal of *Bildung*, the cultivation of character through engagement with the spiritual realm of high culture. At the end of the century, in the face of social dislocation and the growing power of the natural sciences, these elites searched for post-positivist formulations of the social sciences that would salvage both their class ethos and their scientific identity.[22]

According to the philosophers Wilhelm Windelband and Heinrich Rickert, natural science was a nomothetic enterprise that abstracted from concrete experience in order to generate laws that applied universally; thus its thrust was toward the most general aspects of experience. By contrast, the humanistic disciplines were ideographic enterprises that sought to delineate the concrete complexity of experience. Wilhelm Dilthey (1833–1911) emphasized in addition the distinctive method appropriate to the study of human subjects: If the natural objects of nomothetic science were amenable to measurement and causal analysis, human beings, by contrast, were self-conscious and self-motivated; what they did and made could be understood only in the light of their motives and the linguistically and culturally mediated meanings that defined them. For Dilthey, the social sciences were *Geisteswissenschaften*, sciences of spirit or mind; their task, like that of humanistic studies generally, was hermeneutic, to interpret the evidences of human meaning in light of the larger configurations of which they were a part.

In the two decades before World War I, Max Weber (1864–1920) steered a careful middle course through these distinctions, as through the related economic *Methodenstreit*.[23] The sociocultural studies, like history, he argued, aimed to understand concrete reality and used interpretation in order to understand the human purposes that created it. But such studies could be both causal and generalizing sciences. Motives were genuine causes, and historical generalizations, such as "feudalism" and the marginal economist's "economic man," were "ideal types" that permitted scientific analysis; their

---

[21] Theodore M. Porter, "The Death of the Object: *Fin de siecle* Philosophy of Physics," and Dorothy Ross, "Modernism Reconsidered," in *Modernist Impulses*, ed. Ross, pp. 1–25, 128–51.

[22] Fritz K. Ringer, *The Decline of the German Mandarins: The German Academic Community, 1890–1933* (Cambridge, Mass.: Harvard University Press, 1969).

[23] On the debates Weber entered into and his controverted positions, see Thomas Burger, *Max Weber's Theory of Concept Formation*, exp. ed. (Durham, N.C.: Duke University Press, 1987); Manicas, *History and Philosophy*, pt. 1; Max Weber, *Roscher and Knies: The Logical Problems of Historical Economics* (New York: Free Press, 1975); Christopher G. A. Bryant, *Positivism in Social Theory and Research* (New York: St. Martin's Press, 1985), chap. 3. A selective view that distorted the significance of Weber's historicism and value relevance became influential in the United States after World War II; see Max Weber, *The Methodology of the Social Sciences*, trans. and ed. Edward Shils and Henry Finch, Foreword by Edward Shils (New York: Free Press, 1949). See also Chapter 4 in this volume.

usefulness lay in the understandings they could generate about the concrete world. Weber also recognized that values were deeply implicated in the search for social knowledge. To most positivists, scientific knowledge was knowledge of facts only, free of the investigator's subjective values. Weber pointed out that any event or process could be characterized in an infinite number of ways. The characterization that we give to it emerges from the particular set of questions and values that we bring to it: thus our values are built into the questions that frame, and the concepts that constitute, social scientific work.

The turn-of-the-century debates over method, knowledge, and values concerned the relation between science and social activism, as well. Most nineteenth-century practitioners had integrated science, ethics, and social action. The concept of progress allowed evolutionists to embed their values and advice in their stories of historical advance. Even adherents of Mill's chaste positivist separation of science from the art of its application in policy generally found little difficulty in crossing the line. For others, lingering conceptions of natural law and faith in the divine underpinnings of the universe kept alive belief in, as the American sociologist Albion Small put it, the "moral economy of human affairs."[24] When positivism and the limitations of social knowledge came under review at the end of the century, however, just as the disciplines were attempting to secure their scientific authority in a context of heated political debate, the relationship among the social scientists' ideas, values, and prescriptions for action became highly problematical.

Like the positivists, Weber made a sharp distinction between "is" and "ought." The scientific description of reality is not enough by itself, he argued, to dictate how humans should act, and it should be kept separate from evaluative judgments. But unlike many advocates of value-free science, Weber did not erase the values built into social scientific constructs, nor did he free the social scientist from ethical-political judgments. Indeed, Weber's stance was as much a moral response to the disappearance of spiritual meaning in the modern world, an insistence that individuals make their own decisions about what to do and how to live, as it was an effort to clarify the social sciences.

Weber's strategy did not achieve agreement in his own bitterly divided milieu, nor has it since. Although often drawn into escalating political conflict and wartime emotion, academic social scientists have sought by and large to avoid explicit partisanship, which has often meant avoiding more controversial positions. This was especially true in the liberal and neo-positivist American context, where, most often, the debates of this period led social scientists to seek middling, presumably neutral, ground and to define the social scientist's activist role as that of technical expert.

The debates about knowledge and values left a more sharply divided intellectual field. In the United States, the inclination of the social sciences toward

[24] Quoted in Ross, *Origins of American Social Science*, p. 347.

natural science encouraged disciplines engaged with literature, the fine arts, philosophy, and, to a considerable degree, history to construct a counter-identity as "the humanities" around a concern for values.[25] These large oppositions, the inconclusive outcome of the turn-of-the-century debates, and the difficulties inherent in joining social reflection and practice to disciplinary science meant that positions changed and epistemological divisions continually reappeared. As Adam Kuper shows, anthropology largely shifted its domain from biological natural science to the other social sciences in the 1920s, and then, after 1970, to the humanities. Political science, as shown in James Farr's chapter, repeatedly debated the stances appropriate to the discipline's joint commitment to science and democracy. Everywhere, hermeneutic and nomothetic styles of work, value relevance and value freedom continued to claim adherents, even as national and disciplinary trends formed.

THE SOCIAL SCIENCES BETWEEN THE WARS

The social scientists' disciplinary project encountered dramatically different circumstances after 1914, with notably different outcomes in Europe, and North America.[26] In Europe, traditional university arrangements had kept the new disciplines small and scattered, and the war decimated the generation coming of age. Where institutional standing had been achieved – by economics and anthropology in England, for example – strong disciplinary traditions continued into the following decades. At other sites, attrition and postwar politics took a heavy toll. On the Continent especially, the social scientists' confidence and liberal goals were weakened by the rise of socialism, fascism, and virulent nationalism. With the advent of fascist governments in Germany, Italy, and Austria and the Second World War, the European disciplines were in many places disbanded or disrupted, although service to fascist states, as to wartime states of all stripes, provided considerable support for disciplines now offering expertise and practical research.[27]

By contrast, social science disciplines in the United States continued to expand within the growing university system. By 1920, national professional associations numbered 1,000 members in sociology, 1,300 in political science, and 2,300 in economics; and the numbers of teaching and practicing social scientists must have been larger. Each discipline became something of a subculture in its own right and, like a professional guild, provided recruits

[25] John Higham, "The Schism in American Scholarship," in his *Writing American History* (Bloomington: Indiana University Press, 1970).
[26] For basic information on the topics of this section, see the chapters on the disciplines in this volume; Smith, *Norton History*; Ross, *Origins of American Social Science*, to 1930, and Stocking, *After Tylor* and *Race, Culture, and Evolution*.
[27] Stephen P. Turner and Dirk Käsler, eds., *Sociology Responds to Fascism* (London: Routledge, 1992); Ulfried Geuter, *The Professionalization of Psychology in Nazi Germany* (1984) (Cambridge: Cambridge University Press, 1992).

with norms of behavior, patterns of preferment, and hierarchically ordered career tracks that somewhat insulated its members from outside judgment.[28]

In addition, liberal ideologies continued to dominate American politics. Still, the traumas of World War I and the disappointment of pre-war liberal hopes, in America as in Europe, made historical progress seem uncertain and strengthened a sense of historical discontinuity. Explored earlier by modernist artists and intellectuals, this sense of living in a new historical world further eroded the evolutionary systems that had framed nineteenth-century social science and moved all of the social sciences by the 1920s away from historical and toward synchronic forms of explanation. In the decades before and after the First World War, functionalist approaches developed around biological metaphors of adaptation in a wide range of fields: psychologies of learning; studies of politics, parties, and interest groups; empirical studies by institutional economists; interactionist sociologies of the Chicago School; and British functionalist anthropology.

These functionalist approaches were also inspired by scientific ambition and began the move toward statistical techniques. After World War I, and particularly in the United States, they were reshaped by a more rigorous scientism. Historians have attributed this American desire to imitate the natural sciences to the quantitative inclinations bred by individualism and democracy and to the naturalistic bias of American exceptionalism.[29] Blanket national explanations overdetermine the result, however, for considerable variation existed in American social science, as there was considerable like-mindedness in Europe. Moreover, scientism waxed and waned. Scientistic programs were most often launched in response to heated ideological controversies, when science promised to overcome, or to avoid the appearance of, partisanship. Emerging first in the formative decades of the discipline – a period of class politics and professional anxiety – scientism strengthened in the ideological wake of World War I and again following World War II.

Another key to the scientistic stringency that entered American social science during the 1920s lies in the adoption of what might be called an *engineering* conception of science. Since the eighteenth century, the social sciences had had practical purposes – that is implied in their role as guides on the path to modernity. Guidance, however, was often understood as edification or enlightenment, either of the public or, more frequently, of its leaders. The aim of direct social intervention placed social scientists in a

---

[28] Ross, *Origins of American Social Science*, p. 392.
[29] For exceptionalism and its anxieties, see Ross, *Origins of American Social Science*, chap. 2; for democracy and individualism, see Judith Sklar, "Alexander Hamilton and the Language of Political Science," in *Redeeming American Political Thought*, ed. Stanley Hoffmann and Dennis F. Thompson (Chicago: University of Chicago Press, 1998), pp. 3–13. Theodore M. Porter, in *Trust in Numbers: The Pursuit of Objectivity in Science and Public Life* (Princeton, N.J.: Princeton University Press, 1995), links quantification to democracy's suspicion of expert authority, which makes rules of calculation favored for their presumed impersonality.

more active role, and, in a culture that especially valued useful knowledge and that had invented pragmatism, practical intervention had been central to the professional and disciplinary aims of American social science from its inception.[30] But intervention could take a variety of forms. The growing need of government and bureaucratic organizations for procedural rationality, and the testing programs and statistical bureaus of World War I, provided the seedbed for engineering tools that promised prediction and control. At the same time, historical discontinuity and neo-positivism – with its confidence in science's pragmatic relation to reality – encouraged the view that science could reconstruct reality to human purposes.

Together, these attitudes began to reshape scientific language and practice, producing an engineering science that was not only oriented toward technical intervention in the world, but also fundamentally shaped by its technologies of intervention.[31] Seeking predictable manipulation and common disciplinary procedures in research, as well as practical interventions, social scientists tried to remake the "science" of their disciplines in the image of their interventionist techniques. As Theodore Porter suggests in his chapter on statistics, Pearson's neo-positivist program of statistical measurement and correlation, later augmented by regression analysis and the analysis of variance, provided the crucial mathematic tools. Advocates of more rigorous social sciences often argued in these years that construction of "basic" science was an endeavor separate from "application," but, in fact, the continuing practical aims that directed research and the similar intellectual values that governed both scientific and practical strategies erased the separation. We can locate that engineering mentality in American psychology between the wars, and identify its beginnings in sociology, political science, and economics.

The behaviorism of John B. Watson (1878–1958) first asserted the engineering program. An animal psychologist, Watson held that any resort to subjective states was illegitimate, a metaphysical inference from observable behavior. Rather, all behavior was composed of reflex responses to stimuli, conditioned by environmental learning and compounded to form complex

---

[30] David Hollinger, "The Problem of Pragmatism in American History," in his *In the American Province: Studies in the History and Historiography of Ideas* (Baltimore: Johns Hopkins University Press, 1985), pp. 23–43. For a correction of misreadings of Dewey's instrumentalism, see Robert Hollinger and David Depew, eds, *Pragmatism: From Progressivism to Postmodernism* (Westport, Conn.: Praeger, 1995), pp. 78–81, 118.

[31] For the engineering model in Progressive Era social work, see Stephen Turner, "The Pittsburgh Survey and the Survey Movement: An Episode in the History of Expertise," in *Pittsburgh Surveyed: Social Science and Social Reform in the Early Twentieth Century*, ed. Maurine W. Greenwald and Margo Anderson (Pittsburgh: University of Pittsburgh Press, 1996), pp. 35–49; for a variety of usages of the engineering metaphor, see John M. Jordan, *Machine-Age Ideology: Social Engineering and American Liberalism, 1911–1939* (Chapel Hill: University of North Carolina Press, 1994); for the engineering model as "instrumental positivism," see Bryant, *Positivism in Social Theory and Research*, chap. 5; for engineering models in the natural sciences, see Philip J. Pauly, *Controlling Life: Jacques Loeb and the Engineering Ideal in Biology* (Oxford: Oxford University Press, 1987), and Ronald Kline, "Construing 'Technology' as 'Applied Science': Public Rhetoric of Scientists and Engineers in the United States, 1880–1945," *Isis*, 86 (1995), 194–221.

patterns. The behaviorist, he announced, "wants to control man's reactions as physical scientists want to control and manipulate other natural phenomena."[32] Watson's conception of science owed less to physics than to engineering: "What was desired," Kurt Danziger suggests, "was knowledge of individuals as *the objects of intervention rather than as the subjects of experience.*"[33]

The neo-behaviorisms of Watson's leading followers – Edward Tolman, Clark Hull, and B. F. Skinner – were grounded in the same Darwinian naturalism, technocratic utopianism, and stimulus-response technology. However, they developed more sophisticated psychologies and made use, chiefly in their programmatic statements, of new strains of the philosophy of science. One was Percy Bridgman's operationalism, which identified the meanings of all scientific concepts with the concrete operations performed by scientists in verifying them, thus ruling out of science concepts that could not be reduced to such operations. The logical positivism developed by the Vienna Circle of philosophers of science posited science as a single method for all realms of knowledge and restricted science to statable empirical and logical claims. Holding the authoritative high ground, these philosophies of science added legitimacy to the behaviorist effort.[34]

Historians have shown that most psychologists, including the neo-behaviorists, did not adopt Watson's proscription of attention to mental states. But in the instrumentalist context of the 1920s, behaviorism captured the imagination of the discipline. "Behavior" rather than "mind" became the common name for American psychology's object of study.[35] At the same time, behaviorism provided the context in which the stimulus-response model became standard for research practice in psychology, applied to complex as well as reflex behavior. By the 1930s, laboratory practice had also incorporated Francis Galton's method of studying individual variation, the basic technique of the practical psychologists. From the statistical aggregate, experimenters now sought predictions of how the individual, an abstract statistical construct, responded to varied conditions. With their subjects constructed as "singular, non-communicating individuals . . . anything social or cultural could enter this world only in the form of stimuli external to the individual."[36]

While historians have noted the engineering mentality of behaviorist psychology, Mary Morgan's original interpretation in Chapter 16 of economics as an engineering science suggests that this tendency lies deeper in the social

[32] John B. Watson, *Behaviorism* (1924) (New York: Norton, 1970), p. 11.
[33] Danziger, *Constructing the Subject*, p. 67; emphasis in original. See also Kerry W. Buckley, *Mechanical Man: John Broadus Watson and the Beginnings of Behaviorism* (New York: Guilford Press, 1989).
[34] Laurence D. Smith, *Behaviorism and Logical Positivism* (Stanford, Calif.: Stanford University Press, 1986).
[35] Franz Samelson, "Organizing for the Kingdom of Behavior: Academic Battles and Organizational Policies in the Twenties," *Journal of the History of the Behavioral Sciences*, 21 (January 1985), 33–47.
[36] Danziger, *Constructing the Subject*, particularly chaps. 7–8; and Kurt Danziger, *Naming the Mind* (London: Sage, 1997), chap. 6, quotation at p. 99.

science project and in the conditions that encouraged scientism. During the interwar decades, the basis for an engineering approach was forged by both institutional and neoclassical economists, the two main varieties in a still-pluralistic discipline; for both, statistics represented a scientific method that would yield greater practical realism. Mathematics began to make substantial headway in the 1930s; the number of mathematical and econometric studies began to climb as the discipline responded to the urgent need for economic intervention created by the Great Depression and to the influx of a cohort trained in mathematics and physics. It was largely the Second World War, however, that gave economists the opportunity to develop an array of quantitative engineering tools for analyzing practical economic and allocational problems, and it was the postwar political climate that brought these tools to dominance.[37]

During the interwar decades, behaviorism, not economic technologies, was the standard-bearer for an engineering conception of science in the other social sciences and in the new Social Science Research Council. Behaviorist assumptions permeated the quantitative social research that became the hallmark of American social science: Social and political phenomena were cast as the behavior of aggregates of individuals responding to external stimuli, and thereby made subject to statistical analysis, prediction, and control. Sociology and political science mounted urgent efforts to adopt quantitative measurement and statistical procedures in research. Survey techniques were beginning to provide a technology for measuring public opinion by the 1930s, but the engineering program outran the training, tools, and market necessary for its enactment and, as was the case in economics, fully came to fruition only after World War II.[38]

The individualistic methodological assumptions imbedded in engineering science conformed well to the liberal individualism of American society, but fit as well the aims of bureaucratic organizations of various political kinds. The engineering science forged in wartime became a major support of welfare states throughout the world and was usable for managerial tasks in both democratic and authoritarian political contexts. Engineering tools were often inserted into research framed by different conceptions of science. In psychology, for example, Jamesian empiricism and the tradition of natural history provided support for ideographic study of the concrete by a group of social and personality psychologists, who nonetheless cast their research

---

[37] Mary S. Morgan and Malcolm Rutherford, "American Economics: The Character of the Transformation," in *From Interwar Pluralism to Postwar Neoclassicism*, ed. Mary S. Morgan and Malcolm Rutherford (Durham, N.C.: Duke University Press, 1998); Philip Mirowski, "The When, the How and the Why of Mathematical Expression in the History of Economic Analysis," *Journal of Economic Perspectives*, 5 (Winter 1991), 145–57.

[38] Samelson, "Organizing for the Kingdom"; Danziger, *Naming the Mind*; Ross, *Origins of American Social Science*, chap. 10; Bannister, *Sociology and Scientism*, chaps. 11–12. Cf. Jennifer Platt, *A History of Sociological Research Methods in America, 1920–1960* (Cambridge: Cambridge University Press, 1996), a nominalistic argument for the gap between programmatic statements and actual methods.

in engineering terms, thus constricting the richer humanistic frame.[39] As Ellen Herman shows in Chapter 38, the psychologists' therapeutic language sometimes worked against the engineering model and sometimes incorporated it.

Engineering science also began to construct a "mainstream" that after World War II would marginalize alternative forms of reflection. Still, mainstream hegemony was always contested by opponents with substantial support and influence. During the 1930s, a resurgence of political engagement led vocal minorities to question the scientistic style of objectivity. Hermeneutic and normative approaches existed alongside scientific ones, particularly in subfields such as symbolic interactionism and political theory. Historical perspectives especially influenced political science and institutional economics, and institutional economists were conspicuous in government and in the study of labor and business. Historically oriented theory courses were common in all of the disciplines. Particularly important was the discipline of anthropology, where Darwinian functionalism had been reshaped by the historicist and hermeneutic conception of social science of Franz Boas (1858–1942). Between the wars, American anthropologists focused on culture as the defining paradigm for their discipline and developed holistic ethnographic approaches to its study. Sociologists Robert and Helen Lynd could claim the authority of anthropology, for example, to justify their cultural study of *Middletown* (1929) to the skeptical psychologists and sociologists overseeing the project.[40]

If we move now from the United States to Europe, we move to a differently configured historical scene, although the differences were not absolute. The Great Depression set off an international search for economic technologies that were statistically and mathematically based, with notable work done in the Netherlands, Scandinavia, and France. Anglo-American social science inhabited a particularly wide domain, from hereditarian statistical investigations to social surveys and policy studies for the emerging welfare state.

Likewise, the desire to construct the social sciences as sciences remained in Europe, as in America, a defining feature of the social science disciplinary project. However, an engineering conception of science did not take hold. European social scientists developed a greater interest in empirical statistical studies and practical technologies during this period, but these carried less legitimacy in European academic milieux, and outside of economics the market for such technologies was still limited. More deeply grounded in

---

[39] Katherine Pandora, *Rebels within the Ranks: Psychologists' Critique of Scientific Authority and Democratic Realities in New Deal America* (Cambridge: Cambridge University Press, 1997); Danziger, *Naming the Mind*, chap. 9.

[40] Richard Wightman Fox, "Epitaph for Middletown: Robert S. Lynd and the Analysis of Consumer Culture," in *The Culture of Consumption: Critical Essays in American History, 1880–1980* (New York: Pantheon, 1983).

philosophy and history, and with Marxism a forceful presence as inspiration or as opponent, European social scientists occupied a wider philosophical and political range. As a consequence, they worked within conceptions of science that acknowledged greater complexity in the social world and that allowed the combination of holistic styles of analysis with naturalistic assumptions and quantitative methods. Gestalt psychology, a holistic vision of psychology and natural science, and the synthetic *Annales* school of social science history were, in different ways, signal cases in point.

CROSSING BORDERS IN INTERWAR SOCIAL SCIENCE

If one feature of interwar social science was scientism, strident in the United States and muted in Europe, a second was the intensifying movement across disciplinary borders.[41] Much of the creative work during this period was linked to the demands and opportunities produced by such traffic. Ironically, it was the weakness of disciplinary structures in Europe and their strength in the United States that stimulated border crossing.

In France, for example, the collapse of Durkheimian sociology opened the way for the more established synthetic disciplines of history and geography to take up its ambitions through the *Annales* program. Durkheim's successors moved into anthropology and sociology, maintaining strong ties to philosophy. As Claude Lévi-Strauss remarked in 1945, French sociology did not "consider itself as an isolated discipline, working in its own specific field," but rather as a "method" or "attitude" manifested in a number of related disciplines. Ultimately, the weak disciplinary focus encouraged the production of broad social theorists, such as Lévi-Strauss, Louis Dumont, and Pierre Bourdieu.[42] Much the same situation existed in French psychology, where the absence of disciplinary institutionalization encouraged the broader inquiry of the Swiss psychologist Jean Piaget.

In central and eastern Europe, the unstable succession of reactionary and radical regimes forced some academic social science disciplines to move in formalist or reactionary directions and pushed the free-market Austrian economists, who lacked an academic base, into a "bunker mentality" of fervent ideological certainty.[43] But such conditions could also produce openness

[41] For the basic information on which this section is based, see the chapters on disciplines in this volume; Smith, *Norton History*; Ross, *Origins of American Social Science*, to 1930; Stocking, *After Tylor*; Morgan and Rutherford, "American Economics"; Peter Wagner, "Sociology," in *The History of Humanity*, ed. UNESCO, vol. 7: *The Twentieth Century* (London: Routledge, forthcoming).

[42] Victor Karady, "The Prehistory of Present-Day French Sociology 1917–1957," in *French Sociology: Rupture and Renewal Since 1968*, ed. Charles C. Lemert (New York: Columbia University Press, 1981). The quotation is from Claude Lévi-Strauss, "French Sociology," in *Twentieth Century Sociology*, ed. Georges Gurvitch and Wilbert E. Moore (New York: Philosophical Library, 1945), p. 505.

[43] Claus-Dieter Krohn, "Dismissal and Emigration of German-Speaking Economists after 1933," in *Forced Migration and Scientific Change: Emigré German-Speaking Scientists and Scholars after 1933*, ed. Mitchell G. Ash and Alfons Sollner (Cambridge: Cambridge University Press, 1996), p. 188.

to interdisciplinary and heterodox work. The carnage of the war and the communist revolution in Russia stimulated left-wing intellectuals in the Weimar Republic, socialist Vienna, and communist Russia to seek new solutions from social science. The socially oriented developmental psychology of Lev S. Vygotsky and the multidimensional social research of Paul Lazarsfeld (1901–1976) were among the notable products of these distinctive interwar conditions.[44]

Another influential heterodox effort was the Institut für Sozialforschung, founded in Frankfurt in 1923 and staffed by intellectuals of Jewish descent, radical politics, and varied interdisciplinary interests. Max Horkheimer, who became their leader, refused identification with "sociology"; he understood his task to be the construction of social theory, an ongoing process of theoretical critique, supplemented and informed by empirical research. One important strain in this rich milieu was psychoanalysis. Sigmund Freud's emphasis on the determining influence of unconscious mental processes, the radical implications of his sexual theory, and his Jewish identity made psychoanalysis suspect among academic and medical specialists in Europe, but it flourished in modernist cultural circles and in the brief efflorescence of radical social theory and practice during the 1920s and early 1930s. Joining Marx to Freud and cultural analysis, Frankfurt theorists examined how individuals internalize, and culture reproduces, the power relations of capitalist society. The migration of the Frankfurt Institute to New York City in 1933 became one of the many journeys that was to alter intellectual possibilities in the United States.[45]

For American social scientists, the problem of the interwar years was not the weakness of disciplinary structures but their strength. Disciplinary form and scientific aspiration created an expectation of disciplinary unity, but in terms of theory, internal practice, and public voice, the disciplines were fractured. At the same time, they divided the human/social subject into separate and often incompatible pieces. If sociologists in the United States had firmer authority over the "social" than did their European counterparts, they gained it by abandoning large areas of the economic, political, and cultural world that might well have been within their purview. By taking the universal individual as its subject, psychology left the psychological assumptions of the separate disciplines to themselves, producing such disparate creatures as the rational, acquisitive "economic man" and the culture-bearing, norm-laden subject of anthropology. Theoretical approaches and statistical techniques developed in one discipline often crossed into others, but there took on different shapes. Each discipline asked different questions, framed by different assumptions,

[44] Cf. Smith, *Norton History*, pp. 616–22, 783–98; see also the essays on Austria, Hungary, and Poland in *Sociology in Europe*, ed. Birgitta Nedelmann and Piotr Sztompka (Berlin: Walter de Gruyter, 1993).

[45] Martin Jay, *The Dialectical Imagination: A History of the Frankfurt School and the Institute of Social Research, 1923–1950* (Boston: Little, Brown, 1973).

and these differences became traditions of discourse into which practitioners were socialized. The commonalities focused work inward, but could not produce fundamental agreement or halt the proliferation of subfields.

The differences within social science were multiplied during these years by the crossing of national borders. Intellectual exchanges through conferences, fellowships, visiting appointments, and emigration in search of opportunity resumed in the 1920s, taking scholars in both directions across the Atlantic. In the 1930s the traffic moved westward, as hundreds of social scientists emigrated from Germany and central Europe, some previously exiled from Russia, and resettled in England and North America, mostly in the United States. In psychology, perhaps 15 percent of practitioners left Germany, and a third of university professors. In all fields, the exiles were disproportionately Jewish, on the political left, and in positions marginal to the German academic establishment. These characteristics were hardly assets in the search for American academic appointment. Obtrusive left politics was probably less tolerated in the United States than in Germany, and anti-Semitism, if less virulent than in Europe, was still a major factor in hiring and preferment. The refugees' talent and the active concern of some American intellectuals and officials allowed them to make their way.[46]

The full impact of these emigrés on American social science would not be felt until the 1950s. They would have the greatest influence where their native dispositions blended with American interests and styles of work, as in the case of Lazarsfeld; but many emigrés widened the spectrum of intellectual positions available. They added weight to tendencies that had been marginalized by the reigning scientism, as, for example, Gestalt psychologists authorized American psychological "rebels" seeking more holistic perspectives. In political science, Germans substantially strengthened the fields of political theory, international relations, and comparative politics; after World War II began, they took the lead in constructing Americans' understanding of totalitarianism. In economics, a German network of "new classical" economists committed to interventionist state policies found a ready home at the New School for Social Research and in New Deal policy making. A large contingent of Austrian economists – libertarian in outlook, less likely to be Jewish – more easily found appointments in established American universities.[47]

Major support for crossing both disciplinary and national lines came from American philanthropic foundations. In Europe's state university systems, research was supported through special institutes, often organized around leading professors and often parochial in their concerns. Occasionally, private

---

[46] Ash and Sollner, eds., *Forced Migration*, Introduction and pts. 2–3; Earlene Craver, "The Emigration of the Austrian Economists," *History of Political Economy*, 18 (1986), 1–32.

[47] Krohn, "Dismissal and Emigration"; Alfons Sollner, "From Public Law to Political Science? The Emigration of German Scholars after 1933 and Their Influence on the Transformation of a Discipline," in *Forced Migration*, ed. Ash and Sollner, pp. 175–97; Pandora, *Rebels within the Ranks*. John G. Gunnell, *The Descent of Political Theory* (Chicago: University of Chicago Press, 1993), emphasizes the discordant influence of emigré political theorists on American political science.

donors, with specific political or policy aims, supported research institutes or projects, as in Frankfurt and Vienna. In the United States, private foundations became major players in social science research, and their influence reached Europe as well. Rockefeller foundations invested forty million dollars in social science during the 1920s and continued with large sums through the 1930s, while the Russell Sage and Carnegie Foundations contributed lesser amounts. The different aims of foundation trustees and social scientists were brought into working connection by foundation managers such as Beardsley Ruml and Lawrence K. Frank, themselves trained social scientists, who urged the production of scientific knowledge "which in the hands of competent technicians may be expected in time to result in substantial social control." This engineering conception of science was joined to an interest in interdisciplinary research: Disciplinary barriers, it was believed, were a major cause of the failure of the social sciences to achieve the scientific power of the natural sciences.[48]

The most important Rockefeller effort was the Social Science Research Council (SSRC), the first organization to join the range of social science disciplines. A testament more to their secure disciplinary individuality and growing self-consciousness as sciences than to any common identity, the SSRC was a means for distributing Rockefeller funds. Although it left specific research projects to investigators, it promoted new research directions – such as behaviorism – that promised an engineering science, and programs such as the University of Chicago community studies that encouraged social sciences to work together. The Rockefeller Foundation also invested in European social science, bringing individual scholars to American universities and supporting institutions that did empirical and interdisciplinary work that reflected its own conception of science, such as the London School of Economics (LSE), a major beneficiary, and the Deutsche Hochschule für Politik in Berlin, the only German representative of an autonomous political science. The foundations' financial support both consolidated the scientistic "mainstream" that was forming in the United States and contributed substantially to those outside it.[49]

Funding and the crossing of disciplinary and national borders helped to create several innovative projects in interwar social science. Initiated earlier by A. R. Radcliffe-Brown and the Polish emigré Bronislaw Malinowski, British functional anthropology announced itself as a school in the 1920s.

[48] Samelson, "Organizing for the Kingdom," quotation at p. 39; Barry D. Karl and Stanley N. Katz, "The American Private Philanthropic Foundation and the Public Sphere, 1890–1930," *Minerva*, 19 (1981), 236–70; David C. Hammack and Stanton Wheeler, eds., *Social Science in the Making: Essays on the Russell Sage Foundation, 1907–1972* (New York: Russell Sage, 1994); Ellen Condliffe Lagemann, *The Politics of Knowledge: The Carnegie Corporation, Philanthropy, and Public Policy* (Middletown, Conn.: Wesleyan University Press, 1989).

[49] On foundation funding in Europe, see Earlene Craver, "Patronage and the Directions of Research in Economics: The Rockefeller Foundation in Europe, 1924–1938," *Minerva*, 24 (1986), 205–22; Ash and Sollner, eds., *Forced Migration*, passim.

When the charismatic Malinowski, located at the LSE, convinced the Rockefeller Foundation and some colonial administrators of the practical value of functionalism for the policy of "indirect rule," Rockefeller funneled a quarter-million dollars into postdoctoral fellowships in Africa, solidifying the centrality of fieldwork for the discipline and assuring the dominance of functional anthropology in Britain. The Foundation also sent Radcliffe-Brown and Malinowski on tours of American universities, to which they subsequently returned for extended periods, and the exposure encouraged Boasian anthropologists to put greater stress on cultural integration.[50]

As Elizabeth Lunbeck shows in Chapter 39, psychiatry and psychoanalysis were also drawn during these decades toward collaborative efforts with anthropology in the study of "culture and personality," an effort that drew on emigré theorists to show how culture expressed, and was expressed in, personality. For psychologists, major centers for such work were the interdisciplinary child development research institutes funded by Rockefeller during the 1930s, and the study of personality and social psychology pioneered by Gordon Allport and Lois and Gardiner Murphy.[51] The Institute of Human Relations at Yale, funded in 1929 by Rockefeller to develop an integrated behavioral science effective in social intervention, had less success. In the seminars of the neo-behaviorist psychologist Clark Hull, psychoanalysis – like every other psychological theory – was recast into behaviorist and presumably verifiable constructs, but hopes for an integrated social science foundered on theoretical differences.[52]

Two features of these interwar movements are especially noteworthy, because they were new to American social science and because they later became salient in the postwar "American model" of social science. One feature was an interest in the unitary character of both individuals and societies and their systemic interconnections. While the interest in systems may have been rooted in the structural character of interwar political and economic problems, European concepts provided key holistic perspectives, and border crossing in an era of search for disciplinary unity encouraged synoptic views. Border crossing also encouraged confrontation on the level of theory, a focus new to the heavily empiricist, inductivist American social sciences. The reigning philosophies of science added weight to the centrality of theory construction. The logical positivists, many of them now transplanted, brought to the empiricist positivist tradition a concern for science considered as a system of

[50] On the effects of the imperialist context, funding, and cross-national influence on functionalism in anthropology, see Stocking, *After Tylor*, chap. 8.

[51] Jon H. Roberts, "The Human Mind and Personality," in *Encyclopedia of the United States in the Twentieth Century*, ed. Stanley I. Kutler (New York: Scribner's, 1996), vol. 2, pp. 877–98; Pandora, *Rebels within the Ranks*.

[52] Jill G. Morawski, "Organizing Knowledge and Behavior at Yale's Institute of Human Relations," *Isis*, 77 (June 1986), 219–42.

linguistic propositions. From a different perspective, the British philosopher Alfred North Whitehead, also teaching in the United States, urged that all observation be considered a selection in terms of a "scheme of abstraction": In science, that scheme should be a "theory . . . of an ideally isolated system" operating according to "general laws."[53]

Talcott Parsons (1902–1979) drew many of these threads together. Initially inspired by his study of institutional economics to seek a synthetic social view, Parsons chose a European graduate education focused broadly on social theory, moving from the LSE and Malinowski to studies of Weber at Heidelberg. Appointed as an instructor at Harvard, where sociology still occupied a disdained corner of the Economics Department, he looked for a general theory that would carve out a unique sphere for sociology. With no mention of previous American sociology, though adopting its individual voluntarism, Parsons used his European sources and new philosophies of science over the following decades to fashion a functionalist theory of social action that emphasized the way shared norms produce social integration. Institutionally, he moved to establish a Department of Social Relations (1946) that would include the new interdisciplinary work in psychology and anthropology. For Parsons and others, these systemic views and theoretical ambitions came to fruition in the revitalized social sciences of the postwar era.[54]

SOCIAL SCIENCE IN ASCENDANCY, 1945–1970

In the decades following World War II, a renewal of the social science disciplinary project brought the social sciences to their highest point of self-confidence and of intellectual and popular authority in the United States and around the world.[55] The liberal Enlightenment vision of a progressive modern society guided by science gained energy and urgency from the defeat of fascism, the disintegration of colonial empires, and the threat of communism. University systems expanded and democratized, providing a vibrant academic base for the social science disciplines, and the market for

---

[53] Whitehead quoted in Charles Camic, "Introduction: Talcott Parsons before *The Structure of Social Action*," in Talcott Parsons, *The Early Essays*, ed. Charles Camic (Chicago: University of Chicago Press, 1991), p. xxxiv.

[54] On Parsons and his role in American sociology, see Camic, "Introduction"; Howard Brick, "Society," in *Encyclopedia of the United States in the Twentieth Century*, ed. Kutler, vol. 2, pp. 917–40; Howard Brick, "Talcott Parsons's 'Shift away from Economics', 1937–1946," *Journal of American History*, 87 (September 2000), 490–514.

[55] For the basic information on which this section is based, see the chapters on disciplines in this volume; Smith, *Norton History*; Morgan and Rutherford, "American Economics"; Wagner, "Sociology"; *American Academic Culture in Transformation: Fifty Years, Four Disciplines* [philosophy, literary studies, political science, economics], *Daedalus*, 126 (Winter 1997); Wagner, Wittrock, and Whitley, eds., *Discourses on Society*; Meinolf Dierkes and Bernd Biervert, eds., *European Social Science in Transition* (Frankfurt: Campus Verlag; Boulder, Colo.: Westview Press, 1992); A. W. Coats, ed., *The Post-1945 Internationalization of Economics* (Durham, N.C.: Duke University Press, 1996); Burton R. Clark and Guy R. Neave, eds., *Academic Disciplines and Indexes*, vol. 4 of *The Encyclopedia of Higher Education* (Oxford: Pergamon Press, 1992).

social science services, cultivated by wartime governments, grew during post-war reconstruction. As the strongest power to emerge from the war and a society that had escaped fascism and communism, the United States promoted its ideologies and cultural products around the world. United States government agencies, private foundations, universities, and disciplinary organizations supported extensive exchange of social science faculty, students, and books. American models of social science were selectively imported into countries outside the Soviet sphere of influence, while the American model of graduate education that linked teaching and research was often emulated. At the same time, the disciplinary form of American social science was encouraged by UNESCO, which organized international disciplinary organizations, whose members were to be national disciplinary organizations. Many countries where such disciplines had hardly existed formed national societies in response, including L'Association Francaise de Science Politique (1949) and the British Sociological Association (1950).

In time, national governments added massive support for the social sciences. By the late 1930s, the Rockefeller Foundation had become disillusioned with the practical results of social science philanthropy and had begun to withdraw from the field. The new Ford Foundation took up the slack, and in the United States private foundations continued to be the major source of research funds for the social sciences until the late 1950s. At that point, governments became more important. As recovery proceeded, European and Japanese governments extended substantial support as well, much of it through government-organized research institutes. In the United States, the federal government invested, on average, a billion dollars a year during the 1970s, the peak decade of support, shifting the greatest portion of research funds into commissioned research, where the research agenda was set by the granting agency. For university research understood as basic, the National Science Foundation (NSF) became the major player. Given its natural science core, and the intensely anticommunist climate of the Cold War, the NSF established an official standard for social science research of "objectivity, verifiability, and generality," which tended in practice to mean methods modeled on the natural sciences and politically acceptable practical purposes.[56]

The social sciences that developed in the United States – and were exported abroad – thus magnified the scientism of the interwar decades while inscribing American values; but they also assimilated into this American framework some of the migrating European perspectives. The postwar social sciences were cast as theories of integrated systems, now a hallmark of science. A new neoclassical economics absorbed Keynesian theory, loosely joining micro- to

[56] Dean R. Gerstein, R. Duncan Luce, Neil J. Smelser, and Sonja Sperlich, eds., *The Behavioral and Social Sciences* (Washington, D.C.: National Academy Press, 1988), Appendix A: "Trends in Support for Research in the Behavioral and Social Sciences"; Daniel Lee Kleinman and Mark Solovey, "Hot Science/Cold War: The National Science Foundation after World War II," *Radical History Review*, 63 (1995), 110–39, at p. 124.

macroeconomics. Parsons attempted a universal structural-functional theory of social systems and offered to incorporate the other social sciences as subsystems. These other disciplines meanwhile propounded a systems theory of American pluralist politics, a holistic concept of culture, the psychosocial integration of culture and personality, and neo-behaviorism in psychology. As Marie-Claire Robic shows in this volume, even geography shed its identity as a synthetic study of the geographical world and regrouped around abstract theories of spatial interrelations. Retaining a basis in individualistic, voluntarist premises, these theories examined the structures that integrated individuals into systematic wholes, such as personality, role, norm, status, and bureaucracy, as well as the costs of functional failure, such as social strain and deviance. In America's consolidating liberal society, functionalist systems, and the imported concepts of classical sociology, Freudian psychology, and Keynesian economics took on new relevance. The anthropologist Margaret Mead, the psychologist B. F. Skinner, and the sociologist David Riesman reached a wide public by addressing the tensions between social cohesion and individualism.

If theory provided one leg of postwar social scientific authority in the United States, the other was the explosion of engineering technologies that governed empirical research and professional practice. As the chapters in Part IV show, social science techniques for managing, surveying, testing, and evaluating spread through every area of American life during the postwar decades. Psychology, with its multiplying clinical specialties, supplied probably the largest number of practitioners. In line with the era's theories, behavioral social research methodologically endowed individuals with autonomy, while substantively enmeshing them in a world of increasing social complexity.

In the 1950s climate of Cold War scientism and burgeoning professional practice, engineering technologies came to dominate theoretical and applied research in psychology and economics. In psychology, the statistical format of independent and dependent "variables" replaced stimulus-response as the standard model of research practice, perpetuating the behaviorists' individualistic, reductionist, and technocratic style and reaching for a technical unity among the discipline's diverse fields.[57] In economics, modeling became the characteristic feature of the dominant neoclassical paradigm. Simplified models of the neoclassical economy provided both practical tools in mathematical theoretical research and, when fitted to empirical data, the primary tools for policy research, reshaping the discipline into an engineering science. As Morgan notes, "It was the simplistic quality of such models, particularly the smaller ones, with their effective reduction of complexity and their ability to produce answers explainable in terms of rather simple propositions of economic efficiency and rationality," that led to their

---

[57] Danziger, *Naming the Mind*, chap. 9.

widespread use.[58] Research organized around "variables" also dominated the empirical work of sociologists and the behavioral program of political scientists as they moved into survey methods and statistical analysis of voting or of social-psychological "traits." Behavioral science even made some inroads into anthropology, where it supported a statistical program in comparative anthropology begun before the war.

Through translation of structural-functional concepts into behavioral variables, theory was sometimes linked to these methods, notably by Robert Merton and Lazarsfeld at Columbia, thereby offering the promise of interdisciplinary convergence in "behavioral science."[59] Herbert Simon worked to develop a mathematically based behavioral science linked to economics rather than to functionalist social science, drawing on defense and RAND corporation research, including operations research, game theory, organization and systems theory, cybernetics, and artificial intelligence.[60] The social sciences also played an important role in another, looser form of interdisciplinary integration, the new area study programs promoted by Cold War concerns.

The attraction of American theories and methods varied in different European countries and across disciplinary areas, though in all cases American patterns were "translated" rather than imitated, in some places reinforcing historic traditions, in others energizing a deliberate break with the past. Perhaps the best fit between American paradigms and local purposes occurred in Sweden. There, economists had long had strong international ties, chiefly to European countries, although interest in American economics had risen even before the Second World War. The war accelerated a shift away from the German-language orbit, while the rapid growth of English as a second language that began during the 1950s facilitated study, research contacts, and publication in the United States. By 1990, about 90 percent of economics dissertations in Sweden were written in English. Some economists, such as Gunnar Myrdal, had taken up sociological topics before the war and also developed American contacts. With its roots in government investigation, empirical sociology, along with economics, was seen as an important tool of the expanded welfare state. Swedish social democrats found American structural-functionalism serviceable for their own vision of a harmonious egalitarian society.[61]

[58] Morgan, "Economics," p. 301 of this volume. On the importance of the Cold War in this development, see also Craufurd D. Goodwin, "The Patrons of Economics in a Time of Transformation," in *From Interwar Pluralism to Postwar Neoclassicism*, ed. Morgan and Rutherford.

[59] Cf. Bernard Berelson, "Behavioral Sciences," in *International Encyclopedia of the Social Sciences*, ed. David L. Sills (New York: Macmillan, 1968), vol. 2, pp. 41–5.

[60] Hunter Crowther-Heyck, "Herbert Simon: Organization Man" (PhD dissertation, Johns Hopkins University, July 1999).

[61] Coats, ed., *The Post-1945 Internationalization of Economics*, particularly p. 389 and Bo Sandelin and Ann Veiderpass, "The Dissolution of the Swedish Tradition," pp. 142–64; Katrin Fridjonsdottir, "Social Science and the 'Swedish Model': Sociology at the Service of the Welfare State," in *Discourses on Society*, ed. Wagner, Wittrock, and Whitely.

Elsewhere in Europe, as in Sweden, the influence of American sociology peaked during the 1960s with the expansion of universities and welfare states. After the New Left student revolt in 1968 and American intervention in Vietnam, however, American sociology faced a Marxist and New Left backlash as well as more limited institutional growth. In France and Germany, this double dynamic did not prevent sociology from becoming a center of postwar creativity. The German revival of social theory led by Niklas Luhmann, Jürgen Habermas, and the returned Frankfurt school had a distinctly German idiom, though adaptations of American pragmatism and structural-functionalism played some part. France, too, after 1970 forged a distinctive sociological tradition out of its own Durkheimian and philosophical traditions and a variety of international sources. This development, like the strengthening of *Annales* social history, was aided by newly funded postwar research institutions: the Centre National de la Recherche Scientifique, and the "Sixieme section" of the École Pratique des Hautes Études.[62]

Following much the same pattern as sociology, American-style political science also expanded during the 1960s, only to face critiques from the left thereafter. In West Germany, American-style political science was imported as a support for democracy, as against past totalitarianism and the resurgence of conservative, normative theory. In Britain, the long Oxbridge tradition of philosophical and historical study of politics opposed the new import – forcing, for example, the creation of a national Political *Studies* Association.[63]

Italy offered a still less welcome environment for American social sciences. With sharper antagonisms between left, right, and center, and with a system of university governance that allowed little disciplinary autonomy, even economics faced cross-pressures and divisions similar to those elsewhere experienced by the more fragile disciplines of sociology and political science. University expansion, and with it opportunities for the social sciences, did not occur until after the crisis of 1968, and hence in a Marxist climate of impatience for structural change. A mix of native, European, and American varieties of social science in time shared the Italian field.[64]

Even where American social science found a European base, however, exponents often disengaged the empirical approach of American disciplines from

[62] Lemert, ed., *French Sociology*, pt. 1; Richard Munch, "The Contribution of German Social Theory to European Sociology," in *Sociology in Europe*, ed. Birgitta Nedelmann and Piotr Sztompka (Berlin: Walter de Gruyter, 1993). For similar patterns in social psychology, see Klaus R. Scherer, "Social Psychology Evolving: A Progress Report," in *European Social Science*, ed. Dierkes and Biervert; Pieter J. van Strien, "The American 'Colonization' of Northwest European Social Psychology after World War II," *Journal of the History of the Behavioral Sciences*, 33 (Fall 1997), 349–63.

[63] Hans Kastendiek, "Political Development and Political Science in West Germany," in *The Development of Political Science*, ed. David Easton, John G. Gunnell, and Luigi Graziano (London: Routledge, 1991); Malcolm Vout, "Oxford and the Emergence of Political Science in England, 1945–1960" in *Discourses on Society*, ed. Wagner, Wittrock, and Whitley.

[64] Pier Luigi Porta, "Italian Economics through the Postwar Years," in *The Post-1945 Internationalization of Economics*, ed. Coats; Luigi Graziano, "The Development and Institutionalization of Political Science in Italy," in *Development of Political Science*, ed. Easton, Gunnell, and Graziano.

its scientizing pretensions and ideological thrust. As one observer of the development of political science in Europe has noted: "American political science is grounded in liberal individualism.... By contrast, European politics is grounded in collective concerns, whether socialist, conservative or *etatist*. Institutions make the subject of study different."[65] The situation is similar in social psychology: The North American approach highlights "individual functioning with respect to social input or content," while the Europeans focus "on the effects of social embedding on all aspects of human performance."[66] French sociology, too, influenced by its existentialist and Marxist context, developed around a problematic of "structural constraints on practical action," a problematic that led sociologists to study such topics as order, change, structure, practice, power, and class relations, but largely to ignore such American concerns as deviance and values.[67]

THE SOCIAL SCIENCE PROJECT CHALLENGED, 1970–2000

After 1970, the bases not only of the social sciences' postwar authority, but of the disciplinary project itself, were severely challenged.[68] The postwar politics that had renewed the social sciences' liberal goals were shattered by the political crises of the 1960s in the United States and Europe. By the 1970s, the civil rights movement, political conflicts, youth revolt, and the rise of feminism challenged the smooth liberal functionalist premises of sociology and political science and the blinkered individualism of psychology, while postcolonial developments cast doubt on the legitimacy of the anthropologist's gaze. Then, as politics in the United States and parts of Europe moved to the right after 1980, government funding of social science research declined. The social sciences associated with social democratic sympathies and statist policies, already weakened from the left, were challenged by new political and intellectual centers on the right, and the challenge gained momentum as communism collapsed.[69]

The social scientists' project was buffeted not only by political shifts but also by long-accumulated discontents with modern society. The most vocal critics repudiated the liberal Enlightenment vision of modernity guided by science and technocracy, declaring it to be monolithic and coercive, and sought alternative, postmodern bases for individual freedom. Theoretical attacks on positivism and new linguistic critiques of knowledge fueled the postmodern vision and worked more broadly to reopen fundamental questions

[65] Wittrock, "Discourse and Discipline," p. 269.
[66] Scherer, "Social Psychology Evolving," pp. 184–5.
[67] Lemert, ed., *French Sociology*, pp. 26–7, 41.
[68] For basic information, see the sources cited in note 55.
[69] Charles H. Page, "The Decline of Sociology's Constituency," *History of Sociology*, 6 (Fall 1985), 1–10.

about the viability of the social science disciplines and their relationship to science and the humanities.[70]

These challenges increased the divisions within the social sciences. Except in economics, none of the new theories of the 1950s became paradigmatic in their disciplines, and they left behind disciplines divided by schools and subfields, interdisciplinary overlays, and technological practices. The intense ideological and theoretical debate of these decades spawned new fields, such as the study of women and gender, and sharpened discord, so that subfields often went their own way, rarely communicating with each other or contributing to a common matrix. The sheer size of the disciplines encouraged fragmentation. In 1995, there were over 80,000 doctoral psychologists in the United States and 76,000 holders of doctorates in the other social science disciplines.[71]

Not the least source of the postmodern critique and of disciplinary pastiche was the changed social world itself: a hybrid world that seemed to confirm perspectival conceptions of knowledge. At the same time native traditions of social science, themselves the product of cross-national influences, were becoming part of international social science networks. With the rise of English as a world language and of American cultural dominance, fears of Americanization were often voiced. Still, as A. W. Coats said of economics, the most homogenous international discipline, a universal science that obliterates national schools remained a "chimera," and in the other social sciences, national problematics and disciplinary diversity created greater variety.[72] The newly formed European Economic Community competed for attention with the United States, as did continentwide social science institutions in Latin America and Africa. If there were international disciplinary communities, hybrid American, European, and native models of social science jostled within and outside their boundaries.

These challenges to the viability of the social science project produced diverging responses. In the late nineteenth century, the social science disciplines had pulled away from biology, historicism, and economics in an attempt to form autonomous disciplines; in the late twentieth century, these alternative bases returned, mounting transdisciplinary programs to reclaim the social science field. Standing on opposite sides of the divide between scientific and humanistic domains of knowledge, they pulled in very different directions. Social scientists who turned to economics and biology, disputing or ignoring postmodern critics, sought to renew the social science project on the firmer bases of mathematics and natural science, while those

[70] See Richard J. Bernstein, *Beyond Objectivism and Relativism: Science, Hermeneutics, and Praxis* (Philadelphia: University of Pennsylvania Press, 1985) and his *The New Constellation: The Ethical-Political Horizons of Modernity/Postmodernity* (Cambridge, Mass.: MIT Press, 1991).

[71] *Doctoral Scientists and Engineers in the United States: 1995 Profile* (Washington, D.C.: National Science Foundation, 1998).

[72] A. W. Coats, "Conclusion," in *The Post-1945 Internationalization of Economics,* ed. Coats, p. 396.

who turned to historicism, absorbing the critique, sought to reconstitute the project on a more defensible basis. The prevalence of scientism aggravated the crisis and the responses to it in the United States: Both the antipositivist reaction – sharpest in anthropology – and the renewal of scientism were stronger there than in Europe.

Economics was one of the chief claimants to the social science project, especially in the United States, where it emerged from these decades in a stronger disciplinary position than the other social sciences. Neoclassical economics integrated micro- and macroeconomics around a free-market paradigm and expanded the use of mathematics and engineering tools in theory and practice. Through the study of economic growth, path dependency, economic history, and a new institutionalism, some economists attempted to stretch or diversify the paradigm, but most graduate programs succumbed to what Neil DeMarchi called the mathematizing "juggernaut."[73] Moreover, the political shifts that battered the other social sciences served to benefit economics. In the United States, the left radicalism of the 1960s had little influence in economics, while the conservative and libertarian politics of the following decades rewarded rational choice theory and the generally antistatist neoclassical mainstream of the discipline. With its abstract reach into any kind of choice under constrained conditions, economics colonized other social sciences, carrying the authority of scientific advance that, between the wars, had belonged to psychological behaviorism.

Biology extended into the domain of the social sciences through a number of interdisciplinary fields, such as neuroscience, sociobiology, ecological theories, and population genetics. Cognitive psychology and rational choice theory formed links to Darwinian theory, constructing a paradigm for the psychological and social world congruent with that of the economists, but very different from the more social and social-structural focus of the European social sciences.[74]

At the same time, the new authority of hermeneutic and historicist philosophies enabled humanistic psychologies, historical sociology, anthropological textualism, cultural studies, political theory, and the institutional study of politics to take deeper hold. Historicism offered not a paradigm but a philosophy that grounded the social sciences, like other humanistic studies, in their diverse, historically and ethically based problems. As Jacques Revel notes of the *Annales* historian-social scientists, historicism suggested a return to Max Weber's effort to construct sciences of the historical world. On that ground, disciplinary fragmentation might be transformed into a genuine pluralism, where engineering technologies do not preempt critical reflection.

---

[73] A. W. Coats, "Report of Discussions," in *The Post-1945 Internationalization of Economics*, ed. Coats, pp. 383–6.

[74] On Darwinian realignment, see Gerstein, Luce, Smelser, and Sperlich, eds., *The Behavioral and Social Sciences*.

No one transdisciplinary move dislodged the others or shattered entrenched disciplinary institutions, although they did place additional strains on disciplinary boundaries.[75] Unlike the postwar paradigms based on such unitary disciplinary objects as society, state, and economy, the moves toward economics, biology, and historicism shifted the focus to processes that transcend disciplinary boundaries; unlike earlier interdisciplinary efforts, they led to the transfer of methods. From other corners of the academic domain, where postmodern views stressed the intermingling of economic, social, political, and cultural power, sociologists contemplated the "death of the social," and political scientists the dispersion of "the political" into all areas of study. As engineering sciences purveying an array of research and practical technologies, and as collections of diverse, often contradictory, kinds of social knowledge, the social sciences at the end of the twentieth century were not the coherent disciplines or the rational liberal compass for modernity that had been originally projected. They were nonetheless deeply embedded in the modern world and, like it, still very much in transit.

---

[75] See *Open the Social Sciences: Report of the Gulbenkian Commission on the Restructuring of the Social Sciences* (Stanford, Calif.: Stanford University Press, 1996).

14

STATISTICS AND STATISTICAL METHODS

Theodore M. Porter

Statistics assumed its recognizably modern disciplinary form during the period from about 1890 to 1930. These dates are comparable to those for the formation of disciplines in the leading fields of social science. Statistics, however, changed during this period from an empirical science of society, as it had been during the nineteenth century, into a mathematical and methodological field. Although it disappeared as a social science per se, as an area of applied mathematics it became an important source of tools, concepts, and research strategies throughout the social sciences. It also provided legitimacy for, and contributed to a redefinition of what would count as, social knowledge.

In its nineteenth-century incarnation, as itself a social science, statistics was guided by a different set of ideals – not academic detachment, but active involvement in administration and social reform. The social science of statistics was practically indistinguishable from government collection of numbers about population, health, crime, commerce, poverty, and labor.[1] Even its most self-consciously scientific advocates, such as the prominent Belgian astronomer and statistician Adolphe Quetelet (1796–1874), often had administrative responsibility for the organization of official statistics. This alliance of scientific and bureaucratic statistics did not disappear abruptly. But it was gradually subordinated to a new order in which statisticians assumed consulting roles, offering their expertise to statistical agencies but also to many others. At the end of the nineteenth century, it still appeared possible that statistics might succeed in the universities as a quantitative social science. Instead, it was recreated as a mathematical field.

Even as a branch of applied mathematics, statistics remained a close and indispensable ally of social science. In this guise, it was one of the seven constituent fields that in 1925 made up the new (United States) Social Science Research Council. Revealingly, statistics was identified from the beginning as a promising basis for interdisciplinary cooperation. Statistical methods

---

[1] See the introduction to Part I of this volume.

were increasingly perceived by social scientists as bearing the authority of mathematics, while government statistical offices were reduced to sources of data. Statistical mathematics during the twentieth century stood for objectivity and technical rigor – which, paradoxically, were particularly valued in connection with practical and applied research. The development of the new statistics was closely associated with that of professional social science, as well as with the biology of populations and, at their intersection, with eugenics. From about 1930 to 1970, statistical analysis became almost mandatory for empirical or experimental research throughout social science, with the partial exception of ethnographic and clinical work. It was important also for a wide range of applied and professional studies, including agriculture, medical testing, education, engineering, surveys of all kinds, and business administration.[2] Its history is not one of the autonomous development of a mathematical specialty, but rather of a panoply of alliances and interactions that promoted the development of mathematical tools and stimulated new methodological ambitions.

ESTIMATION AND ERROR

The word "statistics" (*Statistik*) was coined during the eighteenth century in Germany to designate a descriptive science of the state. The genre might be compared to that of the modern encyclopedia article on a country or state. The quantitative study of populations and economies went under a different name, "political arithmetic." As background to the methods now called statistical, political arithmetic contributed more than the old *Statistik*. By 1700, demographic numbers and the theory of probability had begun to be brought together for purposes of calculating annuities and life insurance premiums. During the 1770s and 1780s, just before the French Revolution, sophisticated probabilistic mathematics was developed to estimate populations, as of France. There was at the time no census, but the law required a recording of births and deaths. What was then needed to determine the total population was a multiplier, the ratio of population to births (or to deaths). Mathematicians hoped to approximate this number using full counts of a few smaller populations. Pierre-Simon Laplace (1749–1827), the greatest probability mathematician of his age, showed how to calculate the error to be expected for any given sample size, or (turning the problem around) how many people to count in order to attain, with a specified probability, a certain degree of precision. These estimates, however, presumed that the chosen town or towns could be taken as representative of the whole of France. Laplace understood that this was not strictly valid, but, since the problem had no mathematical solution, he said little about it.

[2] Theodore M. Porter, *Trust in Numbers: The Pursuit of Objectivity in Science and Public Life* (Princeton, N.J.: Princeton University Press, 1995), chap. 8.

In Laplace's time, high functionaries were interested in what mathematicians could tell them about population measures.[3] As full censuses became standard during the nineteenth century, probabilistic estimates fell from favor. The new quantitative science of statistics, which developed from about 1820 to 1850, owed much to the tradition of political arithmetic, but it kept its distance from probability theory. A conscientious statistician was one who insisted on a complete count, the only way to insulate statistics from mere speculation. A few mathematicians continued to advocate the use of probability to estimate errors in population counts, but they had little role in the concrete business of (social) statistics.

During the nineteenth century, the use of probability theory in the analysis of data became above all the business of astronomy and the sciences of the observatory. Often the astronomer or surveyor had many measures of the same quantity, or a cloud of observations that had to be reduced to a single line. In 1805, the mathematician Adrien-Marie Legendre proposed the "method of least squares" for solving problems of this kind. A few years later, Laplace and Carl Friedrich Gauss worked out ways to ground the method in probability theory. For the rest of the century and beyond, this problem of data reduction supported a tradition of mathematical investigation and refinement, providing important mathematical background for statistics as it developed beginning in the 1890s. The connections of error theory to nineteenth-century social statistics were modest. Astronomical observation was, however, linked to the form of experimental psychology, called psychophysics, announced by Gustav Theodor Fechner (1801–1877) during the 1850s. Fechner used least squares to process his data and even wrote a treatise on the study of "mass phenomena." His work launched a continuing tradition of statistical analysis in psychology. Another distinguished pioneer of statistical methods at the intersection of astronomy and psychology was the American philosopher and metrologist Charles Sanders Peirce.[4]

STATISTICAL MODELS OF REGULARITY AND VARIATION

The quantitative science of (social) statistics, as it developed during the 1830s and 1840s, was practically oriented to address questions of disease, poverty, and crime. When "statists" spoke about method, they tended to emphasize the solidity of numerical facts, which were often supposed to speak for themselves, rather than tools of analysis. They insisted on complete counts instead of estimates, and rarely mentioned probability theory. Quetelet argued tirelessly that probability was needed to raise the standard of statistical practice, but his examples were highly abstract, and he did not use probability

[3] Eric Brian, *La mesure de l'État: Administrateurs et géomètres au XVIIIe siècle* (Paris: Albin Michel, 1994).
[4] Stephen Stigler, *The History of Statistics: The Measurement of Uncertainty before 1900* (Cambridge, Mass.: Harvard University Press, 1986).

methods to analyze social numbers in his own writing. It was not in order to process data, but rather as a model or theory, that he put probability to work. He supposed that the physical and moral characteristics of the individuals who make up society are shaped by a multitude of small causes, including nutrition, schooling, religion, and laws. The combination of these circumstances accounts for the physical features of men and women, such as height and weight, and also for moral traits, which he expressed as probabilities or "propensities" – to marry, to commit crime or suicide, to perform acts of heroism, to write books, and so on. The random differences among individuals are swamped at the level of society, where it is not variation but the average that prevails. Quetelet personified this as *l'homme moyen*, the average man.

On this foundation he built a statistical order. The life of "man," he argued, is characterized by an unfailing pattern of "statistical laws," "astonishing" regularities from year to year in the numbers of births and deaths, marriages, murders, thefts, and suicides. Regularity in the biological order was unsurprising, because natural, but Quetelet and many of his contemporaries were shocked at the stability revealed by judicial statistics, which the French government began to publish during the late 1820s. He wondered whether immoral and criminal acts were produced by some mysterious fatality rather than by human free will. In the end, he explained them as characteristics of "society" rather than of individuals. In proposing a new version of social science, he also articulated the most fundamental principle of statistical reasoning. It is possible to build a coherent science at the level of the collective by attending only to frequencies or rates without seeking causes of individual behavior.[5]

Quetelet's statistical version of social science posed a series of problems for his successors. The doctrine of statistical law retained its ability to shock for several decades, and indeed became all the more controversial after the English historian Henry Thomas Buckle expressed it in a particularly provocative way in a popular work of 1857, *The History of Civilization in England*. English moralists were bothered by this ostensible challenge to free will and moral responsibility. German statisticians criticized Quetelet for dissolving individuals into society and also, conversely, for reducing society to a sum of individuals. They saw him as ignoring the distinction between human freedom and mechanical law, and even as denying, through his social determinism, the possibility of improving society by reforming laws and institutions. The stability of statistical series thus became a serious issue, whose dimensions were moral and social as well as quantitative. These debates were particularly lively in Germany, where statistics flourished as an academic field – a social

---

[5] Theodore M. Porter, *The Rise of Statistical Thinking, 1820–1900* (Princeton, N.J.: Princeton University Press, 1986), chap. 2; Ian Hacking, *The Taming of Chance* (Cambridge: Cambridge University Press, 1990), chap. 14.

discipline allied to the sciences of state – for several decades after about 1860. These questions of statistical law provided the first occasion for using tools of probability theory for the analysis of social numbers. The defining issue was not inference, measurement, or uncertainty, but the relation of individual action to the collectivity.

Wilhelm Lexis (1837–1914), the most accomplished of the mathematical statisticians in Germany, wrote a series of papers and a short book on this broad topic beginning in 1875. He interpreted Buckle as implying a degree of regularity in social actions greater than could be explained by mere chance – a stabilizing force, or mysterious fatality, governing human moral actions. If suicides or murders were independent random events, like coin tosses or throws of dice, the expected regularity of the numbers from year to year was a purely mathematical problem. A combinatorial formula from Laplace's student Siméon-Dénis Poisson defined for him the standard of statistical regularity, which he called "normal dispersion." Buckle, he supposed, had claimed "subnormal dispersion" for acts like suicide, but the empirical returns gave no example of this, and hence no support for any mysterious fatality. Only the ratio from year to year of male to female births, a result from biology rather than from social science, was consistent with the model of independent chance events. Virtually every statistical series involving moral actions showed annual variability much greater than chance – that is, "supernormal dispersion." Hence, he thought, these series could not be comprehended in terms of the basic laws of probability. Lexis understood society as something complex, no mere sum of its parts, and as composed of fundamentally diverse individuals rather than of Quetelet's average ones. He explained its structure in broadly probabilistic terms, as a system of many groups characterized by inhomogeneous probabilities. But he did not try to assign numbers to these chances, and in the end his program for probability-based statistics was quite limited.

STATISTICAL MATHEMATICS: CORRELATIONS AND REGRESSIONS

Statistics as a field of applied mathematics arose principally out of biological rather than social or economic investigations. Francis Galton (1822–1911), younger cousin of Charles Darwin, questioned the scientific status of ordinary statistical compilations, but he praised Quetelet's mathematical program. Galton's admiration was firmly linked to his use of the astronomer's error law (the distribution he dubbed "normal"), which Quetelet, following Laplace, introduced as the limit of a binomial. (It could be approximated by the probability distribution of the outcomes of a thousand coin tosses.) Just as any astronomical observation was subject to many small errors, which could be positive or negative, so too would humans vary randomly from the mean on account of climate, disease, nutrition, and the health of the mother. As a consequence, most human traits (e.g., height, the circumference of the chest) and even moral characteristics should be distributed within a population in approximate accordance with the bell curve. That is, Quetelet construed human variation as mathematically identical to error. He explained this discovery as proof that the average man was an authentic type. Galton was fascinated by his work for just the opposite reason. Quetelet's curve was a tool for investigating biological variability, which Galton, following Darwin, now recognized as the raw material of evolutionary change. Galton was effectively the founder of eugenics, a program for improving humanity through selective breeding rather than social reform. He was interested not in mean values, but in the tails of the error distribution, where exceptional individuals were gathered. Why are statisticians so often content with averages, he wondered? "Their souls seem as dull to the charm of variety as that of the native of one of our flat English counties, whose retrospect of Switzerland was that, if its mountains could be thrown into its lakes, two nuisances would be got rid of at once."[6]

Having failed in his first efforts, during the 1860s, to launch eugenics as a reform campaign – it would eventually take off during the 1890s – Galton devoted the intervening decades to biological investigations of heredity. He was almost unique in pursuing a statistical approach to these questions, and also in rejecting entirely the Lamarckian mechanism of use inheritance. It was easier to experiment on plants than on people, and Galton, like the still-unknown Gregor Mendel, chose peas. He also assembled family records to provide evidence of heredity in people. He learned that the offspring of exceptional parents tend to "revert" or "regress" toward the mean, and he worked out the elementary mathematics of this relationship. "Regression" thus arose not as a statistical method but as a biological law. Yet Galton regarded statistics also as a set of tools of wide applicability, and this was reinforced when he found that his mathematics of heredity applied also to the problem of "correlation." His prototype for correlation involved relations of bodily measurements, *correlation* as a measure of the tendency for height and length of arm to vary in the same direction.

Galton was delighted when better mathematicians than he, such as the economist Francis Edgeworth and the applied mathematician Karl Pearson (1857–1936), took an interest in his methods. Pearson was also won over to eugenics, and soon was devoting his immense energy to promoting the statistical study of evolution. He formed a "biometric laboratory" and then, with funds from Galton, a eugenic one, at University College London. He attracted students from all over the world who wanted to learn his methods. In 1901, in collaboration with his colleague, the biologist W. F. R. Weldon, and with Galton's support, he formed a journal, *Biometrika*, devoted to the

---

[6] Francis Galton, *Natural Inheritance* (London: Macmillan, 1889), p. 62.

study of life (*bios*) through the use of calculation and measurement (*metron*). Statistics was still very far from being a discipline in 1900, but Pearson gave it intellectual and institutional coherence. His work and that of his students, including W. S. Gosset (1876–1937) (known to history as "Student") and George Udny Yule (1871–1951), provided the crucial point of reference in statistics for more than two decades beginning about 1895.[7]

Pearson's group worked out some very specific mathematical tools, including the chi-square test in 1900 and the Student's t-test in 1908. But this was a program with a mission, not a miscellany of techniques. Pearson had been won over to statistics just as he was finishing an influential work of philosophy, his *Grammar of Science* (1892). This was a radically positivistic work, which presented the world as being full of variability, so that entities like "atom" and even "circle" were valid only as abstractions. He cast doubt on the concept of causality, which he regarded as merely a way of summarizing experience. That is, he seemed already to be interpreting the world through the lens of statistics, and his philosophy was eminently suited to his emerging statistical program. He was not a strong advocate of experimentation, holding rather to the social-statistical project of investigating mass phenomena using very large numbers of observations. He was keenly interested in evolutionary and eugenic issues, especially in the question of nature and nurture, and he deployed his methods to measure their relative contributions to human ability and success. Statistics was for him a measurement discipline, the basis for a new form of expertise, and a crucial resource for "the modern state."

Pearson considered biology to be the proper basis for social science, but his statistical methods were also put to work by many others who did not. Yule, who was at odds with his teacher for much of his life, calculated the correlations of various social factors with poverty, as a contribution to what was by then called sociology. Other students worked on the statistics of public health or of criminology. Pearson's biometric methods, however, were developed and applied most self-consciously in economics and psychology. In the 1930s, scholarly societies were formed around statistical methods in these areas and given names that clearly alluded to Pearson's project: the Econometric Society and the Psychometric Society.

Econometrics, as Mary Morgan has shown, derived above all from statistical investigations of the business cycle. In 1932, the worldwide economic depression gave a crucial impetus to the founding of the Econometric Society. This was a highly international body, and included people trained in physics and mathematics as well as in economics. They aspired, with a commitment that has rarely been matched, to join theory to statistics; they wanted explanations of the economic slump, not mere correlations. Pearson's

---

[7] Stigler, *History of Statistics*; Porter, *Rise of Statistical Thinking*; Donald MacKenzie, *Statistics in Britain, 1865–1930: The Social Construction of Scientific Knowledge* (Edinburgh: Edinburgh University Press, 1981).

positivistic philosophy, very attractive to an earlier generation of American social scientists, was less appealing in these circles. Yet economists relied heavily, as Pearson had, on large data sets, often collected by governments. Much work in economic statistics concerned time series: annual numbers or indexes of growth, production, wages, prices, and unemployment. Here, replication with controls was impossible, or at least very difficult, and every analysis was complicated by changes in the economy over time. Causation remained elusive with regard to purely econometric studies.[8]

In the period after World War II, empirical economics came increasingly to mean running regressions. This became cheaper and easier to do as social scientists gained access to ever more computer power. Typically, an economist might undertake to measure the effects of education on wage levels, for example, by modeling income as a function of years of education and some other clearly relevant factors, such as age, sex, and one or more geographical variables. Solving the equation meant finding weights to assign to these variables in order to "explain" as much of the variance in the data as possible. Our specimen regression might lead to the conclusion that each year of education beyond eighth grade corresponds to an increase in salary of a certain number of dollars. A regression could not, however, distinguish the effects of education from those of prior differences in ambition, intelligence, or opportunity that led some but not others to pursue higher education. A debate between theoretical and statistical economists during this period centered on the charge that these statistical methods were indiscriminate, "measurement without theory." But the alternative, said their defenders, seemed to be "theory without measurement." As a practical matter, this form of econometrics was highly successful. Since the 1950s, such regressions have become standard tools in the social science disciplines, including, eventually, sociology and political science as well as economics. One might even argue that they have reshaped these fields.

Psychometrics arose primarily from educational testing, especially from efforts to measure intelligence. This was mainly an American endeavor, but with important European sources. Alfred Binet (1847–1911), canonized in the United States as the French pioneer of IQ testing, disdained what he regarded as the number fetish of American psychologists, and on occasion even denied that intelligence was the sort of thing that could be measured at all. Yet he also measured skull size as an indicator of mental ability, and his standard set of questions to assess the intelligence of schoolchildren enabled him at least to place them on a scale, if not to assign them a number. He used his tests, though not mechanically, to determine whether poor school performance was due to intellectual retardation; this remained for him a clinical decision.

[8] Mary Morgan, *The History of Econometric Ideas* (Cambridge: Cambridge University Press, 1990); Judy Klein, *Statistical Visions in Time: A History of Time-Series Analysis* (Cambridge: Cambridge University Press, 1997).

Another resource was provided by Charles Spearman (1863–1945), an Englishman. He defined and called attention to *g*, or general intelligence, which was not for him a mere basis for measuring and classifying, but a defense of the unity and integrity of the human mind. His statistics followed the English biometric tradition, though at the time of his first important papers in 1904 he was a student at Wilhelm Wundt's psychological laboratory in Leipzig. Spearman used correlation measures to demonstrate the interconnections of human mental abilities, as revealed by school success in subjects such as Latin and mathematics. Soon he began to develop a new statistical method of "factor analysis" to demonstrate that all of them were dependent on a unified entity behind the particular faculties, his *g*. His critics, notably the Chicago psychologist Louis L. Thurstone (1887–1955), subsequently inverted his method to decompose *g* into its factors, trying in this way to make it disappear.[9] In the United States, Binet's style of questions and Spearman's statistics became elements in a systematic program of mental measurement, which came together during and after the First World War. John Carson's chapter in this volume surveys the development of this program and how it was used to sort students in American schools. The crucial point here is that mental testing was associated with a distinctive set of statistical tools, deriving from English biometry, and also that this practical project contributed immensely to the expanding role of statistical tools in psychology generally.[10]

STATISTICAL MATHEMATICS: SURVEYS AND SAMPLES

The history of survey methods as tools of inquiry, planning, and intervention is introduced here in chapters by Eileen Janes Yeo and Susan Herbst. During the twentieth century, sampling became standard also in academic social science, especially in sociology and political science. While its basic mathematics can be found in Laplace, the practical problems of survey sampling were manifold, and nineteenth-century statisticians generally eschewed it. The strategies of generalizing from part to whole were promoted instead among opponents of statistics, most notably by Frédéric Le Play and his school in France. Beginning in 1895, the Norwegian statistician A. N. Kiaer (1838–1919) began to discuss "representative sampling" at meetings of the International Statistical Institute, and also began to use it in his own country. It was for him a question not of mathematics, but of identifying typical or representative regions. The "purposive" selection of representative individuals or groups remained appealing, and it took many decades to persuade census

[9] Gail A. Hornstein, "Quantifying Psychological Phenomena: Debates, Dilemmas and Implications," in *The Rise of Experimentation in American Psychology*, ed. Jill G. Morawski (New Haven, Conn.: Yale University Press, 1988), pp. 1–34; Olivier Martin, *La Mesure de l'esprit: Origines et développement de la psychométrie, 1900–1950* (Paris: L'Harmattan, 1997).

[10] Kurt Danziger, *Constructing the Subject: Historical Origins of Psychological Research* (Cambridge: Cambridge University Press, 1990).

bureaus, polling companies, and the like of the advantages of probability-based surveys.

Probabilistic sampling was appealing, however, because it provided a well-structured mathematical method of generalizing from hundreds or thousands of interviews to a whole population, and of estimating the range of error. For about two decades after 1906, the probabilistic approach was identified with the British social scientist Arthur L. Bowley (1869–1957), who defended random sampling and introduced its mathematics. Statisticians, at least, came to regard a 1934 paper by Jerzy Neyman (1894–1981) as having settled the matter.[11] Neyman, who came to London from Poland to study with Karl Pearson, is known for importing some of the analytical rigor of Continental mathematics into the more pragmatic biometric tradition. This important paper reveals another dimension of his activity, a close involvement with official statistics that remained more central to the work of statisticians in eastern and southern Europe than in Britain or America. Neyman set up this mathematical consideration of alternative sampling procedures very concretely as a problem of securing unbiased numbers from thousands of variably sized packets of Polish or Italian census forms.[12] Academic social scientists found these arguments convincing. In time, and to a certain degree, so did political pollsters and their ilk. In the postwar social sciences, statistical methods helped to define the standard of practice in social and political surveys.

STATISTICAL MATHEMATICS AND EXPERIMENTAL DESIGN

Beginning early in his career, R. A. Fisher (1890–1962) was at odds with Karl Pearson. Fisher was a strong mathematician, yet he also practiced statistics in a very earthy and pragmatic way, having spent the most productive years of his career, from 1919 to 1933, at an agricultural experiment station in Rothamsted, England. He was also a leading figure in the "evolutionary synthesis" of Mendelian genetics and biometric statistics. As a statistician, Fisher emphasized the importance of performing suitably designed experiments in order to get beyond correlation to the identification of causes. Gosset, in his capacity as an employee of the Guinness Brewery, had already gone some way in the direction of experimental statistics. Pearson was not notably enthusiastic, joking that only naughty brewers would draw conclusions from such small numbers.

[11] Alain Desrosières, *The Politics of Large Numbers: A History of Statistical Reasoning* (Cambridge, Mass.: Harvard University Press, 1998), chap. 7.
[12] Jerzy Neyman, "On the Two Different Aspects of the Representative Method: The Method of Stratified Sampling and the Method of Purposive Selection," *Journal of the Royal Statistical Society*, 97 (1934), 558–606.

Fisher expressed his mature statistical vision in his 1935 book, *The Design of Experiments*. A proper experiment required controls, which must be selected randomly rather than purposively. In a simple Fisherian agricultural experiment, a patch of ground would be divided into similar blocks, and these assigned to experimental and treatment groups at random. The experimental plots might receive a fertilizer, such as bone meal. Since crop yields always vary from one patch to another, often for unknown reasons, the comparison of just one fertilized block to an unfertilized block would be an entirely unreliable guide. Fisher treated the blocks as independent units in a statistical design. Since they had been assigned to treatment groups at random, he could apply the mathematics of chance. His method of analysis took this form: to compare (for example) the actual difference of yields with the difference that might be expected one time in twenty, even if the fertilizer were entirely ineffective. If the observed difference exceeds that standard, one can say that the "null hypothesis" (of no effect) can be rejected at the 0.05 level. The fertilizer has then passed what is called a "test of significance." Fisher preferred where possible to test not just one factor, but many, and he developed analytical methods suitable to experiments with multiple variables.

This was an experimental protocol for dealing with irrepressible variation. The physicist's ideal experiment, by contrast, involved tight control of every factor but one, so that a single result could be decisive. In the human sciences, as in crop studies, this often was not possible. The controlled statistical experiment was created for fields such as experimental psychology and therapeutic testing in medicine. Some aspects of Fisherian experimentation, including the use of randomization, had long since been practiced in some areas of educational psychology, and in parapsychology.[13] Psychologists adopted Fisher's program very quickly. But his methods were not suited to all forms of psychology. Gestalt psychology, for example, appealed to immediate perceptual experience – as in the line drawing of a duck that can also be seen as a rabbit – in a way that had little to do with statistics. In the new statistical regime, such psychologies were marginalized, especially in the United States. If, however, the object was to determine how levels of lighting affected industrial productivity, or whether a new style of reading instruction improved average student performance, then the Fisherian experiment was a perfect model. Psychology textbooks offered new versions of his agricultural methods, leaving out the manure, and quickly reshaped the discipline.[14]

Psychology, perhaps the most enthusiastically statistical of all the social or human disciplines, has responded to most statistical innovations, beginning

[13] Trudy Dehue, "Deception, Efficiency, and Random Groups: Psychology and the Gradual Origination of the Random Group Design," *Isis*, 88 (1997), 653–73; Ian Hacking, "Telepathy: Origins of Randomization in Experimental Design," *Isis*, 79 (1988), 427–51.

[14] Gerd Gigerenzer, Zeno Swijtink, Theodore Porter, Lorraine Daston, John Beatty, and Lorenz Krüger, *The Empire of Chance: How Probability Changed Science and Everyday Life* (Cambridge: Cambridge University Press, 1989), chaps. 3, 6.

with psychophysics and the least squares method. It has also developed important new methods and techniques of its own. In the 1950s, Fisher's "analysis of variance" and "f-tests" rapidly became dominant in experimental psychology, especially (again) in the United States. The controlled experiment might seem less promising for the other social sciences, whose object is a whole society or economy rather than the thinking or behavior of the individual. But the methodology of controlled trials has more recently been applied under the auspices of national governments to questions of social policy. Usually it is children, the poor, or criminals who are investigated in this way. What consequences can we anticipate from a "negative income tax," or from work requirements for welfare recipients, or from state provision of methadone (or even heroin) to addicts? Some of these "experiments" are inadvertent, the result of conflicting policies that social scientists can then investigate. Others have been planned and coordinated by experts in experimental design, often at the level of cities or neighborhoods.[15]

THE STATISTICAL ETHOS IN SOCIAL SCIENCE

The field of mathematical statistics has, since Pearson's time, been marked by serious division and sometimes by bitter controversy. Paradoxically, it also has often been supported as the one right way. Fisher's ideal of experimental design was challenged by Neyman, in alliance with Karl Pearson's son Egon Pearson, and later by Bayesians interested in the subjective dimension of probability. Textbook authors in the social sciences almost never mentioned these differences, but rather introduced a compromise version, usually involving a Fisherian test of significance, and called it simply "statistics." Statistics was, for them, a jealous method that brooked no alternatives. Since the time of Karl Pearson, and even of Laplace, probability and statistics have been associated with the idealization of "scientific method," which was supposed to replace fallible human judgment. Pearson held up science as the subjection of personal interest to what is valid for all, and Fisher exalted statistical tests as a democratic alternative to the fading authority of aristocrats. Social science has construed statistics as something unitary, and valued it as an indispensable tool for producing objectivity.[16]

The history of statistical methods is an international one. Important statistical traditions developed in India, Australia, Russia, Scandinavia, and the Netherlands as well as in the major countries of Europe and North America. The most prominent statisticians from the late nineteenth century to 1935 were British, yet the statistical impulse in social science was consistently

---

[15] Trudy Dehue, "Establishing the Experimenting Society: The Historical Origin of Social Experimentation according to the Randomized Controlled Design," *American Journal of Psychology*, 114(2001), 283–302.

[16] Gigerenzer et al., *Empire of Chance*; Porter, *Trust in Numbers*.

strongest in the United States. Statistics became an important ingredient in what Dorothy Ross calls here the disciplinary project. Like that project itself, statistics was not confined within disciplinary boundaries. Statistical methods provided a degree of unity for social science, even if they assumed distinctive forms within the various disciplines and subdisciplines. They also embodied an ethos, which came to be widely shared across the disciplines. The reverence of social scientists for statistics enshrined a vision of personal renunciation and impersonal authority in the name of higher truths and public values.

15

PSYCHOLOGY

Mitchell G. Ash

Psychology occupies a peculiar place among the sciences, suspended between methodological orientations derived from the physical and biological sciences and a subject matter that extends into the social and human sciences. The struggle to create a science of both subjectivity and behavior, and the related effort to develop professional practices utilizing that science's results, provide interesting examples of both the reach and the limits of such scientific ideals as objectivity, measurability, repeatability, and cumulative knowledge acquisition. In addition, psychologists' struggles to live by such ideals while competing with others to fulfill multiple public demands for their services illuminate both the formative impact of science on modern life, and the effects of technocratic hopes on science.

The aim of this chapter is to sketch the results of a broad shift in the historiography of psychology over the past twenty years, from the achievements of important figures and the history of psychological systems and theories to the social and cultural relations of psychological thought and practice.[1] In the process, I hope to bring out the interrelationships of psychological research and societal practices both with one another and with prevailing cultural values and institutions in different times and places, while at the same time attempting to bring out certain common threads in this varied narrative.

---

[1] For a summary of this shift, see Laurel Furumoto, "The New History of Psychology," in *The G. Stanley Hall Lecture Series*, vol. 9, ed. Ira S. Cohen (Washington, D.C.: American Psychological Association, 1989). For comprehensive overviews, see Roger Smith, *The Fontana History of the Human Sciences* (London: Fontana, 1997) (in the United States, *The Norton History of the Human Sciences*); Kurt Danziger, *Constructing the Subject: Historical Origins of Psychological Research* (Cambridge: Cambridge University Press, 1990) and his *Naming the Mind: How Psychology Found Its Language* (London: Sage, 1997). Informative efforts to incorporate recent work while retaining a more traditional narrative are: Ludy T. Benjamin, Jr., *A History of Psychology: Original Sources and Contemporary Research* (New York: McGraw-Hill, 1988); Thomas H. Leahey, *A History of Psychology*, 3rd ed. (Englewood Cliffs, N.J.: Prentice Hall, 1992); Ernest R. Hilgard, *Psychology in America: A Historical Survey* (New York: Harcourt Brace Jovanovich, 1987).

One of those common threads is that the history of psychology has been a continuous struggle by multiple participants to occupy and define a sharply contested, but never clearly bounded, discursive and practical field. The emergence and institutionalization of both the discipline and the profession called "psychology" are often portrayed as acts of liberation from philosophy or medicine, but these efforts to establish scientific and professional autonomy have never completely succeeded.

A second common thread is that the history of psychology as a science and that of the psychological profession are inseparable, at least in the twentieth century. Scientistic discourse and professional practice have been linked together in the use of metaphors and methods of prediction and control. But in other ways, too, enhanced public attention to particular social problems has led to the development of new methodological instruments, such as intelligence tests and personality inventories, that have had significant feedback effects on research.

A third common thread of psychology's history is that while psychologists have struggled to establish their work as internationally recognized science, they also have drawn upon local traditions. As a result of such efforts, the contents of both the discipline and the profession have varied according to particular social and cultural circumstances in ways that do not easily conform to grand narratives of progressive knowledge acquisition and practical success.

The first two parts of this chapter focus on the creation and contested identity of the scientific discipline called psychology in Europe and the United States from 1850 to 1914. The third and fourth parts outline the multifaceted struggle for dominance within the discipline and the contested professionalization of the field until 1945. The final part examines the impact of American dominance in both scientific and professional psychology during the postwar era.

ROUTES TO INSTITUTIONALIZATION, 1850–1914: ENGLAND AND FRANCE

A fundamental claim of recent research is that the emergence of psychology as a distinct subject of scholarship during the seventeenth and eighteenth centuries did not lead automatically to its institutional separation from philosophy. Some criteria for the existence of a discipline – that it be taught as a subject in schools, with journals and practitioners, a subject matter, and intended methods of study – were indeed met to some extent and in some places during the eighteenth century. In addition, conceptual frameworks from that period, such as the system of psychological faculties (thought, feeling, and will) and associationism, continued to shape psychological discourse through the nineteenth and – in the case of associationism – into the twentieth century.[2] But none of this, including the recognition of psychology as a teaching field within philosophy and pedagogy, led to continuous representation of the subject as a natural science in professorships designated for that purpose, or to the institutionalization of empirical research training in psychology, as opposed to systematic discussion in textbooks.

The widespread introduction of so-called physiological or natural scientific psychological discourse in the middle third of the nineteenth century, and the institutionalization of laboratory instruction on the model recently established in the natural sciences in the last third, had a major impact on the subsequent development of psychology as a discipline. The transition from empirical to experimental psychology was hardly complete or easy. Moreover, as will be shown, even after its establishment experimental psychology never succeeded in dominating the entire discipline.

The institutionalization of scientific psychology took quite different forms in different parts of Europe and the United States. Indeed, the components of now-standard psychological research procedure were drawn from a variety of approaches, each of which was rooted in a particular set of social and cultural circumstances. Nowhere in Europe was academic institutionalization a simple or straightforward affair; and nowhere in Europe or America did the process lead inevitably or directly to occupational professionalization.

England was the home of the statistical research practices pioneered by Francis Galton (1822–1911). These practices targeted not psychological processes assumed to be essentially similar within all individuals, but rather distributions of performances *among* individuals.[3] Galton first presented this approach in two books, *Hereditary Genius* (1869) and *Inquiries into Human Faculty* (1883), where he attempted to show, first, that physical and mental capabilities are quantitatively distributed in the same way, and, second, that both are therefore inherited to the same (large) extent. Charles Spearman (1863–1945) took the approach a step further by distinguishing in 1904 between "general intelligence" – a factor underlying all performances in a test series and presumed to be hereditary – and so-called 's' factors accounting for differential performance on specific tests, presumed to be teachable. In this work, Galton, Spearman, and others identified themselves as members of an educated elite concerned to protect its status in a democratizing society by instantiating the qualities it valued as the ones to be selected for – in eugenical marriages, school grading, and the like. By the second third of the twentieth century, this group-data approach had become the predominant research mode in both academic and applied psychology in the English-speaking world, for reasons to be discussed.

---

[2] Gary Hatfield, "Wundt and Psychology as Science: Disciplinary Transformations," *Perspectives on Science*, 5 (1997), 349–82, and his "Psychology as a Natural Science in the Eighteenth Century," *Revue de Synthese*, 115 (1994), 375–91.

[3] Danziger, *Constructing the Subject*.

But this outcome was by no means clear in 1900. Rather, while Galton and his followers advanced anthropometric testing and other quantitative data-gathering techniques, such as the questionnaire method, as well as the statistical treatment of results, philosophers such as James Ward and G. F. Stout followed their predecessors John Stuart Mill and Alexander Bain in constructing systematic psychologies that continued the theoretical traditions established by English empiricist and Scottish "commonsense" philosophy, while departing from them in certain respects.[4] Neither these initiatives nor the establishment of journals like the *British Journal of Psychology* in 1909 led to academic institutionalization; as late as the 1920s, there were only six university chairs for psychology in England. Psychological practitioners of various kinds far outnumbered academics in the membership of the British Psychological Association at the time of its founding in 1901 and for decades thereafter.[5]

The strongest advocates of scientific psychology in France at this time, the philosophers Hyppolite Taine and Théodule Ribot (1838–1916), shared a coherent vision of the field as a synthesis of medical and philosophical approaches. The coexistence of clinical "exceptional case" studies, based on Claude Bernard's idea that illness is a form of "adduced" natural experiment, alongside controlled or "induced" experimentation remained a distinguishing feature of French research.[6] But institutional fragmentation made it difficult to realize this integrative vision. The first university course in psychology, taught by Ribot at the Sorbonne in 1885, was located in the Faculty of Letters rather than in the Faculty of Sciences or Medicine. Ribot had already introduced the "new" psychology in France in the 1870s with his books on British and German developments, and he continued to argue that scientific psychology belonged to biology, not to philosophy. However, his course included no laboratory instruction, other than demonstrations at laboratories associated with the Faculty of Medicine. His appointment to a chair at the prestigious Collège de France in 1888 brought no change in this situation. Nonetheless, he encouraged younger figures, such as the physicians Pierre Janet and George Dumas as well as the biologist Alfred Binet (1857–1911), to adopt a natural scientific approach.[7]

After studying with the neurologist Jean Martin Charcot, Binet attempted to establish an explicitly biological science of higher mental processes. In 1894, he succeeded the physiologist Henri Beaunis as director of the first

---

[4] G. F. Stout, *Analytic Psychology*, 2 vols. (London: Allen and Unwin, 1909); *Brett's History of Psychology*, ed. and abr. R. S. Peters (London: Allen and Unwin, 1962), pp. 675–86.

[5] Leslie S. Hearnshaw, *A Short History of British Psychology, 1840–1940* (London: Methuen, 1964); Nikolas Rose, *The Psychological Complex: Psychology, Politics and Society in England, 1839–1939* (London: Routledge, 1985).

[6] Jacqueline Carroy and Regine Plas, "The Origins of French Experimental Psychology: Experiment and Experimentalism," *History of the Human Sciences*, 9:1 (1996), 73–84.

[7] John I. Brooks III, "Philosophy and Psychology at the Sorbonne, 1885–1913," *Journal of the History of the Behavioral Sciences*, 29 (1993), 123–45; Laurent Mucchielli, "Aux origines de la psychologie universitaire en France (1870–1900): enjeux intellectuels, contexte politique, réseaux et stratégies d'alliance autor de la 'Revue Philosophique' de Théodule Ribot," *Annals of Science*, 55 (1998), 263–89.

psychological laboratory in France, which had been founded in 1889 and was located in the Sorbonne's Faculty of Sciences. Also in 1894, he established France's first scientific psychological journal, *L'Année Psychologique*, which at first he largely wrote himself. In 1895 he published a research program that he called "individual psychology," the conceptual basis of which was the conversion of mental faculties into biological functions.[8] However, the laboratory attracted few students, and Janet rather than Binet was appointed to succeed Ribot at the Collège de France in 1902.

Lobbying in the Ministry of Education through a group he headed called the Society for the Scientific Study of the Child brought Binet the official commission that led to his publication, with Theodore Simon, of the first intelligence tests in 1905. The tests' purpose was not to measure intelligence directly – Binet doubted that this was possible – but rather to establish practical criteria for separating "subnormal" from "normal" children, in order to provide the former with special education. But this effort to fulfill practical needs by scientific means did not lead to an institutional breakthrough. The tests were not widely used in France because of the opposition of schoolteachers; and there, as in England, extensive academic institutionalization of psychology did not result.[9]

ROUTES TO INSTITUTIONALIZATION, 1850–1914: GERMANY AND THE UNITED STATES

Germany is generally regarded as the homeland of scientific psychology. An often-told scientific success story leads from Johann Heinrich Herbart's program for the measurement of sensations (in response to Kant's claim that mental events, lacking the attribute of space, could not be measured), by way of Hermann Helmholtz's measurement of the speed of nervous impulses and Gustav Theodor Fechner's psychophysics (the measurement of relations between external stimuli and just-noticeable differences in sensation), to Wilhelm Wundt's (1832–1920) "physiological psychology." However, the picture is more complicated than this. In the 1870s, systematic psychologies derived from Herbart, Rudolph Hermann Lotze, Franz Brentano, and others shared a crowded stage with Fechner's psychophysics and the *Völkerpsychologie* launched in 1860 by Moritz Lazarus and Heymann Steinthal, which took an ethnological, linguistic, and historical approach.[10] Wundt is

[8] Alfred Binet and Victor Henri, "La Psychologie individuelle," *L'Année Psychologique*, 2 (1895), 411–65.

[9] Theta Wolf, *Alfred Binet* (Chicago: University of Chicago Press, 1974); William H. Schneider, "After Binet: French Intelligence Testing, 1900–1950," *Journal of the History of the Behavioral Sciences*, 28 (1992), 111–32.

[10] Franz Brentano, *Psychologie vom empirischen Standpunkt* (Leipzig: Duncker & Humblot, 1874); Moritz Lazarus and Haim Steinthal, eds., *Zeitschrift für Völkerpsychologie und Sprachwissenschaft*, 20 vols. (Berlin: Dümmler, 1860–90). Cf. Geroge Eckard, ed., *Völkerpsychologie – Versuch einer Neuentdeckung* (Weinheim: Beltz Psychologie-Verl.-Union, 1997).

celebrated as the founder of the world's first continuously operating psychology laboratory at the University of Leipzig in 1879, but experimenters
such as Georg Elias Müller (in Göttingen) and Carl Stumpf (in Halle, Munich, and Berlin) worked parallel to, not as imitators of, Wundt, pursuing at
times quite different research programs.

More important than these distinctions, however, are the common cultural
assumptions embodied in the organization and content of psychological
research practices in Germany. In Wundt's laboratory, in contrast to the
situation in Britain and France, experimenter and subject were generally equal
in status and often changed roles. They employed mechanical apparatus to
control and thus objectify stimulus presentation, and their knowledge claims
were universal; but Wundt's coworkers and competitors all supplemented
their data charts with extended records of their subjects' self-observations,
thereby showing themselves to be engaged in an instrument-aided version
of the self-discovery traditional to members of the German educated middle
classes. Disputes between Wundt and Stumpf during the 1880s about the
expert status of trained subjects indicated that both the content of, and
control over, such research practices remained disputed terrain.[11]

The establishment of an infrastructure, including journals (such as the
Zeitschrift für Psychologie und Physiologie der Sinnesorgane, founded by
Hermann Ebbinghaus [1850–1909] and others in 1890) and the Society
for Experimental Psychology (founded in 1904, with Müller as its first
chairman), the frequent assertions by Wundt and others that psychology had
finally become an autonomous science, and Ebbinghaus's famous claim that
"psychology has a long past but only a short history" all suggest that the "new
psychology" was on firm ground in Germany by 1905. However, there was no
agreement on the subject matter or method of the discipline. As William Stern
put it in 1900, there were "many psychologies, but no one new psychology."[12]

One reason for this was what Kurt Danziger has called the "positivist repudiation" of Wundt by a younger generation of experimenters, including
Ebbinghaus and Müller, intent on extending apparatus-driven experimental techniques and quantitative presentation of results from sensation and
perception to higher mental processes, such as memory.[13] A second area of
disagreement was the effort, opposed by Wundt, to refashion laboratory
techniques into professional practices, for example, to assess the veracity of
witnesses' testimony in court, to test the performance of schoolchildren at
different times of day, and to assess the skills of industrial workers.[14] The

[11] Adrian Brock, "Was macht den psychologischen Expertenstatus aus?," *Psychologie und Geschichte,*
2 (1991), 109–114.
[12] William Stern, "Die psychologische Arbeit des 19. Jahrhunderts," *Zeitschrift für pädagogische
Psychologie,* 2 (1900), 414.
[13] Danziger, *Constructing the Subject,* chap. 3.
[14] Wolfgang G. Bringmann and Gustav Ungerer, "Experimental versus Educational Psychology:
Wilhelm Wundt's Letters to Ernst Meumann," *Psychological Research,* 42 (1980), 57–74.

third and most important area of dispute was the persistence of an explicitly humanistic philosophical tradition with competing conceptions of the subject matter, methods, and practical uses of psychology. In *Elemente der Völkerpsychologie* (1911) and other works, Wundt himself denied that experimental methods were sufficient to study the higher mental processes and produced his own, explicitly humanistic *Völkerpsychologie*.

The controversy sharpened at the turn of the century, as neo-Kantians such as Heinrich Rickert and Wilhelm Windelband excluded natural scientific methods and explanatory principles from psychology in principle, while the phenomenological philosopher Edmund Husserl and others attacked a (variously defined) epistemological and logical "psychologism."[15] Open conflict broke out in 1912, when over 110 German teachers of philosophy signed a public statement opposing the award of any further professorships in the field to experimental psychologists. But this protest failed, because state officials responsible for funding new positions remained unpersuaded that the discipline had any obvious link to professional or state civil service training.[16] The result was that, until the Nazi era, experimental psychologists in Germany maintained their own laboratories, journals, and association, but generally continued to compete for chairs in philosophy.

Wundt's American students and others rapidly transferred the new "brass instrument psychology" from Germany to the United States during the 1880s and 1890s, but the positivistic concepts employed by Edward Bradford Titchener (1867–1927) and others to justify using such tools were quite different from Wundt's. The sheer size of the country and the decentralized structure of the emerging American university, with its collegially organized departments in place of one-man institutes, facilitated rapid institutionalization. By 1910 there were more psychological laboratories in the United States than there were universities in Germany. The founding of the American Psychological Association in 1892 preceded that of the American Philosophical Association in 1904.[17]

This rapid growth masks continuity with the past as well as disagreement on the scope and methods of the new discipline in America. Indigenous roots included instruction in psychology as part of the required philosophy courses taught by college presidents, such as James McCosh (1811–1894) at Princeton — courses taken by many of those who later advanced the "new" psychology. These courses and their teachers encouraged an orientation toward moral

---

[15] For the varieties of "psychologism," see Martin Kusch, *Psychologism* (London: Routledge, 1995).

[16] Mitchell G. Ash, "Psychology in Twentieth-Century Germany: Science and Profession," in *German Professions, 1800–1950*, ed. Geoffrey Cocks and Konrad H. Jarausch (New York: Oxford University Press, 1990), pp. 289–307.

[17] John M. O'Donnell, *The Origins of Behaviorism: American Psychology, 1870–1920* (New York: New York University Press, 1985), chap. 3; Charles R. Garvey, "List of American Psychological Laboratories," *Psychological Bulletin*, 26 (1929), 652–60; Michael M. Sokal, "Origins and Early Years of the American Psychological Association, 1890–1906," *American Psychologist*, 47 (1992), 111–22.

issues and concentration on useful knowledge, rather than the emphasis on empirical foundations for philosophy of mind prevalent in Germany.[18]

Equally formative of the discipline in America, albeit for different reasons, was the work of Darwin and Spencer. Evolutionary thinking reinforced the emphasis on biological functions versus mental faculties, and also led to an emphasis on development and thus an interest in the psychology of children and animals. These trends also existed in Europe; but in the American context, the evolutionary concept of function made human adjustment appear to be a natural continuation of organic adaptation. Such views supported evolutionary theories of cognition such as those of James Mark Baldwin, while at the same time granting psychologists so inclined the authority to intervene in social practice as agents of species betterment.[19] Education and child study thus came to be of central concern to American psychology; here John Dewey, G. Stanley Hall, and Edward Thorndike were the opinion leaders, though they advanced different research and reform programs.

Disagreements on the proper scope and methods of psychology in the United States were similar in some respects to those in Germany. Thus, the members of Titchener's informal group of "experimentalists," which began to meet apart from the APA in 1904, were not opposed to applied work per se, but insisted on employing rigorous methods both within and outside the lab. Their explicit aim was to standardize the behavior of "normal" experimental subjects; the implicit, not always intended, result was to produce a knowledge instrument prepared for technological use.[20] By contrast, activists such as Hall, who pioneered the use of questionnaires in the United States and was perhaps the best-known public advocate of the "new" psychology in America, were less concerned with laboratory-style rigor than with translating moral issues into scientific ones, motivated by a concern for progressive reform.

The case of intelligence testing combined social reform and technocracy. Though Binet's tests were not widely accepted in France, they quickly became popular in the United States after Henry H. Goddard, the director of a training school for so-called feeble-minded children, propagated them in nearly messianic terms as instruments of human betterment.[21] After the success of Goddard and others, Lewis M. Terman revised the Binet–Simon scale for use in American schools in 1915, and later extended it to studies of

---

[18] Dorothy Ross, *G. Stanley Hall: The Psychologist as Prophet* (Chicago: University of Chicago Press, 1972); Michael M. Sokal, ed., *An Education in Psychology: James McKeen Cattell's Journal and Letters from Germany and England, 1880–1888* (Cambridge, Mass.: MIT Press, 1981); Graham Richards, "'To Know Our Fellow Men to Do Them Good': American Psychology's Enduring Moral Project," *History of the Human Sciences*, 8:3 (1995), 1–24.

[19] O'Donnell, *Origins of Behaviorism*, chaps. 4–5; Robert J. Richards, *Darwin and the Emergence of Evolutionary Concepts of Mind and Behavior* (Chicago: University of Chicago Press, 1987).

[20] Deborah J. Coon, "Standardizing the Subject: Experimental Psychologists, Introspection, and the Quest for a Technoscientific Ideal," *Technology and Culture*, 34 (1993), 757–83.

[21] Leila Zenderland, *Measuring Minds: Henry Herbert Goddard and the Origins of American Mental Testing* (Cambridge: Cambridge University Press, 1998).

the gifted. Terman's linking of "mental age" to another hierarchical, linear order – school class years – proved well suited to American schools in their role as sorters of a socially and ethnically diverse population.[22]

William James (1842–1910) attempted in his own way to combine science and reform. Himself an evolutionist in certain respects – and, as such, a major contributor, along with Dewey and James Rowland Angell, to the creation of a distinctly American functional psychology – James also continued the moralist, pragmatic tradition of indigenous philosophical psychology. Though he published a plea for "psychology as a natural science" in 1894, he also criticized the "psychologist's fallacy" – the tendency to substitute psychologists' scientific conceptions of reality for their subjects' reported experiences – in his classic text, *The Principles of Psychology* (1890). He opposed narrow experimentalism because he favored a more expansive conception of consciousness and thus also of the subject matter of psychology than that of the experimentalists. But his later proposal to study the experiences of psychics and mystics with the same objectivity as those of "normal" adults was not widely accepted. James is thus rightly cited both as a founder of and as a perpetual embarrassment to the "new" psychology.

Central to the establishment of psychology in the United States, as it was in Germany, was a rhetorical strategy aimed at separating the philosophical past from the scientific present. Here, as in the other human sciences in America during this period, an emphasis on social usefulness that actually harked back to Scottish commonsense philosophy now presupposed an engineering model of science; the adoption of that model was central to the "new" psychology's struggle for scientific and expert authority.[23] An inwardly directed counterpart to such rhetorical advocacy was the distinction made in American psychology textbooks of this period between trained psychologists and so-called naive observers; this had the effect of separating psychologists from their own ordinary selves, who would otherwise be representative of commonsense views of mind.[24] Thus, instrumental conventions of objectivity were employed to construct a professional identity that could also serve as a social resource in the public sphere.

---

[22] Paul D. Chapman, *Schools as Sorters: Lewis M. Terman, Applied Psychology and the Intelligence Testing Movement, 1890–1930* (New York: New York University Press, 1988).

[23] David E. Leary, "Telling Likely Stories: The Rhetoric of the New Psychology, 1880–1920," *Journal of the History of the Behavioral Sciences*, 23 (1987), 315–31; Jill G. Morawski and Gail A. Hornstein, "Quandary of the Quacks: The Struggle for Expert Knowledge in American Psychology, 1890–1940," in *The Estate of Social Knowledge*, ed. JoAnne Brown and David K. van Keuren (Baltimore: Johns Hopkins University Press, 1991), pp. 106–33; Jill G. Morawski, ed. *The Rise of Experimentation in American Psychology* (New Haven, Conn.: Yale University Press, 1988). Cf. Ronald Kline, "Constructing 'Technology' as 'Applied Science': Public Rhetoric of Science and Engineering in the United States, 1880–1945," *Isis*, 86 (1995), 194–221; John C. Burnham, *How Superstition Won and Science Lost: Popularizing Science and Health in the United States* (New Brunswick, N.J.: Rutgers University Press, 1987).

[24] Jill G. Morawski, "Self-Regard and Other-Regard: Reflexive Practices in American Psychology, 1890–1940," *Science in Context*, 5 (1992), 281–307.

COMMON FEATURES OF THE "NEW" PSYCHOLOGY

Despite the multiple routes to institutionalization and different research practices, certain common features of the "new" psychology can be identified, all of them part of the field's self-conscious identification with natural science during this period.

One of these common features was a reliance on what Lorraine Daston has called instrumental objectivity to establish scientific standing.[25] With their heavy brass instruments for the controlled presentation of stimuli and for measuring reaction times, the creators of the "new" experimental psychology participated in the culture of precision characteristic of nineteenth-century physics and physiology, and thus acquired scientific respectability. They also reconstituted the object to which their efforts were addressed. What had been mental and moral capacities became psychical functions; and the sensing, perceiving, conscious mind became an instrument that functioned, or failed to function, in a measurably "normal" way.

A second common feature of the "new" psychology was the use of physiological analogies, in turn often based on mechanical physics and technology.[26] The term "inhibition," for example, blended organic and machine metaphors and applied them both to human action and to society; in this case the language was taken in part from the operation of regulative devices in machines.[27] A further example is the metaphor of psychical energy. Soon after scientists and engineers applied the idea of energy conservation to human labor in order to create a science of work intended to make the "human motor" run more efficiently, Emil Kraepelin and others extended the effort to "mental work"; Hugo Münsterberg gave the result the name "psychotechnics."[28]

A third common feature of the "new" psychology was a studied vagueness about the mind–body relationship. Terms like "energy" and "inhibition" effectively linked psychology with the natural sciences and industrial culture, but their use in both the mental and physiological realms implied a solution to the mind–body problem that had not actually been achieved. Many psychologists asserted some version of "psychophysical parallelism" or claimed

[25] Lorraine Daston, "Objectivity and the Escape from Perspective," *Social Studies of Science*, 22 (1992), 597–618; cf. M. Norton Wise, ed., *The Values of Precision* (Princeton, N.J.: Princeton University Press, 1995).

[26] Horst Gundlach, "Zur Verwendung physiologischer Analogien bei der Entstehung der experimentellen Psychologie," *Berichte zur Wissenschaftsgeschichte*, 12 (1989), 167–76.

[27] Roger Smith, *Inhibition: History and Meaning in the Sciences of Mind and Brain* (Berkeley: University of California Press, 1992).

[28] Cf. Siegfried Jaeger, "Zur Herausbildung von Praxisfeldern der Psychologie bis 1933," in *Geschichte der deutschen Psychologie im 20. Jahrhundert*, ed. Mitchell G. Ash and Ulfried Geuter (Opladen: Westdeutscher Verlag, 1985), pp. 83–112; Joan Campbell, *Joy in Work, German Work: The National Debate, 1880–1945* (Princeton, N.J.: Princeton University Press, 1989); Anson Rabinbach, *The Human Motor: Energy, Fatigue and the Origins of Modernity* (New York: Basic Books, 1990).

a more intimate, functional relationship between mind and brain, but few were very precise about the nature of that relationship.[29]

A fourth common feature was the use of the term "experimental" itself. But the term's meaning was contested, and laboratory psychologists shared it with a rather different research community, the spiritualists and psychical researchers. Until quite late in the century, the term *psychologie expérimentale* referred in both France and Germany to seances; alternative designations were *psychologie expérientielle* and the more modest and more common "empirical psychology."[30] The experimentalists actively opposed spiritualism and attempted to expose quack practitioners in Britain, Germany, and the United States; but studies of altered mental states in psychics and mystics conducted by James, Janet, and others also supported a more expansive conception of psychology.[31]

This broader view was not widely accepted at first, due to a fifth common feature of the "new" psychology – a tendency to restrict its subject matter to topics that could be addressed by the natural scientific methods and apparatus then available, such as psychophysics, sensory psychology, attention span, and retention. One result of this self-restriction was an uneasy tension between efforts by Wundt, James, and others to preserve the notion of a volitional, active mind and the actual stuff of experimental research – measurable reactions to external stimuli.[32] Another result was the exclusion of social or "crowd" psychology from experimental psychology; brass instrument methodology was plainly not applicable to groups.[33] Most widely noticed by contemporaries, however, was the gap between the psychological insights into human sensibility and motivations produced by great writers and the dry texts produced by the "new" psychologists.

A sixth common, also contested, feature of late-nineteenth-century psychological science was its gendered dimension. The head–heart dichotomy and the worship of the (female) "beautiful soul" persisted through the nineteenth century; but its role in the "new" psychology was ambiguous.[34] The

---

[29] Anne Harrington, *Medicine, Mind and the Double Brain* (Princeton, N.J.: Princeton University Press, 1987); Mitchell G. Ash, *Gestalt Psychology in German Culture, 1890–1967: Holism and the Quest for Objectivity* (Cambridge: Cambridge University Press, 1995), pp. 96–7.

[30] Danziger, *Constructing the Subject*; Carroy and Plas, "The Origins of French Experimental Psychology."

[31] Marilyn Marshall, "Wundt, Spiritism, and the Assumptions of Science," in *Wundt Studies*, ed. Wolgang Bringmann and Ryan D. Tweney (Toronto: C. J. Hogrefe, 1980), pp. 158–75; Janet Oppenheim, *The Other World: Spiritualism and Psychical Research in England, 1850–1914* (Cambridge: Cambridge University Press, 1985); Deborah J. Coon, "Testing the Limits of Sense and Science: American Experimental Psychologists Combat Spiritualism, 1880–1920," *American Psychologist*, 47 (1992), 143–51.

[32] Lorraine Daston, "The Theory of Will and the Science of Mind," in *The Problematic Science: Psychology in Nineteenth-Century Thought*, ed. William R. Woodward and Mitchell G. Ash (New York: Praeger, 1982), pp. 88–118.

[33] Japp van Ginneken, *Crowds, Psychology and Politics, 1871–1899* (Cambridge: Cambridge University Press, 1992).

[34] Lorraine Daston, "The Naturalized Female Intellect," *Science in Context*, 5 (1992), 209–36.

generalized, "normal" adult mind that the experimentalists usually posited as their subject matter was at least implicitly the common property of both sexes, while the vocabulary and practices of objective science carried unmistakably masculine symbolism.

COMPETING "SCHOOLS" AS CULTURAL CONSTRUCTS, 1910–1945

The struggle for intellectual dominance in early-twentieth-century psychology has been depicted since the 1930s as a battle of competing "schools."[35] This view has its uses, but conveys the false impression that all schools competed on an equal basis everywhere. Behaviorism captured both expert and popular attention in the United States in the 1920s, but the new approach was hardly taken seriously in other countries until after 1945. The "reflexology" of the Russian physiologists Ivan Pavlov and V. M. Bekhterev did not become a dominant approach in psychology even in the Soviet Union until the 1940s. Gestalt psychology and other initiatives from Germany were received with interest but also with considerable skepticism in other countries. Psychoanalysis had established itself as an international movement by the 1920s, but had acquired few academic adherents at that time.[36] Thus, the histories of these competing schools are plainly more complicated and culturally contingent than is often acknowledged in conventional accounts. We can best locate these contingencies by looking more closely at German-speaking Europe and the United States, where the discipline was most fully developed.

In German-speaking Europe, both the "crisis of psychology" announced by the Viennese professor Karl Bühler (1879–1963) in 1927 and ideological battles over holism in psychology reflected the hothouse atmosphere of the interwar years.[37] The most widely received view internationally was that of Gestalt psychology. Developed by Max Wertheimer (1880–1943), Wolfgang Köhler (1887–1967), and Kurt Koffka (1886–1941), Gestalt theory claimed, among other things, that immediately perceived structures (*Gestalten*) and relationships rather than punctiform sensations are the primary constituents of consciousness. Nearly all participants in these debates agreed on the central importance of key words like *Ganzheit* and *Gestalt*, but the actual content of these terms differed across the political spectrum. Felix Krueger, Wundt's successor and head of the so-called Leipzig school of "holistic psychology"

---

[35] This portrayal dates from the period itself. See Robert S. Woodworth, *Contemporary Schools of Psychology* (New York: Ronald Press, 1931); Edna Heidbreder, *Seven Psychologies* (New York: Century, 1933).

[36] Gail A. Hornstein, "The Return of the Repressed: Psychology's Problematic Relations with Psychoanalysis, 1909–1960," *American Psychologist*, 47 (1992), 254–63; Bernd Nitzschke, ed., *Freud und die akademische Psychologie: Beiträge zu einer historischen Kontroverse* (Munich: Psychologie-Verlag-Union, 1989); Graham Richards, "Britain on the Couch: The Popularization of Psychoanalysis in Britain 1918–1940," *Science in Context*, 13 (2000), 183–230.

[37] Karl Bühler, *Die Krise der Psychologie* (Jena: Fischer, 1927).

(*Ganzheitspsychologie*), emphasized the role of feeling in perception and espoused neo-Romantic cultural conservatism. William Stern's personalism focused on the individual as a "psychophysical whole" in a manner congruent with liberal politics, while the Gestalt psychologists, who located themselves mainly to the left of center politically, employed holistic vocabulary to ground a rigorously natural-scientific worldview.[38]

Such controversies were inseparable from the parlous situation of psychology as a profession in Germany in the 1920s. The challenge of philosopher Eduard Spranger's "humanistic" psychology in *Lebensformen* (1922); alternative professional practices, such as handwriting analysis, advocated by Ludwig Klages in *Handschrift und Charakter* (1917); and typological personality diagnostics based on Ernst Kretschmer's *Physique and Character* (1921) strengthened the hand of culturally conservative holists and increased the pressure to develop modern research instruments congruent with German cultural tradition. Similar controversies over the cultural content of research and professional practices took place in other countries.[39]

In Austria, the work of the Vienna Psychological Institute formed a bridge between old and new, theory and practice, Europe and America. The institute was founded in 1922 in part as a way to bring Karl Bühler to Vienna; predominant in his Department of General Psychology was epistemologically oriented cognition research under the direction of Egon Brunswik. However, proponents of the Social Democratic Party's school reform program hoped for scientific support for their child-centered approach to education. In rooms located at the city's adoption center, the institute's Department of Child and Youth Psychology, led by Charlotte Bühler (1893–1974) and her associates Hildegard Hetzer and Lotte Schenck-Danziger, created so-called baby tests – performance measures for assessing the behavioral development of infants. Charlotte Bühler acquired some of her expertise as a Rockefeller Fellow in the United States, and Rockefeller Foundation funding also supported the sociographic and survey research of the institute's Research Center for Economic Psychology under Paul Lazarsfeld during the late 1920s and early 1930s. All of this put the Vienna Institute, along with those in Jena and Hamburg, in the forefront of the transition to practice-oriented basic research in German-speaking psychology.[40]

In the United States, multiple versions of behaviorism competed for attention and adherents during the 1920s. As proclaimed by John B. Watson (1878–1958) in his famous article "Psychology as the Behaviorist Views It" (1913), radical behaviorism excluded consciousness altogether from psychological

[38] Ash, *Gestalt Psychology in German Culture*, pt. 3; Anne Harrington, *Reenchanted Science: Holism and German Science from Wilhelm II to Hitler* (Princeton, N.J.: Princeton University Press, 1996).

[39] Trudy Dehue, *Changing the Rules: Psychology in the Netherlands, 1900–1945* (Cambridge: Cambridge University Press, 1995).

[40] Gerhard Benetka, *Psychologie in Wien: Sozial- und Theoriegeschichte des Wiener Psychologischen Instituts, 1922–1938* (Vienna: Wiener Universitätsverlag, 1995).

science in favor of establishing "prediction and control" of behavior; in his later writings, Watson advocated Pavlovian conditioning as a form of social engineering. However, the often-alleged behaviorist "revolution" has proved difficult to find in retrospect, despite the widespread popularity of Watson's writings. Far more significant within the field at the time were the social science and child development programs generously funded by the Laura Spelman Rockefeller Foundation. The administrators and researchers of these programs were not doctrinaire behaviorists, but they shared a belief in hard facts – in the idea, for example, that measuring children's growth and IQ test scores over time would produce scientific norms of human development – and hoped to utilize this factual knowledge in order to rationalize society.[41] Such scientific and technocratic beliefs were embodied both in Watson's radical behaviorism and in the middle-of-the-road functional psychology that remained the majority approach.

Critics of behaviorism could call on Gestalt psychology for support; the Gestalt theorists Kurt Koffka and Wolfgang Köhler pressed their cases during frequent visits to America and in their writings in English-language publications before they came to the United States permanently themselves.[42] Despite a certain skepticism toward the Gestaltists' holism, the Harvard professor Gordon Allport (1897–1967) and other prominent psychologists, including Gardner Murphy, Lois Barclay Murphy, and Henry Murray, advocated a person-centered conception of psychology. More receptive to European ideas, they were also generally more liberal politically and less technocratic in orientation than the majority of behaviorists.[43] These dissenters were responsible to a great extent for the introduction of "personality" as a psychological subject in America.

One important impact of behaviorism was the project of an experimental social psychology, which emerged in America during the 1920s. Floyd Allport (1890–1971) fought a double battle, differentiating social psychology from sociology and defending individualism against collectivism. The immunization strategy he employed in order to support his claims to expertise and to defend himself against charges of advocating a "group mind," as many crowd and folk psychologists had done, was to limit his research to social influences on the overt behavior of individuals in artificially constructed, short-term situations.[44]

[41] Franz Samelson, "Organizing for the Kingdom of Behavior: Academic Battles and Organizational Policies in the Twenties," *Journal of the History of the Behavioral Sciences*, 21 (1985), 33–47; Hamilton Cravens, *Before Head Start: The Iowa Station and America's Children* (Chapel Hill: University of North Carolina Press, 1993).

[42] Michael M. Sokal, "The Gestalt Psychologists in Behaviorist America," *American Historical Review*, 89 (1984), 1240–63.

[43] Katherine A. Pandora, *Rebels within the Ranks: Psychologists' Critique of Scientific Authority and Democratic Realities in New Deal America* (Cambridge: Cambridge University Press, 1997).

[44] Kurt Danziger, "The Project of an Experimental Social Psychology: Historical Perspectives," *Science in Context*, 5 (1992), 309–28.

The decade of the 1930s was dominated by competing versions of what came to be called neo-behaviorism, which were also alternative approaches toward reintroducing theorizing into psychology. Edward Tolman (1886–1959) tried to integrate purposive motivation and cognitive processes into behavior theory, going so far as to claim that white rats framed "hypotheses" as to which maze route would yield an expected food reward.[45] Clark Hull (1884–1952) developed an elaborate hypothetico-deductive model of learning based on what he took to be Newtonian axiomatics; he then tried to expand the model from the habit hierarchies of classical conditioning to personality theory. Finally, B. F. Skinner (1904–1990) developed operant conditioning in the 1930s. Theoretical influences came in this case from Ernst Mach's positivism and from the physicist Percy Bridgman's operationalist philosophy of science. To Skinner, these inputs justified an approach that yielded little theoretical output, producing careful measurements of the relative likelihood of simple behaviors, such as rats' or pigeons' pressing a bar to obtain a pellet of food under rigorously controlled conditions, and suspending all efforts to explain such behavior. The most prominent nonbehaviorist effort to bring systematic theorizing into psychology was that of the emigré Kurt Lewin (1890–1947). Lewin advocated what he termed the "Galilean" study of ideal-typical behavioral situations, exemplified in his Iowa studies of "democratic" and "authoritarian" leadership in children's groups.[46] In America, Lewin incorporated some aspects of American-style experimentation, for example, the operationalization of variables. But his work remained the search for "pure" forms of group action rather than for social influence on the behavior of individuals.[47] Lewin and his American competitors shared an antipathy for blind fact gathering, an admiration for classical physics, and a willingness to draw upon the philosophy of science, especially operationism and logical positivism, to legitimate their positions.[48] They differed in their basic conceptual foundations and also in the physics they chose to emulate. The competition continued through the 1940s and was resolved, if at all, only by the rapid fragmentation of the discipline in the 1950s.

In Britain and France, psychology remained, in comparison to Germany and the United States, weakly institutionalized from the 1920s to the 1940s. Yet precisely this situation enabled a wide range of theoretical explorations and practical applications, including alternatives to American behaviorism, to flourish. In England, the links to educational practice were as tight as they were in the United States. Cyril Burt (1883–1971), originally a London

---

[45] Edward C. Tolman, *Purposive Behavior in Animals and Men* (New York: Century, 1932).

[46] Kurt Lewin, Ronald Lippitt, and Robert K. White, "Patterns of Aggressive Behavior in Experimentally Created 'Social Climates,'" *Journal of Social Psychology*, 10 (1939), 271–99.

[47] Mitchell G. Ash, "Cultural Contexts and Scientific Change in Psychology: Kurt Lewin in Iowa," *American Psychologist*, 47 (1992), 198–207.

[48] Laurence Smith, *Behaviorism and Logical Positivism: A Reassesment of the Alliance* (Stanford, Calif.: Stanford University Press, 1986).

school official, adapted and expanded Spearman's concept of general and specific intelligence in studies of educational performance, delinquency, and so-called backward children, then developed a mathematical basis for factorial approaches to intelligence and personality testing in *The Factors of the Mind* (1940); he succeeded Spearman as a professor at University College London. The practical impact of his efforts to combine academic and applied work was so great that he was later knighted for his contributions. Controversy over accusations that he manipulated or even invented some of the data on which he based his confident claims did not emerge until after his death.[49] During the same period, the Cambridge professor Frederick Bartlett (1886–1969) published his pioneering study *Remembering* (1932), in which he established the role of learned schemata in retention and laid the foundations for considering memory as a process of active reconstruction rather than of rote recall. Less well remembered is that Bartlett used folktales in this study, and spoke of the "social constructiveness" of cognition in an effort to integrate his research with social and cultural anthropology.[50]

In France, psychology remained divided between medicine and philosophy, as it had been before 1914; there was no separate degree until 1947.[51] One result was that alongside the strictly experimental work of Henri Piéron (1881–1964), Binet's successor as director of the Psychological Laboratory at the Sorbonne, philosophers and sociologists felt free to consider psychological issues in broader and less scientistic or behavioristic ways. The debate over Lucien Lévy-Bruhl's concept of "primitive" mentality, for example, contributed to the emergence of the "mentalities" concept of the *Annales* school of history.[52] In French-speaking Switzerland, the biologist and philosopher Jean Piaget (1896–1980), building upon Eduard Claparède's functional psychology but also hoping to confirm views advanced in contemporary liberal Protestant thought, began his pioneering studies of cognitive development in children.[53] In the 1930s and 1940s, the philosopher Maurice Merleau-Ponty expanded phenomenology by drawing upon Gestalt psychology, as well as

---

[49] L. S. Hearnshaw, *Cyril Burt, Psychologist* (Ithaca, N.Y.: Cornell University Press, 1979); Steven J. Gould, *The Mismeasure of Man* (New York: Norton, 1981), chap. 6; Robert B. Joynson, *The Burt Affair* (London: Routledge, 1989); Nicolas John Mackintosh, ed., *Cyril Burt: Fraud or Framed?* (Oxford: Oxford University Press, 1995). For the broader context, see also Adrian Wooldridge, *Measuring the Mind: Education and Psychology in England, c. 1860–c. 1990* (Cambridge: Cambridge University Press, 1994); Rose, *The Psychological Complex.*

[50] David Bloor, "Whatever Happened to 'Social Constructiveness'?," in *Bartlett, Culture and Cognition,* ed. Akiko Saito (London: Psychology Press, 2000), pp. 194–215.

[51] Francoise Parot and Marc Richelle, *Introduction a la Psychologie. Histoire et méthodes,* 4th ed. (Paris: Presses Universitaires de Paris, 1998).

[52] Cristina Chimisso, "The Mind and the Faculties: The Controversy over Primitive Mentality and the Struggle for Disciplinary Space at the Inter-war Sorbonne," *History of the Human Sciences,* 13 (2000), 47–68; Laurent Mucchielli, "Aux origines de la nouvelle histoire en France: l'évolution intellectuelle et la formation du champ des sciences sociales (1880–1930)," *Revue de synthèse,* 1 (1995), 55–99.

[53] Fernando Vidal, *Piaget before Piaget* (Cambridge, Mass.: Harvard University Press, 1994).

upon studies of children's perception by Henri Wallon and Piaget, in his *The Structure of Behavior* (1942) and *The Phenomenology of Perception* (1943).

DYNAMICS OF PROFESSIONALIZATION TO 1945

The turning point for the public visibility of professional psychology in the United States came with the mass use of intelligence tests by the U.S. Army during the First World War. The remarkable fact here is that the route of application ran not from the "normal" to the "pathological," but rather from socially marginal populations – the so-called feeble-minded and schoolchildren – to "normal" adults. The deep historical significance of this event lives on in the very terminology of psychological testing; a series of psychological tests is still called a "battery," and a collection of therapeutic methods is referred to as an "armamentarium." The interaction of two emerging professions – applied psychology and the professional officer corps – reshaped the aims of intelligence testing, the test instrument itself, and ultimately conceptions of the objects being assessed. Intelligence became not intellectual or problem-solving capacity alone, but a sum of skills and (presumably hereditary) aptitudes for certain kinds of learning.[54]

"Binet testing," as it was then called, continued to fuel the expansion of professional psychology in both the United States and Britain during the 1920s. The use of quantitative assessment or classification instruments and of "Galtonian" group data in basic research and professional practice spread rapidly in both countries, primarily because the products thus created supported the classifying functions required by administrators – initially in schools, and later also in industry and social service agencies.[55] It was during this period that the field became more open to women; but a gender hierarchy emerged, with industrial psychology remaining male-dominated, while female "Binet testers" and social workers took on more people-oriented functions.[56]

The extraordinary variety of psychological applications and the vastly increased numbers of trained psychologists available to carry them out during the Second World War contrasts strongly with the narrow focus on sorting soldiers during World War I. In addition to the use of tests in personnel management, fields of application included the employment of social

---

[54] Michael M. Sokal, ed., *Psychological Testing and American Society* (New Brunswick, N.J.: Rutgers University Press, 1987); Richard von Mayrhauser, "The Practical Language of American Intellect," *History of the Human Sciences*, 4 (1991), 371–94; John Carson, "Army Alpha, Army Brass and the Search for Army Intelligence," *Isis*, 84 (1993), 278–309.

[55] Danziger, *Constructing the Subject*.

[56] Laurel Furumoto, "On the Margins: Women and the Professionalization of Psychology in the United States 1890–1940," in *Psychology in Twentieth-Century Thought and Society*, ed. Mitchell G. Ash and William R. Woodward (Cambridge: Cambridge University Press, 1987), pp. 93–114.

psychology in morale research and applied human relations, incorporation of
psychophysics and experimental psychology into studies of human–machine
interactions – for example, at Harvard's Psychoacoustical Laboratory – and
diagnostic testing in clinical psychology. All of this led in turn to significant basic research programs during the postwar period. Amid this flurry of
activity, initiatives by women psychologists aimed at increasing their representation in the discipline's governing bodies took a back seat, in part due to
differences among the women psychologists themselves.[57]

The professionalization of psychology in Germany took a rather different
course. During the First World War, efforts focused on the use of techniques
from psychophysics to instrumentalize human subjects in a mechanized bat-
tlefield. Examples included the adaptation of psychophysical techniques to
develop sound-ranging devices and to test the visual discrimination ability of drivers and pilots.[58] Under the name "psychotechnics," this approach
continued into the Weimar era, particularly in industry.

After the Nazi takeover, the directors of four of the six leading psychological institutes in Germany were dismissed because they were Jewish; a fifth,
Wolfgang Köhler, the head of the Berlin institute and one of the few German
academics to protest Nazi policies publically, left voluntarily in 1935.[59] The
Marburg professor Erich Rudolf Jaensch and others tried to "Nazify" their
earlier viewpoints; but more important developments were the rapid growth
of military psychology as a result of German rearmament and the resulting
shift from psychotechnical skill testing to "intuitive" character diagnosis.

In contrast to the situation in the United States during World War I, the
primary purpose in Germany was elite officer selection, rather than the sorting of large numbers of average recruits. Though paper-and-pencil and skills
tests were used, these were secondary to the extended observation of officer
candidates in simulated command situations intended to induce expressions
of the candidate's "deeper" self. The personality characteristics sought had
considerable affinity to the traditional virtues of the Prussian officer – the
will to command and the ability to inspire troop loyalty. By contrast, diagnostic efforts based on Nazi "race psychology" could not be translated into
professional practice.[60]

In the United States, too, personality diagnostics ultimately became a royal
road to professionalization. However, in contrast to Germany, quantitative

---

[57] James H. Capshew, *Psychologists on the March: Science, Practice and Professional Identity in America, 1929–1969* (Cambridge: Cambridge University Press, 1999), esp. chaps. 3–7.
[58] Horst Gundlach, "Faktor Mensch im Krieg: Der Eintritt der Psychologie und Psychotechnik in den Krieg," *Berichte zur Wissenschaftsgeschichte*, 19 (1996), 131–43.
[59] Mitchell G. Ash, "Emigré Psychologists after 1933: The Cultural Coding of Scientific and Professional Practices," in *Forced Migration and Scientific Change: Emigré German-Speaking Scientists and Scholars after 1933*, ed. Mitchell G. Ash and Alfons Söllner (Cambridge: Cambridge University Press, 1996), pp. 117–38, at p. 118.
[60] Ulfried Geuter, *The Professionalization of Psychology in Nazi Germany* (1984), trans. Richard Holmes (Cambridge: Cambridge University Press, 1992).

methods based on techniques of factor analysis developed by L. L. Thurstone and others predominated, despite competition from "projective" tests such as the Rohrschach in the 1930s and 1940s. The history acquired a gendered dimension in the construction of "female" and "male" traits in early personality research. In the Terman–Miles Aptitude Interest Analysis of 1936, for example, psychologists assigned "masculine" and "feminine" point values to subjects' responses on a 910-item multiple-choice test. Using such tools, personality researchers acquired authority over the definition and interpretation of culturally selected attributes. In addition, they justified their emerging diagnostic role as screeners authorized to recommend clinical assistance to those who deviated from the tested norms.[61]

THE POSTWAR ERA: "AMERICANIZATION" AND THE ALTERNATIVES

In the United States, the postwar years saw explosive expansion and differentiation in both the scientific and professional realms. The establishment of a divisional structure within the American Psychological Association (APA) in 1947 – already negotiated during the war – reflected this process. Despite the optimism of the time, it proved difficult to subsume all aspects of psychology's protean identity within single university departments or graduate programs.[62] Fragmentation was most obvious in the different research practices institutionalized in experimental, social, and personality psychology.

In experimental psychology, neo-behaviorist learning theory challenged a revival of cognition research by advocates of the so-called New Look and information processing approaches.[63] Common to both neo-behaviorism and the new cognitive psychology, however, were an emphasis on standardizing experimentation by "operationalizing" variables, distinguishing "independent" from "dependent" variables, and using statistical significance testing to evaluate results.[64] An increasingly fragmented field held itself together, if it did so at all, by enforcing such methodological conventions on ever-widening groups of researchers via the increasingly extensive guidelines of the *Publication Manual* of the APA.[65] Among the results were a relative lack of interest in field research and phenomenological exploration and, by

[61] Jill G. Morawski, "Impossible Experiments and Practical Constructions: The Social Bases of Psychologists' Work," in *The Rise of Experimentation in American Psychology*, pp. 72–93.
[62] Capshew, *Psychologists on the March*, pp. 205–8.
[63] Howard Gardner, *The Mind's New Science* (1985) (New York: Basic Books, 1996).
[64] On the postwar triumph of statistics, see Danziger, *Constructing the Subject*; Capshew, *Psychologists on the March*, chap. 10.
[65] Charles Bazerman, "Codifying the Social Scientific Style: The A.P.A. 'Publication Manual' as a Behaviorist Rhetoric," in *The Rhetoric of the Human Sciences: Language and Argument in Scholarship and Public Affairs*, ed. John S. Nelson, Donald McCloskey, and Allen Megill (Madison: University of Wisconsin Press, 1987), pp. 125–43.

implication, the prestructuring even of basic research to suit the needs of an expert society.

The problematic implications of statistical significance testing become clear much later, in the debate over computational models of mind. In this case, psychologists seeking instruments of control via standardized inference provided tools such as Bayesian statistics, which then generated metaphors and concepts, the justification of which was easier because the tools were already in common use. The scientists then found the instruments informing their theorizing, or they found themselves, claiming, quite implausibly, that "normal" subjects, not socialized into the use of these techniques, nonetheless solve problems in the same way, by applying "incomplete" or "naive" versions of statistical inference.[66]

Beneath the loosely forming net of methodological convention, substantial differences existed. In educational psychology, for example, the preferred research tools were the correlational methods pioneered by Galton. In 1957, Lee Cronbach even spoke of the rival research communities as "two disciplines."[67] A comparable methodological split occurred in experimental social psychology and personality theory. In a broad survey of the field, Dorwin Cartwright spoke openly of "hard" and "soft" or "messy" methods to distinguish learning theory from social and personality psychology.[68] Nonetheless, experimental studies of social influence on perception by Solomon Asch, and of prejudice by Gordon Allport and others, captured the imagination of many in the field. At the same time, the *Authoritarian Personality* study, begun during the war and published in 1950, played on widespread worries among American liberals that fascist and anti-Semitic attitudes were not limited to Nazi Germany. The popularity of such studies was symptomatic of a widespread tendency during the period to psychologize, and thus to individualize, social problems.[69] Meanwhile, developmental psychology went its own way, taking the work of Jean Piaget as a touchstone for numerous studies closely related, as the earlier work of Arnold Gesell and others had been, to the practical needs of schools for age-related developmental norms.

By the 1970s, both the sheer number of psychologists (over 70,000; over 100,000 by the end of the century) and the international representation of psychology had reached levels that could not have been imagined fifty years earlier. The growth was worldwide, but more than two-thirds of the total were Americans. The openness of both discipline and profession to

[66] Gerd Gigerenzer, "From Tools to Theories: Discovery in Cognitive Psychology," *Science in Context*, 5 (1992), 329–50.

[67] Lee Cronbach, "The Two Disciplines of Scientific Psychology," *American Psychologist*, 12 (1957), 671–84.

[68] Dorwin Cartwright, "Lewinian Theory as a Contemporary Systematic Framework," in *Psychology: A Study of a Science*, vol. 4: *General Systematic Formulations*, ed. Sigmund Koch (New York: McGraw-Hill, 1959), pp. 7–91.

[69] Franz Samelson, "Authoritarianism from Berlin to Berkeley: On Social Psychology and History," *Journal of Social Issues*, 42 (1986), 191–208.

women continued, and indeed increased from the 1950s onward. According to a National Science Foundation survey for the years 1956–8, for example, 18.49 percent (2,047) of all American psychologists were women; this was the highest percentage for any single discipline. Today more than half of the doctorates awarded in the field go to women. However, the gender concentration that began in the 1920s continued, with women being more numerous in developmental and educational psychology and men in experimental, industrial, and personnel psychology.[70] Such numbers, and the extent of the institutional anchorage of psychology in the United States, were more than sufficient to assure that the research and professional practices institutionalized there would spread throughout the world.

The most important exceptions to the overall trend were the near-worship of Piaget by developmental psychologists, and the positive reception of applications of factor analysis to personality testing and diagnostics by the British psychologists Hans Eysenck and Raymond Cattell. In cognition research, too, British work such as that of F. C. Bartlett and Donald Broadbent, as well as the work of Soviet theorists such as Alexander Luria, were mobilized to lend respectability and theoretical sophistication to the resurgent field in the United States. Nonetheless, in cognitive science, too, the pervasive influence of computer metaphors and associated information-processing models was plainly of Anglo-American origin.

During this period in the two German states psychology itself became a laboratory for Cold War science. In West Germany there was striking continuity with the Nazi period at first; nearly all those who had held professorships in 1943 still did so in 1953. By the 1960s, after an intense controversy that had both nationalistic and generational dimensions, this older generation had been supplanted by younger advocates of American-style, meaning data-driven, research and statistical presentation and assessment of results.[71]

In East Germany, continuity with the past was most clearly evident in the appointment of Kurt Gottschaldt, a former student of the Gestalt psychologists who had carried out extensive twin studies at the Kaiser Wilhelm Institute for Anthropology during the Nazi era, to a full professorship at the Humboldt University in East Berlin. The context here was the decision, for pragmatic reasons, of East German party and state officials to utilize "bourgeois" scientists until a "new intelligentsia" could be trained.[72] By the late 1950s, however, Gottschaldt had come under pressure from proponents

---

[70] Margaret Rossiter, "Which Science? Which Women?," *Osiris*, 2nd ser., 12 (1997), 169–85, data at 170–5.

[71] Alexandre Métraux, "Der Methodenstreit und die 'Amerikanisierung' der Psychologie in der Bundesrepublik 1950–1970," in *Geschichte der deutschen Psychologie im 20. Jahrhundert*, ed. Ash and Geuter, pp. 225–51.

[72] Mitchell G. Ash, "Kurt Gottschldt and Psychological Research in Nazi and Socialist Germany," in *Science under Socialism: East Germany in Comparative Perspective*, ed. Kristie Macrakis and Dieter Hoffmann (Cambridge, Mass.: Harvard University Press, 1999), pp. 286–301, 360–5.

of a "Marxist-Leninist" psychology based, ironically, in Wilhelm Wundt's Leipzig.[73] He departed for the West in 1962, but his successor in Berlin, Friedhart Klix, skillfully presented his own mixture of Soviet-style cognition research and American information-processing approaches as a new "Marxist" psychology in tune with the "scientific-technical revolution."[74]

In Western Europe outside Germany and France, the predominance of American and British work in academic psychology was secure by 1970. For example, citation rates for English-language publications in the leading Dutch psychology journal rose from 20 percent in 1950 to over 70 percent in 1970; by then the citation rate of American publications in social psychology dissertations was well over 90 percent.[75] The work of British researchers such as Bartlett, Broadbent, Eysenck, Cattell, and their students quickly found supporters in America, which led to a merging of traditions. This was also the case in clinical psychology, thanks to the positive reception of research from the Tavistock Institute and elsewhere.

The professional history of psychology after 1945 nonetheless continued to be affected by contingent local circumstances. The rise of clinical psychology in the United States, for example, was originally driven by the need to deal with large numbers of mentally ill veterans after World War II. The initially established division of labor between test-based clinical diagnostics and psychiatric treatment soon became complicated, as clinicians engaged in a wide variety of psychotherapies often, though not always, inspired by psychoanalysis. The new field ultimately brought forth its own basic research in both clinical and academic settings, which led to the emergence of scientific communities based on methodological norms quite different from those of experimental and developmental psychologists. This was the background of the controversy over "clinical versus statistical prediction" during the early 1950s.[76] In addition, an eclectic, so-called humanistic psychology movement arose in opposition to both behaviorism and psychoanalysis, becoming widely popular in psychotherapy, social work, and the emerging field of counseling psychology.

In Germany, as in the rest of Europe, the rise of clinical psychology came approximately ten years later than in the United States. There, however, in contrast to the United States, the supremacy of personality diagnostics and its quantitative tools had already been established in basic research before the professionalization of the clinical field. Another important difference indicative of a persistent European tradition was that clinical training in academic settings was based far more on cognitive and behavioral techniques than on

---

[73] Stefan Busse, "Gab es eine DDR-Psychologie?," *Psychologie und Geschichte*, 5 (1993), 40–62.

[74] Friedhart Klix, *Information und Verhalten* (Berlin: Deutscher Verlag der Wissenschaften, 1966).

[75] Pieter J. van Strien, "The American 'Colonization' of Northwest European Social Psychology after World War II," *Journal of the History of the Behavioral Sciences*, 33 (1997), 349–63.

[76] Paul E. Meehl, *Clinical versus Statistical Prediction* (Minneapolis: University of Minnesota Press, 1954).

psychoanalysis. Barriers to the academic institutionalization of psychoanalytic research and training in the universities proved surmountable only in exceptional cases, such as that of the Sigmund Freud Institute in Frankfurt am Main under Alexander Mitscherlich.

In sum, by the 1980s, if not earlier, what had been at the turn of the century a multifaceted but predominantly European discursive and practical field had become deeply dependent economically, institutionally, and culturally on American research styles and professional practices.[77] When and to what extent the kinds of obsessions with psychological topics typical of American popular culture came to pervade European culture cannot be considered in detail here. But it had become clear even to casual visitors by the 1980s that psychobabble and the associated group workshop culture had become just as firmly anchored there, at least in Western European (and especially German) middle- and upper-middle-class culture, as it had in the United States.

American predominance was contested, though with at best only partial success, by dissident local-language movements, most notably in France and Germany. Most significant in the end, however, was the contrast between American predominance in both academic and professional psychology worldwide and the insecure standing of trained psychologists in America itself. Vagueness and confusion in the use of the term "psychologist" in public discussion have been remarkably consistent over time; the term itself lacks legal protection in any case. All of this, not to mention the omnipresence of self-help books, which are placed on the psychology shelves of many bookstores whether their authors are psychologists or not, indicates that even in the United States, where most of the world's psychologists live and work, trained academics and professionals can hardly claim hegemony over psychological discourse in the public sphere to the degree that physical scientists can in their own fields.

CONCLUSION: SCIENCE, PRACTICE, SUBJECTIVITY

Given this incomplete victory in the century-long struggle for scientific and professional autonomy and authority in psychology, it might well be asked why such a shakily legitimated field has acquired such an important role in twentieth-century culture and society. Roger Smith suggests that the discipline grew in constant interaction with "psychological society," drawing its authority from and simultaneously giving voice to "a significant sense in which everyone in the twentieth century . . . became her or his own psychologist, able and willing to describe life in psychological terms."[78] Nikolas Rose argues that psychological practices make possible particular kinds of social

[77] On the "psychologization" of American postwar culture, see Ellen Herman, *The Romance of American Psychology* (Berkeley: University of California Press, 1996).
[78] Smith, *Norton History*, p. 577.

authority, assembled at first ad hoc, then grafted onto all activities connected with the stewardship of human conduct in liberal-democratic polities, from law and penal administration to education and parenting. No single profession has monopolized the codification and certification of these activities, which are aimed at simplifying the administration of modern life by producing calculable individuals and manageable social relations. Precisely because it is so diffuse and widespread, psychological knowledge shapes the practices of welfare states and justifies them with a rationale according to which individuals are required to be free, and feel obligated to correct or repair defects if they fail to cope on their own.[79] Such a view could explain why reflexive practices, nicely epitomized in the phrase "working on oneself" or "working on a relationship," have become the norm in late modern societies.

A further implication of such views is that psychology's alleged objects themselves – mind, behavior, and personality – are not simply invariant fixtures of the species, but may have cultural as well as natural histories. These histories also require study, in order to understand the historical development of scientific discourse about them. Such questions have only recently received the attention they deserve, despite the focus on "mentalities" in cultural history.[80]

In light of the long view taken in this chapter, the predominance of behaviorism in the American-Saxon cultural region in the middle third of this century becomes an episode in a much larger story. However, it is a characteristic episode, for both the discourse of prediction and control and its associated practices have persisted, even as the so-called cognitive revolution has reintroduced mentalistic vocabularies. One reason for such continuities appears to be that not only the members of the discipline and profession called psychology, but also the modern culture and society in which they function, require, and may even desire, both technocratic discourse and the instruments that embody and enact it.

---

[79] Nikolas Rose, *Governing the Soul* (London: Routledge, 1990) and his *Inventing Our Selves: Psychology, Power and Personhood* (Cambridge: Cambridge University Press, 1996).

[80] For important first steps in this direction, see Norbert Elias, *The Civilizing Process* (1939), 2 vols, trans. E. Jephcott (New York: Urizen, 1978); Gerd Juttemann, ed., *Die Geschichtlichkeit des Seelischen: Der historische Zugang zum Gegenstand der Psychologie* (Weinheim: Psychologie Verl. Union, 1986); Irmingard Staeuble, " 'Psychological Man' and Human Subjectivity in Historical Perspective," *History of the Human Sciences*, 4 (1991), 417–32; Roy Porter, ed., *Rewriting the Self: Histories from the Renaissance to the Present* (London: Routledge, 1997).

16

ECONOMICS

Mary S. Morgan

Economics has always had two related faces in its Western tradition. In Adam Smith's eighteenth century, as in John Stuart Mill's nineteenth, these might be described as the science of political economy and the art of economic governance. The former aimed to describe the workings of the economy and to reveal its governing laws, while the latter was concerned with using that knowledge to fashion economic policy. In the twentieth century, these two aspects were more often contrasted as positive and normative economics. The continuity of these dual interests masked differences in the way economics was both constituted and practiced during the twentieth century, when these two aspects of economics became integrated in a particular way. These two wings of economics, originally a verbally expressed body of scientific lawlike doctrines and associated policy arts, in the twentieth century became more firmly joined together by the use of a set of technologies routinely and widely used within the practice of economics in both its scientific and policy domains.

In the twentieth-century history of economics, tool development and changes in economic theory need to be set alongside demands for advice generated by overwhelming events in the economic history of the times and strong economic ideologies in the political arena. These processes interacted to generate a Western technocratic economics very different in style and content from the economics of previous centuries, one we might characterize as an engineering science.

I thank Malcolm Rutherford for his willingness to let me draw on our joint work in this chapter, and I thank the editors of this volume, Ted Porter and Dorothy Ross, for their incisive comments, their encouragement, and the overwhelming patience they displayed toward a recalcitrant author. Many historians of economics, especially Roger Backhouse, provided suggestions and comments for which I am grateful.

ECONOMICS AS ENGINEERING

To understand twentieth-century economics as a science in the mold of engineering is to see that the economics profession came to rely on a certain precision of representation of the economic world, along with techniques of quantitative investigation and exact analysis that were alien to the experience of nineteenth-century economics, when the extent of such technologies of representation, analysis, and intervention were extremely limited. The engineering metaphor also suggests that twentieth-century economics is best characterized as a science of applications and implies a technical art, one that relies on tacit knowledge and decidedly human input as in the eighteenth-century term "art of manufactures."[1] Because of inherent limitations on the field's ability to access and control its subject matter, even economists' most exact theories had to be explored on a case-by-case basis, and the practical application of quantifying technologies could never be automatic, but always involved human judgment. There are certain parallels here to psychology's effort to "control" the individual, although, perhaps because of the presence of centrally planned Eastern economies for most of the century, Western economics fought shy of the view that direct control is the aim of economic science, either as a way of validating scientific explanation or as a program of social action.

From the point of view of economic policy, the engineering notion embodies elements of both the operation and design of systems, and it is subject to different interpretations at different times in the practice of twentieth-century economics. In terms of operating the economy, notions of control engineering were explicitly discussed during the 1950s experience of the "managed" economy. The way the macroeconomy was pictured implied that the economy was subject to governmental control. At the same time, under the influence of cybernetic thinking, the economic behavior of each individual was pictured as being controlled by personal feedback loops. More flexibly, in the 1960s, governments were thought to have the economic power only to "fine tune" the macroeconomy or to nudge the economy back on course.[2] In the 1920s and 1980s, still less interventionist modes were in favor, and macroeconomic policy was understood to be taking fiscal care and following rules of monetary operation, suggesting the idea of maintaining a smooth-running machine, while at the individual level the issue was one of influencing behavior via incentive systems rather than by mechanisms of control.

---

[1] This contingent and decidedly human element is characterized by Eugene S. Ferguson as an essential part of the engineering mode in his *Engineering and the Mind's Eye* (Cambridge, Mass.: MIT Press, 1992).

[2] An interesting study of these contrasting beliefs can be found in Craufurd D. Goodwin, *Exhortation and Controls: The Search for a Wage-Price Policy* (Washington, D.C.: Brookings Institution, 1975). For a broader picture of the relation between state and economy, see Alain Desrosières's chapter in this volume.

The engineer as designer and constructor was also prevalent in twentieth-century economics. In the 1930s, when it seemed the economic machine was seriously malfunctioning, some economists suggested planning a whole new economy. During the post-1950s period, the goal was more evolutionary and less mechanical, namely, to affect the environment within which people act in order to produce adaptive economic behavior. Western economists were expected to formulate development paths, to design new economic institutions to foster market economies, and to map out transition paths for postcommunist economies. Throughout the century, they were asked to carry out technical assessments of economic decisions and to tinker with, or design anew, incentive structures for all sorts of everyday cases.

Economic technologies were not only policy tools for designing and justifying interventions in the world but also scientific tools, forged for theory development and to find out about the world. These tools were not independent of high theory; rather, they supported its development. They were also critically involved in new ways of making sense of phenomena and constructing facts about the economy.

Around 1900, there was relatively little mathematics, statistics, or modeling contained in any economic work: Economics was a verbal tradition. In the first half of the twentieth century, a massive growth in the collection of economic data and associated empirical investigations built a detailed knowledge base in economics, leading to the development of specialized statistical tools under the label of econometrics. Concurrently, but more slowly, mathematics was adopted, both to express economic theories and to formalize arguments. During the 1930s, the technology of modeling was introduced into theoretical and econometric work. The full dominance of these technologies – measurement methods, mathematics, statistics, and modeling – occurred only after 1940, but by the end of the century economics had become a modeling science in both theoretical and applied work. Economics became, in effect, a tool-based discipline.

These quantitative techniques gave economics the aura of scientific modernity. But while economics portrayed itself as the most scientific of the social sciences, its claim to such a title had less to do with any success in using mathematics to formulate general laws or using statistics to predict economic events – the criteria often applied to the physical sciences – and more to do with turning economics into a discipline whose methodology relies on technical tools to buttress claims for economic knowledge.

This account of twentieth-century Western economics begins with a picture of the economics discipline around 1900, and then analyzes how the tools that economists fashioned, the theories they developed, and the economies they tended mutually shaped one other and changed the discipline. A further important element in this mix was the role of economic ideology, which was critical to the development of tool-based economics and to the increasing dominance of American styles and ideas within Western economics during the latter part of the century.

ECONOMICS FROM THE NINETEENTH TO THE TWENTIETH CENTURY

Considered as a field of study, economics had already gathered sufficient academic respectability to have chairs in many universities by the mid nineteenth century. By 1900, it had its own separate academic societies and journals, and its subject matter had become to a large extent separate from its older ancestors, moral philosophy and politics, and from newer siblings such as sociology. Nevertheless, the creation of separate university departments of economics, the growth of professional positions both inside and outside academia, and the advent of graduate education were subject to considerable national variation in timing and outcomes over the first half of the twentieth century.[3] With independence, economics developed specialized subfields, such as labor economics and international trade, but local demarcation disputes continued as economic history, industrial relations, and business management gained their own disciplinary positions.

During the late nineteenth and early twentieth centuries, economics was characterized by a considerable pluralism of beliefs, theories, and methods. It is difficult to view any one school of economics as being dominant, for while there were clearly national differences – and even some "schools" of economics delineated in national terms, such as Austrianism and American Institutionalism – economics throughout this period remained international in terms of its communication lines.[4]

The earlier nineteenth-century English "classical" emphasis on labor as the source of value and the critical element in the creation of wealth had been challenged by the "marginal revolution" of the 1870s.[5] This new account focused on the consumer as the source of valuation of economic goods: Each consumer experienced an increase in overall satisfaction or utility, but at a declining rate, as they increased their consumption of a good. The marginal (last)

---

[3] There is no overall treatment of the professionalization of the discipline, but see, for example, on Britain, John Maloney, *Marshall, Orthodoxy and the Professionalisation of Economics* (Cambridge: Cambridge University Press, 1985); on the United States, Dorothy Ross, *The Origins of American Social Science* (Cambridge: Cambridge University Press, 1991), and Mary O. Furner, *Advocacy and Objectivity: A Crisis in the Professionalization of American Social Science, 1865–1905* (Lexington: University Press of Kentucky, 1975).

[4] Most histories of economics give an account of the various "schools" in this period: See Roger E. Backhouse, *A History of Modern Economic Analysis* (Oxford: Blackwell, 1985); and Henry Spiegel, *The Growth of Economic Thought*, 3rd ed. (Durham, N.C.: Duke University Press, 1991), which places each school into its intellectual context. See Mark Blaug's *Economic Theory in Retrospect*, 5th ed. (Cambridge: Cambridge University Press, 1996), for an in-depth treatment of the theoretical developments.

[5] For a consideration of classical economics, see Margaret Schabas's chapter in this volume.

unit consumed, the least valuable in terms of utility gained, provided the measure of exchange with other goods and thus determined the price paid for all units. There were four variants of this new theory. The English economist William Stanley Jevons (1835–1882) drew on the Benthamite picture of pleasures and pains, the physiology of satiation, and the physics of his day to provide a mathematical formulation of the consumer's feelings. The French economist at Lausanne, Leon Walras (1834–1910), outlined in mathematical form a general equilibrium theory of the economy, in which all the individual consumers' exchanges were matched at marginal values but in which the psychology of feelings and motivations was less prominent. John Bates Clark (1847–1938), the American historical economist, outlined a more complicated vision of multiple bundles of different kinds of utility associated with each good or service. Carl Menger (1840–1921), the founder of the Austrian school, analyzed how individuals satisfy different needs with the same good and outlined an account of how needs were ordered and choices made.[6]

Accounts differ regarding how revolutionary this movement was and how quickly it spread through the profession.[7] They agree, however, that by the early twentieth century, "neoclassical" economics had established a new research approach by combining the older classical focus on production or supply with the new insights of marginalism on the demand side, in a mathematical account developed from the work of Jevons and Walras. This approach continued to gain credibility through the first half of the twentieth century, as the characteristics of what was to become the full-fledged neoclassical economics of the third quarter of the century – namely, formal treatments of rational, or optimizing, economic agents joined together in an abstractly conceived free-market, general equilibrium world – were worked out. This abstract account became widely adopted to the exclusion of other approaches, however, only during the second half of the twentieth century.[8]

One of the reasons for the slow acceptance of the new neoclassical approach was its narrow and unrealistic portrait of the individual. Nevertheless, economists who found themselves at odds with the project also found some of its formulations useful. Thus the American historical economist Richard T. Ely (1854–1943) could use the concepts and analysis to discuss individual consuming behavior without being committed to the utilitarianism and differential calculus of Jevons. Similarly, in the 1930s, Joan V. Robinson

---

[6] Except for Backhouse, *Modern Economic Analysis*, most histories of this "revolution" omit Clark. On differences between the other three variants of marginalism, see William Jaffé, "Menger, Jevons and Walras De-homogenized," *Economic Inquiry*, 14 (1972), 511–24; see also Keith Tribe's chapter in this volume, for an interesting comparison of the historical sequence with regard to Menger and Walras.

[7] See R. D. Collison Black, A. W. Coats, and Craufurd D. W. Goodwin, *The Marginal Revolution in Economics* (*History of Political Economy*, Supplement) (Durham, N.C.: Duke University Press, 1973).

[8] See, for example, the accounts in Mary S. Morgan and Malcolm Rutherford, *From Interwar Pluralism to Postwar Neoclassicism* (*History of Political Economy*, Volume 30 Supplement) (Durham, N.C.: Duke University Press, 1998); and Yuval P. Yonay, *The Struggle over the Soul of Economics* (Princeton, N.J.: Princeton University Press, 1998).

(1903–1983) could use the neoclassical supply-demand graphic framework of Alfred Marshall (1842–1924) to analyze the various elements of labor exploitation, a Marxian concept, inherent in monopoly power.

Perhaps a more important reason was that neoclassical economics at that time had little to say about aggregate questions – that is, about money, growth, technological change, business cycles, or institutions. In these respects we should look rather to individuals such as J. G. Knut Wicksell (1851–1926) in Stockholm and his account of the cumulative process in economics, or to the monetary theories and measurements of Irving Fisher (1867–1947) in America, or to the strongly competing "schools" of economics of the time.

Historical economics remained the economics of choice for the German academy, and the late nineteenth century saw them locked in a bitter *Methodenstreit* with their Austrian neighbors. Whereas the German historical school, associated with Gustav von Schmoller (1838–1917), favored a holism centered on the national level, posited a clear role for the state, and paid close attention to externally adduced evidence, the Austrian school of Menger began with economic individualism, favored abstraction in theory, and advocated introspection as a source of evidence. Both Marxist and American Institutionalist approaches involved historical elements as a matter of method. Both were interested in the nature of the institutions of capitalism. Karl Marx's (1818–1883) economics drew heavily on the earlier classical tradition in its commitment to the labor theory of value and in its desire to provide an account of growth and stagnation as well as of capital accumulation. American Institutionalism, whose most well-known exponent was Thorstein Veblen (1857–1929), focused on the development of habits of economic thought and behavior at both the individual and social levels and on the evolutionary change these experienced.

Thus, between 1870 and 1940, Western economics cannot be easily characterized, since a number of vibrant intellectual approaches coexisted and neither beliefs nor methods fit easily under one label. Only if we look at the entire twentieth century can we see how the various strands of marginalism played out and how the elements of neoclassical economics developed to form a strong paradigm by the 1950s.[9] When, in the last quarter of the century, these essentially micro accounts became formally linked to the aggregate, or macro, level of economics and to certain elements of the institutionalist

---

[9] See Backhouse, *Modern Economic Analysis*. On the development of three American versions of neoclassical ideas during the period 1930–60, see Philip Mirowski and D. Wade Hands, "A Paradox of Budgets: The Postwar Stabilization of American Neoclassical Demand Theory," in *Interwar Pluralism*, ed. Morgan and Rutherford, pp. 260–92. For the two French traditions, and over a longer period, see Robert B. Ekelund, Jr., and Robert F. Hébert, *The Secret Origins of Modern Microeconomics* (Chicago: University of Chicago Press, 1999); and Bruna Ingrao and Giorgio Israel, *The Invisible Hand: Economic Theory in the History of Science* (Cambridge, Mass.: MIT Press, 1990), which also covers Italian thinking. On British neoclassicism over the longer run, see Maloney, *Marshall, Orthodoxy and the Professionalization of Economics*; and Blaug, *Economic Theory*, which also deals with the broader picture of neoclassical theorizing.

agenda to produce "the mainstream" in Western economics, other accounts, namely the historical and Marxist traditions, were pushed to the margins.[10]

The story of these events advanced inside economics faculties usually makes changes in theory, or theoretical debate, the main focus of the narrative.[11] Thus, the history of twentieth-century economics has usually been portrayed as the early domination and inexorable growth of neoclassical microeconomics. If we suspend belief in the inherent progressiveness of that paradigm, however, the changes portrayed in that story have no convincing dynamic, so that other historical factors need to be considered. The standard treatment also downplays the more obvious changes over the century in the way economics was practiced. This account therefore begins with tools for measuring the economy and for developing theories. Such a beginning enables us to show how the history of economics is intimately linked to the histories of economies and their political contexts, as well as to integrate the history of economic methods with the history of economic theories.

MEASURING THE ECONOMY

The drive to measure economic phenomena can best be understood as a movement dating from the late nineteenth to the mid twentieth century.[12] Despite the fact that many economic elements come ready-numbered, the concepts and entities appearing in economic theories present problems of aggregation and combination of the numbers, or of their representative power. Measuring the output of iron, a basic product of the late nineteenth century, required collecting data from many different firms and deciding on appropriate methods of aggregating them to form one series of measurements. The more complex problem of measuring "the price level," that is, the general level of prices, a measurement needed for applied studies in monetary economics, led to the development of index-number theory. This theory dealt with appropriate ways to combine the data collected on prices and quantities of many different goods into consistent sets of numbers from which a price-level series could be calculated.

---

[10] Few texts go beyond 1945 in their coverage; one introductory text that does is Harry Landreth and David C. Colander, *History of Economic Thought*, 3rd ed. (Boston: Houghton Mifflin, 1994); Backhouse's *Modern Economic Analysis* develops a more detailed account. A wealth of biographical material, and some useful subject histories, are contained in John Eatwell, Murray Milgate, and Peter Newman, eds, *The New Palgrave: A Dictionary of Economics* (London: Macmillan, 1987).

[11] One of the few recent texts to eschew such an approach is R. E. Backhouse, *Economists and the Economy*, 2nd ed. (New Brunswick, N.J.: Transaction, 1994); Backhouse follows an earlier tradition of relating the history of economics to economic history.

[12] There is no overall history of the modern measurement movement, but see Judy L. Klein and Mary S. Morgan, *The Age of Economic Measurement* (*History of Political Economy*, Volume 33 Supplement) (Durham, N.C.: Duke University Press, 2001) for a recent set of essays. See also Paul Studenski, *The Income of Nations: Theory, Measurement, and Analysis: Past and Present* (New York: New York University Press, 1958), for an exhaustive account of one important strand – the history of national income and wealth measurement until the 1950s.

The problem of choosing an appropriate index-number formula turned out to be a generic one for much economic measurement, spawning monographs on measurement formulas and debates over the relevant criteria that continue as a highly specialized part of the economics literature.[13] The arguments are technical and abstruse, but the topic is one with considerable practical relevance. A change in the measurement formula may be equivalent to wiping out the measured inflation or growth of an economy for a year, as happened in the United States during the 1990s.[14] There are also profound philosophical implications, for the choice of weighting schemes depends on different assumptions regarding equality among people.

Arguments also arose about the conditions for measurability of unobservable elements, such as "utilities," and about the appropriateness of measurement formulas for various economic concepts that are not already numbered, such as "capital." One particularly important example was the measurement of business cycles.[15] Most economists agreed that the cycle was a genuine phenomenon, but there was no agreed concept of it, let alone a definition or causal account. The cycle might be sought in data on output, prices, or other elements; its periodic length was unclear, as was its shape and regularity. The measurement procedures, concepts, and causal accounts were constructed hand in hand, in different business cycle institutes ranging from Cambridge, Massachusetts, to Moscow, from Vienna to Berlin, from the 1910s to the 1930s. Measurement was not an end in itself, but a necessary prerequisite for predicting the turning points of the cycles in economic activity that beset all economies, an ability much in demand during the interwar period.

The surge of interest in measurement thus had roots in both professional research and political demands. For economic scientists, it began in the strong institutionalist, historical, and empiricist traditions popular around the end of the nineteenth century. Academic economists, like other social scientists, often initiated and collected their own data sets in order to answer specific research questions. The Progressive movement in America and liberal and welfarist movements in Europe were committed to reforms that often relied on social science research and data, and in the face of these movements, governments increased their collection of economic information. But it was the requirements of war economies, and interwar problems, particularly the Great Depression, that massively increased the collection of data by the state and its agencies. By the 1950s, economists in the Western world had access

---

[13] There is no one history of index number measurement, but a glance at Irving Fisher's classic *The Making of Index Numbers* (Boston: Houghton Mifflin, 1922), which includes a huge number of different formulas, will give some insight into the topic.

[14] See the discussion of the Boskin report in "Symposium on Measuring the CPI," *Journal of Economic Perspectives*, 12:1 (1998), 3–78.

[15] See Mary S. Morgan, *The History of Econometric Ideas* (Cambridge: Cambridge University Press, 1990), pt. I.

to a bewildering variety of "official" data. Rarely since then have economists set out to take their own measurements.

Economists' ambitions in the realm of measurement soon led them, along with other social scientists, to develop mathematical statistics. Measurements that had been valued earlier for their own sake, as sufficient evidence in tables and graphs, were now asked to contribute to causal explanation. The methods of correlation and regression, originally designed for biometric data, were immediately adapted and developed by statisticians operating in the social science community.[16] The first multiple regression analysis ever done is reputed to be that of George Udny Yule (1871–1951), an English statistician cum social scientist, in 1899, on the determinants of why different poor law authorities gave out different amounts of relief payments. Beginning in the early twentieth century, economists used such statistical methods to measure parameters in simple relations. Understanding the law of demand, for example, required statistical analysis of the relations between data on the prices and quantities of a good. Methods of statistical analysis were thus welcomed into economics by those with different theoretical backgrounds and methodological approaches: Both historical and neoclassical economists developed faith in statistical evidence and methods.[17]

MATHEMATIZING ECONOMICS

The use of mathematics in economics began at roughly the same time as the drive to measurement, and, though its adoption was in many ways more gradual, it just as inexorably altered the way in which economics was practiced.[18] The introduction of mathematics was particularly associated with marginal utility economics. While it might seem that mathematics was a natural way to deal with the marginalists' account of utility, only two of the four variants of this thesis adopted mathematics: Jevons's account of individual feelings expressed with the differential calculus and Walras's equations for his general equilibrium exchange economy. Though Clark came to adopt

---

[16] For the general role of social scientists in statistical thinking around the turn of the century, see Donald MacKenzie, *Statistics in Britain, 1865–1930* (Edinburgh: Edinburgh University Press, 1981); Theodore M. Porter, *The Rise of Statistical Thinking* (Princeton, N.J.: Princeton University Press, 1986); Stephen Stigler, *The History of Statistics: The Measurement of Uncertainty before 1900* (Cambridge, Mass: Harvard University Press, 1986). For more specialist material on economics, see Judy L. Klein, *Statistical Visions in Time* (Cambridge: Cambridge University Press, 1997); Mary S. Morgan, "Searching for Causal Relations in Economic Statistics: Reflections from History," in *Causality in Crisis: The New Debate about Causal Structures*, ed. Vaughn McKim and Stephen P. Turner (Notre Dame, Ind.: University of Notre Dame Press, 1997), pp. 47–80.

[17] For the history of early developments in statistical economics up until the 1940s, see Morgan, *History of Econometric Ideas.*

[18] The best account of the range of attitudes toward mathematics and quantification held by late-nineteenth-century economists is Theodore M. Porter's "Rigor and Practicality: Rival Ideals of Quantification in Nineteenth-Century Economics," in *Natural Images in Economic Thought*, ed. Philip Mirowski (Cambridge: Cambridge University Press, 1994), pp. 128–70.

the mathematical formulation, Menger and the later Austrian school stood firmly against the use of mathematics in economics.

The development of marginal economics into neoclassical economics in the following generation began along the joint mathematical trajectories set by Jevons and Walras. It is traditional to understand Jevons's project as being concerned with decisions concerning the marginal utilities of the individual, or of individuals in exchange situations, a project most notably taken up by the Irish economist Francis Ysidro Edgeworth (1845–1926), who excelled in mathematics and statistics. The general equilibrium approach of Walras focused on the combination of all of the individual sellers and buyers, a project of interest to the American economist Irving Fisher, a student of the American physicist Willard Gibbs, who provided mathematical proofs of the equilibrium account in several domains. Vilfredo Pareto (1848–1923), the Italian economist who succeeded Walras in Lausanne, looked closely at the problem of the path to equilibrium. The English economist Alfred Marshall railed against the excessive use of mathematics in economics and stressed the notion of economics as a "moral" science. Nevertheless, the direction Marshall took was at least as important as that of Walras and Pareto for the history of neoclassical thinking, since he incorporated classical insights on the nature of production to explore the partial equilibrium of each market, good by good, and over time .

Questions of welfare, equity, and distribution, such as those raised by Henry George's (1839–1897) single tax movement or by Fabian socialists, were now treated with the new marginal and neoclassical tools. Clark replaced his earlier historical and institutional analyses of fair exchange with a mathematical account of the return paid to each factor of production in equilibrium. Pareto developed his criteria of overall welfare based on possible compensation from gainers to losers from any change in circumstances. Arthur Cecil Pigou (1877–1959) used marginal analysis to understand the divergence between private and social interests and Marshall's neoclassical concepts provided the basis for later tool-based analyses of equity and distributive questions arising from governmental actions. Some of these forms of social engineering based on mathematical formulation and calculation had been developed by French engineers during the nineteenth century, but only became general in public economic decision making during the middle and late twentieth century.[19]

By the early twentieth century, although the mathematizing project still had far to go, some key elements of the wider neoclassical picture had been worked out. The introduction of mathematics not only changed the way

---

[19] The importance of engineers in developing and applying these tools as active economists in France in the nineteenth century and in America in the twentieth century has been treated in Ekelund and Hebert, *The Secret Origins*; and in Theodore M. Porter, *Trust in Numbers: The Pursuit of Objectivity in Science and Public Life* (Princeton, N.J.: Princeton University Press, 1995).

that theorizing was carried out and concepts were defined, but also altered the questions considered relevant for study and the way in which they were formulated. For example, the older classical and verbally descriptive account of "free" competition had depicted a state in which firms were free to enter and leave the marketplace and actively competed within it. Early-twentieth-century inquiries into the nature of competition within the neoclassical framework developed the mathematically described concept of "perfect competition," an abstract situation in which no active competition took place between firms.[20] Replacing Adam Smith's famous "invisible hand" description of how order arose in the real economic world, a small group led by the French and American economists Gerard Debreu (b. 1921) and Kenneth J. Arrow (b. 1921) studied the mathematical questions of the existence and stability of the Walrasian "general equilibrium" economy, an ivory-tower speculation about a highly idealized, complex, and formally abstract economy.[21] Welfare economics, which seemed to have foundered on the impossibility of interpersonal welfare comparisons, found a new lease on life with Arrow's formalizing of theorems about social welfare functions and social choice theory. Mathematical theorizing radically changed the objects of study in economics and the kind of truth economists sought.

The proponents of mathematics in economics originally understood mathematics to be the most truthful way to express economic realities. As the twentieth century proceeded, mathematics became a more common, though still contested, form of expression for theory building in economics, until the 1950s, when neoclassical economics became the dominant paradigm. This growing commitment to the effectiveness of mathematics in economic reasoning was accompanied by a gradual weakening of the view that such mathematical representations could be understood to be, or empirically validated as, descriptively accurate.[22] With the retreat from realism, mathematical form took precedence over economic content, and mathematics was seen primarily as a language or tool for the exact expression of abstract theories. However, as the century wore on, the abstraction and formalism associated with mathematization were tempered by the practice of modeling.

[20] For a history of this transformation of the concept of competition, see K. G. Dennis, *"Competition" in the History of Economic Thought* (New York: Arno, 1977). For additional material on the relation to evolutionary thinking at that time, see Mary S. Morgan, "Competing Notions of Competition in Late-Nineteenth Century American Economics," *History of Political Economy*, 25:4 (1993), 563–604.

[21] See Ingrao and Israel, *The Invisible Hand*, for an account of this work. The formalist revolution, as it has sometimes been called, is also treated by Mark Blaug in "The Formalist Revolution or What Happened to Orthodox Economics after World War II," in *From Classical Economics to the Theory of the Firm: Essays in Honour of D. P. O'Brien*, ed. Roger E. Backhouse and John Creedy (Cheltenham: Edward Elgar, 1999), pp. 257–80. See also E. Roy Weinbraub, *How Economics Became a Mathematical Science* (Durham, N. C.: Duke University Press, forthcoming).

[22] See Ingrao and Israel, *The Invisible Hand*; Weinbraub, *How Economics*.

MODELING AND TOOL-BASED ECONOMICS

The mathematization or formalization typical of neoclassical economics has been interpreted as the replacement of words by geometry and algebra or by other mathematical languages. But historians of the discipline have hardly noticed that, during the 1930s, mathematics became attached to another tool – namely, "modeling" – to create a new style of scientific argument in economics.[23]

The term "model" seems to have migrated into economics with Jan Tinbergen (1903–1994), who used his experience in physics to develop econometric models during the 1930s. His models were special: They provided a simple and mathematical representation of the complexity of the real economy, and at the same time they formed the basis for a statistical description of the actual historical and structural relations embedded in the data of the real economy. Tinbergen was one of the leaders of the econometrics movement, an international movement of the interwar period committed to both statistical and mathematical methods and to their union with economics, so that economic relations could be expressed in a rigorous form and measured. To some extent, we can see this movement paralleled in other social sciences: Psychometrics and sociometrics developed their own particular version of statistical methods at the same time that econometrics emerged in economics. Nevertheless, these parallel movements did not take on the econometricians' commitment to mathematical representations (models) and mathematical methods.

Until 1950 or so, the union was maintained and practiced in economics by a small but enthusiastic band of econometricians. Since then, the fields have split; the term "econometrics" now refers only to the statistical side of tool-based economics.[24] Following the lead of Trygve Haavelmo (1911–1999) in the 1940s, econometrics has developed its own branches of theoretical statistics and several highly sophisticated, competing methodologies of application. Econometric models ranged from those describing time patterns to those picturing underlying behavioral mechanisms, from single equations to large models of several hundred equations, as developed by Lawrence R. Klein (b. 1920) and were often constructed for governments; they have formed the mainstay of econometrics into the late twentieth century. Perhaps because of the heavy reliance on this technology in applied economics, economists have invested much research effort in the area. Meanwhile, mathematical modeling

---

[23] Robert M. Solow's "How Did Economics Get That Way and What Way Did It Get?," *Daedalus*, 129:1 (2000), 39–58, offers a similar characterization of late-twentieth-century economics as a modeling science (in an essay that came out as this chapter was being drafted).

[24] For the history of econometrics before 1950, see Morgan, *The History of Econometric Ideas* (containing chapters on Tinbergen and Haavelmo); for the post-1940 period, see Duo Qin, *The Formation of Econometrics* (Oxford: Clarendon Press, 1993); Roy J. Epstein, *A History of Econometrics* (Amsterdam: North-Holland, 1987).

has provided economists with a tool for building and exploring theory, enabling them to build simple mathematical representations of the complex economy or of particular types of behavior and to analyze the theoretical implications by manipulations of the model. The adoption of the modeling style was indeed the primary way in which economics became a mathematized discipline.

Adopted for both statistical and mathematical reasoning in economics, modeling became, especially after midcentury, a distinctive element of both inductive and deductive economics, in both scientific and policy domains. Models were taken as sufficiently accurate representations of the economic world that they formed the basis for both advice to governments and firms and for normal academic science. Each emerging subfield of economic study acquired its own "theoretical" and "applied" economists. To return to the example of business cycles, models such as Tinbergen's both gave mathematical representation to older verbal theories and served as the basis for attaching data to provide measurements of the parameters involved in the relationships. As a consequence, business cycle work suddenly gained a high degree of specificity and exactitude in its claims. Later, with the sudden deepening of economic cycles in the 1970s and 1980s, new mathematical models, labeled "theories," were developed that bore a family resemblance to those of the 1930s; when connected to econometric models and data, these theories were "applied."

Twentieth-century economists viewed their measurement formulas, mathematical and statistical methods, and modeling tools as more "advanced," more properly scientific, than the words and verbal arguments of the nineteenth century, and regarded them as essential to the scientific claims of twentieth-century economics. Economists at the time, and historians since, have linked the use of such tools with the desire to ape natural science. Some notions were indeed imported from other sciences, although these ideas and methods were first adapted to fit economics and then further developed to become tools for economics specifically.[25] During the late nineteenth century, ideas from physics, physiology, and psychophysics were used in the accounts of the marginalists, and ideas from biometrics and social statistics in statistical economics.[26] In the mid twentieth century, information science and artificial intelligence, the so-called cyborg sciences, were another

---

[25] For example, see Marcel Boumans, "Paul Ehrenfest and Jan Tinbergen: A Case of Limited Physics Transfer," in *Non-Natural Social Science: Reflections on the Enterprise of More Heat than Light*, ed. Neil De Marchi (*History of Political Economy*, Volume 25 Supplement) (Durham, N.C.: Duke University Press, 1993), pp. 131–56.

[26] The physics analogy has been vividly discussed by Philip Mirowski in *More Heat than Light* (Cambridge: Cambridge University Press, 1989), and his account critiqued in De Marchi, *Non-Natural Social Science*. For the concern with psychology, see Margaret Schabas, "Victorian Economics and the Science of the Mind," in *Victorian Science in Context*, ed. Bernard Lightman (Chicago: University of Chicago Press, 1990), pp 72–93.

resource.[27] Very often, tools were carried by scientists themselves migrating between fields: Tinbergen brought the tools and concepts of physics with him in the 1930s; Herbert A. Simon (1916–2001) brought tools and concepts from information theory in the 1940s and 1950s. But larger historical factors were also at work in the adoption of tool-based economics: the historicist concern with evidence in the late nineteenth century, the "modernist" movement's focus on abstraction and formalism in early-twentieth-century science and culture, and the positivist philosophy of midcentury. On an historical scale, between these specific impulses and broad cultural factors, events in politics and in the economies themselves significantly reshaped economics.

THE CONTINGENCIES OF ECONOMIC HISTORY AND ECONOMIC RESPONSIBILITY

One of the things that needs to be explained about the adoption of tool-based economics is its timing. With the exception of measurement methods, these tools spread rather gradually before the 1930s. Demands from the policy domain for economic expertise, and especially for a "usable" economics during the period from 1930 to 1950, were critical for the full-scale adoption of tool-based economics that occurred after the 1950s. It is no accident, for example, that the League of Nations supported Tinbergen's econometric research during the late 1930s as part of its attempt to solve the national and international problems of the Great Depression. Both the historical timing and the nature of policy demands affected the character of the economic science that resulted.

Economists had laid claim to a special public policy expertise throughout the nineteenth century; but at that time the range of economic policy considered to be the responsibility of the state, and thus perhaps requiring economic expertise, was somewhat limited. While this range varied by nation, governments generally were taken to be responsible for trade policy, for keeping their own spending within budget, and for monetary and exchange rate policy. In this last case, the late-nineteenth-century view was that the gold standard, by then widely adopted in the Western world, was the ultimate "governor" maintaining the health of the national and international economies and making monetary/exchange rate policy automatic and self-stabilizing. Governments sometimes initiated legislation to protect vulnerable economic groups, but did not consider themselves to have any general economic responsibility for their citizens.

[27] See the papers in pt. 1 (by Mirowski, Sent, and Boumans) of John Davis, ed., *New Economics and Its History* (*History of Political Economy*, Volume 29 Supplement) (Durham, N.C.: Duke University Press, 1997); Philip Mirowski, *Machine Dreams: Economics Becomes a Cyborg Science* (Cambridge: Cambridge University Press, 2001).

The events of the twentieth century radically altered the balance of economic responsibility between the state and individuals across most Western economies. The economic policy experience of the interwar period, combined with that of two world wars, created the view that governments were responsible for intervening to maintain the health of the domestic economy, and thus for the economic security of their own people, as well as for the health of the international economy.[28]

In the case of the two world wars, economic planning and control were required on a hitherto unmatched scale, perhaps since the days of the Roman empire. The experience of economic planning during the First World War was somewhat more ad hoc and piecemeal, during the Second World War more organized and coherent. Regardless, the state's share in the economy grew rapidly during World War I, declined during the interwar period, rose again during World War II, and did not decline much thereafter. The difference between the wars, of course, was the Great Depression. All countries, developed and underdeveloped, experienced a considerable postwar downturn soon after World War I and severe collapses during the 1929–33 period, unmatched by anything after 1950. In the United States, among the most affected in that second depression, aggregate consumption and income fell by 25 percent. International trade and international financial institutions broke down, and the world economy moved towards autarky.[29]

The Great Depression had a profound effect on both the outlook of economists and on the economic responsibilities assumed by governments in the Western world. In the 1920s, most economists believed that business cycles were a regular and natural phenomenon of the capitalist economic system. But the severity of the Great Depression and its unusual length forced them to reexamine their beliefs about how the aggregate economy worked and forced governments to become proactive in economic affairs, with or without the blessing of their economic advisors.

In 1933, for example, Germany and America instituted wholesale economic interventions to end the Great Depression. In Germany, where one-third of the labor force was unemployed in 1933, massive government spending and investment combined with considerable levels of state control, though not central planning, created virtually full employment by 1936, before the full- scale move towards a war economy.[30] By contrast, the American New

[28] Mary O. Furner and Barry Supple, eds., *The State and Economic Knowledge: The American and British Experiences* (Cambridge: Cambridge University Press, 1990); A.W. Coats, ed., *Economists in Government* (*History of Political Economy*, Volume 13 Supplement) (Durham, N.C.: Duke University Press, 1981); Neil De Marchi, "League of Nations Economists and the Ideal of Peaceful Change in the Decade of the Thirties," in *Economics and National Security* (*History of Political Economy*, Volume 23 Supplement), ed. Craufurd Goodwin (Durham, N.C.: Duke University Press, 1991).

[29] James Foreman-Peck, *A History of the World Economy* (Hemel-Hempstead: Harvester-Wheatsheaf, 1995).

[30] Avraham Barkai, *Nazi Economics: Ideology, Theory and Policy*, trans. Ruth Hadass-Vashitz (New Haven, Conn.: Yale University Press, 1990).

Deal is counted a failure by economic historians. State controls were many
but incomplete; federal government spending was high, but more or less
canceled out by state governments' savings. The policy experiments of the
New Deal failed because each agency was staffed by a mixture of economists
and bureaucrats holding divergent views about both economic aims and
means of intervention.[31]

Despite their only partial success, the generation of economists who were
in their prime at the end of the Second World War felt both committed to
prevention of further depressions and optimistic that they had the tools.[32]
To understand why, we need to look more closely at the developments
within economics during the 1930s and their relation to the arts of economic
engineering.

"SOLVING" THE GREAT DEPRESSION: NEW ECONOMICS, NEW EXPERTISE, AND NEW TECHNOLOGIES

Beginning in the 1930s, economists worked with a general distinction between microeconomics (the behavior of the individual or firm) and macroeconomics (the behavior of the aggregate economy), though the labels themselves emerged only during the postwar period and became largely redundant
in the 1990s. Because of the importance attached to explanations of the Great
Depression, this came to be seen as a critical distinction. The mathematical
neoclassical economics of the first half of the century provided a micro-level
analysis at the level of firms and consumers on both sides of the market and
dealt with a combination of such markets in a general equilibrium account.
But it had nothing much to say about how individuals' different roles in the
economy were *aggregated*, or about the behavior of that aggregate economy,
the macro-level issues that seemed to be relevant for the dislocations of the
1920s and the Depression.

The problems of the aggregate domain were interpreted as questions
of monetary theory and business cycles and were broadly debated within
the extant "schools" of the period: the Austrian tradition, carried on by
Joseph A. Schumpeter (1883–1950) at Harvard and by Friedrich von Hayek
(1899–1992), who after the exodus from Vienna of the early 1930s, was
thriving at the London School of Economics and later at the University of

---

[31] William J. Barber, *Designs within Disorder: Franklin D. Roosevelt, the Economists and the Shaping of American Economic Policy, 1933–1945* (Cambridge: Cambridge University Press, 1996) and his *From New Era to New Deal: Herbert Hoover, the Economists and American Economic Policy, 1921–1933* (Cambridge: Cambridge University Press, 1985); Michael D. Bordo, Claudia Goldin, and Eugene N. White, eds, *The Defining Moment: The Great Depression and the American Economy in the Twentieth Century* (Chicago: University of Chicago Press, 1998).
[32] See the conversations with James Tobin, Franco Modigliani, and Robert Solow in Arjo Klamer, *The New Classical Macroeconomics: Conversations with New Classical Economists and Their Opponents* (Boulder, Colo.: Westview Press, 1984).

Chicago; the Swedish tradition, derived from Knut Wicksell and centered in Stockholm; the Americans, both Institutionalists such as Wesley Clair Mitchell (1874–1948) and orthodox economists such as Irving Fisher; and the Cambridge school in England led by John Maynard Keynes (1883–1946). All were aggregate theorists who assumed some particular beliefs and behavior of individuals, but the precise links between individuals and the aggregate remained unformalized in their accounts. And while they shared the questions posed by events in their economies, they worked with different methods of analysis and proposed different solutions.

In the stereotyped story of policy economics, the category of macroeconomics was put on the map by of the work of one Western economist – John Maynard Keynes. In that story, the importance of Keynes is that his work persuaded governments that they could keep their economies out of depression by adjusting their own spending: By their own actions, they could "manage" the economy. His ideas, which in the main came too late to be responsible for influencing policy during the Depression, were widely adopted after the war.[33]

For the economics profession, the stereotyped story is a different one: The importance of Keynes' work lay not in his solution, but in his analysis of the problem.[34] Keynes suggested that the aggregate level of activity depended on the level of effective demand, which could get stuck at a point at which unemployment remained because markets did not clear. This contrasted with the self-correcting mechanisms, or tendency toward market-clearing equilibrium, assumed in the older orthodox aggregate economics and in much of the newer business cycle economics. In Keynes's account, failures arose because of the ways that, in the aggregate, individuals, firms, and the government – whether as savers, investors or consumers – reacted to current events in the economy in the face of uncertainty about the future.

An adequate history, however, needs to explain why Keynesian economics won out over alternative accounts of the Great Depression, both in the academic domain and as a policy tool. The Stockholm School's analysis shared Keynes's assumption that the world was a disequilibrium world, but their theories involved a much more detailed analysis of the problem of incompatibility of individuals' plans taken together and within each time period.[35]

---

[33] But see the series of papers on "pump-priming" in *History of Political Economy*, 10:4 (1978), 507–48, for an example of tool-based Keynesian style engineering in the 1930s; for later Keynesian influence, see Peter A. Hall, ed., *The Political Power of Economic Ideas: Keynesianism across Nations* (Princeton, N.J.: Princeton University Press, 1989).

[34] Peter Clarke, *The Keynesian Revolution in the Making, 1924–1936* (Oxford: Clarendon Press, 1988), centers on the development of Cambridge analysis; David Laidler, *Fabricating the Keynesian Revolution* (Cambridge: Cambridge University Press, 1999), reviews the debates in aggregate economics and other issues discussed here.

[35] Lars Jonung, ed., *The Stockholm School of Economics Revisited* (Cambridge: Cambridge University Press, 1991); Bjorn A. Hansson, *The Stockholm School and the Development of the Dynamic Method* (London: Croom Helm, 1982); Bo Sanderlin, ed., *The History of Swedish Economic Thought* (London: Routledge, 1991).

Though in many ways attractive as an explanation of what happened at the aggregate level, because it paid full attention to micro behaviors and how these fitted together, it remained largely theoretical and incomplete. The statistical information and mathematical analysis required to make the Stockholm School's approach operational, either as a fully articulated aggregate level theory or as a guide for general advice or government action, did not seem feasible in the 1930s. Ragnar K. Frisch (1895–1973), a Norwegian econometrician of the period, did try to develop a planning model based on consumption requests, with some family resemblances to the Stockholm ideas, and quantified the calculations required. They were of a similar order to those required under socialist planning, another alternative solution to the Depression available in the Marxist tradition. Following the work of Italian economist Enrico Barone (1859–1924), the period from 1920 to 1960 saw a vehement theoretical debate between the Marxist tradition, represented notably by the Polish econometrician Oskar Lange (1904–1965), and the Austrian tradition, represented by Ludwig von Mises (1881–1973) and Hayek. The issue was whether markets were necessary for economic efficiency. It turned out that the socialist planned economy could reach as good an outcome as the free market economy in terms of optimal production and welfare for all individuals, for a given technology and distribution of income – the "Pareto optimum." The information assumed for the necessary calculations did not exist, however, in the absence of a market.[36] "Austrians," who eschewed data and calculations and made their arguments in the traditional manner in words, used the principle of methodological individualism in their scientific accounts and held to a strong belief in the efficacy of the free-market system to solve all economic ills, a stance that became increasingly untenable as the Depression continued.

Keynes's book, *The General Theory of Employment, Interest and Money* (1936), was difficult; like contemporary analyses of business cycles, it was written in the old style, yet with some attempt at formal analysis. But his ideas were very quickly translated by economists in Britain and the United States into simple mathematical models of the macro economy; the longest-lived and flexible, the "IS-LM model," came from John R. Hicks (1904–1989), who was at that time developing a general equilibrium account at a miniature level.[37] These macro models were manipulated to give specific answers to concrete and real policy questions, using the comparative static method, well known among economists and understood from Marshall's microeconomics of the early century. The Keynesian analysis did demand new aggregate data, such as aggregate income and consumption, but once assembled the data could be used to measure parameters of the Keynesian relationships using

---

[36] Don Lavoie, *Rivalry and Central Planning: The Socialist Calculation Debate Reconsidered* (Cambridge: Cambridge University Press, 1985).

[37] William Darity, Jr., and Warren Young, "IS-LM: An Inquest," *History of Political Economy*, 27: 1 (1995), 1–42.

statistical models and methods.[38] The resulting model-based analysis, if not Keynes' book, produced answers that could be explained to governments, and it was deemed more scientifically advanced than the older "commonsense" analysis. The element of surprise in its advice – that governments should spend their way out of depression, not save because times were bad – was also important in making it acceptable in the political domain; in the 1940s and 1950s, politicians wanted new solutions to the old economic problems. Thus, whereas the alternative economic accounts of aggregate economics available in the 1930s relied on general verbal advice or analytical and planning tools that were too complex or too demanding of data or calculation to be feasible, the Keynesian account generated what might be called intermediate technologies, that is, practical ones for governments in need of policy prescriptions and scientists seeking adequate explanations of events.

The exact historical claims about when, where, and from what sources Keynesian economics was put in place are subject to debate.[39] The more important point is that economic expertise and usable technologies were developed together. After 1950, with the aid of new data, new statistical methods, and simple mathematical models of the economy and economic behavior, economists made their advice effective across a wider range of fields – from older domains, such as regulation of natural monopolies and monetary policy, to newer problems, such as the creation of stabilization schemes and the control of war economies and finance. The profession demonstrated its ability to respond to a range of regular problems, such as the design of subsidies for farmers, and to economic emergencies, such as hyperinflation, with new policy prescriptions that turned out to have varying degrees of success and failure. The failures were, perhaps, a more important dynamic for the history of economics than the successes.

THE FEEDBACK FROM ECONOMIC ENGINEERING TO HISTORICAL EVENTS

Economists' engineering and historical contingencies constantly interact, producing new economics, technologies, and expertise. In this interactive context, macro- and microeconomics became formally joined. Keynesian ideas appeared to be reasonably successful during the 1950s and 1960s, when

[38] Studenski, *The Income*; national income accounting also provided a considerable stimulus for such data collection and usage. On the work of the Russian-born Simon Kuznets in the United States, see Carol S. Carson, "The History of the United States National Income and Product Accounts: The Development of an Analytical Tool," *The Review of Income and Wealth*, 21:2 (1975), 153–82; Mark Perlman, "Political Purpose and the National Accounts," in *The Politics of Numbers*, ed. William Alonso and Paul Starr (New York: Russell Sage, 1987); on the work of John Richard N. Stone in the United Kingdom, see the entry on him by Angus Deaton in *The New Palgrave*, ed. Eatwell, Milgate, and Newman, vol. 4, pp. 509–12. See also Ellen O'Brien, "How the 'G' got into the GNP," in *Perspectives in the History of Economic Thought: Method, Competition, Conflict and Measurement in the Twentieth Century*, ed. Karen I. Vaughn (Aldershot: Edward Elgar, 1994).

[39] Hall, *Political Power*.

the analysis was used to design fiscal policy and to "manage" the economy. This was perhaps the high period of the economist as engineer, advising the government on how to set the levers of economic control. Western governments used economists' models and calculations to dampen the economic cycles in their economies and to engineer relatively stable growth, low inflation, low unemployment, and a reasonable balance of payments. In certain open economies, those with a relatively high level of trade compared to their gross national income, there were problems in timing the levers. In retrospect, it appeared that these levers were rather crude tools: They were designed to change incentives for individuals in the system, even though the ultimate aim was to affect the aggregate. In addition, the government itself was an actor, and its own spending and saving was another control lever. Such economic engineering thus did not mean external control over an object, but rather conscious action taken by one of the major components of the machine.

Governments' ability to manage or control their economies suffered a severe breakdown during the 1970s. The most immediate evidence of that failure was the new phenomenon of "stagflation," both high inflation and high unemployment, a combination inconceivable within Keynesian economics, which perceived a trade-off between the two. The problem prompted a number of diagnoses. First, the theory and policy design of Keynesian economics focused on the demand side of the economy, while economists gradually concluded that stagflation resulted from changes on the supply side – in particular, from the large shock given by the 1973 rise in oil prices. A second explanation connected the rising inflation with the neglect of monetary elements in the Keynesian system, a critique led by the monetarist Milton Friedman (b. 1912). Another element in the account was the role of expectations: As people got used to the amount of inflation in the economy, they modified their behavior based on an expected amount of inflation remaining in the system and so exacerbated the stagflation. A fourth element was that the government's actions were being second-guessed, thus invalidating its power to manage the economy while at the same time being an actor in it. This "Lucas critique," named after the Chicago economist Robert E. Lucas (b. 1937) and built on earlier versions of the same insight, was another nail in the coffin of the government as controller of the economy. Economists judged, in effect, that the Keynesian demand management of previous years had helped to create stagflation and that its continuation after the supply-side shock had exacerbated the problem; they represented this in an aggregate supply-demand analysis that became popular at the time. Thus, in a simple domain transfer, a standard neoclassical micro-level tool was applied to the macro context to explain a phenomenon and a policy failure at the aggregate level.

Economists' accounts of stagflation spawned the "rational expectations revolution," an analysis that connected the microeconomics of uncertainty at the individual level with the impact of policy tools at the macro level.

Developed primarily by Lucas, this thesis argued that individuals should be assumed to hold "rational expectations," that is, that they made use of all the information they had and so did not make systematic errors; such expectations might be taken as formally equivalent to those embedded in the economic and econometric model. As a result of the stagflation experience, economists came to hold the view that macroeconomic models should always have adequate micro foundations, that is, that they should be consistent with a set of assumptions, mathematically represented, about the behavior of the individuals in the economy. The technology of new economic models thereby served to underwrite the integration of macroeconomic theory with neoclassical microeconomic theory.[40] The individuals represented in the economy were now also bound tightly into the model by the presumption of rational expectations. Thus the push for micro-macro integration was a result of the practical experience with stagflation, but its particular form was determined by the two postwar disciplinary contexts of an increasing mathematical formalism and, as we will see in the next section, the renewed ideological attraction of individualism.

The most striking case of feedback from economic engineering to economic events and ideas came with the collapse of communism, which Western economists largely blamed on the failures of Eastern block economics. Eastern European economics was the product of firmly held ideologies and strongly based theories of production, along with techniques of central planning; it had delivered growth rates substantially above those of the free capitalist West for much of the early postwar period. When their citizens grew disenchanted with the economic outcomes produced in later years by their own economic experts, they were ready and eager to invite Western economists into their countries to teach them "modern" economics. Western expertise did not prove entirely equal to the task of designing economic institutions for the Eastern countries' transition to capitalism, and that experience challenged Western neoclassical mainstream economists to incorporate the role of institutions into their formal models.

THE IDEOLOGICAL TURN IN AMERICAN ECONOMICS

The day-to-day practice of economics turned technical at midcentury, just as economic ideas became a central and more highly specified element in the ideologies of different world power blocks.[41] Particularly in American economics, the acceleration toward tool-based economics and the

---

[40] Kevin D. Hoover, *The New Classical Macroeconomics: A Sceptical Enquiry* (Oxford: Blackwell, 1988); Backhouse, *Modern Economic Analysis*.

[41] This section draws particularly on my essay "American Economics: The Character of the Transformation," written jointly with Malcolm Rutherford, in *From Interwar Pluralism*, ed. Morgan and Rutherford, pp. 1–26. I thank Malcolm Rutherford for allowing me to draw on that material here.

development of a full-fledged neoclassical economics were intimately connected with the ideological war. These connections are important to an
account of Western economics, for it was during this period that American
economics became dominant in the Western discipline, just as the United
States gained economic and political dominance.

The thesis that American war and Cold War experience were critical for
the turn of American economics to a tool-based discipline in general, and to
neoclassical economics in particular, requires amplification. Tool-based economics had been important in the American experience of fighting the war,
not only in economic policy terms but in other areas as well, for mathematical and statistical techniques and modeling could be turned to many ends,
specifically to direct war aims. Indeed, the economic side of the war effort
was partly determined by businessmen rather than by economists, while the
economists were employed in tasks like the design of bombing raids. The
war experience also produced data and planning experience that were grist
for the mill of statistically minded Institutionalists. Research on such matters
as linear programming, operations research, game theory, and decision theory, involving concepts and mathematical techniques that became mainstays
of later twentieth-century neoclassical economics, were generously funded
as defense expenditures, and such research and funding continued into the
Cold War years.[42]

The economic values enshrined in the Cold War between East and West
are well known. Postwar Western economic values were more clearly defined
in opposition to the centrally planned East. The leader of the "free" West,
the United States, preached a theory of free markets as the most efficient
ones. The Eastern bloc economic ideology began with Marxian production
planning and aimed at fairness, not efficiency. Meanwhile, Western European
ideals marked a middle way, aiming for reasonably free, and thus moderately
efficient, markets and a reasonable level of distributional equity through
welfarism and state intervention. The Western economic ideology bore down
more strongly on the academic community in the United States than on those
in Western Europe, with consequent effects on the views held by economists.

While war work supported tool-based economics, the American political
movement against communism in the later 1940s and the McCarthyism of
the early 1950s decided the issue in favor of neoclassical economics at the
local level. Although the overall picture has yet to be filled in, it is clear
that economists had to be careful in expressing their views.[43] One economist

---

[42] Mirowski, *Machine Dreams*; Robert Leonard, *From Red Vienna to Santa Monica: Von Neumann,
Morgenstern and Social Science, 1925–1960* (Cambridge: Cambridge University Press, forthcoming)
and his "War as a 'Simple Economic Problem': The Rise of an Economics of Defense," in *Economics
and National Security* (*History of Political Economy*, Volume 23 Supplement), ed. Craufurd Goodwin
(Durham, N.C.: Duke University Press, 1991).

[43] Ellen W. Schrecker, *No Ivory Tower: McCarthyism and the Universities* (New York: Oxford University
Press, 1986); Craufurd Goodwin, "The Patrons of Economics in a Time of Transformation," in *From
Interwar Pluralism*, ed. Morgan and Rutherford, pp. 53–84.

writing about this period suggested that moving to tool-based economics was a defensive option against ideological persecution, though this sometimes proved to be an inadequate defense, particularly for those whose values were not aligned with the new ideology. There are examples of American economists of mild left-wing sympathies (including one future Nobel Prize winner, Klein) leaving the United States for the safety of Europe. Others who held such views remained, for the effects of loyalty oaths and McCarthyism were uneven. Nevertheless, economists who preached Keynesianism – viewed by some as close to socialism – or who had advocated postwar socioeconomic planning of the sort associated with Institutionalist positions were particularly at risk from university administrators, local state governments, and research institute trustees, who sought to purge their faculties of "reds" and "pinks" during the late 1940s and early 1950s.

Though neoclassical economics had been slow to spread in the United States during the interwar period, unlike the economics of institutionalism, it was nonetheless one of the forms of economics unambiguous in its support of capitalism. The ideal abstract neoclassical economy takes as its problematic the efficient use of existing resources, and analysis of this model suggests that result is best achieved by minimizing interference in the market. The neoclassical theory of distribution, in part developed by the American economist J. B. Clark around the turn of the century, assumed that the efficient economy would also be characterized by a just distribution for each contributing factor: Labor and capital would earn precisely what was due to them. In this privileging of efficiency and the ideal economy, the important questions of equity arising from the original distribution of wealth in the actual economy are left to one side. The values of neoclassical economics were perfectly aligned with the American position in the ideological war, so that during the postwar years the virtues of free individuals operating in free markets, or "economic democracy," came to seem inseparable from the virtues of political democracy.

In sum, it was neoclassical economists, whose mode of analysis had come to rely most heavily on the adoption of statistics, mathematics, and modeling technologies – those same techniques that had proved so efficacious during the war – who found their economic values most closely aligned with those of postwar society at large. In this context, pressures to conform to the newly (re)established American ideal of free markets and individual capitalism boosted the adoption of neoclassical economics at the expense of the previously dominant Institutionalist approach within the economics profession in America.[44]

Throughout the postwar period, American neoclassical economists claimed that tool-based analysis provided a mantle of scientific neutrality

<hr />

[44] Malcolm Rutherford, "Understanding Institutional Economics: 1918–1929," *Journal of the History of Economic Thought*, 22 (2000), 277–308.

with respect to all ideological positions. This claim could not be made by the free market and libertarian "Austrian" tradition, by then largely domiciled and increasingly naturalized in America, for their methods were old-fashioned words, which no longer held the guarantee of scientific objectivity. Only in the 1980s and 1990s, when the political climate had turned so far to the right as to obscure their ideological tinge, did the Austrian accounts associated with Hayek and Schumpeter of the functioning of free markets, the role of competition as both a creative and a destructive agent, and the self-organizing nature of the market economy feed successfully into American mainstream economics, which then developed their ideas on the role of information and the evolution of competition in formal and technical ways. After the fall of the Eastern communist regimes, some of the ideas and questions associated with the "old" American Institutionalists also found their way back onto the agenda. But these too were now integrated into the mainstream, so it was difficult at first to recognize the congruence between "old" and "new" institutionalists, whose ideas could be found in realms ranging from law and economics, in the work of Ronald H. Coase (b. 1910), to economic history, in the work of Robert H. Fogel (b. 1926) and Douglass C. North (b. 1920). The "old" concerns with economic justice and the inseparability of theory and evidence were lost, but interest in economic habits and institutions reappeared in investigations into the rules and conventions of behavior, the legal and economic arrangements of economic units, and the processes of learning and adaptation.[45]

From this discussion it appears that tools and values cannot be divorced. But in the following sections we will see that tools remained partially independent of values, and that differences in values enabled Western economics as a whole to retain a certain variety. First, however, we need to examine more closely the scientific character and value commitments of tool-based neoclassical economics.

TOOLS AND ECONOMIC SCIENCE

The dependency of later twentieth-century economics on technologies, particularly its concentration on the modeling method, involved a subtle downgrading of economists' scientific ambitions. Published papers and books at the start of the twentieth century tended to treat specific real questions by invoking general claims or laws about how the economy works and discussing them in the context of specific cases. Alternatively, they treated questions empirically, almost as a piece of economic history, rather than invoking any particular explanations or laws. Economics was seldom abstract, and the

---

[45] Malcolm Rutherford, *Institutions in Economics: The Old and the New Institutionalism* (Cambridge: Cambridge University Press, 1994).

distinction between theoretical and applied economics could not easily be made.[46] A century later, economics papers tended to treat specific questions directly, either in abstract terms, by means of mathematical modeling under the heading "economic theory," or empirically, through econometrics. By the late twentieth century, there were no longer any "laws" of economics and few general theories – only models of concrete, but not necessarily real, cases.

We can see this process at work in the twentieth-century mathematical work characterizing individual behavior. From the 1890s to the 1930s, economists of the neoclassical persuasion retreated from the possibility of measuring individuals' underlying utilities and satisfied themselves with representing the situation of the choice between goods in mathematical form. Particularly in the United States, they also turned away from making claims about motivation and psychology.[47] The postulates used to characterize such individual choice behavior were outlined in Britain in the 1930s by Hicks and Roy D. G. Allen (1906–1983) and axiomatized by the American economist Paul Samuelson (b. 1915) in the 1940s, creating the depersonalized "rational economic agent" of the latter half of the twentieth century. This was a highly idealized and abstract representation, not thought to characterize any real person or actual behavior. Neoclassical economics used this model person to explore not the reasons for action, but the consequences of acting rationally, as defined by those economists, in a specified situation.

To its many critics, this portrait of individual self-interested behavior seemed highly restrictive, yet it did not forbid very much: Rationality was narrowly defined, but to behave rationally, an individual had only to prefer more goods to fewer and to maintain a certain consistency in choice situations. This allowed simplified models of behavior to be invoked in concrete and complicated situations. A good example is the postwar development of the economics of the family, a case where other social scientists resented neoclassical economic work as imperialist. In this subfield, developed by the American Gary S. Becker (b. 1930), economists explored the consequences of their general theory of individual behavior for such typical decisions as which parent should go to work and whether or not to have another child. Modeling suggested the "rational" and "efficient" decision for the specific family situation modeled. Such concrete "theoretical" – that is, mathematical – models became attached to real situations when they were reformulated for statistical work. Econometricians added greater realism and complexity to the model of economic rationality by taking other factors into account and by assessing the fit of the model to real-world data.

---

[46] Roger E. Backhouse, "The Transformation of U.S. Economics, 1920–1960, Viewed through a Survey of Journal Articles," in *From Interwar Pluralism*, ed. Morgan and Rutherford, pp. 85–107.

[47] A.W. Coats, "Economics and Psychology," in *Method and Appraisal in Economics*, ed. Spiro J. Latsis (Cambridge: Cambridge University Press, 1980), pp. 43–64.

In such neoclassical modeling, it was the restrictive neoclassical assumptions of self-interest depicted as rationality that enabled the reduction to simplicity necessary for the mathematical models, and it was this that non-neoclassical economists found objectionable. For critics, the effect of the program was to erase whatever did not fit the paradigm. But while it may have seemed otherwise, the neoclassical program did not prove immune to such criticisms, and modeling developed in three new directions during the late twentieth century. First, the dual impact of critiques of the economists' notion of rationality by Herbert Simon and Amartya Sen (b. 1933) in the 1970s and the results reported from laboratory experiments in the 1980s broadened the concept and theoretical characterizations of economic rationality. The "rational economic agent," who had become so pervasive in economics during the third quarter of the twentieth century, came, in the final quarter, to be used more as a benchmark for the exploration of behavior patterns that varied from that ideal. Second, it was no longer assumed that each microeconomic individual acted independent of other individuals; rather, they had to be modeled in situations of interaction. Third, economists found a way within their paradigm to take institutions into account. Despite appearances, the tools of neoclassical economics turned out, by the end of the century, to be adaptable to a wider range of assumptions (and so implicit values) and a greater variety of situations than had earlier been conceived.[48]

We can see this flexibility in the field of "game theory." This was a body of investigation, dating from the classic work by John von Neumann (1903–1957) and Oskar Morgenstern (1902–1977) published in 1944, and later developed primarily in America and Germany, that became dominant in late-twentieth-century economics and was exported to evolutionary biology and political science. In game theory, individual "agents" are placed in situations of interaction with each other called "games."[49] This placement is not usually real, but a thought experiment worked through in a model representation in mathematical form. Since the 1980s, such investigations have been one of the main foci of the growing program of laboratory experimentation in economics, using methods similar to those found in social psychology.[50] This has allowed economists to study the processes of economic interaction and learning in a "controlled" field. The "games" in both thought and real experiments, are defined as situations with rules of interaction or "institutions": who moves first, how many moves there are, what kinds of moves can be made, and so forth. As in the usual modeling method of neoclassical micro

[48] For a discussion of the individualistic values imbedded in marginal and neoclassical economics at the turn of the last century that is compatible with the range of commitments discussed here, see Maloney, *Marshall, Orthodoxy, and the Professionalization of Economics*.

[49] E. Roy Weintraub, *Toward a History of Game Theory* (*History of Political Economy*, Volume 24 Supplement) (Durham, N.C.: Duke University Press, 1992).

[50] Vernon L. Smith, "Experiments in Economics," in *New Palgrave*, ed. Eatwell, Milgate, and Newman, vol. 2, pp. 241–9.

theory, each type of game could be "applied" to concrete situations in which individuals or firms (the economists' "agents") might find themselves. This has enabled game theorists to apply their ideas to specific fields, such as industrial economics, where strategic choice has a natural role in the problem of describing and understanding the behavior of competing firms.

The dominant neoclassical economic theory of the postwar period was in many ways rather general; modeling gave it content because economists used the method to explore what the theory would mean in specific, rather simple, circumstances. By contrast, the larger economic world was seen as incredibly detailed and complex. Modeling, even the more elaborate econometric models maintained by economists in government, made the economy seem open to investigation. It was the simplistic quality of such models, particularly the smaller ones, with their effective reduction of complexity and their ability to produce answers explainable in terms of rather simple propositions of economic efficiency and rationality, that made neoclassical advice ubiquitous in the economic sphere and invasive even in the political and social spheres.[51]

THE NEXUS OF TOOLS, SCIENCE, AND IDEOLOGY

Although the values of neoclassical economics were aligned with those of the general market orientation of Western, and particularly of American, economic ideology, tools and ideology were not fully aligned, especially in the policy domain.[52] Even while relying on economic theory to espouse the benefits of free markets and unfettered capitalism, American economic policies in the domestic arena and those exported abroad remained interventionist and depended on tools. For example, the Marshall Plan required that recipient countries have an overall economic policy constraint conceived in Keynesian aggregate terms, (and this in turn required the local provision of national income accounting systems, based on Richard Stone's design), which, at that time of reconstruction, required some strict domestic policies, even though at the same time commitments to open markets were extracted.[53] Western ideologies and tools figured prominently in the relationships among countries, donors, and international agencies. Through its own Foreign Aid Program and its dominance among economists in international agencies, such as the International Monetary Fund and the World Bank, the United States

---

[51] For a good example, see Jacques J. Polak, "The IMF Model at 40," *Economic Modelling*, 15:3 (1998), 395–410; and for a more general portrait of the insider's view, see William R. Allen, "Economics, Economists, and Economic Policy: Modern American Experiences," *History of Political Economy*, 9:1 (1977), 48–88.

[52] The classic treatment of the interrelations of values and theory development, rather than tools, is K. Gunnar Myrdal, *The Political Element in the Development of Economic Theory* (New York: Simon and Schuster, 1963).

[53] M. J. Hogan, *The Marshall Plan: America, Britain and the Reconstruction of Western Europe, 1947–1952* (Cambridge: Cambridge University Press, 1987); for discussion of national income accounts, see note 38.

exported beliefs in the virtues of free competition and an economy free of
government direction along with a set of tools meant to aid in the design
of economic policy, planning, and project assessment. The economics of
the "free world" seemed to require an arsenal of economic tools of intervention to make sure that it worked "properly" – that is, according to the
donor's design – in new countries. Even economists who had little sympathy with Western economic ideals soon learned to use the tools in order to
maximize the aid their economies received. The ideologies of Marxism and
communism of the Eastern bloc countries also connected their satellites to
economic engineering, for Marxian economies required structural analysis of
the economy and high levels of data collection and calculation for purposes
of production planning.

   Nevertheless, tools were more genuinely autonomous, or detachable, from
values in policy usage than is suggested by these observations: Tools were
neither totally domiciled nor fully independent in either Western or Eastern
ideologies. One tool that was widely thought to represent the task of central planning is the Leontief input-output matrix, developed by Wassily W.
Leontief (1906–1999), a Russian economist who emigrated to America. This
method uses industry-level data, on inputs into and outputs from each industry, to portray the technical interrelations between the sectors of the economy
in matrix form. Such matrices can be used to understand and analyze technical relations and to predict or plan industrial output at various levels, ranging
from the industry level to the national economy. This technique fitted neatly
with the economic theory of Eastern bloc countries that assume labor creates value in production, so growth has to be understood and planned at
the level of production. In fact, however, it was only in the 1960s that the
tool was imported from the United States to play a marginal role in Soviet
central planning, which had relied on the more practical method of material
balances. In any case, the use of such matrices does not necessarily require
the theoretical commitment to a labor theory of value, and input–output
analysis has been by no means confined to Eastern bloc countries. Norway,
for example, has used these methods in conjunction with a form of national
budgeting accounts as a standard part of its economic information and policy
analysis since the Second World War. French indicative planning of the postwar period was also based on a version of the method. Leontief constructed
such matrices for the U.S. economy as part of an academic research initiative
during the 1930s, and they have also provided tools for academic research
into economic performance. Such tables were used by the U.S. government
in the 1940s to predict the probable economic response to the end of the war
within different economic sectors. Thus, although not the main policy tool,
input-output tables have often been constructed and used for policy analysis
in Western countries.

   During the second half of the twentieth century, the tool-based style and
neoclassical content of American economics became the dominant influence

not only in policy terms but also within Western economic science. The disciplinary background helps to explain how this American economics was exported to other countries.[54] One of the main conduits was through the adoption of American economics education, the development of graduate school training based on American lines, and the preference to send students for training in the United States rather than somewhere else. Whereas during the late nineteenth century American economists had typically undertaken training in Europe, mainly in Germany, by the late twentieth century the flow had been reversed; the preferred place of economics study for Europeans became America. The decline of European imperial power during the postwar period meant that economists who had earlier looked to Britain or France as the educational model, as the place to train graduate students, and for leadership in economic science and expertise, began to look elsewhere. For example, Australia became more American-oriented in its economics and began to see American economics as the new role model. India later followed a similar route, initially having imported Soviet planning ideas and found training opportunities in the Eastern bloc. New members of America's informal empire were even better candidates for importing American economics. South Korea soon began sending its brightest students to the United States for economics graduate training; they found homes in university departments and in important positions in government on their return.[55] International agencies such as the International Monetary Fund and the World Bank contributed to the Americanization process. Early repositories of American economics at a technical level, they also exported these ideas directly, by training other nationals and by specifying in their operational and technical manuals how to evaluate policy regimes, design programs, and assess project proposals.

We know most about this process of Americanization of economic science from certain cases in Latin America. Here the record describes specific attempts by a combination of governmental, academic, and charitable American institutions to instil "good" or "modern" – that is, neoclassical tool-based – economics into the academic and political elites of Latin American economies.[56] Latin Americans, both those who approved of the import of American economics and those who disapproved, openly interpreted the

[54] A.W. Coats, ed., *The Post-1945 Internationalization of Economics* (*History of Political Economy*, Volume 28 Supplement) (Durham, N.C.: Duke University Press, 1996); A.W. Coats, ed., *The Development of Economics in Western Europe since 1945* (London: Routledge, 1999).

[55] Young Back Choi, "The Americanization of Economics in Korea" in *Post-1945 Internationalization*, ed. Coats, pp. 97–122.

[56] Veronica Montecinos, "Economists in Political and Policy Elites in Latin America," and Maria Rita Loureiro, "The Professional and Political Impacts of the Internationalization of Economics in Brazil," in *The Post-1945 Internationalization*, ed. Coats, pp. 279–300, 184–210; J. G. Valdes, *Pinochet's Economists: The Chicago School in Chile* (Cambridge: Cambridge University Press, 1989). For earlier attempts to export "good" social science, see Earlene Craver, "Patronage and the Directions of Research in Economics: The Rockefeller Foundation in Europe, 1924–1938," *Minerva*, 24:2–3 (1986), 205–23; Martin Bulmer and Joan Bulmer, "Philanthropy and Social Science in the 1920s: Beardsley Ruml and the Laura Spelman Rockefeller Memorial, 1922–29," *Minerva*, 19:3 (1981), 347–407.

changes in their academic economics as Americanization; but European academics preferred to see the trend as one of "internationalization" or even of "denationalization," for they were never quite so open to channels of American domination.

European academics gradually became more American in their concern with academic credentials and citations and their adoption of American-style graduate training schemes, all of which created mechanisms for conformity. Yet in many respects European economics retained its individuality. This may be because of the wider range of economies and ideologies that co-existed within European democracies, and the greater public service ethos of European economics, which made European economists more likely to spend some of their working time outside the ivory tower of the university and inside government or in politics.[57] For example, in Italy and Japan economics was, for much of the postwar period, home to active groups of Marxist economists.[58] Despite the American role in reconstruction, many Marxists regained their positions at the end of the war, for they had been active in resisting the fascist war regimes in those countries. Dutch economics remained largely wedded to what is known as the Tinbergen legacy, involving technocratic management of the economy and a practical commitment to social justice in analysis and outcomes. Norwegian economics also remained to some extent concerned with the econometric legacy of Frisch, displaying its own brand of commitment to economic planning and policy design. French economics supported a strong group of modernists of high theory in the mathematical and statistical domains, but such economists represented only a small part of the economics profession in France, which seemed, like Germany, to remain relatively immune to the internationalist trend. In Britain, while the Keynesian legacy continued into the 1970s, academic and policy economists were, from that time, more ready to follow American examples in both disciplinary and theoretical respects. In Europe as a whole, the concern for economic security and a relatively equal economic distribution kept issues of political economy firmly on the scientific and policy agendas. In scientific endeavor, as in the sphere of policy advice, tools proved in part autonomous and applicable in circumstances where the values of rationality and efficiency inherent in American neoclassical economics might be taken to be second-order values.

Most late-nineteenth-century Western economists read several languages and often wrote in many. Despite language barriers, communication between members of recognized national schools was effective and active; yet national schools thrived. By contrast, with the domination of American economics

[57] R. L. Frey and Bruno Frey, eds., "Is There a European Economics," *Kyklos*, 48:2 (1995), 185–311.
[58] Pier Luigi Porta, "Italian Economics through the Postwar Years," and Aiko Ikeo, "The Internationalization of Economics in Japan," in *The Post-1945 Internationalization*, ed. Coats, pp. 165–83, 123–41.

during the late twentieth century, the languages of scientific economics have become unambiguously mathematics, statistics, and English. These shared languages have been advanced as another of the reasons why the tool-based style of American economics has proved an effective scientific export. But the existence of shared tools and language, and the partial autonomy of tools from ideology, have also provided an easy entry for challenges to American mainstream ideas. Thus, some of the most interesting developments of late-twentieth-century economic analysis have come from third world economists operating within the first world community, the most notable example being Sen's analyses of famines and poverty.

CONCLUSION: THE DYNAMICS OF
THE ECONOMICS DISCIPLINE

The twentieth-century discipline of economics, its ideas, methods, institutions, "schools," and the shifting of what constitutes the "mainstream," depended not only on the everyday internal dynamics of normal science, but also on the demands of changing historical realities at local, national, and international levels. This is the way "nature" works in economics: The economies throw up unexpected economic events or demands of such magnitude that they exert a strong discipline on the pattern of economics. At the same time, the economic science of the twentieth century has, by means of its engineering interventions in the economy, engendered new economic "events," to be reckoned with by new generations of economists. Thus the use of technological methods of analysis and tools of intervention, a particular feature of Western economics in the twentieth century, created a peculiarly reflexive dynamic for the discipline. The practice of economics over the twentieth century changed from a primarily verbal method to one dependent on mathematics, statistics, and modeling. This move was connected to the growing power of an American-dominated neoclassical economics, but it was also dependent on many other contingencies, generated from inside economics and from outside. The histories of tool-based economic science and of the economies it analyzes cannot easily be separated, nor can they be pulled apart from local ideologies, the foreground within which economics thrives.

17

POLITICAL SCIENCE

James Farr

The idea that politics is, or can be, the subject of science is an ancient one that reaches back to Aristotle. Early modern expressions of the idea can be found in Machiavelli and Hobbes, as well as in Enlightenment thinkers from Hume to the American Founders. "Science" was understood as the systematic knowledge of first principles, whether prudential or philosophical, and "politics" as the public life of a city-state, kingdom, or republic. This old science of politics became remote in time and worldview during the nineteenth century with the fluorishing of the democratic state and the empirical natural sciences. In 1835, Tocqueville foresaw the consequences in *Democracy in America*: "A new political science is needed for a world itself quite new."[1]

The democratization of politics and the scientization of knowledge are two forces of modernity that explain the formation and transformation of the social sciences in general.[2] But these forces are particularly crucial for understanding a "new" political science, given their conscious problematization by those who have styled themselves "political scientists."[3] Political

---

[1] Alexis de Tocqueville, *Democracy in America* (1835), ed. J. P. Mayer (New York: Harper and Row, 1966), p. 12.

[2] Mary O. Furner, *Advocacy and Objectivity: A Crisis in the Professionalization of American Social Science, 1865–1905* (Lexington: University Press of Kentucky, 1975); Thomas L. Haskell, *The Emergence of Social Science: The American Social Science Association and the Nineteenth Century Crisis of Authority* (Urbana: University of Illinois Press, 1977); Peter T. Manicas, *A History and Philosophy of the Social Sciences* (Oxford: Basil Blackwell, 1987).

[3] Bernard Crick, *The American Science of Politics* (Berkeley: University of California Press, 1959); Edward A. Purcell, *The Crisis of Democratic Theory: Scientific Naturalism and the Problem of Value* (Lexington: University Press of Kentucky, 1973); David M. Ricci, *The Tragedy of Political Science: Politics, Scholarship, and Democracy* (New Haven, Conn.: Yale University Press, 1984); Raymond Seidelman with the assistance of Edward J. Harpham, *Disenchanted Realists: Political Science and the American Crisis, 1884–1984* (Albany: State University of New York Press, 1985); James W. Ceaser, *Liberal Democracy and Political Science* (Baltimore: Johns Hopkins University Press, 1990); James Farr, John S. Dryzek, and Stephen T. Leonard, eds., *Political Science in History: Research Programs and Political Traditions* (Cambridge: Cambridge University Press, 1995); Rogers M. Smith, "Still Blowing in the Wind: The American Quest for a Democratic, Scientific Political Science," *Daedalus*, 126 (1997), 253–87.

scientists were unique among the emergent social disciplines in using "science" in their chosen name, and they made the politics of a democratic age their principal inquiry and fundamental problem. Like Tocqueville, they displayed considerable ambivalence about democracy and what they should do about its shape and progress. Democracy needed to be explained and understood. But it also needed to be educated, because citizens wanted improvement, administrators needed training, and officials required statecraft. To satisfy democracy's needs and their own competing goals, political scientists looked to natural science as a model – either to emulate or against which to pattern their own methodologies and cultural authority. Initiating debates over "democracy" and "science" that persist to this day, they abetted the fractiousness of modern democratic discourse and voiced the perennial allure and contentiousness of desiring a natural science of politics. Their debates demonstrate the constructedness and contestedness of modern political science, as well as establishing its story line from 1890 to 1970.

THE DISCIPLINING OF POLITICAL STUDIES, TO 1890

As an academic discipline, political science emerged out of moral philosophy, which had dominated collegiate education in Europe and the United States. Moral philosophy laid out a system of modern natural law that tied political science to ethics, as well as to law, history, and economics. It educated young citizens of gentrified background in their duties, including those that would come with office or public trust. Thus, collegiate political science was essentially the science of public duty, and it was the public duty of moral philosophers to teach this science to new citizens. History was an auxiliary inquiry that charted the development of modern political institutions and portrayed the disturbing spectacles of ancient democracies compared to which new republics and modern nations were evolutionary improvements.[4]

Francis Lieber (1800–1872) was particularly important in the emergence of political science from moral philosophy. A German emigré to the United States, Lieber represented the internationalism of early political science. A friend of Tocqueville, he formed a scientific cloverleaf with two great European figures, Johann K. Bluntschli (1808–1881) in Germany and Edouard Laboulaye (1811–1883) in France. His *Civil Liberty and Self-Government* became the first proper textbook of the new discipline and helped to earn

---

[4] Anna Haddow, *Political Science in American Colleges and Universities, 1636–1900* (New York: Appleton-Century, 1939); Stefan Collini, Donald Winch, and John Burrow, *That Noble Science of Politics: A Study in Nineteenth Century Intellectual History* (Cambridge: Cambridge University Press, 1983); Dorothy Ross, *The Origins of American Social Science* (Cambridge: Cambridge University Press, 1991); James Farr, "From Modern Republic to Administrative State: American Political Science in the United States," in *Regime and Discipline: Democracy and the Development of Political Science*, ed. David Easton, John G. Gunnell, and Michael B. Stein (Ann Arbor: University of Michigan Press, 1991), pp. 131–67.

him the first professorship of political science, at Columbia College in 1857. Lieber raised new methodological issues regarding electoral statistics and textual hermeneutics. Favoring representative government and the authority of trustees, he inveighed against women's suffrage and majoritarian "democratic absolutism,"[5] and encouraged the university reform already under way in Europe, especially in Germany.

The German university transformed scientific education through research and specialization, bestowing on the professorate cultural prestige and political authority. It was emulated by aspiring political scientists in the United States at Johns Hopkins and Columbia Universities, where the disciplinary infrastructure took shape. Johns Hopkins created the first Political Science Association (1877), and Columbia the first graduate School of Political Science (1880). These universities also published the discipline's first journals, *The Johns Hopkins Studies in History and Political Science* and *Political Science Quarterly*. To these university developments may be added the beginning of specialized schools for the training of civil servants, especially at the École Libre des Sciences Politiques (1871) and the London School of Economics and Political Science (1895).

The infrastructure of schools, associations, and journals institutionalized the new discipline by the turn of the century, but it did not definitively establish what made the discipline a distinctive science. At the École Libre, under Émile Boutmy (1835–1906), the plural political *sciences* of law, history, and economics coexisted for the purpose of enlightening civil servants. At Columbia, the School of Political Science, founded by John W. Burgess (1844–1931), also included sociologists such as Franklin Giddings and economists such as E. R. A. Seligman. At the London School, cofounder Beatrice Webb despaired of hiring a lecturer in political science, since it was "a trifle difficult to teach a science which does not yet exist."[6] However, the disciplinary infrastructure was crucial for convening scientific debates, providing public platforms, and giving at least a *nominal* answer to the question, "Who are political scientists?" They were scholars and educators in professional schools and university departments who read journals and joined associations that invoked the name "political science."

In America, disciplinary institutions grew immediately and continuously through the next century. In Europe, significant expansion did not occur until after World War II, and then under American influence.[7] For these reasons, like those that Tocqueville had foreseen regarding democracy, the

---

[5] Francis Lieber, *Civil Liberty and Self-Government* (1853), 3rd ed. (Philadelipha: J. B. Lippincott, 1901), p. 156.

[6] Quoted in Jack Hayward, "Cultural and Contextual Constraints upon the Development of Political Science in Great Britain," in *The Development of Political Science*, ed. David Easton, John G. Gunnell, and Luigi Graziano (London: Routledge, 1991), p. 95.

[7] William G. Andrews, ed., *International Handbook of Political Science* (Westport, Conn.: Greenwood Press, 1982).

turn-of-the-century discipline was largely an American invention and remains so to this day. However, many European thinkers proved very influential. Indeed, the discipline's grand theories would almost invariably be European, beginning with the theory of the state.

STATE AND PLURALISM THEORIZED, 1890–1920

The theory of the state dominated the imagination of the new discipline and was elaborated most fully in texts of German legal philosophy. Bluntschli insisted that political science was *Staatswissenschaft* (science of the state), not "a chapter of moral philosophy." Its teachings concerned the forms, ideals, and development of the state, conceived as the highest political association of a legally constituted nation, with government as its deputed agent. Vaultingly ambitious as political philosophy and empirical history, the theory proved timely when analyzing nineteenth-century developments, especially the rise of representative democracy. This state form, governed by elected officials limited by law, contrasted favorably with direct democracy, whose propensity was "despotism of the mob."[8] German, French, and Italian variants of *Staatswissenschaft* proliferated. The theory also proved useful in Britain, as well as in the United States,[9] the country that had first used the term "state" to name itself. The distinction between state and government was important for American political scientists such as Burgess, who had studied in Germany, especially in the context of post–Civil War nationalism, to which political science contributed. "The national state . . . is the self-conscious democracy," Burgess proclaimed, but "the most advantageous political system" was "a democratic state with an aristocratic government."[10] Other American political scientists, among them Woodrow Wilson (1856–1924) in *The State* (1889), denied the separation of state from government, but for no less elitist purposes. The state was *in* its government, ruled by the elected or appointed few. Wilson expressly agreed with Bluntschli that politics is the special province of the statesman, and administration of the technical official. Whatever the precise status of government, the "state" met the conceptual demands of scholars searching for general theory and the political aspirations of a professional class alarmed by popular excitements in a democratic age. Here theory and institutionalization joined, paradigmatically in America but also in Europe, where democratic demands for expanded suffrage and

---

[8] Johann K. Bluntschli, *The Theory of the State*, 3rd ed. (Oxford: Clarendon Press, 1895), pp. 3, 463.

[9] Peter Wagner, *Sozialwissenschaften und Staat: Frankreich, Italien, Deutschland, 1870–1980* (Frankfurt: Campus Verlag, 1990); Pierre Favre, *Naissance de la Science Politique en France, 1870–1914* (Paris: Fayard, 1989); John G. Gunnell, *The Descent of Political Theory: The Genealogy of an American Vocation* (Chicago: University of Chicago Press, 1993).

[10] John W. Burgess, *Political Science and Comparative Constitutional Law*, vol. 1 (Boston: Ginn, 1891), pp. 3, 72.

responsive government were being heard. The university educated the few in *Staatswissenschaft*, while providing political legitimation for the educators.

The theory of the state provoked a *Methodenstreit*, a debate over methods, concerning the facts, theories, and values of the democratic state. Bluntschli put forward philosophy and history as complementary methods, warning against their respective perversions into "abstract ideology" and "mere empiricism."[11] For others, these methods diverged. Those drawn to philosophy emphasized statist ideals and normative values. Those drawn to historical actualities echoed the aphorism of Sir John Seeley: "History without political science has no fruit; Political science without History has no root." They also codifed a "comparative historical method" that traced the empirical development of different states, joining anthropologists and philologists who sought the comparative histories of different races and languages.[12] Links to biology provided for Burgess the methodological imprimatur of "Natural Science," but this self-legitimating gesture invited further debate. Burgess thought the comparative historical method required studying transformations of constitutional law, whereas Wilson insisted upon study of actual politics, since "Politics is the life of the State." James Bryce (1838–1922), the most insightful commentator since Tocqueville on American democracy, sidestepped the normative "merits" of democracy so as "to present simply the facts of the case," including unsavory ones about corruption, party bosses, and uninformed public opinion.[13] This formative *Methodenstreit* set the pattern for later debates attending the aspiration to science, debates over theory versus fact and fact versus value.

Political scientists were also intent on reforming the government and administration of the state, and thus were actors on democracy's stage as well as scholars. Wilson, for example, thought that the "democratic forces of the age" – including popular education, immigration, and "cheap printing" – created "new problems of organization" that political scientists must solve. He championed legislative leadership and "progressive policy" modeled on the English parliamentary system. He also advanced a reformist "science of administration" – useful even in England, since "England herself is close upon democracy" – to place national civil service above the fray of party politics and public opinion. Reform of city government was urged in works such as *Municipal Problems* (1897), by Frank Goodnow. Crediting English and French models, Goodnow espoused a conservative Progressivism. The city should be substantially independent of state government and absolutely free from domination by political parties. A strong city council with constrained

---

[11] Bluntschli, *Theory of the Modern State*, p. 1.

[12] Seeley and comparative method are discussed in Collini, Burrow, and Winch, *That Noble Science*, chap. 7; for American developments see Ross, *Origin of American Social Science*, chap. 3.

[13] Burgess, *Political Science*, vol. 1, p. vi; Woodrow Wilson, *An Old Master and Other Political Essays* (New York: Scribner's, 1893), p. 51; James Bryce, *The American Commonwealth* (1893), 2nd ed. (New York: Macmillan, 1922), p. 4.

mayors and appointed experts in policy would properly separate "political" from "administrative" functions, a distinction that the discipline's emergent subfield of public administration henceforth took as its premise. Given an un-informed electorate and in some places "a large ignorant negro population," elections should be held only to fill key offices, with infrequent referenda. Political science could thus serve the cause of administrative reform and its own "theory of limited democracy."[14]

Disciplinary institutions expanded during the early twentieth century, especially in America, as democratic and scientific debates morphed into new shapes. Departments of political science followed in the wake of the democratization of higher education at state universities, among them the Universities of California (1903), Illinois (1904), Wisconsin (1904), Michigan (1911), Minnesota (1915), and Kansas (1917), as well as at some private ones, including Northwestern (1915) and Stanford (1919). Harvard created a pro-fessorship in the science of politics housed in its Government Department. The American Political Science Association (APSA) was founded in 1903 by professors, lawyers, and civil servants, many of them members of the American Historical Association (AHA) or the reformist American Social Science Association (ASSA).[15] The APSA delineated the discipline's three principal subfields which remained roughly stable thereafter: political theory or philosophy; public law, including constitutional, international, and ad-ministrative law; and "the general study of government," which encompassed political relations, policy, and administration. Over time, the third field de-volved into separate American, comparative, and international subfields.[16] A new journal, the *American Political Science Review* (*APSR*), was a further sign of the discipline's development and of its attention to both science and politics, as scholarly articles were published alongside "News and Notes" of actual politics. The *APSR* also published the latest mechanism for contin-uing the discipline's *Methodenstreit* and its pronouncements on democracy: the annual presidential address.

The early presidential addresses thematized science and exposed contin-uing anxieties about democracy. For some, science displayed a realistic, em-pirical, and nonpartisan orientation that proceeded above common opinion but beneath the heights of "philosophical speculation." "A habit of mind which checks controversial attitudes and faddish enthusiasms," it sought "to influence the statesman on the one hand and to supply intellectual pabulum to the people on the other." When taught, it "creates in the class which leads

---

[14] Wilson, *Old Master*, pp. 105, 111, 118, 136; Wilson, "The Study of Administration," *Political Science Quarterly*, 2 (1887), 198; Frank J. Goodnow, *Municipal Problems* (New York: Macmillan, 1897), pp. 147, 309–10.

[15] Institutional developments (with slight differences in departmental dates) are discussed in Albert Somit and Joseph Tanenhaus, *The Development of American Political Science: From Burgess to Behavioralism* (Boston: Allyn and Bacon, 1967), chap. 5; Ricci, *Tragedy of Political Science*, chap. 3.

[16] Westel W. Willoughby, "The American Political Science Association," *Political Science Quarterly*, 19 (1904), 108.

a nation the proper temper and attitude" but without "vain hopes of introducing the certitude or the authority of science into politics." This latter note of caution on the part of Bryce, who served as the fourth APSA president while serving as British ambassador to the United States, was shared by Harvard's president, A. Lawrence Lowell. His address in 1910 analogized political science to physiology in its search for "the organic laws of a political system," but without denying that the "ultimate object of political science is moral." Wilson, the new governor of New Jersey, invoked the image of "law and fact" in his address. But, he confessed, "I do not like the term political science," since human relationships require "insight and sympathy."[17] The discipline's identity debates continued to implicate its name.

The theory of the state receded, however, as an explanation of democratic politics. The word "state" itself would not disappear, and political scientists would continue to wrestle with the genuine problems that the theory confronted. As long as there is the least pretense to popular government or to national aspirations, a concept of political association will be needed that persists across changes in regime, or that a people or nation may reappropriate from their delegated governments in times of revolution. But the gestures toward "facts" and "political systems" begun by Wilson, Bryce, and Lowell stimulated strong condemnations of nineteenth-century state theory and its comparative historical method. The first professor of political science at the London School of Economics, Graham Wallas (1858–1932), wished to reverse the trend that would "analyse institutions and avoid the analysis of man." Only via psychology could political science improve its "curiously unsatisfactory state" and assist elites in representative democracies, for "those who have had most experience of its actual workings are often disappointed and apprehensive." Henry Jones Ford, another of the early APSA presidents, retained the term "state," but called for the use of more "data" and "the evidence of behavior" in writing its "Darwinian . . . natural history."[18] The invocation of Darwinism evoked little applause in political science (unlike sociology),[19] but "data" and "behavior" were words with bright shining futures. So too was "process," which figured prominently in Arthur F. Bentley's stunning critique, *The Process of Government*. Bentley (1870–1957) decried the "barren formalisms" of "a dead political science," especially that "spook," the state. The subject of political science should be interests and activities associated

[17] Frank J. Goodnow, "The Work of the American Political Science Association," *Proceedings of the American Political Science Association*, 1 (1904), 43; Albert Shaw, "Presidential Address," *American Political Science Review (APSR)*, 1 (1906), 182; James Bryce, "The Relations of Political Science to History and Practice," *APSR*, 3 (1909), 16; A. Lawrence Lowell, "The Physiology of Politics," *APSR*, 4 (1910), 7, 9; Woodrow Wilson, "The Law and the Facts," *APSR*, 5 (1911), 11.
[18] Graham Wallas, *Human Nature in Politics* (1908), 3rd ed. (New York: Knopf, 1920), pp. ix, 25, 37; Henry Jones Ford, *The Natural History of the State* (Princeton, N.J.: Princeton University Press, 1915), pp. 1, 131.
[19] John S. Dryzek and David Schlosberg, "Disciplining Darwin: Biology in the History of Political Science," in *Political Science in History*, ed. Farr, Dryzek, and Leonard, pp. 123–44.

with "group phenomena." Democracy simply concerned "large group interests," not any "mysticism of 'the people.'" Bentley even took to task those
who paid lip service to factual inquiry:

> Your political scientist thinks he is going a long way afield and that he
> is meritoriously portraying 'actual' government when he inserts in his work
> some remarks on the machine [or] the boss. . . . But the boss himself is almost
> as formal an element in political science as is the president or governor. When
> you state him you have not stated the living society. You must still go behind
> to find what are the real interests that are playing on each other through his
> agency.[20]

The emphasis on interests and activities concerned not only theory but
also an interested, activist view of the discipline's practical purposes. The
allegiance to Progressive reformism persisted, and in Bentley and Charles
A. Beard (1874–1948) it found more radical expression of opposition to entrenched economic interests in politics. The APSA expressed the majority's more conservative, professionalized activism when, in 1913, its premier
committee listed the interests of political science departments for a new
century: "(1) to train for citizenship; (2) to prepare for professions such as
law, journalism, teaching, and public service; (3) to train experts and to
prepare specialists for government positions." A fourth "might be added"
regarding "research."[21] This was not unscientific activism, but a "political"
science that brought its expertise to bear on civic, legal, and administrative
practices.

World War I completed the transformation of the youthful discipline. The
war ended Progressive and Fabian hopes to keep attention focused on the
reform of domestic policy in the United States and England. It also helped
worldwide to pin "the state" on Germany and German political philosophy.
Ernest Barker, who would soon become the first professor of political science at Cambridge, led an assault in 1915 on "the discredited state," which
had, as one American later remarked, "offered so excellent an apology for the
Goose-step."[22] Motivated by nationalism, the assault nonetheless embodied other meanings, including a face-lift for "democracy." Having criticized
democratic politics as a political scientist, Woodrow Wilson as president of
the United States entered the hostilities with the famous declaration that
"the world must be made safe for democracy." Disenchantments persisted
regarding democratic realities, but the term "democracy" reached an historic
elevation that it maintained throughout the twentieth century. Moreover,
British and American political scientists swept clean any remaining thoughts

---

[20] Arthur F. Bentley, *The Process of Government* (1908), ed. Peter H. Odegard (Cambridge: Harvard
University Press, 1967), pp. 162–3, 263, 455.
[21] Quoted in Somit and Tanenhaus, *Development of American Political Science*, p. 82.
[22] Ernest Barker, "The Discredited State," *Political Quarterly*, 2 (1915), 101–26; William Y. Elliott, *The
Pragmatic Revolt in Politics* (New York: Macmillan, 1928), p. 86.

that the state was a monistic sovereignty acting with absolute authority over the lives of citizens. American federalism had never been like that anyway, but the scientists of the state had not attended to the teeming life of group activities that operated beneath or beyond government. Although British political scientists such as Barker faced a more centralized national government, as well as a history of Idealist statism and legal positivism, the loyalties of British subjects to their state were also cleaved by church, association, and trade union. Rule was thus dispersed. "Polyarchy" was Barker's term; others dubbed it "the pluralistic state" or simply "pluralism."

Pluralism replaced state theory as the discipline's primary and contested theoretical preoccupation. The state existed, but it was composed juridically and explained scientifically by group interests and functions. Reminiscent of Tocqueville, pluralism found in modern democracy both a diversity of interests and the danger of majority tyranny. Sociological methods were now required for studying a democracy composed of social groups. Harold Laski (1893–1950), an Englishman at Harvard and later at the London School of Economics, where he succeeded Wallas as professor of political science in 1926, developed the most influential account of the pluralistic state in *Studies in the Problem of Sovereignty* (1917) and *Authority in the Modern State* (1919). Like Bryce before him, Laski continued to make political science a transatlantic discipline. He was joined in Britain by G. D. H. Cole, and both drew upon Otto von Gierke. Their efforts "set the agenda for political theory" in Britain and commanded disciplinary attention worldwide.[23] In France, the pluralist research program was joined by jurists such as Leon Duguit, and in the United States by political scientists and sociologists such as Mary Parker Follett and Ellen Deborah Ellis. Follett placed pluralism in the service of associational democracy and pragmatic individualism. The task was to understand "the method of self-government," which meant understanding the "group process" of political organization and the "psychic process" of self-development. For "democracy *is* group organization," not the polling booth; "it is the bringing forth of a genuine collective will, one to which every single being must contribute the whole of his complex life." The "whole body of citizens" was at stake with pluralism, not just a few "'good' men in office." For her part, Ellis credited pluralists with rejecting German authoritarianism while articulating the group basis of genuine democracy.[24]

Apart from the pluralist debate, Follett and Ellis were notable figures in a discipline that provided virtually no support for academic women and that had fought women's suffrage with tortured arguments and gendered

---

[23] Paul Q. Hirst, *The Pluralist Theory of the State* (London: Routledge, 1989), p. 8.

[24] Mary Parker Follett, *The New State: Group Organization the Solution of Popular Government* (New York: Longmans, 1920), pp. 7, 11, 75, 168; Ellen Deborah Ellis, "The Pluralist State," *APSR*, 14 (1920), 393–407.

imagery.[25] Even then, Follett was without department and as much sociologist as political scientist; and Ellis taught at Smith, one of the historic women's colleges. Things would not change in this regard for several decades. But the discipline was at a crossroads in 1920, with the Nineteenth Amendment enfranchising women in the United States, the end of the war, and a host of new challenges.

A "NEW SCIENCE" OF POLITICS, 1920–1945

Between 1920 and 1945, political science continued its steady pace of institutional development in the United States and a slower though noticeable one elsewhere. Membership in the APSA jumped from 1,300 to 3,300 during those years, and most universities added political science departments to their rosters. The discipline continued its "hole-and-corner existence at British universities," though Cambridge inaugurated a named chair in 1928, joining those at the London School of Economics, Oxford, and University College London. In France, political science – "the daughter of history, law . . . and geography" – remained in the family of subjects considered necessary for educating civil servants. The study of "politics as science" emerged in 1920 at the Deutsches Hochschule für Politik in Berlin and thrived there until placed under the Ministry of Propaganda in 1933. During the liberal-democratic interlude of Weimar, the Hochschule combined "pragmatic and democratic concerns with political education," as against the conservative *Staatswissenschaften* still regnant at universities.[26]

Profound problems faced democracy, fueling the discipline's debates, including mass immigration, expanding enfranchisement, economic depression, authoritarian movements, totalitarian states, and another world war. Three National Conferences on the Science of Politics (1923–5), held in the United States, advocated "actual observations of political processes" in order to solve or control such "problems of politics."[27] "Democracy" was the name for the representative governments and constitutional systems that political science was to study and help guide. While most political scientists had by then come to endorse the broad ideals of democracy, they were at odds regarding the capacity of the people, as opposed to elected representatives

[25] See Beverly B. Cook, "Support for Academic Women in Political Science, 1890–1945," *Women and Politics*, 6 (1986), 75–104; Mary G. Dietz and James Farr, "'Politics Would Undoubtedly Unwoman Her': Gender, Suffrage, and the Origins of American Political Science," in *Gender and American Social Science: The Formative Years*, ed. Helene Silverberg (Princeton, N.J.: Princeton University Press, 1998), pp. 61–85.

[26] Somit and Tanenhaus, *Development of Political Science*, pp. 91–4; entries by Hayward and Favre in *International Handbook of Political Science*, ed. Hayward, quotations at pp. 155, 355; Hans Kastendiek, "Political Development and Political Science in West Germany," in *Development of Political Science*, ed. Easton, Gunnell, and Graziano, pp. 108–26.

[27] Charles E. Merriam, *New Aspects of Politics* (1925, 1931), 3rd ed. (Chicago: University of Chicago Press, 1970) p. 334.

and elites, to participate in it. Francis G. Wilson, who called himself a prag-
matist and pluralist, thought that an "inactive electorate" was a "permanent
factor" in America's "conservative democracy." He was not complaining, for
"it is a tribute to the rational character of the citizen that he is little interested in voting." One APSA president went further, rejecting the "dogma"
of universal suffrage: "[T]he ignorant, the uninformed, and the anti-social"
should not vote; governance should be left to "an aristocracy of intelligence
and character." However, Beard, who had been president of both the APSA
and the AHA, sounded a more optimistic and populist note. "Our democracy rests upon the assumption that all human beings have a moral worth in
themselves."[28] This undergirded his – and he hoped the twin disciplines' –
commitment to popular democratic education.

   The debates over democracy implicated those over science, and the latter
were acrimonious during this period. Some political scientists were attracted
and others repulsed by the more rigorous views of a naturalistic, value-free
science that emerged. These debates premiered in the United States, but they
drew in and upon European thinkers. George E. G. Catlin, a Cornell-trained
Englishman, wrote that there was "no such thing as political science in more
than a barren name." However, such a science was possible, if the natural scientific attitude regarding experimentation and avoidance of "political values"
were copied, and a measurable unit of political life found. Where economics
had money, political science could use "power." William Bennet Munro,
an old-line municipal reformer and skeptic regarding "gigantic campaigns
of civic education," also felt that a "pure science of politics" would remain
"backward" until it embraced an "analogy from the new physics" and discovered the "inexorable laws of . . . politics" that governed phenomena beneath
state, group, and behavior. Munro offered his own "law of the pendulum,"
which ostensibly explained how political attitudes swung from one extreme
to the other, especially in times of ideological extremism. William Yandell
Elliott, a conservative constitutional liberal who had studied at Oxford,
thought that such views "very aptly summed up the prevailing scientistic currents among American Political Scientists." He thereby added Catlin's and
Munro's names to a list that included pluralists such as Laski, Duguit, and
John Dewey, whom he blamed for making political science "positivistic, behavioristic, descriptive – and morally blind." The moral blindness, he opined,
was due to pragmatism, which had proved "impatient with representative
government" and had accommodated fascists like Mussolini, "the prophet
of the pragmatic era." Against pragmatists and "the 'pure' political
scientist," Elliott championed "political theory" with avowedly "normative
principles."[29] Despite considerable differences, it was plain to all participants

[28] Francis G. Wilson, "The Pragmatic Electorate," *APSR*, 24 (1930), 32, 35, 37; Walter J. Shephard,
"Democracy in Transition," *APSR*, 29 (1935), 18, 20; Beard quoted in Ricci, *Tragedy of Political
Science*, p. 95.
[29] George E. G. Catlin, *The Science and Methods of Politics* (New York: Knopf, 1927), pp. 84, 298;
William B. Munro, *The Invisible Government* (New York: Macmillan, 1927), pp. 35–7; Munro,

that the fate of democracy was intertwined with the contested fate of science. For many, the commitment to both science and democracy was problematic, if not contradictory. Value-free science appeared to preclude a rational defense of the value of democracy.

During the interwar years, Charles E. Merriam (1874–1953) stands out as the most important figure in the discipline's history. Merriam left an indelible mark on the institutional shape and intellectual framework of political science.[30] He was an unparalleled organization man, building a department at the new University of Chicago that shifted the discipline's center of gravity away from Columbia. As in sociology, and later in economics and law, a "Chicago School" emerged composed of the colleagues and students Merriam drew together. Merriam established a pattern of collaborative research that, like his Series on the Making of Citizens, was frequently international in composition. He founded the Social Science Research Council (SSRC) in 1923 with donations from the Rockefeller Foundation. The SSRC and corporate philanthropy thereafter played a crucial role in the scale of research, as well as in the kinds of questions social scientists asked or were funded to answer.[31] Merriam, furthermore, exemplified the scholar-activist-expert in political science. On behalf of Progressive causes, he ran successfully for the Chicago City Council and unsuccessfully for mayor. He served as an adviser to presidents from Wilson to Truman, playing especially significant roles in propaganda activities during World War I and in national planning during the New Deal.

Merriam's scholarship was a trail of the discipline's intellectual migration from statism through pluralism to an emergent behavioralism. He wrote extensively about power, parties, and elections, as well as about the history of political thought. In *Non-Voting: Causes and Methods of Control* (1924), with Harold F. Gosnell, he developed survey instruments to investigate why, in the immediate aftermath of women's suffrage, so few citizens registered or voted. He kept *The Making of Citizens* (1931) and *Civic Education in the United States* (1934) on the discipline's list of practical concerns, and he placed public policy *On the Agenda for Democracy* (1941). In answering *What Is Democracy?* (1941) and contrasting *The New Democracy and the New Despotism* (1939), he reinforced the "democratic" allegiances of political science in a world plagued by new totalitarian states.

Merriam was also famous for *New Aspects of Politics* (1925), his methodological manifesto for "the new world made over by modern science." Indeed, he

"Physics and Politics – An Old Analogy Revisited," *APSR*, 22 (1927), 5, 10; William Y. Elliott, "The Possibility of a Science of Politics," in *Methods in Social Science: A Case Book*, ed. Stuart Rice (Chicago: University of Chicago Press, 1931), p. 74; Elliott, *Pragmatic Revolt in Politics*, pp. 53, 84–5, 337.

[30] Barry S. Karl, *Charles E. Merriam and the Study of Politics* (Chicago: University of Chicago Press, 1974).

[31] Donald Fisher, *Fundamental Development of the Social Sciences: Rockefeller Philanthropy and the United States Social Science Research Council* (Ann Arbor: University of Michigan Press, 1993).

narrated the discipline's history in terms of changing scientific methods: "(1) The a priori and deductive method, down to 1850; (2) the historical and comparative method, 1850–1900; (3) the present tendency toward observation, survey, measurement, 1900– ; (4) the beginnings of the psychological treatment of politics." The last stage gestured to everything inhabiting "the borderland between psychology and politics," including psychiatry and behaviorism, but it mainly identified the study and control of attitudes, opinions, and personal character that Merriam christened "the science of political behavior." Like behaviorism, this special political science was control-oriented and ostensibly value-free. Unlike behaviorism, however, it accepted the irreducible reality of mental states such as opinions, including public opinion, and welcomed any contributions psychoanalysis might make to political science or democracy. "But this is fundamental – that politics and social science [including psychology] see face to face; that social science and natural science come together in a common effort and unite their forces in the greatest task that humanity has yet faced – the intelligent understanding and control of human behavior."[32]

Merriam's most important disciplinary legacy was the "Chicago School" itself. His earliest colleagues and students – among them Gosnell, Leonard D. White, and Harold D. Lasswell (1902–1978) – vindicated Merriam's gestures toward a science of political behavior by delivering its methods and substantive research. After World War II, under the banner of behavioralism, the school came to count as members V. O. Key, David Truman, Gabriel Almond, and the future Nobel laureate Herbert A. Simon. On this famous roster, Lasswell stands out, both before and after 1945. Like his teacher Merriam, Lasswell reinforced those disciplinarians who made "science" synonymous with methods. He generated countless "skills" – configurative analysis, elite analysis, cohort analysis, and more – that were unleashed on topics around power and psychology – for example, in *Psychopathology and Politics* (1930) and *World Politics and Personal Insecurity* (1935). His bracing definition of politics in the subtitle of his *Politics: Who Gets What, When, and How* (1936) appealed to a discipline disabused of juridical notions in a world lurching again toward war. At the same time, Lasswell drew his hypotheses about the discontents of democracy from Europe's grandest thinkers – Marx, Weber, and Freud. He also helped to popularize the theories of Gaetano Mosca, Vilfredo Pareto, and Roberto Michels, whose early-twentieth-century analyses of "iron laws," benighted masses, and inevitable elites were further reminders of the fragility of modern mass democracy.

Propaganda lay at the core of politics and political science for Lasswell, who defined it as "the management of collective attitudes by the manipulation of significant symbols," or again as "control over opinion by significant

---

[32] Merriam, *New Aspects of Politics*, pp. 105, 132, 173, 348, 350; on behaviorism, see Kurt Danziger, *Naming the Mind: How Psychology Found Its Language* (London: Sage, 1997).

symbols." Lasswell's dissertation on the topic was followed by numerous studies during the 1930s and 1940s, including *World Revolutionary Propaganda: A Chicago Study* (1939) on the methods of domestic communists. These studies spawned still more methodological "skills" like content analysis. Straddling theory and practice, propaganda, while value-neutral, could yet be instrumental in the service of any cause. It was a "mere tool . . . no more moral or immoral than a pump handle." Democracy needed propaganda as much as fascism or communism, perhaps more, given its emphasis on speech and deliberation. Lasswell saw this in the wake of World War I, choosing a blunt-edged synonym for speech: "Democracy has proclaimed the dictatorship of palaver, and the technique of dictating to the dictator is named propaganda."[33] His views remained the same at the outset of World War II, when he promoted *Democracy through Public Opinion* (1941) and directed the Experimental Division for the Study of War-Time Communications at the Library of Congress.

During the 1930s, the emigration of intellectuals from Germany and Austria, most of them fleeing Nazi persecution, profoundly influenced the discipline. Though political science was "not a familiar vocational category" before they emigrated, save for those at the Deutsches Hochschüle, many emigrés found home in or around departments of political science, including those at Columbia, Chicago, Harvard, and the New School for Social Research. A few, such as Paul Lazarsfeld, Karl Deutsch, and Heinz Eulau, would bring with them a rigorous understanding of scientific methods and an appreciation for positivism and empiricism as philosophies of science. But the majority, who were trained in history, philosophy, and law – among them Carl Friedrich, Franz Neumann, Leo Strauss, Eric Voegelin, and Hannah Arendt – would excoriate positivism and sometimes science itself. Their voices helped stoke the intensity of debate over science and democracy during the ensuing decades. Emigrés on both the left and the right usually judged the liberal democratic experiment under Weimar an evident failure as well as a cause of the rise of totalitarianism. Some, like Voegelin, were bitter toward "the rotten swine who called themselves democrats – meaning the Western democracies" for allowing German expansion in the first place. But even for the less condemnatory, the thirties and wartime forties would renew scrutiny of both the meaning and the prospects of democracy.[34]

The Second World War was itself the immediate subject of *A Study of War* (1942), by the Chicago School's Quincy Wright. Reflections during the war by the emigré Hans Morgenthau led to his masterpiece of political realism, *Scientific Man versus Political Power* (1946). Moreover, the war secured

[33] Harold D. Lasswell, *Propaganda Technique in the World War* (New York: Knopf, 1927), p. 9; "Propaganda," in *Encyclopedia of the Social Sciences*, ed. E. R. A. Seligman (New York: Macmillan, 1934), vol. 12, p. 525; and "The Theory of Political Propaganda," *APSR*, 21 (1927), 627, 631.

[34] Gunnell, *Descent of Political Theory*, chaps. 7–9, at pp. 185, 198.

the democratic self-identification of Allied political scientists in the face of totalitarianism. Out of faith or commitment, political scientists decided to place their skills in the service of democracy, despite continuing reservations about the people's civic capacities and continuing affirmations of science's neutrality in value conflicts. In 1942, Lasswell hailed the discipline's "developing science of democracy" as "an arsenal of implements for the achievement of democratic ideals." Science was conceived now more than ever as technical expertise with a "realistic focus of attention" and a "new feeling of relevance." Democracy was understood as "a pattern of symbol and practice" in need of its own "symbol manipulators" and a "redirection of education."[35] Political scientists enlisted in their respective countries' war efforts, often in propaganda and intelligence. War service in turn had important consequences for judgments about science and democracy. Having "vacated their ivory towers and come to grips with day-to-day political and administrative realities in Washington and elsewhere," many American political scientists of the behavioral persuasion felt "a strong sense of the inadequacies of the conventional approaches of political science for describing reality, much less for predicting."[36] Others, including emigrés and political theorists drawn to European grand theories, felt an equally strong sense of the inadequacies of the newer behavioral approaches for interpreting and criticizing reality.

BEHAVIORALISM AND DEMOCRACY'S CRITICS, 1945–1970

For a quarter-century after the war, the discipline expanded explosively in terms of specialized curricula, scholarly publications, new departments, and association membership. Under the auspices of UNESCO and backed by the United States, an International Political Science Association (IPSA) was formed in Paris in 1949. It brought together the few established national associations from the United States (founded in 1903), Canada (1913), Finland (1935), and India (1938), and helped to create other national associations that would later comprise it. L'Association Française de Science Politique formed simultaneously with the IPSA and was presided over by André Siegfried, a political geographer known for studies of voting and political parties. During the fifties, national associations formed in Germany, Switzerland, Belgium, the Netherlands, Israel, Australia, and Argentina. Such association activity became possible because of the number of scholars in departments and schools of political science that emerged during the period. The Deutsches Hochschule für Politik, for example, was revived in 1948 and incorporated

[35] Harold D. Lasswell, "The Developing Science of Democracy," in *The Future of Government in the United States*, ed. Leonard D. White (Chicago: University of Chicago Press, 1942), pp. 25, 31, 33, 34, 43.
[36] Robert A. Dahl, "The Behavioral Approach in Political Science: An Epitaph for a Monument to a Successful Protest," *APSR*, 55 (1961), 765.

into the Free University of Berlin in 1959 as one of its largest departments. In France, the regionalization of political science outside of Paris and the École Libre des Sciences Politiques was noteworthy, while in Britain the study of politics became more prominent in the expanded curriculum of both established and newly created universities.[37]

American influence on the worldwide development of political science was pervasive and in places hegemonic. This was largely the result of the expansion of United States government activity abroad and government support for domestic research with international implications, through grants or direct employment.[38] Some political scientists, such as Leonard D. White at Chicago, were conscious of having a mission: "We have a practical task of world education in the American way of life and in the spirit of American government, made in its image." American influence was often welcome, but not invariably. When the British association was formed in 1950, for example, the very name "political science" was strongly opposed by Cole, the erstwhile pluralist. The name Political Studies Association was coined instead. In 1951, the London School of Economics bestowed its chair in political science on Michael Oakeshott, known for questioning the rationalism upon which science rested; and Bernard Crick would complete the British reaction to American scientism and liberalism in his unflattering study of *The American Science of Politics* (1959). Resistance to political science came *to* America, as well. Franz Neumann reflected on "the German exile" in 1953: "Bred in the veneration of theory and history, and contempt for empiricism and pragmatism, the German exile entered a diametrically opposed intellectual climate: optimistic, empirically oriented, ahistorical, but also self-righteous."[39] The climate about which he complained, and about which White felt confident, was shaped by behavioralism.

Behavioralism was broadly identified with a plea for quantitative methods of research into contemporary political behavior found paradigmatically in voting, legislatures, and secondary associations outside government. In short, it was about methods, behavior, and American-style liberal democracy. The term "behavior" was quite general, covering subindividual properties like attitudes and supraindividual ones like whole systems. And "method" meant any quantitative or operationalized instrument. Behavioralism was thus not a grand theory in the European style, nor as programmatic as behaviorism in psychology, but more an orientation, an approach, a persuasion expressed in

---

[37] Entries by Trent, Favre, and von Beyme in Hayward, ed., *International Handbook of Political Science*, pp. 34–46, 154–68, 169–76; Kastendiek, "Political Development"; Jean Leca, "French Political Science and Its Subfields," in *Development of Political Science*, ed. Easton, Gunnell, and Graziano, pp. 108–26, 147–86.

[38] Terence Ball, "American Political Science in Its Postwar Political Context," in *Discipline and History: Political Science in the United States*, ed. James Farr and Raymond Seidelman (Ann Arbor: University of Michigan Press, 1993), pp. 207–21.

[39] Leonard D. White, "Political Science, Mid-Century," *Journal of Politics*, 12 (1951), 18; Neumann quoted in Gunnell, *Descent of Political Theory*, p. 186.

a distinctive idiom of empirical research and a metalanguage of "variables."[40] Continuities with the Chicago School and Merriam's science of political behavior were evident, since leading behavioralists were members, descendants, or allies of the School. Most, in turn, remembered vanguard figures such as Wallas and Bentley. Not all political scientists shared the persuasion; many continued to study institutions, ideas, administration, and policy in historical, comparative, and international contexts. Indeed, numerically, nonbehavioralists represented the majority of practicing political scientists and theorists. However, no one could ignore behavioralism's high rhetorical profile or its rapid capturing of institutional bases of power. The SSRC helped to route research in a behavioral direction when in 1951 it formed a Committee on Political Behavior. A network of institutes and centers with behavioral leanings also emerged, including the Survey Research Center at the University of Michigan, the National Opinion Research Center, and the Center for Advanced Study in the Behavioral Sciences. The Rockefeller, Carnegie, Ford, and National Science Foundations – as well as various agencies of the United States government – supported their efforts.[41]

The crucial decade for behavioralism fell between 1951 and 1961, with 1953 serving as a symbolic date. In that year, the APSA president, E. Pendleton Herring, pronounced that "Americans respect technology and science; political scientists envy authority that can be based on experiment, not argument."[42] Most significantly, David Easton published *The Political System: An Inquiry into the State of Political Science*, behavioralism's manifesto, thereby doing for Easton's generation what Merriam's *New Aspects of Politics* had done for his. Educated at Harvard (ironically, under Elliott), and newly ensconced at Chicago in Merriam's old office, Easton criticized not only "traditional" inquiry into the state, but also the "historicism" of political theorists and the "hyperfactualism" of earlier empiricists. The "political system" should orient political research, and "the authoritative allocation of values" should replace "power" in the definition of politics.[43] Rebuffing many emigrés and political theorists who, like Voegelin in *The New Science of Politics* (1952), condemned science in the name of classical philosophy and normative values, behavioral political scientists hailed "scientific method" and set a methodological course for the discipline's mainstream. Statistics and the sample survey established the pattern for what would count as "scientific" study. Methods became more quantitative, and the star of the "methodologists" rose.

[40] On "variables," see Danziger, *Naming the Mind*; on behavioralism, see the works cited in note 3; see also Somit and Tanenhaus, *Development of American Political Science*; James Farr, "Remembering the Revolution: Behavioralism in American Political Science," in *Political Science in History*, ed. Farr, Dryzek, and Leonard, pp. 198–224.

[41] Somit and Tanenhaus, *Development of Political Science*, pp. 167–72.

[42] E. Pendleton Herring, "On the Study of Government," *APSR*, 47 (1953), 961.

[43] David Easton, *The Political System: An Inquiry into the State of Political Science* (1953), 2nd ed. (Chicago: University of Chicago Press, 1971), chaps. 4–5, 10.

Behavioralists produced a prodigious quantity of substantive research, especially about parties, public opinion, legislative behavior, and other constituent features of representative democracy. Investigation into the process of interest-group politics, associated earlier with Bentley, was revitalized in Truman's *The Governmental Process* (1951). Lasswell continued to write on questions of power and elites, and inaugurated *The Policy Sciences* (1951) as the special contribution of political science to postwar democracy. The Chicago School inquiry into propaganda continued in methodological treatises on content analysis, as well as in substantive studies like Almond's *The Appeals of Communism* (1953), in which Cold War politics were now evident. The best and most influential empirical research concerned voting and public opinion, especially in the collaborative studies *Voting* (1954) and *The American Voter* (1960). These "realistic" studies also placed voting at the center of a model of representative democracy that downplayed the individual citizen and highlighted the competition for votes among elites within an overall system of liberal institutions: "*Individual voters* today seem unable to satisfy the requirements for a democratic system of government outlined by political theorists. But the *system of democracy* does." This earned for them such labels as "empirical democratic theory" and "the theory of democratic elitism." In works such as *A Preface to Democratic Theory* (1956) and *Who Governs?* (1961), Robert Dahl emerged as behavioralism's most significant empirical democratic theorist. He revitalized the terms "polyarchy" and "pluralism" in the course of discrediting both Madisonian and populist theories of democracy, the better to underscore the role of secondary associations and the competition of interests. Polyarchal democracy displayed not majority rule but "minorities rule."[44]

The appeal of laws and generalizations about representative democracy was evident in the reception accorded "Duverger's law," which explained the number of parties in terms of the vote-counting system. Winner-take-all districting, for example, produced two major parties. Named for Maurice Duverger, the French legal scholar and political sociologist, "Duverger's law" bolstered the international appeal of behavioralism, evident also in *The Study of Political Behavior* (1958), by the British scholar David Butler, and in *Introduction à la Science Politique* (1959), by Jean Meynaud. While American behavioral research into democracy was never blatantly partisan or nationalistic – indeed, much of it was critical and reformist in spirit – it conveyed the sense that American-style liberal democracy was the best system attainable in a century as turbulent as the twentieth. In the course of explaining European instability, Seymour Martin Lipset

---

[44] Bernard R. Berelson, Paul F. Lazarsfeld, and William N. McPhee, *Voting: A Study of Opinion Formation in a Presidential Campaign* (Chicago: University of Chicago Press, 1954), pp. 306, 312; Angus Campbell, Philip E. Converse, Warren E. Miller, and Donald E. Stokes, *The American Voter* (New York: Wiley, 1960); Robert A. Dahl, *A Preface to Democratic Theory* (Chicago: University of Chicago Press, 1956), chaps. 3–5.

famously remarked that such democracy was "the good society itself in operation."[45]

The late fifties and sixties witnessed a series of strident debates over behavioralism and its liberal-democratic allegiances. Easton figured prominently in these debates, as did Lasswell, Dahl, Truman, Almond, and Eulau. In 1961, Dahl tried to inscribe "An Epitaph for a Monument to a Successful Protest," in memory of the behavioral approach. But rumors of its demise were premature, and Dahl himself demonstrated that "the revolutionary sectarians have found themselves . . . becoming members of the Establishment." Key books continued to be published, not only on behavioral research but also on the "way of life" led by those of "the behavioral persuasion." Some leading figures – such as Truman and Almond – continued through the sixties to speak on behalf of behavioralism from the APSA presidential podium, invoking new philosophies of science, such as Thomas Kuhn's. There was no little irony in making behavioral political science a "paradigm," given Kuhn's antipositivism. But the resilient call for a "science" of politics remained the principal point. And it was put to the test by political scientists, many of them emigrés, who advertised themselves or were vilified as "traditional" or "normative" stalwarts of history and philosophy. Hannah Arendt thought that "the unfortunate truth about behaviorism and the validity of its 'laws' is that the more people there are, the more likely they are to behave and the less likely to tolerate non-behavior. . . . The rise of the 'behavioral sciences' indicates clearly the final stage . . . when mass society has devoured all strata of the nation." Morgenthau thought that Lasswell's turn to policy as applied behavioralism revealed "the tragedy of political science," given that it was "if not hostile, in any case indifferent to the necessary contribution of political philosophy to political science." A volume of *Essays on the Scientific Study of Politics* (1962) by followers of Leo Strauss took direct aim at Bentley, Lasswell, and Simon. Strauss ended the volume by praising "classical political science" since Plato for upholding the reality of "the common good" and by excoriating "the new political science" for scientism, liberal relativism, and blindness to "the most dangerous proclivities of democracy" in the midst of the Cold War. A facetious twist concluded his fiery denunciation of behavioralism: "One may say of it that it fiddles while Rome burns. It is excused by two facts: it does not know that it fiddles, and it does not know that Rome burns."[46]

Behavioralism and liberal democracy faced other critics, both in America and in Europe, who spoke from the political left. Sheldon Wolin criticized

---

[45] Seymour Martin Lipset, *Political Man* (Garden City, N.Y.: Doubleday, 1959), p. 439.

[46] Dahl, "Epitaph," p. 766; Heinz Eulau, *The Behavioral Persuasion in Politics* (New York: Random House, 1963), p. viii; Hannah Arendt, *The Human Condition* (Chicago: University of Chicago Press, 1958), pp. 43, 45; Morgenthau quoted in Crick, *American Science of Politics*, p. 208n; Leo Strauss, "Epilogue," in *Essays on the Scientific Study of Politics*, ed. Herbert J. Storing (New York: Holt, Rinehart and Winston, 1962), pp. 311, 327.

"methodism," hailed the "vocation" of political theory, and praised participatory over representative democracy.[47] C. B. Macpherson, the Canadian political theorist, criticized the myths of liberalism and enumerated non-liberal variants of democracy in *The Real World of Democracy* (1966). Titles like *Apolitical Politics* (1967), *The Bias of Pluralism* (1969), and *The End of Liberalism* (1969) captured the antiestablishment mood. In West Germany, the "Americanization" of political science, long resisted by conservative inheritors of *Staatswissenschaft*, came under fresh attack by self-identified Marxist political scientists and critical theorists including Jürgen Habermas. Revival of interest in Marx and Marxist analyses of politics also characterized French and Italian political science. Even when rejected, Marx(ism) was seriously debated. In France, too, where sixties politics gave an edge to such discussions, "mimicking American empiricism" was not only avoided, but the development of poststructuralist and postmodernist analyses – most strikingly those of Michel Foucault – contributed greatly to the critique of science.[48] These more radical American and European critics thought that behavioralism, science, and method – as well as the attendant values of American-style liberal democracy – were part of the political problem.

Riots in America's cities, assaults against Cold War policies, and protests in the United States and Europe over the Vietnam War exacerbated debates within the discipline. Their ferocity suggested the limits of liberal pluralism, and their unexpectedness dramatized the inability of political scientists to predict actual behavior. A left-leaning Caucus for a New Political Science was formed in 1967, critical of the APSA for having "failed to study, in a radically critical spirit, either the great crises of the day or the inherent weakness of the American political system." A "new" political science must reverse these misfortunes; if not, the Caucus announced *An End to Political Science* (1970). Even erstwhile behavioralists recognized the import of the era for the discipline's identity. Indeed, it was a defining moment and an act of professional bravery when David Easton delivered his APSA presidential address in 1969. "A new revolution is underway in American political science," he began. It was a "post-behavioral revolution," now that "the last revolution – behavioralism – has . . . been overtaken by the increasing social and political crises of our time." These crises suggested the scientific "failure of the current pluralist interpretations of democracy" and the political failure of a discipline that appeared "more as apologists" for than "objective analysts" of American policy. Easton called for a postpositivist conception of method, as well as for a "Credo of Relevance" and a Federation of Social Scientists to revitalize the

[47] Sheldon S. Wolin, "Political Theory as a Vocation," *APSR*, 63 (1969), 1062–82. Also see the journal *democracy* that Wolin edited, which gave voice to participatory democrats.
[48] See Kastendiek, "Political Development"; Leca, "French Political Science"; and Luigi Graziano, "The Development and Institutionalization of Political Science in Italy," in *Development of Political Science*, ed. Easton, Gunnell, and Graziano, pp. 108–26, 147–86, 127–46.

discipline's time-honored affiliation with the cause of reform.[49] The postwar era of political science was at a close.

DEMOCRATIC PROSPECTS AND THE POSTBEHAVIORAL CONDITION, FROM 1970

As the 1970s unfolded, the crises of the discipline and the political world receded, but without foretelling any disciplinary unity or confidence in democratic stability. Some behavioralists would complain that the discipline was still too "pre-behavioral" or that it needed to keep alive a Lasswellian "skill revolution" to lend expert consultation to liberal democracy.[50] Political theorists disengaged from disciplinary debate and reengaged in history and grand theories, as did many continental political scientists. In Britain, "political studies" continued apace, despite some behavioral inroads at the Universities of Essex and Strathclyde supported by a new Social Science Research Council, founded in 1965. Even the history of political thought – so sharply criticized by behavioralists – was renewed at the core of the discipline at Cambridge when Quentin Skinner was made professor of political science.

Not postbehavioral*ism* but a "postbehavioral condition" can appropriately describe political science after 1970. The condition received different diagnoses, mostly about drift, division, and disenchantment. Less dire accounts still pointed out that the discipline's subfields, not to mention the various national associations, were differentiated and frequently isolated from one another. Nonetheless, institutional developments and perennial debates over science and democracy continued. We may end this short history by tentatively sketching some disciplinary trends whose long-term consequences were not yet clear at the end of the century.

The nominal identity of political science as a professional academic discipline with numerous departments, journals, and association members was as securely institutionalized as ever, though some changes are noteworthy. The IPSA counted forty national associations in its membership by 1980, and the number slowly rose over the following years, chiefly because of national reorganizations in central Europe and the former Soviet Union. APSA membership was 11,700 in 1975, falling to 8,400 by 1983, and then climbing to the historic level of 13,900, including 3,600 women, in 1995. The proportion of women in political science was lower than in sociology and history, but higher than in economics. However, the presence of women in the discipline was significant and their contributions more prominent, as was feminism in political theory. The year 1991 witnessed the election of the Harvard political

---

[49] Caucus platform quoted in Seidelman, *Disenchanted Realists*, p. 198; David Easton, "The New Revolution in Political Science," *APSR*, 63 (1969), 1051, 1057, 1061.

[50] Heinz Eulau, "Skill Revolution and the Consultative Commonwealth," *APSR*, 67 (1973), 169–91; John Wahlke, "Pre-Behavioralism in Political Science," *APSR*, 73 (1979), 9–31.

theorist Judith Shklar (1928–1992) as president of the APSA, the first woman to hold the post. The second, Elinor Ostrom, served in 1996. *Women and Politics* began publication in 1980, taking its place among the dozens of professional journals read mainly by self-identified political scientists and theorists.

Both the aspiration to and the resistance to a natural "science" of politics persisted, if less intensely, as was evident in debates surrounding rational choice theory and contestatory philosophies of science. The "rational choice revolution" frames political decisions as rational choices amenable to formal economic models and game theory, thus making the discipline "a genuinely scientific enterprise."[51] The revolution dates from *An Economic Theory of Democracy* (1957), by Anthony Downs, though the general attraction to economics among political scientists dates back to moral philosophy as well as to the work of Lieber, Catlin, and Beard. But it was in the postwar period, during international economic reconstruction, that political science and economics were more tightly drawn together in methods and in substance, as was evident in the works of Charles Lindblom, Dahl, and Simon, who was trained in public administration and won the Nobel Prize in economics. William Riker, for years the standard-bearer of the new rational choice revolution, put economic theorizing to work on time-honored problems in political science. His brief for procedural democracy and limited government took clearest form in *Liberalism against Populism: A Confrontation between the Theory of Democracy and the Theory of Social Choice* (1982). The political as well as methodological stance reflected in this work and others like it evoked considerable reaction in the discipline. Broader contestations of any sort of "science" of politics found support in hermeneutics, critical theory, and postmodernism. These philosophies gained favor among many political theorists and students of international relations in the United States, as well as among political scientists in Europe, who were more likely than their American counterparts to appeal to philosophical discourse when seeking disciplinary self-understanding. To these methodological developments may be added the reinvigoration of European-style grand theory, an appeal to "bring the state back in," and a revival of historical study of democratic institutions and of the discipline itself.[52]

Notwithstanding the importance of debates over science, those over democracy remained most significant for the discipline's identity and sense of place in the public world. It is the issues surrounding democracy that students

[51] Kenneth A. Shepsle, "Studying Institutions: Some Lessons from the Rational Choice Approach," *Journal of Theoretical Politics*, 1 (1989), 148.

[52] See Morgan's chapter in this volume; Kristen R. Monroe, ed., *The Economic Approach to Politics: A Critical Reassessment of the Theory of Rational Action* (New York: Harper Collins, 1991); Ian Shapiro and Donald P. Green, *Pathologies of Rational Choice: A Critique of Applications in Political Science* (New Haven, Conn.: Yale University Press, 1994); Quentin Skinner, ed., *The Return of Grand Theory in the Human Sciences* (Cambridge: Cambridge University Press, 1985); Peter B. Evans, Dietrich Rueschemeyer, and Theda Skocpol, eds., *Bringing the State Back In* (Cambridge: Cambridge University Press, 1985); Farr, Dryzek, and Leonard, eds., *Political Science in History*.

328   *James Farr*

of political science take with them when they join the ranks of educators, administrators, officeholders, and, most often, lawyers. It is also these issues that bring professional political scientists onto the public stage as advisers, pollsters, and commentators. The old quarrel between liberals, pluralists, elitists, and participatory democrats was given new life during the 1980s and 1990s with the entry of communitarians and deliberative democrats, as well as the revival of "civil society" and "the return of the citizen."[53] The problems of democracy took new forms because of increasingly multicultural nation-states in North America, the end of the Cold War, democratic revolutions in the former Soviet bloc, efforts to create a European Union, and the globalization of democratic movements. Such problems require debate and explanation not only in the interest of science, but also for the education of citizens and leaders of states. Political science will continue to be identified as the discipline of democracy – with all of the complexities, ironies, and contradictions of its own historical debates – as long as it takes up these tasks.

---

[53] Ronald Beiner, ed., *Theories of Citizenship* (Albany: State University of New York Press, 1995); Robert D. Putnam, *Making Democracy Work: Civic Traditions in Modern Italy* (Princeton, N.J.: Princeton University Press, 1993); Sidney Verba, Kay Lehman Schlozman, and Henry E. Brady, *Voice and Equality: Civic Voluntarism in American Politics* (Cambridge, Mass.: Harvard University Press, 1995).

18

SOCIOLOGY

Robert C. Bannister

Sociology emerged in response to the problem of social order in modern society in the wake of the American and French Revolutions and the rise of industrialism and market capitalism. A precondition of the project was the recognition of a civil society apart from any particular political form. Combining skepticism and a faith in reason, sociologists insisted that society is not a reflection of a natural or divine order but is nonetheless subject to rational analysis. Whereas Enlightenment theorists had viewed society in terms of a "social contact" and a convergence of individual interests, sociology explored the forms and structures that make "society" possible.[1]

Taking sociality as its subject, sociology differed from the other social sciences in claiming no specific area as its own, such as primitive society, politics, or the economy. While the other social sciences took their subjects as given, the first academic sociologists expended vast energy arguing that there was such a thing as "society" to be studied. As a result, the discipline developed a decade or more later than anthropology, political science, and economics. Strategies to legitimate the new discipline ranged from claims that it was the capstone of the social sciences to more limited proposals to study social relations.

Sociology had its roots in the theories of August Comte and Herbert Spencer and in empirical work previously conducted by census bureaus, state labor boards, and reform organizations. A tension between theory and practical knowledge persisted throughout the various stages of its history: (1) a preacademic era, during which the concept of "sociology" emerged (1830s–1860s); (2) the proliferation of organicist and evolutionist models of society (1870s–1890s); (3) parallel traditions of statistics and social investigation

---

[1] Shmuel N. Eisenstadt, *The Form of Sociology* (New York: Wiley, 1976) and his "Sociology," in *International Encyclopedia of the Social Sciences*, ed. David L. Shils (New York: Macmillan, 1968), vol. 15, pp. 25–36; Peter Wagner, "Science of Society Lost," in *Discourses on Society*, ed. Peter Wagner, Bjorn Wittrock, and Richard Whitley (Dordrecht: Kluwer, 1991), pp. 218–45; Heinz Maus, *A Short History of Sociology* (London: Philosophical Library, 1962).

(1830s–1930s); (4) a "classical period" coinciding with mature industrialization and the formation of modern nation-states, during which sociology became an academic discipline (1890s–1910s); (5) the interwar flowering at the University of Chicago in the United States, paralleled in Europe by a relative decline and virtual disappearance following the rise of fascism; (6) a worldwide revival under United States influence after 1945, when, ironically, American sociological theory was being re-Europeanized; and (7) fragmentation and continuing crisis following the radical assaults of the 1960s.[2]

Sociologists recounted this history in a series of competing narratives. In the *positivist* scenario dating to Comte, the logic of science advances knowledge inexorably, albeit incrementally, as metaphysical speculation yields to empirically grounded social laws. *Pluralist* accounts, responding to the reality of conflicting "schools" during the interwar years, instead stressed the multiplicity of complementary approaches. *Synthetic* histories identified a "true" sociological tradition that took shape between 1890 and 1910 in the work of Émile Durkheim and Max Weber, among others. Despite substantial differences, these accounts shared the assumption that sociology, like natural science, is cumulative, progressive, and entirely cognitive. *Humanist* histories of the 1950s and 1960s focused instead on a "classic" tradition extending back to the eighteenth century, a relatively small group of theorists who addressed the breakup of the European ancien régime and the emergence of modern society with an aesthetic sensibility and a moral passion similar to that expressed in literature and philosophy.[3]

Since the 1960s, *contextual* histories have stressed the role of social, institutional, ideological, and cultural factors in shaping the discipline.[4] Challenging the positivist scenario, contextualists historicize the concept of "science" itself, termed "scientism" or "objectivism." Struggles over theory, and the split between theory and empirical work, appear as contests for social influence and authority rather than as movement toward a unified sociological tradition.

[2] For periodization, see Edward Shils, "Tradition, Ecology, and Institution in the History of Sociology," *Daedalus*, 99 (1970), 760–825; Terry Clark, "The Stages of Scientific Institutionalization," *International Social Science Journal*, 24 (1972), 658–71.

[3] Typology adapted from Donald N. Levine, *Visions of Sociological Tradition* (Chicago: University of Chicago Press, 1995), chaps. 1–5. Examples include: John Madge, *The Origins of Scientific Sociology* (New York: Free Press, 1962) [positivist]; Don Martindale, *The Nature and Types of Sociological Theory* (Boston: Houghton Mifflin, 1960), and Eisenstadt, *Form of Sociology* [pluralist]; Talcott Parsons, *The Structure of Social Action* (New York: McGraw-Hill, 1937) [synthetic]; Robert A. Nisbet, *The Sociological Tradition* (New York: Basic Books, 1966), and Raymond Aron, *Main Currents in Sociological Thought*, 2 vols. (New York: Basic Books, 1965–7) [humanist].

[4] Continuing the typology of note 3, see: Thomas L. Haskell, *The Emergence of Professional Social Science* (Urbana: University of Illinois Press, 1977), Robert C Bannister, *Sociology and Scientism* (Chapel Hill: University of North Carolina Press, 1987), and Stephen and Jonathan H. Turner, *The Impossible Science* (Newbury Park, Calif.: Sage, 1990) [social/institutional]; Irving M. Zeitlin, *Ideology and the Development of Sociological Theory* (Englewood Cliffs, N.J.: Prentice Hall, 1968), and Dorothy Ross, *The Origins of American Social Science* (Cambridge: Cambridge University Press, 1991) [ideology]; Arthur J. Vidich and Stanford W. Lyman, *American Sociology: Worldly Rejections of Religion and Their Direction* (New Haven, Conn.: Yale University Press, 1985) [cultural-religious].

For contextualists, a discipline born of concern for social and moral reconstruction at the start of the modern age appears more often to be a servant rather than a critic of the status quo. This contextualist critique contributed to a crisis that continues three decades later.

THE FOUNDERS, 1830s–1860s

Although social and economic changes during the middle third of the nineteenth century provided a common context for the emergence of preacademic sociology, the work of the first sociologists mirrored important national differences in the timing and intensity of the modernizing process. In France, members of a relatively powerful middle class, remembering the excesses of both the Terror and the Napoleonic dictatorship, vacillated between a wish to fulfill the egalitarian promises of the Revolution and a desire for social and moral order. For Auguste Comte (1792–1857), the burning issue was the French Revolution and its aftermath. Breaking with his Royalist, devoutly Catholic father, Comte embraced republicanism. After collaborating with the early socialist Henri de Saint-Simon from 1814 to 1824, Comte outlined his sociology in the *Cours de philosophie positive* (1830–42) and the *Système de politique positive* (1842). There he announced the "law of three stages" and a hierarchy of the sciences wherein knowledge proceeds from the theological to the metaphysical and finally to the positive or scientific stage. The last science to develop is "sociology," a term he coined in 1839. Sociology was to be the basis of governance in modern society, although after his break with Saint-Simon, Comte increasingly viewed scientists as the least capable of rulers. In his later work, Comte outlined a Religion of Humanity, a normative theory complete with priests and ritual.[5]

In England, removed by a century from Civil War and Glorious Revolution, the promises and perils of the Industrial Revolution took center stage. In *Social Statics* (1850), his first book, Herbert Spencer (1820–1903) defended a "moral sense" philosophy against the utilitarian "expediency doctrine" that looked to government to achieve the greatest good of the greatest number. In *The Study of Sociology* (1873) Spencer produced the first major treatise on sociological method since Comte's *Cours*. In the *Principles of Sociology* (1876–93) he provided a functional analysis of social institutions, using extensive ethnographic and historical materials within a comparative and evolutionary framework and arguing that all societies proceed from the simple to the complex or, in an alternate formulation, from the military to the industrial, a unilinear view for which he was later criticized.

---

[5] On Comte, see Mary Pickering, *Auguste Comte* (Cambridge: Cambridge University Press, 1993); on later positivism, see Christopher G. A. Bryant, *Positivism in Social Theory and Research* (New York: St. Martin's Press, 1985).

In Germany, which had a weaker middle class, modernization was the work of a relatively closed elite, who accepted its technical and economic but not its political and social consequences. Sociology was rooted in traditions of philosophical Idealism and Romanticism, and was shaped by the perceived excesses of Enlightenment rationalism and individualism and by Germany's uncertain sense of national identity. The result was a legacy of holistic analysis, historical consciousness, distrust of reason, and alienation from modernity. Sociological elements surfaced in the works of historians and philosophers from Herder to Hegel to Karl Marx and a host of lesser-known figures. But the vitality of the historical/philosophical tradition impeded the development of "sociology" proper, a term virtually unknown in Germany until the late 1870s.

In the antebellum United States, the anomaly of chattel slavery in a "free" society inspired George Fitzhugh's *Sociology for the South* (1854) and Henry Hughes' *Treatise on Sociology* (1854), critiques of northern industrial society and the first American books to employ the term "sociology." In the North, utopian socialists drew on Comte and others in their quest for alternative social orders. Although these particular trails went cold after the Civil War, the desire for social reconstruction and the presence of a large middle class disposed to embrace modernity made the United States a fertile ground for the new discipline.[6]

ORGANICISM AND EVOLUTIONISM, 1870s–1890s

Comte and Spencer also contributed to the proliferation of organic and evolutionist metaphors that developed from the 1860s onward. A natural rather than a metaphysical object, the social organism for Comte provided an object worthy of human veneration, legitimating the Religion of Humanity.[7] For Spencer society *was* an organism, literally, not simply by analogy. He conceded, however, that social organisms differ from biological ones in the sense that consciousness adheres in the organism's separate parts, not in a centralized "social sensorium," thereby preserving his methodological and political commitment to individualism and laissez-faire.[8] Continental theorists refined and extended organic analogies through the end of the century. In *Bau und Leben des sozialen Körpers* (1875–8), Albert Schäffle drew extensive analogies between the human body and the social body – for example, likening the nuclear family to the basic cell, and the police to epidermal protective tissue. Albion Small and George Vincent introduced Schäffle to American

[6] Eisenstadt, *Form of Sociology*, pp. 15–16.
[7] Donald Levine, "Organism Metaphor," *Social Research*, 62 (1995), 239–65.
[8] On "social Darwinism" and Spencer, see J. D. Y. Peel, *Herbert Spencer* (New York: Basic Books, 1971); Robert C. Bannister, *Social Darwinism* (Philadelphia: Temple University Press, 1979), chaps. 2–3; Jonathan Turner, *Herbert Spencer* (Beverly Hills, Calif.: Sage, 1985), chap. 3.

readers in *An Introduction to Sociology* (1895), one of the first textbooks in the field.

Organic/evolutionist works also emphasized conflict among groups and races. Already developed in Walter Bagehot's *Physics and Politics* (1873), conflict moved center stage in the work of the Austrian "struggle school," represented by Ludwig Gumplowicz (1838–1909) and Gustav Ratzenhofer (1842–1904). They played important roles in shifting sociology's attention from the individual to the group and group interests, an influence evident in Small's later work. In *Social Evolution* (1894), Benjamin Kidd (1858–1916), a British civil servant, produced one of sociology's first best-sellers. Kidd argued that the struggle for existence, although necessary to human progress, could not be justified by reason – since rational calculation was always self-interested – but only by what he termed a "super-rational" sanction, an irrationalist echo of Comte's Religion of Humanity.[9] Meanwhile, racialist ideologies surfaced in the works of Joseph Arthur Gobineau, Georges Vacher de Lapouge, and Otto Amon, each of whom enjoyed an especially wide audience in Germany.

By the 1880s, sociologists had perceived a threat in the alliance with biology: It undercut the need for a separate discipline and, in Spencer's laissez-faire version, tainted the discipline among social reformers and other constituencies crucial to its success. In *Dynamic Sociology* (1883), the American Lester Frank Ward (1839–1913) addressed both issues. Rooted in evolutionary biology, sociology would study the ways in which basic human drives give rise to "social forces." In this process, according to Ward, mind emerges (the "psychic factor"), allowing scientific direction of human affairs, and the creation of a polity he termed "sociocracy." Sociology was "dynamic" in reconciling human desire and social order, an emancipating vision from which Ward and his disciples retreated after the 1890s.

Challenging Ward's reading of evolution, the Yale professor William Graham Sumner (1840–1910) defended laissez-faire in countless essays and in such widely read works as *What Social Classes Owe to Each Other* (1883). An Anglican clergyman turned economist, Sumner looked warily to "sociology" to analyze how the biological "struggle for existence" described by Malthus was conditioned by social rules and norms that governed the "competition for life." Although attracted to Spencer on ideological grounds, Sumner fashioned his own "science of society," the term he preferred to sociology, from anthropology and the historical ethnography of the German Julius Lippert. In his pathbreaking work *Folkways* (1906), he emphasized the power of social mores to shape individual behavior.[10]

Although later branded and dismissed as "social Darwinists," a politically charged pejorative that warned against the alliance between biology

[9] James Alfred Aho, *German Realpolitik and American Sociology* (Lewisburg, Pa.: Bucknell University Press, 1975); D. Paul Crook, *Benjamin Kidd* (Cambridge: Cambridge University Press, 1984).
[10] Bannister, *Sociology and Scientism*, chaps. 1, 6–7.

and sociology, the evolutionists and organicists played an important role in shaping the discipline. Spencer's contributions included an early analysis of sociological method, the use of ethnography and the comparative method, and incisive treatments of religion, the military, the professions, and other social institutions. The organic metaphor also served to legitimate the dispassionate study of society by scientific methods and to advance holistic methodological positions.[11]

STATISTICS AND SOCIAL INVESTIGATION, 1830–1930

Empirical work developed alongside, although separate from, sociological theory in the realm of statistics and the social survey. Statistics was rooted in the work of the Belgian Adolphe Quetelet (1796–1874), the social survey in that of Frédéric Le Play (1806–1888), a conservative French reformer who pioneered studies of the working classes. The earliest investigations were census tabulations to meet the administrative needs of consolidating nation-states; vital statistics, growing from epidemiological and actuarial concerns; and "moral statistics," which reflected anxiety over social problems. To these were later added the practical concerns of charity and settlement house workers in ministering to the poor.

Although modern statistics is a twentieth-century development, its key elements took shape in two stages during the nineteenth century. Viewing variation as accidental, Quetelet assumed that data for any group display a normal distribution around a mean, arguing that the average represents the group's essential "type." Combining the concerns of earlier statisticians with the technical tools of astronomers, Quetelet helped to shape the conviction that regularity in masses does not depend on assumptions regarding the causes of individual behavior, and that social science is the study of laws rather than simply of facts. In the 1870s, drawing on studies of heredity and evolution, Darwin's cousin Francis Galton focused instead on variation. Refined mathematically in the work of Karl Pearson and George Yule, this "new statistics" was concerned not with calculating averages, but with measuring and describing the distribution of traits in any given population. In *The Grammar of Science* (1892), Pearson advised scientists to analyze experience in terms of probabilities rather than "causes." By substituting "correlations" for "causes," statistics provided a way to measure in the absence of theory.[12]

In Britain, early industrialization forced the "social question" to the fore earlier than in other countries, and with it an interest in collecting statistics.

[11] Levine, "Organic Metaphor."
[12] Theodore M. Porter, *The Rise of Statistical Thinking, 1820–1900* (Princeton, N.J.: Princeton University Press, 1986); Bernard J. Norton, "Pearson and Statistics," *Social Studies of Science*, 8 (1978), 5–33.

Initially created during the era of poor law reform of the 1830s, institutions such as the Manchester Statistical Society (1833), were separate from the older universities. Although the economic crisis of the 1870s dampened enthusiasm for this work – since neither statisticians nor economists appeared to have any solutions – statistics, along with the social survey, flourished in Britain, supported by government agencies and professional and reform organizations, until well into the twentieth century.[13]

Sociology elsewhere responded slowly to these developments. Although Lester Ward served a lengthy apprenticeship at the United States Bureau of Statistics, his sociological work contained only the simplest numerical tabulations, as did most other studies, even those by sociologists who touted the value of "statistics" when they meant only counting. In *Suicide* (1897), however, Durkheim pioneered the use of a comparative, quantitative analysis for determining suicide rates, before turning in his later work to questions that could not be addressed statistically. In 1915, the British economist A. L. Bowley developed sampling techniques that transformed later survey work. By the 1920s, the "new statistics" entered American sociology in the work of two of Giddings's students, William Ogburn and Stuart Rice, and in studies by Dorothy Thomas, who had studied with Bowley at the London School of Economics. More than a change in method, the new statistics signaled the rise of a value-free "objectivism," a pejorative term implying that this sociology treated social activities as inert objects and was thus more concerned with the "how" than the "why" of human behavior, with control rather than with amelioration.[14]

Whereas sociologists eventually embraced statistics, the profession turned its back on the survey tradition. The social survey grew out of charity work in Great Britain; the most important British surveys were Charles Booth's *The Life and Labour of the People of London* (1889–1903), B. Seebohm Rowntree's *Poverty* (1902), and the urban studies of Patrick Geddes, a Scottish naturalist who was the first to apply the term "ecology" to social phenomena. The earliest social surveys conducted in the United States focused on race and immigration, both relatively more important there than in Britain, the best-known being W. E. B. DuBois' *The Philadelphia Negro* (1899), Jane Addams and associates' *Hull House Papers* (1895), and the "Pittsburgh survey" (1909–12).[15]

In the 1920s, however, Chicago sociologists disparaged surveys as the work of "social politicians," as compared to sociological research, which

[13] Philip Abrams, *The Origins of British Sociology, 1834–1914* (Chicago: University of Chicago Press, 1968).

[14] Anthony Oberschall, "The Two Empirical Roots of Social Theory," *Knowledge and Society*, 6 (1986), 67–97; Gary Easthope, *A History of Social Research Methods* (London: Longman, 1974), chap. 6.

[15] Milton Gordon, "The Social Survey Movement," *Social Problems*, 21 (1973), 284–98; Martin Bulmer, Kevin Bales, and Kathryn Kish Sklar, eds., *The Social Survey in Historical Perspective, 1880–1940* (Cambridge: Cambridge University Press, 1991).

involved the systematic testing of hypotheses. Although the secretary of the *Recent Social Trends* (1933) project had worked in the survey tradition, this government-sponsored work ignored the tradition entirely. Echoes of the survey lingered in the work of some of Giddings's students, but the new community study exemplified by Robert and Helen Lynd's *Middletown* (1929) and by the sampling surveys of the 1930s had no direct connection to the earlier tradition. This development revealed important changes in the entire field of social investigation: a shift from local to regional and national issues; new sources of funding from foundations, government, and industry rather than from local elites and charity organizations; and, as with the embrace of the new statistics, a shift from alleviating to managing social problems.[16] From a concern of scattered theorists and diverse social investigators, sociology had become a discipline with an institutional base in academia, which brought more focused intellectual discussion and an effort to bring sociological expertise to bear on public polices.

THE "CLASSICAL" ERA, 1890s–1910s

After the 1870s, the problems of modernity assumed new forms. Unification struggles in Germany, the United States, and Italy and the creation of the French Third Republic left in their wake problems of nationhood and national identity. Accelerated industrialization forced attention to providing more adequate state responses to the "social question." Universities emerged as major sites for the organization and dispensing of social knowledge. "Science" assumed new authority, while itself being transformed from theoretical knowledge of the past to practical, instrumental control of the present. National differences continued to affect sociology's fortunes, resulting in resistance in Europe and a relatively quick embrace of a positivistic, instrumentalist sociology within the newly created universities of the United States.[17]

A crisis in classical economics provided the opening, as sociologists sought to explain social forces that were economically important but not strictly economic. Epistemologically, sociology challenged the individualistic assumptions of classical economics; politically, its crude reliance on a self-regulating market based on rational calculation; and institutionally, its prior establishment as *the* science of society within the university. This conflict was played out in the careers of all of the classical European sociologists and

[16] Stephen P. Turner, "The World of the Academic Quantifiers," and Martin Bulmer, "The Decline of the Social Survey Movement," in *Pittsburgh Surveyed*, ed. Maurine W. Greenwald and Margo Anderson (Pittsburgh, Pa.: University of Pittsburgh Press, 1996), chaps. 10, 11.
[17] Peter T. Manicas, *A History and Philosophy of the Social Sciences* (New York: Basil Blackwell, 1987), chap. 10, and his "The Social Science Disciplines," in *Discourses on Society*, ed. Wagner, Wittrock, and Whitley, chap. 3.

also in the United States, where Giddings, for example, began his career as a "marginalist" economist, then turned to sociology to explain what determines economic preferences. The Ward–Sumner confrontation of the early 1880s was but one of a series of battles between sociologists and their economist colleagues.[18]

The outcome of this pre-war project was finally a paradox. In Europe, classical sociology, despite its intellectual brilliance, gained little institutional permanence and left little immediate legacy. In the United States, the institutional success of prewar sociology, despite its intellectual shortcomings, provided a basis for sustained development and also, ironically, for the revival of the European classical tradition after 1945.

Academic sociology in France developed in several stages: a formative period dating from the appointment of Émile Durkheim (1858–1917) to the faculty of letters at the University of Bordeaux in 1887 to the publication of the *Année socioloqique* in 1898; its establishment as a university discipline in 1913, when Durkheim's chair at the Sorbonne was first titled Education and Sociology; and the eventual dominance of a "Durkheim school."

The Durkheimians' ability to establish sociology as a discipline in their own image was the result of an interplay of theory, institutional strategies, and the cultural/ideological milieu.[19] Durkheim and his chief competitors each presented viable theoretical paradigms: Durkheim in the view that society, a reality apart from individuals, must be studied using rigorous, often statistical methods; René Worms in his elaboration of the organic analogy in *Organisme et societé* (1896); and Gabriel Tarde in the view that social life can be reduced to processes of "invention" and "imitation" whereby an elite leads a sheepish mass, a view elaborated in *The Laws of Imitation* (1890, English trans. 1903). Each had an institutional base: Durkheim in Bordeaux and Paris, Worms as editor of the *Revue internationale de sociologie* (1893) and academic entrepreneur par excellence, and Tarde as professor of philosophy at the Collège de France. Each spoke indirectly to the political concerns of the day: Durkheim and Worms to those who wanted more social stability, Tarde to an elitist right still not reconciled to the legacy of the Revolution.

Durkheim's competitors, however, had fatal weaknesses. Worms's theory of the social organism was fast losing ground; his eclectic enterprise lacked the ability to provide career opportunities; and his highly abstract support of social stability offered nothing compelling to any faction within French politics. Tarde proved weaker than either of his rivals, although his theory of

---

[18] Wagner, "Sciences of Society Lost," pp. 226–33; Norbert Wiley, "The Rise and Fall of Dominating Theories in American Sociology," in *Contemporary Issues in Theory and Research*, ed. William E. Snizek, Elizabeth R. Fuhrman, and Michael K. Miller (Westport, Conn.: Greenwood Press, 1979), pp. 52–3.

[19] Philippe Besnard, ed., *The Sociological Domain, the Durkheimians and the Founding of French Sociology* (Cambridge: Cambridge University Press, 1983); Roger L. Geiger, "The Institutionalization of Sociological Paradigms," *Journal of the History of the Behavioral Sciences*, 11 (1975), 235–45.

the central role of the individual in the process of invention and imitation influenced work in crowd psychology and American sociology. The metaphysical cast of concepts such as "imitation" and Tarde's anachronistic view of science failed to distinguish sociology from philosophy, and his aristocratic biases were not compatible with the prevailing republican ideology.

Durkheim, by contrast, had considerable strengths. Concerned with structures rather than with the individual, he argued in *The Rules of Sociological Method* (1895) that "social facts" are the subject of sociology. External to the individual, they exercise a coercive, constraining power and are not reducible to biology or psychology. Society is sui generis, and sociology is a field with its own subject matter. Societies are characterized by two different forms of integration, "mechanical" and "organic," the latter resulting not from forced similarities but from differences created by the division of labor. Modernity thus holds the promise of organic unity. When social integration breaks down, however, the result is "anomie," a state of normlessness the consequences of which Durkheim examined in *Suicide*.[20] In *The Elementary Forms of the Religious Life* (1912, English trans. 1915), he viewed religion – the entire realm of "the sacred" as opposed to the "profane"– as the "collective representation" of a group consciousness that frees the individual from personal interests, regulates behavior, and provides a sense of well-being. Whereas *Suicide* employed comparative statistics, *Elementary Forms* focused on a single case, the totem religion of Australian Aborigines.

Although Durkheim's call for social integration combined spiritual appeal and political relevance, he was widely attacked during his lifetime. Numerous critics objected to his anti-individualistic "social realism," his scientific pretensions, and his analysis of religion. His program nonetheless provided a three-pronged route to disciplinary formation and definition. The notion of society as sui generis provided an ideal platform for disciplinary autonomy; indeed, the extremism of the Durkheimians on this point was a source of strength. Institutionally, Durkheim and his followers cultivated their connection with philosophy, an already established discipline, while also serving such classical disciplines as history and geography. The fact that *Suicide* provided a concrete example was a further source of strength. To this was added carefully orchestrated connections with social science in Germany, Britain, and the United States, using *l'Année Sociologique* as a showcase for their own brand of sociology. Finally, Durkheim's insistence that society provided a basis for civic morality neatly dovetailed with French republicanism, winning the support of key figures in government and education. This ascendance assured that Durkheim's influence

---

[20] On Durkheim, see Steven Lukes, *Emile Durkheim* (New York: Harper and Row, 1972); Kenneth Thompson, *Emile Durkheim* (London: Tavistock, 1982). On his legacy, see Victor Karady, "Prehistory of Present Day French Sociology," in *French Sociology*, ed. Charles C. Lemert (New York: Columbia University Press, 1981), pp. 33–47, and his "Durkheimians in Academe," in *Sociological Domain*, ed. Besnard, pp. 72–89; Albert Salomon, "The Legacy of Durkheim," ibid., pp. 247–66.

would be passed on to later generations, although his disciplinary program was eroded when World War I ushered in an era of fragmentation and stasis.

German classical sociology in the work of Ferdinand Tönnies (1855–1936), Georg Simmel (1858–1918), and Max Weber (1864–1920) made an even sharper break with the traditions of Comte and Spencer than did the French. Each man eventually reshaped sociology elsewhere, notably in the United States, and also helped the discipline to gain grudging acceptance in Germany. But none succeeded in establishing an institutional presence or a sociological tradition comparable to that of their French and American contemporaries.

Of the three, Tönnies in *Gemeinschaft und Gesellschaft* (1887) remained the closest to the nineteenth-century evolutionary tradition, restating as "community" and "society" a distinction that echoed Sir Henry Maine's "status" and "contract" and similar dichotomies between traditional and modern society. Simmel, by comparison, decisively rejected Spencerian organicism as well as the German Idealist distinction between *Natur* (nature) and *Geist* (spirit or mind). He insisted that "society" was real but consisted of the patterned interactions of individuals. Sociology was to focus on the "forms" of this interaction.

Weber rooted his sociology in German historical and legal thought.[21] Whereas Durkheim and Tönnies studied structures, Weber stressed the individual actor. Social structures such as the state and the church, when reduced to their elements, consist of social activity and the repetition of specific actions. Sociology is the study of human activity from the perspective of its meaning to participants, whether or not consciously intended. Weber distinguished "traditional" from "modern" society by virtue of the latter's "goal-oriented rationality," not its "organic" solidarity, his central concern being the process of rationalization that had been transforming Western society for centuries. Transcending the historicist/positivist divide, he denied that the natural and social sciences are identical, but also insisted that it is possible to generalize about the realm of human activity. In his doctrine of "ideal" types, he located a level of abstraction that, by highlighting certain elements of a reality, allows a qualitative comparison of similarities and differences and is not merely a statistical average. The method of "*verstehen*," removed from idealist metaphysics, provided a tool for exploring motivation, the unique causal factor in human activity.

Weber's substantive interests ranged from agrarian society in antiquity to medieval trading associations, religion, politics, and bureaucracy. In

---

[21] On Weber, see Reinhard Bendix, *Max Weber* (Garden City, N.Y.: Doubleday, 1960); Arthur Mitzman, *The Iron Cage: An Historical Interpretation of Max Weber* (New York: Knopf, 1969); Wolfgang J. Mommsen, *The Age of Bureaucracy; Perspectives on the Political Sociology of Max Weber* (New York: Harper and Row, 1974); Dirk Käsler, *Max Weber*, trans. Philippa Hurd (Chicago: University of Chicago Press, 1988); see also the chapters in this volume by Stephen Turner and Dorothy Ross.

The Protestant Ethic and the Spirit of Capitalism (1904–5, English trans. 1930), the work for which he is probably best known among nonspecialists, Weber argued that Calvinism, by making one's work into "calling," cultivated the worldly asceticism necessary for the development of capitalism. Testing this theory, he subsequently analyzed the relation between economic ethics and social life in Confucianism and other non-Western religions. Bureaucracy, pervasive in most industrial countries, was a second example of the rationalization of modern society. Its hallmarks are a ruled-defined division of labor, hierarchical organization, recruitment based on expertise, a separation of official from personal concerns, and an established career line. Bureaucracy for Weber constituted the most efficient mode for organizing and managing tasks on a large scale, especially in government, although he recognized that in practice bureaucracies are often inefficient and pose a threat to the individual.

Although a German sociological community was developing by the turn of the century, several factors continued to impede successful institutionalization within academia. These included pessimism about the future and about sociology's ability to further progress, and a split between the strong historical/philosophical tradition among academically oriented social theorists and the reform-minded empirical studies conducted by nonsociologists. Although some financial support came from the Verein für Sozialpolitik, a research and policy organization founded in 1872, its reformist goals did not interest most sociologists. A sharp distinction between the social and natural sciences, rooted in the work of Wilhelm Dilthey, appeared to deny to sociology the positivist claim to objectivity.

These impediments framed the careers of the German triumvirate. Tönnies remained at Kiel in northern Germany and, oddly, rarely applied his theory in his own research, so his influence never matched the personal esteem in which he was held. Simmel, a Jew of independent means, was unable to obtain a professorship and was also an outsider politically, commenting only occasionally on current events. His influence came through his writing and his sparkling lectures, whose audiences included the American Robert Park and a who's who of European intellectuals.[22]

Weber was a preeminently successful academic, having been appointed at the age of thirty-two to replace the economist Karl Knies at Heidelberg. His wide-ranging interests and brilliant intellect made him a central figure among prominent academicians from a variety of disciplines. Although remaining above partisan wrangling, he was sympathetic to the nationalist program of the Pan-German League in his youth, and in 1918 he accepted an offer to run for nomination to the National Assembly. But his fearless honesty kept him from allying for long with any party, leaving him by the

---

[22] Donald N. Levine, "Simmel's Influence on American Sociology," *American Journal of Sociolology*, 81 (1975–6), 813–45, 1112–32.

end of World War I disillusioned with a Germany he had once loved. Compounding these problems, a nervous breakdown in 1898 cut short his academic career. He published and traveled feverishly until the outbreak of war in 1914, but his diverse interests meant that he did not address a single constituency of would-be sociologists, and he never again held a regular academic position.

European classical sociology, while often brilliant intellectually, thus failed to achieve firm institutional bases. By 1914, the term "sociology" had become widely recognized, sociological journals thrived, and professional societies multiplied. But within the universities the story was different. In France, by 1914 there were only four sociology courses offered by the Paris Faculty of Letters, and only a half-dozen para-sociological offerings elsewhere. Durkheim's greatest influence would be on anthropologists, economists, geographers, and historians, notably the *Annales* group, led by Lucien Febvre and Marc Bloch. But even members of this group were marginal within the universities. Likewise, no chair of sociology appeared in Germany until 1919, or in Italy through the interwar years. Although American sociologists knew the works of Tönnies, Weber, and even Durkheim, into the late 1920s they were as likely to cite Spencer, Tarde, or even Leonard Hobhouse, a British sociologist who continued to work in the evolutionist tradition.[23]

While Europeans attracted disciples, Americans created departments, the first at the University of Chicago under Albion Small in 1892, a second at Columbia under Franklin Giddings two years later. By 1914, other important programs existed at Yale under Sumner, at the University of Wisconsin under Edward A. Ross, and at the University of Michigan under Charles Horton Cooley. In 1895, Small founded the *American Journal of Sociology*, later to become the official organ of the American Sociological Society, organized in 1905. By 1920, American universities had granted approximately 175 PhD degrees in sociology, approximately fifty each at Chicago and Columbia.[24]

This disciplinary success was the result of a decentralized, loosely organized, relatively new university system; a well-educated, reform-minded public; and a relatively clear demarcation in the United States between civil society, church, and state, reinforced by a cultural diversity that made it easy to think of various forms of social life coexisting in a single political order. Although poverty and industrial conflict concerned American sociologists, the problem of national identity and central state authority, settled in principle if not in fact by the Civil War, were relatively less important in the

[23] Werner J. Cahnman, "Tönnies in America," *History and Theory*, 16 (1977), 147–67; Roscoe C. Hinkle, "Durkheim in American Sociology," in *Essays on Sociology and Philosophy*, ed. Kurt H. Wolff (New York: Harper and Row, 1964); Peter Kivisto and William Swatos, "Weber and Interpretive Sociology in America," *Sociological Quarterly*, 31 (1990), 149–63.

[24] Nicholas C. and Carolyn J. Mullins, *Theories and Theory Groups in Contemporary American Sociology* (New York: Harper and Row, 1973); Robin M. Williams, "Sociology in America," in *Social Science in America*, ed. Charles M. Bonjean, Louis Schneider, and Robert L. Lineberry (Austin: University of Texas Press, 1976), pp. 77–111; Wiley, "Dominating Theories," pp. 48–79.

United States than in France and Germany. Attention instead focused on the transformation of a rural, ethnically homogenous country of communities into a heterogeneous, urban nation; on race and immigration and the moral and cultural issues raised by diversity; and on individual interaction, rather than on power or authority.

These factors gave American sociology a distinct coloration, although not always to its advantage by European standards. In the pressure to differentiate itself from competing social sciences, sociology cut itself off from philosophical and historical traditions that strengthened classical European theory, intellectually if not institutionally. The lavish capitalist patronage and local business support that created so varied and open a university system left professors vulnerable to political pressures, as many discovered during a series of academic freedom battles in the 1890s. The tension between "science" and "activism" created a desire to distinguish sociology not only from socialism but also from "Christian sociologists" and other do-gooders, resulting in a separation of academic sociologists from social workers, town planners, and other potential constituencies who created their own professional training schools. By 1920, there were two identifiably "American" strains within academic sociology: the heavily empirical urban ecology and community studies of the Chicago School, and a scientistic, quantitative neo-positivism emanating from Columbia.

Albion Small (1854–1926) and Franklin Giddings (1855–1929), the initial sources of this division, refined their competing systems over several decades. Abandoning the social organism metaphor in *General Sociology* (1905), Small described "association" as a "process" wherein conflicting "interests" compete, converging to form "groups," which are the fundamental units of sociology, a theory derived from Ratzenhofer. In *The Principles of Sociology* (1896), building on Spencer and Tarde, Giddings described social evolution as a threefold process: aggregation; association via "consciousness of kind" and "imitation"; and selection, wherein a social version of natural selection weeds out "ignorant, foolish, and harmful" choices. In *Inductive Sociology* (1901), he began a retreat from concern with the subjective elements in social behavior in favor of a statistical, probabilistic sociology later termed "pluralistic behaviorism."[25]

In sociological theory, the most important developments at Chicago came not from Small but from William I. Thomas.[26] Drawn to sociology by his reading of Spencer, the early Thomas viewed humans as creatures of instinct; for example, he classified men as "katabolic" and women as "anabolic" in *Sex and Society* (1907). In his *Source Book for Social Origins* (1909), which was

[25] Robert W. Wallace, *The Institutionalization of a New Discipline: The Case of Sociology at Columbia University, 1891–1931* (Ann Arbor, Mich.: University Microfilms International, 1989); Charles Camic, "The Statistical Turn in American Social Science: Columbia University, 1890 to 1915," *American Sociological Review*, 59 (1994), 773–805.

[26] Norbert Wiley, "Early American Sociology and the *Polish Peasant*," *Sociological Theory*, 4 (1986), 20–40.

influenced by the anthropology of Franz Boas, he repudiated Spencer's uni-linear evolution and looked at primitive cultures to find the crucial elements in social change: attention or individual response; habit and crisis, whereby attention is alternately relaxed and disturbed; and control, the end of all social interaction. The nature and rate of change depend upon the actions of extraordinary leaders, the level of culture, and experience of previous adjustments. Thomas later outlined an alternative to "instinct" theory in his doctrine of the "four wishes" – recognition, response, new experience, security – while stressing the importance of "attitudes" and the way individuals act on the basis of their "definition of a situation." This reorientation led him to emphasize the importance of "behavior documents" such as biographies, diaries, and medical reports, most notably in *The Polish Peasant in Europe and America* (1918–20), written with Florian Znaniecki, a work that later fueled a debate between advocates of the "case study" and of "statistics."

Like Thomas's study of the "disorganization" of an immigrant community, other major works of the prewar era addressed the problem of social order. In *Folkways* (1906), Sumner argued that the most expeditious social practices first become "folkways," then gain coercive power as "mores," against which there is no appeal. In *Human Nature and the Social Order* (1902) and *Social Organization* (1909), Charles Horton Cooley described the "looking glass self," whereby identity is created by a process of social interaction and natural "primary" groups are replaced by artificially created "secondary" ones. In *The Process of Government* (1907), Small's sometime student and critic Arthur Bentley provided an early statement of "interest group theory," just as Ogburn and others would translate Giddings's teachings into an administrative liberalism. In *Social Control* (1901), E. A. Ross provided a rationale for eugenics and immigration restriction.

British prewar sociology – statistics and survey work aside – was at best a footnote to developments on the Continent and in the United States. Breaking with utilitarian and laissez-faire traditions, British sociological theory combined evolutionism and philosophical Idealism in a "new liberal" faith that the modern social order held the materials for progress and individual self-fulfillment, given some guidance from government. Its chief representative was Leonard Hobhouse, professor of sociology at the London School of Economics (1907–29), the only chair in Britain until after World War II. In 1903, Hobhouse joined a coalition of theorists, social survey workers, and eugenicists – among them Geddes and Francis Galton – to form the London Sociological Association.

Britain nonetheless failed to develop a vital sociological tradition. A powerful force in the late nineteenth century, British philosophical Idealism contained the conceptual materials for a sociological theory that might have paralleled those of Durkheim and Weber. But the Idealist worldview in Britain was hostile to what Spencer and Kidd had defined as "sociology" and remained mired in the Hegelian conviction that the "state" was the basic unit of modern society. The resistance of the older English universities to the new social sciences and the vitality of governmental and extra-university support for empirical studies also impeded the marriage of theory and practice.[27]

INTERWAR YEARS

On the surface, American and European sociology during the interwar decades was a study in contrasts. In the United States, sociology experienced a rebirth in the work of the "Chicago School," while Columbia's influence continued through the work of its graduates. Philanthropic foundations funded much of interwar sociology, including the Lynds' *Middletown* (1929) and Gunnar Myrdal and associates' *An American Dilemma* (1944). In Europe, by contrast, there was a dispersal of energies as the founders' hope of uniting different levels of sociological analysis gave way to a separation of theory and research and to institutional fragmentation. In the end, however, sociology suffered setbacks on both sides of the Atlantic. During the 1930s Chicago declined in output and influence, and the profession as a whole coped with a loss of financial support and often-bitter infighting.

Newcomers won "Chicago sociology" its fame: Robert Park, a former newspaperman who arrived at Chicago in 1913; Ellsworth Faris, a former missionary who succeeded Small as departmental chair in 1925; Ernest Burgess, a sociologist of the family; and William Ogburn, author of *Social Change* (1922).[28] Chicago sociology was actually a mosaic, defined by individuals and generations. The urban ecology of Robert Park (1864–1944) and his students; the Park–Burgess textbook, *An Introduction to the Science of Society* (1921); and their combination of theory and firsthand study of urban settings initially gained the department national attention.[29] Park described cities in terms of a series of concentric zones, "natural" areas such as skid rows and rooming house districts. He described social interaction in terms of competition, conflict, accommodation, and assimilation, a process that yields temporary peace among specific groups, but is constantly repeated as new groups make their own claims and move from the central ghettos to middle-class neighborhoods and suburbs. Since differences are never eliminated, individuals and groups maintain a measurable "social distance" from one another, a concept derived from Simmel.[30] Park garnered funds from the Laura Spelman Rockefeller

[27] Stefan Collini, "Sociology and Idealism in Britain, 1880–1920," *European Journal of Sociology*, 19 (1978), 3–50. Cf. Noel Annan, *The Curious Strength of Positivism in British Political Thought* (Oxford: Oxford University Press, 1959).

[28] Martin Bulmer, *The Chicago School of Sociology* (Chicago: University of Chicago Press, 1984).

[29] Cf. Anthony J. Cortese, "The Rise, Hegemony, and Decline of the Chicago School of Sociology, 1892–1945," *Social Science Journal*, 32 (1995), 235–54; Jennifer Platt, "The Chicago School and Firsthand Data," *History of the Human Sciences*, 7 (1994), 57–80.

[30] Fred H. Matthews, *The Quest for an American Sociology: Robert E. Park and the Chicago School* (Montreal: McGill Queen's University Press, 1977).

Foundation, channeled them through an interdisciplinary committee, and worked with community agencies to further his research program on the city, race, and immigration.

A second element in the Chicago mosaic was the rigorously quantitative statistical sociology of William Ogburn (1886–1959), which gained ascendancy after his appointment in 1927 through the work of such students as Philip Hauser and Samuel Stouffer. For Ogburn, sociology was to be quantitative and value-neutral, a view he put into practice in influential statistical studies of legislation, voting, and social indicators; in his 1929 presidential address to the American Sociological Society, "The Folkways of a Scientific Sociology"; and as an advisor to governmental agencies during and after the New Deal.

A third and most enduring strand of Chicago sociology was "symbolic interactionism," so named by Herbert Blumer (1900–1987) in 1937. In response to Ogburn's objectivism, which he termed "science without concepts," and Park's drift toward what one sociologist has termed "instrumental positivism,"[31] Blumer argued that individuals and groups act on the basis of "meanings" that they attach to objects, creating symbolic systems used to communicate and analyze experience. Drawing on the work of George H. Mead and, by extension, of Thomas, symbolic interactionism was refined in the 1940s and 1950s by Blumer, Arnold Rose, and Erving Goffman, especially in Goffman's widely read *The Presentation of Self in Everyday Life* (1959). Against the structural-functionalist emphasis on the performance of externally defined social roles, symbolic interactionists stressed individual and interpersonal definition, thus providing a counterpoint to the Harvard-based sociology of Talcott Parsons.[32]

Chicago nonetheless declined as a sociological power in the 1930s. The Depression made the work of the Parkians seem less relevant; Rockefeller funding ended; and the world crisis of the late 1930s gave the pessimism of some European social theorists greater appeal.[33] The Columbia department, by contrast, created a second generation network of quantifiers, committed to making sociology rigorously "scientific," the most prominent being James P. Lichtenberger (PhD 1910) and Stuart Rice (1924) at the University of Pennsylvania; Howard Odum (1910) at the University of North Carolina; F. Stuart Chapin (1911) at the University of Minnesota; and Ogburn (1912) at Chicago. The increasing influence of the educational foundations worked

---

[31] Bryant, *Positivism in Social Theory*, chap. 5.

[32] Ross, *Origins of American Social Science*, pp. 428–48. On symbolic interaction, see J. David Lewis and Richard L. Smith, *American Sociology and Pragmatism* (Chicago: University of Chicago Press, 1980); Hans Joas, "Symbolic Interactionism," in *Social Theory Today*, ed. Anthony Giddens and Jonathan H. Turner (Stanford, Calif.: Stanford University Press, 1987), pp. 82–115.

[33] Bulmer, *Chicago School*, pp. 205–6; Ruth Cavan, "Chicago School of Sociology," *Urban Life*, 11 (1982–3), 407–20; Bernard Farber, "The Human Element: Sociology at Chicago," *Sociological Perspectives*, 31 (July 1988), 354.

in the same direction. Science, typically equated with statistical analysis, provided a seemingly absolute standard in place of outworn customs and assumptions, now branded "subjective," while at the same time satisfying the grant givers' preference for "realistic" and politically uncontroversial projects.[34]

These developments occurred against a backdrop of broader changes within American sociology in the 1930s. Although foundation funding dropped sharply, it left a legacy of bitterness among sociologists who felt excluded or marginalized by a foundation-created "establishment" that included some of the leading quantifiers. The rise of fascism, although ignored by most American sociologists, contributed to mounting criticism of "value-free" scientism.Charles Ellwood's *Methods in Sociology* (1933) and Robert Lynd's *Knowledge for What?* (1939) attacked narrowly statistical work, while calls for a revival of "theory" created a climate for the future reception of Talcott Parsons's work.[35]

Internecine struggles and the increasing number of sociology faculty members and undergraduate courses together produced fragmentation. Meanwhile, other developments contributed to a postwar revival: new opportunities for government service; the increased sophistication of interviewing techniques, market research, and public opinion polling; and an influx of German and Austrian refugee scholars.

Continental sociology continued to produce accomplished individual theorists, although the swiftly changing political currents and the divide between academic theory and application-oriented research continued to impede institutional success. In France, chairs of sociology existed only at Bordeaux, the Sorbonne, and Strasbourg, the latter two occupied by Durkheim's principal heirs, Paul Fauconnet and Maurice Halbwachs. The centralized university system continued to withhold official recognition, while the changed political climate made both the conservative LePlayist tradition and Durkheim's non-clerical civic religion appear anachronistic. Yet, as Durkheim's heirs moved in two different directions, they sowed seeds that would eventually flower after World War II. One was the exploration of the collective mind and group morals, as in Fauconnet's study of sanctions and the work of Marcel Mauss, whose social anthropology was a spiritual forerunner of the structuralism of Lévi-Strauss and others. A second was a more positivistic, statistical approach, as in Halbwachs's reexamination of Durkheim's *Suicide* and the work on wages of his colleague François Simiand. Inspired by Halbwachs's struggles with other disciplines and with German and American sociology – a combativeness solidly within the Durkheimean tradition – this emphasis on

[34] Bannister, *Sociology and Scientism*, chaps. 11–12; Turner and Turner, *Impossible Science*, chap. 2.
[35] Turner and Turner, *Impossible Science,* chap. 2; Edward Shils, "The Calling of Sociology," in his *The Calling of Sociology* (Chicago: University of Chicago Press, 1980); Robert C. Bannister, "Principle, Politics, Profession," in *Sociology Responds to Fascism*, ed. Stephen P. Turner and Dirk Käsler (London: Routledge, 1992), pp. 172–213.

quantitative methodology kept alive a tradition that was to flourish again in the 1950s.[36]

In Germany, sociology appeared to thrive during the Weimar years as universities established chairs for distinguished incumbents: at Cologne, Leopold von Wiese, proponent of a "systematic" or "formal" sociology in the tradition of Simmel; at Frankfurt, Franz Oppenheimer, a follower of Gumplowicz who emphasized group processes; at Heidelberg, Karl Mannheim, chief representative of the sociology of knowledge, and the cultural sociologist Alfred Weber, younger brother of Max; and at Leipzig, Hans Freyer, a conservative who made important contributions to the history of sociology. On the eve of the Nazi takeover, Weimar sociology flowered in a number of important books, including Mannheim's *Ideology and Utopia* (1929). In Vienna, a group of researchers under Paul Lazarsfeld revived an empirical tradition that had a long history in German-language sociology, though it was rarely given university support or recognition. The later sociologies of Alfred Schutz and Norbert Elias, although not recognized for several decades, also had roots in the Weimar period.[37]

The creation of the Institute for Social Research at the University of Frankfurt in 1923, meanwhile, launched the peripatetic Frankfurt School, whose members established branches in Paris and Geneva before moving in 1934 to quarters provided by Columbia University in New York. Its members, many of them Jewish, included the director Max Horkheimer, Theodor Adorno, Herbert Marcuse, and Leo Lowenthal. A common denominator was an animus against "positivism," a term used loosely to encompass the French nineteenth-century tradition, the logical positivism of the Vienna Circle of philosophers, and less rigorous American versions. In 1950, Horkheimer and other key members would return the Frankfurt School to Germany, where its tradition continued in the neo-Marxist "critical theory" of Jürgen Habermas. Although Adorno and others made important contributions to sociological research, subsequent battles between "positivists" and "critical theorists" during the 1960s would deepen a divide between theory and empirical work that continued to blight German sociology.[38]

Despite the initiatives of the 1920s, Weimar sociology as a whole remained mostly promise at the time of Nazi ascendancy. Proposals to make sociology the centerpiece of university reform met with vigorous opposition

---

[36] John E. Clark, "Sociology and Related Disciplines between the Wars," in *Sociological Domain*, ed. Besnard, chap. 12.

[37] M. Rainer Lepius, "Sociology in the Interwar Period," in *Modern German Sociology* ed. Volker Meja, Dieter Misgeld, and Nico Stehr (New York: Columbia University Press, 1987); Dirk Käsler, "In Search of Respectability: The Controversy over the Destination of Sociology during the Conventions of the German Sociological Society, 1910–1930," in *Knowledge and Society: Studies in the Sociology of Culture Past and Present. A Research Annual*, ed. Robert Alun Jones and Henrika Kuklick (Greenwich, Conn.: JAI Press, 1983).

[38] Martin Jay, *The Dialectical Imagination* (London: Heinemann, 1973); Rolf Wiggershaus, *The Frankfurt School*, trans. Michael Robertson (Cambridge, Mass.: MIT Press, 1994).

from related disciplines, while the major impetus for sociology came from the "folk school" movement, labor courts, trade unions, and other nonuniversity sources of the sort that supported Lazarsfeld's studies. Although sociology became involved in debates over the imposition of a democratic political culture, sociologists themselves, unlike the Durkheimians of the Third Republic, failed to create an image of their discipline that supported the democratic program, nor could they resist the Nazi takeover. By 1938, two-thirds of all sociology teachers had been expelled from the universities. "Nazi sociology" brought a revival of holistic, idealist, and biologized approaches, with an emphasis on racial theory, the folk, and community – all intellectual dead ends. Meanwhile, a "realistic" sociology of area research, town planning, and labor policy became a branch of state administration.

Ironically, Nazi persecution laid a basis for the international postwar revival in which German emigrés played a major role. To Hans Gerth, Reinhard Bendix, Lewis Coser, and Kurt Wolff was left the task of introducing the international community to the work of Simmel and Weber; to Lazarsfeld that of transmitting the German empirical tradition; and to Erich Fromm, Max Horkheimer, and Theodor Adorno that of explaining fascism and the "German catastrophe."

INTERNATIONAL REVIVAL AND AMERICAN HEGEMONY, 1945–1960

The climate for sociology improved dramatically after 1945. In the major Western countries the discipline established itself solidly in universities, in departments within government and industry, and in public esteem. Contributing to this renaissance was a general enthusiasm for applied science, disillusionment with Stalin-era Marxism, and the rise of the welfare state. As historical and philosophical studies became overly specialized, educated public audiences increasingly turned to the social sciences.

Although these influences operated universally, national differences persisted. In the United States, university departments played a dominant role, creating an American research model that soon influenced work in most other countries. In France and Germany, university teaching and research institutes proceeded on separate paths, although research gained considerable support and sometimes academic status. In Britain, the teaching of sociology spread beyond the London School of Economics, but not until the 1960s to Oxford and Cambridge, as sociology struggled toward a closer relation with empirical work in the survey tradition. Prominent newcomers included the Netherlands, the Nordic countries, Latin America, and Japan.[39]

---

[39] Maus, *Short History*, chaps. 17–19.

Harvard and Columbia followed similarly disjointed paths toward postwar dominance. Appointed at Harvard in 1930, Pitrim Sorokin (1889–1968) was already an established scholar when he emigrated from Russia after the Revolution, but *Social and Cultural Dynamics* (1937–41), his magnum opus, was a sprawling review of 2,500 years of human history, in the tradition of Toynbee and Spengler, that left few openings for development by graduate students. He was also organizationally inept, and by the mid-1930s so disillusioned with Stalinism that he appeared to be soft on fascism.[40]

Talcott Parsons (1902–1979), in contrast to Sorokin, began his academic career inauspiciously with a ten-year stint as an untenured instructor. Although his education at Amherst (1920–4), the London School of Economics, and Heidelberg (1924–6) introduced him to the work of Veblen, Radcliffe-Brown, and Weber, it left him with a foreign doctorate and an uncertain position in a Harvard Economics Department that was less interested in theory than in certain technical issues that he found boring. Yet Parsons finally proved to have strengths that Sorokin lacked.[41]

Developing his theory in stages over several decades, Parsons drew on the classical European theorists most American sociologists had ignored. But, without acknowledging any American influences, he preserved an emphasis on conscious behavior, or voluntarism, that was squarely within the tradition of Cooley, G. H. Mead, and Thomas. During his first two decades at Harvard (1927–47), he elaborated this voluntaristic "action" theory in *The Structure of Social Action* (1937), tracing it to convergence in the work of Alfred Marshall, Émile Durkheim, Max Weber, and Vilfredo Pareto. Attacking the utilitarian, rationalistic conception of the individual, Parsons argued that society is held together by common values that orient individual choices of means and ends in the pursuit of goals. Although biological and environmental constraints limit accomplishment, social action must be understood sociologically, not reduced to biology or psychology.

As head of a new Department of Social Relations (1946), Parsons amplified his position. His "systems theory" in *The Social System* (1951), or "structural-functionalism" (a term Parsons disliked, favoring "structural analysis"), treated social structures – institutions and the norms that sustained them – in terms of the functions they served. From the late 1950s onward, he refined systems theory to deal with the interaction of social subsystems and to develop a cybernetic model of the ways in which the culture controls social change, interests already evident in his earlier work on the professions.

Parsons's voluntarism enhanced his appeal to audiences that might otherwise have been deterred by his opaque prose and muddy definitions.

[40] Barry Johnston, *Pitirim A. Sorokin: An Intellectual Biography* (Lawrence: University Press of Kansas, 1995).
[41] Francois Bourricaud, *The Sociology of Talcott Parsons* (Chicago: University of Chicago Press, 1981); Peter Hamilton, *Talcott Parsons* (London: Tavistock, 1983); Ken Menzies, *Talcott Parsons and the Social Image of Man* (London: Routledge, 1976).

To graduate students, his theory appeared to be original and open, inviting countless future projects. Institutionally, he effectively transcended the Chicago–Columbia divide. Presented in the language of European social theory, his antipositivism effectively positioned him against both classical economics and the statistical objectivism of Giddings's students, while at the same time upstaging the Chicago functionalist tradition. His personal convictions were also suited to the changing political climate. A left-of-center liberal, he attacked laissez faire in the 1930s, supporting the New Deal's social welfare and regulatory measures. At the end of the decade, he warned against the dangers of Nazism and joined anti-isolationist faculty groups supporting mobilization. In the early 1940s, his analysis of fascism was the most penetrating until that time by an American sociologist. His systems theory, so critics later charged, was likewise well suited to the administrative corporate liberalism of the 1950s.[42]

Columbia was soon drawn into Parsons's orbit, although not before Robert MacIver (1882–1970) failed to create a different sociological tradition there, for many of the same reasons as for Sorokin's failure at Harvard. Born in Scotland, MacIver studied classics at Edinburgh before accepting a post in political science in Canada in 1915. Although well received, his major works from *The Community* (1917) to *Society* (1931) straddled the divide between political philosophy and sociology. He had few disciples, a colleague later observed, because he possessed no distinctive method or model of analysis beyond his own genius.[43]

As chair from 1929 to 1950, however, MacIver rebuilt the Columbia department, appointing Robert K. Merton (b. 1910) and Paul F. Lazarsfeld (1901–1976). A student of Parsons, Merton termed his mentor's systematic functionalism "premature," a philosophy rather than a method for testing hypotheses empirically. In *Social Theory and Social Structure* (1949) he argued that earlier functionalism, particularly that of the British anthropologists A. R. Radcliffe-Brown and Bronislaw Malinowski, had overemphasized the degree of integration within societies: No sociocultural item was universally functional within any system, and there were no indispensable requirements for social integration, but rather a range of available alternatives. A corollary was a distinction between recognized and intended (manifest) functions, and unrecognized and unintended (latent) ones. Merton developed "middle range" theories such as "reference groups" and "relative deprivation" to analyze the family, the university, science, and bureaucracy.

The Austrian-born Lazarsfeld, after emigrating to the United States in 1933, founded and directed the Office of Radio Research at Princeton, New Jersey

[42] William Buxton, *Talcott Parsons and the Capitalist Nation-State* (Toronto: University of Toronto Press, 1985).
[43] Mirra Komarovsky, "MacIver," *American Sociologist*, 6 (1971), 51–3; Leon Bramson, ed., *Robert M. MacIver* (Chicago: University of Chicago Press, 1970).

(1937–43), later the Bureau of Applied Social Research. In 1940 he joined Merton in the Columbia Sociology Department. In influential statistical studies of *The People's Choice* (1944) and *Voting* (1954), he analyzed the relation between political and popular culture. After 1945, a Merton–Lazarsfeld collaboration attempted to "operationalize" structural-functional theory, creating a Columbia renaissance and an informal alliance with Harvard, where Parsons and Samuel Stouffer had a similar collaboration.[44]

Wartime issues inspired such major collaborative efforts as Dorothy S. Thomas and colleagues' *The Spoilage* (1946), Samuel Stouffer and colleagues' *The American Soldier* (1949), and Theodor Adorno and colleagues' *The Authoritarian Personality* (1950). Sociology also reached a wider audience in David Riesman's *The Lonely Crowd* (1950), C. Wright Mills's *White Collar* (1951), and William Whyte's *The Organization Man* (1956), as "status," "norm," "role," and countless other sociological terms entered the American vocabulary.

Although Parsonsian/Mertonian functionalism thus dominated American sociology between 1945 and the early 1960s, one should not exaggerate the Parsonsian monopoly or American international influence. The revival of sociology in postwar Europe was greatly influenced by American models, but it also had indigenous roots in the needs of the emerging welfare states and was built on older national traditions.

By 1945, despite Durkheim's influence on related disciplines, France still had no specifically sociological instruction or practitioners, although several professors taught courses or pursued research bearing the label. The Centre d'Etudes Sociologiques in 1946, under the Russian-born George Gurvitch, was a gathering point for historians, geographers, and others interested in "empirical" research in their own areas. The appointment of Raymond Aron and Jean Stoetzel to chairs in sociology at the Sorbonne in the mid-1950s, the provision of state funding, and the creation of a publications system led to an expansion of sociological research and new interest in American models, earlier examples of which had included studies of industrial workers in the late 1940s.[45]

After the disastrous hiatus of the Nazi era, German sociology revived with the reconstitution of the German Sociological Society in 1946, the publication of von Wiese's *Studien zur Soziologie* (1948), the return of prominent exiles, among them René Konig, Horkheimer, and Adorno, and the reestablishment of the Institute in Frankfurt in 1950. University departments of sociology, however, played a relatively minor role. Through the 1950s, sociology continued to be taught under the aegis of other disciplines, in

[44] Allen H. Barton, "Paul Lazarsfeld and Applied Social Research," *Social Science History*, 3 (1979), 4–44; Anthony Oberschall, "Paul F. Lazarsfeld and the History of Empirical Research," *Journal of the History of the Behavioral Sciences*, 14 (1978), 199–206.
[45] Jean-Michel Chapoulie, "The Second Birth of French Sociology," paper presented at the Twelfth World Congress of Sociology, Madrid, 9–13 July 1990.

research institutes, or by faculties outside the university structure funded by
industrial and commercial interests. During the 1960s and after, sociological
authors – Adorno, Horkheimer, Ralf Dahrendorf, and Habermas – gained
international attention, although it was as social theorists or philosophical
anthropologists, not as sociologists per se.[46]

A split persisted between a mainstream sociological practice based on
American examples and a body of theory with roots in German tradition.
The first produced more empirical studies in a decade than German sociology
had produced in its entire history: studies of public opinion, community,
the family, industrial relations, education. Although some of these studies
were guided by theory – industrial studies, for example, by the work of the
British-trained Dahrendorf – the major theoretical debates took place apart
from empirical work. Returning to traditional themes of German sociology,
conservatives in the anti-Enlightenment tradition, such as Arnold Gehlen and
Niklas Luhmann, probed issues of rationalization and modernization,
while "critical theorists" of the revived Frankfurt School subjected the
Enlightenment tradition to critical scrutiny.

In 1976, the sociologist Robin Williams pointed with pride to American
sociology's postwar accomplishments: the accumulation of data in the many
subfields – politics, education, the military, health; the use of new methods –
participant observation, scaling, multivariate analysis; and, above all, the
new ways of thinking about human society – a heightened "awareness of
irony, ambiguity and paradox," a recognition that "good intentions produce
undesired results and vice versa," and "a more complex and steady view of
social realities than can be found in either utopian or cynical orientations."[47]
By this time, however, these values and the assumptions behind them were
already under siege.

THE 1960s AND AFTER

The 1960s spelled the end of "modern" sociology. In the United States,
Parsons's hegemony and Merton's "middle range" compromise gave way to
a politically charged humanist/positivist divide. Conflict theorists attacked
Parsons for ignoring the reality of force and repression, notably C. Wright
Mills in *The Sociological Imagination* (1959) and Alvin Gouldner in *The Coming Crisis of Western Sociology* (1970). Symbolic interactionists, phenomenologists, and exchange theorists took aim at Parsons's rigid model of human
behavior and alleged blindness to the complexities of cognition and reality construction. French neo-Marxists argued that no single, abstract social system is common to all societies, but rather that historically specific

---

[46] Volker Meja, Dieter Misgeld, and Nico Stehr, "The Social and Intellectual Organization of German Sociology since 1945," in *Modern German Sociology*, ed. Meja, Misgeld, and Stehr, pp. 1–56.
[47] Williams, "*Sociology in America*," pp. 91–7.

social systems reflect underlying "forces of production." A new breed of positivists, armed with computers and mathematical sophistication, insisted that Parsons's theories be tested empirically. Feminists charged that functionalism reinforced existing gender roles. Sociobiologists raised the perennial specter of biological determinism.[48]

Contextualist historians meanwhile told a story of cycles rather than of progressive advance, with success a product not of universal truths but of institutional and ideological factors. Appeals to "science" appeared to be at best discipline-legitimating ideology, at worst a cloak for socially reactionary ends. Although sociology's defenders could reply that their discipline had immensely enriched the social vocabulary, amassed information useful to its diverse patrons and constituencies, and refashioned rather than abandoned the liberal tradition, the critics appeared to have carried the day. By the end of the 1970s, ironically, the winners were not the "radical" contenders but workaday methodologists, now armed with computers, backed by a mathematics lobby working through the Social Science Research Council, and dedicated to evaluating governmental programs quantitatively. Within the discipline as a whole, the result was fragmentation and what one observer has termed an "interregnum."[49]

A decade later, this challenge threatened more than a simple repetition of earlier cycles. As conflict and varieties of neo-Marxist sociologies gave way to poststructuralist/postmodernist approaches, critics deepened the challenge to sociology's basic tenets: its claim to provide universal knowledge, its emphasis on order and system, and its privileging of "expert" over lay understandings of society.[50] At issue was not just one or another theory or methodology, but the very concepts of "society" and the "social." "[The] death of the social," wrote the French philosopher Jean Baudrillard of sociology, in an extreme statement of this position, would also be its own death.[51]

At the same time, the divide between positivistic policy research and varieties of postmodernist theory undermined what remained of the cooperation between theory and research that had characterized the discipline during its most productive periods. While some sociologists urged rejection of postmodernism altogether, others noted hopefully that postmodernists address issues that have always engaged the sociological imagination: the major structural transformations in Western society, their impact on social interaction and identity, and the need for new methods and strategies. In this climate, the future of the discipline appeared to be as uncertain as it had been at the start of the academic era.

[48] Mullins and Mullins, *Theory Groups*, chaps. 7–11.
[49] Wiley, "Dominating Theories."
[50] Peter Wagner, "Sociology," in *The History of Humanity*, ed. UNESCO, vol. 7: *The Twentieth Century*, (London: Routledge, forthcoming).
[51] David R. Dickens and Andrea Fontana, eds., *Postmodernism and Social Inquiry* (New York: Guilford Press, 1994).

19

ANTHROPOLOGY

Adam Kuper

Cultural anthropology, social anthropology, ethnology, *Volkskunde* and *Völkerkunde*, anthropology *tout court*: It would be foolhardy to attempt a common definition for these terms, let alone to specify a shared program for what appears to be rather a series of loosely connected, geographically variable, and historically unstable projects. Indeed, a coherent history of the varieties of world anthropology is not a plausible enterprise.

My strategy here is to concentrate on the development of what came to be known in the early twentieth century as social anthropology (the usual term in Europe), or cultural anthropology (the American designation), the dominant traditions throughout this period. Second, I will identify common elements in the trajectories of these traditions, although much of the argument will necessarily concern various national schools. In tracing the modern history of this discourse, I have adopted the conventional, though certainly debatable, division into three stages: the evolutionist debates and the confrontations between evolutionists and diffusionists, roughly, 1860–1920; the social science or behaviorist phase, running from about 1920 to 1970, when the theoretical models were drawn from sociology and psychology or from structural linguistics; and the more recent period, during which the dominant project has been what Clifford Geertz termed "the interpretation of cultures" and the most potent theoretical influences have come from philosophy, semiotics, and literary theory. To be sure, some national schools developed along very different lines; and while this periodization is most apparent in the development of cultural anthropology in the United States, even there, all three orientations – roughly, evolutionist, functionalist or structuralist, and interpretivist – coexisted uneasily throughout the twentieth century. Nevertheless, this framework can be used to organize a preliminary account.

THE EVOLUTION OF CULTURE AND SOCIETY

The Société d'Anthropologie de Paris was established in 1859, swiftly followed by similar initiatives in London in 1863 and in Berlin in 1869 (each, of course, with its own journal). The impetus for the establishment of these anthropological societies, which attracted scientists of the caliber of Paul Broca, Francis Galton, T. H. Huxley, E. B. Tylor, and Rudolf Virchow, was the burgeoning debate on the great question of human origins, stimulated by discoveries in Europe by paleontologists and archaeologists and by the publication of Charles Darwin's *Origin of Species* in 1859. Perhaps precisely because the new horizons seemed so immense, men from all corners of the intellectual world began to come together in the late 1850s to discuss them, and a new field of discourse began to emerge.

Nevertheless, the early history of anthropology was not orderly. Its practitioners straddled the conventional division between the natural sciences and the humanities. Race, culture, and the history of human institutions were subjects in which medical doctors, lawyers, and even missionaries and explorers could all claim some expertise, as could a variety of specialists, including geographers, palaeontologists, geologists, philologists, classicists, and students of the Bible. Moreover, it was not easy to decide which topics were relevant to the new discourse. Papers delivered in the 1860s to the Anthropological Society of London, for example, ranged from "Notes on Scalping" and "Abnormal Distension of the Wrist" to "Danish Kitchen Middens," "The Fossil Man of Abbeville Again," and "The Gypsies in Egypt."[1] There was also a great range of empirical reference. It was generally accepted that anthropologists should concern themselves with "primitive peoples," either because were taken to represent living fossils, or, more pragmatically, because it was thought that they were on the verge of extinction, or at least about to lose their cultures. At the same time, one source of excitement in the new field was the hope that a comparative study of "primitive" societies would throw new light on the texts of classical antiquity, including the Bible.

By the 1880s, courses in anthropology were being offered in universities in Germany, France, Britain, and the United States, but the subject could still seem to be a veritable cabinet of curiosities. E. B.Tylor's textbook, *Anthropology* (1881), covered human origins, "Man and Other Animals," race, language, the arts – which he divided into the "Arts of Life" ("instruments and machines") and the "Arts of Pleasure" (poetry, music, dance, and the plastic arts) – along with science, the spirit world, history and mythology, and "Society." In most of Western Europe, prehistory and human biology had established themselves as distinct fields by the turn of the century. Nevertheless, ethnology (or social anthropology) still embraced philology, folklore, material

[1] From the *Anthropological Review* (1864), quoted in John Burrow, *Evolution and Society* (Cambridge: Cambridge University Press, 1966), p. 125.

culture, and what Tylor called "comparative jurisprudence." Throughout the twentieth century, anthropology in the United States joined in a promiscuous, though seldom passionate, embrace what came to be termed the "four fields" of human biology, archaeology, linguistics, and cultural anthropology.

Notwithstanding this diversity, the various intellectual projects had a certain unity. Indeed, the president of the Anthropological Society of London explained that his society had been formed in 1863 precisely because "the time has arrived when it has become absolutely necessary that all the different branches of science relating to man shall no longer be isolated."[2] Debates began to focus on four emerging research programs that seemed bound to transform the understanding of human origins and human nature.

The first issue concerned the antiquity and origins of humanity. Jacques Boucher de Perthes had published his pathbreaking findings on the French Paleolithic in 1847, and his research was given a firm scientific context after the excavation of Brixham Cave in 1858. In 1863, T. H. Huxley published *Evidences as to Man's Place in Nature*. In the same year, Charles Lyell vouched for the antiquity of the human fossils that had been discovered in Europe in association with stone tools, and he related these finds to Darwin's theory about the origin of species in *The Geological Evidences of the Antiquity of Man, with Remarks on Theories of the Origin of Species by Variation*. In 1865, Darwin's close associate, John Lubbock, published *Prehistoric Times*, which surveyed and synthesized the findings of the new archaeology and palaeontology. A linked set of questions were now addressed. Could the prehistory of humanity be recovered? Did the human species have one common origin, or did different human races have separate origins? Finally, was there a direction – a progression – in human history?

A second project had been defined by Theodor Waitz, professor of philosophy at Marburg, in the first volume of his encyclopedic *Anthropologie der Naturvölker* (1858). Did differences of mentality and culture reflect biological variations? Or, as Waitz himself believed, was cultural variation constructed on a common human basis, the "psychic unity of mankind"? Although the Anthropological Society of London quickly published the first volume of Waitz's book in translation, its members were more inclined to take the view that there were great racial differences in mentality. During the American Civil War, London was torn by conflicts between the "anthropologicals" and the "ethnologicals," believers in human monogenesis confronting those who inclined rather to the view that human races were equivalent to species, and that each had a separate origin.[3] A commitment to monogenesis did not, however, necessarily entail a belief that all races were equal. In the 1880s, even the Darwinians increasingly took the view that cultural differences were rooted in race.

---

[2] Quoted in Burrow, *Evolution and Society*, p. 120.
[3] George W. Stocking, Jr., *Victorian Anthropology* (New York: Free Press, 1987), chap. 7.

The third project took up Auguste Comte's theory that human rationality had progressed through stages, represented by the Victorians as magical, religious, and scientific. Magic was just bad science and ineffectual technology. Religion was a hangover from the irrational fears and dreams of our remote ancestors. These were the established rationalist doctrines, but now Darwinian theory offered a scientific alternative to the biblical account of creation and divine guidance. Utilitarian theorists argued for a rational ethics to replace a revealed morality. The hour of science had surely arrived.

The second volume of Tylor's *Primitive Culture* (1871) was devoted to the development of religion, introducing a series of studies on what came to be termed "totemism" that dominated the work of the next generation of British anthropologists and attracted the attention of Durkheim and Freud. Religions could be ranged in a series according to their intellectual sophistication, but later religions all derived from a primitive system of theology and retained traces of their origins. The fundamental religious ritual was sacrifice, and the origins of sacrifice were to be found in primitive "animist" practices, in which the gods were fed the souls of sacrificed animals. The clear implication was that Christian rituals were suffused with primitive relics. They should no longer fool any rational person. These ideas were developed by J. F. McLennan, William Robertson Smith, and J. G. Frazer, who argued that all religion originated in "totemism" and that the primordial religious practices could still be observed among the Australian aborigines.[4]

The fourth project concerned the origins of civil institutions. The classical philosophical questions about the sources of legitimate government were now reexamined in an evolutionary framework, and with reference to observations of so-called primitive societies. The debate was initiated by three lawyers, Henry Maine (*Ancient Law*, 1861), J. F. McLennan (*Primitive Marriage*, 1865), and an American, Lewis Henry Morgan (*Ancient Society*, 1877); and Friedrich Engels identified its potentially radical implications (*Der Ursprung der Familie, des Privateigenthums und des Staats*, 1884). Despite their polemical differences, largely concerning the priority of matrilineal kinship systems, these authors quickly established a consensus on the central issues. The most primitive human societies were based on kinship and made up of exogamous descent groups. Only after a very long period of savagery and a great revolution – the greatest in human history, according to Henry Maine – did territorially based groups at last replace kinship groups as the basis of society. This revolution was accompanied by the invention of private property and the emergence of marriage and the family.[5]

[4] E. E. Evans-Pritchard, *Theories of Primitive Religion* (Oxford: Oxford University Press, 1965); George W. Stocking, *After Tylor* (Madison: University of Wisconsin Press, 1995), chap. 2.
[5] Adam Kuper, *The Invention of Primitive Society: Transformations of an Illusion* (London: Routledge, 1988).

These four projects, though distinct, had a great deal in common, and there were early attempts at synthesis, notably by Lubbock and Tylor.[6] The shared problems concerned the way in which human institutions and knowledge had developed from a savage to a civilized condition. The authors also generally treated the same body of literature, a mixture of classical sources and the reports of travelers to, and residents of, the tropics. Finally, they confronted the same problems of method. How was the immense history of the human species to be reconstructed? How could the diverse reports on "primitive peoples" be classified and ordered?

In Germany, geographical syntheses were generally preferred, with the classification of language families providing an influential model for the description of cultural areas. In Britain, however, the Enlightenment model of a single, progressive human history was generally accepted. All societies passed through the same stages of development, if not at the same pace. Human societies were ordered into stages of development, their customs stratified like fossil forms of living species. Anthropologists accordingly sought equivalents to the fossils of the palaeontologists, or the stone tools of the archaeologists, or the linguistic roots of the philologists, that would bear witness to ancient customs and beliefs. Perhaps ancient practices were preserved in fossilized form in some modern ceremonies, or in conservative linguistic forms. There might even be living Stone Age societies. Just as ancient rock formations might surface in some parts of the globe, so backward peoples might survive into modern times. Darwin himself had famously reported that on first seeing the Fuegians, "[T]he reflection at once rushed into my mind – such were our ancestors."[7] This line of reasoning placed a premium upon finding the most primitive living populations, and it was widely believed that these were the native peoples of Australia. At the turn of the century, James George Frazer wrote to Baldwin Spencer: "The anthropological work still to be done in Australia is . . . of more importance for the early history of man than anything that can now be done in the world."[8]

The small and disputatious community of anthropologists also shared some theoretical ideas, or at least debated a small set of alternatives. Darwin was a mighty presence in these circles, particularly in Britain, and some of the anthropologists were strongly influenced by him; Darwin in turn closely followed the debates of the anthropologists. Nevertheless, the influence of Darwinism should not be exaggerated.[9] It is necessary to distinguish between the various aspects of Darwin's message, as his major theses were not equally influential. By the early 1870s, the scientific community had generally become

---

Stocking, *Victorian Anthropology*, particularly chaps. 5 and 6.
Charles Darwin, *The Descent of Man, and Selection in Relation to Sex*, 2nd ed. (London: John Murray, 1874), p. 920.
Robert R. Marett and Thomas K. Penniman, eds, *Spencer's Scientific Correspondence with J. G. Frazer* (Oxford: Clarendon Press, 1932), p. 22.
Burrow, *Evolution and Society*, chap. 4; Stocking, *After Tylor*, passim.

persuaded of the antiquity of the human species, and of its primate origin. The most influential anthropologists also accepted monogenesis for scientific or theological reasons, even if some, perhaps most, were inclined to take the view that environmental pressures had shaped racial differences over a long period, producing significant mental differences across populations. But while the theory of common descent was accepted, the theory of natural selection was disputed by leading biologists, including even Huxley. Potentially the most subversive implication of the theory of natural selection was that there was no clear path of progression in evolution, though in practice most of Darwin's contemporaries remained faithful to an Enlightenment view of unilinear progress, and Darwin himself wrote as though there were a clear development in human civilization. British and French writers were generally wedded to a belief in the universal progress of human civilization, a view that could be combined with a vaguely Darwinian discourse or could accommodate a theory of the hierarchy of human races.

Two books published in 1871 gave a coherent intellectual impetus to these projects: Darwin's *Descent of Man* and Tylor's *Primitive Culture*. Darwin's account of human evolution placed special emphasis on the growth of the brain, which he associated with the evolution of bipedalism and the development of technology. While human behavior was a variation of the behavior of other primates, it had diverged as a consequence of intellectual specialization. It had also surely advanced, and, apparently, progress had continued as the species developed. The archaeological evidence indicated that there had been long-term progress in the field of technology. Tylor now argued that it was not only technology that had advanced with the development of the human intellect, but also language, the arts, social institutions, and the understanding of the world. These various accomplishments formed the elements of a single system, "culture or civilization," "that complex whole which includes knowledge, belief, art, morals, law, custom, and any other capabilities and habits acquired by man as a member of society."[10] In short, the specialization of the brain yielded culture; and the brain had continued to develop, and culture with it.

In Germany, the Enlightenment project had influential adherents, but the Counter-Enlightenment reaction to rationalist, progressive, universal histories was a potent force. The Romantic movement had stimulated the study of German and Slavic dialects and the collection of folktales and folk music, in pursuit of an elusive *Volksgeist* that had somehow resisted the corrosive influence of metropolitan civilization. This nationalist project was criticized in liberal circles, but German scholars were inclined to prefer geographical mappings of cultures to evolutionary reconstructions, and they emphasized the environmental or biological determinants of cultural institutions. While Darwinism triumphed in England, in Germany it met with a critical reaction

[10] Edward B. Tylor, *Primitive Culture*, 2 vols. (London: John Murray, 1871), p. 1.

that precipitated the formation of a very different anthropological discourse, under the leadership of Rudolf Virchow (1821–1902), the leading medical scientist in Germany, a prominent politician of liberal views and the guiding spirit behind the Berlin Society of Anthropology. Darwin had advanced a stimulating hypothesis, Virchow allowed, but it was premature. There was still far too much to be learned about the history of particular species to allow a synthesis. Moreover, Virchow denied that race was the key to human history. Germans and Slavs were not pure races. Racial mixing was widespread if not universal. Biological traits cross-cut the conventional racial classifications, which were in any case influenced by local, environmental factors. Race, culture, language, and nationality did not necessarily, or even usually, coincide.

Virchow's associate Adolf Bastian (1826–1905), who became the first director of the great Berlin Museum of Ethnology in 1886, attempted to demonstrate that, like races, all cultures are hybrids – historically diverse, dependent on borrowings, always in flux. All cultures were rooted in a universal human mentality, with its characteristic tendency to generate similar basic ideas; but cultural development was constrained by the natural environment, and it was shaped by contact between human populations. Borrowing was the primary mechanism for cultural change. Since cultural changes were the consequence of chance local processes – environmental pressures, migrations, trade – it followed that history has no fixed pattern of development. Franz Boas (1858–1942), a student of Virchow and Bastian, introduced this approach into American anthropology at the turn of the century.[11]

DIFFUSIONISM

As American anthropology developed into an organized academic discipline, it was defined by the epic struggle between Boas and his school and the evolutionist – perhaps rather Enlightenment – tradition, represented in the United States by the followers of Lewis Henry Morgan. "The history of the human race," Morgan asserted, "is one in source, one in experience, and in progress."[12] Boas eventually found permanent employment in 1899 at Columbia University, where he established the first graduate department of anthropology in the country. Columbia was a center of the anti-Darwinian reaction in biology, and one of Boas's students, Robert Lowie (1883–1957), later recalled how shocking it was to him, as a young Darwinian, to be confronted there with Thomas Hunt Morgan's critique of natural selection,

---

[11] Woodruff D. Smith, *Politics and the Sciences of Culture in Germany, 1840–1920* (New York: Oxford University Press, 1991); George W. Stocking, Jr., ed., *Volksgeist as Method and Ethic: Essays on Boasian Ethnography and the German Anthropological Tradition* (Madison: University of Wisconsin Press, 1996).

[12] Lewis Henry Morgan, *Ancient Society: Research in the Lines of Human Progress from Savagery through Barbarism to Civilization* (New York: Holt, 1877), p. 6.

and with Edward L. Thorndike's dismissal of links between animal and human thinking.[13] Here Boas found a receptive audience for his critique of evolutionism in anthropology, particularly among the largely German-speaking, first-generation Americans from whom his students were initially drawn. Summing up the teaching of Boas, Robert Lowie emphasized the thesis that no "necessity or design appears from the study of culture history. . . . Neither morphologically nor dynamically can social life be said to have progressed from a state of savagery to a state of enlightenment."[14] Following Virchow, Boas also repudiated racial explanations of cultural difference, a matter of enduring political importance in the United States, and insisted on "the psychic unity of mankind": "There is no fundamental difference in the ways of thinking of primitive and civilized man."[15]

The reason that evolutionary models did not hold had been stated by Virchow and Bastian. History was driven by migration, borrowing, the diffusion of traits, due to chance contact.[16] In consequence, one might identify regional patterns of development, but there was no universal history. Boas conceived of anthropology as an historical discipline, but one that dealt with local histories. Once a number of particular histories had been assembled, the premature evolutionist generalizations would be disproved. He and his students carried out a series of studies of native North American populations, and they deployed their material to show that the theses of Tylor, Morgan, and McLennan did not hold. Totemism was a loose amalgam of beliefs that did not necessarily occur together. Matriliny did not everywhere precede patriliny. Patrilineal societies might become matrilineal, and some matrilineal societies were more advanced technologically and politically than some patrilineal societies. Kinship and territorial groupings coexisted even among hunter-gatherers.[17]

This "diffusionist" critique of evolutionism derived from the German tradition of ethnological research, which was prosecuted particularly in Berlin and Vienna. In the German tradition, however, diffusion was often combined with models of progressive development. Local culture complexes resulted from contacts and borrowings, but there was long-term secular progress. The German critique of evolutionist reconstructions in the Pacific converted the influential British scholar W. H. R. Rivers (1864–1922) from his earlier

[13] Robert Lowie, "Reminiscences of Anthropological Currents in America Half a Century Ago," *American Anthropologist*, 58:6 (1956), 995–1016.

[14] Robert Lowie, *Primitive Society* (New York: Boni and Liveright, 1920), p. 427.

[15] Franz Boas, *The Mind of Primitive Man* (1911) (New York: Macmillan, 1938), Preface to the 1938 edition. No doubt the rise of fascism provoked this unusually uncompromising formulation, but it summed up the argument that he had made in more characteristically provisional and guarded language a generation earlier.

[16] See Boas, *The Mind of Primitive Man*.

[17] See Lowie, *Primitive Society*; Robert Lowie, *Primitive Religion* (New York: Boni and Liveright, 1924); George W. Stocking, Jr., ed., *A Franz Boas Reader: The Shaping of American Anthropology, 1883–1911* (Chicago: University of Chicago Press, 1974).

unilinear evolutionism. His masterpiece, *The History of Melanesian Society* (1914), made diffusionism respectable in British anthropology, but, like the Germans, Rivers was inclined to the view that successive cultural complexes represented evolutionary advances.[18]

FIELDWORK

In the 1890s and the early years of the new century there was also a very general shift in ethnographic methods. The evolutionist writers were not, in general, fieldworkers; they relied on reports from missionaries, travelers, and local specialists, whose work they increasingly came to direct, from afar.[19] "The man in the study busily propounded questions," Marett wrote, "which only the man in the field could answer, and in the light of the answers that poured in from the field the study as busily revised its questions."[20] By the 1890s, however, ethnographic expeditions by professional ethnologists had become common, often financed by the proliferating ethnographic museums, hoping to add to their collections. These expeditions, modeled on those of the natural historians, reinforced the impulse given by diffusionist theory to the study of regional cultural traditions. They were an important element in the professionalization of the discipline: The first British expedition, to the Torres Strait in 1898, effectively served to recruit the initial cadre of British anthropologists for the universities.[21]

At their best, these expeditions, and the surveys they published, provided reliable reports on the distribution of myths, customs, marriage laws, art, and technology within a particular region. Boas himself produced the most substantial of these studies, dedicating himself to the documentation of the cultural traditions of the native peoples of the American northwest coast, and he eventually published over 5,000 pages of records, often in the native language. But Boas and Rivers recognized the limitations of this kind of work. Lowie recalled that Boas "was especially appreciative of men who had achieved what he never attempted – an intimate, yet authentic, picture of aboriginal life. I have hardly ever heard him speak with such veritable enthusiasm as when lauding Bogoras's account of the Chukchi, Rasmussen's of the Eskimo, Turi's of the Lapps."[22] In 1913, Rivers suggested that conventional surveys of the distribution of cultural traits should be supplemented, or perhaps replaced, by "intensive work." In this type of research, "the worker lives for a year or more among a community of perhaps four or five hundred people and studies every

[18] Stocking, *After Tylor*, chap. 5.
[19] Stocking, *Victorian Anthropology*, chap. 3.
[20] Robert R. Marett, *The Diffusion of Culture* (Cambridge: Cambridge University Press, 1927), p. 4.
[21] Anita Herle and Sandra Rouse, eds., *Cambridge and the Torres Strait: Centenary Essays on the 1898 Anthropological Expedition* (Cambridge: Cambridge University Press, 1998).
[22] Lowie, "Franz Boas," *Biographical Memoirs, National Academy of Science*, 24 (1947), 303–22, at 311.

detail of their life and culture." He "is not content with generalized information, but studies every feature of life and custom in concrete detail and by means of the vernacular language." And Rivers concluded that such studies will reveal "the incomplete and even misleading character of much of the vast mass of survey work which forms the existing material of anthropology."[23]

Some of Rivers's Cambridge students, notably A. R. Radcliffe-Brown (1881–1955), had begun to undertake "intensive" research projects, but the first fully realized intensive study was undertaken in 1915, when Bronislaw Malinowski (1884–1942) embarked on two years of field research in the Trobriand Islands, a study that became the paradigm for "participant observation." His guiding notion was that one could not rely on what people said about their customs: People said one thing and did another. It was necessary to witness actual behavior in order to grasp how a social system worked. As Rivers had foreseen, the kind of information on which the evolutionists and diffusionists had typically relied came to be discredited.

VARIETIES OF FUNCTIONALISM: ANTHROPOLOGY AS A SOCIAL SCIENCE

Malinowski's Trobriand researches not only began a revolution in field methods, they also signaled a significant shift in the theoretical climate. What came to be called "functionalism" in anthropology was a coming together of a new method of intensive field research and a new theoretical agenda. Its subject was the workings of a social system, rather than its history, and its focus was on collective psychological processes. Its leading practitioners increasingly drew on theories developed within sociology and social psychology. What came to be termed "social anthropology" was recast in the image of these neighboring social sciences. Radcliffe-Brown defined it as "comparative sociology."

The immediate theoretical inspiration for this new perspective was the work of Durkheim and his school. Radcliffe-Brown had carried out a diffusionist field study in the Andaman Islands (1906–8) along the lines favored by his teacher, Rivers, but almost immediately on his return to England had been converted to a Durkheimian position; the monograph he finally published in 1922, the year that Malinowski published his first Trobriand book, was strictly Durkheimian. Malinowski too was influenced by Durkheim during this period, but he had also studied with Wundt in Leipzig and imbibed his social psychology. For both men, it seems, the initial appeal of Durkheim, and particularly of *The Elementary Forms of the Religious Life* (1912), was that he provided a way of thinking about the relationship between social institutions and collective "sentiments" in a "primitive society."

[23] W. H. R. Rivers, "Report on Anthropological Research outside America," in *Reports upon the Present Condition and Future Needs of the Science of Anthropology* (Carnegie Institution Publication No. 200) (Washington, D.C.: Carnegie Institution, 1913), p. 7.

This is not to say that there was a coherent functionalist theory, even within British anthropology. There were significant theoretical differences between the two main theorists, Malinowski and Radcliffe-Brown. Nevertheless, both insisted that a synchronic framework of analysis was more promising than the established evolutionist and diffusionist approaches, derided by Radcliffe-Brown as "conjectural history." The approach they advocated took the current social life of a people as its subject, and explained it not in terms of its putative history but rather as a working whole, as a machine that performed certain tasks, whose parts all contributed to its efficient operation. In the concluding pages of his first Trobriand monograph, *Argonauts of the Western Pacific* (1922), Malinowski wrote: "[I]t seems to me that there is room for a new type of theory" to be set beside the evolutionism of the "classical school of British Anthropology" and the geographical perspectives of the Germans.

> The influence on one another of the various aspects of an institution, the study of the social and psychological mechanism on which the institution is based, are a type of theoretical studies which has been practiced up till now in a tentative way only, but I venture to foretell will come into their own sooner or later. This kind of research will pave the way and provide the material for the others.[24]

And Radcliffe-Brown insisted more strongly that "at this time the really important conflict in anthropological studies is not that between the 'evolutionists' and the 'diffusionists'. . . but between conjectural history on the one side, and the functional study of society on the other."[25]

Radcliffe-Brown aspired to develop classifications of types of societies and institutions and to identify recurrent social relationships and processes, his work parallelling aspects of that of Durkheim's nephew, Marcel Mauss, who continued the more ethnographic side of Durkheim's program in France. Malinowski developed a variety of functionalist arguments, including a form of biological utilitarianism, but in his Trobriand monographs he was more concerned to demonstrate the practical rationality of Trobriand customs. Individuals, he argued, pursued their self-interest behind a smokescreen of formal pieties. And even the most apparently irrational practices could be shown to have a payoff – magic, for instance, stiffened the confidence of gardeners, or helped canoe builders to organize their tasks.[26]

Natural science models hardly impinged on European social anthropology, which had become institutionally quite distinct from physical anthropology (known, confusingly, as "anthropology" in most European countries). But the reshaping of British social anthropology as a synchronic social

---

[24] Bronislaw Malinowski, *Argonauts of the Western Pacific: An Account of Native Enterprise and Adventure in the Archipelagoes of Melanesian New Guinea* (London: Routledge, 1922), pp. 515–16.

[25] A. R. Radcliffe-Brown, "A Further Note on Ambrym," *Man*, 29 (1929), 50–3, at 53.

[26] Raymond Firth, ed., *Man and Culture: An Evaluation of the Work of Bronislaw Malinowski* (London: Routledge, 1957).

science discipline was not repeated elsewhere during the interwar years. Functionalism and other social science models were not decisive in the development of ethnology in France, and were hardly significant in Germany. These were the two most significant centers of social anthropology – or ethnology – in continental Europe, but the discipline developed in a distinctive fashion in each country.

German ethnology had always maintained a close association with geography, and shared a concern with the spatial distribution of cultural traits.[27] The field was also greatly influenced by philological and folklore studies, which led to a distinctive emphasis on mythologies. Established largely in museum settings, the discipline made a sharp distinction between material objects and "spiritual" traits, although attempts were made to define the ideas that animated particular categories of objects, mythologies, and artistic traditions. At the turn of the century, a distinction developed between *Volkskunde*, the study of national folk culture, and *Völkerkunde*, the study of other cultures, the latter tradition losing a certain impetus when Germany was stripped of its colonies after World War I. Both of these disciplines were deeply implicated in the Nazi science of the 1930s and 1940s, including both racial doctrines and eugenics, and in the development of new theories of colonial rule, but these alliances were shattered by the defeat of the Nazi regime. Only after the war was German ethnology strongly influenced by American, British, and French schools of anthropology.[28]

Robert Lowie has remarked on the peculiar fact that in France "the impulse to field research finally emanated from philosophy."[29] There is considerable truth in this observation, particularly if the central project of Durkheim and his school is understood to be fundamentally philosophical, deriving from Kant and Hegel and concerned with the social history of categories of thought, or of reason.[30] The *Année Sociologique* devoted considerable space to the theoretical analysis of ethnography, and Durkheim's right-hand man, his nephew Marcel Mauss (1872–1950), dedicated his career to the elucidation of what he took to be primitive and innate but evolving categories of thought.[31] For Durkheim, these emerged by way of social experience. Mauss preached the doctrine of the "total social fact." Customs and beliefs are forms

[27] Matti Bunzl, "Franz Boas and the Humboldtian Tradition," in *Volksgeist as Method and Ethic*, ed. Stocking, pp. 17–78.

[28] Walter Dostal, "Silence in the Darkness: German Ethnology in the National Socialist Period," *Social Anthropology*, 2:3 (1994), 251–62; W. Hirschberg, ed., *Neues Wörtebuch der Völkerkunde* (Berlin: Reimer, 1988); Karl-Peter Koepping, *Adolf Bastian and the Psychic Unity of Mankind: The Foundation of Anthropology in Nineteenth Century Germany* (St. Lucia: University of Queensland Press, 1983).

[29] Robert H. Lowie, *The History of Ethnological Theory* (New York: Farrar and Rinehart, 1937), p. 196.

[30] Stephen Lukes, *Emile Durkheim, His Life and Work* (London: Harper and Row, 1973); Steve Collins, "Categories, Concepts or Predicaments? Remarks on Mauss's Use of Philosophical Terminology," in *The Category of the Person: Anthropology, Philosophy, History*, ed. Michael Carrithers, Steven Collins, and Stephen Lukes (Cambridge: Cambridge University Press, 1985).

[31] Marcel Fournier, *Marcel Mauss* (Paris: Fayard, 1994).

of social action, embedded in a system that includes collective beliefs and institutions. However, his teaching encouraged the first generation of French anthropologists to focus on the study of collective representations, notably myths and rituals. By contrast, Radcliffe-Brown in Britain developed a version of Durkheimian theory more concerned with institutional forms and collective sentiments than with collective representations.

Ethnographic field research remained relatively undeveloped until the formation of the Institute of Ethnology at the University of Paris in 1925, under the leadership of Mauss, the philosopher Lucien Lévy-Bruhl, and the physical anthropologist Paul Rivet. Supported by the Ministry of Colonies, the Institute established a training program for ethnographers and organized a series of field expeditions, most famously the Mission Dakar-Djibouti (1931–3) led by Marcel Griaule. These expeditions followed the already old-fashioned survey and collection model, largely concerned with collecting material objects and with studying art and mythology.

After World War II, the tradition of Mauss was recast by the structuralist theorists, Claude Lévi-Strauss (b. 1908) and Louis Dumont (1911–1999). A rival, more sociological school developed under the leadership of Georges Balandier and Paul Mercier, devoted to the study of social change in colonial settings and influenced by British social anthropology, particularly the neo-functionalism of Max Gluckman. In the 1960s and 1970s, younger members of this school participated in a revival of Marxist thinking. Research fellowships became widely available, and a new generation of outstanding fieldworkers emerged, collaborating with historians, philosophers, and sociologists to make Paris one of the major world centers of social anthropology.[32]

During the interwar years the development of American anthropology diverged decisively from the European schools. The Boasians were skeptical about Durkheim's theories, and resistant to the influence of Malinowski and Radcliffe-Brown. A synchronic form of analysis gained ground within Boasian anthropology during the 1920s and 1930s, but it drew inspiration from psychology and psychoanalysis rather than from sociology. Its subject was culture, not society; but each local culture was treated as an organic whole, structured by a particular cluster of values and stylistic forms, rather than as a historical deposit of diverse traits. Culture determined cognition and behavior, and shaped personality and identity. By the time a child can talk, Ruth Benedict claimed, "he is the little creature of his culture, and by the time he is grown and able to take part in its

---

[32] Gérald Gaillard, "Chronique de la recherche ethnologique dans son rapport au Centre national de la recherche scientifique 1925–1980," *Cahiers pour l'histoire du CNRS*, no. 3 (1989), 85–127; Gérald Gaillard, *Répertoire de l'ethnologie française, 1950–1970* (Paris: CNRS, 1990); Jean Jamin, "France," in *Dictionnaire de l'ethnologie et de l'anthropologie* (Paris: PUF, 1991), pp. 289–95; Victor Karady, "Le problème de la légitimité dans l'organisation historique de l'ethnologie française," *Revue française de sociologie*, 23:1 (1982), 17–35.

activities, its habits are his habits, its beliefs his beliefs, its impossibilities his impossibilities."[33]

The "culture and personality" work of the 1930s and 1940s was a genuinely innovative project within Boasian anthropology.[34] Edward Sapir (1884–1939) provided the theoretical inspiration. Ruth Benedict (1887–1948) and Margaret Mead (1901–1978) produced the most widely read case studies, aimed with great success at the general reading public, to whom they conveyed the cultural relativism that was a central tenet of the Boasians. According to Margaret Mead, Boas had come to believe at some time during the 1920s that

> sufficient work had gone into demonstrating that peoples borrowed from one another, that no society evolved in isolation, but was continually influenced in its development by other peoples, other cultures, and other, differing, levels of technology. He decided that the time had come to tackle the set of problems that linked the development of individuals to what was distinctive in the culture in which they were reared.[35]

This loyalist view obscures the radical nature of the break, but Boas did at last come to endorse, though always with reservations, the wave of "culture and personality" studies.[36] In 1930, he wrote:

> If we knew the whole biological, geographical and cultural setting of a society completely, and if we understood in detail the ways of reacting of the members of the society as a whole to these conditions, we should not need historical knowledge of the origin of the society to understand its behaviour. . . . An error of modern anthropology, as I see it, lies in the over-emphasisis on historical reconstruction, the importance of which should not be minimized, as against a penetrating study of the individual under the stress of the culture in which he lives.

Writing as one of the few American adherents of Radcliffe-Brown, Fred Eggan cited this passage and commented derisively: "Some of Boas's students came to believe that 'they had known it all the time.'"[37] In any case, despite their particular interest in psychoanalytic ideas, the parallels between this neo-Boasian orientation and contemporary British functionalism are evident.

It was only after World War II, however, that American cultural anthropology allied itself predominantly, though never completely, with the neighboring social sciences. The decisive influence was the establishment by

---

[33] Ruth Benedict, *Patterns of Culture* (Boston: Houghton Mifflin, 1934), p. 3.

[34] George W. Stocking, ed., *Malinowski, Rivers, Benedict and Others: Essays on Culture and Personality* (Madison: University of Wisconsin Press, 1986).

[35] Margaret Mead, *Blackberry Winter: My Earlier Years* (New York: William Morris, 1972), p. 126.

[36] See Regna Darnell, *Edward Sapir, Linguist, Anthropologist, Humanist* (Berkeley: University of California Press, 1990), esp. chap. 9, "Psychologizing Boasian Anthropology."

[37] Citation from Boas, and commentary, in Fred Eggan, "One Hundred Years of Ethnology and Social Anthropology," in *One Hundred Years of Anthropology*, ed. John Otis Brew (Cambridge, Mass.: Harvard University Press, 1968), pp. 136–7.

Talcott Parsons of the interdisciplinary Department of Social Relations at Harvard in 1946. Parsons's project was designed to establish a new and more systematic division of labor. All social scientists were trying to explain what he called social action, but they tended to be reductionists. However, social behavior could not be reduced to biology, or to economic determinants, or to symbols and beliefs. Social action was shaped at once by the biology and the psychology of individuals, by the social and economic institutions of the society, and by ideas and values. These factors constituted distinct systems and should be studied independently, in the first instance at least. In the new dispensation, psychology would deal with the individual. Sociology would concern itself with social relations and the institutionalization of values. There remained culture, which was the name Parsons gave to the realm of ideas and values. A science of culture would focus on the "culture pattern system," leaving the social system and personality systems for the attention of specialists from other disciplines. The nearest thing to a science of culture in the American universities was anthropology, and Parsons proposed that it be recast as a branch of his master social science, with its own subject matter: culture.

In 1952, the year after Parsons published *The Social System*, the leading figures in American anthropology, Alfred Kroeber (1876–1960) and Clyde Kluckhohn (1905–1960), published their theoretical survey, *Culture*, which was their response, on behalf of anthropology, to the challenge of Parsons. They criticized Parsons for writing of culture "in a sense far more restricted than the anthropological usage" and complained that his conception of culture "leaves little place for certain traditional topics of anthropological enquiry: archaeology, historical anthropology in general, diffusion, certain aspects of culture change, and the like.... In particular, we are resistant to his absorbing into 'social systems' abstracted elements which we think are better viewed as part of the totality of culture."[38] They put up only a token resistance, however, and by the early 1960s the Parsonian program had begun to exercise a decisive influence on American cultural anthropology. Two graduates of the Department of Human Relations, Clifford Geertz and David Schneider, instituted the famous "systems course" as the core of graduate education in anthropology at the University of Chicago, introducing the next generation of anthropologists to Parsons's three-pronged approach to action. A parallel movement took place at Yale, where a similarly integrated "behavioral science" program was instituted. Its anthropological leader was George Peter Murdock (1897–1985), who was dedicated to the development of a comparative and scientific social anthropology that would rest on the firm foundation of a worldwide sample of cultures, against which hypotheses could be tested statistically.[39] The anthropologists influenced by Parsons,

---

[38] Alfred Kroeber and Clyde Kluckhohn, *Culture: A Critical Review of Concepts and Definitions*, as cited in *Papers of the Peabody Museum of Archeology and Ethnology, Harvard University*, 17:1 (1952), 136.
[39] George Peter Murdock, *Social Structure* (New York: Free Press, 1949).

Kluckhohn, and Murdock came to define themselves primarily as social scientists, and they were dismissive of the eclectic, historical, and particularistic studies favored by the Boasians.

But the social science program did not attract everyone. Some contemporaries, equally critical of the Boasian tradition, preferred to see cultural anthropology as a branch of the natural sciences and opted instead for a revival of the evolutionist tradition. Leslie White (1900–1975) dusted off the idea of a progressive civilization, essentially technological and scientific in nature. Culture was a machine for controlling nature, and it was progressive, adaptive, and ultimately universal. His colleague, Julian Steward (1902–1972), emphasized regional, ecological constraints on cultural developments. In the hands of some of their students, evolutionism became a functionalist doctrine, but one that emphasized biological rather than social factors. In the 1960s, ecological determinism became influential. This was in effect a functionalist version of evolutionary theory, in which customs and institutions were explained as adaptations to the natural environment.[40]

ANTHROPOLOGY, COLONIALISM, DEVELOPMENT

The emergence of functionalism in British anthropology has often been linked to the development of the policy of indirect rule in British African colonies.[41] Anthropologists regularly claimed that their science would be of service to the colonial project, but it was only during the 1930s that Malinowski and his students began to pay serious attention to African policy. Malinowski then insisted on a new sort of functionalism, one that addressed problems of change, a theoretical turn that was parallelled in the United States, where Herskovits pioneered the study of "acculturation." Malinowski and Radcliffe-Brown were critical of colonialism, and their relations with colonial grandees were generally strained. Moreover, there was very little specific demand for anthropology, functionalist or otherwise, from the colonial governments until the mid-1930s. Only a small number of colonies appointed government anthropologists, and the research of the university-based scholars was seldom funded directly or indirectly by official sources. There were only about two dozen British social anthropologists before World War II, and some of the rising generation were denied admission to the colonies because of their political radicalism. The African colonies provided the field laboratory for British functionalists, but their main source of funds during the 1930s was the Rockefeller Foundation rather than the colonial

[40] Marvin Harris, *Cultural Materialism* (New York: Random House, 1979).
[41] Talal Asad, ed., *Anthropology and the Colonial Encounter* (London: Tavistock, 1973); Henrika Kuklik, *The Savage Within: The Social History of British Anthropology, 1885–1945* (Cambridge: Cambridge University Press, 1991); Kuper, *Anthropology and Anthropologists*, chap. 4; George W. Stocking, ed., *Colonial Situations: Essays on the Contextualization of Anthropological Knowledge* (Madison: University of Wisconsin Press, 1991); Stocking, *After Tylor*, chap. 8.

governments, and colonial officers did not in general find their studies useful. The most innovative and critical development of functionalism, and the first attempt to apply it in urban settings, was the work of Max Gluckman and the Rhodes-Livingstone Institute in Northern Rhodesia during the 1940s and 1950s. Here a politically radical group of young scholars insisted on studying colonial society as a whole, including its endemic conflicts.[42]

Even if connections between "functionalism" and colonial policy could be plausibly established for British social anthropology, this would not account for the parallel shift of the neo-Boasians toward comparable modes of research and description, or for the adoption of a "Malinowskian" approach by ethnographers in the United States concerned with contemporary American and European societies. Moreover, parallel intellectual currents can be discerned in other social sciences. Dorothy Ross suggests that a "movement toward modernist historical consciousness, the growing power of professional specialization, and the sharpening conception of scientific method" together "produced a slow paradigm shift in the social sciences . . . away from historico-evolutionary models . . . to specialized sciences focused on short-term processes."[43] In short, the emergence of functionalist anthropology is best seen as an instance of a broader reorientation, one that affected all of the social sciences in the United States and Britain, although the move followed a different course in other European countries, and was commonly influenced by American developments after World War II.

Although American anthropologists had maintained a long-standing but often strained relationship with the administrators of Indian reservations, anthropologists in general were slower than other social scientists to play a role in government policy making in the United States.[44] During World War II, however, Margaret Mead and some of her colleagues were recruited to develop cultural profiles of allies and enemies, and Murdock supervised the production of social and cultural guidebooks for the navy during the Pacific campaign. After the war, there was a surge of government support for the social sciences. A number of anthropologists had been persuaded by their war experience that the future prosperity of anthropology depended on demonstrating its utility to governments. Murdock himself was invited by the navy to supervise an ambitious series of field studies in Micronesia, where the navy had been appointed to administer the Micronesian colonies captured by the United States from the Japanese. The research was subcontracted to various anthropology departments, and twenty-one expeditions were eventually financed. In the 1950s and 1960s, anthropologists increasingly participated in multidisciplinary social science research projects in the newly independent

---

[42] Richard Werbner, "The Manchester School in South-central Africa," *Annual Review of Anthropology*, 13 (1984), 1257–85.

[43] Dorothy Ross, *The Origins of American Social Science* (Cambridge: Cambridge University Press, 1991), p. 388.

[44] H. G. Barnett, *Anthropology in Administration* (Evanston, Ill.: Rowe, Peterson, 1956).

states, generally funded by the American government; but when the association of some of these projects with counterinsurgency programs created a backlash, they were gradually abandoned.

In Britain and France, the process of decolonization proved to be a much better source of funding for anthropological research than the colonial governments had been. Research institutes were established, particularly in Africa, staffed largely by anthropologists. All of this funding provided a further impetus for anthropologists to assert their identity as social or behavioral scientists, confirming a tendency already established in Britain during the 1930s. Beginning in the mid-1950s, a series of ambitious "development" programs attracted generous international funding and gave employment to some anthropologists. Particularly in the Scandinavian countries and in Germany, though to some degree also in Britain and in North America, the project of an "anthropology of development" was often attractive to the idealist generation of the 1960s. Enthusiasm waned in the 1980s. Donors abandoned "planification" and began to advocate market strategies, which did not require social research. Anthropologists had in any case become increasingly critical of development policies, and so had become less attractive to potential employers.

Perhaps the most important effect of decolonization was a shift in the subject matter of social and cultural anthropology. During the first half of the twentieth century, research was overwhelmingly conducted among populations defined as primitive – the native peoples of the Americas, Africa, and Oceania, and "tribal" populations in Asia. The abandonment of evolutionist models in the 1920s did not change this situation, although "acculturation" to a modern way of life became an accepted subject for research. American anthropologists had also conducted field studies among peasants in South and Central America, and Dutch and French scholars had introduced anthropological perspectives to traditional Orientalist fields of research in Indonesia and China. During World War II, Margaret Mead and Ruth Benedict and their associates produced anthropological accounts of Germany and Japan. However, it was only during the postwar years that anthropologists began to do intensive fieldwork in India, Indonesia, and China, and also in Europe, effectively turning away from the traditional association of anthropology with the primitive, although the category of "primitive society" was only gradually abandoned.[45]

REACTIONS TO FUNCTIONALISM: ANTHROPOLOGY AND THE HUMANITIES

At the same time that the effects of decolonization were being felt, the theoretical paradigm was changing, and the definition of social and cultural

[45] Kuper, *The Invention of Primitive Society*.

anthropology as social sciences was increasingly being called into question. In Britain, E. E. Evans-Pritchard (1902–1973), Radcliffe-Brown's successor as professor of social anthropology at Oxford, signaled his dissatisfaction with Radcliffe-Brown's sociological determinism in a famous lecture in 1950. He argued that the pursuit of scientific laws was mistaken. Anthropology was not, after all, a science, or even a social science. Rather, "social anthropology is a kind of historiography, and therefore ultimately of philosophy or art" that "studies societies as moral systems and not as natural systems . . . seeks patterns and not scientific laws, and interprets rather than explains."[46] Social anthropology, he concluded, was best regarded as a branch of Oriental studies. This did not necessarily entail a complete break with the established tradition. Evans-Pritchard and Louis Dumont (a colleague of both Evans-Pritchard and Lévi-Strauss) increasingly emphasized the moral, ideological thread in the Durkheimian tradition, and, like Lévi-Strauss, tended to elevate the contribution of Marcel Mauss above that of Durkheim himself.

Evans-Pritchard also advocated the introduction into ethnography of an historical dimension, a move that was given fresh impetus by the development of African history under the leadership of the Belgian anthropologist Jan Vansina. Functionalist perspectives notoriously evaded historical questions, and functionalist anthropologists found it difficult to cope with problems of change. The reintroduction of an historical perspective into ethnographic research was, however, generally associated with a turning away from social theory more generally.

Within American anthropology there was a parallel movement away from the neo-Parsonian project for cultural anthropology.[47] A radical shift away from a social science perspective to a more humanistic, cultural approach is apparent in the career of Clifford Geertz (b. 1926). Geertz had begun as an orthodox Parsonian, insisting that social scientists should "distinguish analytically between the cultural and social aspects of human life, and . . . treat them as independently variable yet mutually interdependent factors." In his early writings on Indonesia, he tended to contrast a traditional state of affairs, in which culture and social structure form a single, mutually reinforcing system, with the modern situation, in which old ideas and values become less and less satisfactory as explanations of the world, and as guides to action in it, and are challenged by new ideologies that may, in turn, introduce fresh lines of social conflict. Culture and social structure must be distinguished analytically, but they act upon each other. Anthropology specialized in culture, the realm of ideas, values, and symbols, but within an interdisciplinary

---

[46] E. E. Evans-Pritchard, "Social Anthropology: Past and Present" (Marett Lecture, 1950), in his *Essays in Social Anthropology* (London: Faber and Faber, 1962), p. 26.

[47] Adam Kuper, *Culture: The Anthropologists' Account* (Cambridge, Mass.: Harvard University Press, 1999).

social science project, never forgetting that culture and society were "mutually interdependent factors."

Geertz moved away from the Parsonian project during the late 1960s. This was the time when the Vietnam War divided the campuses. Parsonian sociology was a specific target of the New Left, which hoped to revive Marxist theory. Geertz, however, pressed the claims of a radically relativist form of cultural idealism as an alternative both to materialism and to Parsonian functionalism. In 1973, he welcomed "an enormous increase in interest, not only in anthropology, but in social studies generally, in the role of symbolic forms in human life. Meaning . . . has now come back into the heart of our discipline."[48]

Ten years later, in his second collection of essays, *Local Knowledge*, Geertz announced that a new configuration of disciplines could be discerned. An interpretive, symbolic anthropology would now link up with linguistic philosophy and literary theory. The common subject matter was culture, but culture now appeared as an object to be deciphered, a text to be translated, rather than as an ideology arising from social processes and constraining them. Moreover, although believing that cultural statements may be translated and interpreted, he repudiated the search for laws of collective imagination.

By the 1970s, "symbolic" or "interpretive" anthropology had become the mainstream movement in American cultural anthropology. Its subject was the study of collective symbolic meanings, and these were generally studied in terms of cognitive rather than social processes. This general reorientation did not necessarily entail a repudiation of scientific approaches in favor of interpretation. A number of scholars who came into anthropology immediately after World War II agreed that the object of study was not society but rather culture, or what Lévi-Strauss sometimes called the superstructure, or what Ward Goodenough called knowledge. But they nevertheless proclaimed that cultural anthropology could borrow scientific methods from cognitive psychology and generative linguistics. Lévi-Strauss returned from wartime exile in the United States to France in 1948 and began to promulgate a radically new program for social anthropology that introduced analytical models drawn from phonetics to uncover the regularities underlying systems of thought. In the United States, a formal, cognitive anthropology emerged, allied to the cognitive movement in psychology and linguistics.[49] Its core project was "ethnoscience," which developed particularly within the Yale school, influenced by the linguist Floyd Lounsbury; it promised a more scientific ethnographic procedure that came to be called "the new ethnography." This project had a great deal in common with the structural anthropology

---

[48] Clifford Geertz, *The Interpretation of Cultures* (New York: Basic Books, 1973), pp. 29, 144.

[49] Roy D'Andrade, *The Development of Cognitive Anthropology* (Cambridge: Cambridge University Press, 1995); Stephen A. Tyler, ed., *Cognitive Anthropology* (New York: Holt, Rinehart and Winston, 1969).

that was being developed at the same time in France by Lévi-Strauss, and both contributed to a general shift of attention from social organization to traditions of knowledge and symbolic communication.

The move from function to meaning, or from social structure to culture, may have been connected to the end of the British and French empires, but it is perhaps best seen as a reorientation of cultural and, to a lesser degree, social anthropology toward the humanities and away from the social sciences, at a time when the behavioral approach was losing ground generally and when the scientific character of the social sciences began to be widely questioned.

NEW DIRECTIONS

There were reactions to this redefinition of the subject matter of anthropology as the study of symbolic culture, and also to what opponents tended to dismiss as an antiscientific cultural relativism. Within American cultural anthropology, the established opposition to Boasian and neo-Boasian anthropology came from the evolutionists. During the 1950s and 1960s, White and Steward had led an evolutionist countermovement. In the 1970s, further varieties of evolutionism emerged in opposition to symbolic anthropology: a neo-Darwinian evolutionism, drawing not on ecology but on the prestigious ideas associated with genetics, and the sociobiology of E. O. Wilson, which was enthusiastically taken up by some cultural anthropologists as well as primatologists, though many physical anthropologists were critical. At the same time, there was a renewal of Marxist theory, in both Western Europe and the United States and also, most decisively, in Central and South America.

For a time it seemed as though the new generation of American anthropologists, the cohort that had passed through graduate school during the turbulent sixties, would be committed to the development of a Marxist anthropology, drawing on the new structural Marxism of the French school and influenced by the dependency theorists of Latin America. Alongside the revival of Marxist theory there emerged a new critique of the political uses of science. Particularly relevant was the critique of colonialism and colonial sciences, formulated first by Frantz Fanon and presented most influentially by Edward Said in his *Orientalism* (1978). Orientalism, and by extension anthropology, was "a kind of Western projection onto and will to govern over the Orient."[50] This radicalism was fueled by the civil rights movement, the experience of the Vietnam War, and the associated disturbances on American campuses. Other social movements – feminism, then multiculturalism – inspired fresh critical assaults on anthropological orthodoxies.[51]

---

[50] Edward Said, *Orientalism* (New York: Pantheon, 1978), p. 95.
[51] Sherry Ortner, "Theory in Anthropology since the Sixties," *Comparative Studies in Society and History*, 26 (1984), 126–66.

One consequence was a critique of the objectivity that had been claimed by the social scientists. Anthropologists were accused of subordinating their research to imperialist interests. More radically, some insisted that objectivity was an illusion. Social reality was necessarily constructed in different ways by differently situated interest groups, and certainly by men and women. Attacking the pretensions of the ethnographer's pose as an objective, scientific observer, critics drew on the resources of modern critical theory to reveal the rhetorical tricks of authorship ("authorizing"). Like all authors, ethnographers were writing "fictions." Nor were these innocent fictions. Caught up as they were in the colonial projects of the great powers, the classical ethnographers were all concerned to impose an order on the actual chaos of voices, perspectives, and situations that they confronted in the field – to inscribe one point of view on history. In this way, they served the interests of a political class that wished to impose an alien order on colonial subjects abroad, or on minorities at home.[52]

The logic of the critique implied that there must be a better way to write ethnographies. Because there were no privileged perspectives, no neutral voice-over was to be tolerated. Ethnographers were urged to experiment, to play with genres and models, to speak ironically, revealing and even undermining their own assumptions. Ethnography should represent a variety of discordant voices, never coming to rest, and never (a favorite term of abuse) "essentializing" a people or a way of life by insisting on a static representation of a culture as an integrated whole. In any case, the very object of ethnography was being transformed. Other cultures were no longer insulated from our own. There were no longer conservative, bounded cultures to be described by observers in that timeless tense, the ethnographic present. The West, or perhaps Capitalism, had spread its tentacles into every crevice of the world. Yet the citizens of the postcolonial states had not simply succumbed to westernization; they had rejected anthropological representations of themselves and were answering back. Some ethnographers insisted that there was a responsibility to give a privileged hearing to the muted voices of the downtrodden – to speak for the oppressed.[53]

These were the common themes of the new generation of American cultural anthropologists, but its theorists drew with different degrees of commitment on a range of critical perspectives, including feminism, literary theory, the subaltern critiques of colonial science, Marxism, and world system theory. Although none of these currents of thought represented a single, monolithic body of dogma, they took for granted a common perspective, derived not

---

[52] James Clifford and George Marcus, eds., *Writing Culture: The Poetics and Politics of Ethnography* (Berkeley: University of California Press, 1986); James Clifford, *The Predicament of Culture: Twentieth Century Ethnography, Literature and Art* (Cambridge, Mass.: Harvard University Press, 1988).

[53] George Marcus and Michael Fischer, *Anthropology as Cultural Critique* (Chicago: University of Chicago Press, 1986); Renato Rosaldo, *Culture and Truth: The Remaking of Social Analysis* (Boston: Beacon Press, 1989).

only from the radical critiques that had flourished in the 1970s, but also from Geertz's success in reorienting cultural anthropology in the United States as a discipline within the humanities.

The roots of this discourse lie in German romanticism, which, like other romantic movements before it, rejects the Enlightenment notion of a common human destiny, shaped by a common nature and a common rationality. It opposes the increasingly global, dominant technical civilization and promotes the interests of marginal groups in their struggle to assert their own identities in the face of this juggernaut. Yet notwithstanding these global claims, and despite affinities with European philosophical currents, the critical writers in American cultural anthropology developed what was a specifically American discourse, shaped by the civil rights movement, the trauma of Vietnam, feminism, and the emergence of identity politics in America. It has also been crucially informed by the internal disciplinary politics of American universities.

The postmodernist program in cultural anthropology suffered from a profound internal strain between the extreme relativism that it promoted and the radical political inclinations of many of its main figures. There was the further problem that a reflexive, critical stance could paralyze research. Clifford Geertz remarked in 1988 that students felt themselves to be "harassed . . . by grave inner uncertainties, amounting almost to a sort of epistemological hypochondria, concerning how one can know that anything one says about other forms of life is as a matter of fact so."[54] One option was to merge cultural anthropology with the more fashionable project of cultural criticism represented by cultural studies. However, much like the outcome of Marxist and feminist critiques of the 1970s, the enduring effect of the postmodernist critique of the 1980s was to emphasize the rupture with the social and cultural anthropology of the previous generation.

One significant consequence was to encourage the development of new topics of research and the introduction of methods from other disciplines. Gender became a major focus of interest, taking over many of the topics conventionally treated under the heading of kinship.[55] A series of approaches were developed to study identity, under the heading of the "self," "embodiment," and the "person." Visual anthropology became a major genre of ethnographic representation. In the 1980s, a major new area of "applied anthropology" emerged – medical anthropology, which soon became the largest subdiscipline in American anthropology. Medical anthropology tapped a rich source of funding in the area of public health, but it also provided an arena in which psychological and cultural approaches, biological perspectives, and ideas from "political economy" could be fruitfully combined.

---

[54] Clifford Geertz, *Works and Lives: The Anthropologist as Author* (Stanford, Calif.: Stanford University Press, 1988), p. 71.

[55] Henrietta Moore, *Feminism and Anthropology* (Minneapolis: University of Minnesota Press, 1988); Carol C. Mukhopadhyay and Patricia J. Higgins, "Anthropological Studies of Women's Status Revisited: 1977–1987," *Annual Review of Anthropology*, 17 (1988), 461–95.

Social anthropology in Europe was influenced by the American debates, but it remained oriented toward the social sciences. Indeed, some anthropologists, notably Pierre Bourdieu, Mary Douglas, Louis Dumont, Ernest Gellner, and Claude Lévi-Strauss, became leading figures in European social thought. The mutual influences between anthropology and social and cultural history were also significant, producing widely diffused research models. A small but creative group of scholars began to apply models derived from cognitive psychology.

The tradition of nationalist ethnology survived in central Europe and in Spain, but elsewhere *Volkskunde*, stigmatized by its association with German nationalism, was largely absorbed into social anthropology. The old distinction between social anthropologists, who worked abroad, and the ethnologists who worked at home disappeared, and social anthropologists began to work increasingly within Europe, and particularly in Eastern Europe. Instead of a celebration of national cultures, however, a critical approach to ethnicity and nationalism became very common. The self-conscious development of a modern European social anthropology was stimulated by the establishment of new centers in a number of southern European universities, and by a great surge in the popularity of the subject among students in northern Europe. The anthropologists of the former Soviet bloc were largely drawn into the orbit of European social anthropology. In 1989, a European Association of Social Anthropology was founded; it had over a thousand members at the end of the century, and a journal, *Social Anthropology*.

At the same time, an indigenous anthropology became strongly established in a number of formerly colonial or "underdeveloped" countries, notably Mexico, Brazil, Peru, India, China, Indonesia, and South Africa.[56] These anthropologists generally adopted Marxist theories during the 1960s and 1970s, but, while responsive to theoretical developments in the metropolitan countries, they became increasingly eclectic, perhaps in part because they were also typically focused on applied problems and engaged in debates with specialists from other fields. Primarily concerned with the problems facing their own societies, all of these schools were drawn mainly to sociological modes of analysis. The situation was very different in Japan, where the development of anthropology had been stimulated by colonial expansion during the first half of the twentieth century. A long-standing engagement with the study of Japan itself was combined with a considerable investment in ethnographic research overseas.[57]

The continued divergence between European social anthropology and American cultural anthropology, and the growing significance of other

[56] W. David Hammond-Tooke, *Imperfect Interpreters: South Africa's Anthropologists, 1920–1990* (Johannesburg: Witwatersrand University Press, 1997); Ulf Hannerz and Tomas Gerholm, eds., *The Shaping of National Anthropologies, Ethnos*, 47 (1982), special issue.
[57] Nobuhiro Nagashima, ed., *An Anthropological Profile of Japan, Current Anthropology*, 28:4 (1987), supplement.

regional centers, suggest that the multiple origins of the world's anthro-pologies continued to shape their histories. However, the various national and regional tendencies can still be characterized in terms of their primary orientations: toward social theory, cultural theory, or the natural sciences. The development of schools of social and cultural anthropology was largely determined by broader currents of thought in these three domains. From time to time, however, anthropologists themselves helped to shape the major discourses of the other human sciences, chiefly through their criticisms and reworkings of theories tied too exclusively to Western experience.

20

GEOGRAPHY

Marie-Claire Robic

During the late nineteenth century, geography was institutionalized as a discipline with ties to both nature and culture, but it was divided into several distinct national schools and competing currents of thought.[1] The chronology of ruptures over the past century also differed by country: The notion of a "modern" geography took hold in the early 1900s, while the diffusion of what was called the "new geography" occurred during the 1950s and 1960s in the United States, and during the 1970s in continental Europe. At the dawn of the twenty-first century, the expansion of geography had not altered the segmentation of the discipline.[2] Nonetheless, several general tendencies – geographic, epistemological, and institutional – transformed the discipline during the second half of the twentieth century.[3] Geographically, the center of gravity of the discipline shifted after the Second World War from the countries of "Old Europe," where it had first flourished – in Germany, France, and Great Britain – toward the United States and the Anglophone world. Beginning during the same postwar period, geography was incorporated into the human sciences rather than the earth sciences, to which it had earlier been attached, inaugurating a greater variety of practices and interactions with the social sciences, especially economics, although after 1980 geography also began to explore its links with the humanities. Finally, the development of new markets after 1950 diversified this formerly professorial

---

[1] William Pattison, "The Four Traditions of Geography," *Journal of Geography*, 63 (1964), 211–16; see also David N. Livingstone, *The Geographical Tradition: Episodes in the History of a Contested Enterprise* (Oxford: Basil Blackwell, 1992). On national schools, see Paul Claval, *Histoire de la géographie* (Paris: PUF, 1995); see also the history by country in Preston E. James and Geoffrey J. Martin, *All Possible Worlds: A History of Geographical Ideas*, 2nd ed. (New York: Wiley, 1981).
[2] John Agnew, David N. Livingstone, and Alisdair Rogers, eds., *Human Geography: An Essential Anthology* (Oxford: Basil Blackwell, 1996); Philippe Pinchemel, Marie-Claire Robic, and Jean-Louis Tissier, eds., *Deux siècles de géographie française. Choix de textes* (Paris: Editions du Comité des Travaux Historiques et Scientifiques, 1984).
[3] Ron J. Johnston, *Geography and Geographers: Anglo-American Human Geography since 1945*, 5th ed. (London: Edward Arnold, 1997).

discipline, so that training in geography became oriented toward spatial planning, the environment, geopolitics, and social expertise.

In this chapter we will first emphasize the conditions under which the discipline emerged and its initial form as a science of synthesis, joining nature and society together. We then focus upon the upheavals of the second half of the twentieth century and their consequences for the discipline as the century closed.

THE INSTITUTIONALIZATION OF GEOGRAPHY AND NATIONAL EDUCATION

Elementary and secondary school teaching played a fundamental role in the development of academic geography as it began to take shape in the 1870s in European and American universities.[4] The institutionalization of geography coincided with the implementation of key education policies and especially of secondary school reforms. Following the Prussian example, chairs in geography began to appear in the German Empire in 1874; beginning in 1877, university positions multiplied in France to train the teachers of history and geography required by the educational programs of 1871–2; a School of Geography was created at Oxford in 1899 and at Cambridge in 1903, following a lively campaign by geographical societies for the teaching of geography. In the United States, mobilization around the reform of secondary teaching also catalyzed an interest in geography at the university level: The foundation of the first department of geography, at the University of Chicago (1903), and of the Association of American Geographers (1904) closely followed the recommendations of the Committee of Ten, an influential group of educational reformers who favored the teaching of geography in schools.

The development of academic geography was thus heavily dependent upon an educational demand that concealed multiple and conflicting objectives: valorization of the homeland, knowledge of the world, and scientific initiation. The geography defended at the Royal Geographical Society in 1887 by Halford H. Mackinder (1861–1947) had to "satisfy at once the practical requirements of the statesman and the merchant, the theoretical requirements of the historian and the scientist, and the intellectual requirements of the teacher."[5] In the form of "local geography," or, under its German designation, the *Heimatkunde*, geography lent itself to the concrete educational methods inspired by Rousseau and the Swiss pedagogue Pestalozzi. It also furnished a basic general knowledge of the contemporary world and,

[4] Horacio Capel, "Institutionalization of Geography and Strategies of Change," in *Geography, Ideology and Social Concern*, ed. David R. Stoddart (Oxford: Basil Blackwell, 1981), pp. 37–69; see also Agnew, Livingstone, and Rogers, eds., *Human Geography*, pp. 66–94.

[5] Halford H. Mackinder, "On the Scope and Methods of Geography," *Proceedings of the Royal Geographical Society*, 9 (1887), 141–60.

through economic geography, an understanding of its natural and human resources. Serving as a pioneer of the natural and social sciences, geography would contribute to the modernization of society.

Along with history, geography also fulfilled an important civic function by exalting the value of the territory of the nation and the greatness of its empire. For nearly all reformers, its primary function was to serve as a means of knowing and loving one's country.[6] Great national traumas provided occasions for defending the cause of geography. In some post-1870 analyses, the French concluded that they had lost the Franco-Prussian War because of their ignorance of geography, and the Spanish responded with the same bitter assessment after losing their last American colonies in 1898. Hugh R. Mill's presidential speech in 1901 before the geography section of the British Association for the Advancement of Science characteristically mixed academic and patriotic objectives, expressing pride at the conquest of the poles, but also stressing the need for studying the mother country: It is "absolutely essential for our well-being, and even for the continuance of the nation as a Power amongst the states of the world."[7]

Regular international geography congresses after 1871, the multiplication and expansion of geographical societies from 1860 to 1890, and university policies promoting geography brought these multiple interests together in the new discipline. Under these conditions, partisans of a pure, universal science often opposed supporters of useful geography. In France at the turn of the century, Marcel Dubois, professor of colonial geography at the Sorbonne, attacked the "ivory tower" in which academics, such as his colleague Lucien Gallois, had enclosed themselves. Indeed, many German geographers before the First World War kept their distance from the national interest. However, although they saw themselves as objective and as working toward the production of a universal knowledge, many geographers of the period contributed to the elaboration of nationalist works. Friedrich Ratzel (1844–1904), an ardent nationalist and pan-German expansionist, was the theoretician of the first treatise on human geography, *Anthropogeographie* (1882–91). Paul Vidal de la Blache (1845–1918) built his reputation as much upon the *Tableau de la géographie de la France* (1903), which celebrated his country's exceptional "geographical personality," as upon the elaboration of a general "human geography" or upon his launching of the *Géographie universelle* volumes.

The development of academic geography was thus part of the production of a "legitimate" geography of the world, legitimate in the sense that it was incorporated into teachings controlled more or less by the state and derived from the political form of the nation-state. Nevertheless, the Western powers'

---

[6] David Hooson, ed., *Geography and National Identity* (Oxford: Basil Blackwell, 1994); Anne Godlewska and Neil Smith, eds., *Geography and Empire* (Oxford: Basil Blackwell, 1994); Benedict Anderson, *Imagined Communities* (London: Verso, 1983).

[7] Cited in Livingstone, *The Geographical Tradition*, p. 216.

uneven proclivity toward imperialism, variations in the relative power of the state and private organizations, and varying degrees of political consensus from country to country contributed to considerable differences in how geography was institutionalized.

THE GLOBE, THE COLONIAL DIVIDE, AND THE "FINITE" WORLD

The territorial stake in geographic investigation existed on two levels: integration within national frontiers and participation in the colonial division of the planet. A geopolitical status quo was achieved during the 1880s and 1890s, when, with the Berlin accords, the parceling out of Africa among colonial powers appeared to complete the enclosure of world space. The legitimacy of the imperial network was largely accepted, and little harsh criticism of colonization arose, except among radical geographers such as the Russian and French anarchists Piotr Kropotkin (1842–1921) and Élisée Reclus (1830–1905), who yearned for the day when "the center will be everywhere." Colonization did foster debate among specialists in medical geography and anthropology and in economic circles, where the problems of Europeans' acclimatization to tropical environments and the capacity for work of indigenous peoples were considered crucial. At the international congresses, however, the colonial question was quickly excluded from authorized debate.[8] In this arena, until the First World War, the subject of discussion par excellence was the unification of the system of spatio-temporal coordinates of the globe. Geographers agreed on one objective: the standardization of measures and the cartographic description of the world, including a project for a map of the world accurate to one part in a million, although they diverged in regard to such practical choices as the meridian of origin.

This production of standardized knowledge of the planet was the work of specialists: meteorologists, hydrologists, geologists, cartographers. The geographers who entered the field at the end of the nineteenth century took up the project of building a unified science, founded on reason. They drew on their diverse experiences and training in exploration, geology, journalism, and history. A number of them were guided by the eighteenth-century representation of the Earth as "the dwelling of man," a habitat requiring practical development.[9] By the end of the nineteenth and the beginning of the twentieth century, however, the possibility of scientific innovation in

---

[8] Marie-Claire Robic, Anne-Marie Briend, and Mechtild Rössler, eds., *Géographes face au monde. L'Union géographique internationale et les Congrès internationaux de géographie* (Paris: L'Harmattan, 1996).

[9] See Clarence J. Glacken's great work, *Traces on the Rhodian Shore: Nature and Culture in Western Thought from Ancient Times to the End of the Eighteenth Century* (Berkeley: University of California Press, 1967).

this domain depended upon the completion of the process of globalization. Globalization meant the creation of a unified planet under the control of the West: "The world is quite ready to unify itself: all the way to the scattered islets in the immense ocean, all lands have entered the zone of attraction of general culture, predominantly of a European type," affirmed Reclus.[10] According to Mackinder, the "world" was henceforth synonymous with the "globe," so that Great Britain, situated at "the end of the world" during the pre-Columbian era, was now in a central position, "in the midst of the (globe-wide) world."[11]

Each in his own way, turn-of-the-century geographers dramatized the sudden emergence of a "finite" geographical space, a world map without any blank spaces. The awareness of terrestrial finitude, later sometimes referred to as "closed space," masked a variety of issues: the Western empires' imperialistic division of the world, the completion of exploration, the depletion of colonizable lands, and the creation of a global market. The "limits of our cage" had been reached, according to Jean Brunhes, and there remained only the task of the scholar – to conquer the Earth through man's intelligence and "modern geography."[12] Ratzel in Germany, Mackinder in Great Britain, and Frederick Jackson Turner (1861–1932) in the United States arrived at similar conclusions, but formulated the program differently. Some, in the manner of Turner, proposed a collective search for a national "new frontier." Others worked within the realm of political geography: In *Politische Geographie* (1897), Ratzel considered the state as an organism whose task was to ensure *"Lebensraum,"* vital space. Mackinder (1904) applied his views to blocks of states: How can the European powers control the continental "Heartland" framed by Eurasia? Formulated during the pre-war era, these doctrines would be incorporated into later geopolitical reflections that were openly designed to nourish strategies of domination, particularly the *Geopolitik* developed in Germany between the wars.[13]

The unification of the Earth was seen as presenting either an intellectual or a practical challenge for the future of humanity. Above all, "finitude" had to be transcended by – as geographers termed it – an "intensive" or "vertical" exploitation and management of the planet.

A SYNTHESIS BETWEEN EARTH SCIENCES AND HUMAN SCIENCES

Geography was conceived as a "bridge," a "crossroads," a "hinge," or a "synthesis" between nature and culture: Geography "is neither a natural

[10] Élisée Reclus, *L'Homme et la Terre* (1905–7), vol. 2 (Paris: La Découverte, 1982), p. 139.
[11] Halford J. Mackinder, *Britain and the British Seas* (London: Heinemann, 1902), p. 13.
[12] Jean Brunhes, "Les limites de notre cage," *Le Correspondant*, 309:5 (December 10, 1909), 833–62.
[13] Claude Raffestin, *Géopolitique et histoire* (Lausanne: Payot, 1995).

science, nor a science of the mind but both at once," according to Alfred Hettner (1859–1941).[14] Drawing on its tradition as an encyclopedic science of the Earth and its inhabitants, geography competed with existing disciplines such as geology and history, and with newer disciplines such as ecology and sociology, by playing a balancing game between the natural and human sciences. Different academic locations sometimes produced inclinations in one direction or the other, coloring national differences. Nearly all first-generation French geographers were trained initially in history; American geography was initially dominated by geology, to which it long remained attached in university departments; the first British geographers were trained mainly as naturalists; and German geographers came from diverse backgrounds.

This nature/culture duality pervaded the discipline. Some geographers focused the science on the diversity of the face of the Earth, making it a descriptive science of "geographic regions" or a "chorological" science, following the terminology of Hettner, one of the rare geographers to reflect upon the epistemology of the field.[15] Others, such as the American geomorphologist William M. Davis (1850–1934), who focused geography on "the relation between an inorganic control and a responding organism,"[16] and Harlan H. Barrows, who advocated the study of human ecology,[17] defended a causal conception.

Drawing from several of the natural science disciplines that acted both as allies and as competitors, geographers adapted naturalistic models as a strategy for achieving the status of an autonomous science. The soil was the basis for descriptions of places and the point of departure for geographical explanations: the essential origin of the diversity of landscapes and of causal links tying together the physical and human constituents of a region. In a number of regional monographs, the stereotyped succession from soil to climate, vegetation, mineral resources, population, and economics reflected the preeminence of this geological factor. In global studies, however, the climatic factor primarily determined regional distribution, and it was often associated with a geography of races that upheld a cultural and moral hierarchy of peoples.[18]

Botany played a role similar to that of geology: It guaranteed scientific status and provided a source of analogies. Whether in Europe, with the work of Eugenius Warming and Andreas Franz Wilhelm Schimper, or in the United States, with the work of Frederic Clements, the role of plant ecology during the 1890s and 1900s was decisive. Paul Vidal de la Blache justified human

---

[14] "Ist weder Natur – noch Geisteswissenschaft, sondern beides zugleich": cited in Bertrand Auerbach, "L'Évolution des conceptions et de la méthode en géographie," *Journal des Savants*, 6 (1908), 311.

[15] Alfred Hettner, "Das Wesen und die Methoden der Geographie," *Geographische Zeitschrift*, 11 (1905), 545–64, 615–29, 671–86.

[16] William M. Davis, "An Inductive Study of the Content of Geography," *The Journal of Geography*, 5 (1906), 154.

[17] Harlan H. Barrows, "Geography as Human Ecology," *Annals of the Association of American Geographers*, 13 (1923), 1–14.

[18] Livingstone, *The Geographical Tradition*, chap. 7.

geography by likening it to the burgeoning discipline of ecology. Several authors conceived of the unity of the Earth under the sign of "biogeography." More generally, geographers were inspired by the theory of evolution to use organic analogies and the model of adaptation of living organisms to their environments. In accord with the pervasive neo-Lamarckian ideology, this geography emphasized the plasticity of living forms and their capacity to adjust themselves to the constraints of the environment, thus prolonging, in a secularized fashion, the ancient providentialist conception of the relationship between people and their habitats.[19]

The alignment of geographers with specialists in the natural sciences produced mixed results. It encouraged modes of analysis and techniques of observation that broke away from the classical methods of the disciplines upon which geography had previously been dependent, such as history. In contrast to history's erudition, founded on state archives, geographers' new fieldwork permitted outlooks on social issues and on historicity that were less confined to the strictly political realm. German geographers developed the study of the migration of peoples and ancient civilizations from sites that were discovered in both rural and urban landscapes. French geographers looked for such traces in the vernacular toponymy, and Olinto Marinelli noted the multiple "topographical traces," such as houses, roads, bridges, and fields, that materialize human activity.[20] Geographers' attention to the variety of humanized landscapes, to regional organizations, and to the creative capacity of societies to transform natural settings served as a model for the human sciences that remained close to the sciences of the state. Lucien Febvre supported this geography *"de plein vent,"* on-site, against a traditional history that he deemed *"événementielle,"* rigidly factual, and too strictly political.[21] In conjunction with Durkheimian sociology, this geographic opening to collective phenomena inspired the historians of the *Annales* school during the 1930s. On the other hand, the adoption of biological metaphors and of the ecological model tended to desocialize human achievements by interpreting them as one would a natural species.

Geography as a hinge between nature and society was a failure, for it could not prevent the formation of two specialities that developed along parallel tracks: Physical (or natural) geography mainly studied geomorphological processes in a strictly naturalistic way, while human (or cultural) geography focused on rural settlements, lifestyles, and economic and demographic matters. Nonetheless, research on three issues addressed the synthesis

[19] John A. Campbell and David N. Livingstone, "Neo-Lamarckism and the Development of Geography in the United States and Great Britain," *Transactions of the Institute of British Geographers*, 8 (1983), 267–94.

[20] Olinto Marinelli, "Alcune questioni relative al moderno indirizzo della geografia," *Rivista Geografica Italiana*, 9 (1902), 231.

[21] Lucien Febvre, *La Terre et l'évolution humaine. Introduction géographique à l'histoire* (Paris: La Renaissance du livre, 1922).

of nature and society: human/environmental relations, regionalism, and landscape.

The first issue considered geography as human ecology. The "natural constraint/response" axis of American geographers, or the "environment/ lifestyle" (*milieu/genre de vie*) pairing of the French school, dealt with the tension between two opposing currents and privileged either human freedom or environmental determinism. For Ellen Churchill Semple (1863–1962) at one end of the scale, determinism was absolute: "Man is a product of the earth's surface."[22] At the opposite end was the extreme form of French "possibilism," illustrated by Vidal de la Blache, who saw the physical framework of natural constraints as an ensemble of "possibilities" or "virtualities" among which people make their choices: "All that touches man is struck with contingency," he stated. A few geographers, such as Ellsworth Huntington, developed a "reversed" determinism that magnified the promethean capacity of groups living in extreme conditions, a constraint/challenge scheme.

Both regional and landscape studies aimed at revealing distinctive spatial individualities that combined natural and human phenomena. The Belgian geographer Paul Michotte argued that the chorological position could form the only basis for "geographicity," since only the synthesis of spaces could distinguish the work of the geographer from that of specialists. The French school of geography conformed to this idea, gaining its reputation through a long line of regional monographs, beginning with Albert Demangeon's *Picardie* (1905). This was also the case among British geographers, notably Andrew J. Herbertson and Herbert J. Fleure, and their tradition of detailed land-use studies. In his *The Nature of Geography* (1939), the same doctrinal position was affirmed in American geography by Richard Hartshorne (1899– 1992), drawing upon an historical tradition that harked back to Kant.

Among landscape geographers, who were prominently interested in the visible, concrete physiognomy of places, Germans had a leading role. With the valorization of the "cultural landscape" by Otto Schlüter (1872–1959) during the interwar years and the work of Carl Sauer (1889–1975) at Berkeley, those interested in the ethno-cultural factors governing areal differentiation emphasized the material imprint of human occupancy in the landscapes of central Europe and North America.

GEOGRAPHY: A SOCIAL SCIENCE OF SPATIAL ORGANIZATION

During the 1930s, geographers were tempted toward practices for which most of them were neither technically nor professionally prepared. With the

---

[22] Ellen C. Semple, "The Operation of Geographic Factors in History," *Bulletin of the American Geographical Society*, 41 (1909), 422.

Great Depression, the rise of national planning policies led geographers to propose regional plans and forced them to undertake tasks in prospective surveys, economic and demographic statistics, and evaluation procedures to which their descriptive expertise was poorly adapted. They produced a cartography that was sensitive to disparities of development within a country, and they began to urge the need for new geographic skills focused on the conception of "rational" and "corrective" land use.[23] These trends in national policies encouraged the formulation of spatial models of economic activity, such as the "central place theory"[24] of Walter Christaller (1893–1969), published in 1933. This theory was used by the Nazi regime to plan the Eastern Territories; it was also used to program the urban scheme of Dutch polders and, after the war, to study national urban systems. Owing to its spatial-oriented view, its theoretical aim, and its focus on urban issues, it became during the 1960s the central point of reference for the "new geography."

Wartime activities accelerated the move toward applied science. Unlike the First World War, when only a few individual geographers played an important role at the Peace Conference, the Second World War offered certain communities of geographers, notably American ones, unprecedented opportunities for serving in geographic information operations. Investment increased in several specialized areas, particularly in cartography. By the end of the century, the geographic information systems unit had become the principal employer of geographers and cartographers in the U.S. State Department.

A shift toward a new vision of geography occurred internationally during the immediate postwar period. Geography was baptized as the study of "spatial interaction" by Edward L. Ullman in the United States, and of "*organisation de l'espace*" or "*espace géographique*" in France.[25] As Ullman specified: "By spatial interaction I mean actual, meaningful, human relations between areas on the earth's surface, such as the reciprocal relations and flows of all kinds among industries, raw material, markets, culture, and transportation – not static location as indicated by latitude, longitude, type of climate, etc, nor assumed relations based on inadequate data and a priori assumptions. I do, however, include consideration, testing and refining of various spatial theories and concepts."[26]

The new paradigm pitted against the older science of regional individuality, the scientific perspective of a nomothetic discipline constituted by spatial interrelations. William Bunge in *Theoretical Geography* (1962) and David Harvey in *Explanation in Geography* (1969) argued that geography should uncover the laws of societal spatial organization. A similar perspective

[23] Robic, Briend, and Rössler, eds., *Géographes face au monde*, chaps. 7–8.
[24] Walter Christaller, *Die zentralen Orte in Süddeutschland* (Jena: Gustav Fischer Verlag, 1933).
[25] Jean Gottmann, *La politique des états et leur géographie* (Paris: Colin, 1952).
[26] Edward L. Ullman, "Human Geography and Area Research," *Annals of the Association of American Geographers*, 43 (1953), 56.

developed in Great Britain during the early 1960s – following the methods outlined in *Locational Analysis in Human Geography* (1965), edited by Peter Haggett, and *Models in Geography* (1967), edited by Haggett and Richard J. Chorley – and in the Scandinavian countries, notably around Torsten Hägerstrand; it then spread around the world. The move to a scientific model coincided with the Western countries' renewed experience of globalization, now from the restricted territorial settings enforced by decolonization. With regional science, geography met the demand for a planning orientation permitting prediction and regulation of spatial patterns and processes. "As the population density rises and the land-use intensity increases, the need for efficient management of space will become even more urgent," American geographers reported in 1965.[27] The urban field became one of the new frontiers of geographical research.[28] Increasingly incapable of justifying its synthetic scope while at the same time its specialized branches expanded, geography found in spatial interaction an integrating concept that could also form a basis for practical intervention on the ground, whether administrative, strategic, or economic. Geography realigned itself with the human sciences and defined itself progressively as a social science.

This formalized and quantified spatial analysis, inspired by positivism and physics, was soon contested by two tendencies: one radical, often of Marxist inspiration,[29] and the other humanistic and attentive to the subjective experience of locality.[30] At the end of the century, all three trends still coexisted.

NEW CHALLENGES: THE GLOBAL SYSTEM, THE LOCALITY, THE ENVIRONMENT

Beginning in the 1980s, geographers posed the problem of the spatiality of societies and people in a more complex way by explicitly participating in the debates in the social sciences and humanities. They adopted the model of complexity, debated the place of human agency in the organization of geographical space, developed a geography of the individual inspired by Anglo-American phenomenology, and produced studies centered on "place" or locality instead of "space," in keeping with the widespread postmodern tendency that distrusted "grand narratives" and valorized fragmentation and

[27] Cited in Johnston, *Geography and Geographers*, 1st ed., p. 70.
[28] Brian L. Berry, "Cities as Systems within Systems of Cities," *Papers of the Regional Science Association*, 13 (1964), 147–63.
[29] David Harvey, *Social Justice and the City* (London: Arnold, 1973); Henri Lefebvre, *La production de l'espace* (Paris: Anthropos, 1974).
[30] David Ley and Marwyn S. Samuels, eds., *Humanistic Geography: Prospects and Problems* (London: Croom Helm, 1978); Yi-Fu Tuan, *Topophilia: A Study of Environmental Perception, Attitudes and Values* (Englewood Cliffs, N.J.: Prentice Hall, 1974).

the effects of context.[31] The new trends all bore witness to the sensitivity of geography to international intellectual currents.

At the same time, geographers and their colleagues in related disciplines faced a new spatial reality: The new ease of long-distance communication and the omnipresence of information were producing unprecedented flexibility in human activities. Had geographic questioning become obsolete, or, on the contrary, did the study of locality have to be valorized in a way that was commensurate with the range of choices opened by the new freedom of localization? Geographers chose the second alternative. In their study of economic dynamics, they focused on the structure of spatial networks, analyzing the connections between places, now more important than spatial continuity and distance. They also emphasized the combined effect that is produced by the concentration of various socioeconomic activities on the same site.

A similar shift occurred in the study of identity, at the level both of the individual and of cultural and political groups. Citizens of the world and of diasporas, and such new "nomads" as exiles, displaced persons, and migrant workers experienced an absence of spatiality, or multispatiality, that prevented them from taking root or anchoring a collective identity in an appropriated territory. The standardized "nonplaces" of modernity, as the anthropologist Marc Augé has called them, came to dominate the world, next to the symbolic "high places" constructed by multiple cultures. These issues were studied by a variety of social scientists, with geographers tending to define their cognitive project in terms of the spatial dimension of society.

And nature? Faith in an all-powerful technique, focus on spatial analysis or analysis of place, and the long-maintained division into physical geography and human geography had detracted from the classical question of the role of the natural environment. Nevertheless, throughout the 1960s, a number of geographers had revived an earlier ecological awareness.[32] They carried out research not on the deterioration of the environment at the global level, but on the natural hazards linked to exceptional phenomena, such as droughts, floods, and earthquakes, and became interested in the perception of risk and in the behavior of people facing natural catastrophes.

From the early 1970s, however, the issue of the environment carried with it anxiety about the future of the inhabited Earth: The risk had become planetary, and the harm to the environment was construed as irreversible. Geographers worked on these issues on two interdisciplinary fronts, in alliance with the powerful fields of the geosciences and naturalist ecology, or allied with anthropology and sociology, where geographers participated

---

[31] See Edward Soja, *Postmodern Geographies: The Reassertion of Space in Critical Social Theory* (London: Verso, 1989).

[32] George P. Marsh, *Man and Nature or Physical Geography as Modified by Human Action* (New York: Charles Scribner, 1864); Gilbert White, ed., *Natural Hazards: Local, National, Global* (Oxford: Oxford University Press, 1974).

in research on the "social construction " and "symbolic appropriation" of nature and the redefinition of the modern landscape as a mix of nature and culture.

At the heart of geography's controversies lay a profoundly ecological question, that of the conditions of habitability of the Earth. Geographical research existed within a continuous tension between unity and diversity, universality and locality, cosmopolitanism and patriotism. Relative to the other scientific disciplines, modern geography was marked by its sensitivity to the evolution of the territory of the nation-state. With the weakening of the nation-state, economic globalization, and the extraordinary shrinking of distances, geographers looked for new ways to combine their studies of the effects of modern communication, the organizing capacities of the economic, cultural, and political groups that fragment the world, and the individual representations of the here and the there.

21

HISTORY AND THE SOCIAL SCIENCES

Jacques Revel

History plus the social sciences: This has been a common formula for more than a century. It has produced extensive discussions and an enormous literature, often quite repetitive, seeking to explain what the relationship between history and the social sciences should be, could be, and cannot be. Still, the terms of the debate have not stabilized. At once epistemological and methodological, the debate also involves power struggles among disciplines and the social representations that they nourish and reflect. For this reason, experiences differ from one country to another. This essay will concentrate on three principal experiences, those in Germany, France, and the United States.

THE PROBLEM POSED

Despite some distant precedents, the problem was not attacked directly until the period when the social sciences were recognized as autonomous disciplines and institutionalized in academia. This was the period from the 1870s to the 1880s – the American Gilded Age – for the sciences of politics and economics and to a lesser degree for sociology, and from 1880 to 1900 in France's Third Republic, where university reforms opened the way for the scientific disciplines of geography, sociology, psychology, and economics. In both America and France, these new sciences embodied the demands for objectivity, method, and positive knowledge, and they expressed the dominant ideologies of progress. The German disciplines provided models for many other countries, but the German social sciences developed in the Humboldtian university within a cultural system built around philosophy, and their ascent appeared threatened at the end of the nineteenth century by the unity of the ideal of *Bildung*, or cultivation.

History, as an ancient discipline, enjoyed particular legitimacy at the century's end, for it had served the construction of national identities. Historians

391

had acquired the intellectual and political authority incarnated in ambitious national history projects, including those undertaken by such major figures as Ranke, Michelet, Bancroft, and Taine. A major reorientation took place between 1860 and 1880. Communities of historians moved toward academic professionalization based on the German model of historical science and cemented their identities around a scientific model: the critique of documentary evidence. Establishing "facts" became the first condition and was often considered the essence of the professional historian's work, and the method and objectivity of knowledge remained the dominant terms of reference. The philosophy of history, which had dominated nineteenth-century thought, became suspect, challenged by these new demands on the discipline.

At this point, history and the social sciences first crossed paths, and the problem of the relationship between the two disciplines began to be posed. The problem was addressed on at least two levels. The first was the status of knowledge and the possibility of objective knowledge about society: Can history and the social sciences be conceived on the model of the natural sciences and their rigorous procedures? This question had been posed since the Enlightenment, but was elaborated during this period.[1] Wilhelm Dilthey, in *Einleitung in die Geisteswissenschaften* (1883), sharpened the issue by opposing the nomothetic sciences – those that study phenomena in their generality and are capable of proposing explicative laws – to idiographic knowledge of the particular, which is associated with the *Geisteswissenschaften*, sciences of mind or spirit.

The second level on which the problem was posed was the confrontation between history and the social sciences. The social sciences conceived their project on the model of the natural sciences, particularly those in France and the United States. Émile Durkheim declared his desire to treat social facts as "things," using the model of physics and the natural sciences. His *Règles de la méthode sociologique* (1894) codified epistemological and cognitive procedures that make objective observation of social facts possible and constructed social laws that were not to be distinguished from the laws of nature. Positivism not only dominated Third Republic France, but also was reflected in the set of propositions about the social sciences in most other countries – except, notably, in the German world. How then can we think about the relationships between history and the social sciences?

THREE ANSWERS

Three types of answers were formulated. The first noted the gulf that separates the related yet irreconcilable ambitions of history and the social sciences. This

---

[1] See Johnson Kent Wright's chapter in Part I of this volume, "History and Historicism."

was the answer of historians the world over, and it was expressed most often by rejection. Behind the pretensions of the new social disciplines, many historians detected the barely masked return of the old philosophy of history. In most Western countries, the principal historical journals devoted many pages to debates on this question around 1900. Most reverted to a position of learned erudition, which explains the publication in France of the famous manual by Charles-Victor Langlois and Charles Seignobos, *Introduction aux études historiques* (1898). Less frequently, the rejection was expressed in a more sophisticated way. The Italian historian and philosopher Benedetto Croce (1866–1952) notably strengthened Dilthey's analyses by challenging the model of scientific knowledge from which both history and the social sciences claimed to take inspiration. According to Croce, the social sciences are but ideological constructions. History, like art, is a kind of knowledge attained by imagination and intuition, which alone permit the apprehension of individual facts.[2]

The second answer, the exact opposite of the first, proposed to conform historians' procedures to the ambitions of the new social sciences. This answer was formulated around Durkheim's project for a social science that would obey the methodological rules newly codified by the sociologist. In the eyes of the Durkheimians, nothing justified the existing divisions between disciplines save technical competencies and the weight of professional traditions. They engaged in a series of confrontations with geographers, psychologists, economists – but, first of all, with historians. In 1903, François Simiand (1873–1935), one of Durkheim's youngest and most brilliant disciples, explained to the French historical establishment that if they wanted to adopt a truly scientific method, they must renounce their habitual "idols" – individual character, singular events, and particular facts – because there is no scientific knowledge of the unique. They must instead construct facts of observation in such a way that they can be integrated into series that permit the determination of regularities and formulation of laws.

In a similar vein, Simiand denounced the weakness of models of causality that were habitually used by historians acting as empiricists and rhetoricians. History must redefine its epistemological agenda and impose the same prerequisites as the social sciences. "There is not, on one side, a history of social phenomena and, on the other, a science of these same phenomena. There is a scientific discipline which, to attain the phenomena which are the objects of its study, uses a certain method, the historical method."[3] The task of history was thus redefined and neatly limited, opening the time dimension to sociological experimentation. Simiand's maximalist proposal anticipated

[2] Benedetto Croce, "La storia ridotta sotto il concetto generale dell'arte" (1893), in his *Primi Saggi* (Bari: Laterza, 1918), pp. 1–41.
[3] Francois Simiand, "Méthode historique et science sociale," *Revue de Synthèse historique*, 6 (1903), 1–22, 129–57.

a unification of all of the social sciences around sociology. But, as a new discipline, weakly instituted in the French academy, Durkheim's sociology did not have the means to fulfill its ambitions. As we will see, the *Annales* would recapture the 1903 program a quarter-century later, and would do so around history.

A third response lay between these two. In the final years of the nineteenth century and the first years of the twentieth century, a number of historians wanted their discipline to incorporate the interests and certain procedures of the social sciences. Their efforts, though inscribed in separate national contexts, had some traits in common: They were relatively marginal within the profession, oriented toward positivism and the prevalent evolutionist perspective, and, in large part, empirical.

Karl Lamprecht (1856–1915), an author with an important and recognized oeuvre, began publishing the *Deutsche Geschichte* during the 1890s and provoked a formidable reaction among the established "Young Rankeans." Against their political history of the state, he proposed to study the economic, social, and cultural history of the German nation by means of a scientific method: "Historical science . . . must replace a descriptive method by a genetic one which tries to formulate general laws."[4] The violence of the *Lamprechtsstreit*, the debate over Lamprecht's proposal, arose from its attack on the quasi-sanctified concept of national history. In other respects, Lamprecht's theoretical ambition remained imprecise in the several texts he published around 1900 in defense of his method (e.g., "Was ist Kulturgeschichte?," 1897). His project was to describe the steps of historical development and to widen historians' interests using the social sciences, in particular sociology and the "psychology of the people," and thus to place history on an evolutionist foundation. Lamprecht's prolific work had no real successor, but the debate it opened was pursued the world over – by the Frenchman Gabriel Monod, the Belgian Henri Pirenne, and the American Earle Dow, among others.

At the same time, in France, Henri Berr (1863–1954), who was by training, not a professional historian but a philosopher, was considering the question that would dominate his life: How, once past the age of grand philosophical systems, is one to produce the conditions for a synthesis of knowledge? Berr's thought was eclectic, and, despite an influential book that was often reprinted, *La synthèse en histoire* (1911), his work was largely that of a tireless intellectual entrepreneur. In the *Revue de synthèse historique*, which he founded in 1900, he published articles by all of the important protagonists of the international debate on history and the social sciences, as well as contemporary epistemological reflections on the nature of scientific knowledge. Berr's key idea was itself a reflection of the approaching epistemological

[4] Georg Iggers, *The German Conception of History: The National Tradition of Historical Thought from Herder to the Present* (Middletown, Conn.: Wesleyan University Press, 1968), p. 197.

crisis: the idea that a synthesis of knowledge cannot be conceived except in an historical perspective. In his view, history must be the site of confrontation between different human activities, and it alone permits knowledge of the present, whether scientific, religious, political, social, or economic. As with Lamprecht, his was an evolutionary perspective that sought to orient the great phases of the "evolution of humanity," the title he gave to the immense encyclopedic collection he launched after the war. Once again, these phases resulted from a series of collective psychological transformations. All of the social sciences, indeed all of the sciences, were convened around history.

Berr's moment was well chosen, for some of the most notable historians, such as Monod, founder of the *Revue historique*, were urging the necessary collaboration between their discipline and the new social sciences.[5] But it was Berr who organized sites and forms of reflection between the disciplines for nearly half a century and prepared the ground for the *Annales*.

The third opening of history to the social sciences, the American New History, shared with the others a global evolutionist perspective rooted in analysis of contemporary situations. It was also colored by specifically national concerns: the questioning of the providentialist version of American exceptionalism and the search for guides to concrete action in both history and the social sciences. The preoccupation with exceptionalism began in the 1890s, in Frederick Jackson Turner's (1861–1932) first writings on the frontier. In these works, he sought to examine the economic and social bases of American exceptionalism from a geographic perspective and also, in a larger sense, to locate the American experience in an account of the evolution of the world.

The interest in practical progress was developed by James H. Robinson (1863–1936) in the manifesto that gave its name to the movement, *The New History* (1912). For Robinson, history functioned as a necessary memory for societies in accelerated transformation, but, enriched by the social sciences, it must also provide models for the "understanding of existing conditions and opinions, and those can only be explained... by following more or less carefully the conditions that produced them." Robinson's "historical-mindedness" resulted in a reversal of time order: "The present has hitherto been the willing victim of the past; the time has now come when we should turn on the past and exploit it in the interest of advance."[6] At a moment when historians tended to turn inward and when the social sciences claimed autonomy, Robinson pleaded for their collaboration in the name of a liberal concept of progress. His prophesying and his call for reconciliation between disciplinary practices was nebulous and eclectic, and his all-encompassing

---

[5] Gabriel Monod, "Bulletin historique," *Revue historique*, 61 (1896), 322–7.
[6] J. H. Robinson, *The New History: Essays Illustrating the Modern Historical Outlook* (New York: Macmillan, 1912), p. 24.

vision founded on social psychology evoked Lamprecht and especially Berr. As a result, the precise form of his proposed articulation between the different approaches to the social remained unclear.[7]

THE RISE OF *ANNALES* HISTORY

These three efforts to join history and the social sciences shared an important genealogy and a limited future. In Germany, the situation was most clear-cut. With rare exceptions – Friedrich Meinecke and Otto Hintze were more moderate – the great majority of German historians actively opposed Lamprecht's "positivist" proposal. Nor did the social sciences welcome the project, with Max Weber proving to be particularly critical.

In the United States, the New History movement, although representing a minority of historians, remained very visible for twenty years. Robinson was an influential teacher at Columbia University and active in the profession, and his message was relayed by Charles and Mary Beard and by propagandists such as Harry E. Barnes in *The History and Prospects of the Social Sciences* (1925) and in journals such as *Social Forces*. As a program for the integration of the social sciences into the practice of history, it remained vague and unrealized. The few efforts made, such as the twelve-volume *History of American Life*, edited by Arthur M. Schlesinger and Dixon R. Fox, were judged "shapeless and sprawling," "an attempt to integrate a vast array of phenomena without a unifying idea."[8] What the New History did succeed in doing was authorizing the limited social-historical initiatives of these Progressive historians: the use of geography by Turner and some of his students to analyze regional development, and the efforts of Beard and his followers, influenced by the economist E. R. A. Seligman's *Economic Interpretation of History* (1902), to ground American political history in the conflicts of economic groups and interests. The Beards' *The Rise of American Civilization* (1928) also gave some attention to women, education, and other cultural matters. But these social-historical efforts remained tributary to political history, and when Beard and Carl Becker questioned the possibility of objectivity, the defining aim of the discipline, they added to their colleagues skepticism of the New History program.

For their part, the social sciences were expanding within American universities. In 1923, the Social Science Research Council, concerned with averting isolation, called for "cross fertilization" across disciplines, an effort that the American Historical Association hesitantly joined. But nothing came of this collaboration. The New Historians were uncertain how to cooperate, and

[7] John Higham, *History: Professional Scholarship in America* (Baltimore: Johns Hopkins University Press, 1983), p. 118.
[8] Peter Novick, *That Noble Dream: The "Objectivity Question" and the American Historical Profession* (Cambridge: Cambridge University Press, 1988), p. 178.

the social scientists, attached as they were to a strictly nonethical conception of their project, doubted that the historians could contribute.

The link with these first experiences was made in France, the country where the debate on the community of the social sciences had been most open at the beginning of the century and where Henri Berr had disseminated ideas, work, and people until the early 1950s. While Berr's activities remained outside the university and lacked academic legitimacy, however, the *Annales* movement was born within the university. At the outset, two historians – one an early modernist, Lucien Febvre (1878–1956), and the other a medievalist, Marc Bloch (1886–1944) – announced a simple project: to establish the community of the social sciences around history. They had been influenced by Durkheim and his school, and by Simiand's polemic. They were also close to Berr and had been influenced by Vidal de La Blache's (1845–1918) geography, which, in their eyes, served as a model of an integrative discipline. The space was clear before them: The discipline of history was singularly more powerful than its partners in the French university, for Durkheimian sociology had been weakened by the death of its founder and the disappearance of a number of its members during the First World War. The offensive was launched in 1929 at the University of Strasbourg, a peripheral but particularly brilliant institution, with the journal *Annales d'histoire économique et sociale*.

Like Durkheim and Simiand before them, Bloch and Febvre began by refuting the compartmentalization of disciplines. They wanted the new journal to become a site of welcome, confrontation, and experimentation for the totality of knowledge on society that was developing in the world. Bearing witness to this aim were geographers, economists, and political scientists on the editorial board, and more still in the soon-to-be international network of the *Annales*. Their personal preference was for social history, but, as Febvre would later write, "a word as vague as 'social' seems to have been created . . . to serve as an emblem for a journal which maintained that it should not surround itself with high walls. . . . There is no economic and social history. There is only history, and history alone, in all of its unity."[9] This supple conception contrasted with the imperious rigidity of Durkheimian epistemology. Bloch and Febvre chose not to rally around an orthodoxy or a school, for their decision to organize the exchange between disciplines around history rested on a voluntarist, empiricist, and eclectic conception of social knowledge, a conception that would become the movement's marker over and above the reformulations it would experience. They proposed a double program of confrontation: between history and diverse social scientific approaches to contemporary social realities, enriching the models of intelligibility of the past, and in the opposite direction, between the experience of the past and the interpretation of the present. Thus the complexity of social time and the

---

[9] Lucien Febvre, *Combats pour l'histoire* (Paris: Armand Colin, 1953), p. 20. The essay is from 1941.

diverse modalities of the historical experience served as the main axes for interdisciplinary practice.

This reformulation was coupled with another significant change. For Durkheim and his disciples, it was the *method* alone that could unify the social sciences. But for historians embarked on the *Annales'* adventure and for their colleagues from other disciplines, it was the *object* supposedly common to all of the sciences that would play this role, "man" in society. (It is no accident that for a long time in France the wording *"sciences de l'homme"* prevailed over *"sciences sociales."*) Theirs was a less ambitious model than that of the Durkheimian sociologists, but it was an immediately operational one. It proved to be extraordinarily productive and was recognized outside of France in successive waves from the 1930s to the 1970s.

A minority, and never holding a dominant position, the *Annales* historians established themselves at the heart of the academic system. Their research and teaching institution became the main site of experimentation for the social sciences in France: the Sixth Section of the Ecole Pratique des Hautes Etudes (1948), which became in 1975 the Ecole des Hautes Etudes en Sciences Sociales.

A movement that extended through the century and considered itself eclectic could not remain unchanged. Because of its plasticity, the *Annales* was able to adapt to various intellectual and institutional circumstances. There was never a single *Annales* "paradigm," but rather a series of paradigms.[10] These shared the aim of treating history as a privileged site of intersection between the diverse human sciences. Fernand Braudel (1902–1985) expressed this aim in his most famous essay, the article on "La longue durée." Published in 1958, the article appeared at a time when France was launching the structuralist offensive, a movement that was theoretically antihistorical and practically concerned with emancipating the social sciences from the historians' rule. In this context, Braudel asked again for an "ecumenical" concept of interdisciplinarity.[11]

Interdisciplinarity was essentially a matter of practice. The *Annales* was not a theoretical movement – no doubt its weakness and its strength. The exchange between disciplines was made most often by borrowing. Certain partners were privileged: first geography, which, in its Vidalian version, demonstrated how the multiplication of points of view could enrich study of a social phenomenon; next economics, during the 1950s and 1960s, around the debates on growth; and finally sociology and social and cultural anthropology beginning in the 1970s. For fifty years, until its reformulations in the 1980s, the *Annales* maintained a global and integrative approach, less concerned

[10] Trajan Stoianovich, *French Historical Method: The* Annales *Paradigm* (Ithaca, N.Y.: Cornell University Press, 1976); for a contrary view, see Jacques Revel, "Histoire et sciences sociales. Les paradigmes des *Annales*," *Annales, Economies, Sociétés, Civilisations,* 6 (1979), 1350–76.
[11] Fernand Braudel, "Histoire et sciences sociales: la longue durée," *Annales ESC,* 4 (1958); reprinted in *Ecrits sur l'histoire* (Paris: Flammarion, 1972), at p. 42.

with establishing hierarchies of instances or identifying rigorous mechanisms of causality than with drawing out the multiplicity of correlations between phenomena. For the *Annales*, history and the social sciences work less through simplification and abstraction than through complexification. With their intersecting viewpoints, they should render their objects more complex and enrich them with meanings generated by the indefinite web of interrelations.[12]

With some limitations, the *Annales* experience was unique, because of its flexible yet powerful program, because it benefited from a power struggle between history and the sciences that in France had been exceptionally favorable to history, and because of the facility with which historians tread on the terrain of other disciplines. Braudelian geo-history, the *histoire des mentalités*, and the historical anthropology of the 1970s were some of the hybrid productions born of these initiatives.

AMERICAN EXPERIENCE COMPARED

Let us compare this experience with the American situation during the same period. Even after the New History was exhausted, the links between history and the social sciences were not entirely destroyed. As John Higham noted, cultural anthropology and the sociology of communities exerted some influence at the end of the 1930s. But on the eve of the Second World War, the partnership was seriously challenged. During the interwar decades the social sciences had become more established than ever, and were redefining their agenda and distancing themselves from the weakening evolutionist perspective that had permitted them to keep contact with the historians' perspective. Their project became "to naturalize the historical world," as Dorothy Ross has forcefully shown.[13] During the 1940s and 1950s, they increasingly privileged an analytical approach that was both functionalist and structural. Their goal became the statistical measure of social behavior in the framework of a larger theory of society conceived as an immense integrating machine. Talcott Parsons's work, in particular *The Social System* (1951), is emblematic of this abstract empiricism, which combined extreme theoretical ambition with demanding conceptual and methodological refinement. At a distance, it can be understood as the scientific ideology of an America become the first world power, sure of itself and of its social and political values, lured by isolationism during the Cold War, and confident that its capacity to integrate differences and conflicts should serve as a model for all contemporary

---

[12] The notion of *Zusammenhang*, the interdependence of social facts, as against "sociological abstraction" is credited to Lucien Febvre, "Une question mal posée: les origines de la Réforme française et les causes de la Réforme," *Revue Historique*, 161 (1929), 1–73.

[13] Dorothy Ross, *The Origins of American Social Science* (Cambridge: Cambridge University Press, 1991), p. 470.

societies. This new formulation of exceptionalism, carried by a conviction to live the "American moment," dominated the 1950s and the 1960s.[14] It abandoned history to adopt an essentially static perspective. It also underlined the theoretical and methodological inadequacies of the historians' work, leaving economists, sociologists, and political scientists to appeal to concepts such as "modernization" and "development" that refer to functions internal to society, rather than to processes analyzed in historical terms.

History as a discipline thus found itself in a subordinate position. In the minds of social scientists history could, at best, proffer factual data. But historians themselves tended to withdraw and defend their own modes of investigation; the profession organized to bolster its autonomy, and some of its eminent members, such as Robert Palmer and J. H. Hexter, called into question the scientific pretensions of the social sciences. All bridges were not burned. Such authoritative voices as those of Richard Hofstadter, H. Stuart Hughes, and David M. Potter reminded historians what could be learned from social scientists in terms of conceptual rigor and scientific procedure.[15] In their research, they tested notions borrowed from sociology and psychology, as in Hofstadter's use of the concept of "status anxiety" in his *Age of Reform* (1955). The Social Science Research Council organized continued confrontation through a series of publications, the most controversial of which was *The Social Sciences and Historical Study* (1954). But was a debate realistically possible between protagonists who stood firm on distant positions that were largely incompatible and, especially, unequal?

Those who remained convinced of the necessity of cross-fertilization often adopted the conceptual presuppositions and operating techniques of economics, sociology, demography, and political science. In the 1960s, Lee Benson was one of the first prophets of this movement for a social scientific history. It influenced political and social history, especially urban history and the history of collective behavior, and, most significantly, the New Economic History fashioned by Robert Fogel, Douglass North, and Stanley Engerman. These economic historians utilized econometrics and the counterfactual approach, techniques that were unfamiliar to historians and that enabled the consideration of ideologically charged issues, such as slavery in the United States.[16] The aim was to apply methodologies tried and tested in other disciplines to historical facts, within an instrumental and positivist perspective arising from the concern with quantification. Economist-historian

---

[14] One dissenting voice was C. Wright Mills in *The Sociological Imagination* (New York: Oxford University Press, 1959).

[15] David M. Potter, *People of Plenty: Economic Abundance and the American Character* (Chicago: University of Chicago Press, 1954), Introduction; Richard Hofstadter, "History and the Social Sciences," in *Varieties of History*, ed. Fritz Stern (New York: Meridian, 1956); H. Stuart Hughes, "The Historian and the Social Scientist," *American Historical Review*, 66 (1960), 20–46.

[16] Robert Fogel and Douglass North, *Railroads and Economic Growth* (Baltimore: Johns Hopkins University Press, 1964); Robert Fogel and Stanley Engerman, *Time on the Cross: The Economics of American Negro Slavery* (Boston: Little Brown, 1974).

David Landes and sociologist-historian Charles Tilly deflated history's ambitions, finding that it was "not a unitary discipline" and contributed only
a temporal and globalizing "perspective" to the common task of seeking
knowledge about humanity. It must learn from the social sciences to define
its questions, and it must construct a procedure that moves from hypotheses
to empirical validation, following experimental criteria. "The social science
approach is problem-oriented. It assumes that there is a uniformity of human
behavior that transcends time and space and can be studied as such; and the
historian as social scientist chooses his problems with an eye to discovering,
verifying or illuminating such uniformities. The aim is to produce general
statements of sufficiently specific contents to permit analogy and prediction."[17] Beginning in 1976, the project of disciplinary integration found a
strong institutional home in the Social Science History Association.

SINCE THE 1960s: MARX AND THE SOCIAL SCIENCES

Although Marx's thought offered one of the stronger conceptualizations of
"history plus social science," his influence was not decisive for a number of
reasons: the suspicion weighing upon Marxist problematics in the West, and
particularly in the United States; the defiance with which official Marxism
in the East treated the "bourgeois" social sciences; and the attraction of
many Marxist intellectuals to theoretical propositions more than to concrete
programs. Experimentation in this domain was the work of "loners" such as
Pierre Vilar and Guy Bois in France or of "deviants" such as Witold Kula in
Poland. The most fertile work was stimulated by intellectual traditions that
were either marginal to mainstream historical work – such as the variety of
British Marxisms, whether of E. P. Thompson or E. J. Hobsbawm or Perry
Anderson – or heterodox, such as the complex currents of Gramsci's influence
on Italian historians.

Since the 1960s, Marxism's influence has mainly been indirect, diffuse,
and ideological rather than operational. During the sixties a new social and
political critique developed in Western countries, calling into question the
dominant consensual postwar models. The New Left in the Anglo-American
world, the rediscovery of the Frankfurt School practically everywhere, and
the atmosphere of broad institutional critique emerging from such thinkers
as Erving Goffman and Michel Foucault came together to sensitize historians
to the mass and diversity of collective experiences and to urge a history "seen
from below" by insisting on the phenomena of crisis, rupture, marginality,
and deviation from central models. This sensitivity – at times populist –
discovered echoes of earlier programs, like those of the *Annales* and of the

[17] David S. Landes and Charles Tilly, eds., *History and Social Science* (Englewood Cliffs, N.J.: Prentice Hall, 1991), p. 470.

British journal *Past and Present* (1952), and reoriented them toward dominated social groups, women in particular. Social history became the most favorable site for an encounter between the historical discipline and the social sciences.

In Great Britain and the United States, the alliance was first made with sociology. Historical sociology (re)discovered Tocqueville, Weber, and Karl Polanyi and distanced itself from structural-functionalism, abstract empiricism, and economism. Inspired by the work of Thompson, Hobsbawm, Barrington Moore, Jr., and Anthony Giddens, it sought to integrate the procedures of the two disciplines.[18]

The same rapprochement was found in Germany, where, after a long eclipse, social history established itself during the 1960s and 1970s and mobilized all of the richness of the German sociological tradition: Weber, but also "Marx mediated through Max Weber," and the Frankfurt School.[19] Centered on the singularity of German historical experience, the social history developed by Hans-Ulrich Wehler (b. 1931), Jürgen Kocka (b. 1941), and the Bielefeld group was one of the most sophisticated efforts to integrate both critical conceptualization and an analytical procedure borrowed from the social sciences into the analysis of grand historical processes, such as modernization, bureaucratization, and the formation of social classes during the nineteenth century.[20]

In France and most Western countries, it was anthropology that became the principal partner, creating a different set of products: true and rare ethno-historical experiences; applications of Claude Lévi-Strauss's rigorous models of structural anthropology to the study of myths; and the transfer of technical instruments of analysis, such as those that renovated historical demography and initiated research in the history of the family. During the 1970s and 1980s, historical anthropology enthusiastically turned its attention to the infinitely diverse experiences of "everyday life" – those valorized by the German *Alltagsgeschichte* and those that stocked the international journals of history. Notable works, in addition to those we have already cited, included *Comparative Studies in Society and History* (1958) and the *Journal of Interdisciplinary History* (1970) in the United States, *Quaderni Storici* (1974) in Italy, *Geschichte und Gesellschaft* (1975) in Germany, *Social History* (1976) in Great Britain, and later, *Odysseus* (1991) in Russia and *Historische Anthropologie* (1993) in Germany. Interdisciplinary borrowing presided over the enormous enlargement of the historian's territory, accompanied by an explosion of practice and a dispersion of research aims.

[18] Cf. Theda Skocpol, ed., *Vision and Method in Historical Sociology* (Cambridge: Cambridge University Press, 1984).
[19] Georg Iggers, *Historiography in the Twentieth Century: From Scientific Objectivity to the Postmodern Challenge* (Hanover, N.H.: Wesleyan University Press, 1997), p. 71, and more generally chap. 5.
[20] Hans-Ulrich Wehler, *Historische Sozialwissenschaft und Geschichtsschreibung* (Göttingen: Vandenhoek & Ruprecht, 1980); Jürgen Kocka, *Sozialgeschichte, Begriff-Entwicklung-Probleme*, 2nd ed. (Göttingen: Vandenhoek & Ruprecht, 1986).

THE PROBLEM REASSESSED

Two types of confrontation and collaboration between history and the social sciences have occurred over the twentieth century. The first model sees the social sciences, or one of them, as decreeing a theory or canon of prescriptive rules to which historians are invited to associate themselves. This was the model that inspired the Durkheimian project around 1900, Lévi-Strauss's structuralism of the 1960s, and social science history in the United States. The second model, the more common and far more supple, is the transfer, more or less crude, of conceptualizations and procedures under the regime of borrowing. This was largely the initiative of historians. But in either model, the positions of history and the social sciences were unequal. The social sciences set the terms of engagement in the first mode; and in both, history took the risk of being placed in a subordinate position, with the notable exception of the *Annales* experience.

The situation changed dramatically during the last two decades of the century. A series of reassessments deplored the fact that "the synthesis ... of history and social science has not arrived."[21] Formerly strong, optimistic, and even triumphant programmatic ambitions were replaced by a recognition that collaboration between disciplines was difficult and that interdisciplinarity was not the rule. This reaction returned attention to the heart of the debate: Interdisciplinarity is a question before it is an answer.

At the same time, in a more or less irregular fashion, most social sciences took what has been called "an historical turn."[22] The phenomenon was complex and ambiguous. It expressed the new uncertainties of disciplines that, because they were less sure of their foundations and of their effectiveness, turned to their own history and questioned their conceptual, institutional, and social genealogy. History, as well as the social sciences, experienced a "reflexive moment." The historical turn also corresponded to the failure, or at least the accelerated decline, of the great integrating paradigms underlying the scientific aim of social science and the intelligibility of the social world, producing the desire to take into account the historicity of social phenomena, not only in the sense that they are the product of a history but also in the sense that their actualization is historical. Some economists, for example, raised questions about the irreversibility of economic phenomena and about their inscription in historical time. Certain anthropologists trained for the structural analysis of myths and rituals sought to explicate the context in which they were uttered and became effective.

---

[21] Andrew Abbott, "History and Sociology: The Lost Synthesis," in *Engaging the Past: The Uses of History across the Social Sciences*, ed. E. H. Monkkonen (Durham, N.C.: Duke University Press, 1994), p. 77.

[22] T. J. McDonald, ed., *The Historic Turn in the Human Sciences* (Ann Arbor: University of Michigan Press, 1996).

This ensemble of shifts, none of which were stabilized at century's end, originated from several different propositions, all of which have different implications for our subject. The most visible and influential at the end of the 1980s and the beginning of the 1990s was the program of a postmodern deconstruction of social knowledge. Its greatest success was in the United States, but from there it influenced a considerable part of the Anglo-American world and beyond. In its most radical forms, this new historicism was at the origin of a skeptical relativism that called into question the existence of a specific field of the social sciences and the possibility of any knowledge about the social. Under the postmodern aegis, the collaboration of the social sciences must be of an essentially critical nature. This is a movement that is inspired by such philosophical proposals as those of Derrida and Lyotard, and that tends to reduce social reality to discursive or cultural productions and to the texts through which we apprehend them; it is a movement as well that has prominently affected the diverse peripheries of American academic institutions. Such a conjunction suggests that a sociological approach to the postmodern turn might itself be wanted.

In France, the late-twentieth-century evolution was different. A period of epistemological anarchy was followed by an effort to redefine a space for the social sciences, history included. Two themes were dominant. The first sought to reconsider disciplinary identities – not to compartmentalize the scientific space, but rather to improve circulation among the social sciences, beginning with an account of the distances between the points of view they construct. The second theme, a more global one, rediscovered the old Weberian account of the common historicity of all the social sciences, inviting us to recognize the demands of a particular regime of science, distinct from that of the natural sciences, in which the work of interpretation is constantly associated with the construction of objects through scientific procedures, including specific procedures of description, explanation, and proof. [23] Beginning with this kind of reflection, the difficult dialogue between history and the social sciences might be pursued in relatively new terms.

[23] Jean-Claude Passeron, *Le Raisonnement sociologique: L'espace non-poppérien du raisonnement naturel* (Paris: Nathan, 1991).

Part III

THE INTERNATIONALIZATION OF
THE SOCIAL SCIENCES

22

THE SCIENCES OF MODERNITY IN A DISPARATE WORLD

Andrew E. Barshay

History, Marc Bloch said, is the "science of men in time... [which is] a concrete and living reality with an irreversible onward rush... the very plasma in which events are immersed, and the field in which they become intelligible."[1] Social science, by analogy, is the science of modernity, "an enterprise of the modern world. Its roots lie in the attempt, full-blown since the sixteenth century, and part and parcel of the construction of our modern world, to develop systematic, secular knowledge about reality that is somehow validated empirically."[2] Modernity, in short, stands in the same "plasmic" relation to social science that time does to history.

Perhaps until the end of the 1960s, common sense might have maintained that *modern* equaled *Western*, that Westernization equaled modernization. In the wake of worldwide events since 1989, it has again pleased some segments of public opinion to reassert this commonsense view, particularly in the name of neoliberal economic reform. The chapters that follow, notwithstanding their differences in approach, focus, and argument, suggest that the equation of *modern* with *Western* is (whether for or against) more an ideological than a historical position. For no single, universal modernity lies waiting at the end of all particular histories. Though powerful (and destructive, according to Serge Latouche), the Western-oriented "drive toward global uniformity" cannot succeed.[3] The paths to modernity are many, and those paths lead to different modernities.

At the same time, we can neither accept an evacuated reading of "modern" as "contemporary," nor banish the West from the larger notion of modern. To understand the internationalization of social science, as these chapters seek to do, is to understand that the relation of (alien) West to (indigenous) modern

[1] Marc Bloch, *The Historian's Craft* (New York: Vintage, 1953), pp. 27–8.
[2] Immanuel Wallerstein, Calestous Juma, Evelyn Fox Keller, Jürgen Kocka, Dominique Lecourt, V. Y. Mudimbe, Kinhide Mushakoji, Ilya Prigogine, Peter J. Taylor, and Michel-Rolph Trouillot, *Open the Social Sciences* (Stanford, Calif.: Stanford University Press, 1996), p. 2.
[3] See Serge Latouche, *The Westernization of the World* (Cambridge: Polity Press, 1996).

is not only one of opposition, but also one of part to whole. And it is this historical "whole," the combination, in some cases by force, of indigenous with Western elements, in order to constitute singular national modernities, that has formed the field of activity, and of intelligibility, for much of social science beyond the "Atlantic Rim."

We may hypothesize, then, that the form taken by social science in a given national setting depends substantially upon the institutional path to modernity taken by that nation, and particularly upon the degree of autonomy – or conversely, of heteronomy – in that development. Without question, the major themes, problematics, and institutional forms of internationalized social science have their origin in one or another Western heartland: It is hard to miss the early imprint, for example, of Historical School economics in Japan, evolutionism and Marxism (initially mediated by Japan) in China, utilitarianism and positivism in India, British social anthropology in Africa, Durkheimian sociology in Egypt, and so on. Yet surely the crucial point is that such impulses did not necessarily remain under the *control* of the heartland; adopting the language and methods of social science itself did not amount to a spiritual self-colonization. Rather, what seems to matter is "not indigenous genesis, but autonomous assimilation."[4] The historical question at hand, therefore, is how to account for and assess the distinctiveness (or mutual similarity) of the national and regional forms of social science. How was a network of Western-derived discourses and practices made meaningful outside the West? How were they translated across the "space-time" of modernity? How did they indigenize and become self-replicating, and in what relation to indigenous systems of knowledge? What, in other words, have been the dynamics of internationalization in social science?

The specificities of the process of internationalization are discussed in the case studies that follow, which treat aspects of social science development in China, Japan, India, the Islamic world, Africa, Latin America, eastern Europe, and Russia. But a few common themes, as well as a broadly relevant chronological framework, may perhaps be identified. Whether among the late-developing empires (such as Japan and Russia) or in the former (semi-)colonies, social science has tended to take as its task the promotion of modern *national development*, usually with a sense of urgent competition with, or threat from, the West. In every case, this development has meant confrontation with legacies of the prenational past, frequently as the source of perceived material backwardness and always of cultural difference relative to the West: The issue was the relation between these two. But whatever the cause of backwardness, social science *was* dedicated to its abolition, and herein lies the reason for its frequently self-conflicted character. What was "national progress," and what was its price? Did development mean the

---

[4] Rikki Kersten, *Democracy in Postwar Japan: Maruyama Masao and the Search for Autonomy* (London: Routledge, 1996), p. 131.

rejection of "tradition," or its co-optation? The supersession of indigeneity, or its reconstitution? What in any case were the specific problems – or areas of strength – in a given society, seen from the point of view of autonomous development? In a broader sense, was it possible for social scientists to identify areas for analysis without combining originally foreign categories and (meta-)narratives with the sure touch of lived experience – or at least of experience as mediated by canons of representation that were indigenous?

Depending on the setting, internationalization would mean coming to terms with a variable *colonial* legacy of administrative techniques and higher education (the latter being institutionally significant, for example, in India, Egypt, and Korea, but negligible in Morocco and sub-Saharan Africa); it would mean thinking through and beyond categories of social inquiry ("caste" in India, "tribe" in Africa, etc.) elaborated under imperialist rule: Only in this way could the indigenous culture(s), self-knowledge, or voice be recovered and the nation created; it did not otherwise exist.

Due to its character as an instrument in nation building, whether in transition from colonialism or not, social science outside the West during its formative period developed most often under direct state (or in some cases, where state authority had been fragmented, foreign) sponsorship. Its history, in turn, has been one of *institutional migration*, with greater or lesser success, to academic settings. The question remains, of course, as to how the academy is governed and maintained. But it does seem that universities, research institutes, and museums have tended to be more sustaining of trends toward professionalism and the solidification of disciplinary identities in the social sciences than has direct subordination to agencies of the state. Institutional migration is also important to the extent that it has been accompanied by some degree of intellectual *decolonialization*, or *indigenization*. It is worth repeating that these terms do not imply the wholesale attempt to "cleanse" social science of foreign elements so much as the *translation* of these elements into something locally meaningful.[5]

It may not be too much to say that the permutations of the Slavophile–Westernizer debates – including the advent of populism – that marked Russia's emergence as the first "developing society" have been replicated, and extended, in every other society that has faced the issue of defining its developmental path. The internationalization, the autonomous assimilation, of Western social science has been part of a many-faceted and deeply ambivalent quest for an effective formula for what has been termed "differential usage" – formulas such as "Chinese essence, Western means," "Japanese spirit, Western technique," "Russian socialism," "socialism with Chinese characteristics," "Islamic economics," to name only a few. Through such a process, both the national past, or tradition, *and* the modern present are selectively

[5] See Wallerstein et al., *Open the Social Sciences*, pp. 56–7.

and mutually invented, and turned toward the inevitably and inherently contested end of constituting a national culture.[6]

Social science, however, was about more than culture or identity. The pursuit of progress through "differential usage" was difficult precisely because it entailed the transformation (or attempted shoring up) of real institutions and practices in a world both unified by and divided among powers that lay largely beyond political reach. Imperialism, democracy, and revolution, the world-altering conflict between the Allies and the Axis, the Cold War, the wave of decolonization and the emergence of a three-world (or North-South) global configuration: These have also shaped the social sciences insofar as they have been concerned with bettering the societies in which they are practiced.

Thus, whether through Spencerian or, later, Marxist lenses, received social forms were subject to "scientific" judgments as to their functionality. Spencerian evolutionism posited a great transition from "militarist" to "commercial" (to "industrial") society; this process was to be mediated by the advanced civilizations on behalf of those ostensibly less favored. In those settings, such as China, the state might have to spur and sustain processes that in the West were thought to have been achieved precisely because the state had stepped out of the way. Marxism, for its part, should have fallen on barren ground outside the capitalist West, but, as with evolutionism, import substitution (of peasantry for proletariat) was possible, and its critique of capitalist industrialization proved powerfully persuasive.[7] Marxism differed from Spencerianism, first in the much greater degree to which it was organized as a political force able (for good or ill) to harness and guide the energy ostensibly produced by class conflict; second, in its dedication to realizing a definitive form of human society; and third, in its development of a "traveling" canon of texts that collectively gave it the status of social science par exellence.

Clearly, these systems were avatars of the idea of progress as necessary and guaranteed – no matter how many lives it might cost. So too was that of modernization, which sought, in a certain sense, to split the difference between Spencerian evolutionism and revolutionary Marxism. What is called "modernization theory" renounced the need for violent transitions while retaining the latter's vanguardism. Upholding the general goal of development, it stood for the willed transformation of "traditional" into "modern" societies, largely along lines defined *for* the peoples of the third world by elites of the first world and some of their own fervent modernizers – albeit in sometimes cynical and destructive rivalry with those of the second, socialist, world. In full cry at the beginning of the 1960s, modernization, both as a concept in social

---

[6] "Differential usage" is the phrase of the Japanese historian Masao Maruyama. See his "Nationalism in Japan: Its Theoretical Background and Prospects" (1951), in his *Thought and Behavior in Modern Japanese Politics* (Oxford: Oxford University Press, 1969), pp. 135–56.

[7] See Eric J. Hobsbawm, *The Age of Empire* (New York: Vintage, 1989), pp. 267–9.

science and as a structuring element in the relation between the "worlds," had been seriously discredited by the end of the decade, and has never recovered the authority it once enjoyed. From the perspective of the post–Cold War present, both the naïve confidence – indeed, the hubris – of most modernization "theorists" and their sweeping denunciation by critics on the left may seem to be no more than mirroring totalisms. The later collapse of almost all socialist regimes (not only those of the former Soviet bloc), followed by the adoption of policies of capitalist transition, and the current course of China do indeed force us to look at modernization and its critique with chastened eyes.

The notion of economic development (as paired with economic growth in advanced societies) survived the demise of the modernization approach. Indeed, it thrived: In its complexity and global reach, "development" may have been the greatest of all knowledge-producing industries, and the one that has most deeply and intimately involved social scientists in its various processes. As such, it has come under sharp critique by figures as different as Samir Amin and Serge Latouche, Andre Gunder Frank and Immanuel Wallerstein, Arturo Escobar and Ivan Illich. Yet it remains the – contested – key word in international social science. On the one hand, the *state* as primary agent of development (and therefore as invested with conceptual primacy) has been decentered; agency has been pluralized. The so-called "world-system," as Wallerstein points out, generates antisystems on its peripheries (including its internal margins); these must be theorized and examined historically, not as antidevelopment per se, but as alternative modes. The dispersion of "statedness" among indigenous "communities," globalized capital, and supranational political entities has caused key notions, such as that of political power and of the market and its presumed benefits, to be recast. To a degree, such transformations have paralleled the rise of "green" and feminist approaches within the Western social sciences, and as with these latter, they have been accompanied by a rethinking of large conceptual premises: Interpretive and linguistic concerns have coincided with the "softening" of the lines between the social and natural sciences. As Wallerstein, representing a group of social scientists drawn from both "North" and "South," puts it:

> Is there a deeper universalism which goes beyond the formalistic universalism of modern societies and modern thought, one that accepts contradictions within its universality? Can we promote a pluralistic universalism, on the analogy of the Indian pantheon, wherein a single god has many avatars? . . . Only a pluralistic universalism will permit us to grasp the richness of the social realities in which we live and have lived.[8]

Nevertheless – to return to the "development" issue – it cannot be said that internationalized social science has entirely lost its bearings as

---

[8] Wallerstein et al., *Open the Social Sciences*, pp. 59–60.

"ethico-practical" activity. Amartya Sen's famous query – what has become of the hundred million women in the third world "lost" due to poverty and ill health? – recalls those bearings. Illich's critique of development may have "underestimated the extent to which the new needs created by the modern division of labor actually do correspond to what people desire," but this point only reinforces the strong and enduring imperative that social science rethink its premises and language as the world is reconfigured around it. "Development" as a term may have to be used in conjunction with qualifiers – development of what, by whom, for whom?[9] Sen's question, however, implies a need for a transformation in conditions that are local but pervasive. As long as it is concerned with grasping such conditions, and aiding and hastening such transformations, internationalized social science will have its ongoing identity and task.

Social science, then, is the science of modernity, outside the West no less than within it. *Modernity* denotes a condition or predicament in which sustained contact with the West forms a salient but not the exclusive element. There is no form of modernity to which that contact is irrelevant. This salience is visible institutionally in the increasing concentration of social science practitioners in academic settings, and in the adoption of intellectual categories and disciplinary divisions that to some degree are then *indigenized*; their external origins may still be recognizable but do not determine their fate. A process of intercultural transfer, and of creation, has taken place. Yet contact with the West, almost everywhere, has proceeded on unequal terms. This legacy of asymmetry is ineradicable. The internationalization of the social sciences, therefore, has unfolded in the historical and existential "space" that, willy-nilly, forms between culture and politics in the modern world.

---

[9] The quotation on Illich is from the brief entry by Michael Ignatieff in *The Social Science Encyclopedia*, ed. Adam Kuper and Jessica Kuper (London: Routledge, 1985), pp. 376–7. See also *International Development and the Social Sciences*, ed. Frederick Cooper and Randall Packard (Berkeley: University of California Press, 1997); *The Development Dictionary*, ed. Wolfgang Sachs (London: Zed Books, 1993); *The Post-Development Reader*, ed. Majid Rahnema with Victoria Bawtree (London: Zed Books, 1997).

23

THE SOCIAL SCIENCES IN LATIN AMERICA DURING THE TWENTIETH CENTURY

Jorge Balan

This chapter is a selective overview of the development of sociology, anthropology, political science, and economics as defined in several Latin American countries. After reviewing the liberal heritage and the influence of positivism and social evolutionism at the turn of the century, the chapter discusses the emergence of sociology and cultural anthropology in Argentina, Mexico, and Brazil, with public education as a central theme. A recently established discipline in Europe and North America, sociology was adopted by learned groups for its promise of a scientific synthesis. Anthropology carried the legitimacy of its links with natural science, although it was the discovery of culture that lent it utility. From the 1940s through the 1970s, as development became the panacea, economics was in the ascendancy in Latin America. Modernization was to follow economic growth, although the relationship was understood to be far from necessary. The economic dependency argument, a dominant framework during the 1970s, was the outgrowth of a theoretical movement that relied on a Marxist third world perspective in rejecting the left-of-center reformist policies then in vogue. From the 1980s until the end of the century, the social sciences became associated with the expansion of higher education. The renewal of political science (grounded in a fresh look at state and society issues), the concern about cultural identity in anthropology and among students of communication, and the hegemony of the neoclassical economic framework were outstanding features of this period of theoretical and thematic diversity. The chapter concludes with a reflection on the Latin American social sciences in a globalized world.

PROLOGUE: POSITIVISM AND SOCIAL EVOLUTION IN LATIN AMERICAN THOUGHT

After many decades of civil wars, the Latin American republics experienced from the 1870s through World War I a period of export-led economic expansion and political order, often imposed by force but legitimized by liberal ideologies and recently adopted constitutions. The liberal doctrines supported faith in progress and science and, shaped by the French tradition, most often put science in the service of the secular state and considered the state to be the main agency for change through public education. The liberal program of secularization, free trade, social reform, and public education was heavily influenced by Comtean philosophy, which gave it a technocratic and authoritarian bias. The educational institutions constructed in that context emphasized encyclopedic learning, scientific and practical training, and adherence to secularism and state control.[1]

Comtean philosophy gained influence through the efforts of educational leaders who had privileged access to liberal politics during the last quarter of the nineteenth century. In Mexico, Gabino Barreda (1818–1881) inspired the newly created *Escuela Nacional Preparatoria* and President Juarez's commission for educational reform. The long period of authoritarian rule under Porfirio Diaz (1884–1910) was heavily influenced by the presence of graduates of that school, the so-called *científicos*, whose basic goal was to bring science to state administration, following the teachings of Comte and Saint-Simon. Brazilian positivists were particularly influential in the *Escola Militar* when, after the war with Paraguay (1865–70), the military became increasingly active in politics on the side of the republicans. Many adhered to Freemasonry and thus felt alienated from an emperor closely identified with the Catholic Church. The abolition of slavery in 1888 and the founding of the Republic in the following year were proclaimed under the positivist banner of *Ordem e Progresso*, words inscribed on the Brazilian flag. Benjamin Constant (1836–1891), a positivist and professor of mathematics at the military school, became minister of education in the first republican government. In Chile, Jose Victorino Lastarria relied on Comtean philosophy in establishing a scientific base for politics in his book, *Lecciones de politica positiva*, published in 1875. Valentin Letelier (1852–1919), a second-generation positivist, founded the pedagogical institute at the University of Chile in 1889 and became the intellectual leader of his country's liberal politics until his death.

In the understanding of society, Herbert Spencer was the most often-cited inspiration for Latin American writers; he was also widely read in the Mediterranean countries. Race was the key concept, often used as a biological term referring to a nationality or a people developing over time. The three major racial influences – Indian, African, and Iberian – were found, more often than not, to be deleterious to social progress. Racism led to self-deprecation, tending to propagate another liberal panacea, European

---

[1] Charles A. Hale, "Political Ideas and Ideologies in Latin America, 1870–1930," in *Ideas and Ideologies in Twentieth Century Latin America*, ed. Leslie Bethell (Cambridge: Cambridge University Press, 1996).

immigration. Unlike Comte, Spencer favored limiting the state's power, and his positions were used to support free-market policies.

Social evolutionism rooted contemporary evils in race and miscegenation or in the geographic environment. In large areas of the continent, the slavery of Africans had long been a justified practice; indigenous groups lacked elementary civil and political rights well into the twentieth century. In countries undergoing rapid urbanization and modernization, such as Argentina, materialism, as evidenced by speculation and usury, and individualism were made responsible for the deterioration of social bonds. A biological or medical framework was applied, through the lens of the Italian school of criminology of Cesare Lombroso, to explain phenomena such as prostitution, crime, mental disorder, and anarchy. Eugenics during the early decades of the twentieth century became both a scientific and a social movement, urging control of immigration and fertility.

Despite being framed by this biological conception of race, the social sciences turned to state-sponsored education as a means to build a nation out of disparate and conflicting social groupings. The empirical study of society and culture had a very thin tradition, but the spread of positivism created great respect for statistical data, observation, and classification. Governmental offices, under the leadership of engineers, medical doctors, and other professionals, collected information and published reports on education, crime, housing, and the standard of living of the working classes.

FROM THE TURN OF THE CENTURY TO THE 1930s: EDUCATION AND NATION BUILDING

In Argentina at the turn of the century, sociology became fashionable among intellectuals who were attempting to understand a rapidly changing society. Export-led growth and European immigration were rapidly transforming this vast, sparsely populated country and turning Buenos Aires into a major city. The focus of attention was the social question, rooted in immigration, urban wage work, family breakdown, criminality, and anarchism. Political order had been established only recently, and social integration had not yet been achieved.

Ernesto Quesada and Jose Ingenieros attempted to create an academic sociology within this context. Both had close links to international academic circles, but they differed politically. Ingenieros was an early example of a politically committed intellectual on the left, at the margins of political power; while Quesada, maintaining a more academic profile, formulated an educational nationalist program.

Ernesto Quesada (1858–1934) became the first to hold a chair in sociology, in the fledgling School of Philosophy and Letters at the University of Buenos Aires in 1904. He also taught political economy and was a skilled historian.

Quesada was particularly well read in German historicism and economics. He admired Gustav Schmoller for his attempt to find national roots in the social science enterprise, but he was also familiar with Marshall and British political economy, which he found too abstract and deductive. He became a member of the American Academy of Political and Social Science, which published his essay on Argentine social evolution in the *Annals*. Quesada's essays focused on the integration of workers into the national society and were rooted in a political debate on labor legislation, which had been proposed as an antidote to the anarchist tendencies shown by immigrant workers.[2] Unlike mainstream Argentine thought, his explanatory framework tended toward historicism rather than the predominant social evolutionism, and he recognized the need to strengthen local communities, often eroded by the central state in its efforts to impose national order.

Commissioned to report on the teaching of history in German universities, where it was a central element in a curriculum designed to reinforce national identity, Quesada concluded that education was crucial to the state's attempt to build a nation. Schooling in Spanish and learning about national history and geography would bring homogeneity to a country torn by regional differences and massive immigration from a score of European countries. It would also be the best antidote to the spread of anarchism and nihilism. His work supported the new standards for "patriotic" education that mandated the use of an official language, Spanish, banned the use of other languages, and oriented the teaching of history and geography toward identification with the nation.

Born in Italy, Jose Ingenieros (1877–1925), was a neurologist and psychiatrist and an early promoter of psychotherapy. He embraced sociology as a synthetic science that should apply to human societies the same methods developed for the study of animal societies. Human diversity, including racial differences, was largely a consequence of adaptation to the highly varied physical environment, although nationalities emerged as transitional adaptations arising from historical factors.[3] Ingenieros also adopted Gustave Le Bon's theory of the biological basis for psychological differences among civilizations. Le Bon's ideas were particularly influential in Argentina, a country concerned with erasing the traces of indigenous and African influences, and socialist writers such as Ingenieros shared in the biologically oriented racist terminology. However, these views admitted a role for environment in hereditary processes and led to some optimism about the consequences of social reform.[4]

[2] Eduardo A. Zimmermann, *Los liberales reformistas: La cuestion social en la Argentina, 1890–1916* (Buenos Aires: Sudamericana, 1995); Ernesto Quesada, "La *Epoca de Rosas* y el reformismo institucional del cambio de siglo," in *La historiografía argentina en el siglo XX*, ed. Fernando J. Devoto (Buenos Aires: Centro Editor de America Latina, 1993).

[3] Oscar Teran, *Positivismo y nacion en la Argentina* (Buenos Aires: Puntosur, 1987).

[4] Nancy L. Stepan, *The Hour of Eugenics: Race, Gender and Nation in Latin America* (Ithaca, N.Y.: Cornell University Press, 1991).

As a social reformer, Ingenieros focused on the dysfunction of capitalist society. Workers had to be protected by enactment of appropriate legislation, as well as by the application of hygiene. A teacher at the School of Medicine in Buenos Aires who worked at public mental health institutions and at the Institute of Criminology of the penitentiary, Ingenieros published articles in specialized medical journals in psychiatry, criminology, and legal medicine. His essays, collected in a book entitled *Argentine Sociology*, established the liberal heritage of the discipline and attempted a defense of its scientific status as the study of social organisms, in continuity with biology. However, he failed to indicate how the new science would collect, classify, and analyze social data. As a militant member of the Argentine Socialist Party and writer for party publications, Ingenieros was denied a chair in legal medicine, which caused him to leave the country for several years. Upon his return from Europe in 1914, he became an advocate of university reform and then an admirer of the Russian Revolution. In many ways, his career established a model for many Argentine intellectuals on the left, who found a highly unstable academic space in the public university.

Unlike academics such as Quesada and Ingenieros, a number of professionals oriented toward social reform worked their way through government offices in the fields of labor, agriculture, and immigration. Led by Juan Bialet Masse, who published the two-volume study *El estado de las clases obreras argentinas a comienzos de siglo* (1904), and Juan Alsina, author of *El obrero en la Republica Argentina* (1905), they were responsible for many studies on the social question. City and national censuses, as well as serving as sources of data, were the occasion for monographic studies of immigration, housing, health, and the family. Many such works were presented in 1916 at a congress on the social sciences held in commemoration of one hundred years of Argentine independence. Sessions organized by Quesada and Ingenieros focused on the standard of living of the working classes and the need for social legislation; others dealt with child and female labor, unemployment, strikes, housing, mutual aid societies, and immigration. Most of the sessions, however, had a legal orientation, and their purpose was usually to argue for legal reform. Almost one hundred of the theses presented at the School of Law during the first two decades of the century were focused on labor issues.

The social science disciplines at the university soon fell victim to a reaction against positivism that drew on the persistent traditional scholastic orientation in educational circles. Although formally defeated by the secular policies of liberal governments, Catholic forces remained strong, particularly in the inland provinces. A rejection of "Anglo-Saxon" capitalism and materialism was supported by a new generation of nationalist writers who looked for inspiration to the Spanish spiritual heritage; and the first successful military coup of the century, in 1930, allowed them to gain increasing influence in public education. Their reading of the German philosophers Wilhelm Dilthey and Heinrich Rickert supported a radical distinction between the natural and the cultural sciences, and rejected empirical methods and general laws in the study of society. Schools of law abandoned the study of working conditions and labor risks in favor of more formalistic legal studies. The fragile tradition of statistical reporting and social analysis based on census data was interrupted; no national census was taken between 1914 and 1947. Only the emerging field of economics – in particular, the group around Alejandro Bunge and the *Revista de Economia Argentina* – developed this tradition further, publishing a series of social and demographic studies on early industrialization and urbanization that revealed, and lamented, the end of mass immigration, declining fertility rates, and a slowing rate of population growth.[5]

Nation building that required state activism in the field of education also laid the groundwork for the development of the social sciences in Mexico, a country at the opposite geographical and cultural extreme of Latin America. Manuel Gamio (1883–1960) brought cultural anthropology to Mexico in order to demonstrate cultural continuity between pre-Columbian cultures, contemporary indigenous groups, and the revolutionary ideals of the nation.[6] Gamio started his career in archeology as a student at the Museo Nacional, graduating in 1907. Awarded a fellowship to study cultural anthropology at Columbia University, he learned from Franz Boas a respect for fieldwork and an appreciation of culture as a basic anthropological concept. However, within the context of the Mexican Revolution (1910–17), which raised nationalist and popular banners against the elitist and foreign-oriented policies of Porfirio Diaz's *cientificos*, Gamio positioned his work closer to Mexican politics than to North American academia. His ideas about culture, although strongly influenced by Boas, became increasingly shaped by the political goal of building a unified Mexican nation out of hundreds of local cultural identities.

Gamio's major accomplishment was a three-volume study of the contemporary and historical cultures of Teotihuacan, *La Poblacion del Valle de Teotihuacan*, published in 1921, in which he rejected racism and adopted a theory of cultural development based upon the concept of "culture areas." Unlike Boas, however, Gamio accepted the notion of the progress of civilization and thus rejected cultural relativism. Boas focused on linguistic differentiation as a key tool in establishing borders between cultural areas, while Gamio was more interested in convergence. Linguistic variations and change were to be studied in order to promote bilingual education, a desired means for national integration of the many linguistic groups of Mexico.

Gamio was director of the recently created Anthropology Department at the Secretariat for Agriculture, the first government office in charge of

[5] Juan J. Llach, *La Argentina que no fue* (Buenos Aires: IDES, 1985).
[6] Guillermo de la Pena, "Nacionales y extranjeros en la historia de la antropologia mexicana," in *La historia de la antropologia en Mexico*, ed. Mechthild Rutsch (Mexico City: Plaza y Valdes, 1996).

indigenous affairs; he had gained visibility with the publication of *Forjando Patria* (1916), a nationalist manifesto for the emerging indigenist movement. His plan for the study of cultural areas was adopted as official policy by presidential decree. Jose Vasconcelos, a leading intellectual in shaping cultural policy, promoted Gamio's work in anthropology, along with rural education and mural art. In 1924, the department was transferred to the Secretary of Education, where *indigenismo* became the basis for educational policy. When Gamio left the position for political reasons, Moises Saenz, who had been trained in education at Columbia under John Dewey, became the leading force behind bilingual and rural education.

Because cultural anthropology, linguistics, and archeology developed in Mexico as applied social sciences in the service of the nation and the government, theoretical issues were highly ideological and intertwined with national politics. Gamio's work initiated a long and conflicted line of applied anthropology. Alfonso Caso (1896–1970) founded the Instituto Nacional Indigenista in 1948, where he organized regional programs throughout the country, with a coordinating center in charge of promotion and research activities. Anthropologists were usually in charge of directing the regional centers. This close connection between anthropology and the national government became the focus of criticism by a younger generation trained during the 1960s and influenced by the 1968 student movement, who perceived their teachers as "organic intellectuals" of the Mexican state.

The first continuing institutional home for the social sciences in Latin America was built in Sao Paulo, Brazil, during the 1930s as part of an education-based project for building a new nation. The rise of industry, the impact of mass immigration, and local pride in opposition to centralized power supported the creation of a modern university where the social sciences could occupy a stable, legitimate place.[7]

By the 1920s, Brazil had a number of prestigious professional schools that were the training grounds for the political and cultural elite in Sao Paulo, but still had no university. Responding to the need to train teachers in order to expand the public education system, a group of writers and professional educators, many of whom knew Durkheim's work or had been trained in France, began a crusade for a regional university. Joined by an association of French scholars, led by Georges Dumas, which had created a French *lycee* in Sao Paulo, they developed the idea of a school of philosophy, science, and letters placed at the heart of the new institution for the training of teachers and the development of basic science. The geographer Pierre Monbeig, the historian Fernand Braudel, and the anthropologist Claude Lévi-Strauss designed the curricula and taught the first classes. Other promising young scholars, such as the economist Francois Perroux and the geographer Pierre Daffontaines, followed them. Some, including Roger Bastide, settled in Sao Paulo because

[7] Sergio Miceli, ed., *Historia das ciencias sociais no Brasil*, vol. 1 (Sao Paulo: Vertice, 1989).

of the war. They introduced high standards and an academic style based on respect for monographic work. However, French sociology and ethnology were characterized by reliance on archival work, rather than on field methods. Durkheim's sociology and Marcel Mauss's ethnology were the prevailing examples, and only Claude Lévi-Strauss and Roger Bastide had rebelled against the established masters.[8]

The educational reform of 1931 had established sociology as a mandatory subject for study in secondary education. Students of working-class and immigrant origin were often attracted to the school by scholarships sponsored by the secretary of education, and over 40 percent of the first graduates were women, which was not normally the case in professional schools. The university established teaching and research as a legitimate profession, and, in the absence of better-trained people to fill the new positions, many graduates remained at the university. Doctoral theses and dissertations required for professorial ranking demanded individual research and familiarity with the relevant bibliography.

During the 1930s, another intellectual project in Sao Paulo adopted the social sciences as its center. The Escola Livre de Sociologia e Politica, founded in 1933, emphasized empirical knowledge and greater contact with the immediate social environment. Donald Pierson, an American trained in Chicago by Robert Park, had first done field work on race relations in Bahia from 1935 to 1937. He moved to Sao Paulo to work at the Escola, where he initiated a graduate program. Pierson gave the Escola an academic turn, with a strong emphasis on empirical research. The journal *Sociologia* became the first of its kind in Latin America. Pierson, his colleagues and students carried out community studies of isolated towns in order to describe the "folk" culture before urbanization, inspired by the work of Robert Redfield. As a private graduate school, the Escola had no articulation with the state educational system, and thus no influence on teacher training and the curriculum; as a result, it failed to create an academic style of work or a school of thought. As a research institution, its studies failed to relate to policy, in spite of the school's original aim, since it largely lived in a foreign-oriented academic world. Still, it influenced Florestan Fernandes, a graduate of the Universidade de Sao Paulo who became mentor of a new generation of social scientists during the 1950s.

The better-known social scientists of the 1930s, however, were not linked to either of these educational projects. Gilberto Freyre (1900–1987), trained in cultural anthropology at Columbia under Franz Boas's influence during the early 1920s, upon returning to his native Pernambuco in northeastern Brazil became an intellectual who paid no attention to disciplinary boundaries or identities, a prolific author of essays on historic and cultural matters, an active

---

[8] Fernando Limongi, "Mentores e clientelas da Universidade de Sao Paulo," in *Historia das ciencias sociais no Brasil,* ed. Miceli.

journalist, and a fiction writer. His early work on slavery and racial relations in Brazil, published in the 1930s, has been translated into many languages and is still considered a key point of departure for scholarly debate on these issues.[9] Francisco Jose Oliveira Vianna (1885–1951), a professor of law, was a critic of liberal constitutionalist thought for its failure to come to terms with Brazilian reality. His studies of Brazilian society and culture, presented in terms of race and geography, and his defense of corporatism were well received in the political and intellectual climate of the authoritarian regime established by Getulio Vargas under the so-called *Estado Novo* (1937–45).

BETWEEN WORLD WAR II AND THE 1970s: DEVELOPMENT AND UNDERDEVELOPMENT

Economic development emerged as a concept in Latin America before economics became an established university discipline and then legitimized the efforts to build the discipline. Changes in world trade brought about by the 1930 crisis and the challenges they posed to national economic policy created the need for professional economists. Self-trained economists with considerable practical experience, and some with foreign academic training, provided the leadership to develop specialized schools. After World War II, intergovernmental agencies within the United Nations were the initial conduits to legitimize the new profession in the government apparatus. Particularly close ties were established with United States universities and foundations. The Cold War brought increasing North American attention to Latin America, and considerable United States involvement in governmental policies and in efforts to secure anticommunist governmental regimes. In that context, economic theory and research, with close ties to economic policy and thus to politics, became a dominant profession in the service of the state, a powerful force shaping public opinion among the educated classes, and the strongest discipline among the social sciences.

Economists during the early decades of the twentieth century were generally trained in law or engineering and had practical experience in government, business, or banking. The debate on economic issues such as custom taxes and monetary policy was argued within established doctrine, borrowed from European sources. The majority favored free trade, basing their policies on Ricardo's theory of comparative advantage, but governments in need of the proceeds from customs were more pragmatic. The successful integration of many Latin American countries into expanding world markets before 1930 provided the best argument in favor of free trade. With few exceptions, there was no economic research within government beyond the preparation of

---

[9] His best-known book in English translation is *The Mansion and the Shanties: The Making of Modern Brazil* (New York: Knopf, 1963).

statistical reports. In the universities, schools of commerce and accounting began to flourish after World War I. Although often called schools of economics, they were academically weak and tended to recruit nonelite students for the study of business management. An economist's credentials were largely established through experience in government or in big business, notably in the banks.

Raul Prebisch (1901–1986), arguably the best-known and most influential economist during this period and a prolific writer until his death, was an exception to the rule.[10] He studied economics at the newly created School of Commerce of the University of Buenos Aires during the early 1920s, where he was fortunate to have Roque Gondra as a teacher of mathematical economics. Prebisch initially worked for the Sociedad Rural, the association of large landowners in Argentina, and was sent for further study of agricultural economics to Australia, where he learned then-current theories of international trade.

The October 1929 crash led Argentina to abandon the gold standard and to introduce exchange controls; the country's first successful military coup in 1930 placed extraordinary power in the hands of the central government. Prebisch joined the government in several foreign missions and international conferences as it attempted to confront the decline in value of Argentine exports. In 1934, he published his first influential paper on the decline, a critique of orthodox equilibrium theories. Soon afterward he became director of the newly created Central Bank, where he founded a research department and a journal devoted to economic research. Research focused on monetary policy and its role in the trade cycle. As a key advisor to the minister of finance, Prebisch inspired industrialist policies. By the end of World War II, he had developed a center-periphery theory to explain the economics of underdeveloped areas. The argument was both theoretical and historical: It linked Latin American development to the shifting of the center of the capitalist economy from the United Kingdom to the United States.[11] Through participation in postwar economic agreements, he built a network of like-minded economists who supported the creation of the Economic Commission for Latin America (ECLA) in Santiago, Chile.

Chile had industrialized more fully than most other Latin American nations. It also enjoyed a more solid academic tradition, built on the national university founded in 1839. Modern economics, separated from schools of commerce, was first taught at the University of Chile in 1935. The school had strong ties with the public sector and supported state intervention in the economy. The Chilean government lobbied to have ECLA established

---

[10] Joseph L. Love, "Economic Ideas and Ideologies in Latin America since 1930," in *Ideas and Ideologies*, ed. Bethell.

[11] Joseph L. Love, *Crafting the Third World: Theorizing Underdevelopment in Rumania and Brazil* (Stanford, Calif.: Stanford University Press, 1996).

in Santiago, and its links with the university gave further legitimacy to the economics profession. In 1953, Prebisch organized the first regional conference of schools of economics in Santiago. With the support of ECLA, the university established a graduate program in 1957 for students from all over Latin America. ECLA was a "think tank" that generated theory, research, and doctrine and also trained economists for government.

In 1950, ECLA published *The Economic Development of Latin America and Its Principal Problems*. Termed the "ECLA Manifesto" by Albert Hirschman, it became mandatory reading for economists, policy makers, and social scientists. The center–periphery argument, as well as the thesis on the deterioration of terms of trade for agricultural export countries, became the founding concepts of the "structuralist" school. The work of authors such as Paul Rosenstein-Rodan and Charles Kindleberger during the war, and of Francois Perroux, Jacob Viner, and Hans Singer in the following years was closely related to that of Prebisch and the ECLA.

In Brazil, the first school of economics was created in Rio de Janeiro in 1945 by Eugenio Gudin and Octavio Gouvea de Bulhoes, economic advisors in the Brazilian government. Eugenio Gudin (1886–1986), an engineer by training, published in 1943 the first Brazilian textbook of modern economics. After attending the conference at Bretton Woods in 1944 and visiting Harvard University, he established a curriculum strong in mathematics, statistical methods, and neoclassical economic theory. Gudin also directed the newly established Instituto Brasileiro de Economia, within the Getulio Vargas Foundation, which did applied economic research for the national government, including work on the national accounting system and the price index. The Instituto, rather than the school, became the academic and political center of the project. Publishing one journal devoted to theory and another on statistical information, it favored a strong central bank to control inflation but opposed protectionist policies.[12]

Although ECLA exerted wide influence, other schools with rival theoretical and policy orientations were soon established. In 1955, the Catholic University in Chile and the University of Chicago established a joint research and teaching program, with the U.S. Agency for International Development providing funds for fellowships and visiting professors. The Catholic University program became the first one in Latin America with a full-time staff of foreign-trained economists actively involved in research. The free-market neoclassicism of Milton Friedman and the teaching and advice of his colleague Arnold Harberger were implemented by the "Chicago boys," who designed the economic and social policies of the Pinochet regime (1973–90). In Argentina, where economics was also established as a discipline in several major public universities during the late 1950s, Chicago economics

[12] Maria Rita Loureiro, *Os economistas no governo: Gestao economica e democracia* (Rio de Janeiro: Fundacao Getulio Vargas, 1997).

became the dominant school in two provincial settings, Mendoza and Tucuman, although it was relatively unimportant at the leading University of Buenos Aires. In Brazil, the Getulio Vargas Foundation followed the Chicago orientation.

The University of Chile also obtained foreign support, initially from the Rockefeller Foundation and later from the Ford Foundation, to enlarge its graduate training capacity into a full-fledged doctoral program. Under the leadership of the American economist Joseph Grunwald, and with funding from those foundations, Chileans also received graduate fellowships in the United States. During the early sixties, the university's structuralist and government-centered approach competed for academic and public power with the Catholic University's free-market orientation.

The favorable orientation of ECLA and the University of Chile toward sociological and political analyses led a group of leading Latin American social scientists during the mid-1950s, working with the sponsorship of UNESCO, to create the Latin American School for the Social Sciences (FLACSO) in Santiago. The university also created during the early 1960s a Center for Socioeconomic Studies, offering interdisciplinary research and training, while ECLA developed an institute for social and political planning (ILPES), headed by the Spanish sociologist Jose Medina Echevarria. FLACSO gave fellowships to a small group of students from all over the region for a two-year full-time master's program in sociology. The curriculum was heavily influenced by the North American graduate training programs, although the faculty was largely European. A degree in political science was added a few years later. As a new program supported by leading Latin American sociologists, it was able to recruit bright and relatively experienced students. Upon return to their countries, often after pursuing doctoral studies in the United States or in Europe, they were responsible for the large-scale renovation of the social sciences in the region during the following decades.

The intellectual environment of the 1960s and 1970s was strongly influenced by the Cuban revolution and the spread of guerrilla movements throughout Latin America, as well as by the political repression unleashed by many military takeovers during this period. The social sciences experienced both radicalization and repression. Marxism became a highly regarded theoretical outlook within the Latin American sociology and politics of development. The influence of the Italian Antonio Gramsci was felt earlier and more deeply in this region than elsewhere, with the exception of Italy, and it often blended with the dependency argument, which incorporated freely the notion of hegemony.[13] The structuralist version of Marxism provided by Louis Althusser and his French school was also widely incorporated into

---

[13] Jose Arico, *La cola del diablo: itinerario de Gramsci en America Latina* (Buenos Aires: Siglo XXI, 1988).

Latin American social science during the 1970s and early 1980s, although its influence proved to be short-lived.

During the 1960s, FLACSO and ECLA attracted many social scientists from Brazil and Argentina looking for academic or political exile, including Fernando H. Cardoso. His initial work, carried out within the research program on race relations of his teachers in Sao Paulo, Florestan Fernandes and Roger Bastide, focused on social mobility and color and on the history of slavery in southern Brazil. As a young faculty member at the University of Sao Paulo, he gathered together a group of young social scientists and philosophers around a seminar on Marx. Cardoso also organized a research group on industrial and labor sociology, in close relationship with the French sociologists Georges Friedmann and Alain Touraine. In 1963, he went to Paris to work on his study of Paulista industrial entrepreneurs. Upon his return, he started a comparative study of entrepreneurs and economic policies in Argentina, Mexico, and Brazil, but was interrupted by the military coup of April 1964, which forced him into exile in Chile. The overall Marxist framework, which predicted an active political role for the industrial bourgeoisie in Latin American capitalist development, was used with flexibility to understand the political coalitions that emerged in the three countries.[14]

During his stay in Chile, Cardoso attempted to integrate the various theories that had influenced him during the previous ten years. Collaborating with a young Chilean sociologist, Enzo Faletto, he produced the most influential sociological monograph in the region during those years. *Dependency and Development in Latin America*, published in Spanish in 1969 and translated into English in 1979, was an historically oriented essay inspired by ECLA's perspective and particularly by the work of the Brazilian economist Celso Furtado,[15] who was also in exile in Chile. Economic dependency within the capitalist system was made the paramount explanatory framework, but it incorporated a concern with the political process of economic policy making. Other Marxist scholars then in Chile, notably Andre Gunther Frank, also relied on the concept of dependency, and on the notion of underdevelopment as an historical process, to reject entirely ECLA's reformist economic policies.[16] Cardoso, who became president of the International Sociological Association before devoting himself to Brazilian politics and being elected president of his country in 1994, often acknowledged ECLA's influence in his ideas about Latin American development and underdevelopment.[17]

---

[14] Joseph A. Kahl, *Modernization, Exploitation and Dependency in Latin America: Germani, Gonzalez Casanova and Cardoso* (New Brunswick, N.J.: Transaction, 1976).

[15] Celso Furtado, *Development and Underdevelopment* (Berkeley: University of California Press, 1964).

[16] Andre Gunther Frank, *Capitalism and Underdevelopment in Latin America: Historical Studies of Chile and Brazil* (New York: Monthly Review Press, 1967).

[17] Fernando H. Cardoso, "The Originality of a Copy: CEPAL and the Idea of Development," *CEPAL Review*, 2 (1977), 7–40.

The sociology of development, often within the ECLA tradition, but also under the influence of Unites States modernization theorists, developed a variety of theoretical and political inclinations in Latin America. Gino Germani (1911–1979), a self-trained sociologist, was the most influential modernization theorist in the region. Born in Italy, Germani studied at the Business School of the University of Rome in the early 1930s. Anti-fascist activism then sent him to prison, where he met communist intellectuals. When he was paroled after one year, he attended lectures in history, literature, philosophy, psychology, and sociology; he read Pareto (officially recognized in fascist Italy) and also, by chance, Durkheim.[18] In 1934 he settled in Argentina, and in 1938 he enrolled in philosophy courses at the University of Buenos Aires. Although the teaching was boring and outdated, in the libraries he found more French sociology and a fine collection of recent American sociological literature, including Parsons's *The Structure of Social Action*, the *American Sociological Review*, and the *American Journal of Sociology*. When the sociology professor, the historian Ricardo Levene, decided to initiate studies on contemporary Argentina, Germani became his assistant and published several papers on the middle class in Buenos Aires. The university lost its autonomy to the military in 1945, however, and Germani and many faculty members lost their jobs. The government of Juan D. Peron (1946–55) exercised close political control of university life and had little use for academic social science.

During those years, Germani started a collection in sociology and social psychology, translating books by Erich Fromm, Harold Lasky, Karl Mannheim, and Raymond Aron, among others, and published a book in 1955, *Estructura social de la Argentina*, which established his reputation in sociology. The book approached Argentine social structure from a socio-demographic perspective, focusing on the changes foreign immigration and internal rural-to-urban migration had brought to Argentina and analyzing changes in family size and composition as a consequence of modernization. Chapters on social classes and electoral sociology showed his familiarity with contemporary American sociology.

The overthrow of Peronism allowed his return to the University of Buenos Aires, where he opened a department and started a degree program in sociology in 1957. The program's curriculum and its link between training and research were largely inspired by programs in the United States and received support from the Rockefeller and Ford Foundations. Young faculty members obtained fellowships to pursue graduate studies in European and American universities, or to FLACSO in Chile. Germani conducted a large-scale survey on immigration, social mobility, and urbanization in Buenos Aires, the first of its kind in Buenos Aires, and worked on a comparative project with other Latin American sociologists. At the same time, he produced several essays

---

[18] Kahl, *Modernization, Exploitation and Dependency*.

on modernization and development. His theoretical framework was largely drawn from the structural-functional approach to modernization of Talcott Parsons, Kingsley Davis, and Karl Deutsch, but, inspired by Erich Fromm and David Riesman, he was inclined to study social psychological rather than economic structural phenomena, in contrast with the predominant theories in the region. When he turned his attention to authoritarian trends in his native Italy, Argentina, and elsewhere, he attributed fascism and populism to rapid and uneven changes in social structure brought about by urbanization and modernization.

In Mexico, the renewal of the academic social sciences and humanities during the 1940s, after many years of political upheaval, benefited from the arrival of a large group of Spanish emigré intellectuals who formed the core of the Colegio de Mexico. Most had training in history or philosophy, but some, like Jose Medina Echeverria, identified themselves with the social sciences. They were responsible for the founding of a large-scale publishing house with government support, Fondo de Cultura Economica, which translated European classical and contemporary authors. The Bank of Mexico and, during the 1950s, ECLA's local office became sites for modern economic research. In the early 1960s, the Colegio started its first research and graduate training programs in economics and population studies, under the leadership of the economist Victor Urquidi. At the Universidad Nacional Autonoma de Mexico (UNAM), where teaching and research took place in separate institutions, Pablo Gonzalez Casanova became a leading figure. He had studied history in Mexico and then gone to Paris to work with sociologists Gurvitch and Friedmann, as well as with the historian Fernand Braudel. In Mexico, where the intellectual traditions in the social sciences were stronger in anthropology and history, Gonzalez Casanova developed a program in sociology at UNAM. In 1963, he published his influential book *Democracy in Mexico*, a study of social and cultural dualities inherited from colonialism that, still present in Mexican society, hindered democratic structures and political participation.

Gonzalez Casanova, along with Germani, had helped to found FLACSO in Chile and a center for comparative research in Rio de Janeiro, also under the sponsorship of UNESCO. In a similar vein, they helped to found in 1967 the Latin American Council for the Social Sciences (CLACSO), which became the major center for the building of research networks in the region. CLACSO best represented social science research outside of the public universities and governmental offices, which thrived in the precarious independent centers operating in countries under military regimes, such as Argentina, Brazil, and Chile.[19]

[19] Jose Joaquin Brunner and Alicia Barrios, *Inquisicion, mercado y filantropia: Ciencias sociales y autoritarismo en Argentina, Brasil, Chile y Uruguay* (Santiago: FLACSO, 1987).

THE END OF THE CENTURY: HIGHER EDUCATION
AND THEMATIC DIVERSIFICATION

The period after the 1980s saw a gradual end to the violent upheavals and authoritarian regimes of the previous decades. By the 1990s, all major countries had established constitutionally elected governments with relatively strong parliaments. The end of the Cold War favored peace agreements in the troubled Central American region, ravaged by war and revolution for twenty years. However, the economic stagnation of the 1980s, a politics of economic adjustment that deeply eroded weak local versions of the welfare state, and the failure of market-friendly policies to produce sustained economic growth or to redress extreme inequality tempered the enthusiasm for democracy.

Within this context, two trends emerged in the social sciences, first, their incorporation as university disciplines into expanding higher education systems, and second, their wide theoretical and thematic diversity. The intellectual agenda became considerably more modest as problem-driven, applied investigation dominated empirical research.

By 1950, there were only some seventy-five universities and just over a quarter-million students, or about a two percent enrollment rate, in Latin America. Between 1950 and 1994, the number of universities jumped from 75 to over 800, and enrollment from 2 to 19 percent of the population aged 20 to 24.[20] Although students tended to be part-time and many institutions had low standards, mass higher education was a real phenomenon. Undergraduate social science enrollments were significant, although a large proportion of students remained in business courses and law schools.

From the perspective of graduate programs and research support, the Brazilian system was by far the most advanced in the region after graduate training became a requirement for an academic career during the 1970s. Almost half of all students in the region enrolled in graduate programs were in Brazilian universities – although the overall higher education enrollment rate in Brazil was well below the median – and were supported by strong fellowship programs and generous research funding. Besides Brazil, only Mexico had a developed system of research and graduate training centers in the social sciences. Initially concentrated at the Colegio de Mexico and UNAM, the first national university, graduate programs spread throughout the country during the 1980s. In the rest of Latin America, this trend is more recent and public support for research and graduate training is weaker, although a network of FLACSO schools spread out from the Chilean base after the mid-1970s.

Among the disciplines, a focus on the state, absent since the turn-of-the-century constitutionalist debates, led to a renewal of political science.

---

[20] Carmen Garcia Guadilla, *Situacion y principales dinamicas de transformacion de la educacion superior en America Latina* (Caracas: CRESALC, 1996).

Corporatism and authoritarianism, the analysis of party systems and political representation, and centralism versus local and regional politics became hot topics for research. Above all, economic policy making, the role of interest groups, and the relative autonomy of the state became favorite themes of both historical and contemporary research. Mediated through area studies programs in American universities, strong links emerged in the study of these topics to democratic politics in Latin America, south and central Europe, and, more recently, Asia.[21]

Considerable strengthening and renewal also characterized cultural anthropology, which during the 1960s and 1970s had largely moved away from its initial focus on issues of ethnicity and race to the economy of peasant agriculture and, more recently, to the analysis of popular culture and multiculturalism. With increased access to a globalized mass media, cultural identity has attracted the attention of anthropologists and communication specialists, often placing them in the role of experts on cultural policies.

Sociology as a discipline had lost considerable attraction by the end of the century, as its ambitious development agenda dissipated and the dispersion among diverse theoretical orientations and substantive specialties inhibited a common framework. Economics, on the contrary, turning its back to the other social sciences, became a lucrative and prestigious profession and a basis for economic policy making and political power. With the growing hegemony of neoclassical theory, "monetarism" displaced "structuralism," although Keynes and the neo-Keynesians still had many followers.

With increased access to higher education, women became well represented in all of the social sciences except economics and were often active in the Latin American feminist movement. Beginning in the 1980s, greater awareness of the role of gender in society and politics led to considerable revision of the analysis of traditional topics and to the emergence of new ones. During the same decades, the erosion of national states' power, accelerated in Latin America by the illegal drug trade,[22] allowed greater room for local and ethnic identities to exercise an influence in culture and politics and turned the attention of social scientists to ethnic, racial, and religious movements.

LATIN AMERICAN SOCIAL SCIENCES IN A GLOBALIZED WORLD

The Latin American social sciences initially emerged within the political context of newly created national societies in a world dominated by imperial

---

[21] Arturo Valenzuela, "Political Science and the Study of Latin America," in *Changing Perspectives in Latin American Studies: Insights from Six Disciplines*, ed. Christopher Mitchell (Stanford, Calif.: Stanford University Press, 1988).

[22] Manuel Castells, *End of Millennium* (Oxford: Blackwell, 1998).

powers and expanding world markets. Unlike other regions on the periphery of this world economic system, the Latin American countries had for the most part been free from direct colonial domination since the nineteenth century. In contrast, no other region of the world was so broadly open to Western intellectual influence, having in fact no major alternative intellectual tradition that was not Western in origin, even if adapted to local realities. During this initial stage, national societies and the building of new nations were important foci of intellectual concern, and the social sciences in Latin America attempted to use the available concepts and theoretical frameworks to analyze what was holding back, or promoting, the emergence of modern nations in the region.

This context changed radically with the breakdown of world markets during the 1930s and the relative closing of the Latin American economies. The ideology and policies of import substitution industrialization strengthened nationalism and corporatism, and placed a premium on an autonomous intellectual development in the social sciences despite the continuous flow of ideas from Europe and the United States. This was an inward-looking phase during which Latin American intellectuals adapted and revised Western concepts and theories in an effort to produce their own autonomous version of the national and regional realities. The social sciences were often successful in producing an original outlook. The key products they were concerned with, however, were ideas and instruments for use by the national states in gaining increased autonomy from the centrifugal forces of local power holders and segmented interests within, or from the imperialist powers and multinational corporations outside.

The ideological and theoretical breakdown that accompanied the end of the Cold War has brought to Latin America, as to the rest of the Western world, considerable confusion about the sense and mission of the social sciences in contemporary societies. Epistemological nihilism has gone hand in hand with narrow specialization and blind market-driven research programs in the social sciences. Yet, there is in place now an academic community with roots in large-scale higher education systems at the national and international levels, within a region that still enjoys the advantages of common languages and a sense of a shared culture, even if we find within Latin America much of the diversity and heterogeneity of the social sciences worldwide.

24

PSYCHOLOGY IN RUSSIA AND CENTRAL AND EASTERN EUROPE

Jaromír Janoušek and Irina Sirotkina

If we assume, as many scholars do, that psychology is mainly European in origin and appeared as a special scientific discipline during the second half of the nineteenth century,[1] we may ask how specific national contexts have affected the discipline's development. Have the particular social and cultural contexts of Russia and central and eastern Europe, for instance, fostered any original developments in psychology? Can we say, as one historian has recently claimed, that there was a specific "Russian way" in the human sciences?[2] What are the results for the human sciences of the "communist experiment" in the Soviet Union and other socialist countries? Did the Iron Curtain that separated the communist countries from the West for several decades create a "splendid isolation," in which psychology proliferated unencumbered by controversies that surrounded the discipline elsewhere? Or did this situation produce a cultural and academic enclave that fostered parochial tendencies and hindered the discipline's development?

Unambiguous answers to these questions are impossible to provide. On the one hand, in Russia after 1917 and in some eastern and central European countries after World War II, political pressures gave a particular twist to local psychological research. On the other hand, there were many conceptual developments in these countries that deserve close attention. There also were intense theoretical debates, of relevance for Western psychology, on questions of psychology's relationship to philosophy and to physiology, the interaction between psychological theory and practice, and the remodeling of human nature.

The variety of geographical, historical, and political conditions is so great that we cannot consider this part of the world homogenous, and any

---

[1] Roger Smith, *The Fontana History of the Human Sciences* (London: Fontana Press, 1997) (in the United States, *The Norton History of the Human Sciences*); William R. Woodward and Mitchell G. Ash, eds., *The Problematic Science: Psychology in Nineteenth-Century Thought* (New York: Praeger, 1982).

[2] Mikhail G. Yaroshevsky, *Nauka o povedenii: Russkii put'* [The science of behavior: The Russian way] (Moscow: Akademiia pedagogicheskikh i social'nykh nauk, 1996).

treatment of the central and eastern European countries and Russia as one region will be politically controversial. Yet it will serve the chapter's purpose to reflect on how one historical moment, one common to all of the countries of the region – the period under communist governments – influenced the direction taken by the human sciences.

RUSSIAN PSYCHOLOGY BETWEEN NEUROPHYSIOLOGY AND THE HUMANITIES

During the early 1860s, with the abolition of serfdom and other political reforms, Russia began the process of modernization. Many people believed that science was the key to this process. The natural sciences, especially, attracted many young Russians, and the 1860s and 1870s produced a cohort of successful scientists, such as the chemists Dmitry Ivanovich Mendeleev and Aleksandr Mikhailovich Butlerov, the biologists Ivan Petrovich Pavlov and Ilya Ilich Mechnikov, and the mathematician Sofia Kovalevskaia. Advances in the natural sciences affected the understanding of human nature, which these scientists viewed as being determined by brain and environment. The generation of the 1860s embraced the radical materialism of Ludwig Büchner, Karl Vogt, and Jakob Moleschott; one may recall the character in Fyodor Dostoevsky's novel *The Devils* who erected an altar using the books of this Unholy Trinity instead of the Bible.[3] Later scientists were more cautious in their views on science and politics.

Modernization revived the old confrontation between the Westernizers, who accentuated everything that was common to Europe and Russia, and the Slavophiles, who stressed Russia's distinctiveness, which they rooted in such institutions as the Orthodox Church and the *obshchina*, a village community with shared land. The mind–body controversy and the problem of free will became facets of intense public debates. In 1861, the discussion started with two publications, the translation of G. H Lewes's *The Physiology of Common Life* (two volumes, 1859–60) and the article "The Process of Life," by the young radical critic Dmitrii Ivanovich Pisarev (1840–1868). The latter argued for the reduction of psychological phenomena to more elementary material processes. These publications were followed by a series of articles by Pamfil Danilovich Yurkevich (1826–1874), a professor of theology and Slavophile, who opposed the materialist approach to psychology.

It is not surprising that in this context the physiological work of Ivan Mikhailovich Sechenov (1829–1905) evoked a lively response.[4] After graduating from Moscow University, Sechenov had gone abroad and worked in the

---

[3] Fyodor M. Dostoevsky, *The Devils*, part 2, chap. 6, sect. 2.
[4] Edwin G. Boring, *A History of Experimental Psychology* (New York: Appleton Century Crofts, 1950), p. 636; Mikhail G. Yaroshevsky, *Ivan Mikhailovich Sechenov, 1829–1905* (Leningrad: Nauka, 1968).

laboratories of Hermann Helmholtz, Karl Ludwig, and Claude Bernard. Experimenting with the reflexes of the frog, he believed that he had discovered the center, in the thalamus area of the brain, that inhibits motor activity. This, Sechenov hoped, would provide the much-sought-for explanation of voluntary acts, that is, actions controlled by the organism and not just triggered by a stimulus.[5] In 1863, he published an essay, "Reflexes of the Brain," intended for a wider audience; it was provocatively subtitled "An Attempt to Introduce Physiological Foundations for Psychic Processes." The brain functioned, he explained, to combine all phases of the reflex: the beginnings (stimuli), the central processes, and the end (external action). The central part of the reflex is what is known as voluntary processes and thought; he described the latter as a "reflex with a delayed end." Sechenov later abandoned his experimental research on inhibition, but in his *Elements of Mind* (1878) he continued to adhere to the global concept of the reflex as a three-part explanation of mind.

Sechenov's polemical essay had a lasting effect. Immediately after publication, it was criticized by the a liberal professor of law, Konstantin Dmitirievich Kavelin (1818–1885), who opposed Sechenov's limited view of mind and his exclusion from psychological methods of introspection and the study of cultural artifacts. Later, Nikolai Iakovlevich Grot (1852–1899) and Georgii Ivanovich Chelpanov (1862–1936), followers of the German psychologist Wilhelm Wundt (1832–1920), also opposed the reflex theory of voluntary acts. They explained will as a subjective movement of consciousness, an expression of special mental energy and of the self, and connected it with the mental causality of motives and acts and with creativity.[6] Grot, the editor of the first Russian psychological journal, *Voprosy filosofii i psikhologii* (Questions of philosophy and psychology, 1889–1918), opened the journal's first issue by discussing free will and determinism. Chelpanov, who taught at Kiev and Moscow, developed a powerful critique of the materialist approach to psychology in his book *Brain and Mind* (which appeared in numerous editions between its first publication in 1900 and the Revolution of 1917).

The divide between materialists and their opponents was in part a consequence of academic positions. The former, as a rule, had backgrounds in science and medicine and worked in medical schools; the latter held positions in humanities departments (called "history and philology departments"), which was where philosophy traditionally belonged in Russian universities. There was, however, one more factor contributing to the schism: In pre-Revolutionary Russia, the materialist approach was firmly associated with the radical political opposition, while the alternative views were often

[5] Roger Smith, *Inhibition: History and Meaning in the Sciences of Mind and Brain* (London: Free Association Books and Berkeley: University of California Press, 1992), chap. 3.

[6] Elena A. Budilova, *Bor'ba materializma i idealizma v russkoi psikhologicheskoi nauke, vtoraia polovina XIX – nachalo XX v.* [The struggle between materialism and idealism in Russian psychological science, late nineteenth and early twentieth centuries] (Moscow: Izdatelstvo Akademii nauk SSSR, 1960), chap. 8.

characteristic of a more reformist, and sometimes conservative, part of the intelligentsia. Sechenov was a typical case: He quarreled with the university officials, his work was censored, and radical and liberal public opinion made him a national hero. It became de rigeur for all Russian physiologists, and later psychologists, to locate themselves as members of "Sechenov's school."[7]

Though not his students directly, Vladimir Mikhailovich Bekhterev (1857–1927) and Ivan Petrovich Pavlov (1849–1936) also traced their genealogy to Sechenov, at least on ceremonial occasions. Both studied in Russia and Germany, and both acquired professional status working in the classical areas of physiology and neurology. Pavlov received the Nobel Prize (1904) for his research on the function of the digestive glands; Bekhterev was a neurologist, the first to give a clinical explanation of what became known as Bekhterev's disease, or spondylarthritis. Both then began to study the relation between the organism and the environment. The crucial phenomenon, the learned reaction, was termed by Pavlov the conditioned reflex; Bekhterev called it the associative reflex. At the outset, both believed in the relevance of their work to psychology. Pavlov occasionally called his work "experimental psychology"; Bekhterev used the term "objective psychology." Both changed their terminology later – Bekhterev came to prefer "reflexology," Pavlov the "doctrine of higher nervous activity." Although Bekhterev was an active public figure and Pavlov rather avoided public life, both were highly influential among the liberal intelligentsia.[8]

Although courses in psychology were traditionally taught in humanities departments, the first psychological laboratories and courses on experimental psychology were introduced in medical schools. The first laboratory was founded by Bekhterev in Kazan in 1885. In 1895, the director of the Moscow University Psychiatric Clinic, Sergei Sergeevich Korsakov (1854–1900), sponsored a laboratory headed by his assistant, the psychiatrist Ardalion Ardalionovich Tokarskii (1859–1901). In Odessa in 1896, Nikolai Nikolaevich Lange (1858–1921) founded the first laboratory in a humanities department. The climax of the institutionalization of psychology as part of the humanities was reached with the founding of the Psychological Institute at Moscow University in 1912.[9]

Despite the dominance of the materialist approach, which became overwhelming during the Soviet years, there was always some resistance to the

[7] Mikhail G. Yaroshevsky, "The Logic of Scientific Development and the Scientific School: The Example of Ivan Mikhailovich Sechenov," in *The Problematic Science*, ed. Woodward and Ash, pp. 231–54.

[8] David Joravsky, *Russian Psychology: A Critical History* (Oxford: Blackwell, 1989), chaps. 3, 5; Daniel P. Todes, "Pavlov's Physiology Factory," *Isis*, 88 (1997), 205–46.

[9] Vladimir Umrikhin, "Russian and World Psychology: A Common Origin of Divergent Paths," in *Psychology in Russia: Past, Present, Future*, ed. Elena L. Grigorenko, Patricia Ruzgis, and Robert J. Sternberg (Commack, N.Y.: Nova Science Publishers, 1997), pp. 17–38.

rigid determinism characteristic of mid-nineteenth-century mechanistic materialism. Beginning with Sechenov, physiologists attempted to make the concept of the reflex more sophisticated, able to accommodate spontaneity and will. Sechenov "discovered" the centers of inhibition; Pavlov speculated about the "reflex of freedom" and the "reflex of purpose"; Bekhterev, notorious for his eclecticism, described political events in terms of "collective reflexes." A little later, the physiologist Aleksei Alekseevich Ukhtomskii (1875–1942) introduced the concept of "the dominant" – the part of the brain that at each instant "dominates" all other parts. Nikolai Aleksandrovich Bernstein (1896–1966) overtly criticized the concept of the conditioned reflex as inappropriate for describing human actions and called for a new "physiology of activity" to replace Pavlov's simplistic schemas.[10]

PSYCHOLOGY AND SOCIETY

The oppressive tsarist regime in Russia curtailed the rights of nearly every social group, and certainly those of the liberal professions. Even those who were willing to cooperate with the state felt that, given the political framework, their work could not be efficient. Physicians in the civil service, for instance, blamed the regime for keeping the majority of the population in poverty and thereby hindering any effort to provide public health care. At the turn of the century, the most radical of them declared that the only way to heal the Russian people was to overthrow the monarchy. Psychiatrists claimed that lack of political freedom was conducive to nervous diseases, and they especially insisted that a genuine social hygiene start with political changes.[11] This was an important reason why the majority of practitioners in both medicine and education embraced the socialist revolution and offered their service to the communist state.

The Russian debates on education, as well as public health, stemmed from the ideas of the *narodniki* (the Populists, a 1860s and 1870s movement of the intelligentsia devoted to educating workers and peasants). They were influenced by moral leaders such as Lev Nikolaevich Tolstoy (1828–1910), who opened a school for peasants' children on his estate, taught there himself, and wrote textbooks for his pupils. The pedagogical movement also benefited from women's emancipation in Russia; education became a sphere in which many talented women attempted to fulfill themselves. Psychologists,

[10] Alex Kozulin, *Psychology in Utopia: Toward a Social History of Soviet Psychology* (Cambridge, Mass.: Harvard University Press, 1984), chap. 3; Irina Sirotkina, "N. A. Bernstein: The Years before and after the 'Pavlov Session'," *Russian Studies in History*, 34 (1995), 24–36.

[11] Nancy Mandelker Frieden, *Russian Physicians in an Era of Reform and Revolution, 1856–1905* (Princeton, N.J.: Princeton University Press, 1981); Julie Vail Brown, "The Professionalization of Russian Psychiatry" (PhD dissertation, University of Pennsylvania, 1981); Irina Sirotkina, *Diagnosing Literary Genius: A Cultural History of Russian Psychiatry* (Baltimore: Johns Hopkins University Press, 2002), chap. 5.

philosophers, and physicians, who contributed to this movement, were sensitive to Western innovations, including pedology and German experimental pedagogy. The latter gave the title to six congresses organized by Russian specialists in education and child psychology between 1906 and 1917. In 1901, the psychologist Alexander Petrovich Nechaev (1870–1943) founded the first laboratory of experimental pedagogy, which included a broad program of experiments by both professional psychologists and schoolteachers. In 1906, the Moscow Pedagogic Union announced a course on pedology; two years later, Bekhterev founded the Pedological Institute in St. Petersburg for day-to-day observation and study of institutionalized children from their birth up to the age of three.[12]

These developments continued after 1917. In the early Soviet years, there was an overriding concern to improve public education, to eradicate illiteracy, which was extremely widespread, and a drive to raise new generations of children free of the taint of pre-revolutionary "bourgeois" values and ideals. The ideal of the "new man" became a real goal toward which people worked. Although colored by Soviet rhetoric, the program was not an unusual piece of educational thinking. At the end of the nineteenth century, the psychologist James McKeen Cattell had remarked that in the United States more than $150 million is spent annually on public schools in the attempt to "change human nature."[13] Throughout the 1920s, developmental psychology and educational research were certainly encouraged by the active interest of Lenin's wife, Nadezhda Konstantinovna Krupskaia (1869–1939). Under the government's auspices, Soviet pedologists started a special journal, founded various institutions, including the Central Pedological Laboratory, organized congresses, and introduced the position of pedologist into the schools. Such leading professionals as Lev Semenovich Vygotsky (1896–1934), Pavel Petrovich Blonsky (1884–1941), Mikhail Iakovlevich Basov (1892–1931), and Aaron Borisovich Zalkind (1888–1936) played active roles in the movement.[14]

In his essay "The Historical Meaning of the Crisis in Psychology" (written in 1927, published in 1982), Vygotsky suggested that the way out of the crisis that psychologists everywhere believed they were experiencing would be through practice. He hoped that new areas like pedology, medical psychology, and psychotechnics would eliminate debates between materialist and idealist psychology, restore the discipline's unity, and help it to regain the status of a natural science. Psychotechnicians, who promised both to increase work efficiency and to look after workers' well-being, also received state support.

---

[12] Artur V. Petrovskii, *Voprosy istorii i teorii psikhologii. Izbrannye trudy* [Problems in the history and theory of psychology: Selected works] (Moscow: Pedagogika, 1984), pp. 42–6.

[13] Quoted in John M. O'Donnell, *The Origins of Behaviorism: American Psychology, 1870–1920* (New York: New York University Press, 1985), p. 153.

[14] René van der Veer and Jaan Valsiner, *Understanding Vygotsky: A Quest for Synthesis* (Oxford: Blackwell, 1991), p. 294; Alexander Etkind, *Eros of the Impossible: The History of Psychoanalysis in Russia* (Boulder, Colo.: Westview Press, 1997), chap. 8.

The Central Institute of Labor was founded in 1921 to study vocational selection, vocational orientation, fatigue, methods of professional training, and the organization of labor processes; as in pedology, a substantial effort went into psychological testing. The leading psychotechnicians, Isaak Naftulovich Shpil'rein (1891–1937?), Aleksei Kapitonovich Gastev (1882–1941?), and Solomon Grigorievich Gellershtein (1896–1967), gained international recognition; in 1931, the International Psychotechnical Congress was held in Moscow.[15]

In 1929, many psychological initiatives were brought to a halt by the politics of the "Great Break" – an offensive to overcome the backwardness of the country and to build a socialist society under conditions of capitalist encirclement, accompanied by political repression. Pedology, viewed as a threat by schoolteachers, was criticized for misuse of tests and closed down by a Communist Party decree in July 1936. Between 1934 and 1936, psychotechnics lost its journal and its research centers. Many psychologists, as well as other intellectuals, perished in the Gulag.[16] At the same time, tactical compromises with the ruling ideology and a shift to topics potentially less exposed to criticism sustained the discipline. During World War II, ideological pressure slightly receded. Psychologists worked on rehabilitation of brain and motor functions in wounded soldiers, studied cognitive processes under combat conditions, and generally succeeded in maintaining teaching and research institutions, most of which were relocated to remote parts of the country.[17]

REMAKING HUMAN NATURE

Although the myth of rebirth is probably as old as humankind itself, the first systematic and state-supported effort to change human nature took place in Russia after 1917. Even before that date, Russian intellectuals had shared the Enlightenment belief in the goodness of human nature when not spoiled by bad education and unfortunate conditions of life. Their accounts of human problems and failures often ended with a reference to "environment" or "milieu" (*sreda*) – a euphemism for unjust social conditions, poverty, and an oppressive regime. Like their Western counterparts, liberal-minded Russians invested their hopes for the betterment of humanity in the reformation of society.

This gave a distinctive color to the discussion of biological and social factors in human development, that is, of inherited and acquired characteristics. During the late nineteenth century, the majority of Russian intellectuals

---

[15] Kozulin, *Psychology in Utopia*, pp. 15–16.

[16] Mikhail G. Yaroshevsky, ed., *Repressirovannaia nauka* [The oppressed science], 2 vols. (Leningrad/Petersburg: Nauka, 1991–4).

[17] Andrei V. Brushlinskii, ed., *Psikhologicheskaia nauka v Rossii XX stoletiia* [Psychological science in twentieth-century Russia] (Moscow: Institut psikhologii, 1997), chap. 3.

vehemently opposed Cesare Lombroso's notion of the inborn criminal type. Marxism – a version of social philosophy – also favored social explanations. When, after 1917, Marxism was either voluntarily accepted or imposed on scholars in the human sciences, the view that human behavior is primarily socially determined became official. By the end of the 1920s, as one historian has noted, this position "was accepted at least on a *pro forma* basis by the overwhelming majority of Soviet psychologists, including the biologically oriented reflexologists."[18] Even biological theories were assessed in social terms. Like Bekhterev's reflexology, Pavlov's theory of conditioned reflexes was primarily perceived as a theory of learning in the service of the new society. In 1923, one of the Communist Party spokesmen, Nikolai Ivanovich Bukharin (1888–1938), called Pavlov's theory a "tool out of the iron toolkit of the materialist ideology."[19]

The other side of the coin was that the revolutionaries viewed human nature as endlessly pliable and, finally, passive – subject to unrestrained management and manipulation. The view took a grotesque shape in the works of Emmanuil Semenovich Enchmen (1891–1966), a Communist Party functionary who studied at Bekhterev's Psychoneurological Institute in Petersburg before World War I. During the early 1920s, Enchmen expected a "revolution in human nature" and suggested "physiological passports" for monitoring changes in the organism.[20] However utopian, his ideas were reminiscent of Western eugenics and, for instance, of Ernst Rüdin's project to monitor the population by completing a "*Psycho-biogram*" for each individual.[21] Yet Marxist philosophers, who looked with suspicion on any reference to biology that was not balanced by a social explanation, severely criticized Enchmen.

The fact that in the Soviet Union social determinism became an officially approved position both caused intellectual problems and led to political consequences. It was taken for granted that in a socialist society people would, in many ways, become better; psychologists, teachers, and medical and social practitioners were all expected to contribute to the improvement. But the question of biology's place in the process was a potentially dangerous one. By the mid-1930s, the biological account of human nature had become more and more unacceptable; as one historian of Soviet genetics notes, "no field

---

[18] Raymond A. Bauer, *The New Man in Soviet Psychology* (Cambridge, Mass.: Harvard University Press, 1952), p. 83.

[19] Quoted in Sergei A. Bogdanchikov, *Proiskhozhdenie marksistskoi psikhologii: Diskussiia mezhdu K. N. Kornilovym i G. I. Chelpanovym* [The origins of Marxist psychology: The discussion between K. N. Kornilov and G. I. Chelpanov] (Saratov: Saratovskii iuridicheskii universitet, 2000), p. 7. See also Daniel P. Todes, "Pavlov and the Bolsheviks," *History and Philosophy of the Life Sciences*, 17 (1995), 379–418.

[20] David Joravsky, *Soviet Marxism and Natural Science, 1917–1932* (New York: Columbia University Press, 1961), pp. 93–7; George Windholz, "Emmanuil Enchmen – A Soviet Behaviorist and the Commonality of *Zeitgest*," *The Psychological Record*, 45 (1995), 517–33.

[21] Paul Weindling, *Health, Race and German Politics between National Unification and Nazism, 1870–1945* (Cambridge: Cambridge University Press, 1989), pp. 384–5.

that linked the biological and the social survived the Great Break intact."[22] The public polemics around the idea that nature can be altered at will, an idea defended by Trofim Denisovich Lysenko (1898–1976), greatly complicated the issue.[23] This was particularly reflected in the later nature/nurture debates as they developed in the Soviet Union in the 1960s and 1970s. As one historian of Russian science remarks, because the proponents of "nature" suffered from political repression, independent-minded intellectuals considered them the "good guys" (anti-Stalinists, anti-dogmatics), while regarding their opponents as the "bad guys" (Party functionaries, pro-Lysenkoists).[24]

In psychology, however, understanding human nature to be socially and culturally determined did not seem so objectionable when it was supported by such leaders of the discipline as Vygotsky, Sergei Leonidovich Rubinstein, and Aleksei Nikolaevich Leontiev, whose integrity was not in question. Their conceptions, which existed side by side with Pavlov's theory, left space for research on memory by Anatolii Aleksandrovich Smirnov (1894–1980) and on individual differences by Boris Mikhailovich Teplov (1896–1965), Vladimir Dmitrievich Nebylitsyn (1930–1972), and Inna Vladimirovna Ravich-Shcherbo (b. 1927).[25] With the decline of communist ideology and the collapse of the Soviet Union in 1991, the idea of the New Soviet Man became a target for irony. All the same, the view that human nature is a product of society and culture, in the wider senses of these words, still stands and is widely shared by psychologists. And the belief that the psychology of individuals might be improved and regulated remains at the core of the psychological enterprise.[26]

PSYCHOLOGICAL THEORY AND MARXISM

From the beginning, psychology has been a contentions science. It constituted itself by negotiating boundaries with philosophy and physiology, and psychology in Russia was no exception. At the beginning of the twentieth century, Chelpanov, one of the leading academic psychologists, was as active in discussing these issues as he was in establishing laboratories. His inaugural lecture at Moscow University in 1907 was "On the Relation of Psychology

---

22 Mark B. Adams, "Eugenics as Social Medicine in Revolutionary Russia," in *Health and Society in Revolutionary Russia*, ed. Susan Gross Solomon and John F. Hutchinson (Bloomington: Indiana University Press, 1990), pp. 218–19.

23 David Joravsky, *The Lysenko Affair* (Chicago: University of Chicago Press, 1986); Nikolai Krementsov, *Stalinist Science* (Princeton, N.J.: Princeton University Press, 1997).

24 Loren R. Graham, *Science, Philosophy and Human Behavior in the Soviet Union* (New York: Columbia University Press, 1987), chap. 6.

25 Elena L. Grigorenko and Inna V. Ravich-Shcherbo, "Russian Psychogenetics: Sketches for the Portrait," in *Psychology in Russia*, ed. Grigorenko, Ruzqis, and Sternberg, pp. 83–121.

26 Michel Foucault, "Technologies of the Self," in *Technologies of the Self*, ed. Luther H. Martin, Huck Gutman, and Patrick H. Hutton (London: Tavistock, 1988); Nikolas Rose, *Governing the Self: The Shaping of the Private Self* (London: Free Association Books, 1999).

to Philosophy." Marxist debates, which later deeply affected psychology in the Soviet Union, started as part of such methodological discussions. Early twentieth-century psychology in the West was marked by the proliferation of psychological theories; it appeared to contemporaries that Marxism might suggest one more paradigm.

Even before the Revolution, Marxism, as a modern form of Westernism, had conquered the minds of many Russian intellectuals, although some of them, such as the philosophers Nikolai Aleksandrovich Berdiaev (1874–1948) and Sergei Nikolaevich Bulgakov (1871–1944), later became its ardent opponents. Some Marxist thinkers, including Georgii Vasil'evich Plekhanov (1856–1918) and Vladimir Ilyich Lenin (1870–1924), wrote on such issues as class consciousness, social moods, and the role of personalities in history, though their political priorities hindered a scholarly dialogue. Social psychology, a subject that Marxist sociology could potentially influence, developed independent of Marxism. Bekhterev wrote on mass suggestion and crowd psychology; the Russian and Polish professor of law Lev Iosifovich Petrazhitskii (1867–1931) gave a psychological interpretation of law; and the Russian and Ukrainian philologist Alexander Afanas'evich Potebnia (1835–1891) distinguished between the individual act of speech and language as an historical phenomenon.[27] At the beginning of the twentieth century, mass movements, wars, and revolutions supplied psychology with an abundance of material. In his *Collective Reflexology* (1921), Bekhterev attempted to explain such events in terms of associated reflexes; his social psychology allegedly stimulated later experiments with groups.[28] All the same, with the exception of the lawyer Mikhail Andreevich Reisner (1868–1928), no representative of psychology took part in promoting Marxism as a scientific methodology before 1917.[29]

The revolution had a varied impact upon the behavioral and social sciences. Because Marxism included its own philosophy and sociology, space for the development of non-Marxist philosophy and sociology quickly disappeared; during the early 1920s, leading representatives of these disciplines either emigrated or were banished from the country. At the same time, party spokesmen such as Lenin, Bukharin, and Anatolii Vasil'evich Lunacharsky (1875–1933) proclaimed that the new society should borrow some traditions from the past. This stance allowed for the development of the psychological sciences so long as scientists maintained a positive attitude toward the Soviet regime. Facing demands to take Marxism into account, academic psychologists at

[27] Elena A. Budilova, *Sotsial'no-psikhologicheskie problemy v russkoi nauke* [Problems of social psychology in Russian science] (Moscow: Nauka, 1983).

[28] Gordon W. Allport, "The Historical Background of Modern Social Pychology," in *The Handbook of Social Psychology*, vol. 1, ed. Gardner Lindzey and Elliot Aronson (Reading, Mass.: Addison-Wesley, 1968), p. 65.

[29] Mikhail A. Reisner, *Teoriia L. I. Petrazhitskogo, marksizm i sotsial'naia ideologiia* [L. I. Petrazhitsky's theory, Marxism, and the social ideology] (St. Petersburg: Vol'f, 1909).

first hoped to compromise by allowing Marxism into social psychology while keeping what they considered to be ideology out of the rest of the discipline. The main spokesman for this view was Chelpanov, at that time still director of the Moscow Psychological Institute. By contrast, the younger generation of psychologists enthusiastically accepted the idea of rebuilding their science on a Marxist foundation. In part, the struggle was a matter of competing for limited funds: A group of younger psychologists, some of them Chelpanov's students, sought to control the well-equipped Moscow Institute. In 1923, Chelpanov was forced to leave, displaced by Konstantin Nikolaevich Kornilov (1879–1957). Kornilov claimed that he was constructing a genuinely Marxist psychology under the name of "reactology"; he used the notion of reaction to reconcile consciousness with reflexes. His synthesis was, however, unsuccessful, though the new staff believed, according to one participant, that by shifting furniture and equipment in the old building they had made psychology genuinely scientific.[30]

Another reason that Kornilov, as well as other psychologists after him, embraced dialectical materialism was that they saw it as a "middle way" between physiological conceptions of the organism and introspective psychology.[31] During the 1920s, when what was understood to be scientific psychology was limited mainly to the study of reflexes and behavior, Marxist debates were instrumental in bringing back concepts of consciousness, motives, and personality. Similarly, in the 1950s, after the United Session of the Academy of Sciences and the Academy of Medicine (1950) elevated Pavlov's theory of higher nervous activity into a model for the human sciences, psychologists succeeded in defending their discipline's independence by manipulating philosophical issues.

At times, Marxist reconstruction assumed ugly forms. Thus, during the 1920s and 1930s participants often used reassessments of "bourgeois" conceptions for purposes distant from scholarly discussion; associating someone with an officially criticized psychoanalysis or *Gestaltpsychologie* could result in a loss of academic position and political repression. After Stalin died in 1953, and Stalinism was officially criticized at the Twentieth Communist Party Congress (1956), previously banned areas of psychology experienced a revival. *The Problems of the "Unconscious"* (1968), by Filipp Veniaminovich Bassin (1905–1992), returned to psychoanalytic themes. A decade later, in 1979, an international symposium on the unconscious gathered in Tbilisi, Georgia, the hometown of Dmitrii Nikolaevich Uznadze (1886–1950), whose theory of *ustanovka* (set, or unconscious attitude) was considered a "Soviet alternative to Freud."[32] Another sign of the changing political situation was

[30] Aleksandr R. Luria, *The Making of Mind: A Personal Account of Soviet Psychology*, ed. Michael Cole and Sheila Cole (Cambridge: Cambridge University Press, 1979).

[31] Graham, *Science, Philosophy, and Human Behavior*, chap. 5.

[32] Ibid., the section "Soviet Freudism."

the return of social psychology, previously stigmatized as a bourgeois psychologization of social phenomena.

The Eighteenth International Psychological Congress took place in Moscow in 1966, providing an opportunity for Soviet psychologists to advertise their research, which had been stimulated at least in part by their dialogue with Marxism. The West discovered the work of Leontiev, Alexander Romanovich Luria, Rubinshtein, Smirnov, Teplov, and Vygotsky. As mentioned, the Marxist discourse had helped their research to go beyond reflex theory and to deepen the philosophical discussion, while remaining on materialist grounds. Soviet psychologists constructively used at least two Marxist ideas containing elements of classical German philosophy – one was the significance of practice (practical activity) for the development of consciousness, and the other, social relations as the essence of human being. We will briefly mention four theoreticians.

Sergei Leonidovich Rubinshtein (1889–1960) studied neo-Kantian philosophy in Marburg before World War I. After his return to Odessa, he taught philosophy at the secondary school level and then at the university. There he published a fragment of his manuscript "The Principle of Creative Self-Activity" (1922), in which he argued that individuals constitute themselves through acts of creative and spontaneous activity. In 1930, Rubinshtein accepted a prominent position in Leningrad, where he wrote what then became a standard textbook, *The Foundations of Psychology* (1935). In his article "Psychological Problems in the Works of Karl Marx" (1934), he discussed the Hegelian/Marxian concept of practice, on which he based his theory of the unity of personality, consciousness, and activity. In 1942, his *Foundations of General Psychology* (an enlarged edition of the 1935 work) received the Stalin Prize, and Rubinshtein was appointed director of the Moscow Psychological Institute. In the late 1940s, during Stalin's xenophobic campaign against "cosmopolitanism," Rubinshtein was dismissed from his position; he returned to academia in 1956, not long before his death.[33]

Vygotsky became known as the author of the so-called sociocultural theory of mental development. Influenced in part by the Marxist concept of human nature as a social phenomenon, he claimed that technical and psychological tools and social relations mediate mental operations. His genetic law of cultural development declared that, in the course of the child's growth, each psychological function appears twice, first on the social plane among people, and then on the psychological plane within the child. His rich monograph *Thought and Language* appeared shortly after his premature death from tuberculosis in 1934. Vygotsky's sociocultural theory had a strong impact on

[33] Ksenia A. Abul'khanova and Andrei V. Brushlinskii, *Filosofsko-psikhologicheskaia kontseptsiia S. L. Rubinshteina* [S. L. Rubinshtein's conception of philosophical psychology] (Moscow: Nauka, 1989); Graham, *Science, Philosophy, and Human Behavior*, pp. 176–84.

various areas of psychology; international recognition, however, came to him only much later, after 1956, when his works began to be republished.[34]

Vygotsky's younger colleague and friend Aleksandr Romanovich Luria (1902–1977) was originally interested in psychoanalysis; his early study of emotional reactions appeared in the United States as *The Nature of Human Conflicts* (1932). During the 1930s, Luria and his team studied the development of cognitive processes under changing cultural and social conditions in Soviet Middle Asia, but the results of this study were officially criticized and their publication was delayed for decades. Luria then moved sideways to work in neuropsychology; his research on the localization of psychological functions in the brain received international recognition.[35]

Another close colleague of Vygotsky's, Aleksei Nikolaevich Leontiev (1903–1979), explored the specific forms of human memory in *The Development of Memory* (1931). Memory, he argued, is mediated by psychological tools, such as specially created signs. During the early 1930s, shortly after the Great Break, Leontiev and his team moved to Kharkov in the Ukraine, which was distant from the capital and hence less exposed politically. There he reoriented his research toward the analysis of human activity in both its mental and practical forms. The concept of activity, relevant to the Marxist notion of practice, was relatively safe from ideological criticism; at the same time, it stimulated a long series of fruitful research efforts by Leontiev and his school.[36]

At a very early stage, scholars in the Soviet Union became disillusioned with earlier beliefs that psychology is independent of its social and political context. In Leontiev's words, they became aware that "in the contemporary world, psychology fulfills an ideological function and serves class interests; to disregard it is impossible."[37] Soviet psychologists therefore did not need to "deconstruct" the myth of science in order to argue that knowledge depends on the context in which it is produced. But they also learned from experience, as in the Lysenko Affair, that blatant social constructions cannot be enduring. This, some scholars believe, is the lesson of the Soviet experiment for science: While constructing their discipline, psychologists discovered both the pliability of psychological material and the point of its resistance.[38]

[34] Alex Kozulin, *Vygotsky's Psychology: A Biography of Ideas* (Cambridge, Mass.: Harvard University Press, 1990); Van der Veer and Valsiner, *Understanding Vygotsky*; James Wertsch, *Vygotsky and the Social Formation of Mind* (Cambridge, Mass.: Harvard University Press, 1985); Mikhail G. Yaroshevsky, *Lev Vygotsky* (Moscow: Progress, 1989).
[35] Michael Cole and Sylvia Scribner, *Culture and Thought: A Psychological Introduction* (New York: Wiley, 1974); Elena D. Khomskaia, *Aleksandr Romanovich Luria: Nauchnaia biografiia* [Aleksandr Romanovich Luria: A scientific biography] (Moscow: Voenizdat, 1992).
[36] Vladimir P. Zinchenko, "Krizis ili katastrofa?" [Crisis or catastrophe?], *Voprosy filosofii*, 5 (1993), 4–10.
[37] Quoted in Graham, *Science, Philosophy, and Human Behavior*, epigraph to chap. 5.
[38] Loren R. Graham, *What Have We Learned about Science and Technology from the Russian Experience?* (Stanford, Calif.: Stanford University Press, 1998), chap. 1.

PSYCHOLOGY AND AUSTROMARXISM

Like tsarist Russia, the Austro-Hungarian Empire had one dominant culture at the turn of the twentieth century. Austria's German culture, however, was more cosmopolitan and less ethnocentric, partly because of its links to the powerful German culture outside its borders, partly because of the considerable influence of non-German cultures within the empire.

German philosophy and psychology, particularly the works of Johann Friedrich Herbart (1776–1841), informed the beginnings of psychology in Austria. Herbart inspired the first European monograph to use the term "social psychology" – Lindner's *Ideas about the Psychology of Society as the Foundation of Social Science* (published in Austria in 1871). Its author, the educator Gustav A. Lindner (1828–1887), applied Herbart's idea that individuals behave in society in almost the same way that ideas behave in the individual soul, if social relations are close enough for them to affect each other.

The young Sigmund Freud (1856–1939) allegedly read Lindner's textbook, although it would be an exaggeration to say that the Herbartian metaphysics of consciousness exerted any direct influence on his future theories.[39] Born in Moravia, Freud spent almost all his life in Vienna, and most of his early coworkers were from Austria and Hungary. Psychoanalysis – with its modernist concept of the unconscious and its ties to both medical science and the humanities – can be considered a product of Austria's cosmopolitan culture.

In Austria at the turn of the century, as in Russia, Marxism exerted a considerable impact on the human sciences, including psychoanalysis. But in contrast with its development in the Soviet Union, where Marxism gradually became an oppressive ideology, the Austromarxism of Otto Bauer (1882–1938) and Max Adler (1873–1937) was characterized by epistemological pluralism and humanistic ethics. The encounter between psychoanalysis and Austromarxism began in the 1909 lecture "On Psychology and Marxism," which Alfred Adler (1870–1937) delivered at one of Freud's "Wednesdays." Although Freud himself was skeptical of Marxism, and his own views on social psychology developed independent of Marxist social thought, his followers Paul Federn (1871–1950), Wilhelm Reich (1897–1957), and Otto Fenichel (1898–1946) worked toward the convergence of psychoanalysis and Austromarxism.[40] In their turn, the Austromarxists Bauer and Max Adler borrowed from Alfred Adler's individual psychology, especially his ideas on education and overcompensation.

Despite the shortage of funds and academic appointments after World War I, the period of the First Austrian Republic was especially productive for

---

[39] Henri F. Ellenberger, *The Discovery of the Unconscious: The History and Evolution of Dynamic Psychiatry* (London: Allen Lane, 1970), p. 536.

[40] Ernst Glaser, *Im Umfeld des Austromarxism* [In the area of Austromarxism] (Vienna: Europa Verlag, 1981), p. 260.

the humanities and social sciences. In Vienna, besides the Psychoanalytical Association, the Vienna Circle stimulated discussion in the epistemology and logic of science. The work of the Circle reflected the influence of Ernst Mach (1838–1916), who had taught in Graz, Prague, and Vienna, and whose philosophy had influenced psychological theories of perception.[41] Other intellectual centers formed around the university department of psychology headed by Karl Bühler (1879–1963) and the child guidance clinic of Charlotte Bühler (1893–1974).

In the precarious institutional situation, scholars in the human sciences moved toward socially oriented and practical work, which the socialist Viennese government encouraged. Alfred Adler organized a network of counseling centers for children, and the Bühlers established a managerial research structure that coordinated university and practical research, much of it in clinical settings. Some social scientists, however, such as the economist Joseph Schumpeter (1883–1950) and the psychologist Jacob Moreno (1892–1974), left the country, paralleling the emigration from the Soviet Union.[42]

Although Austromarxism did not produce theoretical results comparable to the Soviet theories of activity and sociocultural development, it advanced research in social psychology. Paul Lazarsfeld (1901–1976), a schoolteacher trained in psychology by Karl Bühler, reported that Austromarxism turned his attention to social issues, so that he began to study young workers' and village communities. Under the influence of Otto Bauer, Lazarsfeld and his colleagues Marie Jahoda and Hans Zeisel conducted innovative research on a community near Vienna with high unemployment, *Die Arbeitslosen von Marienthal* (1933, translated as *Marienthal: Study of an Unemployed Community*).[43]

Contacts between Austrian scholars and the Frankfurt Institut für Sozialforschung reflected the cosmopolitan nature of German-language culture. Karl Grünberg, a representative of Austromarxism, went from Vienna to Frankfurt to become the second director of the institute. Lazarsfeld contributed to a study of workers' mentality, organized at the Frankfurt Institute by Max Horkheimer (1895–1973) and directed by Erich Fromm (1900–1980).[44] The arrival of fascism of the Italian type in 1934 and of National Socialism of the German type in 1938 ended the dialogue between Marxism and psychology in Austria.

[41] Mitchell G. Ash, *Gestalt Psychology in German Culture, 1890–1967: Holism and the Quest for Objectivity* (Cambridge: Cambridge University Press, 1995), pp. 60–7.
[42] Christian Fleck and Helga Nowotny, "A Marginal Discipline in the Making: Austrian Sociology in European Context," in *Sociology in Europe: In Search of Identity*, ed. Birgitta Nedelmann and Piotr Sztompka (Berlin: de Gruyter, 1993), pp. 104–9.
[43] Smith, *Fontana History*, pp. 616–22, 803. See also Peter Wagner's Chapter 34 in this volume.
[44] Paul Lazarsfeld, "An Episode in the History of Social Research: A Memoir," in Talcott Parsons, Edward Shils, and Paul Lazarsfeld, *Soziologie – autobiographisch* (Stuttgart: Enke, 1975), pp. 149–56; Martin Jay, *The Dialectical Imagination: A History of the Frankfurt School and the Institute of Social Research, 1923–1950* (London: Heinemann, 1972), pp. 9–10.

THE SEARCH FOR NATIONAL IDENTITY IN CENTRAL
AND EASTERN EUROPE

Even putting aside Russia and German Austria, the countries of this region varied economically, politically, and culturally. Before World War I, some of them were formally independent (Romania, Serbia, Bulgaria), but most were part of the Austro-Hungarian Empire. During the interwar period, they all became independent states and sought to reach the economic and cultural level of western European countries.

The development of psychology in these countries paralleled, on a smaller scale, what occurred elsewhere in Europe: the beginnings of experimental psychology of the Wundtian type; the introduction of psychology into university curricula and its practical application in the areas of education, work, health, and commerce; the beginnings of social psychology under the influence of German *Völkerpsychologie* and French theories of imitation and the crowd; and the interaction between psychology and Marxism. Although not a particularly salient part of cultural life, psychology in these countries promoted a scientific approach to social questions, and it was instrumental in modernization.[45]

One psychological topic significant in these countries was the problem of national character or mentality, a problem that emerged from aspirations for national development. The conditions of the Austro-Hungarian Empire fostered romantic, populist versions of nationalism. Under the Habsburg monarchy, in Oszkár Jászi's (1875–1957) view, the glory of the monarchy restrained the expression of a special German patriotism, but allowed space for the patriotism of the subordinate nations to express itself locally in cultural nationalism.[46]

The concept of national mentality drew on the notion of *Volksseele*, or the soul of the people, part of the *Völkerpsychologie* of Moritz Lazarus (1824–1903), Heymann Steinthal (1823–1899), and especially of Wilhelm Wundt. *Völkerpsychologie* meant the study of cultural expression – such areas as language, myth, and custom – and it had two parts: one general, unconcerned with the variations of peoples across space and time, and the other culturally specific.[47] Studies of national character were rooted in the latter, more ethnographic approach. In Hungary before World War I, Jászi wrote on ethnopsychology; later, populist writers such as Gyula Illyés (1902–1983) developed a so-called sociography – a special research project to study peasants and

---

[45] Ivo Banac and Katherine Verdery, eds., *National Character and National Ideology in Interwar Eastern Europe* (Yale Russian and East European Publications No. 13) (New Haven, Conn.: Yale Center for International and Area Studies, 1995).

[46] Oscar Jászi, *The Dissolution of the Habsburg Monarchy* (Chicago: University of Chicago Press, 1929), pp. 447–8.

[47] Gustav Jahoda, *Crossroads between Culture and Mind: Continuities and Change in Theories of Human Nature* (New York: Harvester Wheatsheaf, 1992), pp. 150–1.

other social groups. Sándor Karácsony (1891–1952) analyzed the Hungarian mentality in connection with grammar, a history of passive resistance, and the psychology of the villagers.[48]

In Poland, a well-developed sociology and the struggle for national unity and independence were factors underlying studies of the national mentality. The author of the first Polish monograph on social psychology, the distinguished politician Zygmunt Balicki (1858–1916), believed that chivalry, expressed in altruism and readiness to sacrifice for the nation, was the most important trait of his nation's character. Stanislaw M. Studencki (1887–1944) studied the relation between psychological and anthropological typologies of Poles, and he published a comparative psychology of nations based on the contemporary international literature.[49]

In Romania, the sociologist Dumitru Draghicescu (1875–1945) analyzed the psychology of the Romanian nation through its history, while the psychologist Constantin Radulescu-Motru (1868–1957) attempted an empirical study of national character, especially the features of sociability and emotionality. In his research on the psychology of village people, the sociologist Dimitrie Gusti (1880–1955) developed a model of national character that included cosmological, biological, psychological, and historical factors.[50]

In Bulgaria, research on national character was begun by Anton T. Strashimirov (1872–1937) and continued by Ivan M. Khadzhiiski (1907–1944), who studied, from a historical point of view, the way of life and mentality of the Bulgarian peasantry and of shopkeepers. According to Stoian P. Kosturkov (1866–1949), who developed a multidisciplinary approach to national mentality, the Bulgarians are hardworking, sober, suspicious, lively, and persistent.[51]

The national psychology of the Serbs, Croats, and Slovenes was studied by the lawyer Baltazar Bogišic (1834–1908), by the geographer and ethnologist Jovan Cvijic (1865–1927), and by the psychologist Mihailo Rostohar (1878–1966). While Bogišic studied legal norms and their impact on people's

[48] Gyula Illyés, *A Puszták népe* [People from Puszta] (1936), 3rd ed. (Budapest: Szépirodalmi Könyvkiadó, 1969); Sandor Karácsony, *A magyar észjárás* [The Hungarian mentality] (1939), 2nd ed. (Budapest: Magvetö, 1985).

[49] Zygmunt Balicki, *Psykhologia spoleczna* [Social psychology] (Lwow: Altenberg, 1908), and his *Egoizm narodowy wobec etyki* [National egoism and ethics] (Lwow: Towarzystwo Wydawnicze, 1914); Stanislaw M. Studencki, *O typie psycho-fizycznym Polaka* [On the psycho-physical type of the Pole] (Poznan: Chrześcÿańsko-Narodowe Nauczycielstwo, 1931), and his *Psykhologia porownawcza narodow* [Comparative psychology of nations] (Warsaw: Szkól Powszechnych, 1935).

[50] Dumitru Draghicescu, *Din psihologia poporului roman* [On Romanian national psychology] (Bucuresti: Alcalay, 1907); Constantin Radulescu-Motru, *Sufletul neamului nostru* [Soul of our nation] (Bucharest: Lumen, 1910); Dimitrie Gusti, *La Monographie et l'action monographique en Roumanie* [The monograph and monographic activity in Romania] (Paris: Collections de l'Institut de Droit comparé de l'Université de Paris, 1935).

[51] Anton T. Strashimirov, *Kniga za bolgarite* [Book on Bulgarians] (Sofia: Voenen Zhurnal, 1918); Ivan M. Khadzhiiski, *Bit i dushevnost na nashia narod* [The life and mentality of our nation] (1940, 1945), 3rd ed. (Sofia: Bolgarski pisatel, 1966); Stoian P. Kosturkov, *Vrkhu psikhologiiata na bolgarina* [On the psychology of the Bulgarian] (Sofia: Narodna kniga, 1949).

lives using questionnaires, Cvijic combined direct observation of different Balkan peoples with the analysis of historical records, ballads, and monuments. Rostohar analyzed the moral meaning of nationality; he believed that the essence of nationality lies in awareness of belonging and in the feelings that such awareness generates.[52]

Thomas Masaryk (1850–1937), a sociologist and philosopher and the first president of Czechoslovakia, considered religious faith and earnest everyday work the best traits of the Czech national character. At the same time, he criticized the Czechs' lack of will, emotional vacillation, inconsistent ideas, and tendency to false martyrdom. Similarly, the sociologist and social psychologist Emanuel Chalupný (1879–1958) observed that, though active and purposeful when they begin to do something, his compatriots lack the sustained effort necessary to complete what they have started. He linked this trait to the placement of accents in the Czech language and to the instability of Crech national history – a factor that the political writer Ferdinand Peroutka (1895–1978) also emphasized.[53]

The psychologist Anton Jurovský (1908-1985) linked the Slovak national mentality to religion. He made a distinction between internal factors – such as abilities, emotional irritability and impulsiveness, and empathy and sincerity – and external factors, such as industriousness, modesty, belief in justice, and an active attitude toward life.[54]

Writers in these countries generally used the concept of national character as a way to turn inward, for purposes of national self-definition and cultural development. Their studies of the national mentality therefore included drawing historical lessons, finding positive traits, and criticizing negative ones. These studies thus contrasted with contemporaneous conceptualizations of national character in nations where the problem of creating nationhood had largely been resolved. Writing about larger nations, the Frenchman Alfred Fouillée (*Psychological Outline of European People,* 1903), the German Richard Müller-Freienfel (*Psychology of German Man and His Culture,* 1922), and the Russian Nikolai Berdiaev (*The Soul of Russia,* 1915) treated national traits as stable. They paid more attention to the external mission of the nation than to the construction of national identity. In all such studies, however, scholarly aims were intertwined with political ends.

---

[52] Baltazar Bogišic, *Pravni obitchai u Slovena* [Legal customs of the Slavs] (Zagreb: Jugoslavenska akademia znanosti i umjetnosti, 1866); Jovan Cvijic, *La péninsule Balkanique, Géographie humaine* (Paris: Armand Colin, 1918); Mihailo Rostohar, *Národnost a její mravní význam* [Nationality and its moral significance] (Prague: Library of the Minority Museum, 1913).

[53] Thomas G. Masaryk, *The Meaning of Czech History,* ed. Rene Wellek (Chapel Hill: University of North Carolina Press, 1974); Emanuel Chalupný, *Národní povaha eská* [The Czech national character] (Prague: Lesching, 1907); Ferdinand Peroutka, *Jaci jsme* [The kind of people we are] (Praha: Borový, 1924).

[54] Anton Jurovský, "Slovenská národná povaha" [The Slovak national character], in *Slovenská vlastiveda* [Slovak national history and geography], 3 vols. (Bratislava: Slovenská akadémia vied a umení, 1943), vol. 2, pp. 335–98.

Research on national character receded with the appearance of sociological surveys and the emergence of laboratory experiments with groups. After World War II, psychologists became more interested in studying group behavior under controlled conditions and largely stopped writing about groups, including nations, in general terms.[55] Social psychologists in eastern and central Europe, as in Russia, wanted their work to be accepted on the same level as Western research. Joining Western professional organizations, such as the Transnational Social Psychology Committee of the Social Science Research Council and the European Association of Experimental Social Psychology, researchers had fully entered the international network in the human sciences by the end of the twentieth century.

[55] Smith, *Fontana History*, p. 763.

25

SOCIOLOGY IN EGYPT AND MOROCCO

Alain Roussillon

Sociology, and the social sciences in general, made their entry into the area from Turkey to Morocco through a transfer of European theories, concepts, methods, and interrogations during the colonial period. These transfers, at first provided by the French tradition, then rapidly followed by its Anglo-American rivals, allowed societies freshly open to social scientific investigation to enter into the scholarly representations of their worlds, a prelude to the deployment of the "civilizing missions" of their respective metropolises. Social science disciplines were then mobilized by the new indigenous elites to construct a national apparatus and to contest the self-image that had been reflected in the mirror of colonial science.

The relatively precocious development of the social sciences produced an accumulation of knowledge that diverged, both qualitatively and quantitatively, according to country.[1] Yet one principal result of this process was to consolidate a representation of the unity of this part of the world. It is surely problematic to speak of the "Arab world" or the "Arab-Islamic world" as a stage on which the process of internationalization of the social sciences is played out or as a common identity, be it Arab or Muslim, despite the fact that the producers of these disciplines have asserted such an identity through pan-Arab or pan-Islamic social scientific associations, such as the Association of Arab Sociologists (1985). Such an approach erases specific national developments and makes it difficult to locate the role that Western social science maintained long after independence in the production and reproduction of local social sciences. It also threatens to produce a counterpart of the essentialist identities constructed by "Orientalist" and/or nationalist scholars.

Hence this chapter will analyze the way in which sociology, including anthropology and ethnology, emerged and constituted itself in Egypt and

---

[1] Jacques Berque, "Cent vingt-cinq ans de sociologie maghrébine," *Annales Économies, Sociétés, Civilisations*, 11 (July–September 1956), 296–324.

Morocco, two contexts where it developed along contrasting trajectories. In Egypt, sociology developed first as an academic discipline taught to and by Egyptian scholars and also as a philanthropic practice understood as "social work"; it remained after independence an established presence, sometimes subservient to state power but most often submissive to its command and control. In Morocco, the discipline was the exclusive preserve of the colonialists until independence, after which, in Moroccan hands, it took a critical stance. Banished from the universities in 1970, it has since led a fugitive existence.

Nonetheless, in both contexts colonial intervention produced common problematics. Confronted with the colonial Other, sociologists constructed a counter-paradigm of the reform of the Self, inscribing identity at the heart of the emergence of the social sciences. Reform of the Self was conceived in polar ways: reform of society through the construction of a scientistic avant-garde that would create the necessary conditions for "modernity," and reform of society through a return to an original authenticity led astray by history and adulterated by borrowings. To have been confronted with these foundational dilemmas is what constitutes the common hallmark of Moroccan and Egyptian sociology and social science; the specific ways in which these dilemmas have been confronted constitutes their distinctiveness.

THE ACCUMULATION OF KNOWLEDGE FOR THE OTHER

At the moment when the European powers took their first steps toward a systematic accumulation of knowledge about Egypt and Morocco, these societies practiced only limited and routinized modes of self-representation: *rasâ'il sultaniya*, sultanian epistles, containing bits of social or political philosophy; *khitat* literature, local chronicles, most often dominated by hagiography; *bida'* treatises, blamable innovations, that from time to time took into account practical realities. Still, the monumental work of 'Ali Mubârak, *Al-khitat al-tawfîqiya*, written in the late nineteenth century, facilitated the work of later scholars.[2] In the same way, they had access only to limited information about the evolution of European societies, furnished by diplomats, merchants, and adventurers of all sorts. By the end of the nineteenth century, the weakening of the Ottoman Empire and the establishment of European beachheads in a few major cities had opened new horizons, and a greater number of travelers were making their way to Europe.[3] The Western social sciences had no antecedents in either Egypt or Morocco, despite the later

---

[2] See, e.g., Jacques Berque, *Egypte, impérialisme et révolution* (Paris: Gallimard, 1967).

[3] The unequalled model of this genre remains the Egyptian Rifâ'a al Tahtâwi's narrative of his sojourn in Paris between 1826 and 1831, under the title *Taklîs al-ibrîz fi talkhîs bâris* [The extraction of pure gold in the description of Paris] (Cairo, 1834).

appropriation of the fourteenth-century Arabic philosopher of history Ibn Khaldun – for many Egyptian and Moroccan social scientists, the founding act of their discipline.

IN MOROCCO: MUSLIM SOCIOLOGY AND PACIFICATION

After the first conspectuses of Morocco, sponsored by the Algerian colonial party during the 1880s,[4] an enormous corpus of knowledge was amassed under the auspices of the French protectorate, constructing a "Muslim sociology."[5] In 1904 the Mission Scientifique au Maroc set out to compile a body of doctrine that would orient Islamic affairs, challenging metropolitan Orientalist scholars as the principal laboratory of French colonial thought. These self-proclaimed specialists appeared to be less interested in participating in Orientalist discourse than in exploring the infinite diversity of the Moroccan field. The Mission Scientifique was transformed in 1920 into the Section Sociologique des Affaires Indigènes of the Residence, and in 1925 passed its objectives on to the Institut des Hautes Études Marocaines (IHEM) in Rabat. IHEM produced, among other enterprises, two atlases of Morocco: one hagiographic, including a geography of saints, *zâwiya-s*, holy places, and religious organizations – objects spurned by the dominant Islamology; and the other dialectological, also disdained by Arabic philology. The officers of the "*Bureaux arabes*" had fueled this initial accumulation of knowledge with reports, notes, and *memoires*, and their "exoticism" and "sensualism" was noted by Jacques Berque. "They were not far (aided by Saint-Simonism) from constructing a proper oriental humanism worthy of classical humanism, a strange combination of French chauvinism and adherence to the cité arabe" – a stance that did not predispose them to remain faithful to the codified rules of scientific communication.[6] Not surprisingly, this Muslim sociology was marginal to metropolitan academic research and to the Durkheimian disciplinary field. One can find only fleeting interest shown by Frédéric Le Play in 1877 in the ways of life among Moroccan artisans,[7] notes by Émile Durkheim on Berber tribalism, and a few teachings by Charles Le Coeur, a student of Marcel Mauss, at the IHEM.[8]

Edouard Michaux-Bellaire (1857–1930), who began direction of the Mission Scientifique in 1907, emerged as an emblematic figure of this first exploratory and monographic phase. His sociography – as opposed to

---

[4] For example, the *Reconnaissance au Maroc, 1883–84*, published in Paris in 1888.

[5] On the debates around Morocco, see Daniel Rivet, *Lyautey et l'institution du protectorat français au Maroc, 1912–1925,* 3 vols. (Paris: L'Harmattan, 1996).

[6] Berque, *Égypte, impérialisme ed révolution*, pp. 299–300.

[7] J.-P. Buisson, "Frédéric Le Play et l'étude de niveau de vie d'une famille d'artisans marocains," *Bulletin économique et social du Maroc (BESM),* 59 (1953), 71–83.

[8] Lucette Valensi, "Le Maghreb vu du centre: sa place dans l'école sociologique française," in *Connaissances du Maghreb: Sciences sociales et colonisation,* ed. Jean-Claude Vatin (Paris: CNRS, 1984).

sociology – was situated at the convergence of a meticulous ethnology and a history particularly attentive to pre-Islamic elements, "the incontestable left-overs of a paganism that Islam was not powerful enough to destroy."[9] This was the archaism that justified France's civilizing mission. It is Michaux-Bellaire who first clearly articulated the double opposition upon which both the accumulation of colonial knowledge and the politics of the protectorate were constructed. As Renan wrote, this opposition incites the search for "the true depths of Africa in the dispossessed Berber," who was the "true native, stable and industrious possessor of the earth."[10] Islam was represented as being imposed on indigenous populations and their "residual paganism," behind which this sociology glimpsed the likeness of French laicism and republicanism. Colonial authorities played these supposed cleavages off against each other in order to consolidate colonial control.[11] The opposition between territories and populations subject to the fiscal authority of the client Cherifian state, *bled makhzen*, and the spaces of dissidence, *bled siba*, appeared to observers, Michaux-Bellaire included, to be the key to the Moroccan political system that the protectorate had to preserve at all costs. One policy was instituted for the *makhzen* aiming to reform from above; another, "tribal" policy played on local power struggles in order to obtain a clientele that would rely on the colonial power over and against the Cherifian state.

Commissioned to study the rebellious tribes of the Moroccan south with the goal of pacification, the maritime lieutenant Robert Montagne (1893–1954) later systematized and theorized the resulting "surprising contradiction" operative in the Moroccan context.[12] The sociologist's task was to establish laws of composition between identical parts of a segmented checkerboard; the relation between tribes (i.e., Berbers) and the *Makhzen* constituted at once the principal contradiction of the system and the necessary condition for its functioning. Yet, what made Montagne the veritable "mastermind of an epoch and inventor of a system"[13] was the further object he assigned to colonial science: the recomposition of the Moroccan social fabric. Montagne's *Naissance du prolétariat marocain*[14] compiled the results of a vast collective investigation (1948–50) that optimistically represented the increasing rural migration into urban centers and the extension of detribalizing urbanization as the most effective mechanisms for implanting and generalizing the colonial order. This thesis was laced with a primordial,

---

[9] E. Michaux-Bellaire, "La mission scientifique au Maroc," as cited in F. Hourouro, *Sociologie politique au Maroc, le cas de Michaux-Bellaire* (Casablanca: Afrique-Orient, 1988), p. 126.

[10] Cited in Berque, *Egypte, impérialisme et révolution*, p. 306.

[11] Daniel Rivet, *Le Maroc de Lyautey à Mohammed V: le double visage du Protectorat* (Paris: Denoël, 1999), chap. 10.

[12] Robert Montagne, *Les Berbères et le Makhzen dans le sud du Maroc: Essai sur la transformation politique des Berbères sédentaires (groupe Chleuh)* (Paris: Alcan, 1930), pp. vii–viii.

[13] Abdelkebir Khatibi, *Bilan de la sociologie au Maroc* (Rabat: Association pour la Recherche en Sciences Humaines, 1967), p. 16.

[14] Published in Paris by Peyronnet & Cie, 1951.

obsessional antinationalism that even today makes Montagne's writings unpleasant reading for Moroccan social scientists.

Before independence was finally achieved in Morocco on 2 March, 1956, colonial sociology had time to begin a reevaluation of its analytic paradigms. A large part of the Moroccan and Maghrebian oeuvre of Jacques Berque (1910–1995), appearing in the shadow of the colonial regime of which he was a civil administrator, was based on a critical approach.[15] This was especially true after Berque was appointed to the "*Secteur de modernisation du paysannat*," whose objectives he described as "to combine and blend in a single movement the rise of the indigenous communities and the entrenchment of the French."[16] The question that he addressed particularly to Montagne (in "Qu'est-ce qu'une tribu nord-africain?")[17] interrogated a whole corpus of colonial knowledge and emerged as the inaugural question of a new paradigm taken up by Anglo-American anthropology during the 1960s.

Throughout the colonial period, Muslim sociology was designed exclusively for external use. The native Moroccan was an informer and in no case the recipient of such knowledge, and not a single endogenous accumulation of knowledge on Moroccan society was created within its parameters. Al-Hajwi (1874–1956), the most important intellectual of the period, minister of public education of the "*Makhzen* of the French" and one of the best-informed collaborators in the protectorate, delivered reflections on the reform of Moroccan society in his history of *fiqh* (Muslim jurisprudence). Mukhtar al-Soussi (1900–1963), whose oeuvre resembles most closely a Muslim anthropology from the inside, presented his research on *Sous Berber* culture within the project of conserving the *'ilm* (religious science). The nondisclosure of Muslim sociology and its methods had the effect of passing on to the first generation of indigenous sociologists the thorny question of what to do with this conglomeration of knowledge that took hold of their past and imposed itself as a deforming mirror placed between themselves and their society.

IN EGYPT: INTELLECTUAL RENAISSANCE THROUGH SOCIAL SCIENCE

In the late-nineteenth-century dynastic, proto-national state of Egypt, several institutions reactivated the project of a systematic accumulation of knowledge, initiated in 1798 by the scholars of Bonaparte's expedition. The Société Khédiviale de Géographie, founded by the modernizing ruler Isma'il

[15] See Jacques Berques, *Les Seksawa, recherche sur les structures sociales du Haut-Atlas occidental* (Paris: Presses Universitaires de France, 1954).
[16] In *Bulletin d'information du Maroc*, October 1945, as cited in Khatibi, *Bilan, de la sociologie an Maroc*, p. 21.
[17] In his *Hommage à Lucien Febvre* (Paris: Armand Colin, 1953).

in 1875, served as the home base for Western explorers in eastern Africa. In 1910, the future King Fu'ad founded the Société Khédeviale d'Economie Politique, de Statistique et de Législation, whose journal, *L'Egypte contemporaine*, dealt with "all the problems of national life, posed on a daily basis, projects of reform, whether of legislative, economic, or social nature, information necessary for agronomists, banks, merchants, namely official documents, statistics, and bibliographical material concerning Egypt that would be the object of the most objective and well documented studies."[18] In 1918, the same prince reactivated Bonaparte's Institut d'Egypte, which brought together specialists of many disciplines with the announced purpose of pursuing the program of the *Déscription de l'Egypte*.[19]

The activities of these learned societies, whose membership was largely non-Egyptian, was in large part an adjunct of the exploitation of Egypt upon which the European powers had "cordially" agreed in 1904, but they cannot be reduced to their colonial dimension. The Egyptian situation did not resemble the closed-off Franco-Moroccan field, nor was it reserved for the British, who occupied the country in 1882 but never fully obtained the international community's acceptance of its protectorate. From the point of view of the Egyptian dynasty, these institutions constituted – as did the university, founded in 1908, the Egyptian embassies, the zoo, and the botanical gardens – attributes of sovereignty attesting to its civilizational dignity and the integrity of its prerogatives. Furthermore, these learned societies were a privileged terrain for the converging politico-economic interests of Egyptian elites and French and Italian colonial interests and thus provided space for the systematic critique of British politics.

What decisively differentiated the Egyptian experience from the Moroccan was the pedagogical perspective of these learned societies. Following the instructions of Muhammad Ali, Jomard, an eminent member of the Scientific Mission of the French Expedition, organized a mission of Egyptian students to Paris as early as 1826–31. The Société pour le Savoir et la Publication des Livres Utiles published two hundred texts in translation between 1890 and 1925. One of these translations, Edmond Demolin's *A quoi tient la supériorité des Anglo-Saxons* (1899), would later inspire Muhammad 'Umar to invert Demolin's thesis and elaborate, in Arabic, on the *Présent des Egyptians et le secret de leur arriération*, probably the first indigenous contribution to the sociology of Egypt.[20] The Société also published two important works in 1900 by Qasim Amin: *L'émancipation de la femme* and *La femme nouvelle*.

[18] *Egypt contemporain*, 1:1 (1910), 2.
[19] See Alain Roussillon, "Le partage des savoirs: effets d'antériorité du savoir colonial en Egypte," *Annales islamologiques*, 26 (Autumn 1992), 207–49.
[20] See Alain Roussillon, "Réforme sociale et production des classes moyennes: Muhammad 'Umar et l'arrièration des Egyptiens,'" in *Entre réforme sociale et Mouvement national: identité et modernisation en Egypte*, ed. Alain Roussillon (Cairo: CEDEJ, 1994).

The founding of the Egyptian University in 1908 constituted an important turning point. Such famous orientalists as Louis Massignon, George Hoccart, Gaston Wiet, Robert Cresswell, Carlo Nallino, and Giorgio de Santillana taught classes in Islamic civilization before audiences of wide-eyed students, including Taha Husayn, later "*doyen des lettres arabes.*"[21] Although the first doctorate was granted to an Egyptian only in 1913, at the Sorbonne, and the Belgian statistician M. G. Hostelet had to wait until 1925 to teach his first class, sociology played an integral role in the project of the university. The Durkeimian tradition corresponded well with Egyptian demands for social science. Durkheim's sociology was primarily a reflection on social cohesion. If his thought was not revolutionary, it was nonetheless concerned with the crisis of civilization, a crisis, wrote Raymond Aron, determined by "the non-replacement of traditional morals based on religions." Sociology "should serve to re-establish a morality that satisfies the requirements of the scientific mind."[22] This was precisely the reformist project of the university in reverse order: to identify a scientific mind able to fulfill preexisting moral demands. In addition, Durkheimian sociology furnished a model for intellectual intervention in social and state affairs; in the context of the "cultural renaissance" (*nahda*) of the first part of the century, intellectuals sought a societal paradigm that would allow both the integration of modernity and fidelity to the Self and its history. Such concepts as anomie, consensus, mechanical and organic solidarity, and collective conscience permitted the new Egyptian sociology to conceptualize and represent the process through which Egypt was passing. The sociologist was designated as the arbiter of this process, hence, undoubtedly, the confirmed conservatism that characterized Egyptian academic sociology.

Concerned to see their discipline officially recognized and their graduates appointed to the state apparatus, and preoccupied with appropriating the categories of Durkheimian sociology in order to reform Muslim thought, Egyptian sociologists hardly had time to study their own society. From this perspective, the French tradition failed to maintain its position in the academy or to gain access to an Egyptian social field. Other actors, mobilizing different forms of reasoning, recaptured the learned societies and imposed their paradigms and methods upon sociology. From the beginning of the 1930s, in a context of conservative government response to the Great Depression, charitable and assistance-related institutions became increasingly attentive to scientific forms of social knowledge. The Pioneers' Association, founded in 1929, the Social Reform League (1935), the Association for National Renaissance (1939), and above all the Egyptian Association for Social Studies (1937) simultaneously proposed modes of mobilization and heuristic

---

[21] Donald Reid, *Cairo University and the Making of Modern Egypt* (Cambridge: Cambridge University Press, 1990).

[22] Raymond Aron, *Les étapes de la pensée sociologique* (Paris: Gallimard, 1967), pp. 309–10.

investigation.[23] In effect, these associations were recognized by the state and their members permitted access to new professional positions, including posts in the Ministry of Social Affairs, created in 1939 at their instigation, giving the state control over their activities, budgets, and organizations. Hence the structural "apoliticism" shared by these groups, the explicit refusal to participate in partisan political struggles, a stance that justified their preferred role of expert or advisor to political decision makers.

The compromise between the state and these institutions of social work had an impact on the paradigms of Egyptian sociology, directly determining how knowledge was collected and organized. The purpose was explicitly one of "social engineering," designed by "social experts" and focused on "social problems." What I have analyzed as "the paradigm of three plights"[24] constituted the heuristic converse of the apolitical position of the social-work institutions. The "vicious circle of poverty, sickness, and ignorance" was represented as a closed system in which each of these social ills, defined as both object of study and obstacle to social progress, contributed indefinitely to the reproduction of the others. The new object of study was no longer the modes of social cohesion but in a sense its opposite, the symptoms of social dissolution and the means to combat it. Delinquency, protection of women and children, the condition of peasants, workers, and bedouin populations were all treated as homologous questions, identifying those "disinherited" by progress, determining ways of studying tradition, and locating both problems and traditions in the crisis of civilization through which Oriental societies were passing due to the imposition of rational organizational forms. The paradigm was more American than English, introduced chiefly by Wendell Cleland, a founder of the Egyptian Association for Social Studies, advisor to the Ministry for Social Affairs, dean of the External Division of the American University in Cairo, and a man deeply engaged in Protestant activism directed at the Copts. Many Egyptian candidates were sent to the United States to gain such expertise, mostly with scholarships from the ministry; they would eventually put their competencies at the service of the new independent state after 1952.

The focus on social problems, whatever their practical urgency, significantly displaced sociological objectives. What was at stake in the practice of experts was less the accumulation of knowledge about indigenous society than the command of technologies for its transformation. The study of traditional society was only justifiable as an aid or adjunct to this objective. The success of the sociological endeavor no longer resided in the coherence of the models it deployed but in its capacity to reduce the obstacles to modernity that social structures continued to erect.

---

[23] Alain Roussillon, "Réforme sociale et politique en Égypte au tournant des années 1940," *Genèses, sciences sociales et histoire*, 5 (1991), 55–80.

[24] Ibid.

NATIONALIZATION OF THE SOCIAL SCIENCES: THE INVENTION OF THE SOCIOLOGIST

In Egypt, as in Morocco, independence (obtained in 1952 and 1956, respectively) gave a decisive impulse to the establishment of the social sciences and conferred a double task: the redefinition of the mission of these disciplines and the establishment of new frames of institutionalization and professionalization. Sociology made a new beginning, no doubt, but not a radical break, for the new orientations largely replicated previous patterns. In Egypt, universities continued to form cohorts on the same academic basis, progressively adding Marxism and then functionalism to the Durkheimian substrate and leaving social engineering to state-controlled agencies. In Morocco, sociology again failed to establish a firm academic base. Its excessive radicalism ran afoul of a retraditionalizing state, leaving it to find its path informally, mostly abroad or through nonacademic channels.

IN MOROCCO: THE INITIAL PRODUCTION OF A CRITICAL SOCIOLOGY

During the early years of independence, colonial structures continued to exist in full force in many sectors – agriculture, teaching, the liberal professions, journalism – thus prolonging the pertinence and applicability of relevant bodies of knowledge. The abolition of the protectorate under relatively peaceful conditions, however, left the field open for the most progressive elements of colonial society to accept decolonization, and they were soon joined by French *coopérants*, who played a significant role in preparing the "moroccanization" of the administration and the economy. Since France left nothing behind on the level of universities, research institutions, or trained specialists, it remained the task of such individuals as Paul Pascon (1932–1985) to lay the theoretical groundwork for an indigenous scientific field in Morocco. Pascon's group, including a handful of young Moroccan intellectuals, projected a "global vision of Morocco," but, as self-proclaimed sociologists, they also sought to construct a critical, autonomous, and applied social science.[25] A critical social science would permit opposition to both the colonial order and to the "old Morocco"; it would record the loss of traditional regulations and study the changes introduced into the social and economic fabric by transformations of modes and relations of production. Through the category of "composite society" – a category he elaborated from a mix of Marxism, segmentarism, and the theories of Georges Gurvitch and Jacques Berque – Pascon accounted for the stratification of Moroccan society as well as for the articulation of competing social orders – political, juridical,

---

[25] Abdallah Saaf, *Politique du savoir au Maroc* (Rabat: SMER, 1991), p. 20.

social, symbolic – through which voluntary and forced transformations of modes of life and production emerged.[26]

Pascon and his group sought autonomy as much from the state as from private interests, and they formed their own bureau of applied studies, the Equipe Interdisciplinaire de Recherche en Science Humaines (1959). Functioning as an egalitarian self-managed workers' cooperative until 1963, the group worked primarily with the Union National des Forces Populaires and the technical services of certain ministries. Throughout the 1960s and 1970s, Pascon's contribution to the nationalization of Moroccan sociology was effected from his position as director of the Office Régionale de Mise en Valeur Agricole de Haouz de Marrakech as well as by way of his teachings at the Institut Agronomique et Vétérinaire Hassan II. In the research teams that he directed, inquiries into social history converged with studies of the management of territory inherited from colonial sociology. Coming back to the conclusions of Jacques Berque, Pascon sought in rural society the space to reactivate an "historical project offering a credible alternative capable of opposing foreign capitalist domination," a project, Pascon noted, that was as absent during the 1970s as it had been at the turn of the century.[27]

The creation of the Institut de Sociologie in Rabat (1960) within the young state's new institutions of higher learning marked a point of departure for a second, academic mode of institutionalization of the social sciences in Morocco. Academic status did not prevent this sociology from seeing itself as critical and radical, at least in its incantatory mode, on the basis of a project that we may call culturalist by way of an excessive anticulturalism. Abdelkebir Khatibi's programmatic formulation was to remain programmatic: "The essential task of the sociology of the Arab world consists of advancing a double critique, a) a deconstruction of concepts deriving from sociological knowledge and discourses that have spoken for the Arab world, that have come from the West, and are marked by a Eurocentrist ideology and, b) at the same time a critique elaborated by the different societies of the Arab world for their own use."[28] Against orientalist and colonialist forms of knowledge,[29] these intellectuals affirmed what they called the "normality" of "third world" societies. What was needed in order to capture this normality from the inside, in Khatibi's terms, was a "dissolution of ethnology" in favor of history and the reactivation of the Self in order to "institute the conceptual

---

[26] Paul Pascon, "Le Haouz de Marrakech, Histoire sociale et structures agraires" (state doctoral thesis in sociology, published in Tangier, 1977), and *La maison d'Iligh et l'histoire sociale du Tazerwalt*, with Alidelmajid Arrif, D. Shroeter, Mohamed Tozy, and Henzi Van Der Wusten (Rabat: SMER, 1984); see also the eulogies of Pascon in *BESM*, 155–6 (January 1986).

[27] Paul Pascon, "Repenser le cadre théorique de l'étude du phénomène coloniale," *Revue juridique, politique et economique du Maroc*, 5 (1979), 133.

[28] Abdelkebir Khatibi, "Sociologie du monde arabe. Positions," *BESM*, 126 (January 1975), 1.

[29] Including that of Jaques Berques, against whom Khatibi published a virulent article, "L'orientalism désorienté," in his *Maghreb pluriel* (Paris: Denoël, 1983).

apparata conceived according to the code of the Arabic language."[30] Teaching
and debate dominated activity at the Institut de Sociologie, which counted
266 students in the 1965–6 academic year. Research was mostly limited to
the questions of urban life and language, areas in which colonial impact
was the least contestable; given the absence of ministerial or state command,
access to the countryside was practically impossible. Marxism constituted
the space in which this critique was mobilized – a Marxism less epistemological than directly political, aiming to join the social sciences to social
movements.

The leftward drift of the teaching of sociology, amplified by echoes of
the student protests of 1967–8 and the intense politicization of Moroccan
universities, led to the closing of the Institut de Sociologie in 1970 and the
eclipse of university sociology and social science departments. Through the
end of the century, although sociology was taught in law and economics
schools as well as in humanities departments, the professional designation of
"sociologist" could be acquired and confirmed only in foreign universities.
From the late 1970s onward, the Arabization of university curricula, one of the
first demands of the nationalist Istiqlal Party, was progressively implemented.
Social matters were reinscribed in the framework of what was presented as
an authentic "Islamic" worldview, a worldview that was nevertheless mostly
imported, along with textbooks and references, from the Middle East in
general and from Egypt in particular.

Sociology continued to be practiced in Morocco – the *Bulletin économique
et social du Maroc* persisted – and Morocco has produced talented sociologists,
including some publishing in Arabic – but the discipline's principal interrogations have tended to find more fertile ground in other fields – namely, in
history and geography, where solid professional traditions have been consolidated, and in anthropology, which is still shadowed by its past compromises
with the colonial ordering of the world.

Finally, the perspective of the Other returned during the late 1960s
and early 1970s, but in a position radically different from that of the first
producers of Muslim colonial sociology. Ernest Gellner, Clifford Geertz,
William Zartman, David Hart, John Waterbury, and Dale Eickelmann
converged in Morocco to make it their field of debate, a debate that centered
around segmentarist anthropology and political science. Their arrival broke
up the Franco-Moroccan conversation that had survived decolonization by
attracting increasing numbers of Moroccan social scientists to American universities. The historian Abdallah Laroui severely criticized this anthropology:
"Anglo-Saxon research is in reality a simple reformulation of anterior results
that are far from being accepted: the democracy of Montagne is renamed
segmentarity, the *repression* of Berque is renamed marginality; it is in a way

[30] Khatibi, "Sociologie du monde arabe," p. 9.

superfluous."[31] Aside from these reservations regarding the overly systematic character of certain constructions, what seemed least acceptable was the way in which these social scientists associated Morocco with an *"éternel maghrébin,"* just as anthropologists and neo-Orientalist political scientists, and even Marxist political scientists, had used Ibn Khaldun in the same way.[32] Nonetheless, these works, independent of their claims to science and objectivity, were able to consider systematically the available knowledge on Moroccan social formations, a process that had in some ways been prohibited to insiders.[33]

IN EGYPT: TO REVOLUTIONIZE SOCIOLOGY?

Although the Revolution of 1952 was portrayed by interested parties as a fundamental rupture, the instituional network of Egyptian sociology and its theoretical and methodological options confirmed the choices made during the 1930s and 1940s.[34] The Revolution served only to reorient sociological objectives toward economic and social development as well as toward nationalist consolidation, then focused on the construction of "Arab socialism" as defined by Nasserism. The relationship between sociologists and social workers on the one hand, and sociologists and the state on the other, also remained constant, as did the relationship between Egyptian sociology and the dominant sociological fields abroad.

In the universities, the multiplication of sociology departments (at this writing, there is one in each of the twenty Egyptian universities) assured the academic training of specialists in general sociology or in its principle subdivisions, such as urban, rural, industrial, and military sociology. Still, these departments were created haphazardly in connection with different definitions of the sociological vocation and thus in different relationships with other scientific fields: in association with philosophy at the University of Cairo, with anthropology at the Institute for Social Sciences in Alexandria, with psychology at the University of 'Ayn Shams, and with Islamic theology, demography, and gender studies at the Islamic Department for Girls at Al-Azhar. What assured the coherence of the sociologies that emerged was the reproductive logic that linked establishment of authority in the field to specialization of objectives and methodologies. The rate of attrition of Egyptian social scientists nonetheless accelerated at an increasing rate after the mid-1980s. A network of social service departments (*khidma ijtimâ'iya*)

---

[31] Abdallah Laroui, *Les origines sociales et culturelles du nationalism marocain* (Paris: Maspero, 1977), p. 177.

[32] For example, Yves Lacoste, *Ibn Khaldun* (Paris: Maspero, 1966).

[33] See Rémy Leveau, *Le fellah marocain, défenseur du trône* (Paris: FNSP, 1973).

[34] See Alain Roussillon, "Republican Egypt Interpreted: Revolution and Beyond," in *The Cambridge History of Egypt*, vol. 2: *Modern Egypt, from 1517 to the End of the Twentieth Century*, ed. Martin W. Daly (Cambridge, Cambridge University Press, 1998), pp. 334–93.

came to parallel the sociology departments, awarding thousands of new doctorates.

The abundance of such trained professionals renders paradoxical the small contribution of academic sociology to the body of knowledge on Egyptian society. The large bibliography accumulated by the discipline – 2,000 titles annually since the early 1980s – consists for the most part of manuals, textbooks, redundant compilations, and poor translations. Most masters and doctoral dissertations have been limited to such restricted topics as the verification of a sociological concept or the pertinence of the methods of some school of thought to the Egyptian context.

Parallel to the universities, a series of institutions charged with the systematic accumulation of knowledge on Egyptian society was created by the Nasserian state beginning in the late 1950s. Aside from the Central Agency for Public Mobilization and Statistics created in 1964, which produced and published statistics, the most important institutions were the National Institute for Planning (1960), the Cairo Demographic Center (1964), and, most importantly, the National Center for Criminological and Social Research (NCCSR) (1956). These institutions were designed to analyze "the social and cultural problems that began to manifest themselves in Egyptian society since the Revolution of 1952 and that accompany social transformation, particularly the problems associated with social destructuration and cultural underdevelopment."[35] The association of sociology, social psychology, and criminology embodied in the NCCSR is revealing of the heuristic postures that can take form within authoritarian processes of transformation. Vendettas, prostitution, juvenile delinquency, drug abuse, and the like were the first subjects of the NCCSR's investigations, linking the social sciences to the need for new readings and new techniques of repression. On a more positive note, the social sciences were invited to pave the way for state-initiated reforms by identifying the problems involved in, and the obstacles to, state action. Hence the increased importance during the 1960s of the study of mass communication and elites.[36]

The revolutionizing of sociology thus meant the appropriation of the philanthropic model of social services, adapted without difficulty to the new socialist and Arab nationalist positions. Having dissolved the political parties and repudiated the Muslim Brothers and communists, the Nasser regime turned to the network of social service associations, whose experienced workforce and apolitical character suited it perfectly – so well in fact, that it was mainly personnel of this liberal-conservative persuasion that executed the policies of the Revolution, particularly regarding agrarian reform and the implementation of the educational system.

[35] "Evaluation des réalisations scientifiques du Caire, 1957–82," document published by the NCCSR in 1982 on the occasion of its twenty-fifth anniversary (in Arabic).

[36] See Alain Roussillon, *La sociologie égyptienne de l'Egypte: éléments de bibliographie* (Paris: CEDEJ, 1988), p. 264.

On the other hand, Egypt's regaining control of its own destiny allowed an acceleration of the transfer of knowledge originating in Western Europe and the United States. In Egypt, the 1960s and 1970s constituted a sort of Golden Age of Anglo-American functionalism and its multiple variants. Even Marxist-Leninist ideas circulating in Egypt were filtered through the works of such Western interpreters as Nicos Poulantzas, Louis Althusser, and John Lewis. The field of the social sciences resembled an eclectic patchwork of schools, constituted according to the experiences of Egyptian intellectuals and scholars in Western universities, experiences that continued to condition access to positions of power within the discipline.

The nationalization of sociology did not do away with foreign specialists, who were often preferred by the authorities. One of the first large-scale socio-anthropological investigations that the Revolution made necessary – the study of the displacement of Nubian populations by construction of the Aswan Dam – was placed in the hands of the Social Research Center of the American University in Cairo. This responsibility permitted the American University to establish itself as an impartial research institution in the field of Egyptian social science, despite the deterioration of U.S.–Egyptian relations. Foreign research, particularly American research, increased throughout the 1970s, to such an extent that in 1982 the economic supplement of *Al-Ahrâm* mounted a virulent campaign against foreign research under the title *Une description américaine de l'Egypte*, a reference to the colonial enterprise of Bonaparte's scholars. American research was accused of moving the country toward submission and holding back knowledge from Egyptians. Complicitous Egyptian intellectuals were said to "seek fame and fortune through association with American universities, and hide, behind their collaboration, political dreams that are suspect for Egypt's future.[37] At the same time, the results produced in the orbit of foreign research made clear the small contribution of local social scientists to the accumulation of knowledge on their own society. Since the beginning of the 1980s, Egyptian sociology has perceived itself to be in crisis.

SOCIOLOGISTS IN CRISIS IN EGYPT AND MOROCCO

In Egypt, as in Morocco, the crisis of the social sciences is perhaps the other side of the coin of the internationalization of the Western social sciences.[38] If in Egypt the crisis appeared in the ineffectiveness of the sociologists' truncated project, in Morocco it emerged in the gaps between the project to

---

[37] "Une description américaine de l'Egypt: les dimensions politiques et sécuritaires du phénomène," *Al-Ahrâm al-iqtisâdi* (October 4, 1982), 19.

[38] The Egyptian aspect of this crisis is dealt with in Alain Roussillon, "Intellectuels en crise dans l'Egypte contemporaine," in *Intellectuels et militants de l'Islam contemporain*, ed. Gilles Kepel and Yann Richard (Paris: Seuil, 1990).

Arabize the social sciences and the effectiveness of its implementation. The members of the generation trained in French during the 1960s, who had Arabized their classes by the beginning of the 1970s, came to admit that Arabization effectively isolated the new generation from an international body of knowledge to which their Francophone predecessors had had open access. Arabization of the discipline reactivated modes of conceptual organization that were, if not traditional, at least disconnected from contemporary currents in the social sciences.

In Egypt, calls for the implementation of specifically Arab or Islamic paradigms for the social sciences multiplied during the 1980s and 1990s. Social scientists were told to liberate themselves from imported models that were incapable of capturing the essential Self and that dissociated the discipline from social identity. It is important to underscore the structural homology between these two *da'wa* (callings). The project of an Arab or Islamic social science requires demonstration of the unity of the Arabic/Islamic community, identification of foreign elements or *al-wâfid* ("what comes from the outside," but also that which dominates), and designation of acceptable paths for modernization. Any field of inquiry must display the dialectic between authenticity (*asâla*) and adulteration (*tashwîh*, literally, "to disfigure"), between inheritance (*turâth*) and contemporaneity (*mu'âsara*).[39] Such culturalist social sciences are defined not only by their definition of culture as that which is shared by Arabs of Muslims, but also by the way they place *the origin* at the heart of their project. As a result, the Arab/Muslim social sciences, at least among the Egyptians, have contributed little to theory. It was rather in Indonesia, Malaysia, and then in Iran that the most consistent aims of the Islamicization of the social sciences were being defined.

In Morocco, the social sciences began to be reinstitutionalized at the end of the 1980s. Sociology and political science – the two disciplines ostracized by the state – did not regain their lost autonomy, but they began to adopt a disciplinary character. The creation of the Institut Universitaire de Recherches Scientifique contributed to the emergence of professional researchers, and their publications, mainly written in French, found a public.[40] Most important was recognition of the need for, in Negib Bouderbala's terms, "a cold-blooded look at decolonization," a project that preoccupied the Moroccan social sciences beginning in the late 1980s.[41] History and geography, the most legitimized, deeply rooted disciplines, took on the task of mourning the "old Morocco," permitting the social sciences to take the colonial period as an

[39] Cf. Jacques Berque, *Langages arabes du présent* (Paris: Gallimard, 1974). Also see Alain Roussillon, "Les nouveaux fondamentalistes en colloque: 'authenticité' et 'modernité' – les défis de l'identité dans le monde arabe," *Maghreb-Machrek*, 107 (January–March 1985), 5–22.

[40] Abdallah Saaf carefully analyzes this situation in "L'édition en sciences sociales au Maghreb: Aspects marocains," in his *Sciences sociales, sciences morales?* (Tunis: Alif-IRMC, 1995).

[41] Negib Bouderbala, "Pour un regard froid sur la colonisation. La perception de la colonisation dans le champs de la pensée décolonisée," in *Connaissances du Maghreb, sciences sociales et colonisation*, ed. Jean-Claude Vatin (Paris: CNRS, 1984).

object of knowledge. The ensuing debate enabled the selective appropriation of colonial science through a critique of the conditions of its production, thus permitting the distinction between colonial science and the colonial period.[42] Only then was it possible to ask in what terms the colonial fact contributed to the emergence of an independent Morocco.

Under the influence of nationalism during the 1960s and 1970s, only the economic and political dimensions of colonization had been taken into consideration. The relative withdrawal of the nationalist paradigm and the "cold-blooded look" have underscored the cultural aspects of colonization – including the fact that works of Moroccan social science are mostly written in French. Works of Arabic sociology – less often fieldwork, more often theoretical works on the Islamic foundations of the social order or reconstructions of Moroccan history – serve to accentuate the linguistic bipolarization of Moroccan intellectual life.

Above and beyond their differences, what Moroccan and Egyptian sociologists have in common is their constitution within an historical logic that has opposed to colonial reform the counter-paradigm of identitarian reform. The reformist structure of their project has shaped their relation to power and legitimacy as well as to alterity. The confluence of these tensions has created in both countries a crisis in sociology that is not so much an identity crisis as a crisis of the systems of action and representation articulated by the goal of identity formation. It can be halted only by renouncing the privilege of an essential identity and by the reinsertion of Morocco and Egypt into the process of internationlization of the social sciences – no longer only in relation to the West, and no longer playing the role of a field of experimentation, but rather as producers of knowledge on their own societies and on others.

[42] Fanny Colonna and Claude Haïm-Brahimi, "Du bon usage de la science coloniale," in *Le mal de voir* (Paris: Chaiers Jussieu, Université de Paris, 1976).

26

THE SOCIAL SCIENCES IN AFRICA

Owen Sichone

Since European explorers first began studying Africa, the continent has served as a testing ground for theories central to the development of science generally and of social theory in particular. Research conducted by sociocultural anthropologists, economists, and political scientists in Africa has generated concepts and theories of great importance to the disciplines. The impact of the social sciences on Africa has been equally far-reaching. Policy makers and development planners have tried to employ the social sciences as a means of bringing about social transformation since the colonial days. The development of the social science disciplines by Africans – the subject of this chapter – has been crucially shaped by all these efforts.

Writers on the social sciences in Africa can be divided into two polemical camps. The first is made up of Afrocentric scholars who are preoccupied with correcting unequal power relations in knowledge production. They try to show that there are other ways of knowing than those taught by conventional science. They also try to expose the exclusion of African knowledge systems from university curricula and the history of science, as well as the marginalization of African scholarship in the global academic community.[1] The other side downplays the significance of these power relations and advocates a universal system of scientific knowledge production, one that includes an African role.[2] Lacking a common language, the two camps effectively talk at cross purposes and are engaged in polemical nondialogue. Rather than

---

[1] Thandika Mkandawire, "The Social Sciences in Africa: Breaking Local Barriers and Negotiating International Presence," *African Studies Review*, 40 (1997), 15–36; Paul Tiyambe Zeleza, *Manufacturing African Studies and Crises* (Dakar: CODESRIA, 1997); Jacob F. Ade Ajayi, Lameck H.K. Goma, and G. Ampah Johnson, *The African Experience with Higher Education* (London: James Currey; Athens: Ohio State University, 1996), pp. 21–7; Molefi Asante, *The Africentric Idea* (Philadelphia: Temple University Press, 1998).

[2] Robert H. Bates, Valentin Y. Mudimbe, and Jean O'Barr, eds., *Africa and the Disciplines: The Contribution of Research in Africa to the Social Sciences and Humanities* (Chicago: University of Chicago Press, 1993).

taking a position in this debate, I will discuss features of the social sciences in Africa that both sides must take into account.

The discussion is divided into three parts. The first part is an overview of the colonial history of higher education and social science research in Africa, highlighting the social conditions under which teaching and research institutions were established and the long-term effects of the colonial legacy. I will then look at individual social science disciplines, locating some of the achievements of research in Africa and the varied effects that Africa has had on various disciplines. The last section discusses the impact of African economic and political crises on universities after 1980 and suggests that the social sciences continued to flourish in research centers outside the state-controlled universities.

THE COLONIAL LEGACY

Although social science research by Africans in local institutions dates back to the 1960s, it would be wrong to say that the generation of postcolonial African social scientists "started from scratch."[3] African universities are linked historically to overseas universities, and their teaching and research programs are built on foundations laid during the colonial period. In his television documentary *The Africans*, Kenyan social scientist Ali Mazrui spoke of Africa's "triple heritage" of indigenous African, Islamic, and Western traditions. This was not the first time such a view had been expressed. The influential West African scholar Edward Wilmot Blyden (1832–1912) used the same terminology in his discussion of Islam and Christianity in colonial Africa. The triple heritage thesis suggests a faith in the neutrality of the two great religions and a belief that Africans can use them to change their societies for the better.[4] Given the inequality of the three traditions that make up the triple heritage, it might be more correct to refer instead to the two colonialisms. African knowledge systems have been considered backward by both the Islamic and Western traditions, whose educational systems have produced many generations of African leaders. The work of several cohorts of such converts shows clearly that Western social thought has enjoyed greater prestige than Islamic and traditional African philosophy. Even at the end of the twentieth century, when global society was open to non-Western medical, artistic, and philosophical traditions, most African universities did not include Islamic scholars in courses on political, sociological, or economic theory and relied on standard American and European textbooks. Christian mission schools had the

---

[3] Thandika Mkandawire, as quoted in Anders Hjort af Ornäs, and Stefan de Vylder, *Social Science in Africa: The Role of CODESRIA in Pan-African Cooperation* (Stockholm: SAREC, 1991).

[4] Ali A. Mazrui, *The Africans: A Triple Heritage* (New York: Little Brown; London: BBC Publications, 1986); Valentin Y. Mudimbe, *The Invention of Africa: Gnosis, Philosophy, and the Order of Knowledge* (Bloomington: Indiana University Press, 1988), p. 115.

greatest impact on the erasure of African knowledge systems by virtue of their influence over which aspects of local knowledge could be recorded in writing. Not only was local historical, religious, and other knowledge censored and sanitized in various ways, but the missionaries themselves became the custodians of local knowledge. As one historian has noted, mission school textbooks replaced local elders as sources of historical knowledge.[5]

Education was highly politicized in colonial society, and many of the questions around which educational policy was debated and fought were issues that divided colonial society. Education was an elitist enterprise from which the majority of the population was excluded, whereas traditional informal education, aside from initiation into secret societies, remained integrated into the socialization process. Government educators and mission schools in colonial Africa sought to control the nature and level of education that the colonized people received, although the African elite saw education as a means of achieving upward social mobility and equality with Europeans. The social division of labor in colonies was based on race, and opponents of colonial educational policy criticized the denial to African students of higher education as an attempt to limit non-Europeans to manual labor. Such divisions in colonial society replicated those of the imperial system, in which European countries were the industrial and intellectual centers and the colonies supplied the raw materials and raw data from afar. In the production of knowledge, the role of Africa was to provide raw data for processing in Europe.

All imperial powers used their colonies as laboratories for different kinds of scientific research and experimentation. As European explorers and adventurers raced to "discover" rivers, lakes, mountains, and waterfalls in the interior of the continent, anthropometric studies of African people and ethnographic surveys of their societies also grew rapidly. Such scientific work was closely linked to the creation and mapping of political territories and to the study of populations of colonial subjects. The knowledge so obtained was largely ignored by the European powers when they carved up the continent and created arbitrary borders and countries at the Berlin Conference of 1885 and 1886. These colonial boundaries created problems for the colonial administrators, who had the difficult task of establishing bureaucratic administration over areas of great cultural diversity. Many of the identity, nationalist, and proto-nationalist problems that continue to bedevil African development and to stimulate research in political science, anthropology, and economics, including the lack of a local research capacity, are legacies of this colonial history.

During the nineteenth century, the work of geographers, sociocultural anthropologists, economists, and other scientists in colonial Africa was

[5] Andrew Roberts, *A History of the Bemba: Political Growth and Change in North-eastern Zambia before 1900* (London: Longman, 1973), p. 30.

almost exclusively European. There was hardly any African intelligentsia, and researchers in the major American universities did not show much interest in Africa until after the Second World War. Among African intellectuals – the few African scholars from the independent countries of Liberia and Ethiopia, the members of wealthy Creole families in Atlantic-coast African colonies, and scholars in the Islamic tradition – none made any major contribution to the development of the social sciences, which was the special task of Europe's intellectual aristocracy.

Some members of the small African elite traveled to Europe and attended universities there, but their numbers were insignificant. By contrast, E. W. Blyden, the noted critic of mission education, left his West Indian island home of St. Thomas to obtain, in Liberia, the education that he could not acquire in America, where he had been denied entry to Rutgers University.[6] Blyden is credited with being the first writer to espouse ideas of Pan-Africanism, Negritude, and "the African personality." As a particularist, he claimed a special role for indigenous ideas in African development, but he also considered colonization to be a way of elevating Africa to civilization, especially through the medium of the English language.[7] To Blyden and his contemporaries, the alienness of Western education was not as important an issue as colonial racial discrimination. African scholars from the colonies were thus preoccupied with political issues in their lives and work and concerned themselves mainly with the goal of national liberation. After Blyden, many other intellectuals of African ancestry in North America and the West Indies joined the African dialogue, some living and working in Africa and directly influencing younger generations of African scholars. Common ancestral roots, the shared experience of colonialism, and desire for Pan-African unity made them parties to the identity crisis that colonial intellectuals experienced in Western society.

Ironically, it was while they were studying in Europe that some members of the educated elite started to appreciate and value African culture for the first time. Founders of the Negritude cultural movement, such as Leopold Sedar Senghor (b. 1906) from Senegal, learned about Africa from European Africanists. Senghor's doctrine of Negritude, and specifically the contrast he made between the intuitive African and the rational Greek civilization, was influenced by Levy-Bruhl's work on primitive mentality and by French anthropologist Marcel Griaule's lectures on African culture that Senghor attended.[8] By contrast, Jomo Kenyatta, who studied at the University of London, used his own childhood experiences to write about Kikuyu culture, as we will note again later. Senghor tended to write about black culture in

[6] See Ajayi, Goma, and Johnson, *The African Experience*, pp. 18–19.
[7] Mudimbe, *The Invention of Africa*, p. 103.
[8] Andrew D. Roberts, "African Cross-currents," in *The Cambridge History of Africa*, vol. 7: *From 1905–1940*, ed. Andrew D. Roberts (Cambridge: Cambridge University Press, 1986), p. 261.

general and showed no detailed knowledge of any particular African society. The difference between Senghor and Kenyatta may reflect the British and French educational systems and colonial policies, but there was little difference in the discomfort Africans experienced when they encountered European intellectual supremacy in London or Paris.[9] In European cities, they felt marginalized from the process of knowledge production but also came into contact with radical Europeans, some of whom were opposed to colonial domination. Consequently, many African students became radical anticolonial activists, although the majority of Africa's educated elite remained true to the liberal political philosophies of their Western teachers. The journey west to learn about their own societies, and its displacements, is still made by many African scholars, if only because libraries and universities in Africa are not as well endowed as those in the West.

Europeans in African colonies with large settler populations began engaging in scientific research much earlier. They set up teaching and research institutions in the colonies in order to serve the economic and cultural needs of settler society. In South Africa, the importance of scientific research for the modernization of the colonies had long been recognized, and scientists were employed by government to study various problems of colonial society. The University of Cape Town for English-speaking South Africans traces its beginnings to 1829. Its Afrikaans-language counterpart, Stellenbosch University, began in 1866, while Fort Hare College for black students dates back to 1916. Before the rise to power of the National Party in 1948 and the introduction of the discriminatory policy of Bantu Education, Fort Hare was a major center of learning for the educated African elite in Britain's southern and eastern African colonies.

Stellenbosch University not only catered to the cultural, theological, and agricultural science needs of the Afrikaner population, it also became important as the place where National Party (NP) prime minister and theorist of apartheid Hendrik F. Verwoerd (1901–1966) served as professor of applied psychology until 1932, and as professor of sociology and social work until 1937. Born in Holland and educated at German universities, Verwoerd was a committed Afrikaner nationalist who held strong anti-British and anti-Semitic and racist views that were informed by his social science. Verwoed was the architect of Bantu Education, the policy that the NP used to systematically deny black South Africans access to the country's best educational facilities, sending them instead to substandard ethnic institutions in the homelands. Mission-trained African intellectuals found themselves trapped in a system that prevented them from participating fully in scientific research and development. Bantu Education was a core component of apartheid, the belief that each race ought to develop independent of the others, which Afrikaner social scientists like Verwoed tried to justify.

---

[9] Mkwandawire, "The Social Sciences in Africa," p. 103.

Elsewhere in Africa, colonial education developed differently. The absence of European settlers allowed the emergence of an African educated elite, although in all colonies this process was hampered by missionaries and administrators whose educational policies were similar to those of Afrikaner nationalists. After the local African elite fought to establish a secular, African-controlled institution modeled on British universities, Fourah Bay College in Freetown, Sierra Leone, was founded in 1826 as a trades training institute by the Church Missionary Society; it became affiliated with the University of Durham in 1876.[10] Ethiopia, despite its long history of African Christian scholarship, its system of writing, and the fact that it was not a colonial possession of any European power, established its national university only in the 1930s. The university structure and curriculum adopted in Ethiopia were based on European and American traditions and designed to compare favorably with overseas models.

Independent Christian churches, along with universities, sometimes served as sites of anticolonial sentiment that generated proto-nationalist ideas. Based on the rejection of colonial and Eurocentric readings of the Bible and the failure of missionaries to practice communion with their African brethren, both Zionist and Ethiopian churches tended to expand their goals beyond ethnic or national freedom and to espouse Pan-African liberation. Fourah Bay College tutor James Johnson (1835–1917) was honored by the first pan-African Congress for his work promoting the idea of a pan-African Christian state free of European control.[11] Islamic resistance movements such as the Egyptian Muslim Brotherhood, formed in 1927, appear to have been more national and focused on local problems, though Islam itself is also a global religion.[12]

Thus, until the attainment of independence in the 1960s, scientific research was in the hands of Europeans, and the African intelligentsia's contribution to social science research was not significant. With the attainment of independence, the size of the educational sector and of university enrollments grew rapidly. Not only did African governments construct many national universities, but scholarships to study abroad, especially in the United States, also became widely available as the global power blocs competed to gain influence over the young leaders in the new states. African governments saw the universities as training and research institutions to be used in social transformation and prioritized development-oriented research and teaching programs. The natural sciences and engineering were considered more important than the

[10] Ajayi, Goma, and Johnson, *The African Experience*, pp. 21–7.
[11] E. A. Ayandele, "Africa since Independence," in *The Making of Modern Africa, 1800–1960*, vol. 2: *The Late 19th Century to the Present Day*, ed. John D. Omer-Cooper, Emmanuel A. Ayandele, R.G. Gavin, and Adiele E. Afigbo (London: Longman, 1971), p. 379.
[12] Ali A. Mazrui and J. F. Ade Ajayi, "Trends in Philosophy and Science in Africa," in *Africa Since 1935: The UNESCO General History of Africa VIII*, ed. Ali A. Mazrui (Berkeley: University of California Press, 1993), p. 668.

social sciences and attracted better students and greater funding, but the social sciences nevertheless enjoyed a period of fruitful and prolific research by African scholars and Africanists.

BOURGEOIS ECONOMICS, DEVELOPMENT ECONOMICS, AND POLITICAL ECONOMY

After independence, economics was used to plan, implement, and evaluate development programs. This is the use to which economic thinking had been put since the 1940s and 1950s, when European colonial powers started implementing planned programs. It became an essential part of government work in the newly independent African states, motivated by their desire to develop as quickly as possible, and was supported by international agencies. The United Nations development decade (the 1960s) brought international experts to work in Africa, and economists conducted as much research in the universities as in the development agencies, so that government, World Bank, and university researchers interacted closely.

In the newly independent states, Keynesian and Marxian economic theories provided the theoretical foundation for much of the development planning after independence. As many African countries nationalized existing industries or created state-owned import substitution industries, the universities began to train economists in order to plan and implement national development, and to provide business administration skills for managers of state-owned enterprises. These economics graduates were state officials rather than business managers. By contrast, universities in the settler colonies of Rhodesia and South Africa taught a different kind of economics combined with accountancy. European settlers were engaged in building enclaves of capitalist industry and agriculture. Their family farms, mining houses, and manufacturing industries needed the skills of business administrators, marketing managers, and other commercial professionals.

The contribution of Africanists to various branches of economics as a result of involvement with development projects has been substantial, despite the fact that economic development is an area in which African failure has been most profound.[13] Economic thinking was stimulated by ideological battles between socialism and capitalism that affected university teaching and research programs. "Bourgeois economics" was viewed as a key component of the capitalist-oriented modernization school and considered a tool of imperialism by economists on the left, who formulated dependency theory and Marxist political economy. Africa provided opportunities for comparative research on socialist and capitalist development paths, as well as

---

[13] Paul Collier, "Africa and the Study of Economics," in *Africa and the Disciplines*, ed. Bates, Mudimbe, and O'Barr, pp. 58–82.

a site for polemical debates between the advocates of capitalist and socialist approaches to agricultural development, industrialization, and economic development.

The most exciting studies of development to emerge from the ideological battle came out of East Africa. In 1967, Tanzania launched its *Ujamaa* socialist experiment and began to pursue a policy of collective self-reliance. *Ujamaa* was abandoned after more than twenty years and is generally considered a failed experiment. What it did produce, however, was a large volume of political, economic, and sociological studies of development strategies conducted by students and by more experienced researchers.[14] Tanzanian and expatriate researchers not only studied the Tanzanian *Ujamaa* experiment closely, they also compared it to Kenya's model of capitalism. Within Tanzania itself, differences between state-owned firms and private ones, the impact of aid from China and aid from Western countries, were closely studied and fed into the literature on development. Probably the most influential contribution to these research activities came from economic historian Walter Rodney, who used Latin American dependency theory and Chicago-trained economist Andre Gundar Frank's underdevelopment theory to explain why Africa was not developed.[15] Closely related to the work that New Left scholars were producing in Europe and America, the research done in Tanzania in the 1970s and 1980s was an international as well as a national endeavor, and, equally significant, for the first time African social scientists from several countries were part of the process of creative research and debate.

Other lines of thinking in development economics emerged from the work of William Arthur Lewis. A native of St. Lucia, Lewis was advisor to a number of African and Caribbean governments and was awarded a knighthood in 1963 and a shared Nobel Prize in economics in 1979.[16] Lewis's economic theories clearly gained from his work in Africa. His theory of industrial development, based on unlimited supplies of labor, tried to chart a path out of economic backwardness by means of planned industrialization. Lewis suggested that high wages slowed down economic growth and foreign investment, and his cheap labor theory was used to determine the wage policies of African governments, especially those that employed International Labor Office advisors. Suggesting that higher wages would actually reduce the labor supply, the concept of a "backward-slopping supply curve for labor" was used to justify a continuation of colonial wage policies.[17] As

[14] L. Adele Jinadu, *The Social Sciences and Development in Africa: Ethiopia, Mozambique, Tanzania and Zimbabwe* (Stockholm: Swedish Agency for Research Co-operation with Developing Countries, 1985), pp. 96–131.

[15] Walter Rodney, *How Europe Underdeveloped Africa* (Dar es Salaam: Tanzania Publishing House, 1972).

[16] Mazrui and Ajayi, "Trends in Philosophy and Science," pp. 656–7.

[17] C. C. Wrigley, "Aspects of Economic History," in *Cambridge History of Africa*, ed. Roberts, pp. 123–4.

Lewis's critics noted, this approach paid little attention to the noneconomic state interventions that made such industrial development unlikely in African colonial economies. Egyptian economist Samir Amin blamed economic stagnation in the "Africa of the labor reserves" on a combination of low wages for migrant workers and land alienation.[18] Politically, Lewis's approach was difficult for newly independent states to implement, because in many cases nationalist politicians had mobilized support against colonial rule by highlighting the discriminatory wage policies of colonial regimes. The pros and cons of foreign investment, low and high wages, labor-intensive and capital-intensive production methods begun by Lewis were still being debated in the 1990s.

African economists also struggled with the limiting effect that political balkanization had on markets. Economic integration was made a centerpiece of African liberation by many Pan-Africanist politicians. Once they had become presidents, however, they realized that it was quite difficult to implement. Failed attempts to form integrated economies have included the Ghana–Guinea Union, the United Arab Republic, led by Nasser's Egypt, the Mali–Senegal Union of 1959, and the East African Community. Economists on the Addis Ababa–based Economic Commission for Africa (ECA) probably went further than anyone else in trying to design a program for the integration of African economies. The Lagos Plan of Action, the most important blueprint for the economic integration of Africa, was formulated by a team of ECA economists led by Professor Adebayo Adedeji. Although it was supported by the heads of state of the Organization of African Unity (OAU), in the final analysis the Lagos Plan was not implemented because of the limited ability that the ECA had to influence government policy. Their rivals, the economists working for the World Bank, who rejected both state planning and economic union as impractical, had the weight of the international financial institutions behind them and were able to intervene directly in African central banks and ministries of planning. The Lagos Plan of Action was shelved, although the OAU structures for its implementation remained in existence.

With the end of African socialist experiments, state-owned enterprises were dismantled and the research interests of most economists switched to studies of structural adjustment policies. The concurrent decline of radical scholarship provided space for the liberal tradition in economics to flourish. Africans working for international financial institutions and researchers in the Nairobi-based African Economic Research Consortium were particularly influential during the 1990s. Research into the informal sector, household economics, the introduction of feminist theory into economic theory, and the

---

[18] W. Arthur Lewis, "Economic Development with Unlimited Supplies of Labour," *Manchester School of Economic and Social Studies*, 22 (1954), 139–91; Samir Amin, "Underdevelopment and Dependence in Black Africa," *Journal of Modern African Studies*, 10 (1972), 503–24.

economics of land reform and market liberalization predominated. Political economy was in decline after 1980 but not altogether eliminated, as we will see in a later section.

POLITICAL SCIENCE AND THE POSTCOLONIAL STATE

Political science was even more a Cold War social science than economics. Not only was it the area that attracted most American Africanists, it was also the subject most challenged by the rapid pace of political change in the newly independent African states and thus required continuous research. The military regimes and one-party states that replaced the multiparty parliaments that were set up at independence tended to be unstable, and Africa watchers tried to understand the complex mix of socialist, monarchist, and other political regimes that ruled different countries. Studies of issues such as ethnic conflict, corruption, clientelism, and neopatrimonialism have over the years produced more American doctoral theses in political science than have African-related studies in economics or cultural anthropology.[19]

Within Africa itself, the nationalist leaders who led the struggle for independence provided much creative political theory. Doubts about European civilization had been raised in the minds of African soldiers and students in Europe by the barbarism of the two world wars. Due to improved education and communications, they were able to mobilize the population against colonialism when they returned to Africa, and their nationalist ideas trickled down to the trade unions, mass political parties, and the general population. The struggle for independence radicalized and politicized African intellectuals of different backgrounds and turned them into political thinkers. Thus the Cape Verde agronomist Amilcar Cabral (1924–1973) led the armed struggle against Portuguese colonialism in Guinea Bissau. His revolutionary theory about the need for Africa to reclaim its history and his analysis of the rural population's ethnic and cultural resources were important examples of the local application of Marxism to African situations and were taught in university political science courses.

The Martinique-born Algerian psychologist Frantz Fanon (1925–1961) was also a major influence on radical African political scientists. His ideas on the lasting effects of colonial violence on the mentality of oppressed people and on the catharsis of revolutionary violence were closely studied by African American radicals in the Black Panther Party as well. In Africa, his views on the revolutionary potential of the urban lumpen proletariat had a lasting influence on political theory and practice, as reflected in a number of rebellions by young political activists in countries such as Liberia and Sierra Leone.

---

[19] Michael Chege and Goran Hyden, "Research and Knowledge: Social Sciences," in *Encyclopedia of Africa South of the Sahara*, ed. John Middleton (New York: Scribner's, 1997), pp. 596–601.

Fanon's speculative writings on the inherent inability of the African ruling class to develop the continent in imitation of Europe and America continue to stimulate theories of the postcolonial state among African political scientists.

The Afromarxist variant of the single-party state grew out of the radicalization of national liberation movements in the former Portuguese colonies and the pro-Soviet military regimes in Benin, Ethiopia, and Congo (Brazzaville) during the 1970s. There was very little original Marxian analysis of party politics in the African context. Radical African intellectuals who championed Afromarxism tended to analyze African politics using Maoist or Leninist ideas on vanguard parties and the worker–peasant alliance, depending on whether their ideological allegiance was to China or to Russia.

Presidents Kwame Nkrumah (1909–1972) of Ghana, Julius Nyerere (1922–1999) of Tanzania, and Kenneth Kaunda (b. 1924) of Zambia were themselves influential as African socialist thinkers. Nkrumah was a committed Pan-Africanist, and his writings were mainly concerned with the threat of neocolonialism and how Pan-Africanists could oppose it. He was overthrown in a military coup after less than ten years in office, and his influence was somewhat diminished. Nyerere and Kaunda, on the other hand, survived coup plots and ruled their countries for more than twenty years. They tried to implement African socialism and wrote speeches and books explaining their philosophies, but both lived to see their dream of African socialism collapse.

Ironically, these founders of national universities did not pay much attention to the work of their social scientists. Julius Nyerere's African *Ujamaa* was supposed to be based on traditional kinship structures, but his opposition to capitalism on moral grounds bore a closer resemblance to the criticism voiced in Shakespeare's *The Merchant of Venice* – which Nyerere translated into Kiswahili – than to any of the political economists' studies emanating from the University of Dar es Salaam.[20] Similarly, Nyerere's model of a democratic one-party state was not based on his knowledge of African political systems, or even on major studies by African or European scholars, but on a little-known work by Guy Clutton-Brock that simplified democracy as African-style consensus politics: "The elders sit under the big tree and talk until they agree."[21]

The University of Dar es Salaam debated the nature of the postcolonial state that leaders such as Nyerere were creating. Their research can be characterized as the application of Marxist theory to African conditions and is summarized in the Ugandan Yash Tandon's collection of Dar es Salaam University essays, which captures the creative energy that gripped the university

---

[20] Mazrui and Ajayi, "Trends in Philosophy and Science," p. 674.
[21] D. W. Nabudere, "The One Party State in Africa and Its Assumed Philosophical Roots," in *Democracy and the One Party State in Africa*, ed. Peter Meyns and Dani Wadada Nabudere (Hamburg: Institut für Afrika-Kunde, 1989), p. 2.

during the 1970s and that left its mark on African political science and on the political development of East Africa.[22] Ugandan president Yoweri Museveni's uprising against Idi Amin and Milton Obote owes some of its inspiration to Frantz Fanon's analysis of violence, about which Museveni wrote while studying at the University of Dar es Salaam.[23] A number of political leaders in different countries in southern and eastern Africa participated in the university debates during the 1970s, when Dar es Salaam was a place of refuge for political exiles.

Although the Kenyan political scientist Ali Mazrui does not belong to the radical school of African scholars associated with the University of Dar es Salaam, much of his work has focused on the process of decolonization and has raised concerns about African liberation similar to those of the radical political scientists. He has developed models of analysis of Africa in areas as diverse as gender, violence, international relations, and culture.

During the 1980s, the impending collapse of Soviet socialism was preceded by the abandonment of African socialist experiments and, with it, a shift in political science. One-party and military states that had used the Cold War to obtain aid from the two power blocs found themselves under internal and foreign pressure to democratize. In that context, the modernization school, within which many American-trained political scientists worked, became more salient. Modernization theorists themselves shifted ground: Having once viewed army officers as modernizing elites, they now argued that good governance and democratization were crucial for economic development.[24] The modernization school also produced studies of the role of patron–client relationships in neopatrimonial states, the influence of tribalism and ethnicity on party politics, and the relationship between democratization and economic development. Although concern with the postcolonial state did not diminish, this focus on civil society and other processes outside the state has continued into the 1990s. Increasingly, the work of political scientists has overlapped with that of sociocultural anthropologists, sociologists, and economists.

SOCIOLOGY AND SOCIOCULTURAL ANTHROPOLOGY

Probably the most controversy-ridden social science in postcolonial Africa is sociocultural anthropology. Its history has been so closely tied to that of colonialism, and its research and teaching so closely tied to the needs of

[22] Yash Tandon, ed., *University of Dar es Salaam Debate on Class, State and Imperialism* (Dar es Salaam: Tanzania Publishing House, 1982).

[23] Yoweri Museveni, "Fanon's Theory on Violence: Its Verification in Liberated Mozambique," in *Essays on the Liberation of Southern Africa*, ed. Nathan Shamuyarira (Dar es Salaam: Tanzania Publishing House, 1971), pp. 1–24

[24] Goran Hyden, "Political Science and the Study of Africa," in *Encyclopedia of Africa*, ed. Middleton, pp. 429–31.

colonial administration, that many Africans have continued to see it as a tool of European domination. Although many prominent anthropologists were individually quite anticolonial, and although research conducted in Africa has helped social scientists to refine social theory, these facts have not improved anthropology's standing in the eyes of African scholars. Apart from its usefulness as a tool of colonial administration in such key areas as customary law and tribal authority, sociocultural anthropology also irked Africans because it was the study of primitive societies, their diets, religious beliefs, and political systems. It was essentially a tropical sociology, specializing in symbolic analysis of primitive rituals, tribalism in town settings, and persistence of tradition and backwardness in modern settings. The fact that the comparable modernization of European peasant society was studied by sociologists only reinforced the idea of racial discrimination.

During the colonial period, European sociocultural anthropology enjoyed a golden age of research and theory that was centered on Africa. Studies of witchcraft and magic by Evans-Pritchard, of legal systems by Max Gluckman, and of urbanization, modernization, and African Christianity by Georges Balandier, J. Clyde Mitchell, and Bengt Sundkler became classic texts in the discipline. African research was useful not only in helping Europeans to construct "the Other" but also in allowing comparisons to be made between, for example, independent African churches and cargo cults in the Pacific as local responses to Western culture.

Of the few Africans to have studied sociocultural anthropology during the colonial days, a number of key political leaders stand out: Kenya's first president, Jomo Kenyatta, Mozambican nationalist Eduardo Mondlane, and Ghanaian president Kofi Busia. Kenyatta was a student of Bronislaw Malinowski at the London School of Economics; he wrote his book *Facing Mount Kenya* (1938) mainly to express Kikuyu cultural nationalism. The book was not an ethnography in the British tradition of his teacher but rather a nationalist anthropology that, like nationalist languages, played an important role in identity politics.[25]

After independence, anthropology was demoted to a subdiscipline of sociology in many African universities, and some institutions did not include it in their teaching or research programs. Many sociology departments taught undergraduate courses in rural, urban, industrial, and theoretical sociology in tandem with social work and demography. This meant that anthropological research in Africa was mainly conducted by European and American Africanists to a greater extent than was the case in the other social sciences. During the 1970s, the interest in political economy generated some discussion about modes of production that was based on research done by sociocultural anthropologists, but the perception of anthropology remained a negative one. In the 1980s and 1990s, research into issues of globalization continued

[25] Jomo Kenyatta, *Facing Mount Kenya* (London: Heinemann, 1979).

to blur the boundaries between the disciplines, with issues such as power, poverty, democratization, gender and sexuality, ethnic and national identity, epidemics and other health matters being studied by researchers who favored interdisciplinary methods. The special claim of sociocultural anthropology as the social science based on prolonged fieldwork no longer appeared to be valid. The high cost of prolonged fieldwork also meant that alternative research methods were utilized by anthropologists, producing a more personal, speculative, and literary ethnography that was even less attractive to African readers than the old tribal ethnographies.

The prominent South African anthropologist David Hammond-Tooke noted in his 1997 review of the South African school of anthropology that the future of anthropology in South Africa will depend on the degree to which black scholars are drawn to the discipline.[26] Presumably, black scholars will be attracted to an anthropological approach that can answer questions pertinent to their own lives. During the 1990s, they were drawn to anthropological research that addressed national identity questions. In Ethiopia, a new master's degree program in sociocultural anthropology, supported by Norwegian aid, illustrates the changing attitude toward anthropology in Africa.[27] Ethnological sciences were seen as a valuable resource for resolving what the Ethiopian regime still calls "the national question." The desire to create and define identities is also what drives African intellectuals' concern with "studying and reinterpreting classic ethnographies in the quest to construct new theories about quintessential aspects of African thought and society."[28] Given the strongly felt need for self-definition, it cannot be otherwise.

FROM NATIONAL UNIVERSITIES TO REGIONAL RESEARCH NETWORKS

The 1980s have been labeled the "lost decade" in the development literature. The prolonged economic and political crisis in Africa resulting from the fall in prices of traditional colonial economy exports and the lack of democratic regimes took their toll on the universities. Rising costs and diminishing resources affected research funding in Europe and America as well, but in Africa they led to the decline of universities as research and teaching centers. In addition, funding, academic freedom, and even the lives of students and staff members were violated with impunity by the governing class. As a result, the gains won during the first two decades of independence were lost.

[26] William David Hammond-Tooke, *Imperfect Interpreters: South Africa's Anthropologists, 1920–1990* (Johannesburg: Witwatersrand University Press, 1997).
[27] Rene Devisch and Peter Crossman, *Consultants' Report on Endoginisation and African Universities: A Survey of Endoginisation Initiatives at Six African Universities* (Leuven: Africa Research Centre, Department of Social and Cultural Anthropology, 1998), p. 13.
[28] T. O. Beidelman, "Anthropology and the Study of Africa," in *Encyclopedia of Africa*, ed. Middleton, pp. 55–7.

The best professors moved overseas, and universities were frequently closed because of political clashes or lack of resources.

In order to reverse this decline, some of the scholars who had not sought greener pastures abroad continued to do research in national and pan-African research networks funded by overseas donors. Research, publication, and regular conferences – no longer common in universities – were continued by members of groups like CODESRIA (Council for the Development of Social Science Research in Africa), which was based in Senegal and associated with prominent scholars such as the economist Samir Amin (Egypt), also the head of the Third World Forum; the political scientists L. Adele Jinadu and Claude Ake (Nigeria) and Emannuel Hansen (Ghana); the economist Thandika Mkandawire (Malawi); and the sociocultural anthropologist Archie Mafeje (South Africa). Virtually the same people were active in scientific networks and in university faculties, and were thus influential in producing alternatives to the limited social science priorities of their own national universities and the global academy. These networks operated within a Pan-African social science academy still in its formative stages but were effective in preventing the complete collapse of African social science research.

Apart from CODESRIA, the Ethiopian-based OSSREA (Organization for Social Science Research in Eastern Africa); AAPS (African Association of Political Science) and SAPES (Southern African Political Economy Series), both based in Zimbabwe; and the African Economic Research Consortium, based in Nairobi, were dominated by English-speaking countries. Their research and teaching programs were predominantly policy-oriented and dependent on financial support from overseas. Unlike the universities, they could contribute to capacity building and national development and still function relatively independent of state interference. In 1991, the former executive secretary of the ECA, Adebayo Adedeji, set up "Africa's first think-tank." The African Center for Development and Strategic Studies (ACDESS) created yet another possibility for social science research outside the universities.[29]

At the start of the 1990s in Africa, the ending of apartheid and the almost simultaneous democratization of former one-party states and military regimes heightened the feeling that the liberation of the continent was finally complete. This led to a feeling of optimism across the African continent and diaspora based on the hope that attaining genuine development was now possible. As in the 1960s, another decade of independence, much was expected of the second liberation of Africa, but most of the analysis of the processes of regime change suggests that disappointment with the second liberation came much more quickly. Military regimes, one-party-dominated parliaments, and even countries without functioning states continued to proliferate. Economic structural adjustment continued to be driven by pressure

[29] Adebayo Adedeji, ed., *Africa within the World: Beyond Dispossession and Dependence* (London: Zed Books; Ijebu-Ode: ACDESS, 1993).

from Washington-based economists and to be implemented half-heartedly by African governments.

Against the background of these realities, some African leaders launched a political campaign under the slogan of African Renaissance. The driving force behind this campaign was South African deputy president Thabo Mbeki, an economist by training.[30] It is too early to know what effect this campaign will have on the social sciences, but to date it has been driven largely by cultural nationalism.

This chapter has shown how nation building and economic development, the main social engineering projects of the late twentieth century, have stimulated and constrained social science research in Africa. The instrumental attitude of African social science has been accused of harming basic research. However, creating a better modern society has always been a preoccupation of the social scientist, although history has presented different societies with different issues.[31] The quality of teaching and development planning depends on the quality of research, just as the quality of research may gain from the politics of the day, as the experience of Dar es Salaam University shows. An African Renaissance, if it is to avoid the mistakes of previous attempts to transform the continent, must find a way to widen, not restrict, access to the social sciences.

---

[30] Thabo Mbeki, "The African Renaissance: South Africa and the World," speech at the United Nations University, 9 April 1998 (http://www.unu.edu/Hq/unupress/mbeki.html).

[31] Jinadu, *The Social Sciences and Development in Africa*, p. 11.

27

THE SOCIAL SCIENCES IN INDIA

Partha Chatterjee

This chapter traces in outline the history of the modern social sciences in India from the late eighteenth century to the present. It begins with an account of the "discovery" of India by the European Enlightenment, which created the field of Indological studies. It then describes the practices of the modern disciplines of social knowledge in India in their relation to the institutions of governance created under British colonial rule and thereafter to the project of the Indian nationalist movement. The final section deals with the professionalization of the disciplines during the postcolonial period. The focus is on the disciplines of history, economics, sociology, social anthropology, and political science.

COLONIAL ORIGINS

The decisive event in the early institutional history of modern social knowledge in India was the founding of the Asiatic Society of Bengal in Calcutta in 1784 at the initiative of William Jones (1746–1794), an official of the East India Company and a major linguist of his time. For almost a century, the Asiatic Society was the chief institution in India for encouraging, organizing, and propagating knowledge about the country's history, philosophy, religion, language, literature, art, architecture, law, trade, and manufacture. Most European scholars who worked in India were associated with the Society. They helped to establish Indological scholarship as a specialized field in the world of modern learning.[1]

It was above all in the study of language that Indological scholarship became important for the rise of the scientific disciplines of social knowledge in Europe during the nineteenth century. The study of Sanskrit grammar

[1] O. P. Kejariwal, *The Asiatic Society of Bengal and the Discovery of India's Past, 1784–1838* (Delhi: Oxford University Press, 1988).

based on the classical texts of Pāṇini (ca. 400 B.C.) provided the foundation for modern linguistic analysis and, following upon the pioneering work of Friedrich von Schlegel, Franz Bopp, and Eugène Burnouf, led to the growth of the field of comparative philology. The tracing of linguistic relationships established the common properties of an Indo-European family of languages. This in turn produced, during the second half of the nineteenth century, theories of a common Aryan race, two branches of which were supposed to have migrated and settled in India and Europe, respectively.[2]

The collection and study of classical Sanskrit and Pali texts by European Indologists created the idea of India as a civilization of great antiquity and of great philosophical and aesthetic sophistication. The compilation and translation of these texts into European languages, first by the Asiatic Society but later, most famously, in the series *The Sacred Books of the East*, edited by Friedrich Max Müller (1823–1900), made available to the European intellectual world the materials for the construction of a distinct civilizational entity called India. Every major current of social theory in the nineteenth century took account of this entity in its description of the historical emergence and character of the modern world. The English political economists and utilitarians, the French positivists, and the great system builders such as Hegel, Marx, and Weber all devoted considerable attention to defining the place of India in the dynamics of world history.

The image of India created by the Indologists, confined as they were almost entirely to dealing with religious, philosophical, and literary texts of the "high" (and predominantly Brahmanical) tradition, was supremely abstract. On the ground in India, however, the British rulers, following their military conquests, were faced with the task of raising revenues and keeping order on the vast subcontinent. Carrying out this task meant the collection and recording of a body of empirical information about India of astounding range and detail, often shaped by projects of social engineering in which the colony acted as a laboratory for Physiocrats during the eighteenth century, for utilitarians and liberal reformers during the nineteenth, and for welfarists in the twentieth. Taken in its entirety, governmental information of various kinds still remains by far the most important source of factual knowledge about Indian society.

There were four main forms of production and organization of this knowledge. The earliest was the writing of *land revenue histories*. Soon after the conquests in Bengal, British officials began to compile detailed local histories of the claims, titles, rights, and privileges, both formal and customary, of all classes of people to the use and disposal of land. Soon this became a regular series of published materials on revenue history and land settlement, organized district by district and updated every three or four decades.

---

[2] Thomas Trautmann, *Aryans and British India* (Berkeley: University of California Press, 1997).

The second form of official knowledge was that of the *survey*, which began in British India as early as 1765 with the mapping of the conquered territories. The central institution was the Survey of India, but through the nineteenth century nearly a dozen other specialized and permanent organizations were set up to produce a cumulative body of information on India's natural resources and social and cultural features.

The *census* was the third institutional form of colonial knowledge. Following initial local attempts to count the population, the census of India was conducted once every decade beginning in 1871. It compiled basic information on age, occupation, caste, religion, literacy, place of birth, and current residence for the entire population of British India. The census reports not only presented detailed statistical information but also contained many analytical studies of the caste system, religion, fertility and morbidity, domestic organization, and economic structure. It provided the basis for such widely used government publications as the *Imperial Gazetteers* series, which compiled all relevant information for each district of British India, and the *Tribes and Castes* series, in which scholar-officials put together detailed ethnographies of castes and tribal populations for each region of India.

The fourth form was that of the *museum*, in which archaeological and artistic specimens, texts, and manuscripts were collected and preserved for the use of scholars. The first large-scale museum was set up in 1814 at the Asiatic Society. This collection later became the core of the Indian Museum in Calcutta, established in 1866 as the principal imperial museum. In 1874, the Archaeological Survey was set up in order to record archaeological sites, carry out excavations, preserve historical monuments, develop on-site museums, and build collections of archaeological specimens.[3]

The voluminous published official information provided for European scholars the basis for grand theoretical constructions about the nature of Indian society. Three institutions were thought to contain the key to the mystery of unchanging India: the caste system, despotic kingship, and the village community. The caste system was held to have imposed a rigid division of labor that hindered social mobility. Oriental despotism meant a one-way extraction of the surplus from the peasant communities to a ruling elite immersed in luxury consumption. The largely self-governing and self-reproducing village communities ensured a low-level subsistence production. This, it was argued, explained why, despite frequent changes in political regimes at the top, Indian society had remained stagnant and unresponsive to change.

[3] Bernard S. Cohn, *Colonialism and Its Forms of Knowledge: The British in India* (Princeton, N.J.: Princeton University Press, 1996); C. A. Bayly, *Empire and Information: Intelligence Gathering and Social Communication in India, 1780–1870* (Cambridge: Cambridge University Press, 1996).

NATIONALIST CONSTRUCTIONS

The first formal institution of modern Western learning for Indians was the Hindu College established in Calcutta in 1817. Schools and colleges for Western education proliferated all across India during the subsequent decades, and in 1857 three universities were set up – at Bombay, Calcutta, and Madras – to regulate the courses of study and to conduct public examinations. Through the second half of the nineteenth century, secondary and higher education, consisting mainly of courses in the modern Western sciences and humanities and using both English and the modern Indian languages, expanded considerably, chiefly through the efforts of nationalist educationists and social reformers.

By that time, an arena of public discussion on social and political questions had been created, especially in cities such as Calcutta, Bombay, Poona, and Madras. Intellectuals, often belonging to the new learned societies or associated with particular journals and newspapers, would engage in well-informed and theoretically sophisticated debates. Many of these public intellectuals were lawyers or teachers by profession, but, before the formal disciplinization of the social sciences in university departments during the early decades of the twentieth century, they were the pioneers in modern scientific writing on social questions in India.

As far as traditional genres of history writing in India are concerned, there were two main types. One was derived from the cosmic histories of the Puranic or mythological tradition in Sanskrit, in which mythical stories about gods and goddesses merged unproblematically with dynastic histories of earthly kings and queens. The other was the court history tradition, written mainly in Persian for the Muslim rulers of India, which chronicled the deeds of kings and dynasties. By the eighteenth century, the two genres were sometimes combined in regional forms – genealogical histories of prominent landed or trading families written in the vernacular languages.

These forms were rapidly superseded during the late nineteenth century, following the adoption of Western historiographical modes of writing Indian history by the new Indian intellectual elites. The impetus here is typically represented by the exhortation of the Bengali novelist Bankim Chandra Chattopadhyay (1838–1894): "We must have our own history!"[4] Modern historical writing by Indians emerged mainly through an interlocution with British histories of India, of which the three most influential texts were those of James Mill (1773–1836), Mountstuart Elphinstone (1779–1859), and Vincent A. Smith (1848–1920). Indian historians were strongly attracted by the Indologists' idea of the greatness of ancient Indian/Aryan civilization. Much of their effort went into the discovery, authentication, and interpretation of textual and other sources that threw light on early India. Their

---

[4] Bankimchandra Chattopadhyay, *Bankim Rachanabali*, vol. 2 (Calcutta: Sahitya Samsad, 1956), p. 337.

nationalist persuasions also led them to reject the prejudiced generalization about the Oriental despot: A major focus of Indian research was on establishing reliable chronologies and accounts of political dynasties during the pre-Islamic period. R. G. Bhandarkar's *The Early History of the Deccan* (1884) and H. C. Raychaudhuri's *Political History of Ancient India* (1923) were two of the more important examples of such research. Nationalist historians of the early twentieth century were also concerned to show the existence of responsible monarchies and representative institutions of local governance in early India. Influential examples were K. P. Jayaswal's *Hindu Polity* (1918) and Radha Kumud Mookerji's *Fundamental Unity of India* (1914) and *Local Government in Ancient India* (1919).

A common trope in the Indological construction was a narrative in which an ancient period of civilizational greatness was followed by medieval darkness. This idea was bolstered by the works of British historians of India, who portrayed Islamic rulers as intolerant, degenerate, and brutal. Although W. H. Moreland (1868–1938) helped to pioneer a more systematic and reliable use of sources for the Mughal period, his overall narrative was still one of Mughal India as a medieval tyranny relieved by the advent of British rule. Some influential Indian historians of the period of Islamic rule, such as Jadunath Sarkar and Ishwari Prasad, followed the same pattern. Countering this tendency were works by Muhammad Habib (1927) and K. M. Ashraf (1935), who attempted to describe the Sultanate and Mughal periods as one distinct phase in Indian history with its own economic, social, and cultural achievements – a phase in which civilizational elements from the Islamic world mingled creatively with non-Islamic elements to produce a new synthesis. At the same time, historians such as I. H. Qureshi, in a work published in 1942, emphasized the distinctly Islamic character of the Muslim monarchies in India and insisted that they were benevolent, tolerant, and efficient systems of rule.[5]

Nationalist accounts of the period of British rule began to appear beginning in the late nineteenth century, more in the Indian languages than in English. This was accompanied by new efforts, supported by learned societies, literary academies, and princely states in the different regions, to collect, preserve, and disseminate materials of local and regional history. In Bengal, for instance, the first major critical work of nationalist history – by Akshay Kumar Maitreya (1861–1930) – described the British conquest of Bengal in 1757 as the result of corruption and low intrigue. Contrary to British accounts, it portrayed Siraj-ud-daulah, the last ruler of Bengal, as courageous, patriotic, and a victim of treachery. In northern India, Bharatendu Harishchandra (1850–1885) and the Kashi Nagari Pracharini Sabha launched a highly influential series of history books in Hindi that fed the new nationalist sentiments by telling a story of seven centuries of "foreign oppression" in India under Muslim rule. In Maharashtra too, nationalist histories fed into strong revivalist

---

[5] C. H. Philips, ed., *Historians of India, Pakistan and Ceylon* (London: Oxford University Press, 1961).

feelings for the Maratha Empire as a bastion of Hindu rule. At the initiative of such men as V. K. Rajwade, V. S. Khare, and G. S. Sardesai, valuable work in the collection, editing, and publication of historical sources was produced. Tamil, Kannada, and Malayalam were the other languages in which the histories of regional kingdoms were compiled and published, helped by the support given to these efforts by the princely states of Mysore and Travancore-Cochin.

Academic histories produced in university departments during the early twentieth century showed their nationalist affiliations by choosing subjects such as the history of the Maratha and Sikh Empires, but explicitly critical histories of the period of British rule were rare. These came from nonacademic circles – for example, V. D. Savarkar's strongly anti-British history of the 1857 revolt as the first war of Indian independence.

The first modern social philosophies of Europe to have a significant impact on the new Indian intellectuals were English utilitarianism and French positivism. The works of Jeremy Bentham and John Stuart Mill, as well as those of Auguste Comte and later of Herbert Spencer, were avidly discussed in some of the new learned societies set up in Calcutta during the mid nineteenth century. In 1867, the Bengal Social Science Association was founded to "promote the development of social science" in Bengal. James Long, Lal Behary Day, Iswar Chandra Vidyasagar, Abdul Latif, Rajendra Lal Mitra, and Romesh Chandra Dutt, leading lights of the new intellectual resurgence in Bengal, were active in this body. The ideas of these men were disseminated through the new Bengali periodical press. Leading social thinkers such as Bankim Chandra Chattopadhyay and Bhudeb Mukhopadhyay, though not belonging to any particular circle, were deeply familiar with contemporaneous Western social philosophies.[6]

The main institution of Indian society to come under the new sociological gaze was, not surprisingly, that of caste. Armed with the tools of modern sociohistorical analysis, Indian thinkers attempted beginning in the late nineteenth century to write academic treatises on the Indian caste system that were, or so they claimed, better informed and more nuanced and culturally sensitive than the theories put forward by European scholars. Most of these works, such as those of S. V. Ketkar (1909), Benoy Kumar Sarkar (1914), and Bhupendra Nath Dutt (1944), as well as the multivolume *History of Dharmasastras*, by P. V. Kane (1930–62), consisted of sociological interpretations of classical, mostly Brahmanical, texts. Many such works embodied a nationalist desire to discover a rational kernel in the social institution of caste, based on concepts such as the division of labor and the need to maintain a harmonious unity of the social whole in the presence of natural and social differences.

The first university department for the formal study of sociology as an academic discipline was started at the University of Bombay in 1919 at the

---

[6] Bela Dutt Gupta, *Sociology in India* (Calcutta: Centre for Sociological Research, 1972).

initiative of Patrick Geddes, a town planner and geographer who spent most of his career in India. G. S. Ghurye, his student at Bombay, who did his doctoral work at Cambridge and returned to head the department, is often regarded as the pioneer of academic sociology in India. The Bombay department produced a galaxy of students who would, during the 1950s, dominate the field of sociology and social anthropology. The other sociology department that had a significant impact was the one at the University of Lucknow, where Radha Kamal Mukherjee (1889–1968), D. P. Mukerji (1894–1961), and D. N. Majumdar (1903–1960) were the leading lights. It was largely at this time that Indian sociologists began to turn their attention from textual interpretation to the empirical study and analysis of social institutions and practices in contemporary India.

Even during the early decades of the twentieth century, it was not customary in Indian intellectual circles to make a distinction between sociology and anthropology. Of those who are regarded as pioneers of what is now recognizable as anthropological research, Sarat Chandra Roy (1878–1942) is the most distinguished figure. A lawyer living in the small town of Ranchi in southern Bihar, a region inhabited by tribal populations, he wrote several pathbreaking ethnographic studies of the Oraon, the Munda, and other tribal peoples. He also founded in 1921 *Man in India*, one of the first journals of anthropology in India. Another pioneer was Ananthakrishna Iyer, who studied the tribes and castes of Cochin and Mysore during the first decade of the century. In 1929, Nirmal Kumar Bose published his book *Cultural Anthropology*, which set out a functional theory of culture. D. N. Majumdar carried out many anthropological studies of tribal groups such as the Ho, the Kol, the Korwa, and others; Verrier Elwin studied the tribes of central and northeastern India. Until the 1940s, anthropological study in India largely meant the study of tribal peoples.

When the Zoological Survey of India was set up in 1916, it had an anthropological section. In 1945, less than two years before Indian independence, after much pleading from the anthropologists in the section, the government decided to open an Anthropological Survey of India. Research at the Survey has been dominated to an extraordinary degree by physical anthropology and anthropometry.

The first generation of Indians to take part in public debates over economic issues was active around the middle of the nineteenth century. They were well versed in the writings of Adam Smith, David Ricardo, Thomas Malthus, and John Stuart Mill, and most were enthusiastic supporters of the doctrine of free trade. But by the last quarter of the century, leading Indian publicists on economic questions had become admirers of Friedrich List and the German historical school and critics of English political economy.

The most significant Indian writing on economics during the late nineteenth century came from western India, especially from Bombay and Poona.

This was the region where Indian entrepreneurs started the first modern industries; it is thus not surprising that the most articulate nationalist thinking on economic matters should appear there. Dadabhai Naoroji (1825–1917), the most statistically minded writer of the period, is best known for his demonstration of the "economic drain" from India. He interpreted India's recurrent export surplus with Britain as a symptom of the structural imbalance of a colonial economy and of a net transfer of purchasing power from India to Britain.

A more elaborate framework of nationalist economic thinking was erected by G. V. Joshi, Mahadeo Govind Ranade (1842–1901), and Gopal Krishna Gokhale (1866–1915). Their arguments proceeded from a criticism of the colonial policy of repeatedly increasing the tax revenues to balance the budget. They pointed to the intersectoral imbalances that had emerged in India as a result of this despotic policy and argued for a more comprehensive and subtle view of the national economy as a whole. Their perspective was one of industrialization as the path to national economic growth and the elimination of poverty. They also argued, within the limits of their liberal political views, for state protection and support of infant industries in the face of foreign competition. They were supportd by Romesh Chandra Dutt's two-volume *Economic History of India* (1900, 1902), the first academic-historical account of the deindustrialization of the Indian economy after about 1800. These nationalist writings represented the most influential trend in Indian economic thinking, one that would continue into the late twentieth century.[7]

Until the turn of the century, political economy was taught in colleges and universities in India as part of the study of history. In 1909, the first chair in economics was established at the University of Calcutta, and the first undergraduate honors course was opened. Soon other universities followed suit, and by the 1920s the first generation of professionally trained economists had emerged to take up academic positions in university economics departments.

Beginning in the 1920s, there was a surge in the publication of research monographs dealing with the empirical description of and theoretical problems relating to different aspects of the Indian economy. For example, V. K. R. V. Rao gave the first systematic and reliable estimates of India's national income for the periods 1925–9 and 1931–2 (published in 1939 and 1940). C. N. Vakil and S. K. Muranjan, in a work published in 1927, made an elaborate presentation of the nationalist viewpoint on monetary policy in which they argued for holding India's gold and foreign exchange reserves at home and for allowing a mutual adjustment between price levels and the exchange rate of the rupee. B. N. Ganguli published in 1938 the first systematic study of agricultural production in the Ganges valley, one of the largest agricultural regions in the country.

---

[7] B. N. Ganguli, *Indian Economic Thought: Nineteenth-Century Perspectives* (New Delhi: Tata McGraw-Hill, 1977).

Considering the subjects that would be of greatest interest in postcolonial India, the two areas in which significant developments took place during the period between the two world wars were those of tariff protection and planned industrialization. Works by Jehangir Coyajee (1924) and B. P. Adarkar (1941) strongly argued the case for protection of nascent industries that were in danger of being wiped out by unequal foreign competition. The first book on planned industrialization in India was not by an economist but by an engineer-administrator, Mokshagundan Visvesvaraya (1861–1962), who published his *Planned Economy for India* in 1934. It contained the first elaboration of the idea of planning as a technical exercise carried out by experts, with industrialization as the key to rapid growth and the removal of poverty. Ten years later, a group of Indian industrialists led by Purshotamdas Thakurdas produced the first major planning document, which would become known as the Bombay Plan. After independence, planning would be the most important and challenging area to engage the attention of Indian economists.[8]

SOCIAL SCIENCE IN INDEPENDENT INDIA

When India became independent in 1947, there were a total of twenty universities in the country. By the early 1980s, there were over two hundred. This was the result of a huge expansion in higher education directed and financed almost entirely by the federal and state governments. In particular, there was a massive growth in social science teaching and research. In 1969, the Indian Council of Social Science Research (ICSSR) was set up by the government to promote and coordinate advanced research in the social science disciplines. Over the next two decades, the ICSSR established a network of nearly twenty-five research institutes and regional centers across the country. In addition, the Indian Council of Historical Research was founded in 1972.

The base of social science teaching in India has widened enormously since the 1950s. Further, both research and teaching are now much more closely integrated with international, and especially Anglo-American, professional norms, procedures, and styles in each of the disciplines. In contrast to the colonial period, the bulk of teaching at the undergraduate level is now in the Indian languages. There is consequently a social science literature in these languages that is fed by the professional disciplines. Virtually all advanced research, however, is in English, the language of professional communication among Indian social scientists.

Following independence, two main political concerns shaped historical scholarship in India – (1) the assessment of colonial rule and of the anticolonial

---

[8] Bhabatosh Datta, *Indian Economic Thought: Twentieth-Century Perspectives, 1900–1950* (New Delhi: Tata McGraw-Hill, 1978).

struggle, and (2) the shaping of an historical consciousness of modern nation-hood. Both concerns were strongly affected by the fact that independence was accompanied by the partition of the country along religious lines.

For at least three decades after independence, Indian historiography was primarily engaged in presenting to the world of historical scholarship a modern, professionally sophisticated, nationalist history of India. But by the 1950s, it had divided into two trends. One was exemplified most elaborately by the eleven-volume *History and Culture of the Indian People* (1951–80), of which R. C. Majumdar was the general editor. This series, sponsored by the Bharatiya Vidya Bhavan, a private educational trust, was strongly oriented toward what may be called "Hindu nationalism," that is, a celebration of the ancient past as a history of Hindu civilization, the treatment of the centuries of Muslim rule as a period of foreign oppression, and the description of the anticolonial movement as one of Hindu nationalism challenged by Muslim separatism. This orientation was countered by a trend that described itself as "secularist," one that received official sponsorship from various state agencies but that was also carried forward by a group of Marxist historians. It emphasized the plurality of religious and cultural elements that went into the making of ancient and medieval Indian society and described the freedom movement as the anticolonial struggle of a composite Indian nation hemmed in by both Hindu and Muslim communalist politics. The unfinished *Comprehensive History of India*, began in 1957 and sponsored by the Indian History Congress, was meant to contain the full-fledged statement of this position.

In general, however, historical scholarship beginning in the 1950s was marked by increased professionalization, technical sophistication, and the exploration of new fields of research and new historical sources. Work on the early history of India, which had tended to rely heavily on textual sources, could now be based on material evidence from much-expanded archaeological, epigraphic, and numismatic sources. Already, the early history of India had been pushed back several centuries by the discoveries during the 1920s of the Mohenjo-Daro and Harappa sites in the Indus valley. Beginning in the late 1950s, new excavations in western India and Pakistan took the pre-Aryan Harappan culture back to the third millennium B.C. The evidence from these materials raised doubts about the earlier theory of an Aryan invasion from the north and led many historians to think of the transition from the Indus cities to the Vedic social formation as one of gradual change and intermingling over several centuries.[9]

Another question over which there was a prolonged debate was that of the nature of the state in India. The dominant nationalist tendency was to describe the premodern Indian state as unitary, centrally organized, territorially defined, headed by a strong ruler, and administered by a hierarchical bureaucracy. In many accounts, this model of strong "stateness," supposedly

[9] Romila Thapar, *Interpreting Early India* (Delhi: Oxford University Press, 1992).

exemplified by the Maurya (ca. 322–185 B.C.) and Gupta (ca. 320–510 A.D.) Empires, was evidence of the advanced nature of ancient Hindu civilization. In 1956, D. D. Kosambi, a mathematician and Marxist historian, put forward the idea of two processes of feudalism in India, one from above and one from below. In his *Indian Feudalism* (1965), R. S. Sharma argued that a fragmented and decentralized feudal state formation emerged during the post-Gupta period in northern India. The argument was initially challenged mainly on the ground that the Indian evidence did not fit the model of feudalism as it was known from European history. However, through the 1970s and 1980s, as the debate proceeded on how to characterize the premodern Indian state, historians following Sharma's thesis made the argument for a specifically Indian variant of feudalism, although this view too was not widely accepted. Another argument was advanced by Burton Stein, mainly on the basis of the evidence from southern India; Stein proposed a segmentary state somewhere between the stateless tribal forms of government and the bureaucratic state of the Mughal Empire. As a result of these debates, there is now a much greater awareness, summed up especially in the widely known writings of Romila Thapar, of variations over periods and regions and of the emergence of state formations as a changing societal process. The conventional identification, in both colonial and nationalist historiographies, of the ancient and medieval periods with the periods of Hindu and Muslim rule has been strongly questioned. It is now common to speak of an early medieval period starting three or four hundred years before the founding of the Turko-Afghan kingdoms in northern India during the twelfth century.[10]

Research on the Delhi Sultanate (1206–1526) and the Mughal Empire (1526–1858) made great advances in detail, precision, and theoretical sophistication, especially through the contributions of historians from the Aligarh Muslim University. The standard work on the sultanate was produced in the form of volume 5 of the *Comprehensive History of India* (1970), edited by Muhammad Habib and K. A. Nizami. Irfan Habib's *Agrarian System of Mughal India* (1963), a thoroughly researched account of the Mughal Empire as a centralized bureaucratic state crumbling under the weight of its internal contradictions, especially in the form of a series of peasant revolts, became the classic work on the Mughal period. Most of this work concentrated on economic production, land revenue systems, and bureaucratic structure; it largely avoided other social, religious, and cultural issues. During the 1980s and 1990s, however, the orthodox view of the eighteenth century as a period of decline and disorder was challenged by a revisionist history that claimed that it was instead a period of new beginnings in indigenous economic enterprise, state building, and cultural innovation. The debate among historians such as Burton Stein, C. A. Bayly, Muzaffar Alam, Sanjay Subrahmanyam, and others has now shifted attention to the historical significance of

---

[10] Herman Kulke, ed., *The State in India, 1000–1700* (Delhi: Oxford University Press, 1995).

these new possibilities during the early years of European colonialism in India.

Until the 1970s, writing on the colonial period was dominated, on the one hand, by the emergence in India of a nationalist history of colonial exploitation and of the anticolonial struggle of the Indian people against an authoritarian state and, on the other, by new histories written at centers of South Asian studies in Britain and the United States that described Indian nationalism as the scramble for power of self-seeking Indian elites, themselves spawned by British rule. Both sides in this debate made intensive use of the massive colonial archives and also opened up an extensive range of nonofficial records, literary and visual materials, and oral sources. In the process, beginning in the 1980s, a whole new range of issues concerning the histories of subordinate and marginal groups – peasants, lower castes, tribal peoples, women, religious and linguistic minorities – began to be debated. This work, of which the writings of the *Subaltern Studies* group are well-known examples, not only has spoken of distinct histories of such groups that cannot be encompassed within the terms of a history of the "nation" but also has inflected that national history itself with new questions of cultural politics having to do with the specific roles within the "nation" of regions, classes, castes, and genders. A related aspect is the emergence of well-researched regional histories that have strongly questioned the conventional assumption that developments in northern India were somehow the key to the demarcation of the periods and phases of "Indian" history.

One must also mention the degree to which historical writing in India has become entangled with highly sensitive political issues, of which religious communalism is probably the most contentious, but which also include questions of regional, linguistic, caste, and tribal identity. In many of these debates, historical evidence has been marshaled in support of particular political claims. In a situation where the domains of the professional and the popular are clearly separated by language – English for academic research, the Indian languages for popular dissemination – some historians are worried about maintaining the integrity of their professional roles. Others have sought more effective ways to popularize undistorted historical research.

The new contact of Indian sociologists and anthropologists with international trends in the discipline meant a significant change in the style and content of their research.[11] The most influential orientation during the 1950s and 1960s was that of a structural-functional theory of modernization. The preferred area of research was contemporary Indian rural society, especially small communities in the process of change. The village was usually treated as

---

[11] Sociological writings in India during the period after independence are surveyed in two series published by the Indian Council of Social Science Research, both entitled *A Survey of Research in Sociology and Social Anthropology* (1972 and 1985).

a functional whole, with different caste groups constituting its parts. In this framework, the issues investigated were the local caste structure, factionalism, patron–client relations, the relation between caste and class, and the relation between the village and the outside world. The style was clearly marked in the collection of village studies entitled *India's Villages* (1955), edited by M. N. Srinivas. On social change, Srinivas's suggestion that there were two forms of mobility in contemporary Indian society – namely, sanskritization and westernization – was very influential. Sanskritization meant upward mobility attained by adopting the cultural styles of the upper castes – a process seen in Indian history for a long time. Westernization was the recent phenomenon of adopting the cultural styles of the modern West as a sign of social power and prestige.

Indian sociology was now faced with the task of defining the core of Indian tradition in the face of modernization, and caste continued to be the main focus of attention. Around the late 1960s, at least three sociologists attempted systematic statements of the fundamental structure of Indian society and the changes that it was undergoing: Irawati Karve in her *Hindu Society: An Interpretation* (1965), Louis Dumont in *Homo Hierarchicus* (1966), and Milton Singer in *When a Great Tradition Modernizes* (1972). In India, the Department of Sociology at the University of Delhi, under the leadership of M. N. Srinivas, emerged during this decade as the premier center of research and teaching in sociology and social anthropology. Abroad, the University of Chicago became a very important center of research on Indian cultural anthropology.

During the 1970s, the influence of structuralism was felt in several studies of caste structure, kinship structure, ritual, and religious beliefs, most notably those of J. P. S. Oberoi and Veena Das. Alongside, there was considerable interest in the use of Marxian methods, especially for the study of the relation between caste and class and of social movements. M. S. A. Rao and A. R. Desai, in particular, organized major collections of studies on a variety of social movements in India during the colonial and contemporary periods. Another notable collection is the recent forty-three-volume *People of India* series, edited by K. Suresh Singh, in which the Anthropological Survey of India has attempted to present comparative ethnographies of over 4,500 "communities" living in India. The project is reminiscent of the production of colonial knowledge, except that the research has been carried out by an agency of the postcolonial nation-state.

The inauguration of a developmental state carrying out a program of planned industrialization presented Indian economists with a whole range of new theoretical and empirical problems. The key figure during the 1950s was P. C. Mahalanobis, a physicist and statistician, who took charge of drawing up the crucial Second Five-Year Plan for the government of India. From his base at the Indian Statistical Institute in Calcutta, he organized a continuous series of discussions and training courses on economic growth and planning in which virtually every major economist and statistician in the world took part. At the same time, Presidency College in Calcutta became a major center for the teaching of economics and produced a steady supply of accomplished economics graduates for over three decades. Mahalanobis was also instrumental in organizing a huge official network for collecting and publishing statistical information for economic analysis. During the 1960s, the Delhi School of Economics, under V. K. R. V. Rao, emerged as the premier center of postgraduate training and research in the country.

Beginning in the 1960s, Indian economists were participating in professional research and teaching at the most advanced international levels and in all branches of economics. Nevertheless, economic development and planning, in both its theoretical and empirical aspects, occupied center stage. The study of the welfare aspects of economic policy, especially the relation of economic growth to questions of justice and equity, emerged during the late 1960s. Amartya Sen, Sukhamoy Chakravarty, and Jagdish Bhagwati are among the many scholars who made important contributions to the growing literature on economic development during the 1960s.

By the 1970s, when the initial euphoria of the planning experience had passed, major debates emerged over certain specifically Indian themes in development economics. One was over the role in economic development of the vast agricultural sector: Was it a constraint on economic growth, or could it be suitably restructured to make it a contributor to the process of development? This debate was accompanied by numerous empirical studies on forms of bondage, tenancy, and employment in the rural sector, on the relation between farm size and productivity, on product and credit markets, and on many other institutional features of Indian agriculture. The second theme was the role of public investment in promoting industrial growth. On this, the debate has been mainly between those who have questioned the rationale of import substitution strategies and the economic efficiency of state-sponsored industrialization, and those who argue that without sufficient public investment, both growth and equity would suffer. The former group has largely relied on orthodox, neoclassical, mainly microeconomic arguments, whereas the latter group has mostly used macroeconomic reasoning in the tradition of John Maynard Keynes and Michál Kalecki. Since the 1980s, an important dimension has been added to these debates: namely, the role of the external economy and especially that of direct foreign investment. This theme has raised questions not only about the short- and long-term implications for growth but also about distributive justice and national sovereignty. The fourth theme relates to technology – its import and appropriateness, its adoption and diffusion, the sustainability of technological change, and the possibilities for innovation and indigenous development. An important comparative perspective into which Indian discussions have been drawn in recent years is the so-called success story of industrialization in East and Southeast Asia, as well as the problems faced by the economies of several South American countries. The fifth theme, related in many ways to the other four, is that of the revenue and monetary policies of the government and the legal regulation of economic institutions.[12]

Modern political thinking in India during the late colonial period was mainly liberal in spirit and legal-constitutionalist in method. A parallel stream, however, ran alongside the main one. For three decades begining around 1920, Gandhian leaders kept up a critique of industrial capitalism and the modern state and defended what they claimed was a less violent and more tolerant political society – the "traditional" society of the rural communities. The most significant product of modern Indian political thinking, overwhelmingly liberal but incorporating at several points the "traditionalist" view, was the Indian constitution, written from 1946 to 1950.

The dominant framework of Indian political science during the 1950s was that of liberal modernization theory. While several key institutions of the modern state had been built during the period of colonial rule, India was now said to be in the phase of developing its own democratic processes and the practices of modern citizenship. Features such as patronage relations based on caste and religious loyaties and solidarities based on ethnicity were regarded as vestiges of underdevelopment that would go away with greater democratic participation. In time, however, more complex versions of this modernization theory were produced, such as that of Lloyd and Susanne Rudolph published in 1967, which argued that even supposedly traditional elements such as caste and religion could adapt to modern political institutions and, by transforming themselves, become part of political modernity itself.

The most influential account of the new political system was given by Rajni Kothari, whose work published in 1970 identified its "dynamic core" in the dominance of the Congress Party. Using a largely structural-functional model, Kothari described "the Congress system" as one in which the ruling party connected government and party at various levels, from the national capital down to the localities, accommodated dissidence within itself, and secured the legitimacy of the system as a whole through coalitions and consensus. By the mid-1970s, however, with growing authoritarianism, centralization of power in the hands of a small group of Congress leaders, and especially the state of internal emergency from 1975 to 1977, this model of a consensual Congress system became less persuasive.

Marxist accounts were better able to describe conflicts and the repressive use of state power as systemic features of Indian politics. The state, especially its central structures, was seen as the site over which several dominant classes, none able to achieve hegemony on its own, tried both to outmaneuver one another and to work out coalitional arrangements. The Marxist approach was

---

[12] Deepak Nayyar, ed., *Industrial Growth and Stagnation: The Debate in India* (Delhi: Oxford University Press, 1994); Dilip Mukherjee, ed., *Indian Industrialization: Policies and Performance* (Delhi: Oxford University Press, 1995); Prabhat Patnaik, ed., *Macroeconomics* (Delhi: Oxford University Press, 1995).

less successful, however, in connecting the central account to local societal institutions and mico-level political processes.

Here a structural-functional theory was more commonly used. It assumed that the Congress system was primarily a way of pulling together the various dominant groups in the localities into a single ruling structure. The factions within the Congress Party were said to be the main form through which this was accomplished: The conflicts between factions at lower levels were sorted out by the mediating skills of Congress leaders at higher levels. Later, with the centralization of the Congress during the 1970s, this mediatory form gave way to what was called plebiscitary politics, in which the general elections were turned into a referendum on the leadership of Indira Gandhi, the supreme Congress leader. This allowed the Congress leadership to draw electoral support from the poor, the lower castes, and the minorities without going through the locally dominant groups.

The dominant approach in Indian political science tends to accept the role of the developmental state in modernizing Indian society. However, there is also a critique of the developmental state, which fundamentally questions the project of modernization and describes it as one of conflict, violence, and the marginalization of vulnerable groups. Ashis Nandy, for instance, has argued that the modernist state has failed whenever it has tried to impose on Indian society a set of institutions adopted from the modern West that go against the everyday practices of collective living in local communities.[13]

With the professionalization of the disciplines beginning in the 1950s, numerous social science journals have appeared in India. Of these, the *Indian Economic and Social History Review* and *Contributions to Indian Sociology*, in particular, have great prestige. However, the most remarkable institution is that of the *Economic and Political Weekly*, published in Bombay, which uniquely combines the functions of a newsweekly, a journal of commentary on current economic matters, a professional journal of advanced research in all of the social science disciplines, and a bulletin of academic events in India. Prominent social scientists also play a role in India as public intellectuals, intervening in political, economic, and cultural debates in the print news media and on television. While the bulk of social science activities could be said to provide support for the policies and ideologies of the Indian state, there is nonetheless an active critical component that feeds into oppositional positions and movements. Especially prominent in recent years has been the combination of activist concerns with academic professionalism, particularly in the activities of nongovernmental organizations and in the fields of rural development, poverty alleviation, health, literacy, women's issues, and human rights. This is a new and growing area in which social science research has become directly concerned with issues in the public sphere.

[13] Partha Chatterjee, ed., *State and Politics in India* (Delhi: Oxford University Press, 1997).

28

THE SOCIAL SCIENCES IN CHINA

Bettina Gransow

As academic disciplines, the social sciences are quite young in China, having been banned in the People's Republic from 1952 until the early 1980s. At the same time, they trace back to an extensive tradition, especially the sociological research in China that, along with North America and Western Europe, formed a third flourishing center during the 1930s.[1] During the period of its abolition, Chinese social science became a forgotten chapter in the international history of science. Social scientific study of China fell into oblivion not only because it was neglected for so long in China itself and because of language barriers, but also because of the self-reference of American and European research.

This chapter deals mainly with institutionalized social science in China, which, in a strict sense, includes also studies that do not deal specifically with China. It is concerned especially with what I will call "the Chinese social sciences," referring to those approaches that aim to sinicize or indigenize the social sciences, as well as with approaches seen as a distinct "Chinese school" of social science. I emphasize specific aspects of this definition, and describe some central junctures in the development of the social sciences in China. The chapter starts with the native domains of learning of the Chinese scholars during the Qing Dynasty; it aims to describe these domains as the intellectual space that served as the framework for the reception of Western sciences as a new body of knowledge. The second section shows how a social space for scientific development was created after the abolition of the imperial examination system during the 1920s, and how sociology, economics, and political science became academic disciplines. Third, different strategies for sinicizing the social sciences will be presented, as exemplified by some outstanding personalities, in an effort to characterize the heyday of the Chinese social sciences during the 1930s. The sinicization of the social sciences will be

---

[1] Maurice Freedman, *The Study of Chinese Society: Essays by M. Freedman*, ed. G. W. Skinner (Stanford, Calif.: Stanford University Press, 1979), p. 379.

taken up again in the fourth section with respect to Taiwan and Hong Kong since World War II. The fifth section will deal with the abolition and revival of the social sciences in the People's Republic of China, raising questions about breaks and continuities in the history of the Chinese social sciences.

NATIVE DOMAINS OF LEARNING AND THE EARLY RECEPTION OF THE WESTERN SOCIAL SCIENCES

The introduction of the Western social sciences into China was both helped and hindered by socioethical Confucian norms. Imperial power in traditional China was exercised through an elaborate bureaucratic civil service rather than by a hierarchy of enfeoffed barons. Confucian norms serving to protect that power developed into a broad base of writings on social rules and explanations of human relationships, powerful dogmas with a great influence on the daily lives of the Chinese people. In this sense, ancient Chinese social thought in general, and Confucianism in particular, seems to be overwhelmingly "sociologistic."[2] The emergence of the modern social sciences was encouraged by this avid interest in social phenomena. But because traditional social ethics pursued normative goals attempting to shape individual behavior patterns, social thinking tended to be dogmatic rather than critical. Under the conditions of a paternalistic power structure, critique could only be expressed in the form of a reinterpretation of Confucianism. Confucian scholarship helped to give shape to the critical thinking of Chinese intellectuals into the twentieth century.

An important intellectual turn in Confucian discourse took place during the seventeenth and eighteenth centuries. Advocates of evidential research (*kaozheng xue*) expressed doubts regarding the dominant Confucian ideology of the neo-Confucians of the Song period and attempted to replace their moral philosophy with an empirical approach. They used new philological research methods in order to reconstruct the purity of ancient culture and its precision of conception and expression.[3] The antimetaphysical tone of the Qing philologists corresponded to the substitution of mathematical astronomy for cosmology. This development was supported during the early Qing period by the work of the Jesuits and by an enhanced appreciation of mathematical studies under Emperor Kangxi. However, mathematics was seen in conjunction with the classic texts. "There was no distinction between humanistic knowledge and pure scientific knowledge in traditional Chinese learning."[4]

[2] Benjamin Schwartz, "Social Role and Sociologism in China, with Particular Reference to Confucianism," in his *China and Other Matters* (Cambridge, Mass.: Harvard University Press, 1996), p. 76.

[3] Benjamin Elman, *From Philosophy to Philology: Intellectual and Social Aspects of Change in Late Imperial China* (Cambridge, Mass.: Harvard University Press, 1984), p. 254; see also pp. 27–8, 54.

[4] Limin Bai, "Mathematical Study and Intellectual Transition in the Early and Mid-Qing," *Late Imperial China*, 16 (1995), 29, 50.

China's educated circles became increasingly interested in Western social theories shortly before the turn of the century, after China's defeat in 1895 in the first Sino-Japanese War confirmed the technical and political superiority of foreign powers. There was an interest in getting to know the West in order to enable China to assert itself as a nation among nations.[5] In contrast to the self-strengthening strategy of the Yangwu Movement (1860–95) – which was regarded as having failed because it sought to adopt only natural science, logic, and military-technical capabilities from the West – the Reform Movement of 1898 was also interested in foreign social theories, political systems, and economic policies. The intellectual atmosphere in China had developed to the point of adopting, in particular, theories of social evolution, which the reformers hoped would serve as a key to restructuring state and society.

The reception of Western social sciences in China took a detour by way of Japan. Japan was a more popular destination for study abroad than either Europe or the United States, for reasons of geographical proximity, cultural similarities, and a lower cost of living. But more than that, Japan was seen as having been strengthened by reform based on Western models, and it was believed that the Japanese adaptation of Western knowledge facilitated its efficient and simple comprehension in China. Liang Qichao (1873–1929), who spent more than ten years in political exile in Japan after reform attempts failed in 1899, was especially successful in spreading Western social sciences in China through his translations from the Japanese. At the time, this was a common practice in introducing social science terminology; such language usage was considered contemporary and modern, and was generally accepted in early-twentieth-century China.

Direct translations from English were also available, particularly those by Yan Fu (1854–1921), who, like Liang, belonged to the core of reformers. Works translated by Yan Fu include Thomas Huxley's *Evolution and Ethics* (1898), Adam Smith's *The Wealth of Nations* (1901–2), Herbert Spencer's *The Study of Sociology* (1903), and John Stuart Mill's *On Liberty* (1903). Yan Fu wanted to spark the interest of contemporary Chinese scholars in the significance of Western social science.[6] In contrast to translations from the Japanese, which always used the neologism *shehuixue* for the concept of "sociology" in Chinese, Yan Fu, in his translation of Herbert Spencer's *Principles of Sociology* (1903), for example, used the less common term *qunxue* – literally, "teaching of the groups." Yan used this word, which traces back to the Confucian classicist Xunzi, in an effort to express a specific understanding of the relationship between the individual and society. It reflects close ties to the theory of evolution, the subject matter of many early social scientific translations.

---

[5] Hengyü Kuo, *China und die "Barbaren": Eine geistesgeschichtliche Standortbestimmung* (Pfullingen: Neske, 1967), p. 34.

[6] Cf. Benjamin Schwartz, *In Search of Wealth and Power: Yen Fu and the West* (Cambridge, Mass: Belknap Press of Harvard University Press, 1983).

Traditional Chinese culture and the Confucian orthodoxy of the Qing period thus allowed an intellectual space for scientific study. But this could not give rise to an institutionalized social space or to the creative development of scientific research methods. Science was considered above all to be a body of knowledge to be grasped and assimilated. Moreover, the Protestant missionaries in nineteenth-century China promoted scientific education as a component of character formation, not as the basis for new research. This new knowledge was linked to traditional philosophical categories in order to construct an image of science that was to gain legitimacy through its overlap with traditional forms of learning. This approach went so far as to locate the roots of Western knowledge in Chinese antiquity, thus confirming the superiority of Chinese learning.[7]

INSTITUTIONALIZATION OF THE DISCIPLINES

Approaches to social science teaching and research during the 1910s and 1920s in China were highly diverse, as a wide variety of educational institutions had emerged after the imperial examination system was abolished in 1905. The translation and reception of foreign works formed the basis for social science curricula in the universities and colleges. During the first half of the twentieth century, social scientists experimented with ways to use the new knowledge constructively as a tool of modernization. Diverse influences were exerted on this process of institutionalizing social sciences – by the newly founded Chinese institutions of higher learning, by the missionary universities and colleges, by foreign scientific foundations, and, after the founding of the Nationalist government in 1928, increasingly by the Guomindang. Whereas the social sciences in the Chinese universities stood more in the Confucian tradition of preparing for a career in the civil service, American missionary sociologists strove to teach empirical social research within the context of Christian social reform.

Initially, the main type of school was one that offered training in law and administration, a type adopted from Japan. Starting around 1906, classes were held in sociology, law, and political science. Such schools closely linked issues of state administration and social organization, corresponding to conventional Chinese career patterns. In the tradition of Confucianism, students felt that acquiring social scientific knowledge was an appropriate way to qualify themselves for a government position and a career in the civil service.[8] Starting in the mid-1920s, the number of universities and colleges in China grew rapidly, bringing with them an increase in social scientific course

[7] David Reynolds, "Redrawing China's Intellectual Map: Images of Science in Nineteenth Century China," *Late Imperial China*, 12 (1991), 31, 37–8.
[8] Yung-chen Chiang, "Social Engineering and the Social Sciences in China, 1898–1949" (PhD dissertation, Harvard University, 1986), p. 20.

offerings and academic departments. The Chinese institutions generally offered a greater variety of courses in political science and economics than did the Christian colleges. In sociology, however, the Christian colleges offered about twice as many courses.

Under Guomindang rule, the Nationalist government was oriented more toward political science and economics, as these were viewed as useful in building up the country. Sociology, on the other hand, was associated with socialism and radical politics. This link was suggested in part by the still relatively new and similar-sounding terms for sociology *(shehuixue)*, social sciences *(shehui kexue)*, and socialism *(shehui zhuyi)*; the leftist Shanghai University, for example, offered primarily Marxist teachings in the name of a conventional sociology curriculum.[9] Following the break between the Guomindang and the communists, Shanghai University was closed down in 1927. In order to stop the translation and publication of Marxist and other left-wing works, the Guomindang government passed a series of censorship laws, which were repeatedly intensified in the course of the 1930s and 1940s. But many books and journals succeeded in slipping through the fingers of censorship by using disguised titles, such as *Introduction to the Social Sciences.*[10]

Chinese universities with political science, economics, and sociology departments were centered mainly in the major cities, such as Shanghai, Beijing, Nanjing, and Canton. After the Sino-Japanese War broke out in 1937, numerous universities temporarily moved to areas in southern and southwestern China that were not occupied by Japan. This had the side effect of spatially spreading out the social science institutions within China. The relative status and size of the various academic disciplines remained basically unchanged until into the 1940s. Economics had the highest priority, followed by political science; in a somewhat distant third place was sociology.[11]

Under the influence of missionary sociologists from the United States, empirical social research thrived at the Christian universities and colleges. The development of the Christian-oriented social survey movement in the United States and the failure of the earlier missions in China strengthened the social reform wing among Chinese missionaries. Together with the massive U.S. participation in the development of the Chinese system of education and higher learning, this led to a practice at the missionary universities in which empirical social research was closely linked to social work. Preeminent here was the activity of Princeton-in-China and the work of John Stewart Burgess (1883–1949), which led to the founding in 1922 of what was

[9] Wen-hsin Yeh, *The Alienated Academy: Culture and Politics in Republican China, 1919–1937* (Cambridge: Cambridge University Press, 1990), pp. 156–7.
[10] Lee-hsia Hsu Ting, *Government Control of the Press in Modern China, 1900–1949* (Cambridge, Mass.: Harvard University Press, 1974), p. 83.
[11] Chung-Hsing Sun, "The Development of the Social Sciences in China before 1949" (PhD dissertation, Columbia University, 1987), p. 104.

to become the largest sociology department in China, at Yanjing University in Peking. Burgess, like his later comrade-in-arms Sidney D. Gamble (1890–1968), pursued a dual objective. He wanted both, to compile empirical data to back up Christian social work and to inform Chinese students about the social problems in their own society. The 1921 work *Peking: A Social Survey*, by Gamble and Burgess, was the first "classic" study to follow closely the American "Springfield survey" (1914) in its objectives, investigative methods, and presentation. The survey was to serve as the foundation for a community social welfare program for the entire city of Peking. Despite its scientific success, however, all attempts at implementing such a program to train Christian social workers failed, as neither the YMCA nor the Rockefeller Foundation was willing to finance the program. Instead, the Rockefeller Foundation supported the founding of the Peking Institute for Social Research in 1926, a research institution under Chinese directors (Tao Menghe and Li Jinghan). Against the background of an active labor movement and numerous labor disputes at the time, the Institute conducted a series of studies, including family budget analyses, on the living conditions of urban laborers. Aside from the Peking Institute for Social Research, only one other institute in the area of the social sciences was established solely for research purposes, the Institute for Social Sciences at the Academia Sinica, founded in 1928. The latter had two sites, with the Departments of Ethnology and Economics located in Nanjing, the Departments of Sociology and Law in Shanghai.[12] In the area of economic research, the Nankai Institute of Economics, which was affiliated with Tianjin University, assumed a prominent position. The initial research goal of the Institute was to prepare a comprehensive assessment of the extent and impact of industrialization in China. Under the influence of the Rockefeller Foundation, the emphasis was shifted during the early 1930s to agriculture, rural industry, and local administration.

Demands to separate scientific education from a Christian orientation grew increasingly strong from the Chinese side. This went hand in hand with U.S. educational and research policy, which viewed the universalization of their own social model as being closely tied to evidence of the universal character of science. At the core of this policy was the incorporation of a Chinese elite, which the missionary project had obviously not succeeded in achieving. The most important step in this direction was to enable Chinese students to study abroad. While the Rockefeller Foundation saw the establishment of modern institutions in China as a means for scientific transfer, the concept of studying abroad stressed the trained persons themselves as the means for scientific transfer.[13] After the United States waived half of its

---

[12] Tso-liang Ch'en, "Work of Social Sciences Done in China" (unpublished manuscript, Department of Political Science, Yenching University, n.d.), pp. 78–9.

[13] Peter Buck, *American Science and Modern China, 1876–1936* (Cambridge: Cambridge University Press, 1980), p. 48.

claims from the Boxer Indemnity Fund in 1908, the Chinese government used those funds to finance foreign study for Chinese students in the United States. As in China itself, the interest of the students studying abroad was clearly in the social sciences and humanities.

The first generation of social scientists who had studied abroad and returned to China to assume academic positions gave the social sciences in China a push toward professionalization. They campaigned to set up academic associations and became the directors of the newly founded research institutes. Social scientists did not encounter the obstacles faced by students in technical and natural science fields, who, due to China's lagging industrial development, had difficulty applying what they had learned after returning and who also received little social recognition in Chinese society. Instead, the knowledge the social scientists had gained was almost begging to be applied to Chinese society.

STRATEGIES TO SINICIZE THE SOCIAL SCIENCES DURING THE 1930s

Work performed during the late 1920s emphasized empirical studies and the collection of economic and social data about Chinese society. From 1927 to 1935, there was even a social survey movement. More than 9,000 studies were conducted during that time, 1,739 of which were nationwide studies, and the rest regional studies.[14] Two-thirds of them focused on economic subjects, and approximately one-fourth on social topics. Only six percent dealt with politics. There was a significant increase in the number of agricultural and rural studies during the early 1930s, primarily because of the intensification of the agrarian crisis and the decline of rural trades in the face of foreign competition. This trend was also stimulated by the emergence of a reformist rural reconstruction project in response to the communists' new interest in the peasantry following the breakup of their united front with the Guomindang. Agricultural and rural studies received another push after the Sino-Japanese War broke out in 1937 and several universities moved inland.

The decade of the 1930s was the heyday of the Chinese social sciences. Chinese social scientists shared the goal of acquiring knowledge from the West and using it in the service of the development and modernization of Chinese society. Although in this sense one can identify a collective goal among Chinese social scientists at the time, I would hesitate to speak of a "school of Chinese sociology."[15] In this field alone, the research approaches and scientific strategies employed to address China's social crisis were far too diverse for

[14] Yuren Liu, "Zhongguo shehui diaocha yundong" [The Chinese social survey movement] (Master's thesis, Yanjing University, 1936).
[15] Bronislaw Malinowski, Preface to Fei Hsiao-tung, *Peasant Life in China: A Field Study of Country Life in the Yangtze Valley* (New York: Dutton, 1939), p. xxiii.

that. To illustrate this, I present here some examples showing how Chinese social scientists adapted Western knowledge to the local circumstances in China.

Demands for "sinicized social sciences" were first expressed within the context of the mass education movement in Dingxian. One of the largest and best-known reform projects of the time was conducted in Hebei province under the direction of Yan Yangchu (Jimmy Yen). The sociologist Li Jinghan (1894–1987), who became one of the most important advocates of the Chinese social survey movement when he returned from the United States in 1924, carried out social surveys at the county level. The information collected served as key data in developing a self-help program for peasants in Dingxian. The program encompassed the transfer of elementary education and knowledge in the areas of agrarian technology, personal hygiene, and the organization of village self-administration. As an example of applied social science work, Li Jinghan's *Dingxian shehui gaikuang diaocha* (Social survey of Dingxian) (1933) became a model study for the social survey movement in China. In his preface to Li's book, Yan Yangchu characterized the social scientific study of Chinese society using social surveys and aiming for social reforms as "sinicized social sciences."

Chen Hanseng (b. 1897) attempted to sinicize the social sciences using survey methods, but from a Marxist perspective. After studying the history of economics in Chicago and Berlin, he became in 1924 the youngest professor at Peking University. Soon after, he was appointed director of the Institute for Social Sciences at the Academia Sinica. Between 1929 and 1933, Chen conducted comprehensive surveys on land distribution, as well as on the interconnectedness of foreign capital and Chinese bureaucracy, in order to provide empirically grounded arguments for the political practice of radical land reform.[16] He was forced to leave the Academia Sinica in 1934 because he was a communist, and he founded a "society for the study of the Chinese agrarian economy" so that he could continue his studies. Together with his former staff, he criticized the reform approaches of the rural reconstruction movement and the Dingxian experiment, as well as the agrarian surveys conducted by John Lossing Buck (1890–1962) at Jinling University in Nanjing, which Chen believed reduced the question of China's agriculture to one of agrarian technology.

For reasons less political than methodological, Wu Wenzao (1895–1985) and the faculty of the social anthropology wing of the Sociology Department at Yanjing University dissociated themselves explicitly from the social survey approach. Although they too recognized the fundamental need for social reforms for Chinese society, they sharply criticized the quantitative surveys

[16] Chen Hanseng, *Mu de chayi* [Differences in the Mu] (Shanghai, 1930) and his *Industrial Capital and Chinese Peasants* (New York: Garland Publishing, 1946) and *Agrarian Problems in Southernmost China* (Shanghai: Canton Lingnan University, 1936).

for what they viewed as unscientific methodology. They also criticized the dovetailing of science and social work, of university and administration. Wu Wenzao became dean of the Sociology Department at Yanjing University during the mid-1930s, replacing Xu Shilian (b. 1909), who had given the department a social work orientation. During the second half of the 1930s, Wu initiated a sea change in sociological research in China. Influenced by social anthropology in Great Britain and the United States, he criticized the lack of scientific autonomy in social survey methodology and stressed the advantages of social anthropological community studies. Wu Wenzao supported a systematic linking of sociology and anthropology in China and saw the introduction of functionalistic community studies as the appropriate path toward a sinicization of sociology. He worked toward a comparative methodology in order to relate the different development stages within Chinese society to one another.[17] Wu argued that sinicization was possible for sociology only after independent scientific competence had been developed, which could be identified through a scientifically founded hypothesis and verified through field research. Many of his students achieved international acclaim, among them Lin Yuehua (*The Golden Wing*, 1944), Yang Qingkun (C. K. Yang) (*A Chinese Village in Early Communist Transition*, 1959), and Xu Langguang (Francis L. K. Hsu) (*Under the Ancestor's Shadow*, 1948).

Wu's best-known student was Fei Xiaotong (b. 1910). Fei's interest in empirical social research was sparked by Robert E. Park, who took a teaching position at Yanjing University in 1932 on Wu's invitation. S. M. Shirokogoroff, from the neighboring Qinghua University, was a capable teacher who instructed Fei in the methods of anthropological field research. Under the direction of Bronislaw Malinowski, with whom Fei studied from 1936 to 1938, he conducted his study *Peasant Life in China* (1939), today considered a classic Chinese community study. In contrast to the emphasis on stasis of Malinowski's "romantic escapism,"[18] however, Fei viewed social research as a means of controlling and directing social change induced from the outside.[19] He analyzed the collision between the Western industrial system and Chinese agrarian society – in his comparative culture lectures on China's agrarian sociology, for example, published in 1947 under the title *Xiangtu Zhongguo*.[20] This work developed his ideas on the rural industrialization of China as an alternative path toward modernization.

---

[17] Wu's methodology followed Alfred Radcliffe-Brown, "Proposals for a Sociological Survey of Village Life in China," *Shehui yanjiu* [Social research] (Kanton) 1 (1937), 2, 4–5, 9.

[18] Karl-Heinz Kohl, *Abwehr und Verlangen, Zur Geschichte der Ethnologie* (Frankfurt am Main: Qumran, 1987), p. 46.

[19] David Arkush, *Fei Hsiao-tung and Sociology in Revolutionary China* (Cambridge: Cambridge University Press, 1981), pp. 55–6.

[20] An English translation of *Xiangtu Zhongguo* (Earthbound China) appeared in 1992 under the title *From the Soil: The Foundations of Chinese Society*, trans. Gary Hamilton and Wang Zheng (Berkeley: University of California Press). This work is not the same as Fei Hsiao-Tung and Chang Chih-I, *Earthbound China: A Study of Rural Economy in Yunnan* (Chicago: University of Chicago Press, 1945).

In contrast to these empirically oriented approaches, cultural sociologist Sun Benwen (1891–1979) of Nanjing University sought to sinicize sociology at a theoretical level by developing a cultural sociology. Against the background of debate on Eastern and Western cultures, which had a nationalistic orientation after the Japanese invaded Manchuria in 1931, Sun wrote of culture as a "national essence" that constituted the core of a comprehensive plan for a China-centered sociology.[21] Such efforts show how misleading is the often-heard criticism that Chinese sociologists and anthropologists who returned from abroad merely served as mouthpieces for the scientific schools in which they were trained.

THE SOCIAL SCIENCES IN TAIWAN AND HONG KONG

After World War II, Taiwan and Hong Kong emerged as sites of Chinese social science. This movement was greatly advanced by emigrant Chinese, such as the agrarian sociologist Yang Maochun (Martin Yang) (1905–1985), who went to Taiwan in 1958 following a stay in the United States, and Yang Qingkūn (C. K. Yang) of the University of Pittsburgh, who supported the establishment of sociology in Hong Kong. In Taiwan, the Guomindang maintained its pragmatic concept of the social sciences in general, and of sociology in particular. The field remained small due to limited personnel and equipment. There was a lack of teaching materials, and the writings of mainland Chinese social scientists, a majority of whom remained loyal to the newly founded People's Republic of China, were banned for political reasons. In Hong Kong, however, relevant works by Chinese social scientists remained accessible.

Political stability and academic freedom under the British colonial government offered fertile ground for the development of the social sciences in Hong Kong. Nevertheless, they developed very slowly in the academic sphere during the 1950s and 1960s, and sociological research was not resumed until the late 1960s. The main social science institutions in Hong Kong were at Chinese University and Hong Kong University. According to one of its leading proponents, sociology in Hong Kong was characterized first of all by close ties to U.S. and British sociology and, second, by an emphasis on the study of Chinese society.[22] Many scientists had received their postgraduate training at American and British universities, and their publications targeted the Anglo-American scientific community. No sociology journals were published in Hong Kong, for example. Moreover, during the 1960s and 1970s it was in Hong Kong that Western social scientists conducting research on

[21] Benwen Sun, *Dangdai Zhongguo shehuixue* [The development of Chinese sociology] (Nanjing: Shengli Press, 1948), pp. 284–6.
[22] Rance Lee, "Sociology in Hong Kong," in *Sociology in Asia*, ed. Man Singh Das (New Delhi: Prints India, 1989), p. 101.

China tried to compensate for their lack of options to conduct field research on the mainland by conducting interviews with emigrants. An important institution in this regard was the Universities Service Center, first funded by American foundations and later incorporated into the library of the Chinese University. Whereas basic research in sociology predominated in Hong Kong into the 1980s, the government's response to recent modernization has included a new focus on resolving local problems.

Regarding the development of sociology in Taiwan, according to Xiao Xinhuang (Michael Hsiao) there have been four distinct generations of Chinese sociologists:[23]

1. Sociologists who went to Taiwan after the war. Xie Zhengfu, for example, had served as minister for social affairs in the Nationalist government after a period of study in France. He later supported the introduction of a sociological curriculum and the establishment of a Chinese Sociological Society in Taiwan. Yang Maochun taught rural sociology and in 1960 became director of the newly founded Department of Agricultural Extension in the College of Agriculture of National Taiwan University. His studies had a very applied focus and maintained a close link to the China–United States Joint Commission for Rural Reconstruction. This applied research played a significant role in the success of Taiwanese land reform. Another influential sociologist of this first generation was Long Guanhai, whose research focused on the sociology of urban development.
2. Chinese social scientists who were educated on the mainland, but who did not work in the field until moving to Taiwan. Wen Chongyi belonged to this group.
3. Sociologists who were raised in Taiwan after the war, but who were educated primarily overseas. This third generation of sociologists (including Xiao himself) started playing an influential role in Taiwan during the late 1970s.
4. Students of members of the third generation, who received their education in Taiwan.

Into the 1970s, the social sciences in Taiwan reflected the theories and methodologies of American researchers. After the United States broke off diplomatic ties on 1 January 1979, and in the context of Taiwan's economic boom, local intellectuals began a process of self-reflection, part of the so-called indigenization movement among writers and artists.[24] A flurry of conferences, debates, and opinion polls indicated the need for the Chinese social sciences to join in. Chinese colleagues from Hong Kong and the United States, and even from the People's Republic, were also involved. This marked the reintroduction – after a break of approximately forty years – of sinicization.

[23] Xinhuang Xiao, "Sanshi nian lai Taiwan de shehuixue: lishi yu jiegou de tantao" [Thirty years of Taiwanese sociology: Reflections on history and structure] in *Sanshi nian lai woguo renwen ji shehui kexue zhi huigu yu zhanwang* [Review and forecast on thirty years of Taiwanese humanities and social sciences], ed. Zehan Lai (Taibei: Dongda Press, 1987), p. 342.
[24] Bettina Gransow, "Chinese Sociology: Sinicisation and Globalization," *International Sociology*, 8:1 (1993), 103.

Opinion polls among Chinese scholars showed that the notion of siniciz-ing or indigenizing the social sciences enjoyed great popularity during the 1980s and 1990s in Taiwan and Hong Kong. However, they also showed that a large majority of the social scientists questioned wanted to maintain a universalistic approach. Central to the debate were issues such as whether sini-cization of the social and behavioral sciences meant merely adapting them to a Chinese situation or whether it required an independent theoretical framework. Questions were also raised about whether such hypotheses had to be based on empirical research in Chinese societies, and about whether the desire to assign universal validity to social scientific theories was compatible with the cultural rootedness of its developmental path.

As there had been during the 1930s, there were again diverse opinions about what is involved in indigenizing the social sciences. For example, the Taiwanese social psychologist Huang Guangguo mentioned "indigenous the-orizing," a term referring to the ethnic origin of the authors as Chinese and their efforts to avoid intellectual colonization.[25] He thus seemed to in-terchange the notions of indigenous scientific approaches and indigenous scientists.[26] The sociologists Ye Qizheng and Gao Chengshu postulated a sinicization of the social sciences at the level of critique and hermeneutics, along the lines of the Frankfurt School.[27] Regarding the economic boom in eastern Asia, analyses during the 1980s and 1990s increasingly stressed the sociocultural aspects of trade and commerce, based on the significance of family structures and business networks in the organization of Chinese enterprises. However, at this time there was much more reflection on the Chinese scientists' own position in the scientific community than there had been during the debates of the 1930s. During the late 1980s, this discourse spread to the People's Republic, where the process of reestablishing the social sciences was just beginning.

RECONSTITUTION OF THE SOCIAL SCIENCES IN THE PEOPLE'S REPUBLIC OF CHINA

After two unsuccessful attempts to reorganize (1949–52) and reestablish (1956–7) the social sciences, they remained banned as academic disciplines in the People's Republic until the end of the 1970s. The social sciences were

[25] Michael Harris Bond and Kwang-kuo Huang, "The Social Psychology of Chinese People," in *The Psychology of the Chinese People*, ed. M. H. Bond (Hong Kong: Oxford University Press, 1986), p. 217.

[26] See Michael Cernea, "Indigenous Anthropologists and Development-Oriented Research," in *Indige-nous Anthropology in Non-Western Countries*, ed. Hussein Farim (Durham, N.C.: Carolina Academic Press, 1982), pp. 121–37.

[27] Guoshu Yang and Chongyi Wen, eds., *Shehui ji xingwei kexue yanjiu de Zhongguohua* [Sinicization of social and behavioural sciences] (Taibei: Yongyu Press, 1982), pp. 44–6, 139, 147–8.

finally reintroduced into China as part of Deng Xiaoping's modernization efforts. Social science research and teaching quickly gained a firm mooring in the People's Republic through the founding of the Chinese Academy of Social Sciences, the establishment of social science departments at a number of universities, and the creation of research institutes and scientific journals.

Chinese social scientists who remained on the mainland after 1949 generally expected that their new political leaders would implement democratic coalition politics, and that the Communist Party would reach out to them. But the outstretched hand turned out to have such a tight grip that the social science disciplines were unable to survive for more than a few years. The Chinese government's strict conformity to the politics of the Communist Party of the Soviet Union entailed the banning of the social sciences – including sociology, political science, law, and economics with the exception of political economy – in 1952. Most social scientists were transferred to related fields, such as statistics, or assigned to centers such as the Central Institute for Nationalities, founded in 1951, and the Peking Institute for Politics and Law.

Communist leaders justified these actions by claiming that all of the social sciences exhibited class character. They felt that in China's socialist society, historical materialism performed all of the tasks of progressive social science, and that everything else was reprehensible bourgeois science. Furthermore, against the background of China's role in the Korean War in 1951 and 1952, the anti-Americanism of the Communist Party targeted those intellectuals who had studied in the United States prior to 1949 and were thus suspected of having developed more or less bourgeois attitudes. Mao Zedong's party organization feared that such social scientists would challenge the communists' monopoly of power. Their social reform approaches were seen as competing intellectually with the self-image of the Communist Party as a social revolutionary organization, and on many accounts they sympathized politically with the Democratic League, a party of the intelligentsia founded during the 1940s.

Sociologists such as Fei Xiaotong, Pan Guangdan, Li Jinghan, Chen Da, and Wu Jingchao, as well as the economic historian Chen Zhenhan and the demographer Ma Yinchu, spoke out during the Hundred Flowers Movement, a short phase of political liberalization during 1956 and 1957, for the reestablishment of the social sciences. Many were branded as "rightists" and suffered recriminations up until the end of the Cultural Revolution in 1976. Their publications were banned. The "blacklist" of sociological books and articles alone (including, in particular, many outstanding works from before 1949) contains more than, 1,000 titles.[28]

---

[28] Bettina Gransow, *Geschichte der chinesischen Soziologie* (Frankfurt: Campus, 1992), p. 145.

The question of whether social research continued after the social science disciplines were banned has often been raised.[29] Indeed, social inquiries were conducted during the 1950s and 1960s. As a positive model of a "proletarian" form of social research, they were supposed to counter the negative image of "bourgeois" sociology that had emerged, at the latest, during the 1957 Anti-Rightist Movement. Mao Zedong's *The Question of Agricultural Cooperation* (1955) was the prototype for such proletarian social research. Comprehensive materials on the "four histories" (family, village, commune, factory), which had been collected in the course of the Socialist Education Movement (1963–5), were supposed to serve the construction of an idealized social history. The year 1949 was defined as the dividing line between the former society "before liberation" and the new society "after liberation." This form of social research was directly linked to the goal of ideological education; thus one can speak, at best, of a type of "socialist realism" in social research. This was also the case regarding embellishment of the economic statistics of the time.

With the lifting of the ban came a political freedom of movement that manifested itself in the establishment of the social sciences as independent scientific disciplines based on historical materialism as their ideological leit-motif. This represented a compromise that acknowledged the Communist Party's leadership claims and at the same time gave social scientists sufficient autonomy to establish their academic departments. Censorship and political pressure were no longer directed against the social sciences per se, but *within* the individual disciplines.

Because of the decades-long ban, researchers educated prior to 1949 still constituted the backbone of the social sciences in China, even during the early 1980s. Their experience and comprehensive knowledge, although somewhat outdated, was supplemented by a wave of translations of more recent Western literature. These were subject to strict censorship regulations and were not readily available to students and young researchers.

The main focus of sociological research during the 1980s and 1990s – as it had been during the period before 1949 – was on empirical studies. Social problems in general, and marriage and family issues in particular, were initially given highest priority. Social issues such as unemployment, juvenile delinquency, and the housing shortage were especially pressing after the end of the Cultural Revolution. The traditional focus of the social sciences on social work was taken up again, especially by Yuan Fang, longtime dean of the Sociology Department at Peking University. For example, the work of Lei Jieqiong (b. 1905) on family sociology was related to the rigidly implemented birth control policy of the government. Simultaneous reforms in the agrarian

[29] Siu-lun Wong, "Social Enquiries in the People's Republic of China," *Sociology*, 9 (1975), 459–76; Lucy Jen Huang, "The Status of Sociology in People's Republic of China," in *Sociology in Asia*, ed. Man Singh Das (New Delhi: Prints India, 1989), pp. 111–38.

sector served to enhance the status of the peasant household as a fundamental economic unit. This led to a revival of traditional practices such as arranged marriages, elaborate wedding celebrations, concubinage, and the preference for male offspring.

The small-town study by Fei Xiaotong is outstanding among the empirical projects of this initial phase.[30] Following up on his earlier research, Fei refined his theory of rural industrialization as a modernization strategy suitable for China, adding the concept of rural urbanization. Based on a functionalist analysis of local economic traditions, paths of modernization were sought in which (spontaneous) micro developments and (regulated) macro developments were adapted to each other. At the same time, Fei's project represented an implicit attempt to extract elements for a theory of adapted modernization from an empirical analysis of the revival of traditional economic activities. The project was also a starting point for a number of studies on rural industry, conducted in cooperation with the World Bank.

The extension of reforms to the industrial sector and the cities during the mid-1980s was of major significance for the economic and social sciences, as they now faced the challenges of modernization policies. This pertained especially to economics. Within the given political framework, economic debate dealt with strategies to introduce market elements into the Chinese economic system, which had previously been a planned economy. Discussion picked up ideas from the debate on the liberalization of the economy from the 1960s, and foreign theories were also discussed, including Western approaches and, especially, those of eastern European reform economists such as Brus, Sik, Lange, and, in particular, Kornai. Many Chinese economists agreed on the need to separate the state's ownership rights from the enterprise's management rights, in order to increase the efficiency of business. Both industrial reforms and price reforms were seen as necessary, but opinions were divided on the question of which should come first. One side, led by Li Yining, an economist at Peking University who had been an advisor to the former prime minister Zhao Ziyang, felt that property reform of state-owned enterprises, with a preference for conversion into joint-stock companies, had to precede price reform. Another school, under Wu Jinglian of the Chinese Academy of Social Sciences, supported an integrated reform of the price system, the enterprises, and the macro management system.[31]

Against the background of reform politics, attitude surveys and public opinion polls, which experienced an outright boom during the later 1980s, were often understood from an instrumental perspective. While political

[30] Fei Hsiao Tung et al., *Small Towns in China – Functions, Problems and Prospects* (Beijing: New World Press, 1986); Fei Xiaotong, *Xiao chengzhen, da wenti: Jiangsu sheng xiao chengzhen yanjiu lunwen xuan* [Small towns, big problems: Collection of articles containing research on small towns in Jiangsu province] (Nanjing: Jiangsu People's Press, 1984).
[31] Robert Hsu, *Economic Theories in China, 1979–1988* (Cambridge: Cambridge University Press, 1991), p. 19.

leaders initially supported survey methods in order to exploit the positive attitudes of the public for their own purposes, such methods worked against them to the extent that the high expectations of the urban population could no longer be met, and the surveys increasingly revealed critical attitudes.[32] The Chinese Economic System Reform Research Institute, which operated under the auspices of the State Council and functioned as the think tank of then-prime minister Zhao Ziyang, was the prime example of the link between opinion surveys and national policy. It was disbanded after the violent suppression of the protest movement of 1989.

After the economic reforms, a number of new social phenomena and problems emerged during the 1980s and 1990s, including a major wave of rural-to-urban migration, the emergence of the middle class and of new associations, employment problems (especially for women), corruption and prostitution, and the revival of rural clan associations. The so-called Blue Books, a series published beginning in the early 1990s, offer continuous insight into the processes of rapid social change in present-day China.

Much sociological theorizing involved the search for modernization theories appropriate for China. The lessons of endogenous modernization were quickly shown to be inadequate in many respects for the analysis, not to mention the solution, of the problems of a society like China's, which was dealing with exogenous modernization and the process of readapting a highly elaborate, historically developed culture. Attempts to deal with the relationship between traditional culture and modernization benefited from the debate on the relationship between Confucianism and capitalism that was taking place outside the People's Republic of China, against the background of the East Asian economic upswing. Central to these debates were the writings of Max Weber. Weber's writings were received critically in Taiwan and Hong Kong, or were considered the starting point for a theory of a Confucian ethic, analogous to Weber's Protestant ethic, that was viewed as the driving force behind East Asian modernization.[33] In the People's Republic, sociologists asked what could be learned from Weber's writings for the modernization of China and what could be learned about the obstacles to such development.

The history of the social sciences in China is marked by discontinuity. Periods of intense productivity and creativity have alternated with periods of inactivity, suppression, and forced isolation. This was definitely not a case of steady development or accumulation of knowledge. Nevertheless, some

[32] Stanley Rosen, "The Rise (and Fall) of Public Opinion in Post-Mao China," in *Reform and Reaction in Post-Mao China: The Road to Tiananmen*, ed. Richard Baum (New York: Routledge, 1991), p. 60.

[33] Yaoji Jin [Ambrose King], "Rujia lunli yu jingji fazhan: Weibo xueshuo de chongtan" [Confucian ethics and economic development: Max Weber's writings revisited], in *Xiandaihua yu Zhongguohua lunji* [Collected essays on modernization and sinicization], ed. Yiyuan Li, Guoshu Yang, and Chongyi Wen (Taibei: Guiguan, 1985), pp. 29–55.

forms of continuity in the history of the Chinese social sciences can be identified. One is the dependence of the social sciences on strictly defined political conditions in China. Another is the dependence on foreign scientific organizations with respect to research funding. A third is the work of individual scientists, who remained passionate researchers and dedicated intellectuals despite all of the political adversity and abuse. Fourth is the instrumentalization of the social sciences for the modernization of Chinese society. Finally, there is the demand for a sinicized or indigenous form of social science in China, which has been maintained from the 1930s to the present day. Just as the social sciences had served diverse strategies for renewal of the country during the time of the republic, during the 1980s and 1990s they were directed toward the success of the reform politics, though here too individual opinions regarding the means and goals varied greatly.

The demand for the sinicization of the social sciences illustrates the continuing process of self-reflection by Chinese social scientists, especially by sociologists, who have been constantly forced to face the challenge of *transcultural* understanding and transfer. Just as the modernization of Chinese society during the twentieth century cannot be reduced merely to a process of increased westernization, neither does such a reduction apply to the development of the social sciences in China. Sociocultural features specific to Chinese society not only lead away from Western paths of modernization, but also require for their analysis new and more refined social scientific tools.

29

THE SOCIAL SCIENCES IN JAPAN

Andrew E. Barshay

This chapter traces the history of Japanese social science in five successive "moments" or intellectual orientations that defined problems, structured analysis, and drove disciplinary development. Importantly, they also helped to set the terms of collective agency in public discourse: What was Japan? – a nation of imperial subjects? of classes? of "modern" individuals? a single *Volk*? The account is biased toward elite institutions and scholars, largely because social science in Japan, as in many "late developers," grew out of state concerns and developed as an unequal contest between elite and nonelite scholarship; there was no "free market" of ideas. It is also formalist, in the sense that I see "social science" in terms of the self-consciously professional activity of practitioners themselves.

NEO-TRADITIONALISM AND THE HEGEMONY OF THE PARTICULAR

Japan was the first successful modernizer in Asia. Determined to resist Western domination, elites of the Meiji era (1868–1912) had undertaken a forced march to industrialization and military power through the relentless taxation of peasant production. Initially, this effort was supported by a somewhat free-wheeling Anglophilia. Social Darwinism, the theory of progress, and an ethic of individual and national advancement were the keynotes. Meiji's state makers understood, however, that Japan could never be a pioneer in industrialization or empire building. They sought cultural self-preservation and national strength while avoiding the pitfalls that beset the pioneer: in short, effective followership.

From the late 1880s onward, as the Meiji Constitution (1889) and Imperial Rescript on Education (1890) demonstrate, politicians, officials, businessmen, educators, and publicists increasingly defined "success" in terms of the continued viability of the national culture in the face of Western influence.

The feudal values of obedience and the strong collective consciousness found in Japan's still overwhelmingly agrarian society, they claimed, served to bind the people to the government despite the traumas of industrialization. Japan's "unique" tradition would serve as a brake on both individualism and radical ideologies – the characteristic pathologies, in their view, of modern society.[1] Japan, in short, had modernized through, not despite, tradition; a new, neo-traditional *mode of modernization* had emerged on the world historical stage. An organic exceptionalism – a sense that Japan's mode of modernization was ethnically unique and therefore inherently moral – worked powerfully and persistently to shape Japan's collective self-images and external relations.

Success, however, brought frustration and anxiety. Despite two victorious turn-of-the-century wars and the revision of the unequal treaties, the Japanese were repeatedly reminded of (and became obsessed with) their racial, religious, and geographic alienation from the "civilization" of the Atlantic powers. In the decade following the Versailles Conference, Japan emerged as the bitter "have-not," deprived of its rightful sphere of influence, and soon thereafter as the champion of Asia. Meanwhile, the Japanese empire stoked the fires of anticolonial nationalism in Asia, directed both against the West (especially Great Britain) *and against Japan itself.* Japanese officials and makers of opinion, however, never understood the forces they had stirred up; they thought it was enough to urge Asian unity against the West, and they struck out with violence when nationalist hostility was turned against Japan. A brutal myopia – part and parcel of the same neo-traditional exceptionalism – made its way into Japanese thinking. This, more than desperation born of identity loss, ultimately led to the total war and total defeat that remain the pivot of Japan's modern history.

Of late, the massive discontinuities associated with this experience have begun to soften, at least enough to allow Japan's "neo-traditional" values and mode of development to be seen in recent decades as a model for the industrialization of other Asian societies. Familiarity and apparent likeness, however, should not be taken for intimacy; Japan's relations with its neighbors remain sensitive, to say the least. In its origins and development, its tendency toward (and struggles with) exceptionalism, Japanese social science provides a window on why this should be the case.

Neo-traditionalism held that Japan's achievement and experience were essentially not comparable to those of other peoples, and perhaps nurtured the suspicion that deep communication with the world outside might not be possible. Paradoxically, "noncomparability" derived from Japan's ability, ostensibly unique, to adapt strong impulses from materially "superior" cultures

---

[1] Itō Hirobumi, "Some Reminiscences of the Grant of the New Constitution," in *Fifty Years of New Japan*, ed. Ōkuma Shigenobu (New York: Dutton, 1909), vol. 1, pp. 122–32; Hozumi Nobushige, *Ancestor Worship and Japanese Law* (Tokyo: Z.P. Maruya, 1901); Thorstein Veblen, "The Opportunity of Japan" (1915), in his *Essays in Our Changing Order* (New York: Viking Press, 1930), pp. 248–66.

without sacrificing something called the "national essence." "Social science" was one such impulse, imported in the form of discrete disciplines via translation or by foreign experts. In the immediate post-Restoration era, the major areas included most fields of law, administration, political economy, and historical science, particularly as practiced in Britain, France, Germany, and the United States. There was as yet little professional organization or even disciplinary consciousness. The ethos was one of urgent, even aggressive familiarization rather than of critical reflection or synthesis. This in no way diminishes its intellectual interest, but for present purposes this "preprofessional" activity belongs to the realm of "prehistory" of Japanese social science.[2] Its "history" begins with the dual phenomena of professionalization and "statization," with the emergence of a main line in academic social science characterized by one form or another of particularism: Its ultimate values could be "derived from or reduced to the Japanese social nexus."[3] Social science was to serve the cause of neo-traditional integration; it was a weapon in Japan's national progress and "struggle for existence."

The installation of an exceptionalist hegemony was clearly visible to contemporaries. The Kokka Gakkai (Association for State Science), founded in 1887, drew its membership from officialdom and from the imperial universities, as did the Kokka Keizai Kai (Association for State and Economy), established in 1890. By the 1880s, the concomitant privileging of Prussian-style "state science" had drawn comment – some quite critical – from Japanese observers, especially from journalists and writers. It was not just that the state was enshrined at the core of "social" science. The earliest Japanese translation of "State Science" – *kokkagaku* – tended to emphasize its most conservative aspects, to the detriment of more liberal notions, such as Lorenz von Stein's idea of the "social monarchy," that reflected the size and political strength of the German working class. *Kokkagaku* was also more concerned with administrative techniques of "rule by law" than with the metaphysical underpinnings of the state and its legitimacy.[4] The latter issue had been settled by making the monarchy, with its claim to descent from the sun goddess, an analytical taboo for scholars, to chilling effect. In 1892, Kume Kunitake (1839–1931), trained by Ludwig Riess and a founder of modern Japanese historiography, lost his teaching post at Tokyo Imperial University after publishing an article, influenced by then-current anthropology and comparative religion, in which he declared Shintō to be a "vestige of sky-worship."

A product of the same broad movement was the supersession of laissez-faire or classical economics, which had been vigorously disseminated since the early 1870s. Although private universities such as Keiō and later Hitotsubashi

[2] Ishida Takeshi, *Nihon no shakai kagaku* [The social sciences in Japan] (Tokyo: Daigaku Shuppankai, 1984), chap. 1.
[3] Robert N. Bellah, "Japan's Cultural Identity: Some Reflections on the Work of Watsuji Tetsurō," *Journal of Asian Studies*, 24:4 (August 1965), 574–5.
[4] Ishida, *Nihon no shakai kagaku*, p. 40.

(in economics), Chūō and later Waseda (in law) continued to be associated with British approaches, Tokyo and its fellow imperial universities were firmly drawn toward the German Historical School. As with *Staatslehre*, this was no matter of mere copying. Friedrich List's *System der Nationalökonomie* (1841), for example, was well known in Ōshima Sadamasu's translation. Ōshima (1845–1906), a founder of the Kokka Keizai Kai and translator also of Buckle and Malthus, generally upheld Listian principles. But he strongly rejected List's argument that European colonialism represented a natural division of world labor; and, unlike List, he believed that small producer agriculture, not just industry, ought to receive state protection. Ōshima's estimation of classical political economy, moreover, was nuanced. While British liberalism had helped to "smash our obstinacy," he insisted that "liberty in politics and liberty in trade are quite different matters; that while political liberty can make the people of a country free, freedom of trade means freedom for the people of another country at the expense of the country that grants it . . . "[5]

On the tails of the Historical School, German "social policy" thought, its intellectual progeny, entered official and academic discourse, and this with some urgency. The severely deflationary policies of the 1880s had greatly accelerated the rate of both agrarian tenancy and urban migration. Established in 1896, some twenty years after its German model, and operating until 1924, the Nihon Shakai Seisaku Gakkai (Japanese Social Policy Association) had 122 members by 1909, drawn from academic, official, and moderate labor circles. Apart from stimulating the development of academic economics and empirical social research – Japan's first "social," as opposed to "state," science – the Association sought to prevent the conflict of classes that had traumatized Britain and produced radical movements there and across the continent. This stance led it to advocate protective factory legislation, earning it the enmity of elements of the business world, whose resistance to bureaucratic "interference" in the "warm relations" between employers and workers prevented its passage until 1911.[6] The Association also called for state action to hold off the otherwise inevitable radicalization of workers by enacting policies that would allow potential urban migrants to remain in rural villages.

The earliest professional groups of sociologists began to form shortly before the turn of the century, with university courses beginning to be taught at the same time. Like many practitioners of social policy, sociologists too met suspicions that their interest in "society" – the "lower orders" in the cities – was a matter not of study but of political advocacy.[7] In the overheated atmosphere

---

[5] Ōshima Sadamasu, *Jōseiron* [On the current situation] (1896), in *Meiji bunka zenshū* [Compendium of Meiji culture], ed. Yoshino Sakuzō (Tokyo: Nihon Hyōronsha, 1929), vol. 9, pp. 462, 464.

[6] Kenneth Pyle, "The Advantages of Followership: German Economics and Japanese Bureaucrats, 1890–1925," *Journal of Japanese Studies*, 1:1 (Autumn 1974), 127–64; Ishida, *Nihon no shakai kagaku*, pp. 51–71.

[7] Nozomu Kawamura, *Sociology and Society of Japan* (London: Kegan Paul, 1994), pp. 46–50; Ishida, *Nihon no shakai kagaku*, pp. 45–50.

of the Russo-Japanese War era, professional sociologists felt compelled to distinguish their enterprise from the subversive work of Japan's tiny and harrassed band of socialists. Tatebe Tongo (1871–1945), holder of the chair in sociology at Tokyo Imperial, declared that sociology "began with Comte and culminates in Tongo"; he pursued a national organicism – what he later termed a "statist view of society" (*kokka shakaikan*) – that attacked any and all manifestations of the "skeptical, negative, destructive, and transient" notions of individualism and democracy.[8] Such was the price of professionalization. Still, the pressure of new ideas, particularly those of Simmel, and of new problems, such as urban poverty, began to drive the discipline beyond the conservatism that had framed it.

Some sociologists, undoubtedly, were methodologically disinclined to take up the study of the countryside. But there were disincentives as well. The Civil Code of 1898 had defined the *ie*, or household – rather than the individual – as the normative unit of society. Apart from former warrior households (on which the model was based) and certain great merchant houses, the closest real approximations to the ostensibly traditional household were identified with the countryside. Thus the *ie* (albeit less ferociously than the monarchy) was protected from sociological scrutiny by the double mantle of law and ideology, while a highly idealized image of the "solid core" of frugal owner-farmers was held up for the entire country to emulate.

But this protection was deeply problematic on its own terms. While its extent varied region by region, rural capitalist development was irreversible, as was the state's commitment to it. Officials working within the framework of national exceptionalism saw tradition instrumentally, and had little interest in preserving local customs that they could not control. It was therefore a foreboding of loss that drove the formation of the new discipline of folklore studies (*minzokugaku*), of which Yanagita Kunio (1875–1962) was the founder and the overwhelmingly dominant practitioner. The development both of anthropology, sociology, and ethnology *in* Japan, and of a self-consciously indigenous social science *of* Japan, are unimaginable without Yanagita's influence.

Born the sixth son of a destitute rural scholar of the Chinese classics and later adopted into the family of a high-level jurist, Yanagita displayed strong literary gifts while pursuing a career in the Ministry of Agriculture and Commerce and beyond. Beginning with writings that claimed to record local tales and legends related to him "just as he felt them [*kanjitaru mama*], adding or deleting not a word or phrase," Yanagita developed his core notion of an original, natural village inhabited by the families of *jōmin*, the "people who endure." For Yanagita, the *jōmin* were the silent and invisible "real Japanese," whose timeless lifeways – including ancient ties of service to the

---

[8] Kawamura Nozomu, *Nihon shakaigakushi kenkyū* [Studies in the history of Japanese sociology] (Tokyo: Ningen no Kagakusha, 1975), vol. 2, pp. 6–11.

imperial house – were now threatened with "domicide" by the bureaucracy and capitalism of Tokyo's modern order.

In a sense, Yanagita's folklore was a search for the ur-Japanese, pursued first among the "mountain people" of the interior, and eventually along the "paths of the sea" that tied Japan to Okinawa. Invoking the reconstruction of Japanese tradition by eighteenth-century scholars eager to rescue Japan from the "foreign" taint of long-dominant Chinese ideas and ideals, Yanagita ultimately described his project as a "neo-nativism" – *shin kokugaku*. Yet he was firmly committed to making *minzokugaku* an indigenous *"science* of the native place."　Particularly after the 1920s, as Marxism brought the notion of class to the forefront of social thought, Yanagita produced a number of methodological texts stressing the need for the collection, classification, and close comparison of local legends and tales, customs, dialectology, and so on. Their thrust was to demonstrate what might be termed the "varieties of sameness" across classes and regions.[9]

Prior to 1945, Yanagita's themes were taken up and extended by academic sociologists such as Ariga Kizaemon (1897–1979), who placed particular stress on the hierarchical "family" as the structuring principle for Japanese social relationships generally. Applied during the 1930s and 1940s to the study of Japan's colonial subjects, Yanagita's approach was also appropriated during the postwar years: The anthropologist Nakane Chie (b. 1926), in her well-known *Japanese Society* (1970), pursued the "vertical society" notion that Ariga had developed out of Yanagita's work on the family. In another filiation, associated with the antimodernizationist "people's history" school, Yanagita's critique of the modern in the name of *jōmin* communities under threat was taken in a "left-populist" direction as well.[10]

TOWARD PLURALIZATION: THE LIBERAL CHALLENGE

As the twentieth century opened, the first courses in political science (*seijigaku*) to be conceived independent of the *Staatslehre* tradition were taught in what was then the College of Law in the Imperial University of Tokyo. For Onozuka Kiheiji (1870–1944), with whom this new trend was identified, the state was a proper and legitimate object of empirical inquiry rather than a self-activating subject or realized metaphysical principle. While he carefully abjured any political use of scholarship (in rhetoric, if not in fact),

---

[9] Yanagita Kunio, *Kyōdo seikatsu no kenkyūhō* [Research methods for the study of local lifeways] (1935) and *Minkan denshō ron* [An essay on the transmission of folklore] (1934), in his *Teihon Yanagita Kunio shū*, vol. 25 (Chikuma Shobō, 1964). See also the essays by Hashimoto Mitsuru and H. D. Harootunian in *Mirror of Modernity: Invented Traditions of Modern Japan*, ed. Stephen Vlastos (Berkeley: University of California Press, 1998).

[10] Kawada Minoru, *Yanagita Kunio no shisōshiteki kenkyū* [Yanagita Kunio: A study in intellectual history] (Tokyo: Miraisha, 1985); Kajiki Gō, *Yanagita Kunio no shisō* [The thought of Yanagita Kunio] (Tokyo: Keisō Shobō, 1990).

Onozuka did begin to pry loose the "science" of politics from its identification with administrative technique. It should be noted, however, that independent university departments of political economy (*seiji keizaigaku*) were characteristic only of private institutions such as Waseda University; within the imperial universities, political science was taught as a subfield within faculties of law and administration.

By the beginning of the Taishō era (1912–26), with the waning of oligarchic power and the coincident growth of the middle and working classes, there began movements for universal suffrage and the rights of labor. They were buoyed by the rapid expansion and reform of education and the explosion of mass-circulation newspapers and journals of opinion and entertainment. This impingement of society on politics – even when marked by a populist nationalism – presented a broad and diverse challenge to the hegemony of agrarian-based national exceptionalism, and to the authority of "officialized" social science focused on the exaltation of the imperial state.

Significant attempts to conceive a liberal polity were made by Minobe Tatsukichi (1873–1948) and Yoshino Sakuzō (1878–1933) – the former a constitutional scholar, the latter a political scientist in the line of Onozuka. From their positions in the law faculty at Tokyo Imperial, both sought to broaden the capacity of the political system to "represent" the people, focusing on the Diet as the proper forum for such representation. Both were convinced that the liberation of the individual – *from* traditional constraints and *for* broader participation in society and politics – was the moving spirit of modern times, and a trend from which Japan should not, and could not, claim exemption. Yet their more theoretical contributions were strikingly at odds: Minobe's "organ theory" of the emperor was universalistic and formal in character, stressing not the historical role of the imperial institution but its necessary circumscription once Japan had in place the rationalizing instruments of modern statehood. Yoshino, arguing for "people-as-the-base-ism" (*minponshugi*), took the opposite tack of asserting not only that the franchise could be expanded without attacking or undermining imperial sovereignty, but also that such expansion was consonant with the progressive tradition of the imperial institution itself.[11] For the Christian Yoshino, politics was conducted *for* the popular welfare, with the aim of restoring the proper – harmonious – relation between individual and institution.

The ideas of Minobe and Yoshino won wide acceptance. Minobe's approach informed a generation of bureaucratic training, while Yoshino's spurred an ultimately successful popular movement for universal male suffrage. Taken together, they represent the limit of indigenous liberalism in

[11] Minobe Tatsukichi, *Kenpō kōwa* [Lectures on the Constitution] (Tokyo: Yūhikaku, 1912, 1918), *Nihon kenpō* [The Japanese Constitution] (Tokyo: Yūhikaku, 1921), and *Kenpō satsuyō* [The Constitution in essence] (Tokyo: Yūhikaku, 1922); Yoshino Sakuzō, *Yoshino Sakuzō hyōronshū* [Selected critical essays of Yoshino Sakuzō], ed. Oka Yoshitake (Tokyo: Iwanami Bunko, 1975).

prewar political thought and practice. Their significance for social science lies less in any specific conceptual contribution than in their having vindicated the notion that epochal political change could be conceived and carried out by elites working in consensus with a broad social constituency and without resort to violence. But for such a possibility to be realized, a methodologically independent, empirically grounded political science was necessary. A beginning was indeed made, but its impact was limited. Political science was limited by perspectives that would subsume political processes within those of sociology, which was outpacing the former by the 1910s.[12] More consequentially, an independent political science would eventually require a direct "conceptual" confrontation with the imperial institution itself, which effectively meant denationalization. That was unthinkable.

Along with sociology, economics also acquired academic citizenship at this time. Economics became in some ways *primus inter pares*, the most international, quantitative (though primitively so) and academic of the social sciences. After World War I, independent economics departments were created in major universities. Perhaps because of pressures associated with its location at the hub of the imperial state, Tokyo Imperial's department was particularly prone to ideological and factional disputes, while that of Kyoto quietly assumed international stature as the home of the respected *Kyoto University Economic Review*. The faculties of the Tokyo Commercial College (later Hitotsubashi) and Keiō University were and remain notable. Spurred by the emergence of Marxism as a virtual synonym for social science (a development to be discussed later), the key achievement of interwar economics was its pioneering attempt to examine the contemporary Japanese economy within a rigorous theoretical framework. Marginalists and early Keynesians took their place alongside practitioners of social policy. But certainly through the mid 1930s, Marxian economists enjoyed the clearest identity as a school – and suffered for it in due course.

Appropriately, it was an economist, Sōda Kiichirō (1881–1927), who first articulated a philosophy of social, or more precisely "cultural," science in Japan. Resolutely cosmopolitan, Sōda was an independent scholar who moved easily between economics and philosophy after a decade of study in Europe, most notably with Heinrich Rickert. While heading the bank that bore his family name, he introduced the methodological writings of Weber and Simmel to Japanese audiences. Politically, he espoused the elitist liberalism of the Reimeikai – the Dawn Society – whose public forums on occasion drew audiences in the thousands. Sōda is most closely associated with "culturalism," that is, the notion that each personality can and must "preserve its special and unique significance, and in this sense participate in the creation of cultural products, thereby making possible the realization . . . of its absolute

---

[12] Rōyama Masamichi, *Nihon ni okeru kindai seijigaku no hattatsu* [The development of modern political science in Japan] (1949) (Tokyo: Shinsensha, 1971), pp. 82–92, 142–3.

freedom."[13] Sōda's importance here lies in his having articulated the neo-Kantian distinction between the knowledge of nature and of culture. Nature as an actor without self-consciousness was counterposed to culture, in which the subject of knowledge and action – *ninshiki shutai* – was aware of itself as such. Thus identified and delimited, the notion of "culture" could then provide a basis for the methodological autonomy and differentiation of its constituent sciences. At the same time, neo-Kantianism – Sōda's included – was virtually innocent of the idea of society. One result was a certain barrenness in methodological debate. On the other hand, when the generation of young intellectuals who were exposed to those debates came in large numbers to embrace Marxism as social science par excellence, they tended to demand that it, too, display an appropriate rigor in matters of method. In this sense, "neo-Kantianism was the dialectical premise for Marxism."[14]

Sōda's formalism was emulated in the sociology of Yoneda Shōtarō (1873–1945), who followed Simmel in defining society as the process of mental interaction among individuals apart from the state or household, and the task of his discipline as the study of the *forms* of such association. Two developments of particular interest ensued. Yoneda himself had (in 1919) identified a modern "intellectual class," largely synonymous with a new middle class, as one whose income was derived through knowledge or technical expertise. Too numerous to assimilate entirely to older elites, it was therefore forming ties to the proletariat. This "movement," Yoneda argued, was a social problem of major consequence, especially for the political future of the working class. Yoneda's student Takata Yasuma (1883–1972), in turn, stressed not so much the forms of interaction as those of unity and the will to coexist among individuals, particularly in social classes. The unity of society, whether enforced or voluntary, was an objective fact beyond Simmel's "mental interactions" of discrete individuals. Such work, which clearly prescinded from new forms of urban social interaction, involved a quest for "universal" principles and, as such, constituted an implicit critique of the hegemonic "reality" of the rural community at all levels. In this sense, it fit well with the pronounced universalism of Minobe's "organ theory" of the emperor. At the same time, as shown by Yanagita's increasing nativism and by Takata's move to *Gemeinschaft* in the 1930s, the rural "remainder," as articulated through the state, was at odds with this universalism.[15]

Beyond the academy, engaged and empirical social science was being pioneered by Christian and other social reformers, as well as by academics and scholars affiliated with labor unions and institutions such as the private Ōhara Institute for Social Research. Particularly notable was the evangelist

---

[13] Sōda Kiichirō, *Bunka kachi to kyokugen gainen* [Cultural values and the concept of limit] (1922) (Tokyo: Iwanami, 1972), p. 61; Ishida, *Nihon no shakai kagaku*, p. 99; Rōyama, *Hattatsu*, pp. 141–2.

[14] Ishida, *Nihon no shakai kagaku*, pp. 100, 290–1.

[15] Kawamura, *Sociology and Society*, pp. 54–7.

Kagawa Toyohiko (1888–1960) – among the most famous men in the world in his time – who lived in the Shinkawa slums for four and half years, compiling his surveys of Kōbe and Ōsaka. He wrote extensively on the need for "human construction" among the flawed and wounded personalities of those afflicted by poverty. His studies of the lumpen proletariat, including large numbers of outcastes, sought to embrace the *experience* of modern mass poverty, beginning with its social etiology and extending to considerations of (un)employment patterns, family forms, spending habits, diet, vices, and criminality. A few years prior to the Rice Riots of 1918 and the 1923 Kantō earthquake (which had been followed by a massacre of Koreans in Tokyo), Kagawa strikingly predicted that degradation would predispose individuals with a weakened capacity for self-regulation to mass violence, and this sooner rather than later.[16]

Despite his powerful support of labor, tenant, and outcaste movements, Kagawa was ultimately concerned with moral uplift, and held to a brand of optimistic evolutionism in his social analysis. His concrete hopes for "human construction," along with the liberal visions of Minobe, Yoshino, and Sōda, foundered on the rocks of economic depression and political reaction during the 1930s. Liberalism remained interstitial, a vital irritant in a communitarian environment, but lacked an independent institutional base and motivational force. Following on the decimation of both activist and academic Marxism, liberals in the imperial universities were sacrificed to the guardians of the "national polity," sometimes in the name of university autonomy itself.

RADICAL SOCIAL SCIENCE: THE IMPACT AND FATE OF MARXISM

Writing in the aftermath of Japan's defeat, the political scientist Rōyama Masamichi (1895–1980) argued that liberal political science had been trapped between and shredded by the ideological forces of the left and right, but should not be disregarded by postwar scholars.[17] Rōyama's apologia notwithstanding, the power of the state was in fact valenced in favor of the right; the contest with the left was not equal. A natural harmony between individual and state or community was presumed, and any antagonism required some special explanation. Rōyama was correct, therefore, in identifying a new "threat" from the left, as the image and meaning of "society" sharpened into that of class and class struggle, particularly as embodied in Marxism. As Fukuda Tokuzō, a social policy stalwart and author of a pioneering study of

[16] Kagawa Toyohiko, *Hinmin shinri no kenkyū* [A Study in the psychology of the poor] (1915) and *Ningenku to ningen kenchiku* [Human suffering and human construction] (1920), in his *Kagawa Toyohiko zenshū* [Collected works of Kagawa Toyohiko] (Tokyo: Kirisuto Shinbunsha, 1973), vols. 8 and 9, respectively.
[17] Rōyama, *Hattatsu*, passim.

Japan's social development, put it: "Both socialism and social policy – no, in fact everything to which the term 'social' is attached – today at least take [class struggle] as their chief concern."[18]

Marxism had been introduced during the late 1890s, but it took the Russian Revolution, the Rice Riots, and related labor strikes to confirm the validity of conflict-centered notions of social progress, providing the impetus for a prolonged struggle between the anarcho-syndicalist and Marxist elements of the Japanese socialist movement. In the process, Marxism transcended its role as the ideology of a harried revolutionary movement and became virtually synonymous with "social science" – a term now, for the first time, in common use. A full translation of *Capital* appeared between 1919 and 1925; 15,000 sets of the Kaizōsha edition of the complete works of Marx and Engels (1927–9) were sold in the first edition alone. As elsewhere, the spread of Marxism in Japan depended not only on the existence of a party-authorized, *Capital*-centered canon, but also on its popularization in texts by Engels, Kautsky, Lenin, and Bukharin. Kawakami Hajime (1879–1946), who founded Marxian economics in Japan (and eventually joined the Communist Party), while in certain important respects never committing himself to historical material-ism, was perhaps the crucial "apostle" to young intellectuals. His appeal lay behind the proliferation of "social science" research groups in universities and high schools (and even in middle schools), and their prohibition by ed-ucational officials as early as the mid 1920s. More consequentially, numerous young academics traveled to Weimar Germany to study original texts and in-teract with German (and German-speaking) Marxists representing the entire spectrum of positions, from positivist to revisionist to Hegelian.

In whatever variant, Marxism was "the first *Weltanschauung* in modern Japan which compelled one intellectually to explicate the transformation of social systems in a total and coherent fashion."[19] Its power was all the greater because the various social science disciplines had developed "instru-mentally" as discrete sciences, and, in contrast to Europe, Japan had not experienced the crisis and collapse of evolutionary or positivist systems such as those of Spencer and Comte. Each strength in Marxism, however, brought with it a corresponding flaw. Its systematic character could degenerate into dogmatism, its putative universality recalled its foreign origin (and con-firmed Japan's position as a historically backward "object" of knowledge), and its critical modus operandi often provoked in-fighting and organizational fragmentation.

Ultimately, Marxism's claim to be synonymous with social science derived from its *analysis* of Japanese society itself, one that reflected – but in important

[18] Fukuda Tokuzō, *Shakai seisaku to kaikyū tōsō* [Social policy and class struggle] (Tokyo: Ōkura Shoten, 1922), p. 3.
[19] Maruyama Masao, "Kindai Nihon no chishikijin" [Intellectuals in modern Japan], in his *Kōei no ichi kara* [From the rear guard] (Tokyo: Miraisha, 1982), pp. 107–8.

ways transcended – all of the tensions and problems just described. Its chief contribution took the form of the "debate on Japanese capitalism" that ran from the late 1920s to the mid 1930s. Occasioned by political disagreements over revolutionary goals and strategy, its task was the historical characterization of the developmental process of Japanese capitalism and the modern state. The so-called Lectures Faction (Kōza-ha), following the position of the Comintern's 1927 and 1932 "Theses on Japan," focused its analyses on the entrenched and powerful "feudal" forces that controlled the absolutist imperial regime. Japanese capitalism was "special," a kind of hybrid. Bourgeois political institutions were immature or malformed, and the entire state apparatus was underlaid by a vast base of semifeudal production relations among the peasantry that had been little affected by the political events of 1868. The task of social science, therefore, was to clarify the obstacles to the completion of the democratic revolution as the necessary first step in a two-stage drive toward socialism. The dissident Labor-Farmer Faction (Rōnō-ha), while cognizant of time lags vis-à-vis the West, took a more conjunctural view, regarding Japan as one of a number of imperialist systems of financial capitalism. This meant, by corollary, that prior to the Restoration, Japan's agrarian economy had already developed production relations characteristic of incipient bourgeois domination. The Meiji Restoration was Japan's bourgeois revolution; "vestiges" of feudalism, while still powerful, were incidental and would be swept away in a socialist revolution.

By nature, the debate about capitalism could not be resolved; it ended with the arrest or silencing of its participants by 1938. Because it was bound up with the internal politics of the left, sympathetic observers often felt pressed to declare for one side or the other. This largely obscured the true significance of the debate and the split in perception that had triggered it, but it was not an inevitable effect of Marxism as such. The economist Uno Kōzō (1897–1977), for example, argued that Japan's capitalism was a classic case of "late development" in which the capitalist mode of production was mediated via industry rather than via agriculture. This meant that both the Kōza-ha emphasis on the "semifeudal" peasantry, and the Rōnō Faction insistence on finding evidence of rural differentiation, were misplaced.[20] But it was not until the 1960s, with the virtual disappearance of the peasantry and the supersession of the Kōza–Rōnō debate, that the structure of Japan's prewar capitalism as a whole was *politically* permitted to come into view.

Japanese Marxists have been heroized as the primary victims of political persecution. But we must also recognize that the "national community" in crisis exerted a positive appeal – one that essentially disabled the critical impulse

---

[20] Uno Kōzō, "Shihonshugi no seiritsu ni okeru nōson bunkai no katei" [The process of rural differentiation in the establishment of capitalism] (1935), in *Senjika no teikō to jiritsu: Sōzōteki sengo e no taidō* [Resistance and autonomy in wartime: Impetus for a creative postwar era], ed. Furihata Setsuo (Tokyo: Shakai Hyōronsha, 1989), pp. 151–74.

within Japanese social science that had taken systematic form under Marxian aegis. Social thinkers long weaned of "bourgeois" sociology responded with alacrity to the call of community. Faced with a choice between an open break with the national community – imprisonment, exile, coerced silence – and some sort of compromised life, a great many chose to "return to Japan."

"Return" meant engagement with the state, and more concretely with avowedly reformist officials who were just hitting their stride during the late 1920s and early 1930s, and with their counterparts in the military. Particularly after the invasion of Manchuria in 1931, the state and the military as well as segments of academia and journalism were drawn to contemporary Italian, German, and Soviet models of industrial and economic organization. The South Manchurian Railway had long since employed thousands of researchers; at home, the attempt to mold the economy for total war engaged the efforts of many more. Organizations such as the Shōwa Kenkyūkai (Shōwa Research Association) and the Cabinet Planning Board (Naikaku Kikakuin) recruited both academic and government economists, including luminaries-to-be such as Arisawa Hiromi (1896–1988). For such scholars, the disappearance of alternative foci for their expertise and direct pressure to contribute to Japan's war effort made such service extremely difficult to refuse.

The intolerable strain placed on resources by spiraling military demand meant that many of these wartime economic plans came to nothing. But their failure must not be allowed to obscure the patterns of thinking that drove the planners in their efforts. The rhetoric and substance of their critique of capitalism in its "liberal" phase were drawn in large part from Marxism, while the determinants of their politics were the more "Listian" demands of national and bloc self-sufficiency. For this latter reason, at the intellectual level "class" tended to be transmuted into *Volk* (*minzoku*) as the favored agent of historical change; at the political level, the Marxist could lie down with the "fascist." This intersection of bureaucratic and radical thought in Japanese social science was both representative and momentous: It marked the formation of a technical intelligentsia whose expertise in economic policy and planning was later to be mobilized, without interruption, in the pursuit of recovery and growth after 1945.

POSTWAR SOCIAL SCIENCE: MODERNISM AND MODERNIZATION

Social science during the early postwar decades may be equated with what is known as "modernism" (*kindaishugi*), and with an assault on Japan's "negative distinctiveness" as a state and society. Its temporal starting point was defeat and occupation, its critical genesis a drive to expose the causes of Japan's disaster. The war and the process leading up to it betokened a historical pathology: The *tennōsei*, Japan's imperial system, was held to be inimical to a rationally organized national life and had made Japan unable to coexist with its neighbors in the world. Along with Marxism and eventually superseding it, modernism pursued the *completion* of the distorted and failed first phase of Japan's modernization. Not directly political, modernism animated a "democratic enlightenment" which, while it took certain cues from the rhetoric and policies of Japan's occupiers, had its roots in prewar Marxism, some aspects of liberalism, and in the experience of war itself. The term itself was coined by its Marxist critics, but was widely used.

Modernist writings of the immediate postwar period tended to portray the war as an episode of atavistic irrationality. With time, however, the institutional "usefulness" of economic mobilization and various forms of rationalization, along with the indisputable trend toward social leveling, have come to be acknowledged as elements in shaping the postwar economic regime. The importance of planning and of "engaged" social science was only enhanced by the new dispensation, not least owing to the presence of New Deal "cadres" during the early phase of the occupation. The issue now, many social scientists thought (or dreamed), was to determine whether a democratic Japan was to follow a capitalist or a socialist path. Whichever it was to be (as the government itself recognized), the basic work of data collection and problem definition would require the efforts of scores of economists – Marxians newly released from prison or permitted to return to academic positions, Keynesian generalists (such as Nakayama Ichirō), and policy specialists (Ōkōchi Kazuo on labor, Tōhata Seiichi on agriculture) who had chafed under wartime irrationalities, biding their time.[21]

Ultimately, there was to be no departure from the capitalist path. But the strong presence of Marxian economists in government was a striking feature of the early postwar decades, as was the extensive influence of Marxian approaches among academic economists. Though eventually overshadowed by their Americanized confrères, this "economic left" – figures such as Arisawa Hiromi and Tsuru Shigeto (b. 1912) – made a crucial contribution to what has been termed the "soft infrastructure" or "invisible base" of postwar modernization.[22]

Yet modernism was not about effectiveness per se; nor did "postwar" merely denote straitened material conditions requiring sharpened expertise. At its heart lay the perception of, and desire to reinforce and "vivify," the *discontinuity* that marked Japan's recent past. At its most influential, modernism

---

[21] Special Survey Committee, *Postwar Reconstruction of the Japanese Economy* (Japanese Ministry of Foreign Affairs, September 1946) (Tokyo: University of Tokyo Press, 1992).

[22] Takeuchi Kei, "Nihon shakai kagaku no chiteki kankyō" [The intellectual environment of Japanese social science], in *Shakai kagaku no genba* [Social science at work], ed. Yamanouchi Yasushi, Murakami Junichi, Ninomiya Hiroyuki, Sasaki Takeshi, Shiozawa Yūten, Sugiyama Mitsunobu, Kang Sangjung, and Sudō Osamu, volume 4 of *Shakai kagaku no hōhō* [The methods of social science] (Tokyo: Iwanami Shoten, 1993), p. 43.

was as much a moral as a scientific orientation. Modernists were driven by a degree of collective guilt, expressed not so much toward the victims of Japan's aggression in Asia as, more abstractly, toward history itself. With this was combined a sense of victimhood as members of a generation profligately wasted by the state. The intense self-concern of this "community of contrition" draws criticism today, as does the fact that modernism very quickly lost sight of the external empire in its urgent desire to uproot the pathologies of the vanquished regime. But just this moral seriousness gave modernism its long staying power in its collaborative competition with Marxism.[23]

What sort of social science was modernism? We can draw a few indications from the work of Maruyama Masao (1914–1996) and Ōtsuka Hisao (1907–1996): The former was the foremost historian of Japanese political thought and a practicing political scientist, the latter an economic historian of Europe and later concerned with what is now termed the "North-South" disparity. As Maruyama wrote in 1947, the task now is "to accomplish what the Meiji Restoration was unable to carry through: that of completing the democratic revolution," and "to confront the problem of human freedom itself." The bearer of freedom, however, is no longer the "citizen" of classical liberalism "but rather ... the broad working masses with workers and farmers at the core." Moreover, the crucial issue is not the "sensual liberation of the masses, but rather how and how thoroughly the masses are to acquire a new normative consciousness."[24]

Here we can see the heavy debt to Marxism as well as the Kantian – or neo-Kantian and Weberian – overlay that tied the modernists to the intellectual culture of the interwar era. In Maruyama, whose thought includes strong nominalist and social contractarian elements as well, modernism amounted to a drive to create a critical mass citizenry capable of resisting authority: the *homo democraticus*. In Ōtsuka, who sought to combine Marx and Weber while maintaining textual fidelity to both, the grail was an ethical producer, a man of conscience working among social equals. Postwar Japan, Ōtsuka argued, could no longer subsist on feudal "fairness," but had to strive for modern "equality." An interesting recent criticism of Ōtsuka finds him too optimistic, not Nietzschean enough – too committed a believer in the real possibility of reconstituting Japanese society along "ethically individualistic" lines to face Weber's "iron cage" prophecy head-on. It is argued that Ōtsuka's ethics were insufficiently political – and drawn without acknowledgment from his wartime writings on productivity – and therefore too easily co-optable.[25]

---

[23] Maruyama, "Kindai Nihon no chishikijin."

[24] Maruyama Masao, "Nihon ni okeru jiyū ishiki no keisei to tokushitsu" (August 1947), in his *Senchū to sengo no aida* (Tokyo: Misuzu Shobā, 1976), p. 305.

[25] Yamanouchi Yasushi, "Sengo hanseiki no shakai kagaku to rekishi ninshiki" [Social science and historical awareness in the postwar half-century], *Rekishigaku kenkyū*, no. 689 (October 1996), 32–43.

It is one thing to find the sources of co-optation in the thought of an individual; strong modernists such as Maruyama and Ōtsuka will always have their partisans and detractors. However, unprecedented economic growth, along with the political defeat (in 1960) of activist academics seeking to end Japan's diplomatic and military subordination to the United States in favor of nonalignment, opened the way for the "structural" co-optation of modernism into the discourse of *modernization* (*kindaikaron*). As a set of operating concepts, modernization congealed at the beginning of the 1960s, following its energetic propagation in American scholarship and dissemination in "highbrow" Japanese periodicals such as *Shisō* and *Chūō kōron*. This approach is strongly associated with the rehabilitation of the Meiji and eventually the Tokugawa eras as "forerunners" to Japan's startling run of sustained economic growth; "tradition," that is, was rediscovered as having contributed (via widespread literacy and rural commerce, for example) to the later success of industrialization.

Yet Kawashima Takeyoshi (1909–1992), an eminent sociologist of law at Tokyo University and a close collaborator of Ōtsuka Hisao, saw in modernization a tool for the analysis "not only of social change in the so-called 'East' and 'West,' but in the less developed countries and 'new states' as well." Indeed, he envisioned the "possibility in theoretical terms of being able to treat all these [cases] as a world historical movement headed in the same direction via differing processes." Kawashima's hopes went beyond analytical results:

> To foresee in what direction the grand movement of contemporary world history is headed, and by what route; to search out the path by which to bring humanity true happiness more quickly – this is humanity's fervent desire and a task for social scientists of overriding importance. The approach discussed here may be seen as an effort in that direction.[26]

Kawashima himself retained much of the subjectivist agonism of early postwar modernism, seeing the task of social science as that of guiding the struggle to overcome "traditional" society and values via a revolutionary break – especially, a mental revolution. Its external referents were highly idealized representations of modernity abstracted from the history of revolutions in the West. But, as the decade continued, modernization came to place preponderant value in the smoothest possible continuity from stage to stage in a continuous process of national historical development that could be captured in a set of quantifiable processes culminating in the maximization of GNP. *Modernism* was self-consciously ideological, seeing "value freedom" in social science as something to be struggled for in the process of liberating present reality from the distortions of traditionalistic consciousness; *modernization* tended to assume that "value freedom" or objectivity was assured through the

---

[26] Kawashima Takeyoshi, " 'Kindaika' no imi" [The meaning of "modernization"], *Shisō*, no. 473 (November 1963), 8.

identification of measurable social or behavioral indices. Its key external referent was the contemporary United States, a society "always already" modern and ostensibly freed from "ideology," particularly the ideology of class and class conflict.

According to the modernization approach, Japan was more than a "case"; it was with respect to Japan that the term "case" itself first gained credibility in analyzing the process of historical change. Japan was an exemplar, identified as such by 1961, against which the mere "cases" – Turkey, Russia, Iran, Mexico, Korea, and in general the "developing societies" and "new states" – were to be measured. Convergence (guaranteed Americanization) was the promise held out to all "successful" modernizers; all differences of culture were in the end no more than matters of degree along a scale of functionality.[27]

FROM SCIENCE TO CULTURE

Modernization *à la japonaise* was composed of two elements that had begun to separate out by the end of the 1960s: "growthism" and "culturalism." The former combined the mantra of quantification with the valorization of industrial production for its own sake, providing an enormous stimulus to applied neoclassical and Keynesian economics (known as *kindai keizaigaku*, or "modern economics") and econometrics, and beginning to undermine the commanding position of Marxian approaches. Statistical fetishism was inescapable.

Japan's national purpose was to produce; the political question was, for whom? More fundamentally, how had postwar growth occurred, and how could it be sustained? Government and academic economists focused their cyclical and "macro"-structural analyses on the possibly unique circumstances created by the combination of inherited dualism in industrial structure and the impact of so-called "postwar characteristics" – the need to reconstruct, the legacy of Occupation reforms, and so on. But, over time, the role and habitus of firms, their ostensibly traditional modes of organization and interaction, and the capacity to mobilize and motivate labor were recognized as crucial, engaging the efforts of both Japanese and, in increasing numbers, American researchers.

Certainly the rehabilitation of tradition from early postwar condemnation injected a needed concreteness into the definition of modern or civil society in Japan. American social scientists, both specialists and comparativists, were particularly impressed with the power of corporations to gear secondary and higher education to the production of model employees, even when only

[27] Ishida Takeshi, *Shakai kagaku saikō* [Social science reconsidered] (Tokyo: Daigaku Shuppankai, 1995), pp. 28–34, 100–10; Wada Haruki, "Kindaikaron," in *Kōza Nihonshi* [Symposium on Japanese history], ed. Rekishigaku Kenkyūkai and Nihonshi Kenkyūkai, vol. 9: *Nihonshi ronsō* [Debates in Japanese history] (Tokyo: Daigaku Shuppankai, 1971), pp. 255–82.

a (large) minority could hope to gain "lifetime employment" in a high-prestige firm. Attention also turned to the larger economic role of corporate networks, with their interlocking capitals and their ongoing and intimate relations to the state. In place of a generalized (Western) notion of modern society that looked to independent nonstate organizations – the realm of sanction and custom, voluntary organizations, and so forth – operating on the basis of marketized relations of social "equals," modernization in Japan was seen to have maintained and promoted "neo-feudalism" in human relations. "Culture," it turned out, was the key to growth, and culture meant not convergence but profound and significant difference.

The zenith of "growthism" came with the national celebration of the Meiji Centenary and Expo '70. It might seem that growthism and culturalism should have continued to work in tandem. But the Vietnam War issued a severe check to illusions of American omnipotence, including among the casualties the notion of modernization as a measurable (and guidable) process. Japan's government and corporations, as profiteers of that conflict, were also discredited. Domestically, the costs of growthism were coming due, in the form of staggering environmental pollution, urban hypercongestion, and the sense that corporate interest had come to justify unlimited demands for labor. Neither individual nor community life seemed to have any inherent importance.

This situation ought to have represented a golden opportunity for Marxism. To a degree, this was true: By the 1960s, Marxism had been institutionalized as a requirement, alongside "modern" economics, in many Japanese university departments. The most influential school was that associated with Uno Kōzō, whose system of political economy, along with Maruyama's political science and Ōtsuka's historical economics, has been described as one of the three main currents of postwar Japanese social science.[28] Beginning in the late 1930s, Uno had pursued a logical – Hegelian – reconstruction of *Capital*, developing an original framework of "basic principles" of political economy, along with a three-stage historical model of capitalist development that culminated in "contemporary analysis." Marked by its rigorous separation of economic science from ideological activity and its portrayal of capital as a "structure" of all-generative power, the Uno school represented the apotheosis of Marxism as an "objective" science of political economy. Although Uno's motivation in separating science from ideology stemmed from his revulsion against Stalinist politicization, the pitfall of his system was that if Marxism is construed *only* as a science, it is easily superseded by a more "effective" one.

What happened at the beginning of the 1970s, however, was more than a search for "better" science. As "empirical research" was exposed either as an ideology of expertise for hire (not "value-free" but "value-less," its critics

---

[28] Furihata Setsuo's annotations to Uno Kōzō's article cited in note 20 above.

charged) or as quietistic, a shift in orientation from quantity to quality, from "science" to "culture," extended through the social sciences. Japanese writings on the constitution of society – in the 1970s by Nakane Chie, and a decade later by Murakami Yasusuke (1931–1993) – took on a deeper and deeper hue of what may be termed a "supermodern" perspective: the view, that is, that it was precisely the continuity of "premodern" organizational patterns and ways of thinking that had made possible Japan's unprecedented economic growth. For these analysts, the vitality of Japanese social organizations sprang from their cultural underpinnings: rationalized dependency, corporate personalism, and collective instrumental rationality.[29] In Kōza-ha Marxism and postwar modernism, "culture" – and community – stood for "backwardness," a fetter on rationality. Now, no longer requiring the mediation or checking mechanism of noncorporate civil society, or the promise of convergence with other advanced industrial societies, Japan's "culture" defined the vanguard not of "capitalism," which along with democracy was hardly mentioned, but of a new, information- and relation-oriented "system." Japan had come to embody a future that was not only post-Marxist and postsocialist, but postindividualist and postcapitalist as well.

Yet other visions of culture and community have also been at work in the social sciences. In the 1970s, critical economics migrated from systematic Marxism to the ad hoc milieu of the local residents' and antipollution movements. The target, as attacked in the work of Tsuru Shigeto, Miyamoto Ken'ichi, and Uzawa Hirofumi, was less "capitalism" than the concrete pathologies of "growthism" itself: the penchant for massive building projects, spiraling land values, hurried and unsound engineering, environmental and social destructiveness.[30] Implicitly or explicitly, such work argued that the test of social science lay not in its contribution to "growth," but in the intellectual resources it could provide people, in their localities and as a national community, to help them weather the inevitable cycles of growth and decline in an advanced economy fatefully interwined with the world.

In the long run, culturalism could not survive the decline of growthism. The end of the Cold War coincided with the bursting of the economic "bubble," leading to the most serious downturn in a half-century. Amid external pressures for "liberalization," deep political corruption, and signs of decay in the corporate personalism that had marked the upper tier of the industrial economy, the hegemony of "culturalism" has grown tattered. The implications for social science are ambiguous. Striking work by sociologists of religion (spurred by social traumas such as the Aum Shinrikyō incident and the Hanshin earthquake), new modes of historical inquiry into the Tokugawa

[29] Murakami's final work, *An Anticlassical Political-Economic Analysis: A Vision for the Next Century*, trans. Kozo Yamamura (Stanford, Calif.: Stanford University Press, 1996).

[30] Shigeto Tsuru, *Japan's Capitalism: Creative Defeat and Beyond* (Cambridge: Cambridge University Press, 1993); Uzawa Hirofumi, *Chosakushū* [Collected writings] (Tokyo: Iwanami Shoten, 1994), vols. I, 6–8, 10–12.

past, and local and minority ethnographies all suggest underlying vitality. And a promising theoretical breakout seems to be under way in the analyses by Yamanouchi Yasushi, among others, of the wartime origins of contemporary Japan's "system society" that goes beyond both older notions of historical stages and cultural perennialism.[31] On the other hand, Japanese social science has no strong, urgent focus: no fictive national community, no revolutionary quest, no "modern" democratic personality in need of shaping, no growth-above-all. The Japan–West framework seems to have eroded, but it is not clear that a new social science set on a Japan–Asia axis is a realistic possibility. The history of relations in the region would seem to militate against it, to say nothing of the deepening general uncertainties of the late 1990s. The current situation is perhaps best characterized as a plurality – though not a pluralism – of uncertain significance.

[31] Yamanouchi Yasushi, *Shisutemu shakai no gendaiteki isō* [Contemporary aspects of system society] (Tokyo: Iwanami Shoten, 1996); and the twelve-volume collection edited by Yamanouchi Yasushi, *Iwanami kōza shakai kagaku no hōhō* [Iwanami Symposium: The methodology of the social sciences] (Tokyo: Iwanami Shoten, 1993).

Part IV

SOCIAL SCIENCE AS DISCOURSE AND PRACTICE IN PUBLIC AND PRIVATE LIFE

30

THE USES OF THE SOCIAL SCIENCES

Peter Wagner

The idea of developing social knowledge for the purpose of social betterment assumed its modern form during the Enlightenment. In many respects, the American and French Revolutions were a culmination of that development and the first large-scale "application" of modern social and political theory. At the same time, the revolutions were often interpreted as having brought about a situation in which good social knowledge would permit the steady amelioration of social life. The ways of thinking of the social sciences were also created in that context.[1]

The new, postrevolutionary situation altered the epistemic position of the social sciences, even though this was acknowledged only gradually. Any attempt to understand the social and political world now had to deal with the basic condition of liberty; but an emphasis on liberty alone – as in the tradition of early modern political theorizing during the seventeenth and eighteenth centuries – was insufficient to understand the social order. Thus, in the words of Edmund Burke, if "the effect of liberty to individuals is, that they may do what they please [, we] ought to see what it will please them to do, before we risque congratulations."[2] The use orientation of the early social sciences consisted in offering a variety of ways of dealing with this situation. Aiming at finding out what it pleased individuals to do, the emerging social sciences embarked on developing empirical research strategies to provide useful knowledge. On the other hand, the concern for the practical order of the world in those social sciences translated into attempts to identify some theoretical order inherent in the nature of human beings and their ways of

---

[1] Johan Heilbron, Lars Magnusson, and Björn Wittrock, eds, *The Rise of the Social Sciences and the Formation of Modernity* (Sociology of the Sciences Yearbook 20) (Dordrecht: Kluwer, 1998); see also Göran Therborn, *Science, Class and Society* (London: New Left Books, 1976); Geoffrey Hawthorn, *Enlightenment and Despair: A History of Sociology* (Cambridge: Cambridge University Press, 1976).

[2] Edmund Burke, *Reflections on the Revolution in France* (1790), ed. L. G. Mitchell (Oxford: Oxford University Press, 1993), pp. 8–9.

socializing, namely, the predictability and stability of human inclinations and their results.

THE USES OF THE THEORETICAL TRADITIONS

The roots of the theoretical traditions in the social sciences lie not least in this political *problématique*. The concern of social scientists for the predictability of human action and the stability of the collective order entered into the four major forms of reasoning that have characterized the social sciences throughout their two-century history. Some theorists argued that social location determined the orientations and actions of human beings. There are two major variants of such thinking. The first was what one might call a cultural theory, which emphasized proximity of values and orientations based on a common background. The nation as a cultural-linguistic entity was thus seen as a major collectivity of belonging that gave a sense of identity to human beings in Europe; and, mutatis mutandis, cultural anthropology translated this perspective into other parts of the world. Second, interest-based theories placed the accent on a similarity of sociostructural location and thus on a commonality of interest. In this approach, which strongly shaped the discipline of sociology, social stratification and class were the key categories determining interest and, as a derivative, action.

The third strategy to discursively stabilize human activity was directly opposed to culturalist and sociological thinking. In individualist-rationalist theorizing, full reign is given to individual human beings, and no social order constrains their actions. In the tradition that reaches from political economy to neoclassical economics to rational choice theorizing, intelligibility is here achieved by different means: Though they appear to be fully autonomous, individuals are endowed with rationalities such that the uncoordinated pursuit of their interests will lead to overall societal well-being. These three kinds of reasoning make for a very peculiar set, in the sense that this last one locates the determinant of action almost completely inside the human being, the former two almost completely in the outside sociocultural world. In the fourth approach, the behavioral-statistical one, no such assumptions are made, but attitudes and behaviors of individuals are counted, summarized, and treated using mathematical techniques in order to discover empirical regularities. This approach can be, and has been, combined with all the other three.

These four approaches to social life are all well established, and discussions of their strengths and weaknesses have gone on for many years. What is important in our observations on the uses of the social sciences is that they have all been developed not as purely intellectual projects, but rather with a view to identifying and enhancing those elements of social life that bring stability into the social world. The rationalistic-individualistic idea that a society composed of free individuals would maximize wealth has supported arguments both for the dismantling of barriers to action, as in the introduction of the liberty of commerce, and occasionally for prohibitions on collective action, such as actions by trade unions and business cartels. The socioeconomic idea of defining the interests of human beings according to social position has revealed both fundamental conditions for harmony, as in structural-functionalism, and contradictions in society, as in Marxism. The connection between Durkheim's theory of solidarity and the political ideology of solidarism in the French Third Republic is an important instance of such use of basic modes of social theorizing. The cultural-linguistic idea has informed the understanding of the grouping together of larger collectivities; it was at the root of the idea of the nation as the unit polity, and thus of nineteenth-century nationalism. The behavioral-statistical approach has allowed the aggregation of people into collectivities, not unlike the former two, but has rarely been premised on such strong assumptions about the social bond behind the aggregation. It has flourished not only in state-organized statistical institutes aimed at monitoring the population, but also, particularly in Britain and the United States, in private organizations interested in issues such as poverty and deviance.

These modes of reasoning formed the intellectual basis for some of the key disciplines of the social sciences – cultural anthropology, sociology, economics, and statistics – during a period of internal consolidation of the universities as sites of scholarly research, roughly at the end of the nineteenth century (as discussed in detail in Part II of this volume). In the present context, though, it is more important to emphasize that all of these ways of relating social theories to societal issues have also been employed ever since, even though their plausibility and application have varied across space and time. Their current forms of use, however, are hardly ever pure (with the partial exception of neoclassical economics), but rather are blended with forms of positive knowledge as provided by empirical social research.

THE DEMAND FOR EMPIRICAL SOCIAL KNOWLEDGE

In parallel with the elaboration of the basic modes of social theorizing, and having very much the same objective and ambition, attempts to increase positive knowledge about the new social world were increasingly made across the nineteenth century. Whereas theories tried to provide reasons why such a social world could hold together, research explored experiences of its harmonies or, more often and more consistently, of its strains and tensions. A starting point for many empirical research endeavors was indeed the observation that the Enlightenment, or liberal, promise of automatic harmonization of social life had not been kept.[3] The wide-ranging effects of the new urban

---

[3] Peter Wagner, Björn Wittrock, and Hellmut Wollmann, "Social Sciences and Modern States," in *Social Sciences and Modern States: National Experiences and Theoretical Crossroads*, ed. Peter Wagner, Carol H. Weiss, Björn Wittrock, and Hellmut Wollmann (Cambridge: Cambridge University Press, 1991), pp. 28–85.

and industrial civilization that was rapidly changing living and working conditions for ever larger numbers of people in Europe and America during the
nineteenth century caused increasing anxiety. These changes, often summarily referred to as "the social question" (or "the labor question"), were forcing
themselves on to the agendas of parliamentary bodies, governmental commissions, and private reform-minded and scholarly societies. The impetus for
the search for new knowledge often came from modernizing political and
social groupings that favored industrialization but that also advocated more
or less far-reaching social reforms. These groupings gradually came to embrace the notion that political action to address the "social question" should
be based on extensive, systematic, and empirical analysis of the underlying
social problems. The rising awareness of deep social problems shaped the
social sciences during their period of institutionalization.

In France, social research had been encouraged and pursued since the
early nineteenth century by "enlightened administrators" who had grown
up with the intellectual traditions of the Revolution and the institutional
innovations of the Napoleonic period. They were, therefore, inclined toward an active, modernization-oriented view of society and of the state's
role in bringing about reforms. By midcentury, a more conservative alternative arose in the thinking of Frédéric LePlay, who aimed at maintaining
and restoring the traditional structures of society, but who equally relied
on the systematic observation of society. In Britain, reform-minded individuals, often belonging to the establishment of Victorian England, came
together in a number of reform societies, some of which had close links to
the scholarly world.[4] Concern for health mounted, for example, when army
recruitment during the Boer War revealed the appalling conditions under
which much of the British population lived. Among the reform societies, the
Fabian Society came to play a leading role through the establishment of the
London School of Economics and Political Science, a university and research
center that has continued to be distinguished by its double commitment to
academic inquiry and problem-oriented research.[5] In Germany, immediately
after the founding of the Bismarckian state, the Verein für Socialpolitik became the main initiator and organizer of empirical research on the "social
question." In the United States, social science research originally had the
same characteristics of associational organization and ameliorative orientation that it had in the European countries. The American Social Science
Association (ASSA), created in 1865, embraced the notion that the social
scientist was a model citizen helping to improve the life of the community,

---

[4] See, e.g., Sheldon Rothblatt, *The Revolution of the Dons: Cambridge and Society in Victorian England*
(Cambridge: Cambridge University Press, 1981).

[5] Dietrich Rueschemeyer and Ronan van Rossem, "The Verein für Sozialpolitik and the Fabian Society:
A Study in the Sociology of Policy-Relevant Knowledge," in *Social Knowledge and the Origins of
Modern Social Policies*, ed. Dietrich Rueschemeyer and Theda Skocpol (Princeton, N.J.: Princeton
University Press; New York: Russell Sage Foundation, 1996), pp. 117–62.

not a professional, disinterested researcher. By the turn of the nineteenth century, this model was overwhelmed by emerging disciplinary associations, which splintered off from ASSA and soon began to subdivide.[6]

While the range of comparative observations could easily be enlarged, the apparent parallelism in attention to problems cross-nationally must not conceal the fact that the identification of solutions and even the definition of problems were premised on significantly different discourses and institutional constellations. For our purposes, the role of the state in problem solving and the position of knowledge producers in state and society are the key aspects to be considered comparatively.[7]

STATES, PROFESSIONS, AND THE TRANSFORMATION OF LIBERALISM

The emerging variety of forms of social knowledge and of policy intervention can, as a first step, be traced to different ways of transcending the limitations of a liberal conception of society. For France, this change was closely related to the experience of the failed Revolution of 1848. It thus became evident that the mere form of a democratic polity did not yet provide a solution to the question of societal organization. In Italy and Germany, by contrast, liberal-minded revolutionary attempts had failed, and the emergence of the social question tended to coincide with the very foundation of a national polity. The process of nation building during the decade between 1861 and 1871 profoundly changed the terms of political debate and the orientations of political scientists in both countries. The idea of social betterment through social knowledge appeared to have found its agent: the nation-state. The founders of the Verein für Socialpolitik left no doubt about the intimate linkage between the creation of their association and the inauguration of the Reich: "Now that the national question has been solved, it is our foremost duty to contribute to solving the social question."[8]

On the basis of a great variety of social inquiries, the construction of national social policies was widely urged on the European continent toward the end of the nineteenth century. Such policies would in practice extend the idea of a community of responsibility as it had been developed during

[6] Thomas L. Haskell, *The Emergence of Professional Social Science: The American Social Science Association and the Nineteenth-Century Crisis of Authority* (Urbana: University of Illinois Press, 1977); Peter Manicas, "The Social Science Disciplines: The American Model," in *Discourses on Society: The Shaping of the Social Science Disciplines*, ed. Peter Wagner, Björn Wittrock, and Richard Whitley, (Dordrecht: Kluwer, 1991), pp. 45–71.
[7] On the following, see in more detail Björn Wittrock and Peter Wagner, "Social Science and the Building of the Early Welfare State," in *Social Knowledge*, ed. Rueschemeyer and Skocpol, pp. 90–113.
[8] Gustav Schöneberg, as quoted in Ursula Schäfer, *Historische Nationalökonomie und Sozialstatistik als Gesellschaftswissenschaften* (Wien: Böhlau, 1971), p. 286.

that period in collectivist social theories, be they of a social-interest-based or of a cultural-linguistic kind. In the new intellectual and political situation, it could be argued plausibly that the nation was the relevant, responsibility-bearing community and the state its collective actor, the head and hand, as it were, in the design and implementation of social policies. The nation-state was regarded as the "natural" container of rules and resources extending over, and mastering, a defined territory. This, however, was much less the case in the United States, where a strong central state did not as yet exist. In contrast to both France and Germany (disregarding for a moment the intellectual variety in these contexts), social researchers in the United States tended to be reluctant to posit state and society as collective entities over or beside individuals. Even if the case for individualist liberalism as the predominant politico-intellectual tradition throughout U.S. history has been overstated,[9] the counterpart to such thought in the United States, civic republicanism, was still comparatively much more liberal and individualist than the variants of nationalism, socialism, and organicism that had inspired European social reformers. One consequence of the individualistic inflection of U.S. political culture is that psychology and social psychology have been much more important in the social sciences than elsewhere. As Ellen Herman observes in her chapter, many social problems have been dealt with on the level of individual psychology.

This intellectual specificity of the situation in the United States can be connected to an institutional feature that has shaped the strategies of those academic entrepreneurs who have advocated social reform. In the United States, such advocates of reform based on inquiry, while opposed to the politics of corruption and patronage in particular, were also often distrustful of increasing the power of the state in general. Instead, they tended to advocate the complementary strategy of reform and competence, a type of "profession-based" social policy. If, in the United States, as in Continental Europe, the widening of social responsibility was the issue, then professions were designed as a nonstatist way to exert authority over spheres of social-political action. The specific form of academic institutionalization of social science in the United States, namely as disciplinary associations, was the result of such considerations. As Julie Reuben shows in Chapter 36, professional status concerns also limited American social scientists' engagement with research on education.

For professors in high-prestige, state-run academic institutions on the European Continent, particularly in Germany, it was by contrast quite natural – in intellectual, institutional, and social terms – to see the state as the key policy institution and themselves as its brain. U.S. social reformers not only were doubtful about the "rightness" of state interventions in terms of liberal political theory, but also had no strong reason to connect a

---

[9] As made, for instance, by Louis Hartz, in *The Liberal Tradition in America* (New York: Harcourt, Brace and World, 1955).

reputation-seeking strategy to the state. Their authority was to be based on the knowledge claims inherent in the existence of strong autonomous professions rather than, as in Europe, on the intellectual and social status of representatives of the university as a key institution in the process of nation building.

KNOWLEDGE FORMS OF MASS DEMOCRACY AND INDUSTRIAL CAPITALISM (I): THE TRANSFORMATION OF THE EPISTEMIC CONSTELLATION

As the combined result of the processes just described, a variety of ways of theorizing society, empirical research strategies, and organizational forms for the production of social knowledge were available by 1900. During the first half of the twentieth century, these elements were reassembled, both in the form of an epistemic reorientation, to be discussed in this section, and in the form of a major shift in organizational outlook, to be analyzed in the next section. The result of this process was the emergence of knowledge practices that were oriented toward use by organizational oligarchies, be they in the state, businesses, or associations. Such practices redirected the explanatory ambitions of the social sciences and, without abandoning them, deflected the basic theoretical modes of the social sciences.

Many of the detailed analyses that follow can be read in this light. Alain Desrosières observes how economic theorizing enters into a variety of historically changing relations with the concept of a central societal organization, the state. Keynesianism or theories of the welfare state alter neoclassical economics by limiting its reach into the social world or by introducing additional assumptions with a view to changing the societal outcome of economic activities. But they continue to draw on its basic theoretical ideas. The economic way of thinking was also modified when social welfare concerns were introduced, as Ellen Fitzpatrick shows, this time toward a historico-institutional economics that saw the application of economic thinking as dependent on the precise nature of social situations, to be made known through social inquiry. The concern for social welfare, though, also provided for an application of socio-structural thinking, which could identify the social causes of poverty, thus shifting responsibility from the individual to the social situation and permitting the argument that public policies could justifiably intervene in such circumstances.

Because the welfare situation of African American families in the United States was of particular concern, the study of welfare became connected to the concept of race, again as a way to give nonindividual reasons – cultural or biological in this case, rather than socio-structural – for particular social conditions. The argument evolved gradually over the twentieth century. From the late nineteenth century onward, as Elazar Barkan demonstrates, the main use of racial theorizing was as a means to provide arguments for setting boundaries of polities in the era of nationalism and for introducing means to improve a state's population, on the basis of eugenic theorizing. Large-scale emigration, for many European countries, and immigration, for the United States, provided the background for such concern. Even though the origins of modern thinking about differences between human beings emphasized cultural-linguistic features, during the later nineteenth century such thinking increasingly resorted to biological features, allegedly revealed by properly scientific methods. The refutation of those "findings," together with the political discrediting of race-based policies after the defeat of Nazism, led to a return to the cultural approach. Emerging or reemerging in anthropological debate during the interwar years, cultural relativism is the contemporary form of theorizing differences between human beings, as analyzed by David Hollinger. During the past two decades, it has increasingly been linked to political claims for the institutional acknowledgement of, and also the promotion of, diversity. The claim to the right to diversity is not only made on behalf of cultural, linguistic, religious, and ethnic minorities, but also in gender relations, after the earlier emphasis of the women's movement and of feminist scholarship on the right to equality (see Rosalind Rosenberg's chapter).

Finally, the behavioral-statistical mode of reasoning finds one of its most significant use-oriented expressions in the twentieth century in survey research. Statistical reasoning had never been entirely detached from policy purposes, since statistical institutes emerged and inquiries flourished first in the realm of the state, before the claim to become a, or even *the*, science of society had been voiced by statisticians. Methodologically dependent on a new understanding of sampling, survey research developed strongly when political actors in mass democracies needed information about the orientations of the voters, whom they no longer knew, and when producers for mass consumption markets faced the same problem (see Susan Herbst's chapter).[10]

KNOWLEDGE FORMS OF MASS DEMOCRACY AND INDUSTRIAL CAPITALISM (II): THE BREAKTHROUGH OF A POLICY ORIENTATION IN THE SOCIAL SCIENCES

The case of survey research makes particularly clear the emerging policy orientation of the social sciences and its impact on their theory and epistemology. As we have seen, the new policy orientation did not mark any radical rupture; the modes of reasoning developed earlier remained alive. However, it considerably redirected research practices and organizational forms. Significantly, the policy orientation itself was dependent on its relation to a

---

[10] Alain Desrosières, "The Part in Relation to the Whole? How to Generalise: A Prehistory of Representative Sampling," in *The Social Survey in Historical Perspective, 1880–1940*, ed. Martin Bulmer, Kevin Bales, and Kathryn Kish Sklar (Cambridge: Cambridge University Press, 1991), pp. 217–44.

feature of social organization that was to some extent novel, and that had never before been addressed by the empirical social sciences. This was the large-scale bureaucratic-hierarchical social organization in all of its forms, including the central state administration – overarchingly powerful, particularly on the European continent – and the giant business corporation and other forms of private organization, which became an increasingly dominant feature of U.S. society.

In this light, it is important to look briefly at the history of organizational analysis. Particularly from a use-oriented point of view, one could have expected an empirical science of state activities to emerge together with rising interest in welfare and other policies. However, especially in Europe, the state long remained a social actor above all, in the sense that it also was kept hidden from the empirical gaze. Despite several attempts, there was no successful establishment of political science as an academic discipline, at least outside the United States, during the "classical" period of the social sciences, the late nineteenth and early twentieth centuries. Comprising elements as various as public law, half-aborted administrative sciences, election studies, and social policy research, the study of things political had become a rather incoherent remainder after the "modern" disciplines had split off.[11] Such a development can best be understood against the background of the post-Enlightenment ambition to understand the social world through its own laws of motion, as described earlier, rather than through orders from a center.

When bureaucracies in state, business, and political parties rose to ever-increasing importance toward the end of the nineteenth century, it became unmistakably clear that there would be no withering away of the state and no self-organization of society. Such observations were at the root of a political sociology of organizations and bureaucracy, which later turned into an organizational theory that became almost the main paradigm in management studies and the new discipline of political science after the Second World War. As such, the study of organizations with a view to enhancing their functioning became one of the major forms of use-oriented social science during the twentieth century (as discussed from the perspective of accounting needs by Peter Miller, and as touched upon by Peter Wagner). It formed the backbone of much policy-oriented research during the twentieth century, especially after the Second World War.

Organizational concerns were the characteristic feature of the emerging policy orientation in the social sciences. They demanded considerable shifts in orientation in several respects. First, they became increasingly concerned with policy actors in a broad sense, especially the top level of decision makers

---

[11] Peter Wagner, "The Place of the Discourse on Politics among the Social Sciences: Political Science in Turn-of-the-Century Europe," in *Texts, Contexts, Concepts: Studies on Politics and Power in Language*, ed. Sakari Hänninen and Kari Palonen (Helsinki: Finnish Political Science Association, 1990), pp. 262–81.

in public administration and business organizations. Second, the substantive focus of research shifted increasingly toward policy areas as objects of public administration, voters as target objects of political parties, and consumers as analogous targets of market-oriented organizations. Thirdly, the *conceptual* perspective increasingly emphasized the functioning of goal-oriented organizations in the social environment.

In all three respects, significant changes in the mode of operation of the social sciences can be observed. New research institutes, often modeled after the Bureau of Applied Social Research at Columbia University, pursued research on commission. The institutes could be university-based, public, or private, for-profit or not-for-profit; and the differences in organizational setting led to quite a variety of different research orientations. Always, however, the institutes were dependent on the commissioning of research projects, be it through the market or through institutional links. The sponsors were obviously large enough organizations that they could afford to pay for the production of knowledge on demand. Such organizations were mainly public agencies, big business firms – including, importantly, the media – and political parties. New fields of social science inquiry were formed that focused on the interests and activities of such organizations, such as education and social welfare, market and opinion research. The knowledge demanded, naturally, had to address the problems of those who demanded it. In the inclusive mass societies of the twentieth century, organizations increasingly directed their activities toward large numbers of people about whose motivations and orientations they knew very little. Ever-larger shares of social science research went into the production of knowledge about such people, as demanded by these organizations in the pursuit of their objectives.

Even though occasional criticism had also been raised earlier – for example, in Theodor Adorno's analysis of the rise of "administered society" with its concomitant form of social knowledge, such developments were increasingly criticized within social science communities during the 1970s. The expansion of funding and the increase in the number of research institutes and of university departments was widely welcomed, but concern was raised about the undermining of the scholarly base of the social sciences because of the increasing imbalance between demand-driven knowledge production and academic research. Many such statements of concern, however, took the disciplinary constitution of the social sciences in academic institutions for granted and saw such arrangement as the normative baseline against which new developments could be evaluated. A different analysis, in which the knowledge practices and modes of theoretical reasoning are themselves set in the context of the longer-term historical development of the relation between knowledge production and sociopolitical institutions, considerably alters the picture. It does not assume that there can be any pure form of social knowledge, uncontaminated by the situation in which it is created, that could provide the measuring rod needed to assess the "drift of epistemic criteria" as a result of social science policy and research funding activities.[12] Rather, it leads to an historical political sociology that is fully interrelated with a sociology of knowledge and of the (social) sciences.

TRANSFORMATIVE MOMENTS: WARS, EXTERNAL AND INTERNAL

To this end, some key aspects of the twentieth-century developments need to be analyzed in more detail. The first such aspect is provided by the observation that there clearly was no steady rise of the "administrative society," but at the very least leaps and spurts in such a transformation. Several of the following analyses, for instance, emphasize the significance of wars as accelerating or transformative moments in the development of the social sciences.

In the United States, the Civil War marked a first such moment, indeed providing the ground for the development of organized social science. In Europe, the wars of the 1860s, culminating in new Italian and German nation-states and the Third Republic in France, provided social science research with a more significant impetus. In Spain, similarly, early social science grew out of formative events in the history of that nation, specifically the experience of losing imperial status in the wake of the Spanish-American War (1898). The 1870s witnessed thriving social research activities, many of which were indeed devoted to providing the knowledge required for organizing national societies. Theoretical and disciplinary consolidation, by contrast, was of little concern. It became the center of attention only later, broadly from the 1890s onward, the period known in sociology as the "classical era."

For the development of new forms of knowledge utilization, however, the First World War was even more significant than the wars of the late nineteenth century. The war effort itself, much prolonged beyond initial expectations and involving the population and the economy much more than had preceding wars, required deeper and more detailed knowledge about both. Psychology and psychiatry offered means to assess human abilities in order to deploy them most effectively in the war, as in intelligence testing, and to determine the impact of the war experience on them, as in the studies of "shell shock" and other forms of war trauma (see the chapters by Elizabeth Lunbeck, John Carson, and Ellen Herman). Doubts about the viability and the desirability of the market mechanism in the economy had already arisen during the closing decades of the nineteenth century. The shift toward a "managerial economy" or toward "organized capitalism" was well under way, at least in the rapidly growing economies of the United States and Germany (see Peter Miller's chapter). However, it was the need to mobilize

---

[12] Aant Elzinga, "Research, Bureaucracy, and the Drift of Epistemic Criteria," in *The University Research System: Public Policies of the Home of Scientists*, ed. Björn Wittrock and Aant Elzinga (Stockholm: Almqvist and Wiksell, 1985).

all productive forces within a short time span and for a particular purpose –
military production and organization – that led to deliberate state efforts to
increase economic efficiency by public intervention and planning (see the
chapters by Alain Desrosières and Peter Wagner). Economics, statistics, and
organizational knowledge were also mobilized to this end.

One of the most important consequences of the war, and of the peace that
ended it, was the disruption of the internationalizing trends of the prewar
decades. Even more than after 1870, the development of resources within
the national societies themselves was prioritized, and the social sciences were
involved in that effort. In this new era, however, the conviction that the
increase of knowledge would directly translate into enhanced understanding
and more effective action was shaken. If scholarly opinion during the 1920s
still oscillated between the hope that industrial societies would return to a
smooth path of development and despair that the conditions for them to do
that had forever disappeared, during the 1930s the view gained ground that
these societies had embarked on an entirely different trajectory for which
new knowledge and new forms of public intervention were required. But the
responses to such insight varied widely. On the one hand, the techniques for
the observation of mass society, such as survey research and statistical inquiry,
were refined and increasingly used to improve knowledge of the condition of
the people and of the economy, in both democratic and totalitarian societies.[13]
On the other hand, the ongoing societal transformation was taken to spell the
failure of the fragmented and overspecialized social science disciplines and
to require the elaboration of entirely new theoretical and research programs,
such as the one that was later to be called "critical theory," initially proposed
by Max Horkheimer in 1931.[14] As a kind of intermediate view and strategy,
the emerging "soft steering" of the economy, later to be called Keynesianism,
and "democratic planning" tried to adapt to the new circumstances just as
much as was needed to keep the institutions of society and politics intact.[15]

The Second World War had a double effect in this context. As in the case
of the First World War, the war effort itself led to the increasing development
and application of centralized planning. But its outcome seemed to indicate
that the third strategy, Keynesian democratic interventionism, was viable in
principle, even though its application was initially limited to the "first world."
A war of a different kind, namely the Cold War, accompanied domestically
by the War on Poverty in the United States, enlisted the social sciences, called

[13] See, e.g., J. Adam Tooze, "La connaissance de l'activité économique: Réflexions sur l'histoire de
la statistique économique en France et en Allemagne," in *Le travail et la nation: Histoire croisée de
la France et de l'Allemagne*, ed. Bénédicte Zimmermann, Claude Didry, and Peter Wagner (Paris:
Editions de la Maison des Sciences de l'Homme, 1999), pp. 55–79.
[14] Max Horkheimer, *Die gegenwärtige Lage der Sozialphilosophie und die Aufgaben eines Instituts für
Sozialforschung* (Frankfurter Universitätsreden, vol. 37) (Frankfurt am Main: Englert und Schlosser,
1931).
[15] Peter A. Hall, ed., *The Political Power of Economic Ideas: Keynesianism across Nations* (Princeton,
N.J.: Princeton University Press, 1989).

either "modern" or "bourgeois" depending on the perspective, in the attempt to prove the superiority of this model. The most systematic effort since the "classical era" to propose a comprehensive social theory and research strategy for the analysis of contemporary societies and their logic of evolution, the modernization theory of the 1950s and 1960s was elaborated in precisely this context (see Michael Latham's chapter).

To what extent this theory offered a useful understanding of Western societies remains contested. It is certain, however, that social research efforts on an unprecedented scale took place under its umbrella. They were driven not least by the hope and expectation that, since the general concepts were available, only a few knowledge gaps needed to be closed by well-targeted empirical research. At the same time, the idea that good knowledge stands in an entirely unproblematic relation to its usefulness was revived. It was only during the 1970s, after signs of crisis had emerged and accumulated, that the presuppositions of "the rationalistic revolution" came to be doubted even by its proponents. The first response to this crisis was not to question its validity, but to inquire into its mode of operation. Research on "knowledge utilization" – initially geared to detecting the obstacles to the good use of knowledge, with the hope of making it possible to remove them once they were detected – was one of the thriving areas of the social sciences during the 1970s. In the course of this research campaign, however, the very model of knowledge use came increasingly to be questioned. The "reflexive turn" of much of the social sciences during the 1980s has one of its sources in this experience.[16]

THE CRISIS OF USEFUL SOCIAL KNOWLEDGE: CRITIQUE, RETREAT, AND REFINEMENT

Reviewing the twentieth-century experience of the use of the social sciences, two key observations can be made. On the one hand, mass democratic, industrial-capitalist societies have been marked by intense efforts to increase social knowledge about their modes of functioning and about their very members. It seems justifiable even to relate the demand for knowledge to a failure, in a rather specific way, of the Enlightenment project. At least in its most optimistic versions, the latter had assumed that once autonomy was granted to human strivings, the use of reason would lead to a harmonious development of social life, in a self-steered, self-organized way. Forms of economic and political freedom were indeed introduced in mass democratic, industrial-capitalist societies (even though such a statement needs much qualification),

---

[16] Björn Wittrock, "Social Knowledge and Public Policy: Eight Models of Interaction," in *The Difficult Dialogue between Producers and Users of Social Science Research*, ed. Helga Nowotny and Jane Lambiri Dimaki (Vienna: European Centre for Social Welfare Training and Research, 1985), pp. 89–109; Ulrich Beck and Wolfgang Bonß, "Soziologie und Modernisierung," *Soziale Welt*, 35 (1984), 381–406.

but the novel institutional arrangements, far from solving all problems for good, created new social and political issues that required new knowledge and understanding.

On the other hand, however, this very foundation of the search for useful knowledge rules out, as a matter of principle, the idea that any logic of control, with "scientification" of human life as its means, can assert itself in any unequivocal way. Although Adorno and Michel Foucault appeared to assume the contrary, there is no totalizing logic of disciplinization, or of the rise of administrative society, and for several reasons. First, there has been significant resistance to objectification, in the form of a political argument, from the late 1960s onward in Western debates, as well as in what is now known as postcolonial discourse. Second, the methodology of the modernist social sciences itself seems to impose limits to objectification. The "complexity," a key term that would be evoked in such contexts, of modern societies escapes even the most sophisticated research technology. And third, in terms of the philosophy of the social sciences, social life and human agency have increasingly come to appear deeply historical, perpetually creating unique and unpredictable situations. Agentiality and historicity are amenable to interpretation rather than to explanation, and every interpretation takes place in language, with its infinitely open range of possibilities of expression.

As a result of a combination of such arguments, the precise mix of which is impossible to assess, the implications of the use of the social sciences have been effectively criticized during the last three decades of the twentieth century. Two different adjustments to such criticism can be distinguished. More moderately, there has been a move from the mere application of general models or theories toward an increasing sophistication in the design of theory and of research. Various approaches are mixed, and their use is made dependent on the assessment and empirical specification of the situation to which they are being applied. In the chapters in Part IV, this kind of reaction is most clearly evident in analyses of the management of the economy and of accounting practices. More radically, although this may on occasion just be one more step in the same direction, we sometimes see the abandonment of any overarching rationality, with a subsequent conceptualization in terms of varieties of particular and potentially competing rationalities. The most obvious examples for such change may be the way that culturalist-holist theories of society, having been radicalized by biologically based theories of race, turned into cultural relativism, and the move from gender studies that emphasized equality to those that emphasized diversity. Elements of such a radical rethinking of the dominance of any singular rationality can, however, also be found in the areas of modernization, accounting, and planning.

In the United States and the United Kingdom, such critical rethinking was accompanied by a crisis of political demand created by the Thatcher and Reagan governments during the early 1980s. The critique of prevailing models of knowledge utilization, linked to a more deeply ingrained conviction that the social sciences are married to strong and interventionist states, encouraged a reduction and restructuring of funding for basic as well as commissioned research. Neoliberalism as a broad economic ideology indeed revives doctrines of societal self-regulation, in which there is neither place nor need for detailed empirical evidence about social situations. It may be noted in passing that even biologist theories of the social resurface in this context, since with new genetic knowledge they can claim to refer to the individual and can be linked to issues of rational choice.

PERSISTENT VARIATION, PERSISTENT *PROBLÉMATIQUES*

By way of conclusion, it would be tempting to paint a picture in which such a neoliberal understanding of the relation between the state and the economy lives forever in harmonious relation with a "postmodernist" understanding of society and culture. The former would need social science only as an underlying framework for thinking the relation between markets and hierarchies; the latter allows for plurality, diversity, and complexity and thus would need social science of the kind of "cultural studies." However, precisely in the light of the recent criticism of "nonreflexive" social science, one should not succumb to this temptation.

As several authors here point out, there is persistent variation in the use of the social sciences across countries and across areas. Social sciences that orient themselves to state and government and whose practical orientation is one of relevance for public policy and state intervention remain more significant in Europe than in the United States. By contrast, research on individuals and their development, with possible applications by the caring professions, including self-help groups and movements, is better developed in the United States. Most methodological development in research on the ways that large-scale organizations can interact with society, such as opinion and survey research for business and political parties, continues to come from U.S. sources. However, the importance of such knowledge tools has considerably increased in Europe as well. And there has been a proliferation of research institutes tied in various ways to social actors, including trade unions, social movements, and nongovernmental organizations.

More generally, neither the thesis of an increasing penetration of life-worlds by a power/knowledge complex nor the opposite view of a retreat to a self-regulation model of society can be sustained. There are persistent *problématiques* in post-Enlightenment societies that will always sustain the demand for useful social knowledge, and that will never permanently be solved by it. (This observation in itself supports the prior argument about the persistence of differences across the variety of possible interpretations of the sociopolitical situation in which one finds oneself.) The demand for knowledge may be driven by the desire to make organizational strategies more predictable. But it may also be meant to justify existing difference and diversity. In either case, it will not succeed in controlling a sociopolitical situation, since human beings always may act in unknowable ways. Nevertheless, across societies and historical periods there is considerable variation in the degree to which the hope of perfectly knowing the social world is upheld, in the ends toward which this hope is entertained, and in the intellectual, institutional, and political means that are used to realize this ambition.

31

MANAGING THE ECONOMY

Alain Desrosières

Since the eighteenth century, economic science has been punctuated by debates on the relation between state and market. Its history has been marked by a succession of doctrines and political constellations, more or less interrelated. They have usually been understood historically in relation to dominant ideas and institutional practices: mercantilism, planism, liberalism, the welfare state, Keynesianism, and neoliberalism. Whatever their dominant orientations, the various states gradually constructed systems of statistical observation. Yet the development of these statistical systems has generally been presented as a sort of inevitable and univocal progress, having little relation to the evolution of the variegated doctrines and practices of state direction and guidance of the economy. The historiography of economic thought, or more precisely, historical works dealing with the reciprocal interactions between the state and economic knowledge, has placed little emphasis upon the modes of statistical description specific to various historical configurations of state and market.[1] In a word, these two histories, that of political economy and that of statistics, are rarely presented, much less problematized, together.

The reason for this gap in economic historiography is simple. Statistics has historically been perceived as an instrument, a subordinate methodology, a technical tool providing empirical validation for economic research and its political extensions. According to this "Whig" conception of the progress of science and its applications, statistics (understood as the production both of information and of the mathematical tools used to analyze that information) progresses autonomously relative to economic doctrine and practice. It is for this reason that the historical specificity of statistics is neglected in the

---

[1] Mary O. Furner and Barry Supple, *The State and Economic Knowledge: The American and British Experiences* (Cambridge: Cambridge University Press, 1990); Michael J. Lacey and Mary O. Furner, *The State and Social Investigation in Britain and the United States* (Cambridge: Cambridge University Press, 1993).

historiography of economic science, and left unproblematized. "Statistics" is here taken to mean the production, recording, and analysis of quantitative data, in the form of series, indexes, econometric models, and many other tools today available in computerized packages.

The history of conceptualizations of the state's role in economic affairs provides a guiding thread for analyzing the relations between statistical tools and their social and cognitive contexts. In what follows, I will present, in a very simplified fashion, five typical historical configurations. *Direct intervention* encompasses a wide variety of perspectives, from mercantilism and Colbertism to socialist planned economies. The French *Etat ingénieur* (engineering state; also, a state administration by engineers) is one of its modalities. At the other extreme, *classical liberalism* minimizes such intervention and extols the free operation of market forces. The *welfare state* (*l'État providence*) seeks to protect salaried employees from the consequences of the extension of this market logic to their own work. *Keynesianism* assigns responsibility to the state for the macroeconomic guidance of society, without challenging its reliance on the market. Finally, *neoliberalism* conceives of the state as seeking to influence microeconomic dynamics, which it endeavors to affect through systems of incentives based on the theory of rational choice. The five configurations just outlined are not meant to describe successive stages in a historical progression, nor are they historically or logically exclusive. In concrete historical situations, they are often mixed together. They have been idealized in this way only to provide a grid on which to arrange the history of the statistical tools employed by each.[2]

L'ETAT INGÉNIEUR: PRODUCTION AND PEOPLE

This configuration has a long history. According to its logic, the state assumes many responsibilities associated with the domain of private enterprise. In seventeenth-century France, for example, Colbert set up royal installations for shipbuilding and tapestry weaving. Peter the Great likewise established industries in Russia. In France, beginning soon after the Revolution, the *Ecole polytechnique* trained engineers in such fields of interest to the state as mining, bridges and highways, and armaments. Polytechnicians became accustomed to overseeing large segments of the French economy from a technical rather than a market point of view. In the tradition of the engineering state, their function as planners had a legitimacy never attained by public engineers

[2] Margo Anderson, *The American Census: A Social History* (New Haven, Conn.: Yale University Press, 1988); Joseph Duncan and William Shelton, *Revolution in United States Government Statistics, 1926–1976* (Washington, D.C.: U.S. Department of Commerce, 1978); for Great Britain, see Roger Davidson, *Whitehall and the Labour Problem in Late Victorian and Edwardian Britain: A Study in Official Statistics and Social Control* (London: Croom Helm, 1985); for a comparative perspective, see Alain Desrosières, *The Politics of Large Numbers: A History of Statistical Reasoning* (Cambridge, Mass.: Harvard University Press, 1998).

in the United States.[3] The role of state engineers was theorized by Claude Henri de Saint-Simon (1760–1825), whose name is associated with a school of industrial thought based on science and technology. This was an important influence on Marxist economics and on centralized planning in the eastern bloc – though Lenin also admired Frederick W. Taylor's campaign to organize labor in capitalistic industries on the basis of quantified time-and-motion studies.

Certain historical circumstances were particularly favorable to direct state organization of the economy. The two world wars entailed, for all of the belligerent nations, a greater centralization and systematic standardization of resources, especially in the armament industries. The Manhattan Project was typical of such state intervention, especially in a nation noted for its reluctance to intervene directly in the economy. Likewise, the resources allocated to the U.S. space program in the 1960s are comprehensible only in light of the Cold War. Even those countries most disposed to practice market economics have experienced, in certain historical circumstances, direct economic intervention on the part of the state.

The Great Depression of the 1930s was commonly viewed at the time as a crisis in the classical market economy. It occasioned serious reflection leading to new doctrines concerning the role of the state. These may be divided into two groups: central planning and Keynesianism. Economic planning was, of course, pushed to an extreme in the Soviet Union; yet in western Europe in the 1930s it was discussed by economists and political philosophers across the political spectrum, from the corporatist right to the socialist left, and equally by Christian reformers, both Catholic and Protestant. These currents of thought, in other respects very different, were at one in opposing economic liberalism, whether for nationalist, humanist, or Marxist reasons. Keynesianism was a less radical alternative, since it did not aim to replace the market economy. Since 1945, planning and Keynesianism have in practice been mixed, in varying proportions, in countries such as France, the Netherlands, and Norway. The analytic distinction, however, is useful for understanding the development of statistics and economic models utilized during the period from 1940 to 1980.

To this end, it helps to consider a saying that gained currency among the economists who laid the foundations of national accounting during the 1940s.[4] "One may think of the economy of a country as like that of a single large firm." Leaving aside its pedagogical uses, this saying points to a technical conception of economics and of national accounting, whose principal tool was input-output analysis, following Leontief's table of industrial exchanges.

---

[3] Theodore Porter, *Trust in Numbers: The Pursuit of Objectivity in Science and Public Life* (Princeton, N.J.: Princeton University Press, 1995).

[4] John W. Kendrick, "The Historical Development of National Income Accounts," *History of Political Economy*, 2 (1970), 284–315.

Its format was like that of the charts that track the flow of materials between the workshops of a single enterprise. The economists behind national accounting defended their methods as free of ideological association, equally applicable to capitalist and socialist economies. What mattered to them was the production and circulation of goods and services, whose monetary representation, deriving from a system of prices, was nothing but a means for calculating macroeconomic aggregates. The essential quantities for the engineering state were production and consumption of each commodity. Here the state was directly responsible for the satisfaction of human needs, just as the technical manager of an enterprise must keep on hand adequate supplies of components in order to maintain continuous production.

This example reveals the historical specificity of the statistics required by the *Etat ingenieur*, which are comparable to the information needed by the general of an army. One measures quantities produced and consumed, supplies and equipment, and, not least, manpower. Demographic variables, such as rates of birth and immigration, are among the concerns of such a state, of which France, with its long-standing population anxieties, forms an exemplary case. On the other hand, information more directly related to the market aspect of the economy has not been central to this statistical program. This was the aspect criticized most vocally by liberal economists of the 1930s and 1940s who followed Friedrich Hayek's opposition to the planned economy. How was it possible to arrange for the optimal allocation of resources without the information revealed by market prices? Certain socialist economists, such as Oskar Lange, attempted to envision a system of planning capable of "mimicking" the market, thereby combining the presumed advantages of both systems.[5] The organization of statistical knowledge in such a hybrid system would have been tremendously complex. In the case of actual socialist countries before 1989, prices were, in effect, mostly arbitrary. Their statistical systems consisted essentially of accounts measured in units of production, which were transmitted to a central office charged with executing the economic plan. In the French case, on the other hand, planning never existed in a pure form, and beginning in the 1950s had a self-consciously Keynesian program tacked onto it.[6]

In one sense, the form of statistics allied to planning constitutes its historical core. Attached originally to a mercantilist system, this "science of the state" began by producing information of immediate use to the prince for the purpose of raising armies and levying taxes. Questions of population and of agricultural and industrial wealth formed the initial subject matter for the eighteenth-century founders of political arithmetic. During this same period, however, a different conception of state–market relations was developed in

---

[5] Bruce Caldwell, "Hayek and Socialism," *Journal of Economic Literature*, 35 (December 1997), 1856–90.
[6] François Fourquet, *Les comptes de la puissance: Histoire de la comptabilité nationale et du Plan* (Paris: Encres, 1980).

France by Turgot and the physiocrats, and in Great Britain by Adam Smith. In time, a different form of statistics would arise, adapted to the new system of economic liberalism.

THE LIBERAL STATE: EXCHANGE AND PRICES

In its most abstract formulation, the pure theory of the market renders statistics superfluous. Prices, made known through merchant exchange and competition between producers, convey all of the information required by this form of economic organization. Given its doctrinaire rejection of central, directing institutions, liberalism had no use for many kinds of statistical information. Statistical institutions, even a permanent census administration, were long resisted in the United States by opponents of the economic role of the state. Theorists of the market economy, such as Jean-Baptiste Say, Augustin Cournot, and Léon Walras, were reluctant to support their hypothetico-deductive reasoning with economic statistics.[7] While statistical knowledge is central for the engineering state, its very existence is paradoxical for the pure liberal state, if such a thing can even be imagined.

Still, many institutions and their statistical operations have been directly justified by the needs of a merchant economy. The first such statistics pertained to international commerce – customs duties, rates of exchange, and management of the currency. The Bureau of Statistics of the English Board of Trade was created in 1833, at just the moment when a series of political and economic reforms liberated the capitalist market from various impediments handed down from the past (such as the 1795 Speenhamland Act on poor relief). The Corn Laws were vigorously debated during this period: Should grain imports be freed from all duties? Industrialists were generally favorable, because free trade in grain would reduce food prices, thereby permitting a reduction of wages. But landholders and their industrial allies were hostile to repeal of the Corn Laws. Their debates prompted ad hoc statistical inquiries into prices and wages. Thus, in contrast to purely theoretical liberalism, "real" liberalism implied for the state a role as organ of economic intelligence, gathering and disseminating information needed by economic agents in order to act in the market.

Another paradoxical example of the need for state intervention so that the full advantages of competitive markets might be realized is to be found in U.S. debates at the end of the century over the problem of industrial concentration. Here again, precise statistical information on the functioning of markets was necessary in order to compose and then apply antitrust legislation. The legislative philosophy deployed against the cartels was radically

---

[7] Claude Ménard, "Three Forms of Resistance to Statistics: Say, Cournot, Walras," *History of Political Economy*, 12 (1980), 524–41.

different from that of the *Etat ingénieur*. The latter aimed to lower the costs of production through economies of scale resulting from the standardization and concentration of production. The liberal state, on the contrary, anticipated a similar drop in costs of production resulting from competition between enterprises, none of which could dominate the market. To these opposing philosophies corresponded very different statistical systems. The engineering state operated on the basis of technical coefficients and functions of production and, more generally, the internal analysis of firms. The liberal state was centered upon market exchanges themselves, and on the effects of variations in price on the behavior of buyers and sellers. This last example makes plain the *co-construction* of a political economy and a cognitive system of statistical information. Statistical systems must not be seen as purely technical or exogenous in relation to specific questions arising within a precise historical context.

Eventually, the social and economic regulation of markets was judged to be impossible without the regular and intense production, and wide diffusion, of statistical data. Such was the case with agricultural statistics in the United States beginning in the late nineteenth century. This project involved collecting, centralizing, and then diffusing as rapidly as possible the latest information on harvests. The knowledge provided by agricultural statistics, when shared among buyers and sellers, allowed for more homogenous and less erratic agricultural prices to be established across the territory of the nation, so that, as much as possible, the revenues of producers would be guaranteed. Elaborate systems, such as sample inquiries to forecast harvests, were set up during the 1920s and have been developed ever since. As before, the essential objective of statistical information was to make the market transparent. This development, however, may also be read in another way, as aiming to provide economic protection for farmers, especially the weakest ones, against the consequences of blind and savage competition. It displays the rise, toward the end of the nineteenth century, of another modality of state intervention in economic affairs. The *welfare state* (*l'Etat providence*) sought to guarantee, in Karl Polanyi's words, the "self-defense of society"[8] against the ravages brought about by a free market in labor, land, and money.

THE WELFARE STATE: PROTECTING WORKERS

During the 1880s and 1890s, after a century of debates on the proper remedies for poverty, nearly all of the industrial nations of Europe created new offices of labor or "bureaus of labor statistics." Rapid industrial growth led to the concentration in cities of workers of rural origin. In the American case, many were immigrants from Europe. The extreme poverty of urban industrial

[8] Karl Polanyi, *The Great Transformation* (New York: Farrar, 1944).

environments had traditionally been the responsibility of local charity and assistance organizations. By the end of the century, the greatly increased magnitude of urban impoverishment inspired a radical rethinking of the problem and of possible solutions. Spurred by the economic crisis of the years 1873–95, this reconfiguration developed in two very different directions, both of which would have important and irreversible consequences for statistical methods. The first of these currents, drawing inspiration from Darwinian evolution, was the eugenics of Francis Galton and Karl Pearson. They sought the cause and remedy for poverty in a biological theory of individual ability, conceived of as innate and hereditary. The quality of a population could be improved, they thought, through a process of selection, comparable to the breeding of animals. Apart from the thinking of a few marginal groups, these ideas have practically disappeared from public discourse. Yet the first formulation of mathematical statistics, with its correlations, regressions, and tests, took place within the frame of this eugenicist "biometrics."[9] Beginning in the 1910s, these statistical formalisms were taken up by economists, such as Henry Moore in the United States and Marcel Lenoir in France, and used in what would become, by 1930, econometrics.[10]

The second current of thinking on the subject of poverty, by contrast, located its causes and remedies not in biology but in society and law. There was a market for labor, whose price was the wage level. Without specific protections and regulations, the life of labor would continue to exhibit the instability and poverty characteristic of nineteenth-century capitalism. The state alone was deemed capable of protecting workers, through laws guaranteeing pensions and insurance for unemployment, sickness, and accidents. The bureaus of labor created between 1880 and 1900 explored and implemented this new form of the state, what eventually would be called the welfare state or *l'Etat providence*. By 1920, this movement had taken on international dimensions with the creation of the International Labor Office, which gathered and coordinated statistical and juridical information provided by various industrialized states.

During the period between 1880 and 1930, labor statistics drove the regeneration of official statistics, in terms of both values studied and methods of investigation. Wages, employment figures, unemployment rates, levels of prosperity according to trade, worker budgets, and cost of living indices were henceforth matters of public interest and subject to state intervention, especially through legislation. They were placed on the agendas of statistical bureaus, which set about inventing new forms of inquiry based on representative samples so that they could be measured. Previously, exhaustive surveys and administrative records of governments were the only sources of statistical

---

[9] Donald Mackenzie, *Statistics in Britain, 1865–1930: The Social Construction of Scientific Knowledge* (Edinburgh: Edinburgh University Press, 1981).

[10] Mary Morgan, *The History of Econometric Ideas* (Cambridge: Cambridge University Press, 1990).

information. Probabilistic sampling, implying the notion of approximation, had been seen as incompatible with the rigor and certainty of official statistics, and so lacked public legitimacy.

The very idea of the welfare state, however, is based on the notion of *insurance*. Protection against risk was assured by statistical calculations of probabilities (measured in terms of frequency) of the various events described by the new labor statistics. The welfare state was thus bound up with probability. It put to work the central intuition of Adolphe Quetelet (1796–1874): that the statistical mean of aggregate values displays stability and predictability absent at the level of individuals. This is the theoretical foundation of insurance. Its method was applied at the level of the national population, which, by this logic, could be thought of as a probabilistic urn from which samples are drawn. These measures could be extrapolated to the entire population, taking into account the uncertainty, or "confidence interval." Thus, political philosophy and the cognitive schemes of the welfare state were tightly imbricated. This new type of state and the new way of doing statistics were constructed at the same time.

THE KEYNESIAN STATE: DECOMPOSING GLOBAL DEMAND

As a consequence of the economic crisis of the 1880s, the protection of wage labor and the statistical investigations by which it was known were inserted onto the agenda of state power. Thus arose the first forms of the welfare state, most notably in Germany under Bismarck. The crisis of the 1930s had similar consequences for the macroeconomic equilibrium between "global supply" and "global demand," the sum of goods and services. Crucially, the notion of centralized regulation of economic equilibrium by the state not only was formulated in theory (by Keynes in 1936) but also was rapidly made operative through national accounting tables and statistical series describing the relations among various components of supply and demand. Here again, state and statistics were co-constructed. As the state gained this new responsibility to preserve macroeconomic equilibrium without sacrificing the market economy, there arose a new mode of description and analysis – national accounting and macroeconometric modeling, such as the system developed beginning in the 1930s in the Netherlands by Jan Tinbergen (1903–1994).[11]

What was most crucially new in the Keynesian perspective was the presentation of the economy as a whole, developing through several macroeconomic flows that could be measured and joined together within theoretically coherent and exhaustive tables of accounts. Directly associated with a form

---

[11] Don Patinkin, "Keynes and Econometrics: On the Interaction between the Macroeconomic Revolutions of the Interwar Period," *Econometrica*, 44 (1976), 1091–123.

of political economy, this model stimulated a complete reorganization of statistical variables and their modes of production beginning in the 1950s. The coherence of the Keynesian model and its double constraint – that tables of accounts be in equilibrium whether arranged according to agents or to operations – drew attention to gaps and contradictions in existing statistical sources. More profoundly, changing the uses of statistical sources also changed their character. For example, there had been inquiries into family budgets since the nineteenth century. They had aimed above all to describe the needs and expenses of working families in relation to wages. This was typical of the statistics of the welfare state, which was concerned above all with wage labor. During the 1950s, these became statistics of consumption for the entire population. Now they described markets for all goods and services and no longer merely the labor market, as had the smaller-scale surveys carried out before 1940. It should be clear from this example that a statistical inquiry is inseparable from its context of use. This point is often forgotten, obscured by the institutional and cognitive division of labor between the producers and consumers of information.

The distinction here between the welfare state and the Keynesian state is, of course, a simplification. It corresponds to two distinct stages in relation to the history of the state, its role in shaping the economy, and the statistics associated with these interventions. The first stage, the protection of wage labor, took shape between 1880 and 1900. The second, macroeconomic piloting, emerged between 1930 and 1950. But since the 1950s, these two forms of action and of knowledge have been closely linked, at least in Western Europe (France, Germany, and Great Britain). Social benefits such as pensions, medical and unemployment insurance, and family allocations provide a major component of worker income, and thus also of the global demand posited by the Keynesian model. For this reason, the crisis of the 1970s and 1980s had different social consequences from that of the 1930s, and unemployment assumed different forms. This is also why the two crises were interpreted in nearly opposite ways. The Depression of the 1930s, interpreted as a crisis of market economics and of laissez-faire, led to an expansion of the role of the state and of social protection. By contrast, the downturn of the 1980s was interpreted as a failure of the very solutions invented fifty years earlier in the form of Keynesianism and the welfare state. These latter were challenged by the ascent of neoliberal ideologies, symbolized by Ronald Reagan and Margaret Thatcher, each of whom cut funding for official statistics in the name of reducing state direction of the economy.

THE FRENCH AND DUTCH PLANS COMPARED

The distinction between the engineeering state and the Keynesian state is far from absolute. In France from 1950 to 1970, these forms of action and

economic analysis were interwoven. Jean Monnet, later a founder of the
European Common Market, established a Commissariat Général du Plan,
or general planning board, in 1945. The French plan brought together three
elements: forecasts to support large public and private investments in infrastructure and the ad hoc financing that such investments required after
the devastation of the Second World War; procedures for consultation and
dialogue between economic and social actors, in the form of specialized
commissions rather than of parliamentary debates; and, finally, a system of
economic analysis and information based on national accounting. This construction combined the *Etat ingénieur* (many former students of the Ecole
Polytechnique were involved in it), the Keynesian state, with its national accounting and macroeconomic analyses, and, finally, an increasingly socialized
state. This last provided a forum for social groups with a particular interest
in reducing social inequalities and thus also promoted such inquiries and the
use of social indicators to describe them.

It seems surprising that until 1970, this French social and cognitive network
did not include the use of macroeconometric models such as those of Jan
Tinbergen and those of Lawrence Klein and Arthur Goldberger.[12] In the
Netherlands, such models had been in use since the 1930s. Still, France and
the Netherlands had much in common. Each created in 1945 a bureau of
economic planning as a response to the occupation and severe destruction
of the war, an idea rejected by the other Western powers. The Germans,
British, and Americans regarded this idea as contrary to market principles
and contaminated by totalitarian associations, both Nazi and Soviet. Two
charismatic individuals gave shape to these bureaus, Monnet in France and
Jan Tinbergen in the Netherlands. Tinbergen devised the first econometric
model in 1936, and his personality helps to explain the Dutch emphasis on
these models, as well as their prominence in social and political debates. In
electoral campaigns, Dutch parties allowed their economic programs to be
fed into the Tinbergen model and to be judged by its results in terms of
growth, inflation, unemployment, and foreign trade.

In France before 1970, planning discussions took place outside the party
system and were not tested by any econometric model. Instead, decisions
unrolled in negotiations within planning commissions, carried out in the
language of engineers and statisticians, who tended to view the economy as
one vast enterprise rather than as a competitive market. As members of elite
state corps, these engineers were in the position of official experts, and they
spoke quite naturally the language of technical rationality symbolized by
Leontieff's input–output tables. The Dutch planners, by contrast, were often
academics with professional positions outside the state apparatus. Also, their
labors were applied to an economy that for centuries had been oriented toward

---

[12] Ronald Bodkin, Lawrence Klein, and Kanta Marwah, eds., *A History of Macroeconometric Model-
Building* (Aldershot: Edward Elgar, 1991).

international commerce. Market dynamics were, for them, a given. The equations of the Dutch model sought to simulate this dynamic, whereas French procedures blended the vision of the engineer with the Keynesians' "comparative statics" – a contrast between a well-documented past and a desired future that provided a basis for discussions within the planning commissions.

The Dutch applied themselves to the dynamic fluctuations of an autonomous market economy, much as one would attempt to mount a galloping horse. Objectives were defined with close attention to the flow of the economy. The procedures of economic planning implied a close articulation between the modeling of objectives and of means, with an emphasis on actual outcomes. The equations in Dutch economic models were designed to mimic the actual path of the economy. The French adopted a more technical and quantitative picture of the economy, leaving the actual dynamic of prices in the shadows. The economic trajectory was reduced to a planned outcome in the target year. French planning arrangements privileged a social procedure, a complex succession of deliberations by experts, national accountants, commissions, and working groups. The French plan mimicked the movement of the economy through negotiations among social groups within the framework of commissions.

THE NEW LIBERAL STATE: POLYCENTRISM AND INCENTIVES

The state forms described here have in common that they are endowed with a center. This applies even to the liberal state, in which the statistics required for antitrust laws or for transparency in agricultural markets must necessarily be centralized. The neoliberal state, by contrast, is conceived as a collection of administrative nodes or distinct territories whose interrelations are negotiated, contractual, and ordered by law. Federated states, or unions of sovereign states like the European Union, provide disparate examples of such modalities. All are based upon notions of subsidiarity, of procedures, negotiation, and networks.[13] The maximum possible liberty is left to the more local levels of society, retaining for higher levels only those powers that lower levels cannot reliably exercise. Established procedures specify structures of negotiation and decision, but do not produce substantive rules. The sites of action and decision, where information is gathered and put to use, are numerous and interconnected. Issues involving collective responsibility have proliferated: the environment, bioethics, child abuse, drug addiction, the prevention of AIDS and other new diseases, the protection of cultural minorities, equality of the sexes, the safety of domestic and industrial environments, and standards of

---

[13] Robert Nelson, "The Economics Profession and the Making of Public Policy," *Journal of Economic Literature,* 25 (March 1987), 49–91.

quality in consumer goods. Each case involves the simultaneous negotiation of appropriate statistics, of division of responsibility, and of methods of assessment. Information is produced and utilized at every link of this circular chain of description, action, and evaluation.

Public action in the neoliberal state involves incentives more than it does regulation. Fiscal incentives, for example, are thought of in terms of microeconomic theory, using a language of individual rational agents, preferences, utility, optimization, and externalities. A typical example of legislation based on microeconomics is the creation of markets in polluting rights, which are viewed as more efficient than limits set by regulation. These procedures can be evaluated by studying the data or by performing quasi-experiments, which aim to measure and model the behavior of actors, including that of public authorities. This last point defines a crucial difference between the neoliberal state and its predecessors. It is closely related to the modern idea of rational expectations. According to this theory, interventionist policies, such as Keynesianism, will be confounded because actors will modify their behavior in anticipation of public decisions.[14] From this perspective, no actor is outside of the game, certainly not the state. Rather, the situation dissolves into several "centers of direction," themselves agents among others, all acting within the parameters of similar economic and sociological models.

The idea of this chapter, that the tools of statistics have evolved in parallel with new forms of the state, may seem to be consonant with the neoliberal sensibility. The realist understanding of statistics, long dominant, treated it as a simple measuring instrument, unaffected by the reality it studied, just as the state, according to the understanding criticized by rational expectations, was external to society. To the extent that production of statistical knowledge is an essential component of economic direction, it is not surprising that regulatory decentralization and endogenization have been accompanied by a similar restructuring of the "centers of calculation" that produce statistics. These are never mere "data," but rather the result of an expensive social process whose economic and cognitive components are parts of the global society that they are supposed to describe.

[14] D. K. H. Begg, *The Rational Expectations Revolution in Macroeconomics: Theories and Evidence* (Oxford: Oxford University Press, 1982); Albert O. Hirschman, *The Rhetoric of Reaction: Perversity, Futility, Jeopardy* (Cambridge, Mass.: Harvard University Press, 1991).

32

MANAGEMENT AND ACCOUNTING

Peter Miller

Accounting is one of the most influential forms of quantification of the late twentieth century. It creates the apparently objective financial flows to which certain Western societies accord such significance, and it makes possible distinctive ways of administering and coordinating processes and people. For a vast range of occupations, from shop floor workers and divisional managers to doctors and teachers, the calculative practices of accounting seek to affect behavior and to constrain actions in a manner and to an extent unimagined a century ago. Yet accounting is also one of the most neglected and least visible of all the quantifying disciplines. While the concepts and practices of the economist, the statistician, and the actuary have received detailed academic scrutiny, those of the accountant have been left in the shadows or relegated to a subsidiary role within a larger story. Only recently has this begun to change.[1]

When accounting does become the object of public scrutiny, this typically concerns the external face of accounting, the reporting of the financial condition of business enterprises to shareholders and other outside parties, and the auditing of such reports. But accounting also has a "hidden" dimension: the financial monitoring, reporting, and evaluating that takes place inside an organization, and is typically treated as confidential even within the firm. This aspect, called management or cost accounting, is made up of practices such as budgeting, costing, and investment evaluation. It is the focus of the present chapter.

By now, management accounting has become almost synonymous with management. Its rise up the corporate hierarchy is intimately linked with what Alfred Chandler has termed "managerial capitalism," the organizing of processes of production and distribution on the basis of large multiunit

---

[1] See Anthony G. Hopwood and Peter Miller, eds., *Accounting as Social and Institutional Practice* (Cambridge: Cambridge University Press, 1994); see also the journal *Accounting, Organizations, and Society* (founded in 1976). An "outsider's" view on accounting is Theodore M. Porter, *Trust in Numbers: The Pursuit of Objectivity in Science and Public Life* (Princeton, N.J.: Princeton University Press, 1995).

enterprises administered by salaried managers.[2] My historical narrative begins where Chandler's ends, and addresses the transformation in management accounting practices since about 1920. It focuses on the ways in which management accounting has been fashioned to make enterprises governable by making individuals accountable, comparable, and responsible. Accounting is one of the most important of those indirect forms of regulation of conduct through which social authorities seek to administer the lives of individuals in accordance with wider economic or political objectives.[3]

This ascendancy of accounting can be attributed in large part to its perceived ability to translate diverse and incomparable phenomena into a single financial figure, such as return on investment or net present value. It appears to make comparable activities and processes whose physical characteristics and geographical locations may have little or nothing in common, such as the assembly of automobiles, the production of foodstuffs, and the delivery of health care. Faith in the apparently hard reality of financial numbers gives accounting much of its legitimacy. For it is by quantifying and abstracting that accounting lays claim to objectivity and neutrality, to a position set above the fray, apart from disputes and political interests.

Accounting practices for internal control purposes have climbed the corporate hierarchy and gained in legitimacy in the course of the twentieth century. This chapter examines three key moments in this process. The first was from 1900 to 1930, when standard costing and budgeting became part of the repertoire of cost accounting. The second, during the interwar years, involved the introduction of concepts of fixed and variable costs from economics. The third was the two decades following World War II, when practices of investment evaluation within accounting were transformed by the introduction of discounting techniques and the economists' notion of the time value of money.

To understand the rise of management accounting, one needs to attend to the links that have been forged between accounting and other disciplines – in particular, economics and engineering. Also important is its association with business administration, which has given accounting much of its meaning, significance, and legitimacy. Such basic concepts as "efficiency," as articulated within scientific management, and the "decision maker," in the administrative science literature, have provided rationales for management accounting and roles for its calculative practices.[4] Management accounting

[2] Alfred D. Chandler, *The Visible Hand: The Managerial Revolution in American Business* (Cambridge, Mass: Harvard University Press, 1977).
[3] Peter Miller and Nikolas Rose, "Governing Economic Life," *Economy and Society*, 19 (1990), 1–31; Nikolas Rose and Peter Miller, "Political Power beyond the State: Problematics of Government," *British Journal of Sociology*, 43 (1992), 173–205.
[4] On the roles of accounting, see Anthony G. Hopwood, "On Trying to Study Accounting in the Contexts in which It Operates," *Accounting, Organizations, and Society*, 8 (1983), 287–305; Stuart Burchell, Colin Clubb, Anthony Hopwood, John Hughes, and Janine Nahapiet, "The Roles of Accounting in Organizations and Society," *Accounting, Organizations, and Society*, 5 (1980), 5–27.

has a curiously hybrid nature, simultaneously administrative practice and social science.

INDIVIDUALIZING EFFICIENCY

Between 1900 and 1930, cost accounting was transformed, and its domain massively expanded, by the invention of standard costing and budgeting.[5] By 1930, on both sides of the Atlantic, cost accounting could be based on predetermined or standard costs and was no longer limited to ascertaining actual costs after the event. Waste and efficiency began to be assessed using the difference, or "variance," between actual costs incurred and a normal or standard cost established in advance. Cost accounting could bring "preventable inefficiencies" to the attention of management, so that these might be eliminated.[6] This development made possible a new way of governing the factory. Efficiency was now individualized, and employees at all levels were made accountable to prescribed standards or norms of performance.

Standard costing owed much to the movement originating in the United States that came to be called "scientific management." Indeed, F. W. Taylor's paper of 1903 on shop management contains many essential elements of what would later become standard costing. Through the costing framework enunciated by Harrington Emerson,[7] an American efficiency engineer, Taylorism helped to shape Charter Harrison's elaboration of a fully integrated standard costing and budgeting system in 1930.[8] Taylor's celebrated *Principles of Scientific Management* sought to advance national efficiency by attacking what he saw as the vast and largely invisible waste that had secreted itself within the daily actions of every individual.[9] Others, such as Frank and Lilian Gilbreth, joined his crusade.[10] Long-established practices in trades such as bricklaying were to be dissected and analyzed in terms of the waste held to reside in all of the minute components of such an activity.[11]

Taylor and his followers identified inefficiency with lay knowledge and practices. Scientific management required "expert" interventions if

---

[5] Peter Miller and Timothy O'Leary, "Accounting and the Construction of the Governable Person," *Accounting, Organizations, and Society*, 12 (1987), 235–65; Peter Miller and Timothy O'Leary, "Governing the Calculable Person," in *Accounting as Social and Institutional Practice*, ed. Hopwood and Miller, pp. 98–115.

[6] G. Chester Harrison, *Standard Costing* (New York: Ronald Press, 1930), p. 8.

[7] Harrington Emerson, *Efficiency as a Basis for Operation and Wages* (New York: Engineering Magazine Co., 1919).

[8] Ellis M. Sowell, *The Evolution of the Theories and Techniques of Standard Costs* (Tuscaloosa: University of Alabama Press, 1973).

[9] Frederick Winslow Taylor, *The Principles of Scientific Management* (New York: Harper and Brothers, 1913).

[10] Frank B. Gilbreth, *Applied Motion Study* (New York: Sturgis and Walton, 1917); Frank B. Gilbreth and Lilian M. Gilbreth, *Fatigue Study: The Elimination of Humanity's Greatest Unnecessary Waste* (New York: Sturgis and Walton, 1916).

[11] Horace B. Drury, *Scientific Management* (New York: Columbia University Press, 1915).

individuals were to give their best in their work, whether on the shop floor or in the office. Their presumed scientific expertise gave legitimacy to these managerial interventions. Scientific management reflected the almost messianic role for the engineering profession envisaged by some of its leaders in the United States. Emerson remarked that efficiency is "not an ethical or financial or social problem, but an engineering problem." It was, he continued, "to the engineering profession, rather than to any other" that people should look for "salvation from our distinctly human ills."[12]

Still more might be achieved if efficiency norms could be given financial form. As early as 1886, H. R. Towne, then president of the American Society of Mechanical Engineers and a mentor of Taylor's, had wanted to construe the engineer as an economist.[13] Efficiencies, after all, must ultimately take the form of cost savings. Emerson echoed these sentiments, arguing that engineers and accountants needed to collaborate in the task of detecting and analyzing inefficiencies. Finally, G. Charter Harrison, whose career spanned the professional bodies of industrial engineering, chartered accountancy, and cost accountancy, and whose writings provided the first full articulation of standard costing, helped to cement this temporary alliance of engineering and accounting.[14] Standard costing made the engineering concept of scientific management visible and calculable in financial terms.

The individualizing of efficiency was, however, not to be limited to the shop floor. The leaders of the scientific management movement envisaged their principles as ultimately embracing everyone, including managers themselves, and standard costing provided the means by which this might be achieved. Standard costs, Harrison argued, are equally applicable to the "five-dollar-a-day trucker in the factory or a five-thousand-dollar-a-year executive." Without this mode of analysis, "No man can realize his fullest possibilities."[15] Although the engineers had envisaged standard costing as no more than an appendage to scientific management, accounting's facility for expressing in monetary form the standardizing ambitions of the engineers brought about a far-reaching and long-lasting metamorphosis.

An apparently simple technical change in ways of calculating costs thus effected a profound transformation in ways of governing the enterprise. It provided norms and standards of efficiency for everyone in the firm. It interposed between the worker and the boss a calculative apparatus that claimed neutrality and objectivity. Discipline was to be founded on knowledge of facts and on deviations from a norm. Ideas and devices that had hitherto been the province of engineers were to become the cornerstone of a transformed cost

---

[12] Emerson, *Efficiency as a Basis*, p. 5.

[13] Henry R. Towne, "The Engineer as Economist," *Transactions of the American Society of Mechanical Engineers*, 7 (1886), 428–32.

[14] David Solomons, "The Historical Development of Costing," in *Studies in Cost Analysis*, ed. David Solomons, 2nd ed. (London: Sweet and Maxwell, 1968).

[15] Harrison, *Standard Costing*, pp. 27–8.

accounting, which created a new and individualized mode of governing the factory. With the development of standard costing, governing the enterprise became firmly established as part of the territory of accounting.

LINKING COSTS TO DECISIONS

Cost accounting underwent another major transformation during the inter-war years, and in the process was accorded a clearly defined role as the basis for what managers now identified as their primary role, making "decisions." This transformation of accounting drew its ideas and tools from economics rather than from engineering. The notions of fixed and variable costs, and of marginal costs, helped bring about this change. Cost-volume-profit calculations and break-even graphs operationalized and gave visual form to these ideas.

During the early 1920s, students at the School of Commerce and in the Department of Political Economy at the University of Chicago were shown what J. M. Clark described as "an experiment in a type of economic theory which is largely inductive." His object was unused capacity, or costs that do not vary with output. He went so far as to speculate whether "the whole body of economic thought must become an 'economics of overhead costs.'" But economic theory was useful not only for graduate students in economics, according to Clark, but also for accountants, who should be aware of "the meaning of cost from the standpoint of disinterested economic science," For it is this concept of cost that "embodies, in a sense, that impossible goal to which his practical devices serve as approximations."[16]

Clark was not an accountant, but he thought that "the unconventional standpoint of an outsider" might help to "throw a useful light upon the question of what cost accounting can and cannot be expected to do."[17] He argued that the analysis of costs should not be constrained by the rules of financial accounting. Cost accounting might include certain items that would be excluded for purposes of making up the income statement. He noted that railways had brought clearly into view the importance of the notion of overhead costs. For it had been understood as early as the mid nineteenth century that additional traffic could be carried on the railways at little or no additional cost.[18] Price discrimination could be justified on the grounds that added traffic was not responsible for fixed costs, costs that did not increase with traffic volume. In any case, he held that it was impossible to determine the proper share of costs traceable to an individual shipment

---

[16] John M. Clark, *Studies in the Economics of Overhead Costs* (Chicago: University of Chicago Press, 1923), pp. ix–x.
[17] Ibid., p. 234.
[18] Chandler, *Visible Hand*; Porter, *Trust in Numbers*; Solomons, "The Historical Development."

or unit of business.[19] Initially, the railways were seen as different from other industries because so much of their cost was "constant," or fixed, but soon this argument was applied to other industries. The distinction between "constant" and "variable" costs became a general principle for the classification of costs.

Clark was concerned with the "underlying functions" of cost accounting. He argued that these were multiple, and required "an elastic technique" that he described as "cost analysis" or "cost statistics." The functions of cost accounting, according to Clark, were various: to help determine a normal or satisfactory price for goods sold; to help fix a minimum limit on price cutting; to determine which goods are most profitable and which are unprofitable; to control inventory; to set a value on inventory; to test the efficiency of different processes and of different departments; to detect losses, waste, and pilfering; to separate the "cost of idleness" from the cost of producing goods; and to "tie in" with the financial accounts. "The purposes of cost analysis require a number of different conceptions and measures of cost, and the natural result is a plea for the development of a sufficiently varied technique to satisfy these quite independent requirements."[20] These included the "total economic sacrifice of production," including interest on all investment; differential costs; complete records of actual costs and standards of efficient performance against which to compare them; residual costs; and total operating expenses. His purpose was to define a distinctively managerial role for cost analysis, one that would embody the self-evident truths of economic science.

On the other side of the Atlantic, a little over a decade later, others argued in similar terms. Again, economic concepts were held up as corrections to the accountants' perception of cost behavior. Here, the key figures were Ronald Edwards and Ronald Coase, both from the London School of Economics and Political Science (LSE). Edwards, a "Lecturer in Business Administration with special reference to Accounting," had some ten years' experience as a professional accountant. But this did not prevent him from appealing to the language and concepts of economics to formulate what he called the "businessman's entrepreneurial problem." The most important issue vis-à-vis costs, he argued, was "the extent to which they change with output." Thus the avoidability or unavoidability of costs should be the principal concern. He defined "variable cost" as the additional expense incurred by producing the unit to be costed. What remained were then "fixed costs." Cost accounting should be based on "differential" or "marginal" costs, for those are the costs that vary with output.[21] Cost accountants should ignore unalterable expenses, and they should not spend their time calculating arbitrary allocations of fixed

[19] Clark, *Overhead Costs*, p. 10.
[20] Ibid., pp. 236, 257.
[21] Ronald S. Edwards, "The Rationale of Cost Accounting," in *LSE Essays on Cost*, ed. J. M. Buchanan and G. F. Thirlby (New York: New York University Press, 1981), pp. 76, 81. First published in Arnold Plant, ed., *Some Modern Business Problems* (London: Longman, 1937).

expenses to departments. Cost accounting should address the entrepreneurial problem, which implies a focus on marginal revenue and marginal cost.

A further and decisive step was the linking of cost accounting to the category of decision. Ronald Coase, then a lecturer in economics at the LSE, was particularly influential here. For while both Clark and Edwards had sought to alter cost accounting by appealing to the concept of marginal cost, neither had explicitly and firmly linked these concepts to an undifferentiated notion of decision. As Coase remarked in *The Accountant*, in setting out what he regarded as the basic concepts of cost accounting: "[A]ttention must be concentrated on the variations which will result if a particular *decision* is taken, and the variations that are relevant to *business decisions* are those in cost and/or receipts. This reasoning applies to every business *decision*." Business decisions, he continued, should depend on "estimates of the future."[22] His "general rule" was that a firm should expand production so long as marginal revenue is expected to be greater than marginal cost, and the avoidable costs of the total output less than the total receipts. Even if it was "utopian" to think that such a position might be reached in practice, he hoped that "the cost accountant may so refine his technique to take account of variations in cost and thus facilitate the task of the businessman."[23]

In the same year, and back on the other side of the Atlantic, there was emerging a new literature that depicted executives as decision makers. Chester Barnard's *The Functions of the Executive* referred to decision making as a "burdensome task," one that "men generally try to avoid." But they cannot. Interacting decisions distributed throughout organizations make up "the essential process of organization action which continually synthesizes the elements of cooperative systems into concrete systems."[24] Here was a distinctive role for the recently formed class of salaried managers who controlled large multiunit enterprises, one reaffirmed by parallel developments in the literature of business administration. Some influential lectures by William Vatter at the University of Chicago during the mid-1940s linked this function specifically to accounting. Vatter was unequivocal: "The only reason for collecting financial data about a business, from the managerial viewpoint, is that decisions must be made." Accounting records, he continued, "are of use to management only because they provide a basis for decision-making."[25]

Statements such as these, although still relatively novel in 1950, were to become the norm over the following two decades. During the 1950s and 1960s, the idea of decision, and the tools by which academics and managers might model decisions, were to become a core part of financial planning

---

[22] Ronald H. Coase, "Business Organization and the Accountant" (1938), in *LSE Essays on Cost*, ed. Buchanan and Thirlby, pp. 98, 100.

[23] Coase, "Business Organization," p. 102.

[24] Chester I. Barnard, *The Functions of the Executive* (Cambridge, Mass.: Harvard University Press, 1938), pp. 189, 187.

[25] William J. Vatter, *Managerial Accounting* (New York: Prentice Hall, 1950), pp. 102, 506.

and control. The new economic analysis provided accounting with some of its most basic devices, such as the cost-volume-profit graph and break-even analysis. Also, and equally importantly, cost accounting was provided with a route by which it might move further up the corporate hierarchy. By linking costs to choices, by calling these choices decisions, and by establishing linkages between the concept of decision and the practice of management, cost accounting acquired a new managerial significance. Cost accounting gradually metamorphosed into management accounting, with greatly widened scope.

MAKING THE FUTURE CALCULABLE

The concepts and calculations of economics have also contributed to the making of management accounting by suggesting a new way of framing and evaluating large-scale investments. This was grounded in the notion of the time value of money (the idea that a given sum of money is worth more today than at some point in the future) and the related practice of discounting (translating cash flows expected at some point in the future into present values).

Discounting techniques were in existence long before they were recommended as a management tool for the evaluation and comparison of investment proposals. Principles of compound interest were firmly established in actuarial practice as early as the sixteenth century, and annuity tables began to appear in the late seventeenth century. In the decades around 1900, engineering and political economy developed distinctively modern ways of representing and calculating, based on discounting and the use of net present value.[26]

In accounting circles, there was considerable hostility to the use of discounting practices for the purpose of investment evaluation even as late as the 1930s. In the United Kingdom, a series of articles in *The Accountant* in 1938 by Edwards generated a heated exchange of correspondence. Edwards insisted that, when considering alternative investments, "the influence of time must be eliminated and this is effected by discounting all receipts to their worth at a given date, say the date of the investment."[27] This brought forth a sharp riposte from Stanley Rowland, his colleague at the LSE. Rowland announced his "most fundamental disagreement," and his alarm at the prospect that

[26] Ian Hacking, *The Emergence of Probability* (Cambridge: Cambridge University Press, 1975); Robert H. Parker, "Discounted Cash Flow in Historical Perspective," *Journal of Accounting Research*, 6 (1968), 58–71; Theodore M. Porter, *The Rise of Statistical Thinking, 1820–1900* (Princeton, N.J.: Princeton University Press, 1986).

[27] Ronald S. Edwards, "The Nature and Measurement of Income: I–XIII," *The Accountant*, 99 (July–September 1938), 13–15, 45–7, 81–3, 121–4, 153–6, 185–9, 221–4, 253–6, 289–91, 325–7, 361–4, 397–401, 429–32, at p. 14.

"certain of his colleagues at the London School of Economics are associated with the views expressed."[28] It is not just the severity of Rowland's reply that is of interest, but also the way in which he situated accountancy. He thought perhaps that "Mr Edwards has 'gone berserk,'" and he drew the battle lines in terms of an onslaught by economists upon accountants. Referring to the "unanimity which is so characteristic of economists," he suggested that Edwards was "enjoying for its own sake the sport of bludgeoning the heads of accountants with intent that they shall be both bloody and bowed."[29] In a statement almost moving in its evocative appeal to the accounting ledger as a domain of objectivity and security, in contrast to what he saw as the speculative and showy world of the economist, Rowland proposed the outright rejection of Edwards's proposals. Let us, he suggested, "leave these nightmare thoughts and get back to a world in which cool sanity reigns. Let the accountant sit before his ledger and regard it with confidence as the bed rock on which his whole scheme rests. Let him record the present as it flows into the past and let him leave to others the risky business of tearing aside the veil which conceals the future."[30] Edwards replied by pointing out that the increased net worth concept was "neither new nor strange" in economic theory or in actuarial science.[31] Still deeply dissatisfied, Rowland denounced Edwards's theory as "sheer insanity" and "dangerous nonsense," one "in which assumptions are disguised as truths."[32]

More than a decade later, in the United States, principles of discounting were being urged insistently upon managers. Joel Dean, probably the most influential American advocate of discounting techniques, was still uncertain of their importance in his 1951 book, *Capital Budgeting*.[33] Yet three years later, in a highly influential *Harvard Business Review* article, he was unequivocal. Discounting principles, he said, offered a novel theoretical framework for managers seeking to understand investment decisions. A new economic-financial mentality should replace the accounting mentality, represented most typically by the payback method. This meant something more than substituting one technique for another, and it entailed a fundamental change of thinking about investments. Economic reasoning, especially about the time value of money, should be reflected in all investment decisions, he argued. Economic expertise should supplant personal intuition and the rule of thumb in making investment decisions. This would allow the ranking of investment opportunities, their comparison to alternatives, and a consideration of their

[28] Stanley W. Rowland, "The Nature and Measurement of Income," *The Accountant* (24 September 1938), 426.

[29] Stanley W. Rowland, "The Nature and Measurement of Income – II: A Rejoinder," *The Accountant* (15 October 1938), 519.

[30] Rowland, "Nature and Measurement II," p. 522.

[31] Ronald S. Edwards, "The Nature and Measurement of Income," correspondence section of *The Accountant* (22 October 1938), p. 575.

[32] Rowland, "Nature and Measurement" (24 September), pp. 609–10.

[33] Joel Dean, *Capital Budgeting* (New York: Columbia University Press, 1951).

net economic worth to the company. The "productivity of capital" was to be the decisive test, an objective measure of the economic worth of individual investment proposals. For the "discounted-cash-flow method of computing rate of return is demonstrably superior to existing alternatives in accuracy, realism, relevance, and sensitivity." Its range was almost unlimited. Not only investment in plant and machinery, but welfare and prestige investments such as gymnasiums, country clubs, and palatial offices should be analyzed by reference to the "directional beam of capital productivity." Investment decisions should contribute to "an enlightened intellectual environment throughout the company" in which all concerned would understand the economics of capital expenditures.[34] Financial expertise would transform the manager into a calculating individual. Dean was supported in this ambition by a general enhancement of the prestige of economic knowledge within the academy and among business schools.[35]

In the United Kingdom, widespread support for discounting principles in investment evaluation developed several years later. Toward the end of 1959, an article in *Accountancy* emphasized the importance of the time value of money in comparing returns on prospective investments.[36] In 1961, a series of leading articles in the same journal argued strongly, and at considerable length, that present value calculations were superior to both return on investment and the payback method for the control of capital expenditure.[37] Much discussion was to follow. This included an invited response by Reynolds to a series of criticisms of present value methods that had been published in one of the leading professional accounting journals in the United States; an extended tutorial on the principles and calculations of discounted cash flow analysis in *Accountancy*; and an editorial in the same journal commending a television series that showed modern accounting and statistical techniques, such as discounted cash flow, in action.[38] The discussion was not confined to accounting journals; it appeared in the pages of *The Economist* in 1964, and in a rapidly burgeoning literature that extolled the virtues of discounting procedures for a wide range of business decisions.[39]

[34] Joel Dean, "Measuring the Productivity of Capital," *Harvard Business Review*, 32 (1954), 120–30, at pp. 129, 121, 130.
[35] For example, see Jack Hirshleifer, "On the Theory of Optimal Investment Decision," *Journal of Political Economy*, 66 (August 1958), 329–52; James H. Lorie and Leonard J. Savage, "Three Problems in Rationing Capital," *Journal of Business*, 28 (October 1955), 229–39; Franco Modigliani and Marcus H. Miller, "The Cost of Capital, Corporation Finance, and the Theory of Investment," *American Economic Review*, 48 (June 1958), 261–97.
[36] H. J. H. Sisson and C. R. Goodman, "Capital Expenditure Decisions: Measuring the Prospective Return," *Accountancy*, 70 (1959), 597–600.
[37] P. D. Reynolds, "Control of Capital Expenditure," *Accountancy*, 72 (July 1961), 397–404; (August 1961), 471–5; (September 1961), 538–45.
[38] See, for instance, P. D. Reynolds, "Business Mathematics," *Accountancy*, 72 (September 1964) pp. 819–820; (October 1964) pp. 881–2; (November 1964) pp. 1039–40.
[39] For instance A. M. Alfred, "Discounted Cash Flow and Corporate Planning," *Woolwich Economic Papers*, no. 3 (1964), 1–18; A. M. Alfred and J. B. Evans, *Appraisal of Investment Projects by Discounted Cash Flow* (London: Chapman and Hall, 1965); Anthony J. Merrett and Allen Sykes, *The Finance and*

By the second half of the 1960s, the climate in the United Kingdom had changed fundamentally from that of the late 1930s. Interest in discounted cash flow techniques had reached such a point that when the London and District Society of Chartered Accountants organized a talk by the chief economic advisor to Courtaulds on such techniques, they had a capacity turnout, and members had to be refused tickets. Those who attended were not disappointed, for the speaker "held his audience so rapt that the rustle of turning papers, as he occasionally referred them to a section in the notes provided, seemed near thunderous and everyone forgot the beer and sandwiches waiting outside."[40]

The three moments discussed here by no means exhaust the repertoire of cost and management accounting. They serve, however, to mark the process by which accounting has moved up the corporate hierarchy across the twentieth century. They demonstrate also the ways in which management accounting has provided a calculus for governing the conduct of individuals within the formally private sphere of the enterprise. And they illustrate the permeability of accounting to other bodies of expertise, especially economics and engineering.

During the early 1980s, much of this was called into question. The rediscovery of the factory involved a new concern that enterprises in the United States were being administered excessively or exclusively according to financial numbers. In the early decades of the twentieth century, accounting held out the promise of demarcating a financial domain that would be neutral, objective, and calculable, one that would allow long-distance control of persons and processes. But over the past two decades or so, management accounting has been subjected to a barrage of criticism. Precisely the distanced and abstract nature of such expertise, once regarded as a major advantage, is now identified as the problem. The ideal of managing by the financial numbers alone has been increasingly questioned.[41] The financial mentality of large corporations, particularly in the United States, has seemed to entail a preoccupation with short-term cost reduction rather than long-term competitiveness. A pervasive myopia afflicting American industry has been traced back to the calculative practices and mentality of accounting. And this in turn has been traced to the institutions through which such expertise has been promulgated and disseminated, namely the business schools, universities, and management consultancies that for many years have promoted

Analysis of Capital Projects (London: Longmans Green, 1963); Louis W. Robson, "Capital Investment in Relation to Increased Productivity," *Accountancy*, 74 (December 1963), 1068–75.

[40] *Accountancy* (March 1967), 156.
[41] H. Thomas Johnson and Robert S. Kaplan, *Relevance Lost: The Rise and Fall of Management Accounting* (Boston: Harvard Business School Press, 1987); Peter Miller and Timothy O'Leary, "Accounting Expertise and the Politics of the Product: Economic Citizenship and Modes of Corporate Governance," *Accounting, Organizations, and Society*, 18 (1993), 187–206.

the "new management orthodoxy." The implications of this questioning of financial expertise remain to be seen, but to the extent that accounting has come to be viewed as synonymous with managing, its significance is potentially far-reaching. In such a context it is no longer a challenge to just one body of expertise, but to a core component of business administration as currently conceived.

These recent developments sound a note of caution, discouraging any dystopian image of a world ruled totally by accounting numbers. So, too, does a geographical purview that extends beyond North America and the United Kingdom. The events described here are almost exclusively Anglo-American, and although the financial mentality that has characterized these regions is widespread, it is not universal. The history provided here is therefore necessarily partial. The image of enterprises ruled wholly or largely by financial numbers is at odds with management practices in a number of countries. The governing of the enterprise takes many forms, and varies considerably. Although the contrast between Western and non-Western modes of management may be the most striking, there is considerable diversity even among European countries. To this extent, multiple histories of "accounting" are needed. Increasingly, however, such local histories are likely to converge or at least to overlap, as the ideas and practices described here, and the apparatus of business schools, consulting firms, and universities that promote financial expertise, become ever more global.

33

POLLING IN POLITICS AND INDUSTRY

Susan Herbst

Survey research has a relatively short history, since the systematic practice of aggregating preferences dates back only to the nineteenth century. Scholars, statesmen, and businessmen had an interest in the nature of public opinion long before the nineteenth century, of course, but technically sophisticated attempts to quantify popular sentiment trailed far behind theorizing and discussion of it. In the twentieth century, most Western democracies witnessed a tremendous surge in survey research with the emergence of large commercial firms devoted to counting individual opinions, preferences, and attitudes. This chapter will focus on three moments in the development of survey research: the proliferation of the straw poll in mid nineteenth-century America, the vital period between 1930 and 1950 across several national settings, and contemporary debates over the uses of opinion research in a democratic state.

The meaning of the term "public opinion" itself is tied to historical circumstances, as are methods for measuring it. These days, we have all become accustomed to the constant flow of polling data in our mass media, and to their underlying assumption – that public opinion can be defined as the aggregation of individual opinions. But public opinion has not always been conceptualized or measured in an aggregative fashion. For example, Jacques Necker (1732–1804), the finance minister of France, proposed that public opinion was equivalent to the "spirit of society."[1] Public opinion was a wise court, embedded in communication and conversation, which made societies stable, rising up slowly and rationally when necessary in response to important events. Necker viewed the salons of the period (elite drawing-room discussions of politics, art, and religion) as manifestations and indicators of public opinion – a far cry from the polls and surveys of today.

During the nineteenth century, a critical mass of writers, social reformers, political party operatives, marketers, and others began to think about

---

[1] Quoted in Keith Michael Baker, *Inventing the French Revolution: Essays on French Political Culture in the Eighteenth Century* (Cambridge: Cambridge University Press, 1990), p. 193.

systematic opinion aggregation as a tool for enhancing democratic life by giving voice to the views of common citizens. English and American social reformers surveyed the living conditions of urban populations, and polling citizens' opinions about politics became increasingly popular and widespread. The straw poll of the nineteenth century was a pivotal development in the history of polling, since it marked the first time that widespread opinion research – careless though it often was – became fully integrated into electioneering, newspaper journalism, and community life.

POLITICAL POLLING IN NINETEENTH-CENTURY AMERICA

From the mid nineteenth century through the early twentieth, political party workers, journalists, and citizen activists in the United States engaged in survey research at little cost and with great constancy and fervor. Their surveys were political polls, conducted for the express purposes of building party strength, fund raising, campaigning, and election forecasting. Straw polls typically focused on how citizens would vote in upcoming presidential elections. While many polls were conducted by political party operatives, journalists traveling the country by boat or rail also conducted polls while covering political campaigns. Party workers wanted to understand voter preferences so that they could direct resources to constituencies they hoped to persuade. Journalists, the human agents of a new infrastructure of mass communication, had different motives. They hoped to excite readers about the "horse race," to provide interesting reports from the field about the nature of voters' intentions, and to boost circulation as a result. This early journalistic interest in polling is still evident today, when newspapers and news networks are among the major producers of surveys and are also important clients for research suppliers.

Scores of straw polls could be found in local and major regional newspapers at election time. The *Chicago Tribune*, a staunchly Republican paper with the highest circulation in Chicago after its purchase by Joseph Medill in 1855 (it was founded in 1847), was one daily with a great interest in the straw poll, both as election predictor and as rhetorical device. Medill (1832–1899) was a leader in the Republican Party and a great promoter of Abraham Lincoln, so the editors of the paper put the full force of their partisan writing into articles about the 1860 presidential race between Lincoln and his opponent, Stephen Douglas. The following straw vote report from the 7 October edition of the *Tribune* is representative of the sort of polls found in the paper:

> They had an excursion from Hillsdale to Goshen, Indiana on Thursday. A vote for President was taken, with the following result: For Lincoln, gentlemen 368, ladies 433 – total 796 [sic]; for Douglas, gentlemen 156, ladies 60 – total 216; for Breckinridge, gentlemen 5, lady 1. Lincoln over all 574. The

following is a vote taken for the various nominees for the Presidency, on the train leaving Galesburg for Quincy the morning of the 4th inst. Lincoln 110, Douglas 43, Breckinridge 4.[2]

Despite an exclusively male franchise, this poll and many others included female preferences, probably in an effort to attract women readers in their role as consumers.

Polling was not limited to political operatives and professional journalists. In the mid and late nineteenth century, American citizens commonly polled each other and sent those data to political party workers and to the highly partisan newspapers of the day. For example, in the summer of 1856 one ambitious gentleman – most likely a traveling salesman – polled a total of 2,886 people throughout the Northeast about their preferences in the upcoming election and published his report in the *New York Times*.[3] Many citizens conducted smaller-scale polls querying their neighbors, friends, fellow club members, and workplace colleagues about the presidential race. These straw pollsters were always devoted partisans, intent on demonstrating that their favored paper's candidate was winning.

The abundance of polls in nineteenth-century America is reflective of the political culture of that period, a time of flamboyant public displays of partisanship, public debates, and a generalized carnival-like atmosphere surrounding elections.[4] Straw polling was an integral aspect of campaigns and enabled citizens to cast symbolic votes during the exciting anticipatory days before the election. Yet in addition to the increasingly intense rivalry between the major parties and the high level of participation in politics among free men, there are some other specific reasons why this particular form of quantitative opinion assessment emerged during these years. One of the most important was the changing nature of the American news media. During the mid and late nineteenth century, newspapers were working to increase their readerships and to support their operations using advertising revenues. The papers developed a wide range of new story genres and reporting techniques in order to build circulation figures. Publishing straw polls was one of those techniques, since readers were (and still are) interested in the competition among candidates and curious about the preferences of their fellow citizens.[5]

[2] *Chicago Tribune*, 7 October 1860, p. 2.
[3] Susan Herbst, *Numbered Voices: How Opinion Polling Has Shaped American Politics* (Chicago: University of Chicago Press, 1995), pp. 69–87.
[4] Michael McGerr, *The Decline of Popular Politics: The American North, 1865–1928* (New York: Oxford University Press, 1986); Jean Baker, *Affairs of Party: The Political Culture of Northern Democrats in the Mid-Nineteenth Century* (Ithaca, N.Y.: Cornell University Press, 1983).
[5] Elisabeth Noelle-Neumann, *The Spiral of Silence: Public Opinion – Our Social Skin* (Chicago: University of Chicago Press, 1984); Carroll J. Glynn, Ronald E. Ostman, and Daniel G. McDonald, "Opinions, Perception, and Social Reality," in *Public Opinion and the Communication of Consent*, ed. Theodore L. Glasser and Charles T. Salmon (New York: Guilford Press, 1995), pp. 249–77; Michael Schudson, *Discovering the News: A Social History of American Newspapers* (New York: Basic Books, 1978).

In what ways did the great period of straw polling, which faded with the eventual triumph of the sample survey in the 1930s, foreshadow later developments in the measurement of public opinion? First, straw polling provided a means for reducing the complex nature of citizen attitudes to a calculable, easily reported set of figures. This function is especially important in a large democracy, where political parties, journalists, and marketers are seeking ways to communicate efficiently about public opinion and to influence public opinion as well. Second, polling a huge number of citizens – if only for a brief moment – began at least to address the issue of large population size by gathering many opinions as quickly as possible. Straw polling reflects a preference for number of opinions gathered over depth of response. Third, during the nineteenth century the close relationship between opinion assessment and journalism was solidified. Journalists were central actors in the development of polling and the communication of survey research. Decades later, they became the clients of and major source of revenue for pollsters such as George Gallup. Fourth, because polling was so extensive and popular during the nineteenth century, it advertised the fact that people could express informed opinions about politics. If so, then perhaps they could articulate other desires as well. More and more, individual citizens were viewed as knowledgeable, independent actors with a well-developed sense of their own preferences. The information culture had changed incrementally but dramatically. By the late nineteenth century, with literacy on the rise, there was an explosion of specialty magazines and public education programs established to feed a hungry public that demanded information.[6]

BIRTH OF THE SAMPLE SURVEY

In 1936, the *Literary Digest*, a popular political magazine and organizer of the largest straw poll endeavor in United States history, failed to predict the outcome of the presidential race between Franklin Roosevelt and Alf Landon. The extensive system of the *Digest* poll, a mail survey conducted throughout the early decades of the twentieth century before every presidential race, relied on a mass of public records (e.g., automobile registration, telephone ownership) to locate survey respondents. That same year, George Gallup (1901–1984) accurately predicted the race by using sampling methodology, and thereby established himself as the premier pollster of that early period. After the 1930s, straw polls conducted by traveling journalists and common citizens were rarely found in major newspapers, as the professional political surveyors (Gallup, Roper, Crossley) began to assume the task of tabulating voter preferences.

[6] Theodore Morrison, *Chautauqua: A Center for Education, Religion, and the Arts in America* (Chicago: University of Chicago Press, 1974); Morton Keller, *Affairs of State: Public Life in Late Nineteenth Century America* (Cambridge, Mass.: Harvard University Press, 1977).

Sampling became available as a data collection technique only during the early decades of the twentieth century. Gallup was one of the first researchers to recognize its value in journalism and politics, and his success in 1936 was perhaps the single most dramatic use of the method in history. Sampling revolutionized the practice of survey research, making it possible to poll a smaller number of people than a census, and at the same time to achieve greater accuracy in representation of the populace. The earliest advocate of representative sampling was Anders N. Kiaer (1838–1919), a Norwegian statistician who used purposive sampling methods and then validated his results through comparison to census figures. Kiaer contended fearlessly with leading members of the statistical profession, trying to convince them that representative sampling was both possible and necessary, since population coverage is so expensive and difficult to achieve. The random sample was first used in a social study by the economist Arthur L. Bowley (1869–1957), who surveyed several English towns and published his results in a volume entitled *Livelihood and Poverty* (1915). The statistician Jerzy Neyman (1894–1981) explicated the importance of random selection as a component of stratified sampling, and his paper on the subject published in 1934 served as a foundation for most important statistical work on sampling during subsequent decades.[7] Among the central contributions of the paper was its argument for randomization. One must divide the population into strata or sections determined by a control variable of interest, and then randomly select units to survey (e.g., districts in a country, households in a city) within those strata.

While sampling was developed in part for purposes of social surveying, it was simultaneously tested and applied in fields such as agriculture and mining. In fact, the pollster Emil Hurja (1892–1953), one of Franklin Roosevelt's closest advisors during the early 1930s, began his career as a mining analyst on Wall Street. He had learned sampling theory by evaluating ores, and applied this knowledge to the assessment of the political climate.[8]

Among the many entrepreneurs, academics, and government officials working to take advantage of the new methods of sampling, Gallup was the most prominent. His leading role in this practice was due in part to his sheer inventiveness, but also to his previous success as an academic and newspaper researcher, which had gained him the ear of many influential businessmen and scholars. Before he became a political pollster who moved in elite Washington circles, Gallup had been an industry man. He had worked

[7] William Kruskal and Frederick Mosteller, "Representative Sampling IV: The History of the Concept in Statistics, 1895–1939," *International Statistical Review*, 48 (1980), 169–95; Alain Desrosières, "The Part in Relation to the Whole: How to Generalise? The Prehistory of Representative Sampling," in *The Social Survey in Historical Perspective, 1880–1940*, ed. Martin Bulmer, Kevin Bales, and Kathryn Kish Sklar (Cambridge: Cambridge University Press, 1991).

[8] Melvin G. Holli, "Emil E. Hurja: Michigan's Presidential Pollster," *Michigan Historical Review*, 21 (Fall 1995), 125–38.

to help advertisers systematize their knowledge of which sorts of visual and verbal messages persuaded consumers to buy and which did not. Although it is not widely known that Gallup got his start in the consumer milieu before establishing a political research business, much of the history of political polling can in fact be traced to marketing research. Gallup advocated the use of the sample survey in the commercial and political realms, defending the method as distinctively American – a tool for spreading democracy. A fervent advocate of the popular will, he characterized polling as a finger registering the "pulse of democracy."[9] Gallup devoted considerable effort to justifying the development of the polling industry by claiming that surveying was the premier technique for democratic representation. He and other early pollsters dismissed other means for expressing and measuring opinion, such as political rallies, letters to the editor or to representatives, and town meetings – not as worthy competitors of surveys but as biased techniques, useless in furthering democratic goals. This inability to consider other methods of opinion measurement may have been good for business, but it displayed a narrow-mindedness about the nature of public opinion that critics would later find appalling.

Since the 1930s, there has been much communication among market researchers, political pollsters, academic survey researchers, and government surveyors, because all employ a shared methodology. All have been intent on developing the best sampling techniques for generalizing to large populations, as well as the most effective ways to combat low response rates, questionnaire biases, and interviewer-induced errors. While Gallup worked in the spotlight of public attention, reliance on market research was formalized within manufacturing firms during the interwar period. At General Motors, for example, an energetic advocate of consumer research named Henry Weaver (1889–1949) made extensive attempts to understand the needs of car buyers during the early decades of the twentieth century. Automobiles have long ranked with housing as the largest consumer purchases, and finding out what those drivers desire in a car was and still is vital to car manufacturers. By 1939, GM was budgeting between $300,000 and $500,000 annually for Weaver to investigate consumers' preferences. The historian Sally Clarke has noted that this was probably the largest in-house research budget of any corporation of the period.[10] GM realized early on that the more they knew about car buyers, the more strategically they could act in both designing and selling these expensive items.

By the late 1930s, market research was already enormously popular, and major firms either established their own departments for conducting surveys

[9] George Gallup and Saul Rae, *The Pulse of Democracy: The Public Opinion Poll and How It Works* (New York: Greenwood Press, 1940).

[10] Sally Clarke, "Consumers, Information, and Marketing Efficiency at GM, 1921–1940," *Business and Economic History*, 25 (Fall 1996), 186–95.

or commissioned market research firms to do surveys for them. There was high confidence in the sample survey among marketing experts in the commercial world. An early textbook was published in 1937 by a professor and professional market researcher named Lyndon O. Brown (1903–1966). His *Market Research and Analysis*, which went into multiple subsequent editions, argued eloquently for the use of sampling and provided nuts-and-bolts instructions for the student of market research. Whether or not he knew of Booth's work on poverty in London or the maps produced by Florence Kelley of Hull House, Brown also had a great interest in melding the study of geography with the study of population demographics. But, going further than his mapmaking predecessors, he advocated surveying individuals about their preferences as well as their living conditions. This confluence of interests – in geography, demographics, and opinions – has continued to characterize academic, political, and marketing research of the late twentieth century.[11]

While Gallup and others toiled in industry and political polling during the 1930s and 1940s, academics were also hard at work on many similar projects. One of the earliest such researchers was Paul F. Lazarsfeld (1901–1976), an Austrian immigrant who became one of the most prominent theorists and methodologists in sociology. In an autobiographical essay, Lazarsfeld recalled that when he was a socialist student in interwar Vienna, he and his fellow party members were having difficulty persuading people with their propaganda. This challenge drew him toward the formal study of psychology. In his youth he spouted this formula: "A fighting revolution requires economics (Marx); a victorious revolution requires engineers (Russia); a defeated revolution calls for psychology (Vienna)."[12] Lazarsfeld gained a doctorate in applied mathematics and began to collaborate with psychologists at the University of Vienna. In 1930, he commenced the intensive study of Marienthal, a southern Austrian village devastated by unemployment. That study, published with Hans Zeisel and Marie Jahoda, signaled his move to sociology.[13]

Along with Hadley Cantril, Frank Stanton, and others, Lazarsfeld founded the Office of Radio Research at Princeton University in 1937, with funding from the Rockefeller Foundation. The institute aimed to study the effects of radio on listening audiences. How did people process information they received via this new communication technology, and how might its power – used so effectively by Adolf Hitler – be harnessed? Later, Lazarsfeld moved

[11] Lyndon Brown, *Market Research and Analysis* (New York: Ronald Press, 1937).
[12] Paul F. Lazarsfeld, "An Episode in the History of Social Research: A Memoir," in *The Intellectual Migration: Europe and America, 1930–1960*, ed. Donald Fleming and Bernard Bailyn (Cambridge, Mass.: Harvard University Press, 1969); see also Todd Gitlin, "Media Sociology: The Dominant Paradigm," *Theory and Society*, 6 (1978), 205–49.
[13] Marie Jahoda, Paul Lazarsfeld, and Hans Zeisel, *Marienthal: The Sociology of an Unemployed Community* (1932) (Chicago: Aldine, Atherton, 1971).

to the Bureau of Applied Social Research at Columbia University, where he collaborated with Robert Merton, Elihu Katz, and others to study the effects of the communication media on consumer and political behavior. He later wrote a variety of methodological tracts on survey research and statistics, many of which are still cited in the development of questionnaire design and the analysis of data.[14]

During the 1930s and 1940s, the American survey research community was small but growing, and because of its limited membership there was frequent intellectual exchange among political pollsters, market researchers, and academics. This is evident in the journals of the period – the *International Journal of Opinion and Attitude Research* and the *Public Opinion Quarterly*, both founded in 1937 and still published today as forums for survey research across industry, government, and academe. Survey techniques are not tied to particular fields. In theory, one uses the same sampling techniques and rules of thumb for questionnaire design regardless of the topic of the survey. Data analysis, is likewise standardized, relying routinely on cross-tabulation and general linear models for the reporting of findings. Yet surveys take on a different character in each field because each area has its own constraints. In journalism and in political campaigns, for example, surveys must often be conducted quickly in order to have "news value." This time pressure rules out certain techniques of data collection and analysis. In industry, government, and academe there is often more time to experiment with techniques or to conduct multiple surveys on the same narrow question.

A very interesting debate, which highlights these differences across fields, followed the disastrous American political polls of the 1948 election campaign. In the early fall of 1948, the major political pollsters stopped polling, believing that Thomas E. Dewey would beat Harry Truman handily. They were wrong, of course, and were humiliated by the press and by Truman himself, always ready to demean political forecasting based on polls. In the wake of this error, market researchers reflected on their own craft. Was market research implicated in the mistakes the 1948 election? Frank Coutant, a prominent market researcher of the period, argued in November 1948 that the "upset" of 1948 had nothing to do with his field. Marketers, he explained, survey individuals on fact and behavior, not on mere opinion. "There is no real reason to have our faith [in surveys] shaken."[15] Actually, market researchers did and do ask people about their opinions. However, marketers require no more than a general sense of consumer desires, and they need not provide exact forecasts of behavior.

[14] Paul Lazarsfeld, "The Art of Asking Why," *National Marketing Review*, 1 (Summer 1935), 1–13, and his *Qualitative Analysis: Historical and Critical Essays* (Boston: Allyn and Bacon, 1972).
[15] Frank Coutant, "The Difference between Market Research and Election Forecasting," *International Journal of Opinion and Attitude Research*, 2 (1948–9), 569–74.

EUROPEAN DEVELOPMENTS

The 1930s and 1940s were exciting years in the American survey research community as a variety of individuals and institutes worked toward development of attitude theory and survey techniques. Yet the notion of surveying the population was also of great interest in Europe, site of some of the earliest social surveys. As in the United States, the rise of German opinion polling was very much rooted in market research. In 1934, for example, the successful Gesellschaft für Konsumforschung (Society for Consumer Research) was founded in order to collect data for clients on a variety of consumer attitudes and to engage in what was then called "market observation" (*Marktbeobachtung*).[16]

The Italian case is more complex, as the fascist government struggled with the notion of polling. Interestingly, the opinion surveys of George Gallup and other American researchers were known to the fascist regime in Italy during the war, and even were published in magazines such as *Critica Fascista*, although, as Sandro Rinauro points out, nobody asserted the democratic implications of survey research. Published American and European polls in Italy were often accompanied by commentary denying the value of polling. The fascist regime conducted its own systematic spying on suspect political factions and even hired a firm to conduct a crude survey in order to explore the effectiveness of radio propaganda. In a particularly courageous public act in 1942, a statistics professor at the University of Trieste named Pierpaolo Luzzatto Fegiz (1900–1989) implored the regime to utilize scientific survey research in order to build a democracy. These pleas were answered by another prominent statistician, Corrado Gini (1884–1965), who argued in political terms that opinion polling was antidemocratic, since it enabled politicians to manipulate public opinion more effectively. Gini had the credentials necessary to cast doubt on the technique of polling by questioning the quality of data collected. While the fascists refused to use opinion polling in any sustained way, the first Italian institute for public opinion research – the Doxa – was founded by Fegiz in Milan immediately after the war in 1946. Significantly, the Italian case differs from the American one in that the impetus for survey research emerged from statisticians in academe, not from marketing research. Also, Italy had no tradition of political straw polling.[17]

Not until the late 1960s did French elites recognize polling and survey research as valuable. Although France is now a leader in published polls, the broadcast of which before elections has been contested, French public

[16] Christoph Conrad, "On Market Research Conducted by Independent Organizations in Interwar Germany: Between Business, State, and Academic Research." Paper presented to the conference on Opinion Research in the History of Modern Democracies, Free University, Berlin, 1997.
[17] Sandro Rinauro, "The Diffusion of Public Opinion Surveys in Italy Between Fascism and Democracy." Paper presented to the conference on Opinion Research in the History of Modern Democracies, Free University, Berlin, 1997.

officials and intellectuals resisted survey research for decades. Newspapers did not commission surveys, and in politics the notion of polling seemed inappropriate, since the elected parliament and political parties were believed to be the most reliable representatives of public opinion. Interestingly, even social scientists – who were such active proponents of probability sampling and survey research in other nations – rejected polling as unable to capture the unique, textured, and complex nature of French public sentiment. The prominent postwar sociologist Georges Gurvitch spoke of polls as "les procédés dérisoires de Monsieur Gallup [the ridiculous methods of Mr. Gallup]."[18]

AMERICAN ACADEMIC SURVEY INSTITUTES

While the roots of American survey research can be traced to marketing and political straw polling of the late nineteenth century, academic research on attitude formation did not begin in earnest until survey centers were established during and after the Second World War. These enclaves were largely devoted to the study of social and political attitudes as well as to the methodological and epistemological problematics of sampling and survey design. Three important university-based centers for survey research were established during the 1940s. One was Lazarsfeld's Bureau of Applied Social Research at Columbia, officially formed in 1944, where researchers investigated the effects of communication media on voting patterns, attitude formation, and consumer behavior, often through detailed studies of particular communities. The National Opinion Research Center (NORC) was established in 1941 at the University of Denver, then moved to its current home at the University of Chicago. The NORC was founded by Gallup's colleague Harry Field, with the mission to become "the first non-profit, non-commercial organization to measure public opinion in the United States." Field, intent upon broadening the survey research endeavor beyond commercial applications, preferred a public service orientation. Another large survey operation, the Survey Research Center (SRC) at the University of Michigan, engaged in both applied and basic research. Among SRC projects were studies of income dynamics and of industrial workers, as well as large-scale surveys of electoral behavior. The NORC and the SRC still thrive in today's more sophisticated and competitive world of survey research and have been leaders in the development of survey methodology.[19]

Academic survey institutes conduct basic research, but one of their major clients is often the United States government, which funds and sponsors an immense number of surveys each year in a variety of areas such as health,

---

[18] Loïc Blondiaux, "Comment rompre avec Durkheim? Jean Stoetzel et la sociologie française de l'après-guerre (1945–1958)," *Revue française de sociologie*, 32 (1991), 411–42.

[19] Jean Converse, *Survey Research in the United States: Roots and Emergence, 1890–1960* (Berkeley: University of California Press, 1987).

welfare, crime, and finance. Yet the greatest financial expenditures on survey research today are made not by the federal government but by commercial firms. This represents a shift from the earlier period, when universities were especially prominent in survey research. N. H. Engle, president of the American Marketing Association, noted in 1940 that 57 percent of all marketing projects reported to the *Journal of Marketing* were conducted by academics, 30 percent by government researchers, and only 11 percent by business firms.[20]

THE USE OF POLLS TO INFLUENCE PUBLIC OPINION

Surveys, of course, mean the collection of data about public opinion, but this surveillance is typically undertaken in order to reshape attitudes. Gallup always claimed to be working in the name of democracy, but the data he provided to newspapers and other clients were used strategically, as nineteenth-century straw polls had been, to win votes and entice consumers. And while academic surveyors and market researchers of the 1930s and 1940s certainly did have intellectual and methodological concerns, it was clear to them that sharpening the tools used for attitude assessment would ultimately be most useful to opinion persuaders – such as statesmen, public relations firms, political activists, and advertisers.

That opinion research enables more targeted persuasion is an obvious point, but the close relationship of opinion measurement to the diffusion and heterogeneity of the mass media is often overlooked. Without an extensive media infrastructure for manipulation of public opinion – print and broadcast – as well as a keen understanding of that complex infrastructure, survey research data are not particularly useful. In all eras, those who have wished to persuade citizens or consumers based on opinion data have been forced to conduct a simultaneous assessment of the media landscape. Data about public opinion contribute little to the art of cultivating journalists, setting up a competent public relations operation, developing resonant phrases that might move voters, or writing effective advertising copy. Political operatives of the nineteenth century, for example, often collected masses of data through their straw polling. But even with this extensive (albeit unsystematic) knowledge of public opinion, party workers were constrained by the biases of journalists, the difficulty of getting citizens to listen, the cost of printing pamphlets and organizing rallies, the strength of opposing party organizations, and the speed and intensity of the typical election campaign.

In the current period, citizens cannot easily detect the hidden linkages between opinion data collection and public relations efforts. They know they are being cultivated, through direct mail, broadcast advertisements, telemarketing, and other means. Yet the massive opinion collection efforts

---

[20] N. H. Engle, "Gaps in Marketing Research," *Journal of Marketing*, 4 (April 1940), 345–53.

which often drive these campaigns are invisible, since they are embodied in largely proprietary databases owned by political parties, the state, and private firms. Although census data are available to all citizens in libraries, a complex task of mapping political preferences, psychographic data, and purchasing behavior onto census tracts undergirds many attempts to persuade the public.[21]

POLLING, PERSUASION, AND DEMOCRACY

Polling, by now, is practically ubiquitous.[22] It is often justified and indeed lauded because it gives voice to the views of "the people," as a key mechanism for registering public opinion in the form of objective numbers. There is growing concern about the uses and impacts of these data, however, both among scholars and among the journalists and policy makers who use these statistics most often. Opinion polling has been effectively criticized for several reasons, among them the domination of poll data over other forms of opinion expression, the ways in which polling can narrow public debate, and the unsuitability of polls to reflect the social structure itself.

Some critics argue that as our omnipresent surveys have been rationalized, the range of possibilities for communication about public opinion has become more limited. Ideally, opinion polling should be viewed as one among many means for assessing public preferences. Nonetheless, as newspapers and other media fill up with polling data, as internet polls proliferate, and as marketing suppliers continue to serve their clients, expressions of public opinion that are more difficult to quantify are less often noticed. Journalists, for example, have less incentive to highlight a political demonstration by 100 people when a professionally executed random sample survey on the same issue indicates that the demonstrators are a minority. Polling data have many attractions that demonstrations do not. They can be communicated with great efficiency, and they give the appearance of "hard news." This is attractive to journalists looking for solid and objective news in a messy and complex public sphere. And if journalists believe that political rallies, letters to public officials, focus groups, political theater, and radical arts are less "newsworthy" than polls, these forms of citizen involvement tend to lose their effectiveness.[23]

Surveys inevitably narrow public debate by defining public problems in specific ways. Pollsters, including academic researchers, cannot avoid

[21] Lawrence R. Jacobs and Robert Y. Shapiro, *Politicians Don't Pander: Political Manipulation and the Loss of Democratic Responsiveness* (Chicago: University of Chicago Press, 2000).
[22] See the collection of essays on the growth of polling and the changing nature of the industry in the fiftieth anniversary issue of *Public Opinion Quarterly* (Winter 1987).
[23] Herbst, *Numbered Voices*; Benjamin Ginsberg, *The Captive Public: How Mass Opinion Promotes State Power* (New York: Basic Books, 1986).

narrowing problems when they draw up a list of questions and allowed responses. In election campaigns, for example, candidates from third parties normally receive little support in early campaign polling, since they are typically unknown by citizens participating in surveys. In subsequent polls, the names of these candidates begin to disappear from the polling form, and hence also from campaign journalism. Having lost the attention of the press, their voices are almost entirely delegitimized.[24] Journalists, wittingly or not, persuade the public that only the two major parties matter and that challengers to this system are, with only the occasional exception, insignificant. Surveying can also narrow public debates by focusing on certain issues or policy options and excluding others. Some scholars have argued that there is often a disjuncture between what pollsters, policy makers, journalists, and other elite actors view as "political" issues and what citizens think. Those who conduct and rely on surveys can frame social problems as highly political and divisive, or they can ignore their political resonance altogether; and such choices can have powerful effects on citizen cognition and participation.[25]

Another problem in measuring public opinion by means of surveys is, ironically, rooted in the fact that polls assign equal weight to all respondents' opinions. This has the attraction of democratic resonance, as Gallup himself was tireless in pointing out. By the logic of sampling and random selection, we all have an equal chance of being chosen by a pollster to participate in a given survey. Polling therefore ignores the complexity of social structure and power dynamics by overlooking social inequality and missing key aspects of policy formation. Sometimes the "public opinion" that triumphs in a policy debate is not the public opinion represented by surveys at all, but public opinion as constructed by particular interest groups, leaders, or other parties.[26]

Since surveys were introduced in the mid nineteenth century, and as they have been refined throughout the twentieth century, these tools have been attractive and functional – from instrumental and symbolic standpoints – for leaders, marketers, journalists, and a variety of other social actors. Yet no matter how well we refine particular methodologies and indicators, public opinion is a nebulous entity and will be the site of great struggle in any democracy. Walter Lippmann (1889–1974) and others who have written about public opinion have argued that it is a fiction, invented as we

---

[24] Joshua Meyrowitz, "The Problem of Getting on the Media Agenda: A Case Study in Competing Logics of Campaign Coverage," in *Presidential Campaign Discourse*, ed. Kathleen E. Kendall, (Albany: State University of New York Press, 1995).

[25] Pierre Bourdieu, "Public Opinion Does Not Exist," in *Communication and Class Struggle*, ed. Armand Mattelart and Seth Siegelaub (New York: International General, 1979).

[26] Susan Herbst, *Reading Public Opinion: How Political Actors View the Democratic Process* (Chicago: University of Chicago Press, 1998); Herbert Blumer, "Public Opinion and Public Opinion Polling," *American Sociological Review*, 13 (1948), 242–9.

try to accommodate the vox populi in democratic systems.[27] Surveying should be viewed as one ambitious and provocative means for assessing and influencing citizen preferences. But it should also be recognized as an endeavor that narrows public opinion to a sum of atomized, anonymous opinions solicited by interviewers with their own special concerns and motivations.

[27] Walter Lippmann, *The Phantom Public* (New York: Harcourt, Brace, 1925) and his *Public Opinion* (New York: Free Press, 1965).

34

SOCIAL SCIENCE AND SOCIAL PLANNING DURING THE TWENTIETH CENTURY

Peter Wagner

The social sciences, in broadly their contemporary shapes, emerged after the American and French Revolutions. They offered a variety of ways of dealing with the new postrevolutionary political situation, which enabled, and indeed obliged, human beings to create their own rules for social action and political order. It has been a part of the intellectual tradition of the social sciences from their beginnings to contribute to making the social world predictable in the face of modern uncertainties, or, in the stronger version, to reshape it according to a master plan for improvement.[1]

The general idea of providing and using social knowledge for government and policy purposes was certainly not new. The cameral and policy sciences of the seventeenth and eighteenth centuries were designed for use by an absolute ruler; the very name "statistics" reflects the fact that it was considered science for governmental purposes. The postrevolutionary situation, however, was crucially different in two respects. On the one hand, a much more radical uncertainty had been created by the commitment, even if often a reluctant one, to self-determination of the people, which appeared to limit the possibility of predictive knowledge. On the other hand, this radical openness had been accompanied by a hope for the self-organization of society and its rational individuals, so that the search for laws governing society and human actions emerged beyond – and to some extent instead of – the desire for the increase of factual knowledge of the social world.

As a consequence, two competing concepts of social science with different attitudes toward social planning coexisted throughout much of the nineteenth century. Both anticipated a steady increase in valid social knowledge. But not everyone believed that such knowledge should be actively

---

[1] Johan Heilbron, Lars Magnusson, and Björn Wittrock, eds., *The Rise of the Social Sciences and the Formation of Modernity* (Sociology of the Sciences Yearbook 20) (Dordrecht: Kluwer, 1997). See especially Robert Wokler, "The Enlightenment Passage from Political to Social Science," pp. 35–76, and Peter Wagner, "Certainty and Order, Liberty and Contingency: The Birth of Social Science as Empirical Political Philosophy," pp. 241–63.

translated into planned intervention in the social world. Perhaps the interplay of the free actions of reason-endowed human beings would automatically enhance the well-being of all, as the traditions of political economy and, later, neoclassical economics held; or perhaps a progressive evolution of humankind determined the historical course of societies from lower to higher stages, making interference ineffective and unnecessary. Despite many earlier announcements, then, social planning based on social science knowledge was to be a phenomenon more characteristic of the twentieth than of the nineteenth century.

AMELIORIST SOCIAL SCIENCE AND THE SOCIAL QUESTION

From, broadly, the middle of the nineteenth century onward, though, the more optimistic views on societal self-regulation proved increasingly difficult to maintain, in the face of rising criticism of poverty and prostitution and the deteriorating health of the population. These evils, widely regarded as unprecedented, had at first been seen as transitional problems on the way to a new social order, as the birth pangs of modernity. Now they began to be regarded as persistent and potentially dangerous to the social order, because they appeared together with other major social changes, such as industrialization and urbanization, and because they were linked to widespread discontent.

In this context, explicitly policy-oriented – and, in a loose sense, planning-oriented – forms of social science (re-)emerged in a number of countries. Often their starting point was the empirical elucidation of problematic social situations, a strategy employed by activists as diverse as the hygienists and the group around Frédéric LePlay in midcentury France, reformist moralists in Britain, "mugwump" intellectuals in the Gilded Age United States, and factory inspectors in imperial Germany. Often, the reformism was closely linked to a more comprehensive scholarly ambition and to the creation of semischolarly, semipolitical associations, such as the American Social Science Association, the Verein für Socialpolitik of German historical economists, the Fabian Society, and the LePlayist Société D'action Sociale.[2]

Mostly, the approaches taken were straightforwardly empirical and observational in their methodological orientation and committed to political reformism of a conservative, ameliorist kind, focused on the safeguarding of order.[3] Statistics was often seen as a means to reorder a social reality that

---

[2] Dietrich Rueschemeyer and Theda Skocpol, eds., *Social Knowledge and the Origins of Modern Social Policies* (Princeton, N.J.: Princeton University Press; New York: Russell Sage Foundation, 1996); Dorothy Ross, *The Origins of American Social Science* (Cambridge: Cambridge University Press, 1991), chap. 3; Peter T. Manicas, *A History and Philosophy of the Social Sciences* (Oxford: Blackwell, 1987).

[3] Peter Wagner, Björn Wittrock, and Hellmut Wollmann, "Social Sciences and Modern States," in *Social Sciences and Modern States: National Experiences and Theoretical Crossroads*, ed. Peter Wagner,

appeared to have become recalcitrant.[4] This was particularly the case in newly formed states, such as Italy and Germany, in which the cohesion and homogeneity of society could be taken for granted even less than in other, more firmly consolidated states.

One outcome of these efforts was to link the work of social scientists directly to state concerns, to orient social knowledge to policy making in a way that was novel for the postrevolutionary period and to some extent reminiscent of the earlier policy and cameral sciences. Such state-oriented social science defined the major political issue of the time, often called "the social question," in terms of finding a smooth transition from the earlier restrictive liberalism (or even, as in Germany, the old regime) to a fully inclusive order. Politically, the recognition of the salience of "the social question" spelled an end to any idea or ideology of societal self-regulation. However, the growth of state involvement, while necessary, was generally not seen as a radical break with earlier practice. Social elites simply had to be more responsive to the needs of the population than they had previously been. Empirical social analysis was meant both to demonstrate the need for reforms, also against elite resistance, and to develop and propose the kinds of measures that were required.[5]

Toward that end, initially no particular epistemological or ontological issues needed to be confronted. Broadly speaking, a sober empirical realism appeared to be sufficient for such a problem-oriented social science. Accordingly, a soft version of positivism prevailed among policy-oriented social scientists after the middle of the nineteenth century, one committed to the extension of positive knowledge and sometimes even evoking Auguste Comte's name, but without the religious fervor of the original project of a positive science of society.[6]

SOCIAL SCIENCE AND THE CRISIS OF LIBERALISM

A major transformation in the relation between social science and policy making started gradually after 1870 and culminated in debates around the turn of the century. The earlier reformism increasingly came to be regarded

Carol H. Weiss, Björn Wittrock, and Hellmut Wollmann (Cambridge: Cambridge University Press, 1991), pp. 28–85.

[4] Theodore M. Porter, *The Rise of Statistical Thinking, 1820–1900* (Princeton, N.J.: Princeton University Press, 1986).

[5] Michael J. Lacey and Mary O. Furner, eds., *The State and Social Investigation in Britain and the United States* (Cambridge: Woodrow Wilson Center and Cambridge University Press, 1993); Mary O. Furner and Barry Supple, eds., *The State and Economic Knowledge: The American and British Experiences* (New York: Woodrow Wilson Center and Cambridge University Press, 1990), see especially Mary O. Furner, "Knowing Capitalism: Public Investigation and the Labor Question in the Long Progressive Era," pp. 241–86.

[6] Gillis J. Harp, *Positivist Republic: Auguste Comte and the Reconstruction of American Liberalism, 1865–1920* (University Park: Pennsylvania State University Press, 1995); Terence R. Wright, *The Religion of Humanity: The Impact of Comtean Positivism on Victorian Britain* (Cambridge: Cambridge University Press, 1986), esp. pp. 269–70.

as insufficient for the emerging societal constellation, both in its conception of politics and in its conception of social knowledge.

Politically speaking, liberal elites recognized that industrialization, urbanization, the emergence of an organized working class, and the concomitant demand for full inclusion of all members of society on equal terms posed serious, seemingly almost intractable, problems for the liberal conception of political institutions. Much "realist" political sociology of the time, including works by Robert Michels, Vilfredo Pareto, and Max Weber in Europe and by John Dewey in the United States, sought to identify the required institutional adjustment. At least in their European versions, some elitist conclusions appeared to be inevitable. More conservative-minded authors, especially on the European continent, interpreted the same evidence as confirming their view that liberalism was untenable. Even they, however, perceived on the horizon a transition to a new social order, rather than an adjustment of the existing one. In political terms, therefore, what was at stake was an understanding of the transformation of liberalism.[7]

Ultimately the political balance tipped toward a strengthening of collectivist orientation; the autonomy of the individual was deemphasized in favor of a voluntarism of the collectivity. Both socialism and nationalism provided versions of such a collectivist political philosophy; but even former liberals resigned themselves to social changes that had displaced individual responsibility from the center of politics. Progressivism in the United States, and social democracy in Europe, emerged as new and often quite fragile alliances between socialism and liberalism. Along with them came a new group of political elites, favoring professionalism and science as opposed to the feudalism and clientelism in the old elites, but often also technocratic and state-centered, suspicious of the pluralism and democracy of much earlier liberalism.[8]

The shift in political orientation, this declining faith in the viability of liberalism, was paralleled in epistemology by a renewed skepticism about the other central tenet of the Enlightenment tradition, the intelligibility of human action and the social world. The period around the turn of the century is now considered an intellectually extremely fruitful, even a classical era in many fields of social science, most notably in sociology, psychology, and economics. At that time, however, much of the work was driven by a sense of crisis, a feeling that many of the epistemological, ontological, and methodological assumptions of earlier social science were inadequate.

In terms of epistemology, social science saw itself forced largely to abandon the idea of representing social reality and instead accepted the view that

---

[7] Steven Seidman, *Liberalism and the Origins of European Social Theory* (Oxford: Blackwell, 1983); Peter Wagner, *A Sociology of Modernity* (London: Routledge, 1994), chap. 4.

[8] James T. Kloppenberg, *Uncertain Victory: Social Democracy and Progressivism in European and American Thought, 1870–1920* (New York: Oxford University Press, 1986).

conceptual constructions were dependent both on the means and forms of observation and perception, and on the interest of the observer in the social world. American pragmatism was the most explicit example of such a reorientation, but similar, often much more tension-ridden, discussions marked the European debates, a prominent example being Max Weber's methodological writings. Key concepts once taken as self-evident were now scrutinized and reinterpreted – both collective terms, such as "society," "state," "people," and "religion," and those referring to human beings and their sense of continuous existence, terms such as "individual," "action," "self," "psyche." Certainty about these concepts was especially important, because in some form or other they were indispensable for theorizing the political order in terms of some stable relation between collective phenomena and individual human beings. Such epistemological and ontological questioning had repercussions for methodology. Statistical approaches, for instance, always rely on some assumptions about aggregates, mostly states, and their components, mostly individuals or households. If certainty about these concepts is shaken, the ground for any research methodology would appear to be shaken as well.[9]

As a consequence, the turn-of-the century approaches were more doubtful of the determinist course of human history than earlier social science had been, and also less persuaded that empirical observation gave direct insight into any laws of the social world. This uncertainty was expected to restrict the viability of social knowledge for policy and planning purposes. The earlier call had been for better knowledge that would lead to better action; according to such a conception, action based on uncertain knowledge would entail uncertain outcomes. And indeed, the turn-of-the century debates were marked by a chasm between social philosophizing that tried to live up to these insights, on the one hand, and empirical research that continued and even expanded, remaining rather unconcerned by such issues, on the other. Through the early decades of the twentieth century, however, novel conceptions of the relation between knowledge and action were proposed that turned out to lend themselves to a greater involvement in policy. The world political crisis of the First World War had the effect of giving such considerations a sense of urgency and of focusing the debates.

SOCIAL PLANNING IN MASS SOCIETY: THE FIRST ATTEMPT

The First World War was, among many other things, a giant experiment in social planning. Its unexpectedly long duration and the similarly unforeseen involvement of large segments of the population, as well as trade interruptions

[9] Alain Desrosières, *La politique des grands nombres: Histoire de la raison statistique* (Paris: La Découverte, 1993); Peter Wagner, "Sociology and Contingency: Historicizing Epistemology," *Social Science Information*, 34 (1995), 179–204.

and supply shortages, led to increasing government efforts to direct economic and social activities, mostly with, but sometimes without, the consent of employers, unions, and other social groups. At the end of the war, a widespread impression held such planning to be superior to liberal and market forms of regulation. Direct conclusions had been drawn in the aftermath of the Bolshevik Revolution and, less forcefully, by some bureaucrats in the first Weimar administrations, but the impact of World War I was felt throughout the Western world. This enthusiasm for state planning receded as liberal market democracy appeared to recover during the 1920s, but the planning mood revived again after the world economic crisis of 1929. Social science was now directly involved in such planning moves.[10]

The Austrian economist Rudolf Hilferding, who contributed to Austromarxism but who was active in German social democracy during the Weimar period, had developed his concept of "organized capitalism" even before the war. The notion implied that capitalism was organizable, and that such organization could be pursued from a reformist perspective. Related ideas were developed by a group of broadly left-wing economists at the Kiel Institute for the World Economy during the 1920s. Some members of the group were also involved in the economic planning debates of the early 1930s, when such ideas were found attractive by a number of economists and policy makers ranging from American New Deal liberals to Soviet planners. A testimony to the wide range of that international debate was the Amsterdam World Economic and Social Congress of 1932.[11] The founding of economic survey institutes in many countries, including the United States, France, and Germany, during the interwar period supplied empirical information that could lend itself, potentially, to planning intervention.

While much of that debate was confined to economic terms, some broader conceptualizations of social planning were also proposed. The most comprehensive was probably the *Plan de travail*, developed by the Belgian psychologist and socialist Hendrik de Man, a professor at Frankfurt University from 1929 to 1933. Presented to the Belgian Workers' Party in 1933, the plan was widely debated in Belgium, the Netherlands, and France, where it supported a reformist reorientation of the socialist parties. De Man's case demonstrates particularly clearly the specific reformist-socialist inspiration for social planning, and also its sociophilosophical basis. De Man was well versed in Marxism and social democracy. Yet he gave up social determinism during his German years in favor of a psychologically mediated voluntarism that made reaching socialism a matter of "will and representation" rather than

[10] Peter Wagner, *Sozialwissenschaften und Staat: Frankreich, Italien, Deutschland, 1870–1980* (Frankfurt am Main: Campus, 1990), chap. 9.

[11] Matthias von Bergen, *Vor dem Keynesianismus: Die Planwirtschaftsdebatte der frühen dreissiger Jahre im Kontext der "organisierten Moderne"* (Berlin: WZB, 1995); Guy Alchon, *The Invisible Hand of Planning: Capitalism, Social Science, and the State in the 1920s* (Princeton, N.J.: Princeton University Press, 1985).

of developing material forces.[12] In this respect he was in agreement with much other social theory. He did not, however, dwell on the possible conclusion that the predictability of social life had decreased, but emphasized instead that the malleability of the social world increased once determinism was abandoned.

A related development can be discerned in John Maynard Keynes's economic thinking. Early in the 1920s, Keynes was already emphasizing the relevance of uncertainty in economic life, in defiance of straightforward neo-classical assumptions about complete information and rational behavior. His General Theory, while formalized to a considerable degree, relied at crucial points on the identification of "factors" in economic life that were sociohistorically or psychologically variable, and so required specific identification rather than general deduction. In France, the Durkheimian economic sociologist Maurice Halbwachs argued for a connection between theorizing about the conditions for social order and empirical observation of economic life, an approach that showed some affinity of principle to the Keynesian approach. Halbwachs supported the creation of a French institute for economic surveys, founded as the Institut de Conjoncture in 1938, not least with a view to specifying the conditions for effective political intervention.[13]

This critique of determinism and emphasis on the feasibility of goal-directed, planned political action was joined to a fundamental and critical epistemological presupposition that the social world is, in important respects, not found and discovered but made and invented. It constitutes one strand of the interwar planning debate in the social sciences. Representatives of the other strand severed their ties with turn-of-the-century social theory entirely and put social science on completely new – some would say "modern" – foundations. The key element here is the "scientific worldview" of the Vienna Circle and the unified science movement, which created an unprecedented linkage between positivist philosophy, socialist thought, and modern sociological research – or what has also been called a blend of Comte, Marx, and behaviorism.[14] In an intellectual and political context of doubt and uncertainty, its proponents hoped to reaffirm the social project of modernity by reintroducing sociology as a science with an epistemological standing equal to that of the natural sciences. They made it part of the very same undertaking, the generation of reliable knowledge that lent itself to prediction and planning.

In both intellectual and political terms, the sources of this approach can be traced to the particular situation of turn-of-the-century and interwar Austria, and particularly to Vienna, the capital and major city of the Habsburg

[12] Hendrik de Man, *Zur Psychologie des Sozialismus* (Jena: E. Diedrichs, 1926).

[13] Alain Desrosières, "Histoire de formes: statistiques et sciences sociales avant 1940," *Revue française de sociologie*, 26 (1985), 307.

[14] John Torrance, "The Emergence of Sociology in Austria, 1885–1935," *Archives européennes de sociologie*, 17 (1976), 459; Laurence D. Smith, *Behaviorism and Logical Positivism* (Stanford, Calif.: Stanford University Press, 1986).

Empire as well as of the new Austrian Republic after the First World War.[15] The Austrian socialists and "Austromarxists," who had been confined to theorizing during the stagnant years of the Habsburg Empire, gained and held a comfortable electoral majority in Vienna during the Republic and transformed it into an experimental space for social planning. One of the leading activists and theoreticians of social planning was Otto Neurath, the author of *The Scientific World-View, Socialism and Logical Empiricism* as well as of *Empirical Sociology*. A young member of the same movement was the mathematician Paul Felix Lazarsfeld. The examples of Neurath and Lazarsfeld can serve to demonstrate the particular connection between politics and social philosophy during this period.

Neurath's conviction that scientific rationality and political improvement went hand in hand was conditioned by his perception of himself and others as united in a struggle against both metaphysical worldviews and illegitimate power, a similarly inseparable couple. He witnessed this scientific-political rationality at work in the war economy and participated in the attempts of the postwar revolutionary governments in Saxony and Bavaria to socialize the means of production. Expelled from Germany, he became a leading re-former in Vienna, trying to put rational schemes to work in city politics. In writings on planning, statistics, and socialism, he elaborated the view that individual reason, once given the space to develop, becomes essentially identical to scientific reason. As a consequence, "social technology" could be developed on the basis of an empirical and positive sociology that re-jected all metaphysics, and the sociologist could become a "social engineer." Crude as that view may now appear, Neurath saw his politics as being fully in line with the most rational, and thus the most advanced, science and philosophy of science of his time – the positivism of the Vienna Circle, to which he contributed. As one observer put it, we may see Neurath's relation to Wittgenstein as broadly similar to Hans Eisler's relation to Arnold Schoenberg.[16]

The young Lazarsfeld, who also had clear socialist leanings, was drawn into statistical work at the Psychological Institute of the University of Vienna by Charlotte and Karl Bühler, who were involved in research for the city administration. He founded the Research Unit of Economic Psychology at Vienna University, which acquired research contracts both from the Austrian Radio Company and from the Frankfurt Institute for Social Research. In this way, Lazarsfeld inaugurated the institutional and operational model of social research for which he and the Bureau of Applied Social Research at Columbia

---

[15] Carl Schorske, *Fin-de-siècle Vienna: Politics and Culture* (London: Weidenfeld and Nicolson, 1980); Allan Janik and Stephen Toulmin, *Wittgenstein's Vienna* (London: Weidenfeld and Nicolson, 1973); Michael Pollak, *Vienne 1900: Une identité blessée* (Paris: Gallimard, 1992); Helmut Gruber, *Red Vienna* (New York: Oxford University Press, 1991).

[16] Elisabeth Nemeth, *Otto Neurath und der Wiener Kreis: Revolutionäre Wissenschaftlichkeit als Anspruch* (Frankfurt am Main: Campus, 1981), 77.

University in New York would later become famous. This was the beginning of survey research on commission at university-based but commercially operated research institutes, a model that soon spread from the United States to Europe and then to other parts of the world after the Second World War. Called "administrative research" by Lazarsfeld himself, this research served the planning purposes of the funder without being involved in setting the objectives.

Lazarsfeld's intellectual biography illustrates again the simultaneously political and epistemological nature of the transition from classical social theory to applied social research.[17] Close to Austromarxism himself, Lazarsfeld experienced the difficulties of putting reformist ideas into practice in "Red Vienna." It was in particular the conception of a preconceived unity between the political actors and the people for whom reformist policies were developed – a socialist version of Enlightenment ideas – that proved illusory. In political practice, no such harmonious alignment arose. Indeed, the will of the people was not even known to the policy makers who claimed to serve them. Empirical social research was designed as a way of transmitting knowledge from the people to the elites, always bearing in mind that the kind of knowledge that was called for was shaped by elite views of political feasibility, as implied by the conditions of the research contract. After his move to the United States, Lazarsfeld regretfully accepted the unavoidable decoupling of his political motivations from a research conception that remained otherwise unchanged.

This empirical positivist sociology was one specific, and highly articulate, response to the increased demand for social knowledge brought about by the crisis of liberalism. It found a number of other, much more loosely formulated expressions elsewhere. In the Netherlands, social planning emerged in connection with the draining of the Zuiderzee polders, wetlands that could be made usable for agriculture and settlements. Dutch sociology, known as "sociography" between the wars, had developed a very empirical and applied orientation. H. N. ter Veen, one of its main spokespersons at the time, elaborated proposals for the Zuiderzee colonization and used them to demonstrate the possibilities of sociologically guided social planning. In the United States, a federal report on Recent Social Trends, commissioned by President Hoover and delivered by William F. Ogburn in 1929, was a major example of a social-statistical attempt to grasp the main lines of social development as a guide to government action. And the New Deal, with the foundation of the National Resources Planning Board and the longer-lived Tennessee Valley Authority, made attempts to base planning on social knowledge.

From the late 1920s onward, we can recognize the contours of an empirical positivist social science, oriented toward application and developing

[17] Michael Pollak, "Paul F. Lazarsfeld – fondateur d'une multinationale scientifique," *Actes de la recherche en sciences sociales*, no. 25 (1979), 45–59.

in special institutions, which subsequently shaped the image of the social sciences during the second postwar era. This social science liberated itself from the doubts of the "classical period." Its particular form of empirical social research circumvented the problem of relating the individual to a mass society. Doubts about epistemological and conceptual issues could not be entirely removed, but they could, it was supposed, be contained by starting from the most secure elements one could find, the empirical observation and collection of data on the preferences and behaviors of individual human beings. Conclusions relating to the wider realm of society and politics were reached by aggregation of such data; and the organizing questions were derived from policy needs for "social control." Thus, a "soft" behaviorism became aligned with a similarly "soft" pragmatism.[18]

Such behavioral social research recognizes individual human beings and their doings as a methodological starting point. It mostly rejects any prior assumptions about behaviors as "unfounded" or, in Vienna Circle terminology, "metaphysical." Thus it may be seen as drawing one crucial premise, and not an unproblematical one, from a basic tenet of political modernity – the primacy of individual autonomy.[19] However, this is a very different kind of individualism from that assumed in either liberal political theory or neoclassical economics, where individual rationalities are postulated. In behavioral social research, social regularities can be discovered only through the study of the utterances and behaviors of individuals, and cannot in any way be derived. But after such regularities are identified, they may be reshaped by altering the possibilities of action – for example, in terms of the products advertised or the political party programs advanced.

Neoclassical economics is a post-Enlightenment doctrine – a doctrine of liberal modernity – in the sense that it assumes the self-regulation of a society of reason-endowed (read: rationalistic) individuals. Behavioral social research is a postliberal technology – a tool of the organizers of modernity, of the planners of "modern" society – in the sense that it constructs individuals in order to make them amenable to policy action and planning. The basic cognitive move of this approach was to isolate individuals from each other, ignore whatever social relations they may have, and then to counterpose this atomized mass to the state. "The underlying assumption of social statistics and social research . . . is that singular human beings can be treated as externally related individuals. The State and its individuals are notions from which both social statistics and social research derive." Here sociology echoed Balzac's novelistic social analysis: "Society isolates everyone, the

[18] Ross, *Origins of American Social Science*, chap. 9.
[19] Wagner, "Certainty and Order"; Judith N. Shklar, "Alexander Hamilton and the Language of Political Science," in *The Languages of Political Theory in Early-Modern Europe*, ed. Anthony Pagden (Cambridge: Cambridge University Press, 1987), p. 346.

better to dominate them, divides everything up to weaken it. It reigns over the units, over numerical figures."[20]

PLANNING AND FREEDOM: THE SOCIAL PHILOSOPHY OF PLANNING

While the implementation of this form of social knowledge and planning was yet far from certain, doubts were already voiced as to whether the longing for planning and organization, as well as the forms of social knowledge that accompanied it, were indeed compatible with liberal democracy. Especially in light of experiences with totalitarian regimes, the earliest and strongest promoters of social planning, the model of a direct transmission between masses and elites, mediated by empirical social science, was unpersuasive. Reservations were particularly strong in the United States, where the values of individualism were firmly rooted, and where the necessity for planning was accepted most reluctantly.

By the mid-1930s, whatever enthusiasm there had been about entering a new age had waned, giving way to a more reflective debate about the social and political implications of the move toward planning. The meeting of the American Sociological Society in 1935, devoted to the theme of the Human Side of Social Planning, provided an occasion to review recent developments in the political philosophy of pragmatism and to rethink "social control." There was widespread agreement that a more interventionist state had emerged since the First World War, one heavily involved in planning. Sociologists debated the issue of how a commitment to autonomy and democracy could nevertheless be maintained. Ernest W. Burgess could not resist the conclusion of a Carnegie Foundation report on schools that "American society during the past hundred years has been moving from an individualistic and frontier economy to a collective and social economy." He insisted, however, that any planning in the United States had "to accord with mores of indvidualism, democracy and humanitarianism," the moral bases of American society.[21] By contrast, William F. Ogburn, in a much more technocratic vein, asserted that "some loss of liberty under the predicted conditions is to be expected, for such is the implication of any high degree of organization."[22] Lewis Lorwin put the recent developments into the long-term

---

[20] Dag Österberg, *Metasociology: An Inquiry into the Origins and Validity of Social Thought* (Oslo: Norwegian University Press, 1988), p. 44; Desrosières, *La politique des grands nombres*; Wagner, *Sociology of Modernity*, chap. 7; Honoré de Balzac, *Le curé de village* (1841), as quoted in Gerd Gigerenzer, Zeno Swijtink, Theodore Porter, Lorraine Daston, John Beatty, and Lorenz Krüger, *The Empire of Chance* (Cambridge: Cambridge University Press, 1989), chap. 2.

[21] Ernest W. Burgess, "Social Planning and the Mores," in *Human Side of Social Planning: Selected Papers from the Proceedings of the American Sociological Society 1935*, ed. Ernest W. Burgess and Herbert Blumer (Chicago: American Sociological Society, 1935), p. 33.

[22] William F. Ogburn, "Man and His Institutions," in *Human Side*, ed. Burgess and Blumer, p. 37.

perspective of a "continuous enlargement of organized groups through which the individual has to act in order to shape public policy." But he saw those changes less as an unequivocal loss or gain than as a transformation of the political issue: "Not regimentation versus freedom, therefore, but social control versus unlimited economic power of individuals and minorities is the issue." In theoretical terms, he identified a fundamental transformation of the concept of rights:

> As planning develops it will shift emphasis in our political thinking from the idea of formal rights to the concept of 'real rights' based on capacity; from the notion of the state as a protector of property to that of a leader in the utilisation of our natural and economic resources; from the concept of law as a balancing of individual rights to that of a process of adjusting social relations; from theories of atomistic individualism to those of social solidarity and cooperative action; and from reliance on an assumed metaphysical benevolence of self-interest to a demonstrable hypothesis of the potentialities of scientific guidance of economic and social forces.[23]

By framing the issue as an historical shift between formal rights and substantive commitments, Lorwin captured a basic ambivalence of liberal political modernity that, though often much more implicitly, had characterized sociopolitical debates since the French Revolution, and that is currently again in the forefront of discussion.[24]

While the move towards social planning did not develop the same momentum in the United States that it did in some European countries, and met more principled criticism because of the American tradition of individualism, the emergency situation of the Second World War saw the social sciences nevertheless drawn into planning activities of a large scale, not least with a view to improving the efficacy of military operations and to limiting or mitigating their social implications and "side effects." The volume *The Policy Sciences* (1951), edited by Harold Lasswell and Daniel Lerner, gives testimony both to the involvement of social scientists from all disciplines in war planning and to a willingness to reconsider the possibility of using social knowledge for planning and policy purposes in the light of those – generally deemed to be successful – experiences.

In Europe, the most profound reflections on social planning, in terms both of the underlying conception of social knowledge and of the related political philosophy, were expressed by Karl Mannheim. In his early works, while

[23] Lewis L. Lorwin, "Planning in a Democracy," in *Human Side*, ed. Burgess and Blumer, pp. 42, 44, 47–8.
[24] William H. Sewell, Jr., "Artisans, Factory Workers, and the Formation of the French Working Class, 1789–1848," in *Working-Class Formation: Nineteenth-Century Patterns in Europe and the United States*, ed. Ira Katznelson and Aristide R. Zolberg (Princeton, N.J.: Princeton University Press, 1986), p. 60; Jacques Donzelot, "The Mobilization of Society," in *The Foucault Effect: Studies in Governmentality*, ed. Graham Burchell, Colin Gordon, and Peter Miller (Chicago: University of Chicago Press, 1991), p. 171.

living on the European Continent, he developed a sociology of knowledge and a theory of the role of intellectuals that aimed at a principled reformulation of those issues for an emerging mass society. The earliest version of *Man and Society in an Age of Reconstruction*, which appeared in German in 1935, characterized the major transformation of Western societies in familiar terms as a "crisis of liberalism and democracy" in a highly organized mass society. To rely on laissez faire would lead necessarily to "maladjustment." When the considerably enlarged English version appeared in 1940, Mannheim thought he had sufficient experience with planning under democracy while living in English exile to conclude that "freedom and planning" might possibly be made compatible through some "synthesis of democratic planning."[25]

A SYNTHESIS OF SORTS: THE SECOND ATTEMPT AT SOCIAL PLANNING

After the war, the operating modes of those liberal democracies that appeared to have most successfully transformed themselves into inclusive mass societies – above all the United States, the United Kingdom, and Sweden – were (re-)imported to continental Europe. "Democratic planning" and "modern social science" were two key elements in their modes of operation, and strong efforts were made to implant them firmly on continental soil.[26]

The United Nations' cultural organization, UNESCO, and U.S.-based private foundations were active in promoting a social science oriented toward the empirical study of contemporary policy problems with a view to applications. Paul Lazarsfeld himself was involved in building social research institutes in Austria. *The Policy Sciences* was translated into French by leftist reformers and published with a Preface by Raymond Aron, who, having earlier made arguments for a more "inductive" rather than a philosophical sociology, thus lent his reputation to the development of an applied and planning-oriented social science.

Key areas of the social sciences acquired a cognitive affinity to social planning. In economics, Keynesian theorizing stimulated research on those economic indicators that were seen to be the key variables of macroeconomic steering. In sociology, theories of modernization and development were elaborated on the basis of functionalism and systems theory and were "applied" to those societies allegedly in need of development. Quantitative social research flourished. Although academic sociology, economics, and political

---

[25] Karl Mannheim, *Man and Society in an Age of Reconstruction* (London: Routledge, 1940); *Diagnosis of Our Time: Wartime Essays of a Sociologist* (London: Routledge, 1943); *Freedom, Power and Democratic Planning* (London: Routledge, 1951); Colin Loader, *The Intellectual Development of Karl Mannheim: Culture, Politics, and Planning* (Cambridge: Cambridge University Press, 1985); David Kettler and Volker Meja, *Karl Mannheim and the Crisis of Liberalism: The Secret of These New Times* (New Brunswick, N.J.: Transaction, 1995).

[26] Wagner, *Sozialwissenschaften*, pt. 4.

science also took a "quantitative turn," this kind of social knowledge was increasingly produced on demand on behalf of government agencies, business organizations, and political parties with a view to their own policy and organizational planning needs. Specific methodologies of policy analysis, such as cost-benefit analysis and planning, programming and budgeting systems (PPBS), were developed.

Such efforts also met resistance. Theodor W. Adorno, for instance, criticized the transformation of sociology into statistics and administrative science as the knowledge form of an "administered society." Hannah Arendt's comprehensive study of *The Human Condition* included a fundamental critique of statistics and behaviorism as undermining the conceptualization and understanding of human action. Between the 1950s and the 1970s, however, American and European social science became increasingly policy- and planning-oriented. This perspective can be characterized by its substantive focus on issues of policy, strategy, and administration, and its conceptual focus on the functioning of goal-oriented organizations, both public and private, and their leaders.

In terms of the philosophy of social science, Karl Popper's neo-positivism offered a softer version of the interwar proposals for what was often a quite technocratic social science. His conception of "empirical social technology" that could be used in "piecemeal social experiments" was explicitly based on "trial and error" and directed against "Utopian social engineering." While Popper offered a new linkage of epistemology and politics, his approach showed much greater modesty and hesitancy than some interwar proposals.[27] Popper and Adorno debated their views in 1961 at a meeting of German sociologists. By that time, the reflective social philosophizing of both had been overtaken by the flourishing of empirical, often application-oriented, social research.

In the public realm, the new shift toward policy-oriented social sciences was brought about by reformist discourse coalitions between the younger generation of social scientists and modernization-oriented politicians aspiring to power. In the United States during the 1960s, the reformist drive of the Kennedy administration was translated into the Great Society and War on Poverty programs during the Johnson administration. While these social planning initiatives were soft in the sense of being based on incentives and encouragement rather than on command and restriction, they represented a major program of planned social change.[28]

---

[27] Karl R. Popper, *The Open Society and Its Enemies* (London: Routledge, 1945), pp. 162–3, 291.

[28] Gareth Davies, *From Opportunity to Entitlement: The Transformation and Decline of Great Society Liberalism* (Lawrence: University Press of Kansas, 1996); Lance deHaven-Smith, *Philosophical Critiques of Policy Analysis: Lindblom, Habermas and the Great Society* (Gainesville: University of Florida Press, 1988); Henry J. Aaron, *The Great Society in Perspective* (Washington, D.C.: Brookings, 1978); Herman van Gunsteren, *The Quest of Control: A Critique of the Rational-Central-Rule Approach in Public Affairs* (London: Wiley, 1976).

Similar discourse coalitions formed in many European countries during the 1960s and 1970s, often also in the context of governments moving toward more reform-minded majorities. In many respects, these coalitions were not unlike those that socialist-leaning scholars and reformist administrators had entered into after the First World War. While the more recent alliances were somewhat tempered by recent historical experience, they had much more sustained effects, both on the social sciences and on policy making.

Again, prevailing convictions held that potentially violent strife and conflict, in which one group could gain only at the expense of others, could be transformed into cooperative positive-sum games with the help of social scientific knowledge.[29] In comparison to the earlier effort, this second broad movement for social planning was shaped by the historical experience of totalitarianism, whose recurrence it aimed to avoid by emphasizing democratic consensus. This would prevent planning from becoming the enemy of freedom.[30] On the other hand, the new generation of planners had a markedly higher opinion of advances in social science than had the first planning movement. Intellectual progress, especially methodology, was deemed to have allowed a much firmer cognitive grasp of social reality. Together with the apparent "end of ideology," this meant that social science–based social planning appeared to be ultimately achievable.

Looking back from the early 1980s to the 1960s, a French research administrator, Robert Fraisse, spoke of a pervasive "optimism with regard to the exhaustive cognitive mastery of society." He continued:

> This research is led to endow itself with an aura of the all-comprehensive, owing to the functional use which administration wants to make of its results – and without doubt owing also to the optimism which gives responsible administrators the idea of a strong and continuous growth [of knowledge]. One speaks in terms of knowledge gaps, which are now to be closed. In a certain sense, the objective is the exhaustion of the real, as is evidenced in the requests for proposals of the time which underline the relevance of comprehensive inquiries about consumption, income, life-styles; about regional and national economic accounting; about global modeling of public action systems etc."[31]

This optimism about planning reached its zenith when it made the social sciences themselves one of its objects. During the 1970s, the Organization for Economic Cooperation and Development (OECD) proposed a "social

[29] Pierre Massé, *Le Plan ou l'Anti-Hasard* (Paris: Gallimard, 1965), p. 18; Pierre Massé, *Autocritique des années soixante par un Commissaire au Plan* (Bulletin de l'Institut d'histoire du temps présent, Supplément no. 1, série "Politique économique," no. 1) (Paris: l'Institut d'histoire du temps présent, 1981), p. 38.

[30] Firmin Oulès, *Economic Planning and Democracy* (Harmondsworth: Penguin, 1966).

[31] Robert Fraisse, "Les sciences sociales: utilisation, dépendance, autonomie," *Sociologie du Travail*, 23 (1981), 372.

science policy" intended to optimize its contributions to policy making. OECD also commissioned country analyses of the state of the social sciences in various countries (France, Norway, Finland, Japan) to detect deficiencies and to enhance their efficiency. Some observers spoke of an outright "planification of the social sciences."[32]

AFTER THE PLANNING EUPHORIA

From the mid-1970s onward, it became increasingly evident that social planning had fallen short of these high expectations. The master example of the crisis of planning was the arrival of unmanageable economic downturns, with the simultaneous occurrence of rising inflation and unemployment that served to discredit Keynesianism, which had seemed to rule out such "stagflation." In economics, this experience led to intellectual shifts toward monetarist and "supply-side" approaches, with a much-reduced emphasis on public intervention and a return to market regulation. A similar reorientation also occurred in the other policy- and planning-oriented social sciences. In part as a response to the results of applying social science knowledge to policy practice, attention was directed to such apparently novel phenomena as "unintended consequences," "*effets pervers*," and "implementation problems." Social reality proved recalcitrant to planned intervention.

Though the precise relation between these two phenomena needs further exploration, the crisis of the policy sciences seems to have been deepened by a turn away from objectivist epistemology and by an overemphasis on quantitative methodology in the academic social sciences.[33] The "interpretive turn" or, more broadly, the "linguistic turn" in the human sciences has had strong, though unequal, effects on the sciences devoted to the study of contemporary society. Emphasis on the linguistic constitution of the social world and on the interpretive openness of social representations has brought the social sciences back to a period of epistemological, ontological, and methodological reconsideration that shows many parallels to the "classical era" at the beginning of the twentieth century.

A century ago, such doubts were temporarily resolved by a faith that the social world was indeterminate and for that reason malleable, allied to the search for an actor powerful enough to transform the social world according to a conscious will. This combination made possible the emergence of the planning-oriented social sciences. Under current conditions, the argument for the indeterminateness, or contingency, of the social world is possibly even more strongly voiced than it was a century ago. But the belief in the

---

[32] Michael Pollak, "La planification des sciences sociales," *Actes de la recherche en sciences sociales*, no. 2/3 (1976), 105–21.
[33] Frank Fischer, *Technocracy and the Politics of Expertise* (Newbury Park, Calif.: Sage, 1990); John S. Dryzek and Douglas Torgerson, eds., *Democracy and the Policy Sciences* (Dordrecht: Kluwer, 1993).

existence of a strong actor appears to have been much more decisively shaken. For the time being, the double reorientation in both the planning-oriented and the academic social sciences has entailed the abandonment of the idea of comprehensive social planning. With it, the strong figure of the state as an all-pervasive power center, and of the intellectual committed to universalist values and the search for generally valid knowledge, have almost disappeared from public debate. If one looks for conceptions that may have replaced these strong views, the only contender seems to be the neoliberal and rationalist belief in the optimization of human interaction without a conscious planning subject. A version of such thinking, with almost opposite starting assumptions, can be found in theories of autopoietic systems. Over the long run, however, "weaker" versions of the traditional concepts may emerge, with the state as a "moderator" and, as Bauman suggests, the intellectual as an "interpreter."[34]

Nevertheless, the current lack of fully convincing sociological representations of society and the absence of societal planning do not imply that cognitive representations of society are no longer put forward or that planning has been abandoned. On the contrary, business and other organizations depend on strategic planning, for which they commission expertise on an unprecedented scale. The current proliferation of market assessments and opinion polling attest to the continued vitality of planning. The relative weakening of nation-states and national markets has created this growing need for planning-relevant knowledge. And in the course of such knowledge production, representations of the social world are constantly being produced. However, such plans are less comprehensive and coordinated than the social planning of the 1960s, and they are mostly produced in contexts lacking the commitment to public validation that, despite all criticism and the ongoing transformations of those institutions, still characterizes universities and academia. It is not the idea of planning that is currently in crisis but the possibility of achieving comprehensive social planning under conditions of public exposure and validation.

[34] Jean-François Lyotard, *Le tombeau de l'intellectuel, et autres papiers* (Paris: Galilée, 1984); Zygmunt Bauman, *Legislators and Interpreters: On Modernity, Post-Modernity and Intellectuals* (Cambridge: Polity, 1987).

35

SOCIAL WELFARE

Ellen Fitzpatrick

The persistence of poverty amid economic growth has provoked debate about
the state's responsibility for social welfare since the beginnings of industrialization. Claiming that empirical study was a sine qua non for effective reform,
social investigators played a leading role in formulating policies toward the
poor. Later, academic social scientists developed theoretical models, statistical data, and a language for defining social problems that were placed in
the service of the state and of social and political movements. For better or
worse, the ideas of modern social scientists have helped to write the history
of twentieth century social welfare policy and affected the life fortunes of
millions of people.

SYSTEMATIZING SOCIAL INQUIRY

The relationship between social inquiry and social welfare policy is as old as
efforts to redress human misery. In both Europe and America, notions of the
"deserving" and "undeserving" poor surfaced in some of the earliest measures
devised to lessen the scourge of indigence, such as the Elizabethan poor laws
and early American strategies for "bidding out" and "warning out" the poor.
Data collection, analysis, and the regulation of those who received assistance
were mandated by the logic of policies that distinguished between those who
truly needed help and malingerers, and by a desire always to minimize the
burden of dependency on the state.[1]

---

[1] Theda Skocpol, "Government Structures, Social Science, and the Development of Economic and So-
cial Policies," in *Social Science Research and Government*, ed. Martin Bulmer (Cambridge: Cambridge
University Press, 1987), pp. 40–50; Walter Trattner, *From Poor Law to Welfare State* (New York: Free
Press, 1984); James Leiby, *A History of Social Welfare and Social Work in the United States* (New
York: Columbia University Press, 1978); James Patterson, *America's Struggle against Poverty, 1900–1980*
(Cambridge, Mass.: Harvard University Press, 1981); Michael Katz, *Poverty and Policy in American
History* (New York: Academic Press, 1983); Dietrich Rueschemeyer and Theda Skocpol, eds., *States,*

During the eighteenth and nineteenth centuries, classical political economy bolstered the long-held prejudice that poverty resulted from moral failings. In this view, not only did individuals pursuing their private interests require no assistance from the state, they flourished only within a system of natural liberty. According to Thomas Malthus, poor laws undermined the necessary recognition that the poor "are themselves the cause of their own poverty." In England, opponents of labor and factory legislation invoked classical political economy to justify disregard for the fate of the working class and the poor. They met a powerful challenge as early as the 1840s from Friedrich Engels's examination of the condition of the working class in England and from Karl Marx and Engels's call for revolutionary change in *The Communist Manifesto* (1848). Liberal movements to enlist the state in ameliorating poverty, and thereby to avoid revolution, also stressed social inquiry in achieving that aim.[2]

To uncover the nature and extent of deprivation, investigators of working-class communities counted the poor and detailed their living habits at the local level. Such efforts culminated in the extraordinarily detailed social surveys conducted by Charles Booth and his successors during the 1880s. They not only served as a method of data collection but also realized important political purposes: As local and national governments in Britain abandoned piecemeal their earlier laissez faire stance, surveys monitored the poor, provided fodder for liberal reform constituencies, and laid an intellectual foundation for the development of systems of social provision. Pursued largely *outside* of universities and hence free of any pretense to pure research, such studies permitted British investigators to carry lessons learned in the field into their roles as British civil servants and reform activists.[3]

In Germany, social investigation and government policy followed separate tracks. Beginning in the 1860s, young German social scientists urged attention to workers' struggles, and in 1872, academics, other professionals, trade unionists, and social activists formed the Verein für Sozialpolitik (Association for Social Policy) to conduct research and to prod the new Reich to improve the conditions of the German working class.

Social Knowledge, and the Origins of Modern Social Policies (Princeton, N.J.: Princeton University Press, 1996); Peter Flora and Arnold Heidenheimer, eds., *The Development of Welfare States in Europe and America* (New Brunswick, N.J.: Transaction, 1981).

[2] Sidney Fine, *Laissez Faire and the General Welfare State* (Ann Arbor: University of Michigan Press, 1967), pp. 5–10.

[3] Martin Bulmer, Kevin Bales, and Kathryn Kish Sklar, "The Social Survey in Historical Perspective," in *The Social Survey in Historical Perspective, 1880–1940*, ed. Martin Bulmer, Kevin Bales, and Kathryn Kish Sklar (Cambridge: Cambridge University Press, 1991), pp. 1–48; Martin Bulmer, "National Contexts for the Development of Social Policy Research: British and American Research on Poverty and Social Welfare Compared," in *Social Sciences and Modern States*, ed. Peter Wagner, Carol Hirschon Weiss, Björn Wittrock, and Helmut Wollman (Cambridge: Cambridge University Press, 1991), pp. 148–67; Eileen Yeo, "The Social Survey in Social Perspective, 1830–1930," in *Social Survey*, ed. Bulmer, Bales, and Sklar, pp. 49–65.

Action came under Chancellor Bismarck, who enacted sweeping social insurance measures during the 1880s – including health, accident, disability, and old age insurance – in an effort to forestall class conflict and to consolidate state power. Now less sympathetic toward the Reich and reluctant to continue an alliance with the state, the Verein, under the influence of members such as Max Weber, Werner Sombart, and Ferdinand Tönnies, sought to separate the study of social science from the imperatives of social policy. A similar rupture between the dictates of applied and pure research would occur later in the United States, though under different circumstances.[4]

In America, social provision remained largely private and local. Though widely admired and emulated in the United States, the British style of social investigation did not lead to the creation of national planning agencies or to broad programs of social insurance in late-nineteenth and early-twentieth-century America. A shared tradition of social research and investigators' best wishes yielded no single solution to social problems. Social, economic, and political forces determined the uses made of social scientific research by modern societies. In the United States, the absence of labor parties and resistance to the exercise of federal power curtailed policies of social provision.[5]

The Civil War played an early, if largely temporary, role in centralizing social welfare policy and advancing social inquiry. It produced the United States Sanitary Commission (1861), the first national public health program and one with an investigative and educative bent. It led to the creation of the Bureau of Refugees, Freedmen, and Abandoned Lands (1865), the first federal social welfare agency and one that compiled extensive records. It sought to meet the needs of the South's freed people. The war likewise resulted in the enactment of the Civil War pension system for union veterans and their dependents, the largest social spending program to that point in American history.[6]

Industrialization, urbanization, immigration, and the rise of modern research universities ultimately proved to be far more decisive in advancing social inquiry and social welfare policy. The impetus came partly from middle-class Americans alarmed at the new immigrants, and at urban poverty and political corruption. But workingmen's parties, labor organizations such as the Knights of Labor, trade unionists, socialists, and populist leaders also powerfully shaped public debate with attacks on political and economic inequality and demands for far-reaching reforms. Even the most moderate labor

[4] Irmela Gorges, "The Social Survey in Germany before 1933," and Bulmer, Bales, and Sklar, "The Social Survey in Historical Perspective," in *Social Survey*, ed. Bulmer, Bales, and Sklar, pp. 316–39, 16–17; Anthony Oberschall, *Empirical Social Research in Germany, 1848–1914* (The Hague: Mouton, 1965).
[5] Bulmer, "National Contexts"; Trattner, *From Poor Law*; Skocpol, "Government Structures."
[6] Trattner, *From Poor Law*, chap. 5; Theda Skocpol, *Protecting Soldiers and Mothers: The Political Origins of Social Policy in the United States* (Cambridge, Mass.: Harvard University Press, 1992); Phyllis Day, *A New History of Social Welfare* (New York: Prentice Hall, 1989), chap. 7; Ira Berlin, Barbara J. Fields, Steven F. Miller, Joseph P. Reidy, and Leslie S. Rowland, *Free at Last* (New York: New Press, 1992); Donald Nieman, *To Set the Law in Motion: The Freedmen's Bureau and the Legal Rights of Blacks, 1865 to 1868* (Millwood, N.Y.: KTO Press, 1979).

activists advanced concrete, and often expansive, legislative reforms that out-stripped the halting proposals of middle-class activists and that left out the painstaking oversight provisions endorsed by charity officials and elites.[7]

As early as the 1860s, middle-class reformers had sought more routinized, bureaucratic, and informed measures for addressing poor relief and other so-cial needs. The first state boards of charities, founded in the 1860s, attempted to coordinate state-funded public welfare institutions and to economize by rationalizing procedures and standardizing care. In undertaking research and preparing extensive reports, they made social inquiry an inherent feature of social welfare administration and joined fact finding tightly to the cause of efficient poor relief.[8]

The first clear institutional expression of the belief that effective social welfare policy depended upon the advance of social scientific knowledge was the founding of the American Social Science Association (ASSA) in 1865. Modeled on European groups such as Britain's Nation Association for the Promotion of Social Science, the ASSA joined academic social scientists, char-ity workers, and business and professional elites. Their program to advance knowledge, promote sound social legislation, prevent crime, and uphold public morality could be achieved, the ASSA insisted, only through data collection on social problems such as poverty, crime, intemperance, and prostitution.

This alliance of interests soon fractured. During the 1870s, state charity workers, intent on "practical" as opposed to abstract or theoretical mat-ters, formed their own professional organization. Academic social scientists withdrew to professional organizations as early as the 1880s, partly out of concern that their standing as intellectuals and scientific experts was com-promised by their role as advocates. Economists nonetheless continued to be concerned with ameliorating poverty. Richard T. Ely and other left-leaning ethical economists challenged laissez-faire in the charter statement of their American Economic Association (1883), and some entertained socialist ideals. By the 1890s, the political climate of radical protest and a growing insistence on orthodox professional standards within the academy had led many schol-ars to retreat. Still, a mix of historical-institutionalist and neoclassical theories continued to focus attention on the "realistic" forces that could modify eco-nomic law. Despite such institutional changes, considerable overlap between the realms of social investigation, social work, and social science remained.[9]

---

7 Leon Fink, *Workingmen's Democracy: The Knights of Labor and American Politics* (Urbana: University of Illinois Press, 1983); John L. Thomas, *Alternative America* (Cambridge, Mass.: Belknap Press, 1983); David Montgomery, *The Fall of the House of Labor* (Cambridge: Cambridge University Press, 1987).
8 Trattner, *From Poor Law*, chap. 5; Leiby, *History of Social Welfare*, chaps. 6–8.
9 Thomas Haskell, *The Emergence of American Social Science: The American Social Science Association and the Nineteenth-Century Crisis of Authority* (Urbana: University of Illinois Press, 1977); Dorothy Ross, *Origins of American Social Science* (Cambridge: Cambridge University Press, 1991), chap. 6 and passim; Mary Furner, *Advocacy and Objectivity: A Crisis in the Professionalization of American Social Science, 1865–1905* (Lexington: University of Kentucky Press, 1975); James T. Kloppenberg, *Uncertain*

SOCIAL WORK AS SOCIAL SCIENCE

The interconnections were especially evident in England's and America's Charity Organization Societies (COS). United in their wish to coordinate each nation's flourishing network of state and local charitable organizations, the Charity Organization Societies sent cadres of agents, many of them women, into working-class communities, where they sought to determine "scientifically" who deserved poor relief. In American cities alone, there were over 100 such charity organizations by 1900.

"Science," to these charity workers, meant the attempt to systematize knowledge by keeping detailed records of the poor in central registries – who had sought aid, how much they had received (and from whom), and who was potentially "redeemable." Personal observation and interviews with relief applicants were also essential to the social workers' conception of scientific charity. Finally, research and investigation into the causes of poverty, both in the aggregate and in the case of specific individuals, formed a cornerstone of the entire charity society edifice.

Despite these intentions, or perhaps because of them, the Charity Organization Societies rarely moved beyond a conception of poverty and social welfare that focused on individuals. The tendency to ascribe personal culpability for destitution to moral failings such as sloth, excessive alcohol consumption, drug abuse, and a general inaptitude for thrift and clean living was widespread. In England, the majority report of the 1909 Royal Commission on the Poor Laws, which represented the thinking of many COS members, reflected considerable resistance to the broad state welfare measures proposed by the former COS worker and Fabian Socialist Beatrice Webb in her minority finding. British charity workers found support for their views in the works of Herbert Spencer, who invoked natural law to expand upon and justify his hostility to state action. While Spencer argued that public poor relief simply encouraged the unfit to survive and even thrive, he allowed private aid to the impoverished as long as it did nothing to enable the most wretched of society to multiply. Even influential British COS officials such as Helen Bosanquet, who placed great emphasis on compassion and education, engaged the debate over poverty in terms that never escaped Spencer's dire predictions or the deeply moralistic Victorian conceptions of indigence. More importantly, even the humanitarian commitments of the Charity Organization Societies could not be separated from the dominant social and political purposes they were intended to serve – most notably, the goal of minimizing dependency.[10]

In America, many charity organizations echoed these views. Loathe to contribute to the proliferation of relief systems and viewing themselves as above

---

Victory: Social Democracy and Progressivism (New York: Oxford University Press, 1986); Fine, *Laissez Faire and the General Welfare State*.

[10] Jane Lewis, "The Place of Social Investigation, Social Theory, and Social Work in the Approach to Late Victorian and Edwardian Social Problems," in *Social Survey*, ed. Bulmer, Bales, and Sklar, pp. 148–69; Leiby, *History of Social Welfare*, chap. 8.

almsgiving, they established referral agencies and refused to grant direct aid. Over time, however, many American charity societies began to dispense direct cash assistance to the needy. Nonetheless, when early-twentieth-century American reformers attempted to institutionalize and nationalize social provision through programs such as mothers' pensions, some influential charity officials balked. Among them was Mary Richmond, general secretary of the Baltimore Charity Organization and the founder of the modern casework method. Richmond considered federal and state social welfare programs to be a potential trough of corruption, a likely means of nourishing crippling dependence on government support, and an inadequate substitute for individualized casework with needy clients.

She was far from alone in such sentiments, as the slow progress of social welfare spending in the United States made apparent. Even as a growing body of literature, much of it compiled by charity workers, documented the social and economic roots of poverty, many opposed the enactment of broad social welfare measures. Nothing in the work or the orientation of the Charity Organization Societies, focused as they were on specific cases of indigence rather than on poverty in the aggregate, challenged that resistance. The fusion of social investigation and social work did not in this case provide either an intellectual or a political foundation for generous systems of social provision. The existence of a parliamentary political system, a Labor Party, and active administrative structures in the United Kingdom created greater possibilities for influencing government policy and enacting broad systems of social insurance. This was also the case in Scandinavia, where the existence of government statistical bureaus was reinforced by elite interest in expansive state action, political engagement among academic economists, and a national receptivity to German models of social insurance.[11]

The role of women in American and European charitable organizations deserves special comment, for work in such groups helped to cement women's domination of the field of social work. During the nineteenth century, middle-class women invoked the moral degradation that accompanied poverty and the threat to women and children posed by urban industrial society as a rationale for women's involvement in public life. In 1893, the German Girls' and Women's Groups for Social Assistance Work echoed this sentiment when it called upon bourgeois women to serve the "impoverished classes." This group departed from traditional German charitable societies in its emphasis on acquiring systematic knowledge about social problems and in its careful training in social practice. From these origins, social work education

---

[11] Lewis, "The Place of Social Investigation"; Trattner, *From Poor Law*, chap. 5; Skocpol, *Protecting Soldiers and Mothers*, pp. 424–6; Peter Flora and Jens Alber, "Modernization, Democratization, and the Development of Welfare States in Western Europe," in *Development of Welfare States*, ed. Flora and Heidenheimer, pp. 37–80; Stein Kuhnle, "International Modeling, States, and Statistics: Scandinavian Social Security Solutions in the 1890s," in *States, Social Knowledge*, ed. Rueschemeyer and Skocpol, pp. 233–63.

in Germany flourished, and its trainees staffed the web of municipal jobs created by expansive German social welfare programs. Thus, while many women joined social service groups as volunteers in Europe and America, their long experience led increasingly to paid employment as social workers as charitable institutions became more professionalized and bureaucratized at the turn of the century.[12]

Women were similarly active in the settlement house movement, a second powerful force linking social science to social welfare during the late nineteenth century, and one that sometimes encompassed a broader vision of social provision. The settlement house movement had its origins in Britain, where Toynbee Hall was established in 1884 to bring college-educated men into depressed communities in order to transmit cultural values and to narrow the divide between rich and poor. In America, the settlements quickly became the province of college-educated women, who found in them a place for their talents, professional ambitions, and their wish for community.

Although designed to provide direct, practical service to the poor, the settlements emphasized the common plight of entire groups of poor people rather than privileging work with individual clients, as did the Charity Organization Societies. Such a stance drew the settlements quickly into the larger public arena, where their members contributed to an ongoing conversation about the roots of poverty and social inequality. Among the houses' most important contributions were extensive social investigations into the lives of the poor.

No settlement became more famous in this regard than Jane Addams's Hull House, founded in Chicago in 1889. Its research projects, notably *Hull House Maps and Papers* (1895), were impressive for their methodological sophistication and pathbreaking analyses of urban neighborhoods, work that paved the way for the University of Chicago's urban sociology. Addams (1860–1935) largely endorsed an organic vision of society in which individuals' self-fulfillment was linked to the well-being of the community and depended upon the whole to advance the public good. The environmental explanations of social problems she and other settlement leaders advanced helped to remove the moral stigma from the poor and served as an important rebuttal to social Darwinist explanations of persistent poverty.[13]

---

[12] Linda Gordon, *Pitied but Not Entitled* (New York: Free Press, 1994); Kathryn Kish Sklar, "The Historical Foundations of Women's Power in the Creation of the American Welfare State, 1830–1930," in *Mothers of a New World*, ed. Seth Koven and Sonya Michel (New York: Routledge, 1993), pp. 49–93; Christoph Sachsse, "Social Mothers: The Bourgeois Women's Movement and German Welfare State Formation, 1890–1920," in *Mothers of a New World*, ed. Koven and Michel, pp. 142–9; Jean Quataert, "Woman's Work and the Early Welfare State in Germany," in *Mothers of a New World*, ed. Koven and Michel, pp. 159–87.

[13] Allen Davis, *Spearheads for Social Reform* (New York: Oxford University Press, 1967); Jane Addams, *Twenty Years at Hull House* (New York: New American Library, 1981); Kathryn Kish Sklar, "Hull House Maps and Papers," in *Social Survey*, ed. Bulmer, Bales, and Sklar, pp. 111–47.

In the United States, settlements produced leading figures in early-twentieth century campaigns to draw the federal government into social provision. Florence Kelley, head of the National Consumers' League, Julia Lathrop, the first chief of the Children's Bureau, her assistant and successor Grace Abbott, as well as female voluntary associations such as the General Federation of Women's Clubs were among the activists whose aggressive lobbying on behalf of women and children resulted in some pivotal successes. Most notable was the enactment of mothers' pensions, a program of modest monthly payments to impoverished women (usually widows) with children; by 1920, mothers' pensions had been enacted by forty states over the objections of most charity officials. Only single mothers who met specified economic and moral criteria qualified – a fact that, ironically, gave social workers a central administrative role in a program many of them opposed.

Then, in 1921, Congress passed the Sheppard–Towner Act, the first explicit federal social welfare program in American history. It provided the states with matching funds to construct programs to prevent infant mortality and to improve maternal and child health. Nearly three thousand centers across forty-five states received assistance. The Children's Bureau, as the primary administrator of Sheppard–Towner, became a central authority within the federal government for social workers across the country. Notable as a non-means-tested program, Sheppard–Towner, some hoped, would provide an opening wedge for broader systems of social welfare that would assist all needy citizens.[14]

The "maternalist" drive for more generous social welfare policies was evident in many European nations. Norwegian women in the labor movement advanced proposals for mothers' pensions during the early twentieth century, and their cause was successfully taken up by the Labor Party. In Sweden, female Social Democrats played a central role in developing that nation's early "family policy." British women active in the Labor Party also aggressively sought more generous social welfare programs that responded to the special needs of working women. The Women's Labor League, for instance, pressed for day nurseries, equal wages, and fair employment opportunities.

In the United States, only the transitory Progressive Party in 1912 supported the full range of social welfare ideals advocated by the liberal reformers of the early twentieth century. Yet even among the most liberal reformers, resistance to "universalist" European social insurance programs continued to run deep. Leading figures at the Children's Bureau, for example, endorsed the notion that women with children belonged at home; as we shall see, their vision of the healthy family influenced the later direction of social welfare

[14] Skocpol, *Protecting Mothers and Soldiers*, chaps. 8–9; Joanne Goodwin, *Gender and the Politics of Welfare Reform* (Chicago: University of Chicago Press, 1997); Robyn Muncy, *Creating a Female Dominion in American Reform* (New York: Oxford University Press, 1991); Molly Ladd-Taylor, *Mother-Work: Women, Child Welfare and the State, 1890–1930* (Urbana: University of Illinois Press, 1994).

policy in the United States.[15] There was no single "social work" perspective on social provision, to be sure, but the growing professionalization of social work during the late nineteenth century unquestionably privileged the casework approach. Efforts to incorporate social science into social work training did not alter the focus on individuals, especially as psychology was increasingly emphasized. In England, the Charity Organization established a School of Sociology, which became subsumed by the London School of Economics and Political Science in 1912. But the effect of the change, ironically, was to accentuate the division between social work and social scientific inquiry. In the United States, the University of Chicago's School of Social Service Administration, founded by two Chicago-trained social scientists, Sophonisba Breckinridge and Edith Abbott, resisted vociferously the trend toward psychiatric social work during the 1920s. It emphasized training in political economy, political science, jurisprudence, sociology, and historical research as essential to the study and advancement of social welfare policy. Nevertheless, casework remained at the heart of British and American social work training. Such an approach borrowed theoretical models from and skimmed off the empirical thrust of modern social science, but did little to challenge the individualized approach to dependency that was rooted in the charity tradition and increasingly supported by Freudian psychology. [16]

FROM SOCIAL INSURANCE TO WELFARE

Although academic social scientists in the United States valorized theory, objectivity, and "pure" research, they remained a vigorous force in shaping public debates about social welfare policy during the early twentieth century. In essays and books, and through the lobbying of the American Association for Labor Legislation (AALL), the economist John R. Commons, his students at the University of Wisconsin, and other like-minded scholars explained how shifting conditions in an industrial capitalist economy could plunge even the

[15] Gordon, *Pitied but Not Entitled*; Blanche Coll, *Safety Net: Welfare and Social Security, 1929–1979* (New Brunswick, N.J.: Rutgers University Press, 1995); Day, *New History of Social Welfare*, pp. 292–3; Anne-Lise Seip and Hilde Ibsen, "Family Welfare, Which Policy? Norway's Road to Child Allowances," pp. 40–59; Ann-Sofie Ohlander, "The Invisible Child? The Struggle for a Social Democratic Family Policy in Sweden, 1900–1960s," pp. 60–72; and Pat Thane, "Visions of Gender in the Making of the British Welfare State," pp. 93–118; all in *Maternity and Gender Policies*, ed. Gisela Bock and Pat Thane (New York: Routledge, 1991); Pat Thane, "Women in the British Labour Party and the Construction of the Welfare State, 1906–1939," in *Mothers of a New World*, ed. Koven and Michel, pp. 343–77; Skocpol, *Protecting Soldiers and Mothers*; Ulla Wikander, Alice Kessler-Harris, and Jane Lewis, *Protecting Women: Labor Legislation in Europe, the United States, and Australia, 1880–1920* (Urbana: University of Illinois Press, 1995).

[16] Leiby, *History of Social Welfare*, chaps. 8–9 and passim; Lewis, "The Place of Social Investigation," p. 150; Ellen Fitzpatrick, *Endless Crusade: Women Social Scientists and Progressive Reform* (New York: Oxford University Press, 1991); Steven J. Diner, "Scholarship in Quest of Social Welfare," *Social Service Review*, 51 (March 1977), 1–68; Steven Diner, "Department and Discipline," *Minerva*, 13, no. 4 (Winter 1975), 514–53; Bulmer, "National Contexts," pp. 152–3.

most thrifty and dedicated workers into poverty. Though their solutions differed in details, these social scientists emphasized the need for "insurance" against catastrophe rather than charity or relief once disaster struck. In seeking to place a floor under workers, such scholars sought to remove the taint of "charity" from social provision. Their vigorous advocacy of "workingmen's insurance," including workmen's compensation, unemployment insurance, government regulation of the labor market, and, in the case of the socialist Isaac Rubinow, health and old age insurance, invigorated public debate.

An alliance of activist social scientists, state officials, politicians, trade unionists, and businessmen advanced limited reforms through the 1920s. Workmen's compensation was enacted in forty-two states during the 1910s. Other states passed legislation governing hours and establishing a minimum wage. But the AALL's 1915–20 campaign for health insurance failed, and even in Wisconsin, where Commons and his colleagues spearheaded a drive for unemployment insurance, that legislation was not enacted until 1932.[17]

The Depression marked the turning point in the history of American social welfare policy, setting the parameters of American attitudes and policies toward poor relief for the rest of the century. Academic social scientists and social workers both figured in that outcome, notably in the construction of the most vital social welfare program to come out of the Great Depression, the Social Security Act of 1935.

Social workers, economists, and other experts, many from the Wisconsin circle, were brought into the creation of the Social Security Act at the invitation of Franklin Roosevelt's secretary of labor, Frances Perkins, herself a former settlement worker. Under the direction of the University of Wisconsin economist Edwin Witte, they helped to compile the wide range of social insurance options from which the Social Security Act was crafted. The legislation that emerged was a hybrid. It incorporated the universalist goals of economists and others by enacting a national contributory program of old age and unemployment insurance that would cover all workers. And it addressed the family- and child-centered concerns of women activists and social workers, especially those at the Children's Bureau who wanted to see federal subsidies akin to mothers pensions, by enacting the Aid to Dependent Children (ADC) program.

Although some social workers inside and beyond the Children's Bureau wanted all poor children covered, when ADC was enacted by Congress only poor children deprived of "parental support" were included. Not until the 1950s did the caretakers of these children begin to receive assistance. The states were given considerable discretion to establish eligibility, and the levels of support provided were extremely modest. In a final defeat for women reformers, administration of the program was shifted from the Children's Bureau, as the

---

[17] Skocpol, *Protecting Soldiers and Mothers*, chaps. 3–4; Edward Berkowitz, *Mr. Social Security: The Life of Wilbur J. Cohen* (Lawrence: University Press of Kansas, 1995), chap. 1.

activists had hoped, to the Department of Labor. Social workers were left with the task of carefully assessing a child's home environment to determine eligibility when ADC was administered at the local and state levels.[18]

The Social Security Act institutionalized long-standing categories of the "deserving" and "undeserving" poor. The social insurance programs created by the Act established a category of aid recipients who deserved the government's help because they had worked and contributed to the program. But those who qualified for ADC were merely dependent upon government support. During the Great Depression, public assistance was not as severely stigmatized as it had been in the past, but in the 1950s the split between social insurance and public assistance deepened. Since ADC supported only those children without parental support, it inadvertently rewarded family dislocation, making it still less popular. The fact that, as a means tested program, ADC was also to a large extent a discretionary policy likewise insured that it would remain a hotly contested and politicized subject.[19]

The American system of social provision established during the 1930s diverged markedly from those of many European nations by incorporating market principles of individual responsibility. Social insurance was subordinated to an opportunity structure that stressed employment rather than handouts, and contributory programs were favored. The historical trajectory was different too. Not only Britain and Germany, but also the Netherlands, Austria, Switzerland, France, Sweden, and Italy had all introduced programs of old age, sickness, and unemployment insurance by 1935.[20]

These differences may help to explain why the debate about public assistance remained fractious in the United States through the second half of the twentieth century, even though most Western nations with broad systems of social provision bore larger burdens. Discomfort with "welfare," as the public assistance programs organized by Social Security became known, was keen among both conservatives and liberals even during the 1960s, when liberalism was surging. Although new middle-class entitlement programs such as Medicare, which provided medical insurance to the elderly, were costly, welfare was viewed by many as *the* undesirable drain on the nation's resources, a program of last resort that was being exploited by those who would not work. Although most welfare recipients were white, race became, in the minds of many, inextricably linked to welfare programs when African Americans gained greater access to them in the 1950s and 1960s.

Social scientists played a prominent role in these debates. The rediscovery of persistent poverty amid wealth, exemplified by the publication of

---

[18] Coll, *Safety Net*, chap. 4 and passim; Martha Derthick, *Policymaking for Social Security* (Washington, D.C.: Brookings Institution, 1979); Gordon, *Pitied but Not Entitled*.

[19] Coll, *Safety-Net*; Gordon, *Pitied but Not Entitled*; Michael Katz, *The Undeserving Poor* (New York: Pantheon, 1989).

[20] Flora and Alber, "Modernization, Democratization"; Skocpol, "Government Structures"; Wittrock and Wagner, "Social Science."

Michael Harrington's *The Other America* in 1962, and the ensuing War on Poverty during the Kennedy and Johnson administrations, prompted some social scientists to revisit theories of deprivation and to spell out the connection between poverty and race. In ethnographic studies, Oscar Lewis had theorized that a "culture of poverty" had kept Puerto Ricans and Mexicans on the margins of society, reproducing generation after generation a debased standard of living and meager prospects. The sociologist Daniel Patrick Moynihan, then a member of the Johnson administration and seeking an explanation for the persistence of poverty despite new antipoverty measures, reconfigured the "culture of poverty" idea in 1964 in his report *The Negro Family: The Case for National Action*. African Americans, Moynihan argued, were locked in a cycle of poverty in part because of weaknesses in the black family structure. From slavery onward, social and political conditions had undermined the strength of black men and pushed the black family toward matriarchy; this pathological situation reproduced urban poverty and welfare dependence.

Although Moynihan's ideas were attacked vociferously during the 1960s, the notion that the characteristics of poor people explained the persistence of poverty gained ground as the political center of gravity moved to the right. The conservative George Gilder claimed, in *Wealth and Poverty* (1981), that the poor refused to work and were rewarded for their indolence by the welfare system. Charles Murray insisted in *Losing Ground* (1984) that poverty had actually *increased* because of antipoverty measures. Out-of-wedlock births among African Americans and a reliance on welfare rather than on work and self-initiative had fueled the growth of poverty.[21]

Although neither Gilder nor Murray was a university-based social scientist, both expropriated social scientific research to buttress their sharp criticisms of welfare. With the financial support of conservative interest groups and attention from conservative politicians, they trumped academic social scientists in their ability to reach the American public. Their arguments appealed to middle-class Americans who blamed the social welfare initiatives of the 1960s for their eroding incomes and the decline of American economic power during the 1970s, and provided ammunition for critics of the swollen federal deficit.

In the 1980s, journalists described an "underclass" of young black people, out of work, on the dole, involved in juvenile delinquency or more serious crime, addicted to drugs, and, if a female teenager, pregnant more often than not. Lacking moral values and existing at the margins of American society, they were, one mass circulation magazine reported, "unreachable." In an effort to advance more accurate understanding of the urban poor, academic sociologists such as William Julius Wilson published studies that pointed to male joblessness as the key factor in persistent poverty. The erosion

---

[21] Katz, *The Undeserving Poor*, chaps. 1, 3, and passim.

of good jobs, he stressed, mattered far more than a lack of values in explaining the lives of the "truly disadvantaged." Such arguments complicated political assertions that "workfare," the plan to tie eligibility for public assistance to employment, would easily solve the welfare problem. Yet enthusiasm for workfare grew despite the absence of adequately paid jobs in the inner cities.[22]

By the close of the twentieth century, debate about "welfare" had spread to Europe as well. The trends of postindustrial society, with its globalized economy, proliferation of low-wage service sector jobs, and rising government costs diminished support for expansive social welfare policies in America and in many European nations. Social workers were often left with the task of mediating between the needs of the poor and the shortage of resources. Many were forced to operate within bureaucratic structures that limited their options and eroded their skills. For their part, social scientists had yet to construct the coherent and consistent plan of action to relieve dependency that their early-twentieth-century predecessors had hoped for. The goal itself, perhaps, is a will o' the wisp, given the fundamental political disagreements that lie at the heart of judgments about the nature of poverty and the best ways to address social needs. The marriage of social science and social welfare policy has not ended, nor has the political factionalism and the discouraging reality of persistent poverty, however it is understood.

[22] Katz, *The Undeserving Poor*, chap. 5; William Julius Wilson, *The Truly Disadvantaged* (Chicago: University of Chicago Press, 1987); William Julius Wilson, *When Work Disappears* (New York: Knopf, 1996). For the conceptions of poverty formulated by American social scientists throughout the century, see Alice O'Connor, *Poverty Knowledge: Social Science, Social Policy, and the Poor in Twentieth-Century U. S. History* (Princeton, N.J.: Princeton University Press, 2001).

36

EDUCATION

Julie A. Reuben

The social sciences developed from a philosophic tradition with a keen interest in education. Philosophers looked to education for evidence about human nature and viewed education as an arena for the shaping of individual character and the strengthening of social bonds. As a consequence, education was intertwined with the broad questions that social scientists inherited from philosophy: What is human nature? How is it formed? Can it be changed? How can we explain differences among humans? How and why are societies formed? What are the best forms of social relations? How can social ties be created and maintained?

Despite the centrality of education to the concerns of the social sciences, social scientists' interest in education has waxed and waned. The first generations of social scientists viewed education as a laboratory in which to explore social and psychological theories, an outlet for the practical application of their new knowledge, and an instrument for social and political reform. This early enthusiasm faded, however, as the social sciences became professionalized over the twentieth century. Since the 1920s, social scientists' association with education has been haphazard. At times, education has been at the forefront of the disciplines' research agendas, but more often it has receded to the periphery.

This inconstant relation has been influenced by a number of factors, including social scientists' integration into universities, their changing attitudes towards activism, the relation of universities to schools and the training of teachers, and the development of education as an independent field of research. Because these conditions vary from nation to nation, the relation between the social sciences and education has developed differently across countries. This chapter will focus on the United States, but will compare developments in England and Germany in order to highlight the impact of unique institutional arrangements as well as to point to conditions that crossed national borders.

EDUCATION AND THE PHILOSOPHIC TRADITION

Enlightenment thinkers emphasized the importance of education to the development of free, rational men and a stable civil society. John Locke's (1636–1704) view of children as a tabula rasa, born without knowledge, placed great importance on environment and education. In *Thoughts Concerning Education* (1693), Locke criticized then-current educational practices, particularly the prominence of the classical languages and the reliance on rote learning and physical punishment: Education, to be effective, must build on a person's interests and experience. With gentlemen in mind, he favored home education and viewed individualized education as the only effective way to achieve its central goal – the training of virtue. Locke's work spawned additional treatises on education aimed at inculcating morality independent of church dogma. The most famous, Jean-Jacques Rousseau's *Emile* (1761), advocated "natural" education and elaborated a series of developmental stages, each requiring its own methods of learning, shaped to the interests and capacities of each pupil.[1]

In the second half of the eighteenth century, the emphasis on individual education gave way to an interest in schools as a source of intellectual and moral reform. Johann Heinrich Pestalozzi (1746–1827), an advocate of schooling for all children, believed that successful teaching could be based on psychological knowledge. He pioneered efforts to apply knowledge about how children at various ages learn to the design of classroom lessons. Others, such as Friedrich Froebel and Johann Friedrich Herbart, followed his example and wrote psychologically oriented pedagogical treatises. In addition, philosophers such as Adam Smith, Thomas Jefferson, and Johann Fichte addressed the political and economic importance of education.[2]

This widespread interest in education was closely related to the growth of state-sponsored education. Beginning in the mid eighteenth century, Prussia led the way by developing a universal system of state-licensed schools and normal schools. Pedagogy was introduced as a subject of instruction in both normal schools and universities. A statute requiring instruction in pedagogy at the University of Königsberg led to Immanuel Kant's important lectures on pedagogy. By the nineteenth century, pedagogy was established as a branch of philosophy and a university subject in Germany.[3]

Inspired by the Prussian success, educational reformers in the United States advocated public schools aimed at developing moral character and sustaining public virtue in the new republic. The common school movement, led by Horace Mann (1746–1859) and Henry Barnard (1811–1900), pressed the states

---

[1] John Locke, *Some Thoughts Concerning Education*, ed. John W. Yolton and Jean S. Yolton (New York: Oxford University Press, 1989); Jean-Jacques Rousseau, *Emile; or, On Education*, trans. Allan Bloom (New York: Basic Books, 1979).

[2] Robert B. Downs, *Heinrich Pestalozzi, Father of Modern Pedagogy* (Boston: Twayne, 1975).

[3] Andy Green, *Education and State Formation* (New York: St. Martin's Press, 1990); Immanuel Kant, *Education*, trans. Annette Churton (Ann Arbor: University of Michigan Press, 1960).

to establish inclusive elementary schools that would enroll all children, rich and poor, and provide them with a uniform curriculum. Although in practice there were racial and class limits to this inclusiveness, common school reformers had largely achieved their goals in the northeastern and midwestern states by the 1860s.[4]

Common school advocates also called for a "science of education," which would derive the best principles of teaching from mental philosophy, and introduced the ideas of European writers, such as Pestalozzi, in a number of new educational journals. They also successfully lobbied for the establishment of normal schools to train teachers in modern educational methods. In practice, however, early normal schools mainly taught future teachers the common school curriculum. Hence, pedagogy did not develop in normal schools, nor did it become established as a university subject during the nineteenth century.[5]

In nineteenth-century England, proponents of educational reform were not as successful as their counterparts in Prussia or even in the United States. Instead of establishing a universal system of state-run schools, England expanded education by channeling government funds to existing voluntary schools, reinforcing existing class and religious divisions. Although England had established a number of teacher training institutes by the middle of the century, they were not required for teacher qualification and were too weak to emphasize educational theory.[6]

EDUCATION AND THE DEVELOPMENT OF THE SOCIAL SCIENCES

After the middle of the nineteenth century, the philosophic tradition of which pedagogy was a part began to come under harsh scrutiny. Critics claimed that philosophy was abstract and dogmatic and could not meet the intellectual and practical needs of modern society. They wanted to recast philosophy as a "social science" by basing it on empirical study. The reformist orientation of supporters of social science in the United States and other nations insured that education would become a central part of their research agenda.

In the United States, proponents of social science formed the American Social Science Association (ASSA) in 1865. Assuming that the biggest obstacle

[4] Carl F. Kaestle, *Pillars of the Republic: Common Schools and American Society, 1780–1860* (New York: Hill and Wang, 1983).
[5] Jason R. Robarts, "The Quest for a Science of Education in the Nineteenth Century," *History of Education Quarterly*, 8 (1968), 431–46; Jurgen Herbst, *And Sadly Teach: Teacher Education and Professionalization in American Culture* (Madison: University of Wisconsin Press, 1989).
[6] Harry Judge, Michel Lemosse, Lynn Paine, and Michael Sedlak, *The University and the Teachers: France, the United States, England* (Wallingford: Triangle Journals, 1994), pp. 160–4; Brian Simon, "Why No Pedagogy in England?," in *Education in the Eighties*, ed. Brian Simon and William Taylor (London: Batsford, 1981), pp. 129–33.

to effective social policies was inadequate information, the ASSA encouraged the collection and dissemination of statistics as a tool of social reform. Its Department of Education undertook this task in regard to all forms of schooling and encouraged debate on educational reform.[7]

Prominent members of the ASSA included university leaders, such as Charles W. Eliot and Daniel Coit Gilman, who also worked to reform public schools. They saw such reforms as essential to the changes they were instituting at their universities and viewed education as a natural arena in which to demonstrate their institutions' service to society. Eliot, the president of Harvard University, was chairman of the National Education Association's influential "Committee of Ten" on the curriculum of secondary schools. University presidents Nicholas Murray Butler of Columbia University and David Starr Jordan of Stanford University also headed the NEA, while William Rainey Harper of the University of Chicago led efforts to reform the public schools of Chicago, and Gilman and his successors at Johns Hopkins University, Ira Remsen and Frank Goodnow, all sat on the Baltimore Board of Education. Presidents of state universities, led by the example of the University of Michigan, which set up a system for "accrediting" the state's high schools, tried to position themselves and the institutions they led as leaders of their states' educational systems. These university presidents supported faculty members who joined them in their efforts to guide the development of American schools.[8]

Given their reformist orientation, social scientists did not need much prodding from university administrators to take an interest in education. John Dewey's (1859–1952) work exemplified the links between social science, social reform, and education during the Progressive Era. Dewey, who had studied philosophy and psychology at Johns Hopkins University, was encouraged to examine education while teaching psychology at the University of Michigan. When he moved to the University of Chicago in 1894, Dewey founded the Laboratory School to provide educators with a site to test their theories and social scientists an environment in which to explore the intersection of the individual and society. While at Chicago, Dewey's associations with social reformers such as Jane Addams and with teachers such as Ella Flagg Young stimulated the development of his progressive theory of education, which linked the creation of a child-centered curriculum and a democratic classroom pedagogy with the fulfillment of the hopes of American democracy.[9]

---

[7] Thomas L. Haskell, *The Emergence of Professional Social Science: The American Social Science Association and the Nineteenth-Century Crisis of Authority* (Urbana: University of Illinois Press, 1977).

[8] Hugh Hawkins, *Between Harvard and America: The Educational Leadership of Charles W. Eliot* (New York: Oxford University Press, 1972), chap. 8; Steven J. Diner, *A City and Its Universities: Public Policy in Chicago, 1892–1919* (Chapel Hill: University of North Carolina Press, 1980), chap. 4.

[9] For a mature statement of Dewey's educational theory, see *Democracy and Education*, vol. 9 of *John Dewey: The Middle Works* (1916), ed. Jo Ann Boydston (Carbondale: Southern Illinois University Press, 1980).

Dewey was one of several academics who saw the value of tying education to psychology. G. Stanley Hall (1844–1924), one of the most active promoters of psychology, led the child study movement in the United States, creating a network of teachers and parents to distribute questionnaires on all aspects of children's lives. The goal of the movement was to compile statistics that would establish the stages of cognitive and emotional development, thereby providing the knowledge base for a developmentally appropriate pedagogy. Other prominent psychologists, such as James Cattell at Columbia, shared Hall's enthusiasm and encouraged their students to pursue research related to education. In 1910, over a third of American psychologists reported an interest in education. This interest influenced research practices among American psychologists and left a deep impact on American education. Psychologists established a legacy of research focused on group and individual mental differences, and they developed a program of testing aimed at allowing schools to differentiate students according to their "abilities." Testing and sorting shaped the organization of most American public schools during the twentieth century.[10]

Psychologists, led by Edward L. Thorndike (1874–1949), dominated American educational thought during the early twentieth century. A prolific researcher, Thorndike influenced the methods of educational research, theories of learning, and classroom practice. His book *An Introduction to the Theory of Mental and Social Measurements* (1904) helped to define a distinctive style of psychological research based on group rather than individual data. His animal studies formed the basis for a behaviorist theory of learning emphasizing stimulus-response reinforcement. He also designed a series of experiments that challenged the older idea of mental discipline by demonstrating that there was little transfer of learning from one subject to another. Maintaining that intelligence was largely hereditary, he supported a differentiated curriculum so that students of "lesser intelligence" would not waste their time learning academic subjects for which they were ill-suited. His studies of the psychology of school subjects and his numerous textbooks influenced how teachers were trained, and how they taught students.[11]

Although psychologists had the greatest impact on educational research, social scientists in other disciplines played important roles as well. The sociologist Lester Frank Ward (1841–1913), for example, rejected laissez faire and argued that intelligent intervention in society would speed evolution.

[10] Dorothy Ross, *G. Stanley Hall: The Psychologist as Prophet* (Chicago: University of Chicago Press, 1972); John M. O'Donnell, *The Origins of Behaviorism: American Psychology, 1870–1920* (New York: New York University Press, 1985); Kurt Danziger, *Constructing the Subject: Historical Origins of Psychological Research* (Cambridge: Cambridge University Press, 1990), chaps. 7–8.

[11] Geraldine Jonçich, *The Sane Positivist: A Biography of Edward L. Thorndike* (Middletown, Conn.: Wesleyan University Press, 1968); Ellen Condliff Lagemann, "The Plural Worlds of Educational Research," *History of Education Quarterly*, 29 (1989), 2; Kurt Danziger, "Social Context and Investigative Practice in Early Twentieth-Century Practice," in *Psychology in Twentieth-Century Thought and Society*, ed. Mitchell G. Ash and William R. Woodward (Cambridge: Cambridge University Press, 1987), pp.13–34.

Progress, he argued, depended on universal access to education, because an activist government needed the intellectual talent of all social classes. Albion Small (1854–1926), a sociologist at the University of Chicago, also emphasized education as a potential tool of social reform, arguing that schools should abandon their traditional curriculum, reformulating it around the individual's experience in society.[12]

The work of other sociologists, such as Edward A. Ross and Charles Horton Cooley, helped to shape the debates about curriculum reform in the early twentieth century. Adopting the idea of social control, educators sought to make public education a more effective means of social integration. They thought that a functionally oriented curriculum organized around life situations, such as the family and work, would help individuals to adjust to society. This new curriculum drew on the subject matter of the social sciences, and many social scientists helped to design "model" courses for the public schools. The American Political Science Association, for example, helped to create and promote "community civics," which became one of the classes recommended by the NEA's 1916 committee on the reorganization of secondary education.[13]

Social scientists also participated in efforts to restructure school administration. Frank J. Goodnow (1859–1939), a political scientist at Columbia University, applied to school systems his discipline's prescriptions for administrative efficiency and effective urban government – reduction in the number of elected officials, centralization of authority, and the development of administrative expertise separate from politics. Between 1893 and 1913, the average size of school boards in cities with populations over 100,000 dropped from 21.5 to 10.2 members. Ward elections were eliminated and replaced with citywide election of board members. In addition, power within the district was increasingly concentrated in the office of the superintendent, rather than in the elected board and/or the individual schools. These changes were achieved through a series of political battles in which social scientists marshaled evidence in support of reform.[14]

In England, as in the United States, the development of the social sciences was linked to education. The agenda of the British National Association for the Promotion of Social Science, founded in 1856 to promote empirical research on issues relevant to social reform, included education. Pioneering the social survey, Charles Booth's massive *Life and Labour of the People*

[12] Herbert M. Kliebard, *The Struggle for the American Curriculum, 1893–1958*, 2nd ed. (New York: Routledge, 1995), pp. 21–2, 52–4.

[13] Barry M. Franklin, *Building the American Community: The School Curriculum and the Search for Social Control* (Philadelphia: Falmer Press, 1986); Julie A. Reuben, "Beyond Politics: Community Civics and the Redefinition of Citizenship in the Progressive Era," *History of Education Quarterly*, 37 (1997), 399–420.

[14] Frank J. Goodnow, *Municipal Problems* (New York: Macmillan, 1897); David B. Tyack, *The One Best System: A History of American Urban Education* (Cambridge, Mass.: Harvard University Press, 1974), pt. 4.

of London (1892–97) addressed the relation between class and educational attainment, among other subjects. Michael Sadler (1861–1943) applied the social survey method explicitly to education in a series of surveys of English cities published between 1903 and 1906. These surveys were used to help implement the Education Act of 1902, which created a national system of local education authorities with greater supervision over voluntary schools and with more responsibility for secondary and technical education and teacher training.[15]

English psychologists also supported educational reform. Several of the first generation of psychologists came from teaching, including John Adams and Percy Nunn. Their interests in progressive pedagogy made psychology a natural subject for further study and research. English psychologists such as Charles Spearman (1863–1945) and Cyril Burt (1883–1971) were interested in intelligence and IQ testing. Like their American counterparts, they helped to identify mental differences among individuals and advised educators about programs appropriate to different groups, such as mentally "defective" and "gifted" students.[16]

The connections between social scientists and education followed a different institutional pattern in England. Unlike their counterparts in the United States, aristocratic English universities largely ignored the social sciences and education before World War II. Without opportunities in universities, many psychologists found jobs at teacher training colleges. Cyril Burt, who eventually was appointed to the prestigious professorship of psychology at University College London, spent two decades teaching educational psychology at the London Day Training College and served as an adviser to the London County Council. Other social scientists worked with government ministries that took responsibility for gathering data about education.[17]

In Germany, educational issues also attracted the attention of social scientists. Wilhelm Preyer (1841–1897), who influenced G. Stanley Hall, developed an empirical program of child study based on evolutionary theory. He applied his ideas about child development to education, arguing against rote learning and for curriculum reform in the classical gymnasium. Ernst Meumann and Wilhelm Lay promoted educational psychology in Germany, founding the journal *Die Experimentelle Pädagogik* in 1905. The sociologists Johannes Conrad and Franz Eulenburg completed a number of statistical studies of students at German universities. Since the strong position of pedagogy in German universities insured that philosophical approaches to educational issues continued to be important, German social scientists did not dominate

[15] Raymond A. Kent, *A History of British Empirical Sociology* (Aldershot: Gower, 1981), pp. 52–63; W. F. Connell, *A History of Education in the Twentieth Century World* (New York: Teachers College Press, 1980), p. 98.
[16] Adrian Wooldridge, *Measuring the Mind: Education and Psychology in England, c. 1860–1990* (Cambridge: Cambridge University Press, 1994).
[17] Wooldridge, *Measuring the Mind*, chap. 6.

educational thought to the same degree that their counterparts in the United States did.[18]

DECLINING INTEREST IN EDUCATION

Social scientists' ties to education loosened after World War I. As their disciplines became better established, academic social scientists came to favor theoretical rather than applied research. Disagreements between educators and social scientists over the value of certain types of research also encouraged this shift, as the professionalization of educational researchers distanced social scientists from education.

The increasing prestige of "basic" research within the university contributed to social scientists' declining interest in education. In the United States, social scientists at elite universities distinguished themselves from their colleagues at "lesser" institutions by their research on theoretical rather than practical questions. This pattern was replicated in England, as social scientists gained greater acceptance in universities. After his appointment as professor of psychology at University College London, Cyril Burt abandoned educational research and devoted his attention to psychometric theory.[19]

Social scientists' efforts to increase the "objectivity" of their research intensified the move away from education. During the 1920s, many American sociologists, distressed by what they perceived to be a lack of progress in establishing certain knowledge in their field, rejected the reformism of their predecessors. While involvement in social reform had fueled sociologists' earlier interest in education, the new orientation discouraged it. The change was evident in the career of F. Stuart Chapin, one of the leading proponents of objectivism in American sociology. Chapin's dissertation, published in 1911 as *Education and the Mores*, reflected his social evolutionary views and his desire to transform traditional education. After World War I, Chapin became a proponent of descriptive sociology based on statistical analysis of social indices, abandoned his social evolutionism, and stopped investigating education.[20]

Although psychologists also began to privilege "basic" over "applied" work, they sustained more interest in education than their colleagues in other disciplines. The "scientistic" movement in psychology – behaviorism – was

---

[18] Siegfried Jaeger, "Origins of Child Psychology: William Preyer," in *The Problematic Science: Psychology in Nineteenth-Century Thought*, ed. William R. Woodward and Mitchell G. Ash (New York: Praeger, 1982), pp. 300–21; Marc Depaepe, "Differences and Similarities in the Development of Educational Psychology in Germany and the United States before 1945," *Paedagogica Historica*, 23 (1997), 45–68; Anthony Obershall, *Empirical Social Research in Germany, 1848–1914* (The Hague: Mouton, 1965), pp. 92–5.
[19] Julie A. Reuben, *The Making of the Modern University: Intellectual Transformation and the Marginalization of Morality* (Chicago: University of Chicago Press, 1996), p. 181; Wooldridge, *Measuring the Mind*, 94–6.
[20] Robert C. Bannister, *Sociology and Scientism: The American Quest for Objectivity, 1880–1940* (Chapel Hill: University of North Carolina Press, 1987), chap. 10.

forged as an educational tool and was easily applicable to the sorting of individuals required by mass public schooling. Mental testing, building on the work of Lewis Terman (1877–1956), continued to be an important area of psychological research throughout the twentieth century. John B. Watson's behaviorism also stimulated a spate of child rearing guides during the 1920s. B. F. Skinner (1904–1990) drew on his experiments on animal behavior in the 1930s and 1940s to develop a program of educational reform during the 1950s. He argued that classroom settings did not allow for the variation of stimuli, the attention to the elements of complex behavior involved in mastering school subjects, or the frequency of reinforcement necessary for effective learning; he advocated the use of teaching machines to rectify these limitations. Even in psychology, however, education would never be as central to the discipline as it was at the turn of the twentieth century. After World War I, psychologists found themselves with many more areas in which to apply their expertise. As advertising, industry, the military, and medicine became consumers of psychological expertise, education lost its preeminent place within applied psychology.[21]

Tensions between the needs of practicing educators and the research interests of social scientists also lessened the appeal of applied educational research. Practitioners wanted solutions to children's behavioral problems, but were not necessarily interested in general maturation theory. On the other hand, psychologists studied learning in order to understand mental processes, but were not necessarily interested in teaching children to read. These differences emerged when social scientists and educators worked together and were not limited to the United States. For example, similar conflicts developed between the psychologists Karl and Charlotte Bühler and the Vienna school board, which provided them funds and laboratory space during the 1920s.[22]

The growth of independent educational research in the United States sharpened the divide between social scientists and education. Beginning in the late nineteenth century, universities hired faculty members to teach courses in education. To achieve greater autonomy and status, they successfully pressed for separate departments or schools of education, and sought to influence educational policy by training school administrators and conducting educational research. Faculty members in these new university-based schools of education led efforts to professionalize educational research, forming the American Educational Research Association (AERA) in 1916 and the *Journal of Educational Research* in 1920.[23]

---

[21] Skinner quoted in Robert Glaser, "The Contributions of B. F. Skinner to Education and Some Counterinfluences," in *Impact of Research on Education: Some Case Studies*, ed. Patrick Suppes (Washington, D.C.: National Academy of Education, 1978), p. 219.

[22] Mitchell G. Ash, "Psychology and Politics in Interwar Vienna: The Vienna Psychological Institute, 1922–1942," in *Psychology in Twentieth-Century Thought and Society*, ed. Ash and Woodward, pp. 143–64.

[23] Geraldine Jončich Clifford and James W. Guthrie, *Ed School: A Brief for Professional Education* (Chicago: University of Chicago Press, 1988).

Although the founders of the AERA shared the social scientists' methods of research, they felt that their work needed to be immediately applicable in practice in order to retain the support of public school leaders. They agreed that much social scientific research was too abstract and began training their own doctoral students, who would conduct the kind of research needed by schools. But many social scientists and university administrators did not respect this style of research. The low status of teaching as a feminized profession and the poor reputation of teacher training colleges also made social scientists less willing to study education for fear of being tainted by the field's poor image.[24]

The different pattern of professionalization in England did not produce as sharp a split between the social sciences and education. After a 1925 committee report recommended that universities become more involved in teacher training, some universities founded departments of education, but with the exception of the Institute of Education at the University of London, they were too weak to promote programs of research. The Institute of Education was led by psychologists and retained close relations with the country's premier psychology department at University College London through the 1960s. When education departments began to expand during the late 1950s, the field of education was conceived of as a practical activity that developed its principles from four scholarly disciplines – psychology, sociology, history, and philosophy. This model was institutionalized in university departments, insuring that social scientists continued to have a role in educational research.[25]

RENEWED INTEREST IN EDUCATION

During the 1960s, social scientists in the United States were drawn back to educational research because public policy began to link education to issues of concern to social scientists, such as racism, poverty, and economic development. The 1954 Supreme Court case *Brown v. Board of Education* can, in retrospect, be seen as the beginning of this renewed interest. The decision, which outlawed racially segregated schools, referred to the work of two African-American psychologists, Kenneth Clark (b. 1914) and Mamie Phipps Clark (1917–1983). Their research showed that black children became aware of their racial identity around three years of age and at the same age developed a negative self-image because of "society's negative and rejecting definition of them." In addition, the Clarks' Northside Center for Child

---

[24] Arthur G. Powell, *The Uncertain Profession: Harvard and the Search for Educational Authority* (Cambridge, Mass.: Harvard University Press, 1980).

[25] John B. Thomas, "Day Training College to Department of Education," in *British Universities and Teacher Education: A Century of Change*, ed. John B. Thomas (London: Falmer Press, 1990), pp. 30–2; Brian Simon, "The Study of Education as University Subject in Britain," *Studies in Higher Education*, 8 (1983), 1–13.

Development in Harlem criticized therapeutic models that treated children as divorced from their social context. The center developed innovative forms of "therapy," including a remedial reading program, which demonstrated that children's school performance could be significantly improved by relatively small interventions. This initiated the Center's active involvement in public schools. Although the Clarks' work was unusual during the mid-fifties, within a decade their interest in the links between schooling, self-perception, and social inequality would be at the center of American public policy and social science research.[26]

During the late 1950s, a number of small urban programs challenged views of intelligence as predetermined and raised hopes that intellectual retardation caused by deprivations of various kinds could be reversed. Their findings coincided with the work of psychologists who were attacking the concept of fixed intelligence and emphasizing environmental factors instead. During the early 1960s, sociologists joined psychologists who were researching issues of environment, poverty, and education. This work addressed the effects of preschool conditioning on children's low motivation and inadequate preparedness for schooling. It also focused attention on the ways in which schools discriminated and perpetuated inequalities based on class and race. "Cultural deprivation" became one of the key concepts used to explain "the cycle of poverty."[27]

This research found a receptive audience in the Kennedy and Johnson administrations and helped to shape their economic and educational policies. The Head Start program reflected social scientists' focus on early childhood experiences and the importance of compensating for cultural deprivation. Title I of the Elementary and Secondary Education Act of 1965 was supposed to reverse the inequities of American public schooling by providing special resources and programs for poor children. The prominent role of education in Johnson's antipoverty program increased social scientists' interest in education. Indeed, it was a federally commissioned report evaluating efforts to equalize and desegregate schools that became one of the most debated social science studies of all time. *Equality of Educational Opportunity* (1966), known as the Coleman Report after the sociologist James S. Coleman (1926–1995), who headed the commission, concluded that equalizing resources did not equalize the academic performance of black and white students. It suggested that students' class backgrounds and aspects of their homes and neighborhoods, rather than resources, explained differences in academic achievement. Some black educators were angered by the implication that schools with large numbers of black children would never be successful. Many commentators

[26] Gerald Markowitz and David Rosner, *Children, Race, and Power: Kenneth and Mamie Clark's Northside Center* (Charlottesville: University Press of Virginia, 1996); the Clarks are quoted on p. 35.
[27] Harold Silver and Pamela Silver, *An Educational War on Poverty: American and British Policy-Making, 1960–1980* (Cambridge: Cambridge University Press, 1991).

perceived the report to be another attempt to "blame the victim" and to relieve government of its responsibility to address social injustice. Others argued that the study pointed to the limits of schooling, independent of other economic and social reforms, to overcome deep-seated inequalities. The report encouraged many other studies of social class, race, and schooling.[28]

The veritable renaissance of social science research on education created new fields within the established disciplines. The sociology of education experienced the most dramatic growth, rivaling the influence of psychology in American educational thought by the end of the decade. The American Sociological Society's decision to sponsor the journal *Sociology of Education* in 1963 symbolized the subjects' new legitimacy within the discipline. The economics of education also emerged as a new subdisicpline during this period. Radical economists, led by Samuel Bowles and Herbert Gintis, studied the relation between education and the class structure of the United States. Within more mainstream economics, human capital theory drew economists to research on the relation between education and economic productivity, and school finance became a significant area of interest. The founding of the journal *Anthropology and Education Quarterly* in 1969 signaled anthropologists' new interest in the culture of schools and educational researchers' new interest in ethnographic methods.

Social scientists' increased interest in education reflected their growing acceptance of social activism. The social movements of the 1960s forced academics to reexamine their role in society. A significant number concluded that they had a responsibility not only to produce knowledge but also to work to insure that knowledge was used for good. Their interest was also reinforced by a number of institutional factors. During the late 1950s, the U.S. commissioner of education created a cooperative research program that funneled federal funds for research to faculty members outside schools of education. The Ford Foundation also tried to use its money to encourage social scientists to undertake research on education. At the same time, schools of education at elite universities started to hire more social scientists. Social scientists found that education offered not only a socially significant topic of research, but also a promising career path.[29]

The explosion of interest in education was not limited to the United States. In England, the sociology of education blossomed during the late 1950s and early 1960s. Sociologists began to study the impact of the Education Act of 1944, which created a tripartite system of free secondary schools in which students' placement was determined by their performance on a standard examination. In their classic work of 1956, *Social Class and Educational*

[28] See, for example, Frederick Mosteller and Daniel P. Moynihan, *On Equality of Educational Opportunity: Papers Deriving from the Harvard University Faculty Seminar on the Coleman Report* (New York: Random House, 1972).

[29] Ellen Condliffe Lagemann, *Contested Terrain: A History of Education Research in the United States, 1890–1990* (Chicago: Spencer Foundation, 1996), p. 13.

Opportunity, Jean Floud, A. H. Halsey, and F. M. Martin showed that the Act had not significantly expanded working-class children's educational opportunities or created a system of educational advancement based on merit rather than social standing. The authors challenged the validity of IQ tests, which in their view measured the impact of environment rather than true intelligence. This work, along with Basil Bernstein's study of the different "linguistic codes" of working-class and middle-class children, heightened interest in how economic and cultural differences affected schooling. So-ciologists argued that working-class children's failure in school should be seen as resistance to cultural imposition of middle-class values demanded by the educational system. These ideas found a receptive audience in government when Anthony Crosland became minister of education and science under Labour Prime Minister Harold Wilson in 1965. Crosland advocated comprehensive schools as an alternative to the class-based tripartite system.[30]

As in the United States, public policy and professional opportunities strengthened British social scientists' interest in education. The *Newsom Report on Half Our Future* (1963) and the *Plowden Report on Children and Their Primary Schools* (1967) focused national attention on educational reform and echoed many of the issues raised by social scientific studies of education. A number of policy changes during the 1960s, particularly the establishment of the bachelor's degree in education, integrated teacher training into university education. The new "campus" universities served to accommodate the rapid growth of university departments of education and university centers dedicated to the study of education that hired social scientists. The Educational Research Board of Britain's Social Science Research Council, formed in 1965, provided leadership and money for research on education. Teachers of education marked their new professional success with the creation of the British Educational Research Association in 1973.

CONTINUING TIES?

The surge of interest in education had subsided by the mid-1970s. Although important remnants remained – some social scientists continued to pursue research agendas initiated during the 1960s, and that decade's developments in sociology, anthropology, and economics had a major impact on the cur-ricula and research interests of schools of education – education once again fell toward the bottom of the social science disciplines' research agendas. In part, this was due to a backlash against the social policies that had excited so much interest in education. In the United States, as in other countries, attempts to use education as an instrument of social change came to be seen

---

[30] Wooldridge, *Measuring the Mind*, chaps. 10–12; Silver and Silver, *An Educational War on Poverty*.

both as ineffective and as detrimental to education itself. This change was evident in the report *A Nation at Risk* (1983), which warned of the consequences of declining educational "excellence." The push to achieve educational excellence did not excite the same level of interest among social scientists as efforts to end social inequality.

In addition, tensions between "professional" approaches to educational research and the more theoretical orientation of the social sciences reemerged. Educators felt that schools of education had become too removed from the concerns of schools, and many attributed this problem to the influence of social scientists who had joined their faculties. The shift back toward applied research oriented to the needs of public schools is evident in the recent report *Tomorrow's Schools of Education* (1995), which called on schools of education to devote more of their resources to professional development.

At the same time, schools of education and educational research continued to struggle with their low status in the university. In 1997, the University of Chicago decided to close its Department of Education, claiming that its research did not meet the standards of the Division of Social Sciences. In the hundred years since John Dewey had founded the Chicago department, complex changes in the nature of social science research, the professionalization of educational research, and the relation of universities to primary and secondary education had widened the distance between the social sciences and education. The surge of research during the 1960s demonstrates both the possibility and the promise of closer ties between them, but it also signals the persistent tensions that make those ties difficult to maintain.

37

THE CULTURE OF INTELLIGENCE

John Carson

Over the past two centuries, the concept of human mental ability has undergone three important transformations: from a concept referring to a general faculty to one primarily referring to an individual attribute; from a focus on talents in the plural to one on intelligence in the singular; and from a position of relatively limited cultural significance to one of considerable weight within the United States and, to a lesser extent, within various European countries.[1] These shifts in meaning and emphasis have rendered intelligence a tool available to government, business, and the "helping professions" for the purpose of sorting, classifying, diagnosing, and justifying. Starting in the early part of the twentieth century, determinations of degree of intelligence have been used as aids in the placement of army recruits, in determing the kind of schooling a child will receive, in the hiring of job applicants, and in the decision to allow a person legal immigration. This chapter explores how intelligence has come to play these various social roles. It focuses especially on how experts in the human sciences have both created new meanings for the concept of intelligence and developed technologies that could make those meanings available and useful to a wider public.

FROM TALENTS TO INTELLIGENCE

During much of the nineteenth century, two distinct languages flourished in scientific and intellectual circles to describe the operations of the human

---

[1] On the history of intelligence and its uses, see John Carson, "Talents, Intelligence, and the Constructions of Human Difference in France and America, 1750–1920" (PhD dissertation, Princeton University, 1994); Kurt Danziger, *Naming the Mind: How Psychology Found Its Language* (London: Sage, 1997); Raymond E. Fancher, *The Intelligence Men: Makers of the IQ Controversy* (New York: Norton, 1985); Stephen Jay Gould, *The Mismeasure of Man* (New York: Norton, 1981); Nikolas Rose, *The Psychological Complex: Psychology, Politics and Society in England, 1869–1939* (London: Routledge, 1985); Roger Smith, *The Norton History of the Human Sciences* (New York: Norton, 1997)(in England, *The Fontana History of the Human Sciences*).

mind. Mental philosophers and others interested primarily in what would later be called the "normal" employed a language of character and talents, emphasizing the diversity of the mental faculties and the operation of the *individual* mind. Whether one subscribed to Scottish commonsense realism, to phrenological theories, or to the eclecticism of Victor Cousin (1792–1867), mind was represented as full of active powers, the faculties. Each, when triggered by external sensations, could act relatively independently on the ideas derived from those sensations and could add elements to them not present originally. Two characteristics stand out as particularly salient: first, the sheer number and variety of the faculties; and second, their malleability in response to external influences and the amount of effort exerted to develop them.[2] Thus what talents one possessed and what one became – successful or failed, knowledgeable or ignorant, moral or evil – could be seen to depend largely on early education and moral choices.

This emphasis on the variety of the mental powers and on individual responsibility for nurturing them gave the language of talents much of its political import.[3] Whether employed by those who responded to the postrevolutionary era by emphasizing order and character, or by those who insisted on equality of rights and opportunities, the language of talents provided one means to justify conceptions of the political and social order through recourse to widely accepted beliefs about the nature of human beings. Thus the French refashioned their educational system around the *concours* (competitive examinations), *lycées* (secondary schools), and *grandes écoles* (elite institutions at the apex of the French educational pyramid) in order, among other tasks, to identify and develop talent within the French citizenry and to enlist it in service to the state.[4] Americans did not produce such a thoroughgoing structural connection between talent and merit. Instead, individuals were thought to prosper or fail in the marketplace, as enshrined by American liberal ideology, according to their own hard work and their wise cultivation of particular capabilities. The power of the language of talents was that it conjoined and justified notions of both equal opportunity and unequal success.[5]

In contrast to the plasticity of notions of talent, scientific writers on notions of race and gender deployed the concept of intelligence in order to fix and explore human differences. Created by transforming reason from an absolute attribute into a characteristic that could be manifested in degrees, intelligence and its cognates imposed simple linear order on the animal and human worlds and suggested that this order, because naturalized, could scarcely be altered by human effort. Intelligence provided one way of accounting for what seemed

[2] Francis Wayland, *The Elements of Intellectual Philosophy* (New York: Sheldon, 1864).
[3] Stefan Collini, "The Idea of 'Character' in Victorian Political Thought," *Transactions of the Royal Historical Society*, 5th ser., 35 (1985), 29–50.
[4] Joseph N. Moody, *French Education since Napoleon* (Syracuse, N.Y.: Syracuse University Press, 1978).
[5] Daniel Walker Howe, *The Political Culture of the American Whigs* (Chicago: University of Chicago Press, 1979); Gordon S. Wood, *The Radicalism of the American Revolution* (New York: Vintage, 1991).

the obvious inferiority of certain peoples.[6] Often invoking the concept of the great chain of being – a linear scale stretching from the least intelligent organisms through humans and up to God – naturalists suggested that the human races could themselves be arrayed along this chain, with the"inferior" Africans placed closer to the rest of the animal world and the "superior" Europeans closer to the angelic. When applied specifically to human beings, intelligence as a general mental power varying by degrees and related to the physical nature of the brain allowed measurable external characteristics, such as cranial capacity, to signify power of mind, the measure of a people's place in the hierarchy of races.

Early-nineteenth-century racial anthropologists, then, replaced talents with intelligence, suggesting that human difference might be fundamentally biological in origin, and no more alterable by human effort than the difference in mental power between a monkey and a mouse. At midcentury, these ideas, at odds with both orthodox Christianity and the theories of mental philosophy, remained the province of only a few enthusiasts for racial science. But later in the century, notions of difference built on intelligence were disseminated widely. They were aided by the reduced influence of the more evangelical or conservative forms of Christianity; by the success of evolutionary theory, with its elimination of the gap between humans and other animals; by the adoption of craniometric ideas and techniques by a new generation of scientific psychologists; by the fear of democracy, sparked by working-class and feminist claims to equality; and by the spread of empire.

IQ: MAKING INTELLIGENCE A THING

In the broadest terms, the various conceptualizations of intelligence developed in Britain, France, and the United States during the late nineteenth and early twentieth centuries were produced in tandem with and in response to transformations occurring in most parts of the industrializing West. Three such developments merit particular attention. First, the immense reshaping of social life (urbanization, immigration, and colonization) and of the nature of work (industrialization, bureaucratization, and assembly-line production) undermined traditional methods of assessing, organizing, and managing human beings, and provided both the opportunity and the need for new ways of understanding and ordering the social world.[7] Second, the unprecedented technological innovations of the nineteenth century (railroad, telegraph, telephone, steam engine), and the notion of material progress that they were

---

[6] Carson, "Talents, Intelligence, and the Construction of Human Difference"; Gould, *The Mismeasure of Man*; William Stanton, *The Leopard's Spots: Scientific Attitudes toward Race in America, 1815–59* (Chicago: University of Chicago Press, 1960).

[7] James T. Kloppenberg, *Uncertain Victory: Social Democracy and Progressivism in European and American Thought, 1870–1920* (New York: Oxford University Press, 1986).

taken to confirm, increased the authority accorded to science as the principal means for comprehending the world and its inhabitants. Practices that could be associated with experiment, the laboratory, and science in general were especially favored. With them came an increasing commitment to technocratic solutions to social problems, especially among the managerial, bureaucratic, and professional classes – such as the American Progressives and the French Third Republic positivists – who saw science as one element in their own cultural ascendancy.[8]

Third, the impact of the evolutionary theories of Herbert Spencer (1820–1903) and Charles Darwin (1808–1882), the belief that physical stigmata could indicate mental disorders and criminal tendencies, and the rise of scientific racism all served to strengthen the plausibility of biological understandings of human behaviors as well as of human bodies. Such moves toward biological explanations raised up a host of concerns about the future of society that constituted the other side of the era's obsession with progress. Worries about degeneration, which pervaded Europe during the second half of the nineteenth century, and the reinterpretation of social problems in terms of medical pathology derived from the idea that nations possessed better and worse biological stocks and that the weaker, for a variety of reasons, might be winning.[9] In many respects the culmination of this wedding of anxiety and biology lay in the turn-of-the-century articulations of eugenics. A term coined by Francis Galton (1822–1911) to describe the need for active intervention into the breeding patterns of a population, eugenics implied that biology determined quality and that the success of a civilization depended on enhancing the reproduction of the "best" elements and retarding the reproduction of the least desirable. Widely accepted in Britain and the United States, eugenics had a significant impact throughout Europe and the Americas, and served to confirm for many the legitimacy of searching for biological markers of superiority or inferiority. These were viewed as characteristics not just of individuals, but of definable groups, be they a society's elite or those deemed to be socially marginal.[10]

The rise of "intelligence" is inseparable from these processes of modernity. In Britain, the major impetus came from Darwin's cousin Francis Galton. Thoroughly committed to the biological understanding of human behavior and to the power of statistics, Galton concluded in *Hereditary Genius* (1869) that eminence ran in families and thus that its cause must be inherited,

---

[8] Theodore M. Porter, *Trust in Numbers: The Pursuit of Objectivity in Science and Public Life* (Princeton, N.J.: Princeton University Press, 1995); Dorothy Ross, *The Origins of American Social Science* (Cambridge: Cambridge University Press, 1991); Robert H. Wiebe, *The Search for Order, 1877–1920* (New York: Hill and Wang, 1967).

[9] Mike Hawkins, *Social Darwinism in European and American Thought, 1860–1945: Nature as Model and Nature as Threat* (Cambridge: Cambridge University Press, 1997).

[10] Mark B. Adams, ed., *The Wellborn Science: Eugenics in Germany, France, Brazil, and Russia* (New York: Oxford University Press, 1990); Daniel J. Kevles, *In the Name of Eugenics: Genetics and the Uses of Heredity* (Berkeley: University of California Press, 1985).

like height or any other physical trait.[11] The association of intelligence with Galton's commitment to a biological marker of difference critical to success in life was cemented by the work of Charles Spearman (1865–1945) early in the twentieth century.[12] Employing his new statistical tool, factor analysis, Spearman analyzed the results of some basic psychological tests and concluded that test performance could be explained on the basis of two factors: task specific ability (*s*) and general intelligence (*g*).[13] His mathematical "demonstration" of the existence of general intelligence, although based on methods of assessing ability that would soon be discarded, helped give reality to intelligence as a global, quantifiable, biological object that different individuals possessed to different degrees and that could be used to link notions of biological fitness with success in the world.

In 1904, the same year that Spearman proposed his theory of *g*, the French psychologist Alfred Binet (1857–1911) was asked to serve on a ministerial commission studying abnormal children. Binet had already spent a number of years investigating the higher mental processes from a variety of perspectives. Well versed in the pathological approach to psychological questions that was the hallmark of French scientific psychology, Binet and his colleague Théodore Simon (1873–1961) sought to develop an instrument that could identify subjects of impaired intellectual ability.[14] What resulted was the Binet–Simon Intelligence Scale (1905), a series of thirty tests, mostly verbal, arranged from simplest to most difficult and designed to differentiate the four major classifications of intelligence within mental pathology: idiocy, imbecility, weakmindedness (*débilité*), and normality.

Three features differentiated the Binet–Simon scale from most other psychological instruments. First, it was oriented toward the higher mental abilities and a holistic assessment of their power. Second, the intelligence scale was relational and statistical: Rather than measuring mental ability directly, it used a series of seemingly arbitrary tasks, such as identifying the difference between a fly and a butterfly, that would allow a relative ranking of individual performances calibrated against a standard of what normal children of a given age could accomplish. And third, the primary product of the Binet–Simon scale was a diagnosis allowing classification into a medical/administrative category, not insight into the workings of mind in general or even of an individual mind in particular. Intelligence as defined by the Binet–Simon intelligence scale, especially as the scale was further modified

[11] Francis Galton, *Hereditary Genius: An Inquiry into Its Laws and Consequences* (1869) (Gloucester: Peter Smith, 1972).

[12] Fancher, *The Intelligence Men*; Rose, *The Psychological Complex*; Gillian Sutherland, *Ability, Merit and Measurement* (Oxford: Oxford University Press, 1984); Adrian Wooldridge, *Measuring the Mind: Education and Psychology in England, c. 1860–1990* (Cambridge: Cambridge University Press, 1994).

[13] Charles Spearman, "'General Intelligence,' Objectively Determined and Measured," *American Journal of Psychology*, 15 (1904), 201–93.

[14] Carson, "Talents, Intelligence, and the Construction of Human Difference"; Theta H. Wolf, *Alfred Binet* (Chicago: University of Chicago Press, 1973).

in 1908 and 1911, was something discrete, quantifiable, relative, statistical, developmental, practical, and defined most clearly by its pathological manifestations.[15] It was also an object of little interest to other French psychologists, who were less oriented toward the needs of school and asylum and felt little pressure to identify either the least or the most biologically desirable types of citizens.[16] Indeed, it was only when the test was exported, especially to the United States, that it and "intelligence" itself found true homes.

The Binet–Simon Intelligence Scale came to America in 1908, having been "discovered" by Henry H. Goddard (1866–1957), a psychologist at the Vineland Training School for Feebleminded Girls and Boys.[17] Goddard was interested in methods for the ready and accurate diagnosis of the mental states of residents and potential residents of the school. By the early 1910s, alternative versions of the Binet–Simon scale were being produced around the country, each adapted to the specific needs and conceptions of the investigator. A standard emerged in 1916, when Lewis M. Terman (1877–1956) published his Stanford Revision and Extension of the Binet–Simon Intelligence Scale (the Stanford–Binet scale), a version technically superior in almost every respect and one that quickly became the benchmark for work within the developing field of psychometrics.[18]

Like the Binet–Simon, the Stanford–Binet scale was an individually administered examination in which an examinee was asked questions by a trained psychologist, marked right or wrong on the answers, and then evaluated against a standard determined by his or her age peers. The result was summarized in the calculation of the individual's Intelligence Quotient (IQ) – mental age divided by chronological age times one hundred – a quantity proposed by the German psychologist William Stern. Designed to be constant over time, IQ was described as a measure of the examinee's biologically determined intellectual potential. Opaque to differences in intellectual performance that did not translate into raw score differences, Stanford–Binet homogenized intellect into a linear scale of relative brightness that could encompass not only those already categorized by their degree of intelligence – idiots and geniuses – but everyone in between as well, whatever their age or background or degree of education. And in part through the characterization Terman provided along with the scale, intelligence came to be seen as something biological, quantifiable,

---

[15] Alfred Binet and Théodore Simon, *The Development of Intelligence in Children*, trans. Elizabeth S. Kite (Baltimore: Williams and Wilkins, 1916).

[16] William H. Schneider, "After Binet: French Intelligence Testing, 1900–1950," *Journal of the History of the Behavioral Sciences*, 28 (1992), 111–32.

[17] Leila Zenderland, *Measuring Minds: Henry Herbert Goddard and the Origins of American Mental Testing* (Cambridge: Cambridge University Press, 1998).

[18] Lorraine Daston, "The Naturalized Female Intellect," *Science in Context*, 5 (1992), 209–35; Michael M. Sokal, ed., *Psychological Testing and American Society, 1890–1930* (New Brunswick, N.J.: Rutgers University Press, 1987).

and heritable, and as a decisive influence on behavior and status in life.[19]

Before World War I, dissemination of this psychometric version of intelligence remained limited. The successful introduction of intelligence testing into the American military, as part of the mobilization for the war, changed the situation decisively.[20] Confronted with the immediate need to sort and classify hundreds of thousands of new soldiers, military leaders were open to arguments that intelligence tests might prove of practical wartime value. Although the military itself remained ambivalent about the usefulness of large-scale testing, for American psychologists it proved an extraordinary boon. Forced to construct a new type of intelligence test that could be administered in groups, they succeeded in developing methods that allowed them to assess almost 1.75 million recruits. And in the process, intelligence itself became something familiar to everyone, a quantitative characteristic shown to be as applicable to the average person as to someone manifesting intellectual difficulties, and one that produced clear differentiations across the intellectual spectrum.

INTELLIGENCE AS A TOOL

In the aftermath of World War I, the place of intelligence and its tests in the topography of American culture seemed fairly secure. Many psychologists were involved in the study of intelligence and the development of new means of assessing it; companies specializing in the production of mental tests flourished; and testing was beginning to be used on a large scale at all levels of education and in industry. Finding employment primarily in academe, public education, and industry, American psychological testers constituted a growing interest group whose livelihoods were linked to the promotion of notions of intelligence and its importance. Their successes were particularly noteworthy in two areas: industry and education. Within industry, intelligence testing proved especially popular during the early 1920s, as managers looked to assessments of intellectual ability as one component of their evaluation of applicants for various white-collar positions. The enthusiasm for intelligence testing faded, however, later in the decade, to be replaced by a growing interest in personality as the key to business success.[21]

[19] Lewis M. Terman, *The Measurement of Intelligence: An Explanation of and a Complete Guide for the Use of the Stanford Revision and Extension of the Binet–Simon Intelligence Scale* (Boston: Houghton Mifflin, 1916); Henry L. Minton, *Lewis M. Terman: Pioneer in Psychological Testing* (New York: New York University Press, 1988).
[20] John Carson, "Army Alpha, Army Brass, and the Search for Army Intelligence," *Isis*, 84 (1993), 278–309.
[21] Loren Baritz, *Servants of Power: A History of the Use of the Social Sciences in American Industry* (Middletown, Conn.: Wesleyan University Press, 1960).

In education, by contrast, intelligence remained of central concern throughout the interwar period. During the 1920s, the number of positions for educational psychologists grew rapidly, as modernizing school districts sought guidance in organizing and administering the increasingly diverse student bodies characteristic of urban systems. One role for psychologists was diagnostic: to examine individual children who were manifesting educational problems. Their other major role, however, was more structural: to supervise large-scale intelligence testing as part of the process of placing students on the appropriate academic track. For certain individuals and groups, such testing proved to be of enormous benefit. Potential that may have been ignored because of various forms of prejudice, such as anti-Semitism, often stood out sharply thanks to the mechanical objectivity of the tests. Doors opened for some, however, also proved to be doors closed for others, as individuals and groups who performed poorly, such as African Americans and eastern Europeans, were often shunted away from opportunities that other modes of assessment might have made available.[22]

In large measure, the dissemination of the psychological approach to intelligence in America during the 1920s was based on the belief, fostered by the testers themselves, that intelligence played a critical role in determining an individual's place in society and success in life, and that mental tests were its authoritative gauge. Studies of intelligence carried out using the army data – widely trumpeted by Carl C. Brigham (1890–1943) in *A Study of American Intelligence* (1923) – and analyses of the results of postwar testing served to legitimate both the optimism and the anxieties of the American middle class.[23] Buoyed by the "discovery" that individuals of northern European descent were superior in intelligence to all other groups and that the American occupational hierarchy correlated highly with IQ, middle-class Americans – already fearful about "reds," immigrants, workers, and other seeming threats from within – were at the same time unsettled by the determination that a large percentage of adult American males were feebleminded or worse. Notions of a nation in biological and cultural peril abounded, reflected not only in the vogue for eugenics but also in Supreme Court Justice Oliver Wendell Holmes's famous opinion in *Buck v. Bell* (1927) upholding enforced sterilization of the feebleminded, and in the Immigration Act of 1924, which sought virtually to eliminate the immigration of southern and eastern Europeans, in part on the grounds of their biological unfitness.

In all of these debates, not to mention in schools and prisons and other institutions for the administration of the dependent or marginal, the language of intelligence played an important role, serving to link a perceived social

[22] Paul D. Chapman, *Schools as Sorters: Lewis M. Terman, Applied Psychology, and the Intelligence Testing Movement, 1890–1930* (New York: New York University Press, 1988).
[23] Carl C. Brigham, *A Study of American Intelligence* (Princeton, N.J.: Princeton University Press, 1923).

problem with a biological identity. Some, however, challenged this way of evaluating individuals. William C. Bagley (1874–1946), a psychologist at Teacher's College, Columbia University, worried about the antidemocratic implications of an intelligence whose level was presumed to be set from birth and determinative of an individual's future possibilities; and Walter Lippmann (1889–1974) carried on an extensive debate with Terman about the results of the army testing program and what they meant about the intelligence of the American population.[24] More prosaically, those who were subjects of the tests adopted a range of attitudes, from compliance to indifference to hostility, and public culture as often ridiculed the notion of testing in order to determine one's inborn potential as it supported such an idea. Nonetheless, what emerged and persisted throughout American culture was the belief that intelligence was something real, measurable, and able to influence, if not necessarily to decide, an individual's fate.

In Britain, intelligence and its tests provoked a much more ambiguous response. Following in the footsteps of Galton and Spearman, psychologists such as Cyril Burt (1883–1971) and Godfrey Thomson (1881–1955) worked energetically to establish the science of psychometrics and the practice of intelligence measurement, especially as part of the emerging field of educational psychology. Public interest in intelligence grew substantially after the war, and Thomson was particularly successful in promoting the adoption of intelligence testing by Scottish educational authorities. In England and Wales, however, the results were decidedly mixed. Little interest was evinced in large-scale mental testing, and in the main intelligence assessment, when it occurred, was conducted either for diagnostic reasons or by local educational authorities in the course of the 11+ examination, a test designed to determine which students would enter the university-preparatory curriculum. Many advocates saw standardized intelligence testing as representing a commitment to merit over privilege, a way of opening the class system to infusions of talent from below, and thus as central to ensuring the nation's progress by nurturing the biologically most able, regardless of origin. Ranged against these claims for merit, however, were not only those members of the elite who saw their privileges threatened, but also various groups who argued for a more complex calculus for determining what individual merit might mean.[25] These debates continued well after World War II, resulting in both broad cultural familiarity with the notion of quantified intelligence and piecemeal application of the technology designed to make it visible.

---

[24] The Lippmann–Terman debate is reprinted in N. J. Block and Gerald Dworkin, eds., *The IQ Controversy* (New York: Pantheon, 1976); see also William C. Bagley, "Educational Determinism; or Democracy and the I.Q.," *School and Society*, 15 (1922), 373–84.

[25] Sutherland, *Ability, Merit and Measurement*; Wooldridge, *Measuring the Mind*.

INTELLIGENCE IN AN ENVIRONMENTALIST CONTEXT

During the period from 1930 to 1970, two issues dominated discussions of intelligence in both the professional and, increasingly, the popular literature: the number of primary intellectual abilities, and the degree to which intelligence was inheritable. The question of whether intelligence is one thing or many arose early in the construction of the modern understanding of the term and has persisted up to the present. Spearman's demonstration of the unitary nature of intelligence, his *g*, was adopted by Burt and Terman and by most champions of IQ, and it became the dominant way in which intelligence was understood, both within the profession and popularly. Nonetheless, it did not go unchallenged. Diametrically opposed stood the American educational psychologist Edward L. Thorndike (1874–1949), who argued that the mind was composed of a vast array of specific and intrinsically independent abilities, with no underlying unity. Between them could be found, among others, L. L. Thurstone (1887–1955) and Thomson, who concluded on the basis of factor analysis that the primary mental abilities, though more than one, were few in number. During the postwar period, attempts were made to mediate this disagreement by, among others, Philip E. Vernon (1905–1987), who posited a pyramidical version of intelligence, with specific skills at the base and general intelligence at the apex. He was soon challenged, however, by Joy P. Guilford (1897–1987), whose model of mind eventually embraced 150 independent factors.

While the theoretical disagreement was profound, and the debate among these factions, especially during the 1920s and 1930s, was often sharp, the commitment of each to the existence of intelligence as a real entity with well-defined characteristics never wavered. David Wechsler (1896–1981), for example, began during the late 1930s to develop new instruments for measuring intelligence – the now-dominant Wechsler Intelligence Scale for Children (WISC) and Wechsler Adult Intelligence Scale (WAIS) – out of dissatisfaction with the Stanford–Binet scale and the concept of unitary intelligence. Unable to escape the practical demand for an overall measure of intelligence, however, he also provided an IQ score as well as assessments of verbal and nonverbal ability.[26] Both intelligence itself as a singular, quantifiable entity and its technologies of display were by that time so well established and so thoroughly incorporated into the operating structures of schools and asylums that they had taken on lives of their own, independent of the worries of psychometricians.

If the representation of intelligence as a unitary entity largely persisted from the 1930s until well into the postwar period, the same cannot be said of its characterization as a biological potential genetically determined from birth. As early as the 1910s, questions were raised about the "nature" interpretation

[26] David Wechsler, *The Measurement of Adult Intelligence* (Baltimore: Williams and Wilkins, 1939).

of intelligence, most significantly in the research of the anthropologist Franz Boas (1858–1942) on migration and changes in skull size among native peoples of northwest North America.[27] With enthusiasm for wholly biological and especially eugenic explanations of social phenomena themselves waning (at least in the United States) by the end of the 1920s, a number of psychologists advanced more decidedly environmentalist interpretations of IQ at the level of race and ethnicity.[28] In 1930, Brigham dramatically recanted his 1923 study, which had argued for the existence of a biological hierarchy of European groups (Nordic, Alpine, Mediterranean).[29] At about the same time, Boas's student Otto Klineberg (1899–1992) undertook research on the mean IQs of these European peoples and demonstrated that Brigham's initial findings had been the result of specific environmental conditions and not of underlying biological differences. Klineberg went on to challenge assertions about the innate intellectual inferiority of African Americans, showing that African-American migration to northern cities produced IQ gains that could best be explained in terms of the different educational environments of the North and the South.[30] This shift to "nurture" explanations of group-level differences was given official sanction after World War II, when UNESCO responded to Nazi eugenic policies by convening a conference on race, which concluded that race was a meaningless biological category and that suppositions of natural intergroup differences were unwarranted.[31]

The nearly unanimous rejection of hereditarian explanations for racial and group differences by the 1940s and 1950s was not matched, however, when researchers turned to explaining individual differences in measures such as IQ. There the commitment to biological conceptions of intelligence was much stronger, and the evidence more ambiguous. A number of studies conducted during the 1930s and 1940s, especially at the Iowa Child Welfare Research Station, buttressed the nurture side of the argument. Data on foster child placement, for example, indicated that IQ could change, often dramatically, when children were placed in different social and educational environments.[32] At the same time, research on identical twins by Burt, among others, suggested that a high percentage of an individual's IQ derived from his or her genetic inheritance. Although it is now clear that Burt's results were fraudulent, other studies continued to show the

---

[27] Franz Boas, *The Mind of Primitive Man* (New York: Macmillan, 1911); Carl N. Degler, *In Search of Human Nature: The Decline and Revival of Darwinism in American Social Thought* (New York: Oxford University Press, 1991).

[28] Degler, *In Search of Human Nature.*

[29] Brigham, "Intelligence Tests of Immigrant Groups," *Psychological Review*, 37 (1930), 158–65.

[30] Degler, *In Search of Human Nature*; Kevles, *In the Name of Eugenics*; Otto Klineberg, *Race Differences* (New York: Harper and Brothers, 1935).

[31] Elazar Barkan, *The Retreat of Scientific Racism: Changing Concepts of Race in Britain and the United States between the World Wars* (Cambridge: Cambridge University Press, 1992).

[32] Hamilton Cravens, *Before Head Start: The Iowa Station and America's Children* (Chapel Hill: University of North Carolina Press, 1993).

important influence of hereditary factors on an individual's measured level of intelligence.[33]

What is perhaps most striking about these debates is that they did not seriously upset continued reliance either on the notion of intelligence or on its technology of measurement. The Scholastic Aptitude Test (SAT), developed by Brigham in 1926 as an alternative to content-based tests for college admission, was widely adopted by American universities during World War II as part of the process of accelerating the production of trained individuals for the war. With the end of hostilities, however, instead of being abandoned, the SAT became institutionalized; it was represented as a way to make elite education available to all who were able, regardless of social background or schooling. For those during the 1950s and especially the 1960s who were interested in applying psychology directly to social policy – as did the American Great Society program and the British welfare state – the possibility that intelligence could be increased by improving a child's social environment could be used to justify a range of social programs, from neonatal care to school lunches to early childhood education.[34] And in the mundane tasks of diagnosing learning difficulties and assigning students to educational tracks, intelligence continued to serve as a significant source of both legitimation and guidance in the decision-making process.

CONCLUSION: THE IQ DEBATES, SOCIAL POLICY, AND THE RETURN OF BIOLOGY

A new round of controversies about the nature of intelligence and its measures erupted in 1969, initiated by the work of Arthur R. Jensen (b. 1923), a professor of education at the University of California, Berkeley.[35] Questioning the basis of programs such as Head Start, Jensen contended that environmentalist claims about intelligence were overstated and that both individuals and groups differed in terms of native abilities in ways that had significant social and economic consequences. His views sparked passionate responses from friends and critics alike. Coincident with a perceptible shift away from nurture explanations in a number of the human and biological sciences – epitomized by E. O. Wilson's articulation of sociobiology and by what would soon become the ascendancy of molecular genetics – and away from social interventionism in the realm of politics, Jensen's claims were supported on

---

[33] Leslie S. Hearnshaw, *Cyril Burt, Psychologist* (London: Hodder and Stoughton, 1979); Wooldridge, *Measuring the Mind*.

[34] Ellen Herman, *The Romance of American Psychology* (Berkeley: University of California Press, 1995).

[35] Arthur R. Jensen, "How Much Can We Boost IQ and Scholastic Achievement?," *Harvard Educational Review*, 39 (1969), 1–123.

both sides of the Atlantic by such psychologists as Richard J. Herrnstein
(1930–1994) and Hans J. Eysenck (1916–1997), as well as by policy makers
intent on tempering, if not dismantling, affirmative action programs and the
welfare state.[36]

Jensen's article also appeared in the wake of the social upheavals of 1968,
a time when disenchantment with the Vietnam War and with Western capitalist culture in general had helped to produce a serious, even pervasive,
skepticism about experts and their claims to authority. The biological meritocracy envisioned by Jensen, Herrnstein, and Eysenck, especially given its
highly racialized component, provoked a firestorm of criticism. Biologists and
psychologists including Richard C. Lewontin (b. 1929), Stephen Jay Gould
(1941–2002), and Leon Kamin (b. 1927) joined New Left college students and
social critics in organized opposition, at both the technical and policy levels,
to the hereditarian conception of intelligence being elaborated.[37] The data
on IQ and race and the results of identical twin studies received particular
attention, with charges of racist and antiscientific bias mingling freely with
arguments over the techniques used to measure heritability and the validity
of intergroup comparisons.

What resulted was not so much a victory for one side or the other
as a kind of institutionalized stalemate, marked by periodic skirmishes
over the ensuing twenty-five years. James Q. Wilson's (b. 1931) work on
the connections between criminality and IQ during the mid-1980s, and
then the publication in 1994 by Herrnstein and Charles Murray (b. 1943)
of *The Bell Curve* – in which they argue that America's socioeconomic
stratification is a meritocratic reflection of differences in innate levels of
intelligence – generated strong responses in both the popular and the
professional press.[38] There has been little open public support for claims
that innate biological inequalities in intelligence exist between races or
groups. Nonetheless, a dissatisfaction with the politics of pluralism among
certain segments of the middle and working classes at the end of the
1990s may have made the meritocratic individualism inherent in arguments
such as those contained in *The Bell Curve* more attractive than was publicly articulated. What is certainly clear is that the idea of unitary intelligence remains sufficiently vital in popular as well as scientific culture
to continue to provoke discussion, even when challenged by the theories
of multiple intelligence put forward by Howard Gardner (b. 1943) and

---

[36] Block and Dworkin, *The IQ Controversy*; Degler, *In Search of Human Nature*; Kevles, *In the Name
of Eugenics*.

[37] Block and Dworkin, *The IQ Controversy*; Steven Rose, R. C. Lewontin, and Leon J. Kamin, *Not in
Our Genes: Biology, Ideology and Human Nature* (London: Penguin, 1984).

[38] James Q. Wilson and Richard J. Herrnstein, *Crime and Human Nature* (New York: Simon and
Schuster, 1985); Richard J. Herrnstein and Charles Murray, *The Bell Curve: Intelligence and Class
Structure in American Life* (New York: Free Press, 1994).

Robert J. Sternberg (b. 1949).[39] Skepticism about the validity of intelligence tests and postmodern notions of the fractured self notwithstanding, intelligence in its various guises has become an institutionalized and deeply rooted aspect of culture, and especially of Anglo-American culture, one that is integral to the ways in which resources are allocated and democracy is discussed.

[39] Howard Gardner, *Frames of Mind: The Theory of Multiple Intelligences* (New York: Basic Books, 1983); Robert J. Sternberg, *Beyond IQ: The Triarchic Theory of Human Intelligence* (Cambridge: Cambridge University Press, 1985).

38

PSYCHOLOGISM AND THE CHILD

Ellen Herman

Psychologism is an elusive phenomenon in modern Western culture, located everywhere and nowhere, meaning everything and nothing at all. It refers to the discursive practice of using psychological explanations to make sense of individual and collective experience, and especially to link the two together. Because this explanatory resource is located in the slippery space between academic social science, clinical professionalism in psychology and medicine, and popular culture, psychologism is not fully at home in either the history of science or intellectual history. It has nevertheless often elicited sweeping cultural interpretations.[1]

Psychologism was initially championed by psychological professionals and their enterprising partners in Progressive Era reform before World War I. During this early phase, it was linked to the administration of subjectivity by means of normalizing technologies – standardized tests given to individuals by schools or by the military, for instance. More than a handy toolbox for the managers of mass society, psychological discursive practices were imported into individual projects of self-fashioning. By 1945, they had migrated to popular audiences who began as objects but soon became avid consumers of disciplinary knowledges and practices, inspiring individuals to embrace

---

[1] Robert N. Bellah, Richard Madsen, William M. Sullivan, Ann Swidler, and Steven M. Tipton, *Habits of the Heart: Individualism and Commitment in American Life* (New York: Harper and Row, 1985); Robert Castel, Francoise Castel, and Anne Lovell, *The Psychiatric Society*, trans. Arthur Goldhammer (New York: Columbia University Press, 1982); Ellen Herman, *The Romance of American Psychology: Political Culture in the Age of Experts* (Berkeley: University of California Press, 1995); Philip Rieff, *The Triumph of the Therapeutic: Uses of Faith after Freud* (New York: Harper Torchbooks, 1966); Nikolas Rose, *Governing the Soul: The Shaping of the Private Self* (London: Routledge, 1990); Peter N. Stearns, *Battleground of Desire: The Struggle for Self-Control in Modern America* (New York: New York University Press, 1999); Eva S. Moskowitz, *In Therapy We Trust: America's Obsession with Self-Fulfillment* (Baltimore: Johns Hopkins University Press, 2001). Philosophers used "psychologism" more narrowly, to denote the replacement of logical by psychological statements. An early instance of the broader cultural usage discussed here is Kingsley Davis, "Mental Hygiene and the Class Structure," *Psychiatry*, 1 (February 1938), 55–65.

the varieties of therapeutic experience as the surest path to mental well-being and happiness.

Among the many American figures who promoted a psychological world-view were the Progressives Henry Herbert Goddard and William Healy. At midcentury, professional experts Margaret Mead and Benjamin Spock achieved the status of cultural icons. At the end of the century, Oprah Winfrey was psychologism's most visible proponent, and her massive and enthusiastic audience indicated how thoroughly the therapeutic sensibility had trickled down to the grass roots. Although therapeutic solutions to personal and so-cial dilemmas were periodically ridiculed as psychobabble and accused of contributing to moral relativism and decline, they proved to be as politically flexible as they were ubiquitous. From reducing crime and boosting children's self-esteem to curing the apathy of the American electorate and mending the torn fabric of community life, psychological discourse has been considered practical and profound.

Psychologism has been a meaning-making system adapted to the scientific, bureaucratic, and democratic conditions of modernity.[2] A key component in the transformation of modern selfhood, "the psychological" mapped both a private interior *and* a public exterior geography, serving as both cause and effect of events from human development to group conflict. Society colonized the psychological terrain within individuals, while psychology seeped into the social space between them. By pointing out that persons and populations could not be effectively altered in isolation, psychologism highlighted the precarious balance between individual subjectivity and so-cial reality, and thereby asserted its relevance to the future of public life. With accumulating evidence of irrationalism ranged against the require-ments of the social order for a modicum of predictability and control, ad-justment became an indispensable social asset. The result was the cultural revolution sometimes called the psychiatric, psychological, or therapeutic society.

IN THE AMERICAN GRAIN

In 1957, *Life* boasted that "the science of human behavior permeates our whole way of life" and called it a "brand-new and strictly American" phenomenon.[3] Why did this happen so conspicuously in the United States?

Other nations possessed the requisite infrastructure and deployed psycho-logical knowledge for a variety of administrative and therapeutic purposes. In 1900, French spas were treating hundreds of thousands of cases of nervous

---

[2] Psychologism is especially, but not exclusively, visible in the United States. See Trudy Dehue, *Changing the Rules: Psychology in the Netherlands, 1900–1985* (New York: Cambridge University Press, 1995).

[3] Ernest Havemann, "The Age of Psychology in the U.S.," *Life*, 42 (January 7, 1957), 68.

disease annually.[4] During the first half of the twentieth century, German psychologists established a considerable practical record, even during the Nazi period, when the discipline succeeded in achieving military goals with state support.[5] Britain's Tavistock Institute, a self-described "out-patient clinic for social disorders," coordinated campaigns to place industrial management, consumption, child rearing, and education under the practical guidance of psychoanalytic experts after World War II.[6] From Wilhelm Wundt to Sigmund Freud and Jean Piaget, Europe was the author of psychologies that Americans idealized, imported, and imitated.[7]

But it was in America that individualism and Protestant self-regulation converged with scientific professionalism and consumer culture so completely that discrete innovations were transformed into a veritable worldview. Differences between American and European welfare systems – with European states offering comparatively earlier and more comprehensive social provision – must also be counted as factors. Governments that treated illness, unemployment, and poverty more as matters of structural distribution and less as evidence of personal defect deprived psychologism of cultural nourishment. Becoming psychological required that individuals feel personally damned by failure, vindicated by success, and willing to bear the anxious burden of making themselves up from scratch.

Insistent individualism is nothing new in America. For Tocqueville, who placed it at the heart of Jacksonian society, individualism typified the paradoxical process of universal democratization. The social condition of equality allowed Americans to indulge in the selfishness of privacy, relaxing their vigilance over public virtue. In the short run, individualism made people feel free, optimistic, and arrogant enough to think that they controlled their own destinies. In the long run, Tocqueville predicted that individualism would lock democratic citizens up in a sort of political solitary confinement, corroding the delicate but necessary threads linking the person to the public interest, while simultaneously increasing pressures to conform to mass opinion. As a critic of individualism, Tocqueville highlighted the relationship between the individual, the polity, and society as a major challenge for all who cared about America's future. To monitor the transaction between self and society was an obligation shared by democratic peoples and their governments.

---

[4] Edward Shorter, *A History of Psychiatry: From the Era of the Asylum to the Age of Prozac* (New York: Wiley, 1997).

[5] Ulfried Geuter, *The Professionalization of Psychology in Nazi Germany*, trans. Richard J. Holmes (Cambridge: Cambridge University Press, 1992).

[6] Jaques Elliott, "Some Principles of Organization of a Social Therapeutic Institution," *The Journal of Social Issues*, 3 (Spring 1947), 5. See also Peter Miller and Nikolas Rose, "On Therapeutic Authority: Psychoanalytical Expertise under Advanced Liberalism," *History of the Human Sciences*, 7 (August 1994), 29–64.

[7] A 1981 APA survey ranked only one American, B. F. Skinner, as a leading figure in postwar American psychology. See Albert R. Gilgen, *American Psychology since World War II: A Profile of the Discipline* (Westport, Conn.: Greenwood Press, 1982), Appendix A.

The Protestant mandate to account for one's life provided a morally charged context for psychologism in America, even as the relevant task shifted from the salvation of souls to the cultivation of character to the perfection of personality. New England's Puritan founders admonished their followers to regulate the self, since only disciplined displays of faith and obedience could make tolerable the intolerable unknowability of their eternal fate. As Protestants increasingly shared culture-shaping tasks with Americans of diverse faiths and nationalities during the nineteenth and twentieth centuries, theological dilemmas diminished, but ingrained habits endured. As Max Weber pointed out in *The Protestant Ethic and the Spirit of Capitalism* (1904; English trans., 1930), methodically ordering conduct to cope with spiritual terror had inaugurated an economically revolutionary but spiritually arid cycle of accumulation and investment. With the market transition from production to consumption, self-improvement generated through devotion to a calling gave way to the self-absorption required and encouraged by activities like shopping. By the early twentieth century, advertisers, whose industry both presumed and capitalized on the religiously inspired cult of inwardness, had ushered personality itself into the marketplace, making cosmetics, cars, and other possessions the markers of a selfhood that was as voracious as it was alluring.[8]

From nineteenth-century vogues like mind cure, to the Emmanuel movement of the early twentieth century, to post–World War II inspirational literature, popular Protestantism urged a gospel of individual improvement in which mental health never strayed far from the health of the body or the bank account. "You Can If You Think You Can," intoned minister Norman Vincent Peale, whose early fling with psychoanalysis and flair for popular psychology produced a series of best-sellers equating self-determination and material success with "applied Christianity."[9] Protestantism and psychology both attended to selfhood. That seventeenth-century selves determined by almighty God were so energetic despite a paralyzing theology was the psychological paradox at the heart of Weber's analysis. That twentieth-century selves ruled largely by personal inclination and desire have been so fragile despite their alleged "empowerment" is surely a measure of modern disenchantment.[10]

Psychologism cannot be reduced to the rise of the psychological disciplines, but the special appeal of professional authority in a meritocracy contributed to America's deference to psychological expertise. In America, intelligence

[8] Jackson Lears, *Fables of Abundance: A Cultural History of Advertising in America* (New York: Basic Books, 1994).

[9] Donald Meyer, *The Positive Thinkers: Religion as Pop Psychology from Mary Baker Eddy to Oral Roberts* (New York: Pantheon, 1980).

[10] Thomas L. Haskell, "Persons as Uncaused Causes: John Stuart Mill, the Spirit of Capitalism, and the 'Invention' of Formalism," in *The Culture of the Market: Historical Essays*, ed. Thomas L. Haskell and Richard F. Teichgraeber (Cambridge: Cambridge University Press, 1993), 441–502.

deserved respect because it represented a defensible rationale for social superiority: It produced hierarchy legitimately. Professionalism appeared to be founded on sturdy premises, including educational qualifications, autonomous practice, internal monitoring, and tangible results.

After the late nineteenth century, universities served as the gatekeepers of professional status, and in the academic milieu psychology came to be imbued with the prestige of science. The experimental methods and objective claims of psychological professionals made specialized knowledge about human beings and behavior seem not only plausible, but politically acceptable. The emergence of an industrial order in the late nineteenth century and the Progressive Era effort to tame its worst abuses made psychologism socially necessary as well. Embraced by reformers whose causes ranged from factory labor to pure food to municipal administration, scientific professionalism animated psychologism during its early phases. This gave cultural mavericks a great strategic advantage, and they seized every opportunity to discuss designs for social change as questions of neutral technique.

Only during and after the Vietnam war, a disastrous conflict managed by "the best and the brightest," was expertise recast as something other than a benevolent tool of legitimate democratic authority. Perhaps it was not an oasis of reason in the struggle for power after all, but a new guise for politics and the maintenance of social inequality.

FROM ELITE PATRONAGE TO STATE SUPPORT

Before 1940, private philanthropies like the Rockefeller Foundation patronized psychological practices. After 1945, social psychological management was more likely to be federally funded, an indication that "the psychological" had migrated to the center of government; it figured in the uncomfortable convergence between private and public spheres. Although therapies tended to be associated with humanitarian aims and tests with administrative imperatives, the distinction had been blurred in practice. Helping, supervising, and coercing mingled. Efficiency and enlightenment progressed in tandem, making power's exercise an element of personal experience. This accounts in part for psychologism's ethical quandary and historical importance.

Psychological regulation was firmly linked to war and military institutions throughout the century. During World War I, military officials were persuaded that massive testing programs would lubricate selection and classification procedures, and psychologists toiled to set officers apart from privates and privates apart from men unfit to serve. Robert Yerkes' Committee on Methods of Psychological Examining of Recruits produced novel group intelligence tests known as Alpha and Beta; approximately two million of these were administered by an army of psychologists. (Although later discredited, their most dramatic discovery was demoralizing: The average soldier had a mental age of thirteen, and a substantial number of the nation's defenders were classifiable as morons.) If wartime testing failed to enhance military efficiency, it succeeded in turning the subjectivity of soldiers – and psychology itself – into a key public resource.[11] What soldiers thought and how they felt about the military environment, military authority, and military roles mattered. Because those feelings were potentially positive or negative, military managers who adopted the therapeutic approach would conserve human resources humanely.

World War II was another watershed. It inaugurated a period during which regulatory probing motivated by administrative aspirations – whether in the realm of criminal justice or military operations – gave way to widespread faith in psychology's potential to promote self-understanding and happiness. Wartime mobilization produced millions of standardized tests, supplemented by clinical strategies to "make morale."[12] Easy-to-learn methods of mental control were prescribed. Psychiatrists operated in combat. Soldiers participated in group therapy and were treated with hypnosis and drugs. Initiated to conserve manpower and eliminate troublemakers, contact between individuals and clinicians had enormous consequences for the future trajectory of professional help. The National Mental Health Act of 1946 was a direct outgrowth of anxiety about soldiers' mental well-being – a staggering 1.8 million recruits had been rejected from the armed forces and another 550,000 discharged for neuropsychiatric reasons. The war years also laid the foundation for the National Institute of Mental Health (1949), which pumped millions of research dollars into cementing the linkage between clinical efficacy and psychological science.

The country's governmental and professional elites intended to aid the psychological casualties of war, but veterans and their kin also demanded assistance. They lobbied for a menu of helping options applicable to everyday problems – marital tension and parenting difficulty were frequent complaints – and clients welcomed expanded Veterans Administration outpatient services after the war. The remarkable expansion of this "growth industry" was blanketed in the language of responsible fiscal management. Mental health was a commodity that "could be purchased" for substantially less money than the nation was already spending on mentally disabled veterans, argued William Menninger, the first psychiatrist ever made a brigadier general.[13] Personal well-being and social regulation advanced together.

As an extended illustration, let us consider the domain of childhood.

---

[11] Franz Samelson, "World War I Intelligence Testing and the Development of Psychology," *Journal of the History of the Behavioral Sciences*, 13 (July 1977), 274–82.

[12] Edward A. Strecker and Kenneth E. Appel, "Morale," *American Journal of Psychiatry*, 99 (September 1942), 159–63.

[13] William C. Menninger, *Psychiatry in a Troubled World: Yesterday's War and Today's Challenge* (New York: Macmillan, 1948), p. 410.

CHILDHOOD BECOMES PSYCHOLOGICAL

Children's lives have always attracted enthusiastic interest from those charged with their care. Systematizing the scrutiny of children's lives by turning human development into a laboratory and its study into a science has brought childhood into the psychological orbit. Because this project required specialized training and skill, it necessarily demoted the authority of parents, whose direct, experiential knowledge did not equip them to recognize the signs of normality or to guide their children's growth unassisted. By enveloping childhood with expertise, psychological discursive practices signaled parental incompetence and subjected the developmental process to new forms of government. Childhood was also a public resource that cried out for scientific stewardship and state intervention. Amateurs, however well-meaning, regularly bungled the job; to rely on them was to squander social assets.

Sigmund Freud (1856–1939) was the most famous author of psychological childhood, and his warm American reception publicized the notions that early life was momentous, that regularities were discernable in development, and that unconscious psychological events were pivotal. In Freud's view, painstaking, psychoanalytic retrieval of repressed childhood traumas connected formative experiences to adult outcomes. Determined in childhood, mental life abided no accidents.

But American childhood had been psychologized even before Freud, at virtually the same moment that a psychological discipline emerged. G. Stanley Hall (1844–1924), a founder of American psychology best known for his role in bringing Freud to Clark University in 1909, marshaled evolutionism to claim that individual development recapitulated society's halting steps from savagery to civilization. Although the particulars of Hall's theory were discarded by later generations, the genetic premise of Hall's approach remained intact. Children were pieces of nature and their development a natural, patterned phenomenon. Not eccentric or unpredictable, children had become manipulable objects whose use or abuse would decisively shape their entire lives. The twin mandates of psychologism – to comprehend and to change, to help and to control – were joined in childhood's new history.

To tackle the "nature" of childhood methodically required the tools of science rather than those of sentiment. Psychological and sentimental childhood are not identical, but both presumed to transcend the material imperatives of household subsistence and child labor by attending to the world of emotion.[14] Hall's influence was visible in the rise of the child study movement in the 1890s. Members of the Society for the Study of Child Nature, later renamed the Child Study Association of America (CSAA), armed themselves with cameras, measuring instruments, and orderly habits of observation, interviewing,

[14] Viviana Z. Zelizer, *Pricing the Priceless Child: The Changing Social Value of Children* (New York: Basic Books, 1985).

and documentary record keeping. Directed by professionals in psychology and education, the movement was galvanized by reformist zeal as well as by the desire to invent a novel science. Infusing parenthood with experimentalism would enhance children's welfare. "We need to find a psychological equivalent for the kind of knowledge which the technician has," declared one reformer, as she struggled to articulate the point of parent education. "We have to make all our parents co-partners in research enterprises."[15]

Study groups filled with mothers (fathers were rare) met regularly to conduct "research" by analyzing their children. As vehicles for disseminating authoritative child-rearing advice, members were expected to spread the gospel of popular science. Mothers did not always play this submissive role, especially when expert knowledge contradicted the evidence of experience, but many sought expert knowledge nonetheless.[16] "Here were specialists with a vital grip on the realities of human relationships," wrote the CSAA director Sidonie Gruenberg, in a special issue of *Child Study* devoted to "Parents and the New Psychologies."[17] Obliged to reeducate themselves as pseudo-professionals, parents soon realized that the cost of steering their children's development was high. Parents were to blame for how children turned out. Guilt was the legacy of power wielded by psychological parents over psychological children.

Psychologizing childhood reconfigured the norms of suitable child rearing and worthy parenthood. Older conceptions that made material provision – food, clothing, shelter – the mark of responsible nurture fell by the wayside, and instrumental attitudes toward children – lingering notions, for instance, that their labor was a legitimate family resource, or that corporal punishment was a parental entitlement – were recast as inimical to children's well-being.

Making childhood psychological therefore incorporated powerful class biases into kinship ideals.[18] Even Sidonie Gruenberg, born into a wealthy German Jewish family, found the upper-class tenor of child study groups in turn-of-the-century New York insufferable. She tried to spread the psychologistic message to working-class parents, without much success. A 1913 experiment to organize an entire public school district into study groups excited many teachers, but it failed to interest working parents, who doubted the merit of such a scholarly approach to their kids.

---

[15] Miriam Van Waters, as quoted in Roberta Lyn Wollons, "Educating Mothers: Sidonie Matsner Gruenberg and the Child Study Association of America" (PhD dissertation, University of Chicago, 1983), p. 231.

[16] Julia Grant, *Raising Baby by the Book: The Education of American Mothers* (New Haven, Conn.: Yale University Press, 1998), chap. 5.

[17] Sidonie Matsner Gruenberg, "How New Psychologies Affect Parental Practices," *Child Study*, 6 (October 1928), 11.

[18] Linda Gordon, *Heroes of Their Own Lives: The Politics and History of Family Violence* (New York: Viking, 1988); Ellen Ross, *Love and Toil: Motherhood in Outcast London, 1870–1918* (New York: Oxford University Press, 1993); Kathleen W. Jones, *Taming the Troublesome Child: American Families, Child Guidance, and the Limits of Psychiatric Authority* (Cambridge, Mass.: Harvard University Press, 1999).

Initially, then, psychologism was the ideology of those middle strata whose standard of living had moved them beyond subsistence struggles. Interestingly, these social origins were later codified in the personality theory of Abraham Maslow, who posited a hierarchy of motivations in which survival needs were lower and less human than higher needs for love and achievement.[19] In the era of psychological childhood, parents who merely ensured physical survival, leaving their children's psychological development to luck or fate, were likely to be regarded as incompetent, if not neglectful or worse. Since infant mortality remained high in overcrowded neighborhoods whose residents often lacked decent food and shelter, many Progressive Era reformers knew that moving beyond material deprivation was a luxury most people could not afford. Nevertheless, middle-class medicalizers made their norms into universal edicts. Safeguarding the health of young bodies and minds meant educating mothers as surely as it meant improving housing, sanitation, and milk.[20]

Nowhere was the fit between top-down reform and psychologism's promise to reveal the true forces underlying social problems tighter than in the field of juvenile crime, whose leaders had early on embraced therapeutic justice. Inspired by Jane Addams and John Dewey, novel approaches to young people in trouble severed childhood from crime, discarded legal procedures, and reinvented judges as kindly helpers rather than arbiters of guilt or innocence. These approaches were pioneered in the Chicago Juvenile Court, where the psychiatrist William Healy introduced mental tests in 1909. According to Sophonisba Breckenridge and Edith Abbott, who studied every case that passed through the court in its first decade, only "more exact knowledge" of troublemakers' social worlds and "the researches in biological and psychological laboratories" portended genuine help for children.[21] The Progressive Era sociology of crime produced mountains of statistical social research about urban life. From the 1920s onward, reformers gravitated toward psychological explanations. Studying the individual juvenile delinquent justified the professional role, even if alarming recidivism rates dampened reformers' optimism considerably.[22]

The Chicago model was championed by a host of authorities on delinquency and child psychology, most of them women, who believed in its scientific promises and saw in them a way to advance a liberal welfare state committed to maternalism. Collectively, they issued a call to individualize rather than to criticize, to sympathize rather than to judge. Jessie Taft

[19] Abraham Maslow, *Motivation and Personality* (New York: Harper and Row, 1954).

[20] U.S. Children's Bureau, *Baby-Saving Campaigns* (Infant Mortality Series No. 1, Bureau Publication No. 3) (Washington, D.C.: U.S. Government Printing Office, 1913).

[21] Sophonisba P. Breckinridge and Edith Abbott, *The Delinquent Child and the Home* (New York: Russell Sage Foundation, 1912), pp. 11, 173.

[22] Ellen Ryerson, *The Best-Laid Plans: America's Juvenile Court Experiment* (New York: Hill and Wang, 1978).

(1882–1960), who completed her doctorate in sociology at the University of Chicago in 1913 and became an important social work theorist and educator, helped to mediate the exchange between professional help and social science. In a 1919 manifesto for psychological childhood, Taft urged her professional colleagues to scrutinize individual personalities and to relinquish traditional attitudes about bad children, immoral parents, and their wicked ways. "The self is a very complex, elusive, changing phenomenon and we should approach it with an humble spirit, an open mind and a desire not so much to judge as to understand."[23] Miriam Van Waters (1887–1974), one of the country's best-known prison reformers by midcentury, similarly championed psychological childhood. She surrounded herself with professionals who shared her belief that treatment was always preferable to punishment and that such child-caring institutions as schools, reformatories, and courts should be knowledge-producing laboratories issuing blueprints for individual adjustment.

Between the world wars, Van Waters, Taft, and other champions of psychological childhood promoted an approach to social welfare that stressed therapeutic work as the only effective basis for social work. Improved techniques of data gathering and record keeping had revealed that nostrums about purely material provision were fatally flawed. Taft dismissed "indiscriminate alms-giving" by do-gooders and called for reform that married inner change to the improvement of external circumstances.[24] Psychologically informed home studies suggested that purely rational modes of help were outdated, that impoverished families were plagued by emotional troubles as well as economic crises.

Van Waters, who began her doctorate under G. Stanley Hall but switched to anthropology because of Hall's sexism and autocratic style, never lost her taste for psychology's constructive application. Schooled in a Progressive Era climate "electric with newly discovered complexes which were going to account for all problem children," Van Waters consistently employed mental testers and psychiatric consultants for the delinquents she hoped to redeem, even when it meant raising private funds in the era before such services were routinely provided by public agencies.[25] Her first book, the influential *Youth in Conflict* (1926), advanced an emphatically psychological approach to delinquency, perhaps in part because it concentrated on girls. Throughout her career, Van Waters favored case conferences, which brought to bear intensive professional teamwork, IQ scores, sexual histories, family genealogies, Rorschach tests, and other data on the psyche of the person in question.

---

[23] Jessie Taft, "Relation of Personality Study to Child Placing," *Proceedings of the National Conference of Social Work*, 46 (1919), 67.
[24] Jessie Taft, "The Spirit of Social Work," *The Family*, 9 (June 1928), 104–5.
[25] Freedman, *Maternal Justice*, p. 60.

Taft and Van Waters believed that any child or parent whose misdeeds came to the attention of authorities was "abnormal" enough to merit systematic psychological scrutiny. But psychological discursive practices encompassed all children, however "normal." The habit of routinely measuring children's developmental achievements against established norms was given force by such scientists as Arnold Gesell (1880–1961) of Yale University, a student of G. Stanley Hall's, whose laboratory studies during the 1920s and 1930s yielded scales widely adopted by clinicians and child-rearing experts. The spread of Freudian thought eroded the belief that normality was fixed and unambiguous. In 1930, one essayist concluded that "family life, when it is not a palpable study in mental deficiency, is obviously enough a study in lunacy. Family life doth make idiots and lunatics of us all."[26]

The public reach of these attitudes was evident everywhere from child guidance clinics to Head Start. "We begin to see the emergence of a fairly new conception of the task of social welfare," wrote Lawrence K. Frank prophetically in 1931. "The difficulties and shortcomings of our social life, in economic, political and family affairs, are to be viewed as the products of maladjustment and distortion of personality rather than the operation of large impersonal 'systems' and 'forces'."[27] By midcentury, "nonjudgmental" was a key word in all work with children and other dependents, so rampant that one knowledgeable observer found it indistinguishable from the helping impulse.[28] From child abuse to poverty to bad grades, social problems could be kept at bay by perfecting the skills of professional helpers, guaranteeing that their practices rested on sound research, and insuring that their services would be widely consumed and their worldview gratefully adopted.

To adjust the developing personalities of children through techniques saturated with knowledge about them required tests to determine intelligence, scales to gauge developmental progress, therapies to probe emotional makeup, interviews to compile life histories, and investigations to take the pulse of domestic life. These practices were favored by a wide range of agencies, from those remediating educational failures and placing children in new families to others preventing disease and rehabilitating youthful offenders.

For parents, there were Freudian popularizers like Benjamin Spock, whose *Baby and Child Care* offered advice synchronized with psychologism. The book went through 208 printings and sold 28 million copies in the thirty years following its publication in 1946, making it the country's best-selling book after the Bible. The psychoanalytic developmental narrative also made childhood sexuality and "object relations" bear inordinate meaning for adult

---

[26] Samuel D. Schmalhausen, "Family Life: A Study in Pathology," in *The New Generation: The Intimate Problems of Modern Parents and Children*, ed. V. F. Calverton and Samuel D. Schmalhausen (New York: Macaulay, 1930), p. 275.

[27] Lawrence K. Frank, *American Journal of Sociology*, 36 (July 1931), 156.

[28] Dorothy Hutchinson, "Some Thoughts on Being Non-Judgmental," *Child Welfare League of America Bulletin*, 21 (February 1942), 3–4.

life. Anxieties about sexual abuse, the emotional consequences of day care for young children, and the curious spread of "attachment disorders" that surfaced later in the century were all loosely traceable to Freudian thought.

After 1960, critiques of psychological childhood emerged on both the left and the right. Angry about paternalistic schemes that sacrificed children's due process rights, radicals charged child welfare professionals with enforcing the distorting discipline at the heart of liberal state authority and called for procedural safeguards. At the other end of the political spectrum were conservatives who suspected that psychological childhood winked at bad behavior, reinforced moral relativism, and diluted accountability. For these critics, stern lessons in right and wrong were the answer. Schools, families, and courts should be guardians of virtue, not dispensers of therapeutic excuses. One consequence was the increasing tendency at the end of the century to treat youthful offenders as adults, who deserved punishment rather than counseling.

Even under attack, psychologism remained the inescapable reference point for the cultural conversation about childhood. It altered the experience of childhood and the government of children because it revolutionized thinking about who human beings were, how they developed, and what they needed most, in or out of trouble.

FROM SCIENCE TO HELP: THE GENDER OF PSYCHOLOGISM

The emergence of a "psychological society" after 1945 inflated the ranks of psychological personnel in virtually all areas of American life. Loosely grouped around clinical psychology, psychiatry, and other helping professions whose members stretched from social work to education and theology, psychologism thoroughly permeated the movement that took mental health as its goal. As psychology moved from elite reform early in the century to popular culture after 1945, its occupational prototype shifted from scientist to helper.

Consider the following statistics. Established by twenty-six pioneers in 1892, the American Psychological Association (APA) had grown to over 83,000 members a century later. With fifty-three divisions, forty-two journals, and a large staff engaged in legislation, advocacy, and research, the APA is the largest psychology organization in the world.[29] By century's end, the APA was dominated by clinicians, yet before 1940 few psychologists did therapeutic work. Expansion of the helping trades proceeded rapidly after the war, when healers in clinical psychology and a host of new counseling fields – humanistic therapy and marriage counseling, for instance – institutionalized a market for therapeutic services with promotional flair and generous state funding.

[29] www.apa.org

Psychiatry also was transformed. Traditionally associated with madness and asylums, psychiatry in the late nineteenth and early twentieth centuries set its sights on prevention and treatment of the normal anguish of normal people.[30] With a membership of 42,000 at the end of the twentieth century, the American Psychiatric Association had tilted toward private practice for decades – the decisive shift occurred during the 1940s – with asylum doctors becoming a marginalized minority.[31]

Allied professionals worked under the close supervision of psychiatrists and psychologists, absorbing their ambitions along with their practical technologies. Some historians argue that social work had to borrow psychological theory and technique in order to establish professional credentials during the decade around World War I, while others date the "psychiatric deluge" to 1940.[32] In either case, the adoption of therapeutic ideals was responsible for much of social work's growth during the last half of the century. Between 1975 and 1990, the ranks of clinical social workers grew from 25,000 to 80,000, and membership in the National Association of Social Workers grew to 155,000.[33] Pastoral counseling also exploded after 1945, with new patterns of service delivery and consumption reconstructing the historic nexus between religion and psychology.

While social work was always women's work, the male-dominated psychological profession grew progressively more feminized as psychology gravitated toward helping. The first generation of female psychologists consisted of only twenty-five individuals, nineteen with doctorates, whose professional contributions spanned all the specialties of their day.[34] By 1950, the percentage of psychology doctorates awarded to women was 14.8 percent, and by 1987 it was 53.3 percent, an upward trend coinciding with the sharp turn toward clinical work.[35]

Helping exists on the moral plane of emotion and partiality rather than of reason and disinterestedness, a plane of American culture located below that of science, in part because both helping and science have been historically gendered. Periodic anxieties about feminization illustrate not only that helping was defined as women's work but also that science was not. Ironically, the experimentalism and objectivity associated with the rise of psychologism were threatened by its success as an applied helping science.

[30] Elizabeth Lunbeck, *The Psychiatric Persuasion: Knowledge, Gender, and Power in Modern America* (Princeton, N.J.: Princeton University Press, 1994).

[31] www.psych.org; Jack Pressman, *Last Resort: Psychosurgery and the Limits of Medicine* (Cambridge: Cambridge University Press, 1997), p. 363.

[32] John H. Ehrenreich, *The Altruistic Imagination: A History of Social Work and Social Policy in the United States* (Ithaca, N.Y.: Cornell University Press, 1985); Martha Heineman Field, "Social Casework Practice during the 'Psychiatric Deluge,'" *Social Service Review*, 54 (December 1980), 482–507.

[33] www.naswdc.org

[34] Elizabeth Scarborough and Laurel Furumoto, *Untold Lives: The First Generation of American Women Psychologists* (New York: Columbia University Press, 1987), pp. 134, 142, 168.

[35] National Science Foundation, *Profiles – Psychology: Human Resources and Funding* (Washington, D.C.: National Science Foundation, 1988), pp. 33, 34.

Thus the gendered appeal of nurture both encouraged the spread of psychologism in the culturally feminized sphere of love and family and jeopardized its scientific stature. It is no accident that most vocal critics of psychologism have been men.

Among the men who gave psychologism a bad name, Kingsley Davis, C. Wright Mills, Christopher Lasch, and Russell Jacoby expressed nothing but disdain for the language of "social pathology."[36] Their views were more than sexist reflexes. They complained that psychological discursive practices camouflaged social engineering as science, breathed new life into community-destroying individualism, and promoted a retreat into privatism that subordinated private experiences – such as childhood – to the unforgiving discipline of market values and state power. This powerful critique reminds us that human science often sacrificed the humane for the dispassionate, that reducing the social to the mental obstructed positive change, and that the welfare state established constraints while proclaiming the dawn of enlightened government.

Their protest is tinged with nostalgia, however. Critics fail to appreciate how cultural transformation has complicated the moral terms of modern life. Americans were not simply talked into psychologism. Many veterans, parents, and others found in it a resource that made emotional and cultural sense of how modernity felt, and that helped them to negotiate modern society's treacherous trade between individual and group, inner and outer, self and society, liberation and domination. It will not do simply to reassert the primacy of "the social," insisting that private life stay that way, as if it were possible to sever help from authority, knowledge from power, and psychological from social welfare. Observers may long for a clearer, more comforting moral map than the one constituted around psychologism. It is unlikely that they will find one.

[36] Davis, "Mental Hygiene and the Class Structure"; Russell Jacoby, *Social Amnesia: A Critique of Contemporary Psychology from Adler to Laing* (Boston: Beacon Press, 1975); Christopher Lasch, *Haven in a Heartless World: The Family Besieged* (New York: Norton, 1977); C. Wright Mills, "The Professional Ideology of Social Pathologists," *American Journal of Sociology*, 49 (September 1943), 165–80.

39

PSYCHIATRY

Elizabeth Lunbeck

Psychiatry, a branch of medicine, is a discipline that historically has taken severe mental illness as its object. In the course of the twentieth century, it dramatically expanded its purview, bringing the full range of human behaviors, both normal and pathological, within its domain. By the end of the century, psychiatry dealt with problems in everyday living as well as with schizophrenia and depression. It had become, as psychiatrists sometimes put it, "as broad as life."[1] In the process, it had also moved into the domain of the social and behavioral sciences. This essay will examine that expansion, focusing primarily on American psychiatry.

Unstably situated between genetics and biology on the one hand and the behavioral sciences – psychology, sociology, anthropology – on the other, psychiatry uniquely bridges medicine and the disciplines. Its practitioners take into account "everything from the molecular level to the most basic social issues."[2] They disagree, sometimes vehemently, on its goals, practices, and fundamental truths. Equally important to its instability, psychiatry has been thoroughly remade during the twentieth century. This remaking has fostered an eclecticism that is often interpreted as fragmentation, with psychiatrists advocating a broad range of conflicting models of behavior and disease. The remaking has also fostered a curious relationship between psychiatrists and their discipline's history, with some taking pride in its ancient roots and others willing to jettison them and proclaim their discipline altogether new. "What is psychiatry?" has seemed to many a pressing question throughout the twentieth century.

From its origins as a distinct specialty in the 1820s and 1830s, American psychiatry existed somewhat apart from the rest of medicine. Its practitioners,

[1] Roy R. Grinker, Sr., "Psychiatry: The Field," *International Encyclopedia of the Social Sciences*, ed. David L. Sills (New York: Macmillan, 1968), vol. 12, pp. 607–13, at p. 608.
[2] David R. Hawkins, "The Role of Psychiatry in Society: Introduction," in *American Psychiatry: Past, Present, and Future*, ed. George Kriegman, Robert D. Gardner, and D. Wilfred Abse (Charlottesville: University Press of Virginia, 1975), pp. 131–4, at p. 133.

known as "alienists," labored in large, custodial asylums, isolated from the nation's centers of population and from medicine's main currents, concerned largely with issues of institutional management; its science was an amalgam of moral and religious speculation that no medical school deigned to teach; its punitive practices brought it repeatedly under the scrutiny of an outraged public. When medicine, later in the century, began to embrace science, psychiatry's status became even more marginal. The development of the scientifically identified specialty of neurology, from physicians' experience with gunshot wounds during the Civil War, highlighted the alienists' scientific incapacity. In 1894, the eminent neurologist S. Weir Mitchell berated his psychiatric colleagues, in an address before their professional organization, the American Medico-Psychological Association, for, among other failings, their isolation from medicine's main institutional and scientific currents and their lack of an inquiring spirit. "Want of competent original work is to my mind the worst symptom of torpor the asylums now present," he noted, asking where "are your careful scientific reports?" That his listeners merely concurred with his critique was a measure of their demoralization.[3]

Within two decades, however, the efforts of a number of psychiatrists who were dissatisfied with their discipline's social and scientific marginality had thoroughly transformed it. Nineteenth-century psychiatry had been concerned primarily with insanity. Replacing sharp distinctions between the insane and everyone else, early-twentieth-century psychiatrists adopted a metric mode of thinking about symptoms and persons, arraying the human population on a scale from normal to abnormal, with only a loosely defined demarcation between illness and health. This new brand of psychiatry, turning its attention to normal persons and everyday life, was to have applicability to everyone, from the insane to the worried well. Sigmund Freud's new science of psychoanalysis, which was also a discipline of the everyday, highlighting the significance of life's routine aspects and blurring the distinction between normality and abnormality, also figured in psychiatry's transformation. Freud's well-publicized visit to Clark University in 1909 introduced him to the American psychiatric community. Beginning in the 1920s, mainstream American psychiatry warmed to psychoanalysis, fashioning it into a peculiarly American dynamic psychiatry organized around the operation of mental forces identified by Freud. In the long run, psychiatrists' strategy of focusing their discipline's attention on normal behavior proved successful. Nearly half of those who sought psychotherapeutic treatment during the 1990s did not meet the diagnostic criteria for any defined mental disorder, and a portion suffered only from "problems in living" – the annoyances and anxieties of negotiating day-to-day life. That such were considered therapeutic

---

[3] S. Weir Mitchell, "Address before the Fiftieth Annual Meeting of the American Medico-Psychological Association," *Journal of Nervous and Mental Disease*, 21 (1894), 413–37.

issues rather than, say, moral or religious concerns is a measure of psychiatry's influence.[4]

While dynamic psychiatry forms one part of this story of psychiatry's growing influence, the development of biological psychiatry forms another. Throughout the twentieth century, American psychiatrists debated whether their specialty's province should be brain disease or the more nebulous but also more common "problems in living."[5] Although some have argued the distinction between brain and mind does not hold up in practice, for the most part psychiatrists divided themselves into competing biological and psychodynamic camps. The rest of this essay briefly chronicles psychiatry's twentieth-century fortunes by examining two narratives that tell of its rising influence. One is organized around psychoanalysis and psychodynamic psychiatry, the other around biological psychiatry, that branch of psychiatry that has attempted, from the late nineteenth century to the present, to locate the causes of mental illness in the structure and chemistry of the brain. The first tells of the discovery of psychoanalysis; its American debut in the 1910s; its spreading popular and institutional manifestations from the 1920s through the 1950s, the high point of psychoanalytic hegemony within psychiatry; and its decline thereafter, first as biological psychiatry appeared to be providing answers to enigmas of behavior that psychoanalysis could not solve, and later as the nature of psychiatric practice changed, rendering psychoanalysis an expensive, indulgent, and outmoded form of practice. The second narrative is often cast as a history of discredited treatments (electroshock and lobotomy, for example) – discredited both from within, by psychiatrists skeptical of the treatments' efficacy, and from without, by the popular media and, from the 1960s on, by the antipsychiatry movement. It can also be cast more optimistically, however, by interpreting the early interventions, however misguided they might appear in historical retrospect, as foreshadowings of biological psychiatry's later, less disputed triumphs. Until the 1970s and 1980s, it was dynamic psychiatry that was a participant in the discourse and practices of the human sciences. Then biological psychiatry moved onto that terrain as well.

THE RISE OF DYNAMIC PSYCHIATRY

In the early years of the century, a number of progressive psychiatrists set out to transform their profession and to remake their professional selves. They realized their aims through both institutional and conceptual innovations,

---

[4] William E. Narrow, Darrell A. Reiger, Donald S. Rae, Ronald Mander Schied, and Ben Z. Locke, "Use of Services by Persons with Mental and Addictive Disorders: Findings from the National Institute of Mental Health Epidemiologic Catchment Area Program," *Archives of General Psychiatry*, 50 (1993), 95–107.

[5] Leon Eisenberg, "Mindlessness and Brainlessness in Psychiatry," *British Journal of Psychiatry*, 148 (1986), 497–509, at 500.

bringing psychiatry from the margins into the cultural mainstream, where it has remained since. Institutionally, they abandoned the asylum for the clinic and the consulting room, founding new, urban, and university-based institutions such as the Psychopathic Hospitals in Ann Arbor (1906) and Boston (1912) and the Henry Phipps Psychiatric Clinic in Baltimore (1913). Their aim in these institutions was to treat patients quickly, study them scientifically, and release them expeditiously – in all, to distance themselves from their do-nothing, asylum-based predecessors. Reform-minded psychiatrists also lobbied successfully for new laws that would yield them patients who were not insane but nearly normal. As they envisioned it, patients would no longer be committed to asylums, involuntarily deprived of their liberties in legal proceedings, but instead be admitted to hospitals – voluntarily, if possible, for short stays and without recourse to the courts. Nationwide, one-third of the states adopted new laws that simplified admissions procedures and brought a wider range of persons under psychiatric scrutiny. Through these means, psychiatrists enhanced their specialty's authority, both by aligning it more closely with medicine and by claiming the right to pronounce on a range of social issues, from the domestic to the political.

Their conceptual innovations were no less important. Most significant was their abandonment of the symptom and their turn to the personality as the psychiatric unit of interest. The symptoms around which nineteenth-century psychiatry was organized were relatively rare, displayed only by a disturbed few – the hallucinations and delusions indicative of schizophrenia, for example. These psychiatrists constituted "the psychopathic personality" as a new diagnostic rubric that brought a range of nearly normal behavior – instability, impulsiveness, irritability – into the psychiatric domain and at the same time delineated the personality as an object of analysis. Personality, to the psychiatrist, denoted the entire individual and all of his or her attributes. In the words of William Menninger (1899–1966), one of the century's great practitioners, it covered "all that a person has been, all that he is, and all that he is trying to become";[6] as a standard textbook defined it, personality referred to "the person as he is known to his friends."[7] In contrast to the symptom, it offered psychiatrists a broad investigatory field, for, as they saw it, both its ubiquity (everybody had one) and its separability from the core of the self (its malleability) invited their intervention. Advanced by psychiatric leaders with increasing fervor as the century progressed, the notion that everyone would benefit from psychiatric help was premised on the adoption of the "total personality" as the focus of psychotherapeutic intervention.[8]

[6] William C. Menninger, *Psychiatry: Its Evolution and Present Status* (Ithaca, N.Y.: Cornell University Press, 1948), p. 4.
[7] David Henderson and R. D. Gillespie, *A Text-Book of Psychiatry for Students and Practitioners* (London: Oxford University Press, 1956), p. 131.
[8] Elizabeth Lunbeck, *The Psychiatric Persuasion: Knowledge, Gender, and Power in Modern America* (Princeton, N.J.: Princeton University Press, 1994).

Psychiatrists' turn to the personality was underwritten by the spread of a psychoanalytic point of view, both within psychiatry and in the culture at large. In psychoanalytic theory, symptoms mattered little. The focus was rather, in the words of Karl Menninger (1893–1990), William's equally prominent brother, on "men's motives and inner resources, the intensity of partially buried conflicts, the unknown and unplumbed depths of and heights of our nature"[9] – the total personality that many argued was becoming "the legitimate object of psychotherapeutic attempt."[10] Psychoanalysis was not only a highly specialized therapeutic practice but also a general theory of human behavior. It appealed both to physicians and to the public, particularly in the United States. Younger hospital-based psychiatrists were especially drawn to Freudianism; by 1918, nearly two hundred articles on psychoanalysis – mostly favorable – had been published in medical journals; two small professional organizations had been founded; and psychoanalysis was appearing as a mode of treatment in psychiatric textbooks. Although many psychiatrists bridled at what they saw as the psychoanalyst's excessive focus on sex, branding Freud a decadent, atheistic pessimist, others adapted it to American culture. The public eagerly embraced the new science, reading of Freud's scientific ingenuity and courage, as well as of his radical stance on sexuality, in the mass circulation magazine *Cosmopolitan,* the *New York Times,* and the *New Republic.* Optimistic versions of Freudian concepts such as the unconscious, repression, and displacement entered popular discourse. By 1921, more than forty popular books explaining psychoanalysis had appeared, and a few psychiatrists and psychologists were beginning to offer psychoanalytic treatment for neurotic and hysterical conditions.

Freud, for his part, distrusted the Americans' enthusiasm. As he saw it, psychoanalysis in America was in danger of being watered down, mixed with other elements and made into "a kind of hodge podge" reflective of the public's "lack of judgement" and emotional understanding.[11] He wanted psychoanalysis to remain an independent discipline. In Europe, lay persons as well as physicians were trained as analysts. In the United States, psychoanalysis became a subspecialty of medicine, and only psychiatrists were admitted to training.

The experience of World War I showed that trauma could result in mental symptoms amenable to psychological treatment, and it was critical in smoothing the way for a broader acceptance of psychoanalysis and psychodynamic

---

[9] Karl Menninger, *The Vital Balance: The Life Process in Mental Health and Illness* (1963) (New York: Penguin, 1979), p. 399.

[10] F. J. Hacker, "The Concept of Normality and Its Practical Significance," *American Journal of Orthopsychiatry,* 15 (1945), 47–64, at 49.

[11] Nathan G. Hale, Jr., *Freud and the Americans: The Beginnings of Psychoanalysis in the United States* (New York: Oxford University Press, 1971), pp. 397–400; Hale, *The Rise and Crisis of Psychoanalysis in the United States: Freud and the Americans, 1917–1985* (New York: Oxford University Press, 1995), p. 6.

psychiatry. Thousands of soldiers incapacitated by paralysis, convulsions, fits of trembling, senseless screaming, speechlessness, and terror had to be removed from combat. Neurologists sought, in vain, organic correlates for their symptoms. Many accused the soldiers of malingering. Punitive treatments – electric shock, solitary confinement – proved ineffective. What did work was psychotherapy, administered by psychiatrists at newly established clinics and hospitals.[12]

After the war, the Tavistock Clinic, established in London in 1920, promulgated a psychological approach to childhood and insisted on the importance of the family as a source of personal fulfillment, making it the center of psychodynamic psychotherapy in interwar Britain. For all the recognition accorded it in the United States as well as in Britain, the Tavistock had no academic affiliation.[13] In Britain, as one practitioner observed, psychoanalysis was not part of academic medicine but rather in contact with it.[14] By the 1950s, for example, fully half of university departments of psychiatry in the United States were headed by psychoanalysts – psychiatrists who had undergone years of training in the specialty. Few British psychoanalysts, by contrast, held teaching positions; in 1963, for example, at the Institute of Psychiatry at London's Maudsley Hospital, only 20 percent of the teaching staff members were psychoanalysts.[15] In addition, while there were a number of psychoanalytic institutes located all across the United States, from Boston to Topeka to San Francisco, in England little psychoanalysis was practiced outside London.[16]

American psychiatry's focus on normality and its move from the asylum to the private office enhanced its status and visibility, but it also spurred the formation of allied disciplines that laid claim to the same nearly normal patients so prized by psychiatrists. Psychiatric social work, a largely female profession organized in the 1910s, initially articulated a broad social mission, focused on adjusting patients to their social environments. But, by the 1920s, social work leaders had abandoned social meliorism in favor of exploring patients' inner lives. The personality became their unit of interest, psychotherapy – the psychiatrists' metier – their technique. Early collaboration gave way to sustained conflict, with both psychiatry and social work claiming the same field of expertise. Likewise, by the 1920s, PhD-holding psychologists were working the same therapeutic terrain as psychiatrists, prompting worries among the latter that the former would "grab the whole

[12] Karl Menninger, *Vital Balance*, pp. 62–3; Hale, *Rise and Crisis of Psychoanalysis*, pp. 13–24.
[13] Malcolm Pines, "The Development of the Psychodynamic Movement," in *150 Years of British Psychiatry, 1841–1991*, ed. German E. Berrios and Hugh Freeman (London: Gaskell, 1991), pp. 206–31.
[14] Michael Shepherd, "An English View of American Psychiatry," *American Journal of Psychiatry*, 114 (1957), 417–20.
[15] Aubrey Lewis, "Letter from Britain," *American Journal of Psychiatry*, 110 (1953), 404.
[16] On enthusiasm for Freud and psychoanalysis in Cambridge, among university-based scientists, see Laura Cameron and John Forrester, "Tansley's Psychoanalytic Network: An Episode out of the Early History of Psychoanalysis in England," *Psychoanalysis and History*, 2 (2000), 189–256.

field." Some psychiatrists argued that clinical psychologists should work under medical supervision; others advised collaboration. Psychiatrists' ambitions beyond the asylums and hospitals resulted in battles for control over the highly prized realm of normality.[17] The extraordinary influence of psychoanalysis in the United States was underwritten by the hegemony of dynamic psychiatry, which flourished in the private-practice setting of midcentury psychiatry. Office-practice psychiatry grew dramatically in the 1930s and 1940s. In the early 1950s, 40 percent of American psychiatrists practiced in private settings, and 25 percent of them practiced psychotherapy exclusively.[18] Office-practice psychiatry was largely dynamic, focused on patients' life experiences, mental conflicts, and social environments. Dynamicists were interested in their patients' "predominant modes of behavior," not in specific disease entities. They held that all patients differed from what was normal "only in the degree, persistence and relative unadaptability" of their behavior patterns.[19] They argued that disease was a disturbance in the total economics of the personality rather than something external to the patient. And they constructed personality as a series of adjustments between internal and external relations.

Like their early-twentieth-century counterparts, many of midcentury American psychiatry's most prominent spokespersons were dynamically inclined and proponents of a more activist, socially responsive psychiatry. Their expansive ambitions were underwritten by the experience of World War II. From 1942 to 1945, 1.8 million American recruits were rejected from the services on neuropsychiatric grounds, and the experience of combat produced more than one million psychiatric casualties – young men suffering from combat neuroses and other conditions that were the dynamicists' domain. Psychoanalysts were overrepresented in the service positions. Only 100 of the nation's 3,000 psychiatrists were psychoanalysts, yet they were appointed to many of the top posts in the army and air force. William Menninger, for example, was made chief psychiatrist to the army in 1943, and he appointed four psychoanalysts to his staff. These analysts successfully fashioned psychoanalytic, psychodynamic psychiatry into a medical mold, using medical terms and analogies to characterize the conditions they treated. In addition, they delineated a new, heavily psychoanalytic scientific terminology for psychiatry that would form the basis of the first *Diagnostic and Statistical Manual*, published in 1952.[20] Psychodynamic psychiatrists dominated the field for two decades following the war. As Roy Grinker, Sr., ruefully noted, "psychodynamics is purported to be *the* basic science of psychiatry," a situation of which he and many others psychoanalysts, as well as more biologically oriented

[17] Gerald N. Grob, *Mental Illness and American Society, 1875–1940* (Princeton, N.J.: Princeton University Press, 1983), pp. 235–6, 260–4, at p. 263.
[18] Shepherd, "English View of American Psychiatry," p. 418.
[19] Jules H. Masserman, *The Practice of Dynamic Psychiatry* (Philadelphia: Saunders, 1955), pp. 121–2.
[20] Hale, *Rise and Crisis*, pp. 188–90.

psychiatrists, were highly critical.[21] For as psychoanalysis was "medicalized," psychiatry was, from the opposite point of view, "demedicalized."[22]

BIOLOGICAL PSYCHIATRY

Even as psychoanalysis and the psychodynamic orientation gained in prominence within psychiatry, a number of psychiatrists held that mental illness was at bottom a disease of the brain. Throughout the century, they strove to break free of culture and to create a natural-scientific psychiatry. They promoted therapies, such as electroconvulsive shock and lobotomy, that were widely hailed by the public but later, for the most part, discredited. At the least, such therapies turned psychiatrists' attention from the mind to the brain, anticipating the development of the more successful and less controversial antipsychotic drugs. They imbued psychiatrists with hope and broke the lock of therapeutic nihilism that periodically descended on the profession.

Psychiatrists claimed – and some recent historians agree – that desperation led them to the earliest of the twentieth-century biological therapies.[23] Despite the success of their program of the everyday and their preference for hospitals over asylums, in 1920 most of those in psychiatric treatment were institutional inmates, not private or clinic out-patients. The numbers of persons admitted nationwide to state hospitals (asylums, renamed) increased by 67 percent between 1922 and 1944, from 52,000 to 79,000. Nearly all of these patients were seriously disturbed, diagnosed as psychotic. In addition, the proportion of the patient population classified as chronic increased from 1900 to the 1920s, by which time over half of those in institutions had been there for more than four years, and 13 percent for more than twenty years. Increasing numbers and increasing lengths of stay filled 366,000 psychiatric beds by 1933. Critics charged psychiatrists with incompetence, neglect, callousness, and abuse.[24]

The discovery of an apparent cure for syphilis spurred the psychiatrists' biological project. Late-stage neurosyphilis, termed "general paralysis of the insane" or "general paresis" and characterized by impaired locomotion and speech, partial paralysis, delusions of grandeur, and dementia, accounted for perhaps 20 percent of male admissions to psychiatric institutions during the second decade of the century.[25] The syphilitic origins of general paresis were definitively established between 1905 and 1913, when the causative agent of

---

[21] Grinker, "Psychiatry: The Field," p. 610.
[22] Melvin Sabshin, "Turning Points in Twentieth-Century American Psychiatry," *American Journal of Psychiatry*, 147 (1990), 1267–74.
[23] Most notably, Jack Pressman, *Last Resort: Psychosurgery and the Limits of Medicine* (Cambridge: Cambridge University Press, 1998).
[24] Grob, *Mental Illness and American Society*, pp. 187–98.
[25] Ibid., p. 188.

syphilis – the spirochete *Treponema pallidum* – was discovered and then identified postmortem in the brains of persons who had suffered from the disease. With the development of the Wassermann diagnostic test for syphilitic infection in 1906 and the discovery of an effective treatment for the disease in the drug Salvarsan, psychiatrists claimed that one of the most common of mental afflictions had assumed the shape of an organic medical problem. Almost immediately, they assembled these findings into a paradigm of disease caused by an identifiable agent and amenable to a specific chemotherapeutic agent – the "paradigm of general paresis," to which they were certain other mental disorders would conform. Although attempts to replicate this disease paradigm in psychiatry would prove fruitless, it continued to support the belief that mental illness was a disease of the brain, caused by an organic condition and amenable to somatic interventions, which could be employed to treat large numbers of patients efficaciously and inexpensively.

The best-known of the twentieth-century somatic therapies were electroconvulsive shock therapy and lobotomy, both of which were attended by panegyrics on their introduction.[26] Electroconvulsive therapy was developed from experiments with insulin shock therapy performed by Manfred Sakel, an Austrian Jew practicing in 1920s Berlin. Sakel noted symptomological improvements in the diabetic drug addicts under his care when they were given accidental overdoses of insulin, which withdrew glucose from the blood and induced a hypoglycemic coma or state of "shock." After conducting animal experiments, Sakel began testing this procedure on schizophrenics in 1933, deliberately inducing comas in them and then administering sugar to bring them out of the shock. Sakel reported to an astounded psychiatric community that 88 percent of patients so treated had improved.[27] Psychiatrists in Britain and the United States eagerly embraced insulin therapy, which promised to cure the most intractable of the mental afflictions. More than one hundred American hospitals established special insulin therapy units, as did a number of well-known British hospitals, although the procedure was dangerous – 1 to 2 percent of patients subjected to it died – and difficult to administer.

At almost the same time, the Hungarian physician Joseph Ladislas von Meduna began to consider the notion of inducing convulsions in schizophrenic patients. Arguing that epilepsy and schizophrenia were biological antagonists, he hypothesized that convulsions might work as a treatment for schizophrenia. In 1934, he injected twenty-six patients with metrazol, a powerful cardiac stimulant that could cause convulsions, and reported that the condition of most of them had improved dramatically, with ten fully recovered. By the late 1930s, metrazol shock therapy was in use at many of

---

[26] Percival Bailey, "The Great Psychiatric Revolution," *American Journal of Psychiatry*, 113 (1956), 387–406.

[27] Eliot S. Valenstein, *Great and Desperate Cures: The Rise and Decline of Psychosurgery and Other Radical Treatments for Mental Illness* (New York: Basic Books, 1986), pp. 46–7; Edward Shorter, *A History of Psychiatry: From the Era of the Asylum to the Age of Prozac* (New York: Wiley, 1997), pp. 208–14.

the major psychiatric institutions in the United States. That no one could explain how convulsions worked to improve schizophrenic conditions did not
dampen professional and public enthusiasm for the new therapy. A somatic
treatment that exemplified medical thinking within psychiatry, it offered
hope to the hopeless. Yet, as much as psychiatrists promoted it, patients
feared and resisted it, as the convulsions could be agonizingly painful and
sometimes resulted in fractures. Still, it was safer for patients and easier and
less costly to administer than insulin shock therapy, an important consideration in an era characterized, in both Europe and the United States, by large
institutional populations.[28]

While psychiatrists have consigned both insulin and metrazol therapy to
the dustbin of failed treatments, electroshock therapy is still in use, primarily as a treatment for severe depression. First tried in 1938, electroshock
therapy was developed by two Italian psychiatrists, Ugo Cerletti and Lucio
Bini. Throughout the 1930s, they experimented with dogs, eventually finding that placing electrodes on the dogs' temples allowed them to induce
convulsions without killing the animals. After they claimed success with a
human patient in 1938, news of the newly named "electroshock" treatment
spread rapidly, and psychiatrists in France, Britain, and the United States
began to use it on their own patients. By 1941, more than 40 percent of
American psychiatric hospitals were using electroshock; between 1935 and
1941, 75,000 patients were treated using one of the three shock therapies.
Articles in the popular press – *Time, Newsweek, Reader's Digest* – extolled
their benefits. Many psychoanalytically oriented psychiatrists opposed what
they argued was their inappropriate use, but some of the most prominent
thought that the treatments merited attention. From the vantage point of
1950, the shock therapies seemed efficacious and modern, supported by research at major universities and research centers worldwide. Although no one
could yet claim to understand the causes of mental disease, these therapies
promised to enhance psychiatrists' standing. As one psychiatrist put it, "one
may question whether shock treatments do any good to the patients but
there can be no doubt that they have done an enormous amount of good to
psychiatry."[29]

Prefrontal lobotomy, first performed by the Portuguese neurologist Egas
Moniz in 1935, was supported by the same organicist optimism that underwrote electroshock therapy. For years, physicians had observed intellectual
impairments and dramatic emotional changes in patients with injuries to the
frontal lobes of the brain. Yet expert opinion was divided on whether psychic
activity was located in the prefrontal area. Despite his lack of a consistent
theory to justify what he was doing, Moniz performed twenty lobotomies in

---

[28] Valenstein, *Great and Desperate Cures*, pp. 48–50; Shorter, *History of Psychiatry*, pp. 214–17.
[29] Valenstein, *Great and Desperate Cures*, pp. 52, 61; Shorter, *History of Psychiatry*, p. 222, quotation at
p. 224.

1935 and 1936, drilling holes in patients' heads and severing the connections between the prefrontal lobes and other parts of the brain. In his estimation, 35 percent were cured by the procedure, 35 percent improved, and the rest were unchanged. Moniz reported his results widely. Within months, the operation was being performed in a number of countries – Italy, Rumania, Brazil, and Cuba. Nowhere was it adopted as enthusiastically as in the United States, however, where the neurologist Walter Freeman proselytized on its behalf, hoping to make it into a standard office procedure. More than 18,000 patients were lobotomized in the United States between 1936 and 1957; by 1954, more than 10,000 had been lobotomized in Britain.[30] In both countries the surgery was performed at leading as well as marginal institutions. Like electroshock therapy, psychosurgery was hailed by the popular press, which uncritically exaggerated its effectiveness.

The biological psychiatrists' dream of transforming psychiatry into a fully medical specialty fueled two decades of largely unexamined enthusiasm for psychosurgery, an enthusiasm that was tempered only by the introduction of the psychoactive drugs during the mid-1950s. By 1970, only 300 operations were being performed each year, a fraction of the 5,000 performed yearly during the procedure's halcyon days from 1948 through 1952. Psychosurgery promised to bring psychiatrists the status and effectiveness they desperately sought, offering hopes for cure to the 500,000 chronic patients housed in overcrowded, dilapidated institutions. And with it, they remade their discipline as thoroughly as their early-twentieth-century predecessors who had envisioned a prominent role for psychiatry.

CULTURE AND PERSONALITY

The dynamicists' emphasis on assessing the person in his or her culture and environment first underwrote psychiatrists' ventures into social science. From the 1930s through the 1950s, prominent psychiatrists and social scientists worked to bridge the divide between the psychiatrists' focus on the individual and the social scientists' focus on broader cultural and social processes, meeting on the ground of the personality. In dynamic psychiatry, symptoms were seen not as discrete and limited but as expressions of a person's total personality, the sum of his or her adjustments to the immediate social, cultural, and interpersonal environment. Psychiatry could then plausibly "become an essential part of an over-all science of man – of general anthropology."[31]

[30] Valenstein, *Great and Desperate Cures*, pp. 80–100, 121, 178; Gerald N. Grob, *From Asylum to Community: Mental Health Policy in Modern America* (Princeton, N.J.: Princeton University Press, 1991), p. 130.

[31] Frances J. Braceland, "Psychiatry and the Science of Man," *American Journal of Psychiatry*, 114 (1957), 1–9, at 3.

This science of man is commonly identified with the "culture and personality" school of anthropology that flourished in the United States during the 1930s and 1940s. Anthropologists credit Edward Sapir with founding the field; psychiatrists bestow the same honor on Sapir's collaborator, the eclectic psychoanalyst Harry Stack Sullivan (1892–1949). Sullivan, who held that psychiatry was as much a social as a mental science, was involved in founding in 1936 the Washington School of Psychiatry, an interdisciplinary enterprise for training physicians and social scientists that counted the psychoanalyst Erich Fromm and the anthropologist Ruth Benedict among its faculty. Lines of influence between psychiatry and anthropology flowed both ways. From psychoanalysis, anthropologists learned of the importance of the child's early years and of recognizing unconscious motivations and mechanisms in adult behaviors. From anthropologists, psychiatrists learned of the dangers in assuming that customs were universal and ahistorical; rather, patterns of child rearing, for example, were to be seen as culturally specific and variable. Sullivan hypothesized that mental disorder was culturally specific, proposing that schizophrenia be conceived of in terms of individuals' deviations from group norms. Similarly, studies of psychotic African American patients showed that race prejudice was implicated in their delusional systems.[32]

During the 1930s and 1940s, Sullivan, Fromm, and Karen Horney, among others, turned to exploring individuals' relationships with social reality, focusing on cultural experiences and not, as in classical analysis, on instinctual drives. Using Fromm's concept of "social character," they analyzed patterns of personality structure that, they argued, were typical of groups or classes of people.[33] In *Escape from Freedom*, for example, Fromm argued that Nazi ideology appealed especially to petty bourgeois Germans; characterized by a "love of the strong [and] hatred of the weak," they were submissive to authority and as such vulnerable to the appeal of a strong father figure.[34] The concept of social character, which bridged the divide between individual and group psychology, was then taken up in the social sciences. In *The Lonely Crowd*, the most cited – and arguably most important – work of twentieth-century American sociology, David Riesman, who had undergone analysis with Fromm and who had been heavily influenced by his work, employed the concept of social character to delineate a new American type, the other-directed person, whose conformity to social norms was ensured by his or her "tendency to be sensitized to the expectations and preferences of others." Character, he proposed, was socially conditioned, "*learned* in the lifelong

---

[32] Hale, *Rise and Crisis of Psychoanalysis*, p. 177; Charles S. Johnson, "The Influence of Social Science on Psychiatry," in *Mid-Century Psychiatry: An Overview*, ed. Roy R. Grinker (Springfield, Ill.: Thomas, 1953), pp. 144–56.

[33] Louise E. Hoffman, "From Instinct to Identity: Implications of Changing Psychoanalytic Concepts of Social Life from Freud to Erikson," *Journal of the History of the Behavioral Sciences*, 18 (1982), 130–46.

[34] Erich Fromm, *Escape from Freedom* (1941) (New York: Holt, Rinehart and Winston, 1969), p. 210.

process of socialization." A "confining strait jacket," it limited individual's choices. But it also made social life possible.[35]

Most psychoanalysts preferred a more individualistic approach to the psyche. Psychoanalytic ego psychology, which stressed not – like classical analysis – the ego's defensive aspects but its integrating and adaptive functions, represented a compromise between the ideologically tinged, group-oriented concept of "social character" and the orthodox psychoanalytic focus on individual pathology.[36] During the immediate postwar years, Erik Erikson (1902–1994), ego psychology's best-known spokesman, formulated the concept of psychosocial identity, an elusive concept that encompassed both individual and social characteristics and, more importantly, had implications for everyone. Psychosocial identity depended, he explained, "on the complementarity of an inner (ego) synthesis in the individual and of role integration in his group." His notion of an "identity crisis," an expected stage of adolescence and young adulthood, entered the popular language as a designation for a period of "growth, recovery, and further differentiation."[37] Although Erikson decried the faddish equation of identity with the question, "Who am I?," he was powerless to prevent its conscription in the quest for identity that seemed to be sweeping America. Orthodox psychoanalysts were reporting that their patients suffered not from the classical neuroses but from vague conditions of futility and discontent, not from repression but from not knowing who they were or what they could become; and a spate of popular books, with titles representing variations on *The Search for Identity*, testified to the same phenomenon. By the end of the 1960s, identity crises were everyday events, "finding oneself" a cultural imperative.

Psychiatrists' turn to the social and behavioral sciences produced not only the culture and personality school but also a social psychiatry that stressed the cultural determinants of behavior. During the 1950s, studies of the hospital milieu and its effect on patients' behavior began to appear. In 1954, Alfred H. Stanton, a psychiatrist, and Morris S. Schwartz, a sociologist, published *The Mental Hospital*; other studies followed, including *Human Problems of a State Mental Hospital* and *The Psychiatric Hospital as a Small Society*.[38] The best-known work in this genre was Erving Goffman's *Asylums* (1961), which portrayed the asylum as a dehumanizing "total institution" that stripped inmates of their dignity. Other studies of the incidence and prevalence of mental disorder in selected communities, and of how such were related to social and

---

[35] David Riesman in collaboration with Reuel Denney and Nathan Glazer, *The Lonely Crowd: A Study of the Changing American Character* (New Haven, Conn.: Yale University Press, 1950), pp. 9, 4–5.

[36] Hoffman, "From Instinct to Identity," pp. 138–9; Erik H. Erikson, *Childhood and Society* (1953) (New York: Norton, 1978), pp. 193–4.

[37] Erik Erikson, "Identity, Psychosocial," in *International Encyclopedia of the Social Sciences*, ed. David L. Sills (New York: Macmillan and Free Press, 1968), vol. 7, pp. 61–5, and his *Identity: Youth and Crisis* (New York: Norton, 1968), pp. 16–17.

[38] Grob, *Asylum to Community*, pp. 142–6, discusses these works.

environmental factors, resulted from the same collaborative impulse. Both sorts of studies appealed to popular as well as professional audiences.

Ironically, these critical studies of mental hospitals may have worked to undermine social psychiatry. In any event, by the mid-1960s the medical model in opposition to which social psychiatry had been fashioned was again ascendant, buoyed by the discovery of the first of the antipsychotic drugs, chlorpromazine, in 1952. For the first time, psychiatrists had a means to treat the debilitating symptoms of schizophrenia – hallucinations, delusions, and thought disorders. Pharmacological treatments for mania and depression soon followed, and again psychiatrists were heralding the dawn of a new – this time "psychopharmacological" – era. At a time of renewed scientism within psychiatry, not only social psychiatry but also psychoanalysis was vulnerable to attack. Psychiatrists began to admit that they had overpromised on their ability to treat "the unhappiness of the human condition," and called upon their confreres to narrow their focus to disease and its treatment.[39] The psychosocial model that had dominated American psychiatry from the end of World War II through the 1960s fell into disrepute; the profession was ideologically divided into warring biological and psychodynamic camps. Only with publication in 1980 of DSM-III, the third edition of psychiatry's official *Diagnostic and Statistical Manual*, did the profession unite around a descriptive, nondynamic orientation that signaled its "remedicalization."[40]

Psychiatry has often been characterized as divided between psychodynamic and biological camps, each holding radically different conceptions of the root causes of mental illness. It is not the case, however, that psychodynamic psychotherapy is for the "worried well" and biological psychiatry for the seriously ill. Psychoanalytically oriented psychiatrists treated schizophrenics in institutional settings, especially during the 1940s and 1950s, and, for much of the century, and during the 1950s in particular, psychoanalysis bore the stamp of science, with psychoanalysts seeing themselves – and being seen by others – as the scientists of unconscious and irrational behavior. Likewise, although the history of biological psychiatry is largely a history of the treatment of the severe psychoses, various pharmacological treatments for depression have verged close to the territory of the worried well – notably Prozac, which is used to treat both depression and the personality disorders. Additionally, proponents and practitioners of biological psychiatry have argued that their treatments – from insulin therapy to Prozac – fundamentally change their patients' personalities, yielding not just cures but new persons.[41] In advancing this claim, they have strayed onto the ground created

---

[39] Betram S. Brown, "The Life of Psychiatry," *American Journal of Psychiatry*, 133 (1976), 489–95, at 495.
[40] Michael Wilson, "DSM-III and the Transformation of American Psychiatry: A History," *American Journal of Psychiatry*, 150 (1993), 399–410, at 399.
[41] Shorter, *History of Psychiatry*, p. 209.

by twentieth-century psychiatry that sustains the work of both camps of psychiatrists.

The question of how much of individuals' behavior is traceable to brain abnormalities and genetic makeup and how much to their experiences in life has continued to divide the profession. Claims to have found physical correlates, in the brain, of eccentric behaviors as well as to have traced higher proportions of disturbed – but not quite mentally ill – individuals among relatives of the mentally ill, have suggested genetic causation. Researchers have also isolated a range of genes that they argue are linked to addictive behaviors, obsessive-compulsive disorders, neuroticism, and mania, although many stop short of claiming that the genes cause the conditions, aware that the brain's structure and activity may result from individual life experiences. Despite advances in neuroscience, that is, no firm line can be drawn between mental illness and normal behavior, between the quirks that make life interesting and the illness that makes it difficult. Further, despite a range of scathing attacks during the 1990s on Freud's ethics and practices, psychoanalysis and psychodynamic psychotherapy have remained strong cultural presences, and countless articles and books have debated Freud's claims to truth as well as the efficacy of the science he created.[42] The neuroscientists' foray into what they term "shadow syndromes," eccentricities that were once considered normal but are now classed as illnesses – mild attention deficit disorder, mild obsessive-compulsive disorder – may paradoxically have underwritten this turn, for they were describing the sort of behaviors for which individuals throughout the twentieth century have sought the aid of psychotherapy. If anything, the renewed biological orientation in psychiatry has focused attention on the middle ground of behaviors. At the end of the century, psychiatry's cultural capital remained premised as much on its interventions into normality as on its treatments of abnormality.

---

[42] See, for example, Frederick Crews, ed., *Unauthorized Freud: Doubters Confront a Legend* (New York: Penguin, 1998); John Forrester, *Dispatches from the Freud Wars* (Cambridge, Mass.: Harvard University Press, 1997), pp. 208–48.

40

GENDER

Rosalind Rosenberg

As a social scientific term, gender came into common use only in the final quarter of the twentieth century. But its core idea, that biological sex and its cultural expression are separable, had been evolving for over a hundred years. As rapid urbanization fostered greater sexual freedom and spurred a vibrant women's movement at the end of the nineteenth century, a disparate group of sex reformers, feminists, and university-trained researchers began to question a number of conventional beliefs. Does effeminacy in men signal biological abnormality? Is politics, by nature, a masculine enterprise? Are geniuses disproportionately male? Do females lack sexual drive? At the turn of the century, most social theorists answered yes to these questions. But by the 1970s, even as researchers were mapping the human brain with ever-greater precision, scholars had ceased treating the cultural expression of sex as a direct product of physiology. Symbolic of this dramatic shift, social scientists abandoned "sex" in favor of "gender" when discussing human behavior. Long used exclusively as a grammatical category, "gender" appealed to those who found the biological associations of "sex" too limiting. Here was a term that freed investigators to explore with new intensity the multiple ways in which cultures distinguish males from females, structure sexual experience, and deploy power.[1]

THE AGE OF EVOLUTION: THE LATE NINETEENTH CENTURY

Scientific interest in sex has a long history, but the science that flourished in the late nineteenth century differed in fundamental ways from what had gone before. It was more precise and empirical than any previous work. It

[1] Judith Butler, *Gender Trouble: Feminism and the Subversion of Identity* (New York: Routledge, 1990), pp. 7, 30–1. See also Sandra Lipsitz Bem, *The Lenses of Gender: Transforming the Debate on Sexual Inequality* (New Haven, Conn.: Yale University Press, 1993), pp. 192–3.

enjoyed unprecedented prestige because of the perceived connection between science and the technological accomplishments of the nineteenth century. Most importantly, it spoke to the burning issues of race and sex. Abolitionist movements in both England and the United States had raised the issue of black emancipation earlier in the century, and women's rights movements had opened debates about women's place in society that intensified toward century's end. These debates spilled over into controversies about the nature of femininity and masculinity, which in turn raised questions about sexuality, especially as sharply falling birth rates in Western industrializing countries and the emergence of homosexual communities in cities from New York to Berlin triggered disputes over the legitimate aim of sexual feeling. In this atmosphere, science became a weapon, both for those who sought to legitimate, as well as those who sought to discount, the claims of blacks, of women, and of homosexuals to enjoy social and political equity.[2]

For most of the final third of the nineteenth century, those who wielded the weapon of science to limit the expansion of women's rights and to restrict sexual freedom held the upper hand. Natural and social scientists agreed that women were inherently different from men, physically, temperamentally, and intellectually. Just as "primitive people" had lagged behind Europeans in the course of evolution, so too women had lagged behind men. Both were more childlike, less fully evolved. In *The Descent of Man* (1871), Charles Darwin (1809–1882) declared that males were naturally more variable in their physical and mental characteristics than were females. Through evolution, they had grown stronger and more intelligent because, as a group, they faced the forces of natural selection with a greater wealth of attributes than females enjoyed. Motherhood compounded females' metabolic disadvantage by forcing them into dependence on males. Standing at one remove from the full force of natural selection, they were in some small part immune from its progressive tendencies.[3] The sociologist Herbert Spencer (1820–1903) extended evolutionary thinking to cover society as a whole in his 1874 work *The Study of Sociology*. Over the course of history, Spencer contended, society had shifted from a condition of relative homogeneity to one of heterogeneity, marked by an extensive division of labor. Whereas males and females had once engaged in the same tasks of self-preservation, the success of modern civilization depended on their assuming highly differentiated, complementary roles.[4]

Sexuality played a critical role in this evolutionary tale. At the earliest, most primitive stage of life, evolution proceeded by parthenogenesis, a form

---

[2] Cynthia Eagle Russett, *Sexual Science: The Victorian Construction of Womanhood* (Cambridge, Mass.: Harvard University Press, 1989), p. 2; Jonathan Katz, *The Invention of Heterosexuality* (New York: Penguin, 1996), p. 10.

[3] Charles Darwin, *The Descent of Man, and Selection in Relation to Sex*, 2 vols. (London: John Murray, 1871), vol. 1, pp. 35–8, 111, 273–9; vol. 2, pp. 326–9, 368–75.

[4] Herbert Spencer, *The Study of Sociology* (New York: Appleton, 1874), pp. 315, 373–83.

of asexual reproduction. But the rate of change accelerated greatly with the advent of two sexes and the multiple possibilities that sexual selection added to natural selection. In evolutionary terms, men and women who engaged in sex for reproductive purposes claimed the scientific high ground as agents of evolutionary progress. Homosexuals, by contrast, represented a reversion to an earlier stage of development. The German sexologist Richard von Krafft-Ebing achieved widespread influence during the late nineteenth century with his argument that homosexuality represented a form of biological degeneration.[5]

Biological determinism permeated the work of the male pioneers who established social science departments in late-nineteenth-century universities. Economists and political scientists depicted men as aggressive creatures, driven by self-interest to dominate the marketplace and by evolutionary advantage to guide the state. Women, by contrast, were seen as moral creatures whose natural ties to family placed them outside economic and political analysis. Anthropologists and sociologists, for their part, scorned evidence that matriarchies existed in some cultures by reasoning that such female power could exist only at earlier stages of evolutionary development. Psychologists explained the greater number of scientific discoveries made by men as a sign of men's more highly evolved brain. In all of the social scientific disciplines, naturalism shaped the questions asked, as well as the investigative methods employed in answering them. Administrative efficiency and business productivity attracted sustained attention from male investigators; household labor, infant mortality, and social welfare – topics of greater interest to women – did not.[6]

SEEDS OF DOUBT

Not all those who accepted evolutionary thinking came to the same conclusions. Sigmund Freud (1856–1939) challenged the image of the highly sexed male and relatively less passionate female of evolutionary theory with the argument that women, like men, have sexual needs that must be met to insure health. It was not that Freud had abandoned the belief that women are more passive than men. But whereas medical science viewed psychology as a direct expression of physiology, Freud regarded it as shaped by children's social history within the family. According to Freud, children are born bisexual. Only when they confront the Oedipus complex, through which boys resolve their fears of castration and girls come to terms with the fact that castration has already occurred, do their adult personalities take form, and only

[5] Richard von Krafft-Ebing, *Psychopathia Sexualis* (1902), 12th ed., trans. Franklin S. Klaf (New York: Stein and Day, 1965), pp. 222–368.
[6] Helene Silverberg, ed., *Gender and American Social Science: The Formative Years* (Princeton, N.J.: Princeton University Press, 1998), pp. 3–32.

then do passivity in women and aggression in men gain an all-but-universal ascendance.[7]

Freud's thinking about passivity and aggression also led him to challenge traditional classificatory schemes with respect to homosexuality. Many earlier theories had restricted the meaning of sexual inversion in the case of men to refer only to those who wished to play a passive role in sex. Men who played an active role, whether with men or with women, were deemed to be heterosexual. Freud, however, in common with the British sexologist Havelock Ellis (1859–1939), distinguished feminine and masculine behavior from sexual desire in classifying homosexuals. According to Freud, "The most complete mental masculinity can be combined with male inversion [same-sex desire]." Ellis, less consistent than Freud, tended to refer to female inverts as masculine, but both theorists contributed to a trend in medical discourse toward defining homosexuality as same-sex desire alone, having nothing to do with the conventions of masculine and feminine behavior.[8]

In 1909, Freud introduced his views on sexuality to an American audience in lectures at Clark University. For an intellectual community chafing against puritanical constraints, his ideas helped spark an explosion of creative effort. In the decentralized, modernized, and rapidly expanding universities of America, reform-minded young scholars – especially women – began to challenge biological determinism in general and conventional ideas about gender in particular. The first important work took place in psychology, a new field that created possibilities for women – at least as graduate students – that barely existed elsewhere. Inspired by a burgeoning women's movement and aided by new statistical techniques, female graduate students Helen Thompson Woolley (1874–1947) at the University of Chicago and Leta Hollingworth (1886–1939) at Columbia University tested the mental abilities of male and female college students during the years before World War I. In contrast to the prevailing belief in men's mental superiority, Woolley found greater variation within groups of males and females than between them. At Columbia, Hollingworth confirmed Woolley's findings and went on to challenge two conventional beliefs: that females were statistically less likely than men to be either geniuses or idiots, and that menstruation impaired women's mental acuity.[9]

In the 1920s, as women flooded colleges and universities, entered new professions, won the right to vote, and enjoyed greater sexual freedom, a new

[7] Sigmund Freud, "Femininity," in *The Complete Introductory Lectures on Psychoanalysis*, trans. and ed. James Strachey (New York: Norton, 1966), pp. 576–99.

[8] Sigmund Freud, *Three Essays on the Theory of Sexuality* (1905), as quoted in Charles Chauncey, *Gay New York: Gender, Urban Culture, and the Making of the Gay Male World, 1890–1940* (New York: Basic Books, 1994), p. 124; Havelock Ellis, *The Sexual Impulse in Women*, pp. 191–6, as cited in Paul Robinson, *The Modernization of Sex* (New York: Harper and Row, 1976), p. 17.

[9] Rosalind Rosenberg, *Beyond Separate Spheres: Intellectual Roots of Modern Feminism* (New Haven, Conn.: Yale University Press, 1982), pp. 54–113.

consensus formed around the belief that men and women differed little, if at all, from one another in their mental traits and sexual drives. In 1927, a male reviewer of the psychological literature on sex differences concluded, "few, if any, of the so-called 'sex-differences' are due solely to sex. Individual differences often are greater than differences determined on the basis of sex." The differences between men and women that remained, he added, in a major departure from the biological determinism of the past, were due simply to social causes: "The social training of the two sexes is and always has been, different, producing different selective factors, interests, standards, etc."[10]

Attacks on hereditarian thinking in sociology and anthropology paralleled the work in psychology. As early as the 1890s, urbanization, immigration, and reform efforts inspired innovative work, much of it by women trained in university settings but forced to work outside academia after earning their degrees. The Columbia-trained sociologist and anthropologist Elsie Clews Parsons (1875–1941) relied on family wealth to finance work that questioned the ways in which society classified people by sex. In *Social Rule: A Study of the Will to Power* (1916), for instance, she argued that this classification stemmed from the greater power that men enjoyed in society rather than from anything inherent in male or female physiology. The Chicago-trained economist Edith Abbott (1876–1957) pieced together fellowship support and employment at a school for training social workers as she challenged the evolutionists' depiction of women as absent from the marketplace in her path-breaking study of women's employment and wages, *Women in Industry: A Study of American Economic History* (1910). The Chicago-trained sociologist Katharine Bement Davis (1860–1935) relied on funding from the Rockefeller Foundation to conduct a pioneering study of female sexual behavior. In *Factors in the Sex Life of Twenty-Two Hundred Women* (1929), she showed that white, middle-class, educated, heterosexual women who had grown up before 1900 had enjoyed much more active sex lives – including more homosexual experiences – than conventional opinion had supposed.

The mostly male academicians of the early twentieth century – eager to se-cure the standing of their fledgling enterprises, fearful of political controversy, and largely uninterested in the issues that women investigators considered important – proved cautious in challenging conventional beliefs about gen-der. The few who did so were men preoccupied by race and ethnicity. Tem-porarily established at the University of Pennsylvania, the African American sociologist W. E. B. DuBois (1868–1963) traced the unusually high incidence of female-headed households among urban Negroes to poverty and racism in *The Philadelphia Negro* (1899). At the University of Chicago, the sociolo-gist W. I. Thomas (1863–1947) pointed to the relaxation of sexual inhibitions

---

[10] Chauncey N. Allen, "Studies in Sex Difference," *Psychological Bulletin*, 24 (1927), 299, as quoted in Carl Degler, *In Search of Human Nature: The Decline and Revival of Darwinism in American Social Thought* (New York: Oxford University Press, 1991), p. 132.

among female Polish immigrants as they were exposed to the relatively greater freedom of Chicago in *The Polish Peasant in Europe and America* (1918–20). And at Columbia University, the anthropologist Franz Boas (1858–1942), an immigrant Jew and author of *The Mind of Primitive Man* (1911), trained Ruth Benedict (1887–1948), Zora Neale Hurston (1901?–1960), and Margaret Mead (1901–1978). Each extended his work on race to challenge prevailing ideas about the naturalness of sex differences and the inevitability of male dominance.

In 1934, Ruth Benedict published her immensely popular *Patterns of Culture*. In a chapter entitled "The Pueblos of New Mexico," she not only presented one of the first positive depictions of Native Americans to the nation, but also offered a sharp critique of male-dominated American culture. Isolated on a reservation, the Zuni tribe lived in a matrilineal culture that scorned individual power and violence and granted women far greater authority in family life than white, middle-class women could claim. That a neighboring people could structure their lives so differently, Benedict wrote, proved that the "dominant traits of our civilization ... are compulsive, not in proportion as they are basic and essential in human behavior, but rather in the degree to which they are local and overgrown in our own culture."[11]

In the following year, Benedict's protégé, Margaret Mead, published *Sex and Temperament* (1935), a study of the women and men of three widely differing South Pacific cultures. Their attitudes and behavior, she maintained, were the product of particular cultural conventions, not stages on an evolutionary scale culminating in the American ideal of feminine selflessness and masculine competitiveness. The peoples she studied patterned their behaviors in widely varying ways, following none of the expectations of Western culture. In fact, the men who in her view seemed the most contented, the mountain Arapesh of New Guinea, exhibited characteristics that in America would be deemed feminine because of their association with cooperation and nurturing behavior.[12]

Two years later, Zora Neale Hurston underscored this mounting skepticism of the naturalness of sexual characteristics in her novel *Their Eyes Were Watching God* (1937). Based on the folklore studies she had done in northern Florida under the direction of Boas, Hurston's novel depicted a black culture where happiness came not from an imitation of white society, which led men to measure themselves solely in terms of material success and women to marry for security rather than for love, but from the ability to remain true to one's heart, to work cooperatively with others, and to treat all people as equals. The critique of American culture and gender relations offered by Benedict, Mead, and Hurston contributed to a growing respect

---

[11] Ruth Benedict, *Patterns of Culture* (New York: Houghton Mifflin, 1934), p. 101.
[12] Margaret Mead, *Sex and Temperament in Three Primitive Societies* (New York: Morrow, 1935), pp. 279–80.

for difference in America and to a greater tendency among social scientists to see sex differences as a function of culture and power rather than of biology.

Influenced by this work in anthropology, the psychiatrists Karen Horney (1885–1952) and Clara Thompson (1893–1958) broke ranks with fellow Freudians to develop a cultural approach to the psychiatric study of sex differences. Freud argued that femininity evolved out of childhood experiences. But his contention that female development occurred within the limits set by the "anatomical distinctions between the sexes," the most important of which were the sex organs, made the particular course of female development he described appear to be anatomically determined.[13] No expansion of educational, economic, or political opportunity could overcome the fact that males possessed penises and females did not. In Berlin, influenced by the sociologist George Simmel, Horney challenged Freud, first in her article "The Flight from Womanhood" (1926). Psychological differences between men and women derived from anatomical conditions, she agreed. But the critical organ was not the penis, but rather the womb. Men became masculine because of womb-envy, that is, the power of women to produce children. Moreover, she noted, it was time to discard Freud's belief that the vagina was the sole site of adult sexual feeling in females and to recognize the clitoris as a natural part of the female genitalia.[14] Emigrating to the United Sates, Horney came under the influence of American cultural anthropologists during the 1930s; and by the 1940s, one of her colleagues, the American-born psychiatrist Clara Thompson, had abandoned anatomically based thinking about sex-differences altogether. Women envied the penis merely as the symbol of male power. Vanquish the power, and the envy would disappear.[15]

HEREDITARIAN REJOINDERS

Hereditarian thinking faded in the early twentieth century as broad popular forces inspired social scientists to establish the independence of their disciplines from biology, but those ideas never disappeared. Indeed, they gained new prominence, first in Europe under fascism and later in the United States during the conservative reaction to cultural relativism in the social sciences that accompanied and followed World War II. The return to hereditarian thinking was never as marked in the United States as it was in Europe, but echoes of earlier biological thinking were evident, especially in research on

---

[13] Sigmund Freud, "Some Psychical Consequences of the Anatomical Distinction Between the Sexes" (1933), in *The Standard Edition of the Complete Psychological Works of Sigmund Freud*, 24 vols. (London: Hogarth Press and the Institute of Psychoanalysis, 1953–66), vol. 22, pp. 14–18.

[14] Karen Horney, "Flight From Womanhood: The Masculinity-Complex in Women as Viewed by Men and Women," *International Journal of Psychoanalysis*, 7 (1926), pp. 324–39; Mary Jo Buhle, *Feminism and Its Discontents: A Century of Struggle with Psychoanalysis* (Cambridge, Mass.: Harvard University Press, 1998).

[15] Clara Thompson, *On Women* (New York: Mentor, 1964), pp. 111–41.

women. As the Great Depression drew attention to the economic travails of the so-called "forgotten man," it became increasingly difficult to sustain concerns about the well-educated, often professional "new woman." Moreover, the theory that sex differences were merely the product of socialization lost force when masculinity and femininity persisted, despite the expansion of women's educational and political opportunities.

Hereditarian echoes derived also from the gendering of academic structures. By the 1930s, universities had established themselves as key institutions, and women, who had occasionally found places for themselves on the periphery when these institutions were struggling for acceptance, lost out to men in competition for coveted positions in the midst of the Great Depression. In addition, as research techniques became more sophisticated and philanthropic foundations made it possible to conduct more elaborate research projects, women found it difficult to compete for the necessary grants or to be given the authority needed to run large projects.

When Helen Thompson Woolley opened the field of sex differences in psychology, she did everything herself. A generation later, Lewis Terman (1877–1956) dominated the field on the basis of his access to research funds and research assistants, many of them women. A major grant from the National Research Council, and a team of psychologists, enabled him to write *Sex and Personality* in 1936. In his study, Terman freely granted a "growing tendency" among social scientists "to concede equality or near equality [between the sexes] with respect to general intelligence and the majority of special talents." But he insisted that the personality tests he had developed proved that the sexes differ fundamentally in "their instinctive and emotional equipment." He therefore took strong exception to the work of social scientists such as Margaret Mead who suggested that human nature was almost infinitely malleable.[16]

Sociologists stopped short of Terman's biological determinism in their work, but echoes of the functionalism that had characterized Herbert Spencer's social Darwinism assumed new prominence. At the beginning of World War II, Talcott Parsons (1902–1979) declared that differentiation and specialization within family life were an essential part of the development of complex industrial societies. According to Parsons, by providing emotional support at home, women freed men to play "instrumental" roles in the economic world beyond, making possible increased productivity and abundance for all. A woman who failed to play her proper role threatened not only the happiness of her family but also the economic well-being of the Western world.[17]

---

[16] Lewis Terman, *Sex and Personality: Studies in Masculinity and Femininity* (New York: McGraw-Hill, 1936), p. 461.
[17] Talcott Parsons, "Age and Sex in the Social Structure of the United States," *American Sociological Review*, 7 (1942), 613.

Early in the century, scientific challenges to the perceived naturalness of female subordination had provided important inspiration to those working on race differences, and vice versa. But the Depression and the rise of Nazi power drove a wedge between the two. When the Swedish economist Gunner Myrdal (1898–1987) was writing his influential book *An American Dilemma* (1944), which drew on the expertise of social scientists from America's major research universities, he wrote (perhaps at the urging of his wife, Alma) a chapter on the parallels between race and sex discrimination. Myrdal tucked the chapter away in the Appendix, however, after Frederick Keppel, who had commissioned Myrdal's study for the Carnegie Corporation, warned upon reading the manuscript that both blacks and women would be startled by the comparison.[18]

The trauma of the Great Depression and World War II, followed quickly by the tensions of the Cold War, injected a sober realism into popular faith in the power of social engineering to make a better world. In this context, psychoanalysis, with the emphasis of its orthodox adherents on the limits of the human condition, experienced a sharp increase in popularity, in both the United States and Europe. In 1945, Helene Deutsch (1884–1982), one of the large number of psychoanalysts who had sought refuge from fascism in the United States, wrote that the normal, feminine woman accepted her distinctive sexuality and lived through her husband and children. Women who, due to some unfortunate turn in their psychic development, did not follow this pattern developed a "masculinity complex," in which the "cold, unproductive thinking" of manhood overwhelmed the "warm, intuitive knowledge" of womanhood. Ferdinand Lundberg and Marynia Farnham made the case for natural femininity even more starkly in their 1947 best-seller, *Modern Woman: The Lost Sex.* Tracing feminism to the neurotic impulses of women who had been abused by their fathers in childhood and were seeking revenge by claiming a share of masculine power, Lundberg and Farnham urged that women accept their femininity through subordination to their husbands and the joyful acceptance of motherhood.[19]

Even Margaret Mead, who in *Sex and Temperament* (1935) had taken the argument for the cultural roots of all sex differences further than any other scholar, was trimming sail by the time she published *Male and Female* (1949). In her earlier work, Mead had portrayed motherhood as an incident in the life cycle, a positive experience but not a significant one for the culture at large. By the time she wrote *Male and Female*, at the peak of the postwar baby boom, she had herself borne a child, fallen under the influence of psychoanalysis, and begun to think about the ways in which biology might work dialectically

---

[18] Walter Jackson, *Gunnar Myrdal and America's Conscience: Social Engineering and Racial Liberalism, 1938–1987* (Chapel Hill: University of North Carolina Press, 1990), p. 168.

[19] Helene Deutsch, *The Psychology of Women*, 2 vols. (New York: Grune and Stratton, 1945), vol. 2, pp. 1–55; Ferdinand Lundberg and Marynia Farnham, *Modern Woman: The Lost Sex* (New York: Harper and Brothers, 1947), pp. 140–67.

with environmental forces to shape culture. Maternity became the central feature of this dialectic, the one great problem that all cultures must confront in organizing gender roles.[20]

This new emphasis on the naturalness of sex differences led Freud's successors to reject his assumption that both heterosexuality and homosexuality were specialized derivatives of an earlier bisexuality. Even revisionist analysts such as Clara Thompson, who viewed femininity as a product of cultural forces, treated homosexuality as the sexualizing of a nonsexual, neurotic conflict. The effect of this approach to homosexuality was to pathologize it as a mental illness and make it theoretically amenable to psychiatric treatment. Homosexuality was included in the first official listing of mental disorders published by the American Psychiatric Association in 1952, just as Senator Joseph McCarthy was capitalizing on Cold War fears to drive gays and lesbians from government service, as well as from many private jobs, for posing a threat to national security.[21]

THE REBIRTH OF FEMINISM: ERASING COLOR AND SEX IN THE 1950s AND 1960s

Even as the Cold War and McCarthyism reinforced conservative views of gender and sexuality, social scientists and a few biologists began to build on earlier work in order to question them. These dissidents succeeded because of important changes taking place within higher education. The economic growth of the postwar years led to a renewed expansion of universities and a consequent democratization of higher education in the United States. Ironically, the same Cold War that reinforced traditional beliefs about feminine subordination and domesticity also generated fears that the Soviets' more effective use of women in science would enable it to win the "space race." Experts warned of a shortage of "manpower" in the sciences and urged the government to build "woman power" by encouraging women to seek advanced degrees, especially in the sciences. In war-torn Europe, the democratization of education proceeded more slowly but just as surely, as growing numbers of women sought higher education to support themselves and their families.

With the social sciences in retreat from their pre-war skepticism about sexual differences, the two most important attacks on conservative thinking during the 1940s and 1950s came not from the social sciences, but from philosophy and biology. In France, Simone de Beauvoir (1908–1986) published *The Second Sex* (1949), with its existentialist attack on biological determinism.

---

[20] Margaret Mead, *Male and Female* (New York: Morrow, 1949), pp. 143–60.
[21] Thompson, *On Women*, pp. 98–110; Hale, *The Rise and Crisis of Psychoanalysis*, pp. 298–9; Bem, *The Lenses of Gender*, pp. 92–3.

Woman is the Other, she declared, against which man defines himself in an act of psychic oppression. In a long introductory chapter on "The Data of Biology," de Beauvoir detailed the multiple ways in which physiology affected women's lives. "Woman is weaker than man; she has less muscular strength, fewer red corpuscles, less lung capacity.... In other words her life is less rich than man's." But, she continued, in themselves, these facts "have no significance." What counts is not the body, but what the mind makes of it. For de Beauvoir, as for Clara Thompson, the penis is important not in itself, but as a symbol of power.[22]

The American biologist Alfred Kinsey (1894–1956) accorded far more importance to physical being than did de Beauvoir, yet he did even more than she to undermine conventional views about the fixed nature of sexual experience, first in *Sexual Behavior in the Human Male* (1948) and then in *Sexual Behavior in the Human Female* (1953). A relentless empiricist, Kinsey demonstrated that over time American women had become more sexually active and that the incidence of homosexual contact in American society was much greater than had been assumed. His work helped to inspire the later investigations of William Masters and Virginia Johnson, who concluded in *Human Sexual Response* (1966) that women's sexual response is stronger than men's and that Freud's distinction between the clitoral and vaginal orgasm was unjustified.

Though less well known, work critical of hereditarian thinking continued in the American social sciences, as a result of increased opportunities for women in both the work force and graduate education. By 1965, over a third of all women were in the workforce, and a substantial number of women continued in graduate school and academic positions. Anne Anastasi, Helen Hacker, and Mirra Komarovsky were just a few of the critics of hereditarian thinking whose work provided a language for a new generation of feminists during in the 1960s.

The journalist Betty Friedan led the way with the publication in 1963 of *The Feminine Mystique.* That same year, the Presidential Commission on the Status of Women issued a report that called for state and federal action to improve the status of American women. In its most significant step, the Commission recommended that lawyers seek to expand the reach of the equal protection clause of the Fourteenth Amendment. The idea for doing so came from a civil rights lawyer, Pauli Murray (1910–1986), who believed that the social scientific literature challenging sex differences was just as compelling as the literature challenging race differences. Since the Court had been willing to rely on social scientific evidence to rule in *Brown v. Board of Education* (1954) that race is an arbitrary and therefore illegal basis on which to deny citizens equal treatment, Murray argued, then so too should the Court look to the

[22] Simone de Beauvoir, *The Second Sex* (1952), trans. H. M. Parshley (New York: Vintage, 1989), pp. 32–7.

social sciences for evidence that sex is an arbitrary form of classification.[23] Between 1963 and 1973, arguments first fashioned by researchers during the early twentieth century and updated by their heirs at midcentury provided ammunition for feminists who were struggling to win equal rights for women. Emblematic of social scientists' influence was Ruth Bader Ginsburg's victory in the 1973 landmark case *Frontiero v. Richardson*, in which Ginsburg, relying on Murray's earlier work, successfully likened sex to race as an arbitrary basis for treating women employees less favorably than men in awarding them benefits.

FROM SEX TO GENDER IN THE SOCIAL AND BEHAVIORAL SCIENCES, 1970s TO THE PRESENT

In 1970, women earned only 13 percent of all PhDs in the United States, a lower proportion than in 1930. But during the next decade that figure more than doubled to 30 percent. Caught up in a sexual revolution, inspired by feminism, and fortified by their growing numbers, a new generation of women students turned en masse to the study of what they increasingly termed gender. A growing gay rights movement brought added energy to these investigations. In the year 1973, only twelve articles on gender appeared in journals devoted to anthropology, sociology, psychology, psychiatry, political science, and history. In 1983, the total for the year rose to 210; and in 1993, it soared to 2,607.[24] In the social sciences, the shift from "sex" to "gender" underscored a renewed emphasis on the power of culture to shape human behavior and thinking. Studies of gender emphasized the essential sameness of men and women beneath a malleable overlay of culturally induced differences. This emphasis paralleled continuing efforts by feminist activists to eradicate laws that treated women and men differently. By 1980, not only had most laws that disadvantaged women been struck down, but most laws aimed at benefiting women had also been revoked.[25]

To treat women as though they were men was one way to separate biological sex from its cultural expression. But doing so meant accepting qualities that society had come to value as male as the norm for all of society. Many scholars objected to this approach. The historian of science Donna Haraway warned that feminist scholars "have become antinatural . . . in a way which leaves the life sciences untouched by feminist needs." To turn away from the study of nature, and of human beings' place in it, was to ignore the ways in which experiences and principles associated with females had been systematically

[23] President's Commission on the Status of Women, *American Women: The Report of the President's Commission on the Status of Women* (New York: Scribners, 1965), p. 149.
[24] See the *Social Sciences Citation Index*, which references journals in sociology, psychology, psychiatry, political science, anthropology, sociology, and history.
[25] Barbara Allen Babcock, Ann E. Freedman, Susan Deller Ross, Wendy Webster Williams, Rhonda Copelan, Deborah L. Rhode, and Nadine Taub, *Sex Discrimination and the Law*, 2nd ed. (Boston: Little Brown, 1996), p. 489.

excluded from scientific observation. Only as feminist researchers entered primatology, for instance, had scientists begun to observe matrifocal groups as well as dominance hierarchies; long-term social cooperation as well as short-term spectacular aggression; and flexible process as well as strict structures. Given the power of science in modern society as a legitimizing authority, the scientist and historian Evelyn Fox Keller urged scholars to work "to legitimate those elements of scientific culture that have been denied precisely because they are defined as female."[26]

Feminist scholars throughout the social sciences, from history to economics, followed this lead. Through detailed empirical studies they demonstrated how changing fertility and child-rearing practices had altered the economy, how female social reformers working outside government had shaped the modern state, and how discrimination within the social sciences had affected social thought.[27] Social theorists added to this empirical work. Revising Freudian theory, Nancy Chodorow stressed the centrality of the pre-Oedipal mother–infant relationship in producing gender differences in *The Reproduction of Mothering* (1978). The problem of separating from the mother posed a greater challenge to sons than to daughters, Chodorow argued. To become men, sons had not only to separate from but also to become unlike their mothers, a task that led men to value detachment, independence, and rationality. Daughters, by contrast, could separate from their mothers without losing the connectedness of a common gender identity, and thus had less reason to distinguish between self and other in their mental and moral lives. The educational psychologist Carol Gilligan, following Chodorow, argued that, whereas men often seem preoccupied with achieving independence and avoiding dependence, women see "a world comprised of relationships rather than of people standing alone, a world that coheres through human connection rather than through systems of rules."[28]

In France, postmodern feminist philosophers, literary critics, and psychoanalysts stressed the importance of understanding the dialectical interplay of biology and culture even more strongly. Inspired by the psychoanalytic writings of Jacques Lacan, they attached far more importance to the way in which language and the unconscious shape femininity than did other feminists. Steeped in Marxist culture, trained in dialectical reasoning, and inspired by the uprisings of 1968, writers such as Helene Cixous focused their attention on

[26] Donna Haraway, "Animal Sociology and a Natural Economy of the Body Politic, Part I: A Political Physiology of Dominance," and Evelyn Fox Keller, "Feminism and Science," in *The Signs Reader*, ed. Elizabeth Abel and Emily Abel (Chicago: University of Chicago Press, 1983), pp. 125, 137, 113–14.
[27] See, for example, Mary P. Ryan, *Cradle of the Middle Class: The Family in Oneida County, New York, 1790–1865* (New York: Cambridge University Press, 1981); Theda Skocpol, *Protecting Soldiers and Mothers: The Political Origins of Social Policy in the United States* (Cambridge, Mass.: Harvard University Press, 1992); Silverberg, *Gender and the Social Sciences*.
[28] Carol Gilligan, *In a Different Voice: Psychological Theory and Women's Development* (Cambridge, Mass.: Harvard University Press, 1982), p. 29.

women's exclusion from history and dedicated themselves to wrestling with a male-constructed language that they felt, was alien to them.[29] By calling attention to the ways in which women differed from men, feminist scholars such as Chodorow in America and Cixous in France hoped to undermine what they saw as a tendency in their societies to accept masculine qualities as the norm for all of humanity. The French feminists attacked the problem in writing and in theory, while the Americans, drawing on a more broadly based feminist movement, sought change at an institutional level as well.

According to American feminist scholars who stressed gender differences, the theory of gender sameness and the legal practice of gender neutrality helped only those women who were similarly situated to men, while doing little if anything for those women who were locked into low-paying jobs by their gendered experience as wives and mothers. True equality, they maintained, required preferential treatment for women, including paid maternity leaves and pay scales based on comparable worth , which would raise salaries in positions filled mostly by women. Preferential treatment gained support as minority groups joined in emphasizing the importance of pluralism and the need for special assistance to those who were disadvantaged. But minorities also raised objections to the universalizing tendencies of gender, whether expressed as sameness or as difference. Minority women frequently argued that race, ethnicity, religion, or poverty figured more prominently in shaping and limiting their lives than did gender.

Other challenges to prevailing approaches in the study of gender soon followed. Gay and lesbian scholars questioned the tendency of straight scholars to see the world in dichotomous terms. Foremost among them was the postmodernist French philosopher Michel Foucault (1926–1984), who brought new attention, in his *History of Sexuality* (1978), to the ways in which power has shaped the meaning of sex. Sexuality, in Foucault's view, must not be seen as a biological given, but rather as "an especially dense transfer point for relations of power: between men and women, young people and old people, parents and offspring, teachers and students, priests and laity, and administration and population."[30] The French philosopher Monique Wittig argued, in turn, that heterosexuality, more than patriarchy, was the source of women's oppression. Lesbians, she maintained, were not "women," since they were outside the symbolic order of heterosexual relationships.[31]

As French postmodernism began to influence work in the American social sciences, the historian Joan Scott pointed to the ways in which gender served not only as an element in social relationships based on perceived differences between the sexes, but also as a primary way of signifying relationships of

---

[29] Elaine Marks and Isabelle de Courtivron, eds., *New French Feminisms: An Anthology* (New York: Schocken, 1981), pp. ix–xiii.

[30] Michel Foucault, *The History of Sexuality*, vol. 2: *An Introduction* (New York: Random House, 1978), p. 103.

[31] Joan Wallach Scott, ed., *Feminism and History* (New York: Oxford University Press, 1996), pp. 6–7.

power. Even where women are absent, as in many discussions of political theory, writers have often deployed sexual metaphors to express the relationship
between the ruler (masculine/father) and the ruled (feminine/daughter).[32]
The legal scholar Catharine MacKinnon, among others, built on this expansion of the meaning of gender to encompass the multiple ways in which
societies use sex to structure power relations. By calling attention to rape,
domestic violence, sexual harassment, and pornography as examples of men's
use of sex to achieve dominance, she and others worked to pass laws that
would protect women from such abuses.[33]

Debate over the meaning and significance of gender continued to the end
of the century, but the emphasis was very different from what it had been
a century earlier. Whereas biological determinism had informed virtually all
thinking about what it was to be a woman or a man in the 1890s, by the
1990s gender had become the term of choice in discussing sex differences,
sexual identity, and sexuality in its multiple manifestations, reflecting how
thoroughly theories of social construction had come to dominate work in
the social and behavioral sciences, as well as in everyday speech. The term
"gender," used in reference to sex-based categories, appeared first in the popular press in 1970, and throughout the following decade journalists employed
the term only about three dozen times. But by the end of the century, "gender"
was appearing in the press more than three dozen times every week.[34] In the
first half of the twentieth century, those concerned with the concept of gender were marginal figures – a few women and an even smaller number of
men – who worked in secondary branches of the social science disciplines or
outside universities altogether. By the final decade of the century, social scientists interested in gender were much more securely situated within academia.
More importantly, as gender studies came to focus on the construction of
masculinity as well as of femininity, even those not centrally concerned with
women's experience or sexuality were beginning to use gender theory in their
own work. No longer could one explore the problems of the welfare state, the
failures of conservation efforts, or even the origins of the Cold War without
taking gender into account.[35]

[32] Scott, "Gender a Useful Category of Historical Analysis," *American Historical Review*, 91 (December 1986), 1053–75.
[33] Catharine MacKinnon, *Towards a Feminist Theory of the State* (Cambridge, Mass.: Harvard University Press, 1989), pp. 171–249.
[34] These figures come from a Lexis-Nexis search of the fifty highest-circulation English language newspapers.
[35] Paul Kennedy, *Preparing for the Twenty-First Century* (New York: Random House, 1993), 329–43; Frank Costigliola, "'Unceasing Pressure for Penetration': Gender, Pathology, and Emotion in George, Kennan's Formation of the Cold War," *Journal of American History*, 83 (March 1997), 1309–39. See also Terrell Carver, *Gender Is Not a Synonym for Women* (Boulder, Colo.: L. Rienner, 1996).

41

RACE AND THE SOCIAL SCIENCES

Elazar Barkan

Over the last two centuries, race has carried contradictory meanings to members of different racial and ethnic groups and conveyed distinct and separate symbols even within such groups. Unlike the distinction between gender (social) and sex (biological), race connotes both categories. It conveys a cultural political entity that has certain, if not specific, relations to a group's image of its own primordial characteristics. The mid nineteenth-century belief that "race is everything" was capacious and ill-defined, yet it provided an overarching concept that included meanings both natural and cultural, scientific and popular. Race has long played a powerful popular role in explaining social and cultural traits, often in ostensibly scientific terms. Furthermore, the confusion about race is heightened by the popular illusion, often shared even by scientists, that in premodern times racial distinctions were more orderly and clear, as communities and identities were coherent. This romantic view assumes a stable racial antiquity in contrast to the dynamic, hybrid racial anarchy of modern times. While the idea of race implies a permanent biological entity, an historical overview shows that the meaning of race is provisional and has changed according to political and social circumstances. A close relative of the concept of "race" is "racism," and the two are often confused. Racism, in contrast to the specific and changing content of theories of race, is an ideology of hatred of the Other, and is used as a derogatory term. It was introduced into English from German in 1938, and replaced the word racialism, which had been used to denote a hierarchical view of races but which lacked the stigma of "racism."[1]

---

[1] Elazar Barkan, *The Retreat of Scientific Racism: Changing Concepts of Race in Britain and the United States between the World Wars* (Cambridge: Cambridge University Press, 1992).

THE INVENTION OF RACE

The modern classification of racial groups developed primarily during the nineteenth century. It had several important antecedents before 1800, most notably in the writing of Johann Friedrich Blumenbach (1752–1840). Blumenbach's typology, known as the pentagon – Caucasian, Mongolian, Ethiopian, American, and Malay – was an Enlightenment concept, underscoring the unity and rationality of all humans. He recognized but minimized the importance of racial divisions. His relatively tolerant notion of race, compared to nineteenth-century views, was shaped by debates over slavery. Blumenbach emphasized that races were not distinct, but overlapped, and he underscored the continuity among racial varieties as part of the great unity of humans. He invented the term "Caucasian" to represent the Europeans, and ranked the races along a scale of beauty.[2] During the next century, scientists reified race and portrayed it as a unit, thus taking it beyond its initial taxonomic purpose.[3]

In the first half of the nineteenth century, theorists of race investigated not only human physical features but also cultural characteristics such as language, geography, and history as constitutive racial elements. The connection between race and language came to play a major role in social science and politics. The early search for racial origins through linguistic history was conducted by William Jones (1746–1794), who traced the genealogy of the European languages to India, and by Friedrich von Schlegel (1772–1829), who originated the term and developed the notion of "Indo-European." This became a particularly popular source for racial theorizing within Romanticism, led by philologists and historians. It was then that racial theories became a tool of national competition. In Germany, Johann Gottfried von Herder (1744–1803) was the first to describe the genius of the race. Following Blumenbach in recognizing the obfuscation of demarcations among the races, Herder shifted his attention increasingly to "the physico-geographical history of man."[4] This shortened dramatically the time scale of racial history, to encompass what today we would term national histories. This shift was made possible by nineteenth-century Romantic writings on race, which focused on the racial composition of Europe. They emphasized the racial configuration of the ancient Greeks and of the Roman Empire, and above all the detailed composition of the modern European nations, primarily based on their Germanic tribal genealogy. The search for racial antiquity motivated Teutonic physical anthropology and Aryan (Indo-European) linguistics, which were always

---

[2] Johann Friedrich Blumenbach, "On the Natural Variety of Mankind" (1775), in his *Anthropological Treatise*, trans. T. Bendyshe (London: Anthropological Society, 1865), pp. 98–9, 100.

[3] Ashley Montagu, *Man's Most Dangerous Myth: The Fallacy of Race*, 6th ed. (Walnut Creek, Calif.: Altamira Press, 1998).

[4] Johann Herder, *Outlines of a Philosophy of the History of Man*, trans. T. Churchill (London: J. Johnson, 1803), vol. I, p. 298.

closely related but never quite the same. European tribal genealogy and the search for antiquity became a mainstay of historical writings throughout the century, led by the works of Barthold Georg Niebuhr (1776–1831). In France, Jules Michelet (1798–1874) searched for the most racially pure French people as the unifiers of the nation, while later Ernest Joseph Renan (1823–92) emphasized the racial composition of the nation as a way to celebrate the Celts and to draw distinctions between Aryans and Semites. In England, this trend was exemplified by Walter Scott's novels and by Thomas Carlyle's outlandish political racism, which became representative of the romantic excesses of the search for English genius. Robert Knox (*The Races of Man*, 1850) for example, sought to use race as an explanation for, among other things, the inevitable extinction of non-Europeans. He dismissed British imperialist efforts to protect various aboriginal peoples as being against nature.[5]

The deepest racial distinctions were assumed by polygenism, which became a popular theory especially during the 1840s. It postulated that humans originated as separate races, and it viewed racial differences as fundamental. In its own way, polygenism advocated a revolutionary perspective, for it challenged the biblical creation story that traced all human ancestry to Adam and Eve. Polygenism, however, lost much of its appeal when evolution replaced the Bible as the main explanatory account of the natural world, providing an alternative time scale and mechanism for racial differentiation. Nonetheless, polygenism remained attractive for racial theorists who preferred to accentuate racial distinctions, sometimes classifying humans into species rather than races. The underlying theme was that if races originated separately, they had little or nothing in common, and it was unrealistic to expect them to display equal qualities. The legacy of polygenism is evident in the speculative elements of modern racial thinking. For the last 150 years, one's hypothesis of the antiquity of races has served as a likely indicator of one's racial politics. As a rule of thumb, the further into the past a writer has traced racial differentiation, the more extreme his racism. Champions of early racial separation have generally, though not always, defended policies that have aggravated discrimination and prejudice and have been considered racists by their contemporaries.[6]

SCIENTIFIC RACISM

During the mid nineteenth century, race became an all-consuming passion and was to prove especially alluring to the evolving social sciences. The

[5] Ivan Hannaford, *Race: The History of an Idea in the West* (Baltimore: Johns Hopkins University Press, 1996).
[6] William Stanton, *The Leopard's Spots: Scientific Attitudes toward Race in America, 1815–59* (Chicago: University of Chicago Press, 1982); Stephen Jay Gould, *The Mismeasure of Man* (New York: Norton, 1996); Thomas F. Gossett, *Race: The History of an Idea in America* (New York: Oxford University Press, 1997).

popularity of racial thinking was founded on a positivist belief that racial theories conveyed facts and were empirically verifiable. That is, racial theories were justified as scientific. The Victorians generally believed that race could explain much of life. Their concept of race was not only biological but also included the social, political, cultural, and psychological circumstances of individuals and groups. Politically, the growth of nationalism and of imperial competition provided the context for this intense engagement with racial classification. In order for us to understand, a century later, the Victorian category of race, we may think of it as an identity marker, which included concepts such as ethnicity, religion, nationalism, and class. Disraeli's declaration that "Race is all, there is no other truth" captured the Victorian consensus on race. It underscored the virtues of unmixed pure races, especially the Anglo-Saxons, who were destined to conquer the world.[7] In hindsight, it may be puzzling that the Victorians did not recognize the self-contradictory nature of their multiple assertions about race. From a contemporary vantage point, Victorian society may be seen as a foreign country in its fixation on race. But for more than a century, the theorists of race amounted almost to a who's who of European scholars and authors.

The study of race was institutionalized during the last third of the nineteenth century through a number of anthropological societies established in Paris, London, New York, Moscow, Florence, Berlin, and Vienna. These testified to – and provided the locus for – the growing interest in the scientific study of race, which enhanced notions of hierarchy, antiquity, and the immutability of human races. As the study of race, and racial data, proliferated, racial classification became more problematic. Even before midcentury, the great racial variety made color differentiation inadequate, especially when it came to internal European distinctions. Scientists proposed numerous alternative systems to describe even minute racial distinctions. Especially popular was a racial taxonomy based upon measurements of the skull, known as craniometry. The skull was attractive to physical anthropologists both because its shape was presumed to remain stable over numerous generations and because of the presumed connection between the skull and intelligence. Craniometry classified races according to the mean ratio of breadth to length of the skull, the so-called cephalic index, invented in 1842 by the Swedish anatomist Anders Retzius. Based on the cephalic index, a tripartite European racial division of Nordic (Teutons, Aryans), Alpine, and Mediterranean became the accepted norm. The difficulty was to show that racial categorization by cephalic index – which was, at best, statistical – stood for real racial divisions. There were multiple attempts to fit various other physical characteristics – hair, eye color, nasal indexes, as well as sociolinguistic traits – into racial categories. While the cephalic index was a preevolutionist measure, and was

---

[7] Nancy Stepan, *The Idea of Race in Science: Great Britain, 1800–1960* (London: Macmillan, 1982); George W. Stocking, Jr., *Victorian Anthropology* (New York: Free Press, 1987).

meant only for Europe, it later assumed a far wider significance and was extended to the entire world.

Natural history provided a shared language for writers who traced racial superiority, even though their ultimate focus was on culture as an explanation of racial traits. This was especially true after the publication of Darwin's theory of evolution in 1859. Darwin's main contribution was a conceptual shift from viewing nature as a static entity (creation) to seeing it as a living organism (evolution), from seeing the world as composed of essentially unchanging types to viewing it as a system of interacting and changing populations (species). Darwin's full title, *The Origin of Species by Means of Natural Selection, or the Preservation of Favored Races in the Struggle for Life* (1859), illuminates the centrality of race in Victorian understandings of the struggle for life. Yet Darwin himself wrote relatively little on human races, a meaningful omission in his world, which was saturated with racial thinking. His evasion was even clearer in *The Descent of Man* (1871). There he pointed to the disagreements "among capable judges" (who divided humans into groups varying in number from two to sixty-three) and emphasized that races "graduate into each other, and that it is hardly possible to discover clear distinctive characters between them." Darwin concluded that racial differences were not of great evolutionary importance, but this was mostly ignored by his contemporaries. Other occasional statements in his writings suggest that his views on non-European peoples were not radically different from those of his peers. Racial theorists found enough in his formulation of "natural selection" to support the idea of permanent racial types, despite the evolutionary framework.

It was during this era that racial theories, which would later constitute modern racism, received their modern formulation. Comte Joseph Arthur de Gobineau (1816–1882), who became known as the "father of racist ideology," combined aristocratic pessimism, Romanticism, and biblical metaphysics with biology to transform the European value system into a racial system. Gobineau's views on race were a culmination of pre-Darwinian ideas. He focused on civilization as a racial marker, rather than on physical anthropology, and he viewed the rise of civilization as well as its degeneration as being a result of racial hybridization.[8] He described a history of permanent racial types as a moral genealogy of distinct epochs separated by catastrophes. In this system, the Aryan, Germanic races were the pioneers of modern civilization, and indeed of most other civilizations. Gobineau focused on the disappearance of the primordial race of mythical ancestors, which symbolized the pessimistic turn of racial theories. At the end of the century, Houston Stewart

---

[8] Arthur De Gobineau, *The Inequality of Human Races* (New York: H. Fertig, 1967); Michael D. Biddiss, *Father of Racist Ideology: The Social and Political Thought of Count Gobineau* (London: Weidenfeld and Nicolson, 1970); Michael Banton, *The Idea of Race* (London: Tavistock, 1977), pp. 41–6.

Chamberlain (1855–1927) and G. Vacher de Lapouge (1854–1936) advanced updated versions of the theory of Aryan superiority. These were the most direct scientific sources of Nazi racism.[9]

In its anti-Semitic version, European racial thinking coincided with traditional Christian anti-Semitism. Modern racism postulated an opposition between the virtuous Aryan and the evil Jew. Anchored in mythical thinking, but relying on scientific theories, anti-Semites postulated the Jew as the ultimate Other. Anti-Semitism was part of a class of theories connecting blood, nationalism, and the struggle for survival that later came to be known as social Darwinism.[10] The widespread belief in these theories meant that even the victims of such beliefs often shared in them and accepted the racial essence of the group. For example, Jews often viewed themselves as a separate race, and in extreme cases internalized their proclaimed inferiority into self-hate. (A notorious example was Otto Weininger.)

The theory of evolution was viewed as explaining not only the natural world, but also society and politics. Thus Herbert Spencer's original formulation of "survival of the fittest" became a widely quoted political principle that represented the new social significance of evolution. The mechanism of natural selection in the days before population genetics was misunderstood. "Fittest" was understood as "best" according to the social standards of the day. The aim was to control one's destiny. Thus, those who defeated others showed their superiority. Success was viewed not as chance survival but as a result of effort and genius, and extinction was a mark of inferiority. Most writers continued to believe in the Lamarckian inheritance of acquired characteristics, which especially contributed to the proliferation of social Darwinism.

Social Darwinism was not an ordered theory with leaders and core texts, as were, for example, Utilitarianism and Marxism. Instead, the label "social Darwinian" was a family name describing various ideologies of superiority. The theories were self-contradictory in asserting that society is naturally ruled by superior races, but at risk due to expanding populations of inferior races. Societies exhibited rich structures of biological hierarchy; the rich were superior to the poor, men to women, Europeans to non-Europeans, Germans to Jews, northern Italians to southern Italians, and French aristocrats to peasants. Among the most influential writers of this genre were Herbert Spencer and Benjamin Kidd in England, John Fiske and William Graham Sumner in the United States, Cesare Lombroso in Italy, and Ernst Haeckel in Germany. Here belonged also the popular literature on decline and decadence, represented perhaps most clearly by Max Nordau's *Degeneration*.

---

[9] George L. Mosse, *Toward the Final Solution: A History of European Racism* (New York: H. Fertig, 1985).

[10] Robert C. Bannister, *Social Darwinism: Science and Myth in British–American Social Thought* (Philadelphia: Temple University Press, 1979); Donald C. Bellomy, "'Social Darwinism' Revisited," *Perspectives in American History*, n.s. 1 (1984), 1–129.

The paradox was that for social Darwinists, this successful natural system was on the verge of collapse. Industrialization, population growth, and greater geographical mobility had placed the good old order in danger. If in the past the "fittest" indeed could be expected to reach the top, there was now a danger that the inferior masses would topple this "natural" hierarchy.

Social Darwinian racial theorists never clarified whether the time scale for racial changes and adaptations was to be historical or zoological. Classical archaeology (which studies a history of thousands of years) and paleontology (which studies a history of tens of thousands to millions of years) were often utilized by race theorists indiscriminately to support contemporaneous sociological claims. A similar vagueness was manifested by the theory of recapitulation, which provided another popular anchor for racial hierarchies. The theory held that individual development from the embryonic stage to adulthood (ontogeny) passes through all of the previous evolutionary stages of the species (phylogeny). The younger the embryo, the earlier the evolutionary stage it represents. When applied to races, especially by Haeckel, the theory claimed that the lower races represented earlier evolutionary stages of Europeans, perhaps roughly the developmental stage of children. Not surprisingly, the methodological pluralism led to disparate results and perplexing racial descriptions, rarely distinguishing between culture and biology. Thus few were willing to listen at the end of the century when Thomas Henry Huxley underscored the potential separation between biology (race) and language (culture).

Perhaps the most popular manifestation of racial theories used to advocate policies of biological superiority came in the form of eugenics. Francis Galton coined the term "eugenics" and began a new scientific discipline and a popular social movement. His studies sought to show the correlation between heredity and genius, and in the process contributed to the development of statistics. The movement became especially popular in England, Germany, and the United States, and had many followers, from all across the political spectrum, in other countries. Eugenics claimed that society had to improve its biological composition if it wanted to overcome poverty and social ills. This was translated into anti-immigration, sterilization, and euthanasia policies and received its most diabolic manifestation under Nazism. The movement, which was most widespread before and after World War I, combined scientific investigations with popular culture to advocate widespread xenophobia. From regression analysis of the correlation between anthropomorphic measurements and intelligence, to family competitions in rural fairs and widespread sterilization legislation, eugenics was supposed to provide solutions to many of society's ills. Many eugenic policies have come to be viewed in hindsight as racist.[11]

---

[11] Daniel J. Kevles, *In the Name of Eugenics: Genetics and the Uses of Human Heredity* (New York: Knopf, 1985).

Prominent scientists, such as Charles Davenport in the United States and Bauer and Lenz in Germany, were leaders of the xenophobic application of genetics to eugenics, as was Ronald Fisher, the distinguished English population genetcist and statistician. The Nazis manipulated science most notoriously to achieve racist ends. The ongoing debate among historians over which parts in the Nazi ideology resulted from traditional popular and Christian anti-Semitism, and which were based on scientific theories, is unresolved. The predicament is partially a result of the complementarity among these sources. The Nazi ideology postulated the struggle for global control between the Aryans and the Jews, which only one race would survive, as the core of their belief. Despite the contradictions of racist ideology (for example, claims for a global Jewish conspiracy based on both the superiority and the inferiority of Jews), scientific claims gave credibility to Nazi policies. This was particularly true of their turning the advocacy of racial hygiene into a racist Aryan superiority.[12]

Beyond Nazism, mainstream racial classifications, which began as a local pursuit, went particularly astray when applied globally. Anthropologists found that the extensive data could provide any number of alternative classifications. During the 1920s, Roland Dixon of Harvard University proposed perhaps the most counterintuitive racial classification, a system based on characteristics such as skull shape that placed Australian Aborigines and northern Europeans in the same racial group. The flood of racial systems blurred the scientific definition of race irreversibly. Criticism of these racial theories began in the 1920s, but it was only after eugenics became associated with Nazism that it lost its popular appeal, and antiracism in science became influential. The objections of social scientists to the xenophobic and malicious policies that buttressed these unsubstantiated theories led to a relatively rapid demise of the science of race.

FROM BIOLOGY TO CULTURE

This rise of antiracism occurred in three stages, with some chronological overlap. The first period, from the turn of the century to the 1920s, created an alternative agenda. This included a growing sophistication about mechanisms of biological heredity and a shift among anthropologists, psychologists, and sociologists to studying culture and the social attributes of racial groups instead of racial-biological typology. The second stage, during the 1920s and 1930s, included an initial explicit rejection of racial typology as racism by individual scientists, and the popularization of antiracist ideas

[12] Robert N. Proctor, *Racial Hygiene: Medicine under the Nazis* (Cambridge, Mass.: Harvard University Press, 1988); Paul Weindling, *Health, Race, and German Politics between National Unification and Nazism, 1870–1945* (Cambridge: Cambridge University Press, 1989).

in both interpretive and theoretical work. The third stage had more political manifestations, especially during the period from the 1930s to the 1950s. A group of scientists accepted a professional responsibility to oppose the aberration of racism in the name of science, and undertook to oppose racial topologies. Several became political activists and were ready to address the issue of race. However, most racial theorists among scientists viewed race as too controversial politically and avoided it altogether.[13]

Franz Boas (1858–1942), more than any other scientist, delegitimized racial biological claims as cultural explanations. Beginning in the 1890s, Boas separated race as biology from language and culture and progressively reduced the inferiority attributed to "primitive" or "savage" races in the language of the time. Boas argued for a basic mental equality among the races and between the sexes. This was a radical claim for his period. One of his most important studies showed malleability of the shape of the skull over time, thereby undermining the mainstay of physical anthropology and of the claim for racial stability. Boas had two other roles in fighting racism: as a teacher and as an activist. As the most prominent anthropologist in the United States, Boas taught and educated the most important American anthropologists of the first half of the century, including Margaret Mead and Ruth Benedict. He was the leader in the development of cultural relativism and later, during the 1930s, became a moving force in organizing scientists to oppose lynching and racism in the United States and Nazism and fascism in Europe.[14]

The science of race belonged above all to physical anthropology, which became popular during the mid nineteenth century, then reached its zenith at the turn of the twentieth century and its nadir during World War Two. Boas and others were able to press successfully the objection that anthropological data on race could not support meaningful comparison beyond local variations. As physical anthropology lost ground, race did not become the kernel of any of the new sciences: genetics, social or cultural anthropology, sociology, or psychology. Many scientists in those disciplines who continued to address racial questions were part of the eugenics movement, but race as a scientific vocation vanished from the dominant academic disciplines. During the first half of the twentieth century, while genetics was growing in reputation as the science of heredity, human genetics offered no pertinent knowledge on any of the vital physical distinctions concerning racial classification.

One may safely say that the greatest contribution that the life scientists and physical anthropologists have made to the study of race has been to renounce biologism and their own ability to speak on the subject. Biological definitions of race were discredited by politically active, antiracist scientists, including Lancelot Hogben, Julian Huxley, J. B. S. Haldane, and Lionel Penrose.

[13] Barkan, *The Retreat of Scientific Racism.*
[14] George W. Stocking, Jr., *Race, Culture and Evolution: Essays in the History of Anthropology* (New York: Free Press, 1968).

They persuaded most scientists that biological race had no clear standing in either the cultural or the social arena. As the biological sciences became more agnostic over the place of race, the social sciences came to play an increasing role in studying and interpreting race as a cultural and social phenomenon. Sociologists, for professional reasons, withdrew rather early from attributing social differences to racial taxonomy, but they could not critique its biological foundation. This had to be done by biologists and anthropologists. Once the road was clear, sociologists stepped in. Similarly, psychologists seized on the quandary about race rather early and, by the mid-1920s, began to criticize the notion of racial mental differences. In 1929, Carl Brigham, who had written the standard text on psychological racial differences only a decade earlier, recanted. Otto Klineberg's *Race Differences* (1935) was still more definitive, indicating that there was no vibrant science of race. While race remained a topic in science, its importance had become decidedly secondary.

By and large, the scientists who rejected race seem as a group to have belonged to the genteel milieu of the ivory tower. Most, however, were partial outsiders because of their ethnic marginality, gender, geography, politics, or ideology. This outsider status seems to have made them more aware of the inequalities inflicted by racism and motivated them to oppose the conventions of their day.

The shift in scientific views of race between 1920 and 1950, and the replacement of biological explanations with cultural analysis, contributed to a revolution in race relations far beyond the universities. The forces that shaped this shift are still very much part of the contemporary discourse of race. Race was no longer widely confused with religion or class. With the defeat of Nazism, internal European classifications were no longer viewed as racial. Before World War II, racial studies focused on Europe; after the war, the core of the social science of race moved to the United States. In America, the most significant impact of the shift in racial theory was to facilitate the work of the civil rights movement, which, while it expressed sociopolitical grievances, was able to challenge the intellectual legitimacy of the racist tradition by appealing to science.

The first attempt to assess the impact of this shift in racial thinking on the wider society was made at the outbreak of World War II by the Carnegie Corporation, which funded a study headed by the Swedish economist Gunnar Myrdal that was published as *An American Dilemma* (1944). This study was significant both for its politics and for its substance. Myrdal was an outsider, and he built a wide coalition of contributors, which included white southerners and northerners as well as black social scientists. The study emphasized the good intentions of people aiming to improve race relations and chronicled the general shift away from the rigid scientific racism of the past. America faced a choice, concluded Myrdal; if it wanted to live up to its ideals, it had to overcome its racial prejudice. As race relations and the scholarship of race shifted in the years immediately following the war, Myrdal's study provided the most significant scientific statement for equality.[15]

In the international arena, a new organization, UNESCO, took the lead in disseminating the theory of equality. Its first study by a "world panel of experts" was published in 1950 and declared that there was no scientific basis for racial prejudice or bias. The study highlighted the dramatic transformation in the scientific and public understanding of the concept of race.[16] UNESCO's statement claimed that human equality is based on the similarity of mental capacities among all races; that no evidence for biological deterioration as a result of hybridization existed; that there was no correlation between national or religious groups and race; and that race was less a "biological fact than a social myth." Most significantly, it declared that the brotherhood of man was a scientific fact. The statement was controversial, especially its last component, and necessitated further clarification, which in time became a series of antiracist publications by UNESCO on the status of racial theories. Despite certain persistent opposition, this campaign contributed to the discrediting in the public domain of the scientific credentials of racial theories.

The most direct impact of the new social science on race was its acceptance by the United States Supreme Court in the 1954 case *Brown v. Board of Education*, which declared segregation unconstitutional. The decision transformed antiracism from a respectable intellectual predilection among the liberal intelligentsia to the law of the land. It was substantiated in part by contemporary social scientific research, and it drew especially on *An American Dilemma*. The refutation of the scientific foundations of racism, which occurred between 1920 and 1950, caused a dramatic reversal in the view of race in Western culture. Since the 1950s, those who claim a biological causality for cultural differences along racial (or other group) classifications have generally found themselves outside the mainstream and subject to criticism.[17]

Over the last eighty years, the discourse of race has moved from an "objectivist," hierarchical representation to a pluralist subjectivism. Before the 1930s, dominant groups denied victims of racism the very legitimacy of participating in the discourse about race that determined their inferior social status. Between the 1930s and the 1960s, victims of racism won the opportunity to contribute to these debates. During the 1960s and 1970s, ethnicity studies proliferated. Instead of race, the concept of ethnicity evolved as a competing term. The word "ethnic" in the sense of "racial" first appeared in *We Europeans* (1935) by Julian S. Huxley and A. C. Haddon. Initially, ethnicity sought to replace race, as a cleaner term without the discrimination and racist

[15] Walter A. Jackson, *Gunnar Mydral and America's Conscience: Social Engineering and Racial Liberalism, 1938–1987* (Chapel Hill: University of North Carolina Press, 1990).

[16] *New York Times*, July 18, 1950; UNESCO, *The Race Concept: Results of an Inquiry* (New York: UNESCO, 1952).

[17] Richard Kluger, *Simple Justice: The History of Brown v. Board of Education and Black America's Struggle for Equality* (New York: Knopf, 1975), p. 706.

baggage, and with a greater focus on culture. Sociological and anthropological studies offered extensive data and narratives but relatively little theory on how to differentiate between the two concepts. The prevailing sense among writers on race and ethnicity has been that race as an entity is more stable or biologically significant than ethnicity. Yet efforts to translate general principles of racial classification into specific studies have shown that every proposed system leaves many exceptions. Since the 1970s, a third phase has become dominant in the social studies of race. Contemporary writers tend to privilege plural perspectives on race and to legitimize agency as a way to achieve a richer discussion on racial diversity. The discourse of race is determined not only by what is said, but also by who says it. Race, it is agreed, can no longer be reduced epistemologically to a unidimensional reality. Instead, the subjective perspective of the group has become a constitutive element.

The refutation of racial typology in human classification has led to a parallel elimination of the concept even from such fields as zoology. This contradicts the intuition of many people who continue to treat racial differences as biological, and who have maintained the centrality of race in social and political discourses. The scientific and scholarly debunking of race has itself been politically engaged with questions of immigration and xenophobia, and has reflected scientists' political and ideological commitments, as well as their religious and ethnic affinities. Yet it was only when racial typology ceased to enjoy the status of an objective scientific fact backed by the most advanced biological knowledge that racial factors became a rallying point for minorities' self-identification.

There have been exceptions. Scientists from Carlton Coon to Charles Murray and Richard Herrnstein have maintained that racial differences point to profound biological distinctions.[18] Advocates of such deep racial divergences are frequently viewed as right-wing and racist. Often they capture the public attention, but soon their arguments are effectively criticized, and they are relegated to a footnote. The repeated initial popularity of such studies suggests, however, a public demand for theories of deep racial distinctions, for "real" biological explanations, not merely cultural or social distinctions. At present, there is no credible biological theory of race differences, and there is no reason to expect that one will emerge, given that "internal" racial differences are much more pronounced than distinctions between races.

The biological discourse of race has historically lent itself to racist interpretation, but it was not inherently racist. Nor are cultural readings necessarily egalitarian. For certain racial theories to be viewed as racist, they have to be explicated as such. The genius/inferiority of a race can after all be mirror images. In the last fifty years, those who have strayed from the egalitarian and antibiological consensus have been tagged, often justly, as "racists." However,

---

[18] Carlton S. Coon, *The Origins of Races* (New York: Knopf, 1963); Richard J. Herrnstein and Charles Murray, *The Bell Curve: Intelligence and Class Structure in American Life* (New York: Free Press, 1994).

the instability of such labels may be inferred from recent, politically validated claims that homosexuality may be biologically determined, or from
the new respectability in some circles of the claims of sociobiology, or even
from Leonard Jeffries's theories of "melanin," which seek to establish black
superiority.

A different perspective on race emerges from looking at the intersection of
economics and politics. Since the civil rights movement, race has become a
complex social and economic category. The debate about race as an identity,
like so many other political issues, is articulated in the rhetoric of individual
and group rights and entitlements. Thus, the largest and most comprehensive
antiracist alliance coalesced around an economic agenda. But the identity
issues become even more complicated when the classification of individual
identity comes into conflict with group identity, as in the case of the U.S.
census. In such cases, an individual may define herself differently from the
preference of the group's leadership. The census is the main governmental
tool used to define races and thereby determines the political and economic
resources directed to minorities. These include everything from employment,
education, and health, to enterprise zones and voting rights. The rhetoric of
rivalry, characteristic in the politics of shaping the census, is about political
power and entitlements, or a share of the national wealth. This debate over
the census during the 1990s revealed profound disagreements, about what
constitutes a race and about the method of counting. Each group sought
a system that would result in a larger share of the pie for itself, one that
would "overcome" the barriers of any internal identity differences. Yet the
malleability of racial classification is evident in the growth of the multiracial
phenomenon. The 2000 census was the first official recognition in the United
States that individuals may belong to more than one race.

Race as a social construction remains pervasive and shapes many aspects
of modern existence. It has become a fragmented category. Racial identity
now competes with and complements other group configurations, including gender, ethnicity, nationality, and class. In addition, the idea of group
victimization, which is the alterity of group identity, has become prevalent.
Group victimization, of which racism is the prototype, has been recognized
as a significant experience for many individuals.[19]

THE POLITICS OF RACE

Today social scientists mostly shy away from biology as a racial explanation.
Instead, the focus is on the impact of perceived biological characteristics
on group identity. Racial reality is at once more and less than culture, or
than ethnicity: possibly more visible (depending on the definition), yet more

[19] Dominick LaCapra, *The Bounds of Race* (Ithaca, N.Y.: Cornell University Press, 1991).

diverse, more fragmented, more diffuse. In certain circumstances it is "deeper" than ethnicity, while in other respects it is a part of ethnicity. A fragmented concept of race – meaning different things to different people – is a very contemporary, postmodern configuration. Race in a postracial society is the subject of local (geographic and specific) political activism, not of universal commitment.[20]

Today there are two separate conversations about race in the social sciences. The first, conducted largely but not exclusively by conservatives, engages the old-fashioned question of inherent racial differences. It treats races as biologically stable units while overlooking the fluid margins of race mixture. The second conversation focuses on economic and social conditions and the overlap between the underclass and racial minorities. It responds to the attribution of poverty and crime to racial groups and the confusion of problems of the underclass with characteristics of racial minorities, including the growing black and Latino middle class. This liberal discourse of race involves an attempt to focus attention on the economic and social conditions of the underclass. The conservative discourse seeks to attribute poverty to racial inferiority in order to fend off such liberal social policies. This creates an unholy political alliance between liberals and conservatives that maintains the centrality of race in political discussions of welfare. Consequently, the confusion between economic and social deprivation and assumptions about biological inferiority continue to shape public discourse.[21]

The direct ethical ramifications of the study of race raise the question of how scientific discussion of race differs from traditional prejudices and racism. The science of race faces the predicament of the social sciences' struggle to discover the truth beyond social construction while being constrained by perceived social "reality." Scientists believe that their racial theories are not determined by customary phobias but by rational and empirical examination. Indeed, although views of race have changed over the last two centuries, due largely to the replacement of one theory by another, before 1930 all race theorists (with just a few partial exceptions) subscribed to the ranking of races in a hierarchy. These rankings lent support to, and were closely implicated in, racial discrimination. Yet the scientists and writers who advocated these theories believed that they were explicating valid theories devoid of ethical considerations. It was only during the 1920s and 1930s that theories of race began to be rejected as false and racist. Especially after witnessing the horrors of Nazism, scientists came to recognize the social and political ramifications of their work. For the first time, racial equality became scientifically imaginable.

---

[20] Paul Gilroy, *There Ain't No Black in the Union Jack* (Chicago: University of Chicago Press, 1987) and *Black Atlantic* (Cambridge, Mass.: Harvard University Press, 1993).

[21] James Q. Wilson and Richard J. Herrnstein, *Crime and Human Nature: The Definitive Study of the Causes of Crime* (New York: Free Press, 1998).

In the twenty-first century, the wide acceptance of the notion that race is a social category, together with the growing validation of the idea that individuals have multiple racial and ethnic identities, may lead to a further shift in social scientists' views on race. It is not unreasonable to expect that race will become an even more malleable category, and that individuals who see themselves as belonging to two or more racial categories will become unexceptional, as race continues to fascinate and elude social scientists.

42

CULTURAL RELATIVISM

David A. Hollinger

Of prominent concepts that owe their credibility and popularity to social science, "cultural relativism" is unusual for having received so little clarification from social scientists. The concept is properly associated with a group of anthropologists who flourished in the United States during the second quarter of the twentieth century and who argued, first, that culture rather than biology explains the range of human behavior, and, second, that the sheer diversity of this behavior as seen throughout the world should inspire respect and tolerance rather than invidious judgments. But these anthropologists tried only episodically to fix the meaning of "cultural relativism," and their successors have proved impatient with the terms on which it has been implicated in later debates over moral philosophy, human rights, multiculturalism, and postmodernism. A phrase that had become familiar as an affirmation of liberal values and cosmopolitan tolerance came to be associated instead with the defense of parochial cultures that sanction the abuse of women, and with the dismissal of the ideal of a common humanity. References to "cultural relativism" were more abundant during the 1980s and 1990s than ever before, but the meaning of the term and its relation to the anthropological movement said to be responsible for it were more elusive than ever. Hence "cultural relativism" is a topic without an agreed-upon referent. Indeed, the debate over just what cultural relativism is constitutes a vital part of its history.

The central idea in cultural relativism, said Melville J. Herskovits, is that *"Judgments are based on experience, and experience is interpreted by each individual in terms of his own enculturation."*[1] This emphatic statement of 1955 may imply that questions of right and wrong, and of truth and falsity, demand different answers depending on one's particular culture. But neither Herskovits nor the other anthropologists of his cohort were eager to put

---

[1] Melville J. Herskovits, *Cultural Relativism: Perspectives on Cultural Pluralism* (New York: Random House, 1973), p. 15; emphasis in original; see also the review by I. C. Jarvie, "Cultural Relativism Again," *Philosophy of Social Sciences*, 5 (1975), 343.

it so starkly. Indeed, crisp definitions of cultural relativism – that it means "truth and goodness are relative to your culture," or "one culture is as good as another" – have been the staple of its critics, who have found themselves ridiculed, in turn, for failing to understand it.

"There is not one of its critics in a hundred" who has got "right" what cultural relativism originally was, Clifford Geertz exploded in 1984, while complaining that critics had reduced it to "regarding Hitler as just a fellow with unstandard tastes." Geertz, whose views gain significance from his standing as one of the world's most accomplished anthropologists, was more incisive in explaining what cultural relativism was not than in declaring its positive essence. Getting cultural relativism right, according to Geertz, was not a matter of sharpening the formulations of philosophical doctrine, or of providing dictionary writers with a sounder list of synonyms. Rather, it was a matter of recognizing one "way of thinking" in relation to the rival ways of thinking against which it had been developed in the late nineteenth and early twentieth centuries.[2]

This resolutely historical approach to the meaning of "cultural relativism" has much to recommend it. The anthropological exploration of cultural diversity, especially among distant peoples, unsettled inherited assumptions about the North Atlantic West. Were the moral codes produced by European civilization really the best ones, worthy of being foisted upon the rest of the globe? Anthropologists were the Westerners who looked the longest and hardest at the "Zunis and Dohomeys," Geertz observed, and who came to grips the most honestly with the sheer diversity of the arrangements by which people manage somehow to live, and even to flourish. No wonder it was these explorers of alterity who were the most inclined to insist that "rushing to judgment" about the practices of foreign peoples "is more than a mistake, it's a crime."[3]

This attitude of reserved judgment toward the prodigious expanse of human life is the core of the "relativistic" way of thinking that Geertz urged us to thank for revealing "that the world does not divide into the pious and the superstitious; that there are sculptures in jungles and paintings in deserts; that political order is possible without centralized power and principled justice without codified rules; that the norms of reason were not fixed in Greece, the evolution of morality not consummated in England." Another contribution made by these ethnographic pioneers, Geertz added, was their insistence that "we see the lives of others through lenses of our own grinding and that they look back on ours through ones of their own." To this string of hard-won, cosmopolitan insights Geertz contrasted not only the old ethnocentric, parochial confidence that one's own tribe was a model for the species, but also an ancient faith that often fosters arrogance: the faith that we can locate

[2] Clifford Geertz, "Anti Anti-Relativism," *American Anthropologist*, 86 (1984), 263–4.
[3] Ibid., p. 265.

morality "beyond culture" and that we can attain knowledge that is unmediated by the morality and culture of the knower. This extreme antirelativist faith animates critics of cultural relativism, according to Geertz, just as the rejection of this faith is what most defines it.[4]

The active center of the historic movement in twentieth-century social science that goes by the name of cultural relativism was *a principled doubt that "our" people are right while groups who do things differently are wrong.* But behind the debate over the meaning of the term flagging the movement, and behind Geertz's own intervention in this debate, is an uneasy relationship between two applications of this principled doubt. One application is methodological, the other is ideological. Only by recognizing both, and by appreciating the dynamic interaction of the two, can we achieve a history of cultural relativism that avoids the mistakes against which Geertz warns, and that explains the quarrel over its meaning.

Methodologically, cultural relativism has been a *social scientific* device designed to enable anthropologists to confront, and to attain reliable knowledge about, aspects of *foreign* cultures they might not grasp adequately were they always caught up in judging how the people they study measure up to the standards of the inquirer's home culture. It is this method that Geertz praises for having liberated us from a number of provincial conceits, and that he strives to separate from the dubious, sweepingly philosophical claims attributed to cultural relativism by its detractors. But cultural relativism has been more than a scholarly will to see someone else's culture from the inside. The second, ideological use of moral and epistemic humility has brought cultural relativism directly into the realm of philosophy and "culture wars."

Ideologically, cultural relativism has been a *critical* device fashioned for the purpose of undermining the authority of aspects of a *home* culture. The climax of Margaret Mead's (1901–1978) significantly subtitled *Coming of Age in Samoa: A Psychological Study of Primitive Youth for Western Civilization* (1928) was a breezy, homiletic commentary on the customs of the United States and Samoa, suggesting that middle-class Americans of the 1920s might improve the rearing of their adolescent girls by taking some cues from the sexually relaxed life of the South Pacific.[5] Reflection on the possible implications for "Western civilization" of what had been discovered about other cultures, especially about "primitive" ones, is what made anthropology in the cultural relativist mode a major episode in the intellectual history of the twentieth century, rather than simply another movement within a discipline.

The scientific virtues of cultural relativism were advanced most vocally, and with the most public notice, by men and women who invited attention as social critics, not simply as practitioners of an esoteric *Wissenschaft.* Some

---

[4] Ibid., pp. 275–6.

[5] Margaret Mead, *Coming of Age in Samoa: A Psychological Study of Primitive Youth for Western Civilization* (New York: Morrow, 1928), pp. 195–248.

of the relativizing anthropologists went on to offer cultural relativism as a "philosophy." Herskovits did this frequently, including it in his widely used college textbook of 1948, *Man and His Works*, which invited rebuttals of exactly the kind that struck Geertz, a generation later, as misunderstandings of the epistemic and moral humility that anthropologists had contributed.[6] When cultural relativism's detractors address it as a perspective on value disputes within the North Atlantic West, and evaluate it in its capacity as a theory of truth and value, they do so in an atmosphere created by some of Geertz's predecessors.

FRANZ BOAS AND THE REACTION AGAINST EVOLUTIONARY ANTHROPOLOGY

Cultural relativism has its immediate origins in an argument within the ranks of scholars who faced the enormous panorama of human diversity during the late nineteenth and very early twentieth centuries. Most of the pioneers in the scientific study of human diversity were "evolutionists" who envisioned a gradual, progressive development of the human species. E. B. Tylor in England and Lewis Henry Morgan in the United States were among the most creative of these early anthropologists, who looked upon the "primitive" peoples of their contemporary world as living at an earlier stage of an evolutionary process that culminated in the "civilized" societies of the North Atlantic West. Even scholars who took a relatively generous view of the varieties of humankind and found the scope of human life humbling adopted this hierarchical perspective, which served in many cases to diminish the threat to assumptions of Western superiority that was potentially posed by the discovery of more and more peoples in the Amazon basin, Africa, and elsewhere. The mere presence in the contemporary world of large numbers of such people proved how small a percentage of the human species had ever lived according to the rules and tastes prevailing in modern Europe. The notion of social evolution enabled Westerners to categorize each new discovery of a past or present society as occupying a certain stage in the general progress of humanity. As Thomas Trautmann has explained, these nineteenth-century thinkers in effect laid "the great chain of being" on its side, so that the traditional hierarchies supported by this ancient construct – according to which life forms were distributed from lower to higher on an ascending, vertical scale – were rendered temporal, taking the form of a horizontal time line.[7] This hierarchical, evolutionary analysis of human diversity was subject to great variation, ranging, on the eve of the revolt against it, from

---

[6] Melville J. Herskovits, *Man and His Works: The Science of Cultural Anthropology* (New York: Knopf, 1948), p. 655.

[7] Thomas R. Trautmann, *Lewis Henry Morgan and the Invention of Kinship* (Berkeley: University of California Press, 1987), pp. 20, 222.

the conservatism displayed in William Graham Sumner's *Folkways* (1906) to the savage attack on progress-retarding elites in Thorstein Veblen's *Theory of the Leisure Class* (1899).

The normative assumptions of the evolutionists were not challenged systematically and effectively until the publication by Franz Boas (1858–1942) of *The Mind of Primitive Man* in 1911. This book, which consolidated and elaborated upon arguments Boas had been developing for more than two decades in scientific journals, held that the gap between "primitive" and "civilized" peoples was not nearly so great as had been supposed, and that patterns in human behavior should be understood less in terms of progressive stages than in terms of coexisting, autonomous, distinctive cultures. Boas's critique of evolutionary anthropology was directed especially against its emphasis on biology and race, as opposed to culture; but as Boas explained the workings of culture, he stressed the capacity of cultural conditioning to establish standards for conduct and rationality. The minds of primitive peoples might seem irrational from the perspective of the North Atlantic West in 1911, but those minds, including those of the Indians in the Pacific Northwest studied by Boas, worked according to a rationality of their own. The "general effect of Boas's argument," George W. Stocking, Jr., has observed, "was to show that the behavior" of all humans, "regardless of race or cultural stage, was determined by a traditional body of habitual behavior patterns" passed on from generation to generation. Once "the multiplicity of *cultures* took the place of the cultural *stages* of savagery, barbarism, and civilization," Stocking continued, the varieties of culture "were no more easily brought within one standard of evaluation than they were within one system of explanation."[8] Boas's emphasis on the integrity and autonomy of each culture's value system had much in common with the rudimentary philosophical anthropology that had been sketched a century before by the German theorist Johann Gottfried von Herder.

Yet Boas stopped well short of the inference that no standards could be defended over others. Stocking points out that Boas matched his advocacy of tolerance and respect for cultural difference with a hope that more defensible standards for human conduct could be developed, if one took into account the experience of cultures beyond the one in which he and his fellow Westerners had grown up. Even in the context of Boas's "relativistic, pluralistic critique of evolutionism," Stocking cautioned, "Boas still found in the general development of human culture at least qualified affirmation of the specific values most central to his personal world view: reason, freedom, and human fellowship."[9]

---

[8] George W. Stocking, Jr., *Race, Culture, and Evolution: Essays in the History of Anthropology* (New York: Free Press, 1968), pp. 222, 229; emphasis in original. See also Julia Liss, "Patterns of Strangeness: Franz Boas, Modernism, and the Origins of Anthropology," in *Prehistories of the Future: The Primitivist Project and the Culture of Modernism*, ed. Elazar Barkan and Ronald Bush (Stanford: Stanford University Press, 1995), pp. 114–30.

[9] Stocking, *Race, Culture, and Evolution*, p. 231.

Boas understood that these values were not fully institutionalized in the society in which he lived. He argued against the racist characterizations of Jews, Asians, Slavs, and other groups put forth by advocates of immigration restriction. Boas was the only prominent American scientist to try to expose the scientific inadequacy of many of these racist assertions prior to the mid-1920s. Yet even apart from his activities in trying to reduce racism in public life, Boas's scientific work was certainly directed against aspects of his home culture of which he was critical. To suggest that among Boas's motives was a desire to counteract racism and invidious hierarchies is not to diminish his achievements as a scientist, but only to recognize the ideological as well as the methodological functions of Boas's work.

The ideological as well as the methodological appeal of a relativistic approach to the study of human diversity was visible in the disciples Boas won through his published work and through his teaching at Columbia University in New York City from the turn of the century through the 1920s. Columbia was then an important setting for the interaction of Anglo-Protestant students and faculty with intellectuals of Jewish origin, both from Eastern Europe and from Germany. In this milieu, commentators on the American scene developed a noninvidious, respectful attitude toward immigrant groups that paralleled the outlook toward foreign peoples espoused by Boas and his students. No promoter of this cosmopolitan perspective on the United States was more eloquent, nor in the long run more warmly remembered, than Randolph Bourne, who was inspired by Boas's lectures at Columbia and who reviewed *The Mind of Primitive Man* appreciatively.[10] Many who admired Boas were either of immigrant Jewish stock – like Boas himself, who had come from Germany at the age of twenty-nine – or were, like Bourne and Mead, Anglo-Protestants reacting against what they perceived to be the narrowness of their ancestral culture.

Boas's students and proteges included, in addition to Mead and Herskovits, Ruth Benedict (1887–1948), Alexander Goldenweiser (1883–1953), Alfred Kroeber (1876–1960), Robert Lowie (1883–1957), Elsie Clews Parsons (1875–1941), Paul Radin (1883–1959), Edward Sapir (1884–1939), and Leslie Speir (1893–1961). Of this group of ten leading anthropologists, half were Jewish, four had been born in Europe, and three were women. The leadership of no other social scientific or humanistic discipline of that generation in the United States displayed this demographic mix. American-born Anglo-Protestant males were dominant in the academic profession, but few became anthropologists in the Boasian mode. The early constituency of cultural relativism thus lends credibility to the common speculation that marginality is conducive to the development of relativistic perspectives on culture.

---

[10] Bourne's review of Boas appeared in *Columbia Monthly*, 9 (1911), 27–8. See David A. Hollinger, *Science, Jews, and Secular Culture: Studies in Mid-Twentieth-Century American Intellectual History* (Princeton, N.J.: Princeton University Press, 1996).

BOAS'S STUDENTS AND THE DEVELOPMENT
OF CULTURAL RELATIVISM

The title of Lowie's book of 1929 – *Are We Civilized?* – was well chosen.
The point of reference is "we," and the question to be asked is whether the
distinction between the civilized and the uncivilized stands up to the scrutiny
of anthropologists. Here in full flower is the principled doubt that our ways
are the culmination of the progressive development of the species. And here,
at its most confident, is the classic technique of juxtaposing the norms of
readers and the norms operative elsewhere. "If you saw a man spitting at
another, you would infer that he was expressing contempt," Lowie began
this book, but "in East Africa among the Jagga Negroes," spitting "is a kind
of blessing." There is nothing "natural," Lowie explained, in our conventional
wisdom that spitting is done "in order to show loathing." Folks who spit to
show goodwill are not any less civilized, anthropology tells us, than folks who
spit to show the opposite. But Lowie's relativistic perspective did not preclude
judgments, or a sense of progress. He favored a tolerant cosmopolitanism and
a "united humanity" that struggled against the endless sequence of sectarian,
parochial enclosures erected by his fellows.[11]

Lowie's *Are We Civilized?* represented the cultural relativist movement at
its most glib. Mead was more earnest, not only in *Coming of Age in Samoa*,
but also in *Growing Up in New Guinea: A Study in Comparative Education*
(1930) as well as in her thirty-two other books. Lowie, Mead, Herskovits,
and the other Boasians made substantial, specific ethnographic contributions that have a place in the history of science quite apart from the cultural
relativism that helped to propel their ethnographies and that, in turn, provided the raw materials for its articulation. But it was primarily in their
capacity as public intellectuals that Boas's students developed cultural relativism. The Boasian anthropologists of the 1920s, 1930s, and 1940s, Richard
Handler has summarized, "took seriously the duty of the scholar and scientist to make specialized knowledge accessible to the citizens of a modern
society."[12]

It was therefore fitting that the single book that did the most to define
cultural relativism in the mind of the general reader, and for anthropologists
themselves, was written by a theorist and a gifted writer who did no fieldwork.
Benedict's legendary book of 1934, *Patterns of Culture*, explicitly espoused
"cultural relativity" as a theoretical doctrine. It was recognized, at the end of
the twentieth century, as one of the most widely read books ever produced
by a social scientist in any discipline. It is the central document in the history
of cultural relativism.

---

[11] Robert Lowie, *Are We Civilized?* (New York: Harcourt Brace, 1929), pp. 3, 296.
[12] Richard Handler, "Boasian Anthropology and the Critique of American Culture," *American
Quarterly*, 42 (1990), 253.

The core chapters of *Patterns of Culture* offered descriptions of Dobu, Kwakiutl, and Zuni ways of life. The variations in customs from one society to another illustrated how differently the peoples of the world had selected traits from what Benedict described idealistically as the "great arc of culture." Each selection from this virtually infinite inventory of human possibilities went into creating a distinctive culture, best understood as analogous to the personality of an individual. Each culture was thus a coherent whole, a system that worked on its own terms, deserving, of course, the respect that honorable people give to their individual neighbors. In trying to convey the integrity and dignity of the various cultures of the world, Benedict invoked a powerful, romantic image that became an icon for cultural relativism's benign vision of human difference: She quoted an aged California Indian as saying that God, "in the beginning," had provided every people with "a cup of clay, and from this cup they drank their life." Speaking of his own people, the Indian added, "our cup is broken now."[13] Here, as often in the pages of *Patterns of Culture*, it was Benedict's literary skills, not the scientific findings of her fieldworking colleagues, that most advanced the cause of cultural relativism.

That cause was, for Benedict, indisoluably bound up with a desire for the reform of Benedict's home culture through the reduction of intolerance, prejudice, violence, and greed. If Benedict projected a nonjudgmental attitude in general, her descriptions of the Dobu, the Kwakiutl, and the Zuni were subtly moralistic, and were interspersed with rueful observations about the culture of her readers. Benedict emphasized the cruelty that was routine among the Dobu of New Guinea, and concluded her account by noting that other societies had largely eliminated the "extreme forms of animosity and malignancy" still seen among the Dobu. As Stocking has observed, Benedict presented foreign cultures almost as "pathological parodies of the worst aspects of the puritan and robberbarron traditions" of her own society. The point of learning about cultural diversity was eventually one of judgment at home: "We may train ourselves," wrote Benedict, "to pass judgment upon the dominant traits of our own civilization," among which she counted "capitalism" and "war." Benedict did not specify the criteria by which "we can evaluate objectively" these and other traits of our "Western civilization," but the chief point of reference throughout *Patterns of Culture* was her own society. She explicitly prescribed John Dewey's conception of "social engineering," guided by "rationally selected goals" that could be chosen once the full range of possibilities for human life were gleaned from the results of anthropological study.[14] Benedict's plea for tolerance resonated deeply with readers who saw a globe filled with examples of intolerance – Nazism, imperialism, Stalinism,

---

[13] Ruth Benedict, *Patterns of Culture* (Boston: Houghton Mifflin, 1934), pp. 21–2. Benedict also used this quotation as the book's epigraph. Benedict introduces the concept of the "arc" on p. 24.

[14] Ibid., pp. 172, 249, 271–2; George W. Stocking, Jr., "Ideas and Institutions in American Anthropology: Thoughts Toward a History of the Interwar Years," in *Selected Papers from the American Anthropologist, 1921–1945* (Washington, D.C.: American Anthropological Association, 1976), p. 33.

racism – that were offensive to liberal ideas of freedom and human dignity. Benedict pitted liberal hope against a world of challenging ethnocentrisms and repressions. Although Benedict did not eschew judgment, what *Patterns of Culture* conveyed most compellingly was a sense that negative judgments about human conduct were very likely to be wrong.

What perspective should one take toward cultures that were, themselves, intolerant? Did not the logic of cultural relativism require that "we" respect other cultures that work on the basis of values antithetical to the liberal cosmopolitanism espoused in *Patterns of Culture*? Benedict and her allies did not find this concern nearly as pressing as did discussants of cultural relativism during the 1980s and 1990s. By then, a host of intellectual and political transformations within and beyond the North Atlantic West – to which this chapter will attend later – had given urgency to questions that had struck Benedict, Herskovits, and their associates as nit-picking distractions from the big issues. But the classical cultural relativists did respond occasionally to the "small" issues, as when Herskovits, in a call to arms against the Axis powers, in 1942, suggested tentatively that "the concept of freedom should be realistically redefined as the right to be exploited in terms of the patterns of one's own culture."[15]

Herskovits offered this provocative thought as an aside. He alluded to the damage done to African autonomy by well-meaning European efforts to end "slavery" and "human sacrifice," but the ease with which Herskovits inserted this comment within a justification for warfare against the Nazis revealed the depth of two assumptions found frequently in the writings of cultural relativists. First, boundaries between cultures were sharp and clear. Second, the apparent evils internal to a given culture could be distinguished from the bona fide evil of an assault by agents of one culture upon another culture. Herskovits had no doubt that his culture was being assaulted, but in urging its defense he was extremely cautious. In a formulation implying that some anti-Axis Westerners influenced by cultural relativism might indeed wonder if their own culture had a warrant deep enough to justify warfare, Herskovits said that we must "clearly understand that it is possible to reaffirm in positive terms the fundamental tenets by which we live."[16] But was Germany, one might ask, part of "our" culture, enabling us to take issue with practices internal to Germany even had the Germans kept within their own internationally recognized borders? Or was Germany a different culture, deserving of the same liberty to abuse its own members that Herskovits asked for Africans?

This question is not one the classical cultural relativists were eager to explore. Their own culture was sometimes as large as "the West" and at other times constricted enough to exclude the Nazis. Herskovits often complained of the tendency of discussants to harp on the Nazi case, which he seemed to regard as a wedge that threatened to divide cultural relativists, who were able to agree except on the most extreme and difficult of cases. "Most frequently," Herskovits noted wearily in 1956, the question is asked, "granting the validity of each people's way of life, and the respect to be accorded these ways by peoples whose values are different, what should we do in the face, let us say, of the Nazi policy of the extermination of the German Jews?" Such questions are "not easy to answer," he said, but he went on quickly to make two points. First, "a philosophy based on the scientific findings of cross-cultural study does not imply unilateral tolerance of ideas other than one's own," and, second, "cultural relativism in and of itself does not provide all the rules or all the answers for living in a modern world."[17]

Herskovits's colleague Lowie approached the Nazi question more directly in "Empathy, or 'Seeing from Within,'" an essay that Herskovits admired for what he called its "tough-mindedness." Lowie, writing shortly before his death in 1957, declared that "Nazism furnishes a capital test of ethnological maturity." This admirable quality, Lowie explained, is the ability to avoid passing "moral judgment" on the basis of standards alien to a given culture. He credited Herder with the sound insight that "every people and every epoch must be judged in accordance with local and temporal conditions, not by any extraneous standards." But Lowie's exemplification of this "ethnological maturity" turned out not to display "empathy" toward super-aliens who carried out atrocities. There were "good and bad Germans," explained Lowie, who gave numerous examples of the good. "Aryan friends helped Mrs. Rosenfeld to escape across the Swiss border, at the risk of their own lives," Lowie said, citing one of many cases he used to show how mistaken it would be to think of all Germans as cruel and anti-Semitic. "It is absurd to suppose that any considerable number" of Germans "approved the pogroms of 1938 and later," he added. What appeared to be an ethnography of Nazism, enabling his readers to understand the behavior patterns of a truly foreign culture "as the native sees them," quickly became, instead, an argument that the culture of the Germans during the 1930s and 1940s was highly variable, that it sustained many "good" Germans who behaved as his readers would like to think of themselves as behaving had they been Germans during the Third Reich.[18]

Lowie and Herskovits had difficulty taking seriously the possibility that there could be too much tolerance. Yet this possibility gained credibility as

[17] Ibid., pp. 93–4. These quotations are from a chapter entitled "Cultural Diversity and World Peace," which the editors explain was written in 1956.

[18] Robert Lowie, "Empathy, or 'Seeing from Within,'" in *Culture and History: Essays in Honor of Paul Radin*, ed. Stanley Diamond (New York: Columbia University Press, 1960), pp. 145–6, 152–6. For Herskovits's appreciation of this essay of Lowie's, see James W. Fernandez, "Tolerance in a Repugnant World and Other Dilemmas in the Cultural Relativism of Melville J. Herskovits," *Ethos*, 18 (1990), 162.

the campaign for tolerance achieved a large measure of success, and as many prominent features of the world confronted by the classical cultural relativists were transformed.

THE UNCERTAIN LEGACY OF CULTURAL RELATIVISM

Four major transformations affected the setting in which cultural relativism was discussed during the later decades of the twentieth century. The first, which altered the power relationships within which culture was embedded, was geopolitical. The European colonial empires were replaced by dozens of new nation-states during the two decades immediately following World War II. Many of the peoples studied by anthropologists were located in parts of Africa and Asia where political authority was transferred from the French, British, and Dutch to a variety of indigenous and creole elites. The new governments, while close enough in culture to the "natives" of anthropological lore to bring upon themselves some of the legitimacy conferred by a cultural relativist perspective, usually inherited political borders drawn by the old colonial powers. As a result, the new states embraced several different peoples. Where did one culture end and another begin, and how did these cultures come to be constituted, in part, by their modes of interacting with a civil authority that was ostensibly "theirs," yet in many cases was the tool of an alien ethnic group? These complications were later compounded by the consolidation of a world capitalist economy managed by multinational corporations and serviced by an international labor force on multiple sites of production: This economy simultaneously universalized and particularized the culture it touched. The mechanisms of commercial exchange and the technology of rapid communications promoted the use of the English language and the dissemination of American popular culture throughout the world, while the targeting of particular markets for certain consumer goods reinforced some traditional cultural patterns. In the meantime, in reaction against the capacity of the global economy to spread the culture of the North Atlantic West, many groups, especially in heavily Muslim countries, promoted "intolerant" varieties of religious and ethnic particularism that were articulated in open opposition to Western cosmpolitanism. Cultures became harder to see as cups of clay inherited from time immemorial.

The second transformation entailed a shift in the winds of doctrine within the academic and literary elites of the North Atlantic West. From the 1960s onward, the arrogance and invidiousness against which the cultural relativists had fought was placed more sharply on the defensive by a host of thinkers who emphasized the "situated" character of scientific and ethical judgments. The first and by far the most influential of these new "relativistic" works was Thomas S. Kuhn's (1922–1996) *The Structure of Scientific Revolutions* (1962). Kuhn argued that even the most warranted of truth-claims in the most developed of the sciences depended on the workings of contingent, historically specific human communities. Although Kuhn did not adopt the label "relativist," he was routinely called one, and his work stimulated an enormous controversy over "cognitive relativism." The self-styled "Kuhnian" Richard Rorty, in *Philosophy and the Mirror of Nature* (1979), went further in the direction of denying the role of nondiscursive constraints on knowledge, and in later writings defended ethnocentrism as a basis for cognitive progress and political commonwealth. Geertz's two most widely appreciated books, *The Interpretation of Cultures* (1973) and especially *Local Knowledge* (1984), contributed substantially to the mood of epistemic and moral humility. Hence the intelligentsia underwent a transition "from species to ethnos," in which an earlier generation's struggle to overcome parochialism in the interest of a more truly universal perspective was replaced by a new generation's doubts that any universalist perspective could be anything more than a false front for this or that ethnos-specific frame of reference. Although these thinkers of the later decades of the twentieth century rarely invoked "cultural relativism," the continuity between the new relativism and the old was sufficient to win an audience for commentators who lumped them together, and even for some who appropriated the term for polemical purposes without attending at all to the classical cultural relativists. Under the sign of "cultural relativism," the British philosopher Christopher Norris in 1996 wrote a vigorous critique of Kuhn, Rorty, Walzer, and Geertz – as well as of Ludwig Wittgenstein, and of Michel Foucault, Jacques Derrida, François Lyotard, and a number of other recent French theorists – without even mentioning any of the Boasian anthropologists.[19] "Cultural relativism" had entered a phase of its history altogether detached from its origins, from its early development, and even from its most popular expressions.

The third transformation was the development within the United States of the movement that came eventually to be called "multiculturalism," which called into question the monolithic, integrated, and bounded character of the "home" culture. Multiculturalism began in the late 1960s and early 1970s as an effort to recognize and appreciate the cultural diversity of a national community that had been represented more holistically than could be sustained by the empirical work of social scientists and humanists. The notion of a single "American culture" was also politically suspect in an era when a variety of ethno-racial groups, especially African Americans, were protesting the virtual erasure, by Anglo-Protestants, of their contributions to the cultural life of the United States.[20] From the late 1970s onward, multiculturalism was increasingly defined by an alliance with affirmative action programs

---

[19] Christopher Norris, *Reclaiming Truth: Contributions to a Critique of Cultural Relativism* (Durham, N.C.: Duke University Press, 1996).

[20] For a history and critique of the multiculturalist movement of the late twentieth century, see David A. Hollinger, *Postethnic America: Beyond Multiculturalism* (New York: Basic Books, 1995).

for nonwhites, resulting in greater emphasis on the cultures associated with African Americans, Asian Americans, Latinos, and Native Americans in distinction from European Americans. Although "culture" was omnipresent in each of these demographic blocs, the blocs were identified less by cultural patterns than by skin color: black, yellow, brown, red, and white. Hence a movement that was broadly cultural relativist in orientation developed in ways that cut deeply against the Boasian emphasis on culture as distinct from genetically transferred physical characteristics.

The fourth transformation was the growth of an international feminist movement that challenged the ethical autonomy of particular cultures with a new version of human rights universalism. Although the genital cutting of young females in Muslim-controlled states was the most widely discussed example of a practice that created tension between the rights of women and the rights of cultures, during the 1980s and 1990s feminists identified a wide range of practices in many parts of the world that invited external pressure for reform. The devaluation of female babies in China was another example. Within the United States, feminists criticized the "cultural defense" against criminal prosecution in cases of wife beating and child abuse. Eager to distance themselves from imperialist and missionary interventions in indigenous cultures, these feminists usually tried to work as closely as possible with groups of women within the societies where human rights violations were believed to be taking place. It was in relation to the human rights of females that the term "cultural relativism" was most often mentioned in the 1980s and 1990s.

These four major transformations have rendered less defensible several of the ideas advanced by the classical cultural relativists. It is increasingly difficult to locate and maintain the clear boundaries around which the original cultural relativist program was organized. The conversation about cultural difference is too well developed to allow one to make philosophical claims and then refuse to support them with arguments. Social scientists and other intellectuals who struggle against intolerance and ethnocentrism now require tools sharper than those bequeathed by the classical cultural relativists. Yet, at the end of the twentieth century, the principled doubt that "our" people are right while groups who do things differently are wrong is more widely and deeply entrenched in the intelligentsia of the North Atlantic West than ever before in recorded history. If this fact renders obsolete the emphases and tone of the classical cultural relativists, it also marks the extent of their victory.

43

MODERNIZATION

Michael E. Latham

In 1979, Immanuel Wallerstein proclaimed the death of modernization theory. The concept of modernization, he argued, had finally been recognized as a "cul-de-sac," an intellectual obstruction that had confined decades of social scientific inquiry. Drawing scholars away from questions about the essential nature, historical construction, and lasting power of a capitalist world system, it had only encouraged "comparative measurements of non-comparable and non-autonomous entities." As social scientists invoked objectivity, employed structural-functional indices, and ordered nation-states in terms of relative "progress," they ignored the power that structured global flows of resources. The concept had been, perhaps, a "worthy parable" for its time. By inventing "development" and the "Third World," well-intentioned liberal social scientists had offered "new hope" that destitute peoples might emerge into twentieth-century light. If the "underdeveloped were clever enough to invent an indigenous version of Calvinism . . . or if transistors were placed in remote villages, or if farsighted elites mobilized benighted masses with the aid of altruistic outsiders," then the "underdeveloped too would cross the river Jordan and come into a land flowing with milk and honey." The time had come, however, to reject modernization, "to put away childish things, and look reality in its face."[1]

While perhaps the most striking, Wallerstein's was not the only unflattering epitaph for modernization theory. Starting in the mid-1960s and continuing on through the 1970s, a broad range of scholars generated a massive literature criticizing the idea that all of the world's nations followed the same essential trajectory of growth, a pattern most clearly identified in the history

---

[1] On modernization theory generally, see Michael E. Latham, *Modernization as Ideology: American Social Science and "Nation Building" in the Kennedy Era* (Chapel Hill: University of North Carolina Press, 2000), chap. 2. Daniel Lerner, James S. Coleman, and Ronald P. Dore, "Modernization," in *The International Encyclopedia of the Social Sciences*, ed. David L. Sills (New York: Macmillan, 1968), vol. 10, pp. 386–409, provides a useful overview by several proponents. Wallerstein's critique appeared in his *The Capitalist World Economy* (Cambridge: Cambridge University Press, 1979), pp. 132–7.

of Western accomplishment. From a variety of perspectives, they came to abandon the theory of a common developmental passage marked by factors derived from Western experience.[2] By the mid-1970s, few seemed to mourn modernization's apparent passing.

Such assessments, however, leave a number of important issues unexamined – matters that, two decades later, merit serious attention. Beyond the question of the intellectual validity of the model remain profound questions about the function and historical context of its powerful, influential narrative. Why did modernization theory emerge when it did? Was it actually a new approach, a conceptual breakthrough, as some claimed, or did it simply reframe much older patterns of analysis? What made its poetry so appealing during the 1950s and early 1960s? Why did it decline so swiftly after reaching its peak? And finally, at the beginning turn of a new century, might rumors of modernization's demise be greatly exaggerated?

SOCIAL THEORY AND THE COLD WAR CONTEXT

Much of modernization theory's conceptual power, I believe, came from the way it crystallized and articulated a set of widely shared cultural assumptions derived from liberal, internationalist confidence as well as from earlier Enlightenment models of social change. The carnage of the First World War, the rise of fascism, and the stunning destruction brought about by a second global conflict dashed the hopes of liberal European social thinkers. In the United States, however, social scientists crafted an analysis that explained their country's apparent triumph in World War II and outlined a series of practices for managing a "decolonizing" world. Yet this vision was also shaped by the experience, demands, and anxieties of an expanding Cold War struggle. Modernization, as both an intellectual theory and a political practice, defined a liberal, linear path to "progress" in contrast to dialectical and revolutionary frameworks. It presented America's past as a blueprint for the world's future and put history on America's side.

By 1945, the defeat of fascist Germany and imperial Japan left the United States in a position of unprecedented affluence and geopolitical strength. Even at that moment of victory, however, many perceived a coming crisis. As Harry Truman warned the U.S. Congress in 1947, Soviet actions had divided the postwar world into two "alternative ways of life."[3] Stalin had rejected international economic agreements and brutally abandoned promises to allow independent democratic regimes along Soviet borders. As the United States embraced a global policy of containment, the drive to promote a liberal,

---

[2] "Editorial Foreword," *Comparative Studies in Society and History*, 20 (1978), 175–6.
[3] Thomas G. Paterson and Dennis Merrill, eds., *Major Problems in American Foreign Relations*, vol. 2 (Lexington, Mass.: Heath, 1995), pp. 260–1.

democratic world soon made "development" a paramount concern. Though welcomed by some observers, the erosion of European imperialism also seemed to accelerate a "revolution of rising expectations" in impoverished "new states." Destitution, violence, and political volatility, many feared, only opened the door for expanded communist aggression. As the Truman administration strategist Paul Nitze (b. 1907) explained in the influential document known as NSC-68, the United States faced a decisive struggle and issues that were "momentous, involving the fulfillment or destruction not only of this Republic but of civilization itself."[4]

In that charged atmosphere, many American social scientists turned their attention to creating models useful for understanding, and possibly directing, the course of global change. Over the fifteen years following the conclusion of World War II, approximately forty nations rebelled against former colonial rulers, and the United Nations' original membership of fifty-one increased to one hundred.[5] Fascinated by the rapid "emergence" of those countries and concerned with their strategic import, theorists across several disciplines began to turn toward a most ambitious enterprise. Thinking in terms of comprehensive systems, they considered "development" as far more than a matter of increasing gross national product. In order to analyze the problem of change, they argued, social science needed to undertake a more holistic, comparative evaluation of overall historical patterns. "Modernization," as it came to be called during the mid-1950s, involved closely related, mutually reinforcing transformations in the forms of economic organization, political institutions, and central values that held societies together. A modernizing society passed from one point of equilibrium to another, moving farther up an ordered scale toward the kind of rational economy, participatory democracy, and liberal society that characterized nations like the United States. Although few theorists would have accepted the following list without some modification, most increasingly framed arguments encompassing four central assumptions: (1) "tradition" and "modernity" mark endpoints of a common historical course; (2) political, social, and economic changes are integrated; (3) development advances toward the "modern" state along an incremental, linear path; and (4) the transition of "developing" societies can be significantly accelerated through contact with the knowledge and resources of modern ones.

A good deal of the most influential thinking about modernization, which was soon to become an interdisciplinary concern, first appeared in American functionalist sociology. As early as 1937, Talcott Parsons (1902–1979), the postwar chair of Harvard University's new Department of Social Relations,

---

[4] United States, Department of State, Historical office, *Foreign Relations of the United States, 1950*, vol. I: *National Security Affairs; Foreign Economic Policy* (Washington, D.C.: U.S. Government Printing Office, 1977), pp. 237–8.

[5] Walter LaFeber, *The American Age* (New York: Norton, 1994), p. 563.

had suggested that human agency in all societies was mediated and balanced by a functioning social structure. Values and cultural norms, transmitted through different institutions, played vital roles by regulating behavior and ensuring individual action consistent with the social order.[6] In the Cold War context, Parsons and the University of Chicago sociologist Edward Shils pushed that earlier work to a loftier analytical plane. If social values conditioned individual behavior, the two asked, how might they contribute to the preservation of overall stability and order? How might they be included in a larger map of the social system itself? Thinking of society in terms of an integrated, functioning unit, Parsons and Shils argued that if institutions and cultural ideals served to allocate resources and resolve disputes in harmony with individual needs, the social system would rest at a perfect, consensual point, just as it seemed to in the United States. If, however, it were thrown out of balance by forces of demagoguery, ideology, and repression, as in the case of interwar Germany, order and reason would give way to chaos, brutal efforts at social control, and destructive violence.[7]

Their structural analysis, Parsons and Shils also suggested, could include a dynamic component. Defining a set of "pattern variables," the two outlined a dichotomy between "primitive" societies and "advanced" ones. While the former were based on such qualities as ascribed status, particularism, role diffuseness, and orientation toward the collective, the latter revealed values of achievement, universalism, role specificity, and orientation toward the self. Their framework, based on two static points of equilibrium placed in temporal order, did not specify the causes of change. It did, however, suggest that rapid demographic or technological change might generate new demands upon the functioning social system and require the creation of new values and structures to meet them. If placed under enough strain, a society would make the dramatic shift from one end of the binary divide to the other.[8]

For scholars interested in mapping an integrated, common process of "modernization," that sweeping structural-functional approach proved most appealing. Marion J. Levy, a doctoral student under Parsons at Harvard, published his dissertation in 1949 as *The Family Revolution in Modern China*. Heavily influenced by his mentor's model, Levy claimed that "transitional" China was in the process of confronting new forces transmitted by the impact of "modern industrialized society." Mass communications, manufacturing, and capitalist demands, he argued, would favor "universalistic criteria" over older particularism and break down "traditional institutional patterns." Family relationships and filial identification would necessarily weaken as people absorbed a new set of values and entered contractual relationships with

---

[6] Talcott Parsons, *The Structure of Social Action* (New York: McGraw-Hill, 1937).

[7] Talcott Parsons and Edward Shils, eds., *Towards a General Theory of Action* (Cambridge, Mass.: Harvard University Press, 1951); Talcott Parsons, *The Social System* (New York: Free Press, 1951).

[8] Talcott Parsons, *Structure and Process in Modern Societies* (New York: Free Press, 1960) and his *Societies: Evolutionary and Comparative Perspectives* (Englewood Cliffs, N.J.: Prentice Hall, 1966).

total strangers. "Modern industry," he explained, "is not concerned with who a person is." Its primary interest is "whether or not he can perform certain specific technical functions with a specific level of skill." Though some "traditional" ways of life might endure, the new demands placed upon Chinese society would require that the allocative and socializing functions of the family be fulfilled by new educational institutions and bureaucratic structures. The nation, he argued, would slowly become modern.[9] Although Mao's agrarian revolution drove China down a path that Levy had not anticipated, structural-functional analyses of modernization had become firmly entrenched by the late 1950s. As the sociologist Daniel Lerner (1917–1980) argued, modernization was an authentic global phenomenon: "[T]he same basic model reappears on all continents of the world, regardless of variations in race, color or creed." Mass communications, geographic mobility, increasing literacy, and above all the "empathy" produced when village dwellers entered cities and imagined themselves in the shoes of their social and economic betters, all produced rising expectations, eroded "traditional" fatalism, and engendered a new "participant" society. "The model evolved in the West," moreover, was not the product of ethnocentrism. It was simply "an historical fact."[10] By the middle of the 1960s, many American sociologists had come to agree. Modernization theory made the foreign familiar, it made sense of the dramatic transformations occurring before their eyes.

Modernization proved equally attractive to those searching for a general theory applicable to the massive political changes of the postwar era. A brief look at the Social Science Research Council's (SSRC) Committee on Comparative Politics, an influential group bringing together sociologists, economists, and political scientists, is particularly instructive. In a major article prepared for the SSRC in 1955, George Kahin, Guy Pauker, and Lucian Pye (b. 1921) argued that "profound social and cultural changes are taking place as traditional societies have been exposed to the ideas and the ways of the West," changes raising issues ripe for "empirical investigation and not just speculation."[11] Committee members Gabriel Almond and Myron Weiner concurred and proposed that specialists studying separate geographic regions focus on common, interdisciplinary problems involving "function and the interrelationships between political, cultural, and social processes."[12] As Lucian Pye later reflected, that systemic approach led to a "new stage in the intellectual history of the committee." Adapting the functionalist assumptions of Parsons, the cohort concluded that "to achieve order and a basis for comparison it

[9] Marion J. Levy, *The Family Revolution in Modern China* (Cambridge, Mass.: Harvard University Press, 1949), p. 281 and passim.

[10] Daniel Lerner, *The Passing of Traditional Society: Modernizing the Middle East* (New York: Free Press, 1964), pp. viii, 45–8.

[11] George Kahin, Guy Pauker, and Lucian Pye, "Comparative Politics of Non-Western Countries," *American Political Science Review*, 49 (1955), 1022.

[12] Gabriel Almond, "The Seminar on Comparative Politics, June 1956," *Social Science Research Council Items*, 10 (October 1956), 47.

was necessary to posit certain universal functions of all political processes, some of which must be performed by one structure or another."[13] In the first volume of a series the Committee produced for Princeton University Press, Almond heralded that step as the "intimation of a major step forward in the nature of political science as science." Exploring "primitive" and "non-Western systems," he maintained, would allow scholars to "break through the barriers of culture and language and show that what may seem strange at first sight is strange by virtue of its costume or its name, but not by virtue of its functions." This would allow for an analysis of "political systems as whole systems; and particularly for comparing the modern Western ones with the transitional and traditional." A "formal theory of modernization" would penetrate the opaque layers of history and culture, order the universal process of change, and, when enough data had been gathered, provide genuine predictive power.[14]

Among many American economists, modernization did not have the same type of initial appeal. Macroeconomic theory, it seemed, already provided a larger-scale, integrative model of change. As Albert Hirschman reflected, orthodox thinkers held that "economics consists of a number of simple, yet 'powerful' theorems of universal validity: there is only one economics ('just as there is only one physics')."[15] Gradually, however, by the early 1960s a growing number of scholars concerned with "development" had come to argue that lack of infrastructure, impoverished workforces, slow communications, and rapid population growth made "emerging" countries so different from "advanced" ones that standard models had to be altered. Though "underdeveloped" actors still engaged in voluntary economic exchange, their behavior did not always maximize utility. Serious structural obstacles, moreover, prevented the free flow of resources and assets. There did indeed appear to be a profound gap between "tradition" and "modernity," one heavily shaped by technical, educational, and cultural factors.

For ambitious scholars like W. W. Rostow (b. 1916), the task became one of defining change, of determining how the divide was bridged. Specifically rejecting a Marxist analysis of base and superstructure, Rostow's landmark work *Stages of Economic Growth: A Non-Communist Manifesto* (1960) described societies as "interacting organisms." Nations moved from a "traditional" starting point characterized by "pre-Newtonian science" and "long-run fatalism" through a dramatic "take-off" as a result of improvements in education, transportation, and the rise of new activist attitudes regarding

[13] Lucian Pye, "Political Modernization and Research on the Process of Political Socialization," *Social Science Research Council Items*, 13 (September 1959), 26.

[14] Gabriel Almond and James Coleman, eds., *The Politics of Developing Areas* (Princeton, N.J.: Princeton University Press, 1960), pp. 4, 10, 16, 22, 63.

[15] Albert O. Hirschman, "The Rise and Decline of Development Economics," in *The Theory and Experience of Economic Development: Essays in Honor of Sir W. Arthur Lewis*, ed. Mark Gersovitz and W. Arthur Lewis (London: Allen and Unwin, 1982), p. 374.

the natural world. As in mainstream economics, compound interest and increasing rates of savings and investment were still the engines for economic growth, but success also depended on fundamental changes in "habits and institutional structure." Once "old blocks and resistances" were overcome, a society could "drive toward maturity" and enter an "age of high mass consumption." Like many American social scientists, Rostow also described the process as one typically put in motion by contact with more "advanced" peoples. Activist thinking, present in the United States from the start, could inspire foreign elites to "complete the preconditions and launch themselves into self-sustained growth."[16]

Invoked across disciplines, modernization seemed to promise a kind of "unified field theory" for social science. Where the SSRC addressed the problem of comparative politics by bringing together scholars from several disciplines, other institutions soon amplified that project to consider an even broader spectrum of social change. Started with a planning grant from the Carnegie Corporation in 1959, the University of Chicago's Committee on the Comparative Study of New Nations included sociologists, political scientists, economists, law professors, and specialists in education. Clifford Geertz, another graduate of Parsons's Department of Social Relations, provided a strong voice for the role of anthropology as well. Whereas several prominent figures in his field lamented a model that analyzed culture only in terms of the ideas and values that facilitated or impeded social equilibrium, Geertz embraced that approach and helped to pull American cultural anthropology toward it. He investigated the common "primordial sentiments" of kinship, race, language, region, and religion in Asia, Africa, and the Middle East as barriers to the "integrative revolution" that modernization demanded. The study of cultural patterns, in this sense, did not tell a counter-Enlightenment story about the worlds that the West had destroyed. Culture was instead yet another factor to be included in a much larger attempt to "probe more deeply those facets of human experience that, although they occur in nations that are new, relate to the common destiny of us all."[17]

By the early 1960s, modernization had defined a brave new world – a liberal, internationalist era in which the universal process of change could be mapped, defined, and even accelerated. Hailing rigorous, scientific forms of analysis, theorists claimed conceptual breakthroughs and pivotal accomplishments. A closer look, however, reveals a more complex story. Parsons, Pye, Rostow, and the rest of their interdisciplinary cohort did produce a model suited to their era, but they asked very old questions and produced few original answers. Rather than a conceptual revolution, modernization might be better

---

[16] Walt W. Rostow, *The Stages of Economic Growth: A Non-Communist Manifesto* (Cambridge: Cambridge University Press, 1960), pp. 2, 4–11, 26–7, 144.

[17] Clifford Geertz, "The Integrative Revolution: Primordial Sentiments and Civil Politics in the New States," in *Old Societies and New States: The Quest for Modernity in Asia and Africa* (London: Free Press, 1963), pp. 105–57; David E. Apter, "Preface," ibid., p. viii.

understood as the reformulation of an older discourse, a process whereby older models of social change were reworked and blended together into an holistic pattern.

Long before the American Century, German, French, and Scottish social theorists had posited developmental stages and singled out various causal factors to explain the gap between European achievement and the apparent stagnation of the rest of the world. In the nineteenth century, Auguste Comte defined progress in terms of a consensual, evolutionary transition from a "theological" worldview to a "positivistic" one. Focusing on political economy, Adam Smith held that it was the division of labor that made the poorest English worker's living standard higher than that of an African king. Max Weber proposed that Calvinist asceticism and the pursuit of a divine calling explained the rise of the secular and rational behaviors that promoted capitalism in Western Europe's Protestant countries. Describing the shift from "tradition" to "modernity" as a complex of integrated economic practices and social values, American theorists echoed such explanations to create their own version of Ferdinand Tönnies's passage from *Gemeinschaft* (community) to *Gesellschaft* (impersonal modern society).

Theorists also gave those earlier arguments a more celebratory, less ambivalent tone. Weber believed that instrumental rationality built an "iron cage," and Smith commented that the pursuit of wealth could increase anxiety, fear, and sorrow, but American modernizers displayed no such reservations. Where Tönnies lamented the loss of affectionate, familiar, communal relationships, most of the modernization cohort described unqualified progress. Though they worried about the need for vigilance against "complacency" and "softness," they remained convinced that the United States was a clear success story. The danger facing the republic, as many of them understood it, now came from the possible subversion of its accomplishments by communism, a "disease of the transition" and a "peculiarly inhumane form of political organization" that, while producing economic growth and an expansive state, denied the "possibilities of progressive, democratic development."[18]

MODERNIZERS AND THE STATE

Articulated at the height of Cold War anxiety, modernization theory came to stand at the intersection between intellectual inquiry and the political functioning of the state. As an ideology, a way of looking at the world and the position of the United States within it, modernization brought theory and practice together in striking, powerful ways. Since the end of the First World War, Wilsonians had dreamed of a new world order based on free trade, capitalist growth, and international political cooperation. After a second

[18] Rostow, *Stages*, pp. 162, 164.

global conflict, liberal internationalists finally held the field. They agreed that the global economy should be based on a gold standard, and they created a new International Monetary Fund to lend resources so that countries would not resort to protectionism or devalue currencies. An International Bank for Reconstruction and Development, later called the World Bank, was designed to promote stability by providing funds to stimulate investment, build markets, and integrate both war-torn European nations and poorer, "non-Western" countries into the international market. Finally, the United Nations was created to provide for collective action by resolving disputes, countering aggression, and serving as a forum for debate and discussion. International cooperation, with strong U.S. leadership, was intended to usher in a profoundly democratic and liberal era.[19]

Modernization, in that context, became closely tied to the state's effort to manage and direct a world in flux. Beyond defining the process of global change, many theorists sought to produce "policy-relevant" knowledge in order to identify the key social and political levers that the United States might use to accelerate the process of "development" before a hostile ideology could derail it. Speaking a discourse of scientific objectivity, modernizers derived and naturalized valuable "lessons" they found in the American past. Inspired by the results of comprehensive planning during the New Deal, the Second World War, and the economic reconstruction of Europe, many of them also benefited from the attempt to enlist academia in the Cold War struggle. In 1948, the Central Intelligence Agency and the Carnegie Corporation helped to fund Harvard University's Russian Research Center, an organization that sent analyses of the Soviet social system to the U.S. State Department well before publication. Between 1951 and 1954, the Ford Foundation spent fifty-four million dollars supporting research on human behavior and launched programs designed to answer questions about how the United States might promote democracy and improve stability in the world economy. After the Soviet launch of Sputnik in 1957, the National Defense Education Act also provided federal support for university area studies and language programs. At MIT, the Ford Foundation and the CIA even helped to establish the Center for International Studies, a place where scholars like Rostow, Lerner, and Pye produced knowledge that they hoped would "promote the evolution of a world in which threats to our security and, more broadly, our way of life are less likely to arise."[20]

[19] Akira Iriye, *The Globalizing of America* (Cambridge: Cambridge University Press, 1993), pp. 208–10.
[20] Sigmund Diamond, *Compromised Campus: The Collaboration of Universities with the Intelligence Community, 1945–1955* (New York: Oxford University Press, 1992), pp. 50–110; Roger L. Geiger, *Research and Relevant Knowledge: American Research Universities since World War II* (New York: Oxford University Press, 1993), pp. 50–2, 165; William Buxton, *Talcott Parsons and the Capitalist Nation-State: Political Sociology as a Strategic Vocation* (Toronto: University of Toronto Press, 1985), pp. 168–9, 175; Alan Needell, "'Truth Is Our Weapon': Project TROY, Political Warfare, and Government-Academic Relations in the National Security State," *Diplomatic History*, 17 (1993), 399–420;

Modernization also became firmly established as a policy goal. Interested in using foreign aid to shape the course of economic growth, and worried about a possible loss of credibility in the face of communist gains, Eisenhower created the International Cooperation Administration (ICA) in 1955 to provide development loans. Kennedy and Johnson, searching for a "flexible response" to communism, also called for long-term lending authority, contributions from other "free industrialized nations," and special attention to countries ready to drive toward the "stage of self-sustaining growth." In 1961, Kennedy established the Agency for International Development (AID), an organization designed to promote technical aid, lending programs, and development projects. Crossing from academia into government, many theorists became directly involved in international social engineering. Rostow became a White House national security advisor and chairman of the State Department Policy Planning Council. The Harvard economist Lincoln Gordon joined Kennedy's Latin American Task Force and later became ambassador to Brazil. Pye advised AID, and the Stanford Research Institute economist Eugene Staley led a development mission to Vietnam. Part of a most confident, self-assured cohort, they eagerly mapped out plans for transforming a world ready to learn the lessons that America could teach.

Leaders of the world's impoverished nations often responded positively to those overtures and made their own appeals for modernization. Seeking U.S. assistance and support for his five-year industrial plans, India's Jawaharlal Nehru convinced Washington policy makers that his country was an ideal candidate for American foreign aid. In late 1957, India faced a foreign exchange crisis, rising costs of imported steel, and shortfalls in food production. Substantial American assistance to the world's largest democracy, Nehru argued, would turn that situation around and demonstrate powerfully to other developing and "nonaligned" nations that rapid gains in productivity and industrialization could be achieved along liberal lines. As Nehru put it, borrowing Rostow's formulation, the United States could help push India "out of the morass of poverty" and into "the stage of what is called the take-off into sustained growth."[21] Reformist, democratic leaders in Latin America made similar appeals. In 1958, after decades of being told that the United States' highest strategic and economic priorities lay in Europe, Brazilian President Juscelino Kubitschek proposed "Operation Pan America," a program calling upon the United States, Western European nations, and international lending agencies to provide Latin American governments with forty billion dollars in aid over twenty years. Though foreign assistance never reached those levels, Kubitscheck and democratic leaders in nations such as Colombia

Max Millikan and Walt W. Rostow, *A Proposal: Key to an Effective Foreign Policy* (New York: Harper and Brothers, 1957), p. 3.

[21] Dennis Merrill, *Bread and the Ballot: The United States and India's Economic Development, 1947–1963* (Chapel Hill: University of North Carolina Press, 1990), pp. 161–2.

and Venezuela successfully campaigned for increased aid and supported the Kennedy administration's plan for an "Alliance for Progress" to promote industrialization, improve living standards, and engineer Latin America's advance up the universal developmental ladder.[22]

MODERNIZATION THEORY UNDER FIRE

By the early 1960s, many social scientists from outside the United States, and a growing number in America as well, were beginning to give modernization theory more mixed reviews. For some theorists, such as the Argentine sociologist Gino Germani, the "pattern variables" defined by Talcott Parsons and Edward Shils provided a useful framework. Acknowledging that "tradition" and "modernity" were merely "ideal types," Germani still found them useful reference points for his observations about the "institutionalization of change," the emergence of more specialized professions, and a shift in loyalty toward larger, national entities and away from smaller, localized communities. Social scientists such as Samir Amin also found the concept of modernization useful. Though the Egyptian economist provided a bleak assessment of North Africa's development potential, and asked far more penetrating questions about the impact of colonization than Walt Rostow ever did, he too centered his analysis around the problem of achieving "accelerated growth" and a " 'take-off' to industrial development."[23]

Other social scientists, however, began to identify serious problems with the modernization model. Though originally working within a convergence framework, the Israeli sociologist S. N. Eisenstadt argued that modernization often involved serious conflict instead of consensus. As he warned, erosion of "traditional" loyalties could produce the "alienation of wide groups from the central political and social system." Rather than a harmonious equilibrium, sudden change often resulted in "the development of feelings of anonymity and anomic estrangement." Dependency theorists such as André Gunder Frank launched another critical volley by invoking Marx to reveal what modernization had obscured. The "past" of the "developed world," they explained, looked very little like the "present" of impoverished areas. Because industrialized metropoles had enriched themselves at the expense of Southern Hemisphere satellites, the history of regions like Latin America reflected consistent oppression, not movement along a progressive path. The

---

[22] Stephen G. Rabe, *Eisenhower and Latin America: The Foreign Policy of Anticommunism* (Chapel Hill: University of North Carolina Press, 1988); Michael E. Latham, "Ideology, Social Science, and Destiny: Modernization and the Kennedy-Era Alliance for Progress," *Diplomatic History*, 22 (Spring 1998), 199–229.

[23] Joseph A. Kahl, *Three Latin American Sociologists: Gino Germani, Pablo Gonzales Casanova, Fernando Henrique Cardoso* (New Brunswick, N.J.: Transaction, 1988), pp. 23–73; Samir Amin, *The Maghreb in the Modern World: Algeria, Tunisia, Morocco* (Middlesex: Penguin, 1970), pp. 233–4.

Kenyan political scientist Ali Mazrui delivered his own critique by lamenting the deeply ethnocentric tone of modernization and attacking its emphasis on the diffusion of "rational" ideas. Much like older imperial thought, modernization reflected assumptions that "the more backward of the races," while lacking their own creative potential, could at least imitate their social superiors. In the United States, the Columbia University sociologist C. Wright Mills took Parsons and his cohort to task for assuming that integrative values provided necessary social equilibrium and desirable, harmonious consensus. Their approach, he maintained, was profoundly conservative. It legitimated an existing order and left no space to explore the realities of class conflict, exploitation, and power relations.[24]

Some thinkers even reconceptualized the notion of "tradition" itself. As Indian scholars such as M. N. Srinivas and Rajni Kothari argued, the course of development in their country revealed that "tradition" and "modernity" could hardly be considered discrete and dichotomous opposites. Western printing, film, and radio promoted the transmission of Hindu epics, mythology, and scripture as much as it did scientific knowledge and Western business ethics. Enduring structures of caste and religion, never stagnant or totally rigid in the first place, were not impediments to modern forms of social organization. In providing a necessary element of cohesion, they could actually accommodate and facilitate sweeping change by allowing the absorption of new, potentially disruptive ideas and occupations into a well-defined set of personal roles and social relationships. Appropriating modernization theory and testing its central tenets, such scholars redefined tradition and modernity as mutually transforming, not diametrically opposed. The developing world, they argued, was not converging on a single utopian end point. It was instead changing in ways that reflected a range of varied trajectories and complex interactions of culture, religion, technology, and science.[25]

Such criticism had seriously undermined some of the central tenets of modernization theory by the early 1970s. The greatest damage to the modernization ideal, however, may have been done in the realm of practice. Poverty proved to be far more intractable than modernizers had imagined, and neither American foreign aid nor World Bank loans had engineered a decisive social, political, and economic "take-off" during the 1960s "decade of development." Though sometimes increasing aggregate economic growth, aid programs in Africa and Latin America did little to avert military coups,

---

[24] S. N. Eisenstadt, *Modernization: Protest and Change* (Englewood Cliffs, N.J.: Prentice Hall, 1966), 21–2; André Frank, *Latin America: Underdevelopment or Revolution* (New York: Monthly Review Press, 1969); Ali Mazrui, "From Social Darwinism to Current Theories of Modernization: A Tradition of Analysis," *World Politics*, 21 (1968), 82, 76; C. Wright Mills, *The Sociological Imagination* (New York: Oxford University Press, 1959), pp. 35–42.

[25] M. N. Srinivas, *Social Change in Modern India* (Berkeley: University of California Press, 1966); Rajni Kothari, *Politics in India* (Boston: Little Brown, 1970). See also S. N. Eisenstadt, "Studies of Modernization and Sociological Theory," *History and Theory*, 13 (1974), 252.

weaken oligarchic control, or insure that the benefits of wealth might reach impoverished populations. Advances in the international market did not necessarily go hand in hand with integrated progress toward democracy and social reform.

The most public, prolonged, and painful failure of modernization, however, took place in Southeast Asia. From the decision to support the regime of Ngo Dinh Diem in the mid-1950s through the final United States withdrawal, South Vietnam was understood by many American planners as an urgent and necessary target for modernization. Defining South Vietnam as an "emerging" nation moving through the unavoidable yet destabilizing passage from "tradition" to "modernity," American strategists hoped to accelerate the economic growth, liberal reform, and social progress necessary to win the loyalty of a people torn between their ancient past and competing visions of the future. Using experts from Michigan State University to train a new civil service bureaucracy in Saigon, sending AID workers out to fund rural "self-help" projects, and even promising to build what Lyndon Johnson called a TVA on the Mekong Delta, the United States made modernization an integral part of its strategy during the Vietnam War. Kennedy, Johnson, and others maintained that the destruction of an enemy through military means was only part of a larger task: an effective demonstration of what the United States could do to meet newly awakened hopes and to guide a troubled society toward liberal, capitalist progress.

The promise went unfulfilled. South Vietnamese leaders, resenting U.S. pressure, proved to be far more interested in social control than in popular appeal. American advisors, equating authority with security, often backed repression. Rural peasants, moved off their ancestral lands and regrouped into strategic hamlets, came to lament American intervention far more than they appreciated the provision of livestock, fertilizer, and medical care. Modernization in Vietnam won very few "hearts and minds." The historically rooted vision of a united, independent country presented by the National Liberation Front proved far more attractive than any of the schemes backed by the United States, a foreign power that, despite its rhetoric, seemed much like the imperial invaders that had preceded it. Before long, devastating bombing, the invasion of Cambodia, a massacre at My Lai, and the sharply rising death toll among both Vietnamese and Americans drove questions about the fallacies and the sheer arrogance of American "nation building" to the center of public debate.

By the early 1970s, modernization as both a theory and a practice had been seriously wounded. Intellectuals, testing its assumptions, found its linear schema and its fundamental opposition between "tradition" and "modernity" increasingly suspect. Rather than the purveyor of rationalism, democracy, and transformative, activist values in a "developing" world, the United States often appeared to be an oppressive force denying the very kind of self-determination it called for. Convulsive rioting in U.S. cities, racial violence, and the sheer brutality of the war in Southeast Asia also led many to conclude that America was hardly an example for all to emulate.

Wallerstein's requiem, however, may have been premature. Though some theorists rejected their earlier faith in convergence, the Soviet Union's striking collapse and the fading of state socialism around the world led others to sound hauntingly familiar notes. An ideology closely tied to the enduring Enlightenment vision of progress, modernization has also drawn strength from a continuing confidence in the redemptive power of liberal internationalism. Indeed, even in its reified, social scientific forms, modernization theory reflects a set of tightly held, widely shared beliefs about the United States and its potential to transform a world through open trade, democratic politics, financial reform, and education. From Wilson's vision of world order through the Bretton Woods agreements, the Marshall Plan, and the international aid regime at the close of the twentieth century, those aspirations have endured. Hopes for universal movement toward a democratic, market-oriented, utopian end point seem to be very much alive.

INDEX

Abbott, Edith, 97, 616, 657, 682
Abbott, Grace, 615
accounting. *See* management accounting
Adams, Herbert Baxter, 125–6
Adams, John, 627
Adams, John Quincy, 152
Adarkar, B. P., 490
Addams, Jane, 98, 335, 614, 624, 657
Adedeji, Adebayo, 474, 480
Adler, Alfred, 444, 445
Adler, Max, 190, 444
Adorno, Theodor W., 347, 348, 351, 352, 546, 550, 604
Africa: anthropology and colonialism in, 369; and Negritude, 469; and Pan-Africanism, 469, 471, 474, 476; in relation to modernity, 407–12; social sciences in, 466–81. *See also* indigenization; *specific countries; specific disciplines*
African Americans: and studies of social welfare, 543–4, 619; and theories of intelligence, 645. *See also* race
African Economic Research Consortium, 474
agency: and geography, 383–6; and history, 119; and planning, 594–607. *See also* free will
Agency for International Development (AID), 730
Age of Reform (Hofstadter, 1955), 400
Age of Revolution (1789–1830), and development of social science, 22–6. *See also* American Revolution; French Revolution
agriculture: and political economy, 158, 159, 160; statistics and regulation of markets, 558
Aid to Dependent Children (ADC) program, 617–18
Ake, Claude, 480
Alam, Muzaffar, 492

Al-Hajwi, 454
'Ali Mubârak, 451
Allen, Roy D. G., 299
Allport, Floyd, 264
Allport, Gordon, 228, 264, 270
Almond, Gabriel, 318, 323, 725, 726
Alsina, Juan, 417
Althusser, Louis, 200, 424, 463
American Association for Labor Legislation (AALL), 616–17
American Association for the Promotion of Social Science, 93
American Civil War, 547, 610, 664
American Dilemma, An (Myrdal, 1944), 686, 702–3
American Economic Association, 611
American Educational Research Association (AERA), 629–30
American Historical Association, 311, 396
Americanization: of economics, 302–5; of political science, 325; of psychology, 269–73; of social sciences, 230–5
American Journal of Sociology, 341
American Philosophical Society, 102
American Political Science Association, 311, 313, 315, 324, 325, 326–7, 626
American Political Science Review (journal), 311
American Psychiatric Association, 661
American Psychological Association (APA), 7, 257, 269, 660
American Revolution, 136, 537
American Social Science Association (ASSA), 30, 81, 311, 540–1, 611, 623–4
American Sociological Society, 345, 601, 632
American University (Egypt), 463
Amin, Samir, 411, 474, 480, 731
Amon, Otto, 333

Becker, Gary S., 299

behavioral-statistical approaches, to social science theory, 538–52

behaviorism: and education, 628–9; and engineering concept of science, 220–1, 231–2; and political science, 320–6; and psychology, 262–5, 269, 274

Bekhterev, Vladimir Mikhailovich, 262, 434, 436

Belgium, and Revolution of 1830, 26

Bell Curve, The (Herrnstein & Murray, 1994), 647

Bendix, Reinhard, 348

Benedict, Ruth, 366–7, 371, 674, 683–4, 701, 713, 714–16

Bengal Social Science Association, 487

Benjamin, Walter, 82

Benson, Lee, 400

Bentham, Jeremy, 25–6, 47, 146–8, 181, 487

Bentley, Arthur F., 312–13

Benwen, Sun, 507

Berdiaev, Nikolai Aleksandrovich, 440, 448

Berkeley, George, 173

Berlin Conference of 1885–6, 468

Berlin Museum of Ethnology, 360

Bernard, Claude, 66, 254, 433

Bernstein, Basil, 633

Bernstein, Eduard, 189, 194

Bernstein, Nikolai Aleksandrovich, 435

Berque, Jacques, 452, 454, 458, 459

Berr, Henri, 394–5, 397

Bertillon, Adolphe, 94

Bertillon, Jacques, 95

Bhagwati, Jagdish, 495

Bhandarkar, R. G., 486

Bichat, Xavier, 34, 54

Binet, Alfred, 245, 254–5, 639

Bini, Lucio, 672

biological determinism. *See* determinism

biology: and anthropology, 34, 37; and gender, 678–80, 684–7; and geography, 385; influence of on social sciences, 34, 47–8, 52, 211, 236; and intelligence, 646–8; and psychiatry, 665, 670–3, 676–7; and psychology, 254; and race, 700–2, 704–5; and sociology, 332–3; and statistics, 243, 559. *See also* eugenics; evolution and evolutionary theory; natural sciences; sociobiology

Biometrika (journal), 243–4

Bismarck, Otto von, 32

Bloch, Marc, 341, 397, 407

Blonsky, Pavel Petrovich, 436

Blumenbach, Johann Friedrich, 102, 109, 694

Blumer, Herbert, 345

Bluntschli, Johann K., 307, 309, 310

Blyden, Edward Wilmot, 467, 469

Boas, Franz: concept of social science, 223; and gender studies, 683; and Humboldt, 110; influence of on anthropology in U.S., 360, 361, 362, 367; influence of on sociology, 343; linguistic differentiation and cultural areas, 418; and reaction against evolutionary anthropology, 711–13; and studies of intelligence, 645; and studies of race, 38, 701

Body and Mind (Chelpanov, 1900), 433

Boerhaave, Hermann, 48

Bogisic, Baltazar, 447–8

Böhm-Bawerk, Eugen von, 189–90

Bois, Guy, 401

Boisguilbert, Pierre de, 157

Booth, Charles, 95, 335, 609, 626–7

Bopp, Franz, 483

Boring, Edwin G., 7

Bosanquet, Helen, 612

Bose, Nirmal Kumar, 488

Bossuet, Jacques Bénigne, 114

botany, and geography, 384

Boucher de Perthes, Jacques, 356

Bouderbala, Negib, 464

Bougainville, Louis Antoine de, 100, 104, 105

Bourdieu, Pierre, 224, 377

Bourne, Randolph, 713

Boutmy, Émile, 308

Boutroux, Emile, 66

Bowles, Samuel, 632

Bowley, Arthur L., 96, 247, 335, 581

Braid, James, 141

Braudel, Fernand, 398, 419

Brazil: and economics, 423; and institutionalization of social sciences, 419–21; and modernization, 730; and positivism, 414

Breckinridge, Sophonisba, 616, 657

Brentano, Franz, 255

Bridgman, Percy, 221, 265

Brigham, Carl C., 642, 645, 646, 702

British Association for the Advancement of Science, 28, 178

British Journal of Psychology, 254

British Sociological Association, 230

Broadbent, Donald, 271

Brook Farm community, 80

Broussais, F.-J.-V., 149, 150

Brown, Lyndon O., 583

Brown v. Board of Education (1954), 630, 688, 703

Brunhes, Jean, 383

Brunswik, Egon, 263

Bryce, James, 310, 312

Büchner, Ludwig, 432

Buck, John Lossing, 505

Buck v. Bell (1927), 642
Buckle, Henry Thomas, 47, 51, 125, 126, 241, 242
Buffon, Georges-Louis, 50–1
Bühler, Charlotte & Karl, 262, 263, 445, 598, 629
Bukharin, Nikolai Ivanovich, 190, 196, 438
Bulgakov, Sergei Nikolaevich, 440
Bulgaria, and psychology, 447
Bunge, William, 387, 418
Burckhardt, Jacob, 128
Bureau of Refugees, Freedmen, and Abandoned Lands (U.S.), 610
Burgess, Ernest W., 344, 601
Burgess, John Stewart, 502
Burgess, John W., 308, 309, 310
Burke, Edmund, 23, 537
Burnouf, Eugéne, 483
Burt, Cyril, 265–6, 627, 628, 643, 644
Bury, J. B., 126–7
Busia, Kofi, 478
Butler, David, 323
Butler, Nicholas Murray, 624
Butlerov, Aleksandr Mikhailovich, 432

Cabanis, Georges, 34, 48–9
Cabet, Etienne, 76, 186
Cabral, Amilcar, 475
Cambodia, and Vietnam War, 733
Cambridge University, 178
Cameralism, and political economy, 19, 164
Canguilhem, Georges, 2
Cantillon, Richard, 157, 173
Cantril, Hadley, 583
capitalism: as context for social science knowledge, 543–9; critics of in 1830s and 1840s, 31; critique of in Japan, 526, 527, 528; and influence of Gramsci on social theory, 199; Marxism and political programs, 191; Marx and political economy, 187; Marx's critique of, 32, 189–90, 195; and neoclassical economics, 297; Rudolf and concept of organized, 596; and Weber, 728. *See also* industrialization; poverty; social question
Cardoso, Fernando H., 425
Carlyle, Thomas, 30–1, 55, 695
Carnegie Corporation, 702
Carson, John, 246
Cartesianism, and historicism, 118
Cartwright, Dorwin, 270
Carver, Terrell, 32
Caso, Alfonso, 419
caste system, in India, 92, 484, 487
Catlin, George E. G., 103, 316
Cattell, James M., 436, 625
Cattell, Raymond, 271
Caucus for a New Political Science, 325

censuses: and Europe in 1830s, 28–9; and India, 92, 484; and race, 705; and social surveys, 84–7, 92; statistics and estimation of population, 240. *See also* surveys and survey research; population
Central Intelligence Agency (CIA), 729
central place theory, and geography, 387
Centre d'Etudes Sociologiques, 351
Centre National de la Recherche Scientifique, 233
Cerletti, Ugo, 672
Chadwick, Edwin, 26, 91–2
Chakravarty, Sukhamoy, 495
Chalmers, Thomas, 90
Chalupný, Emanuel, 448
Chamberlain, Houston Stewart, 697–8
Chamisso, Adelbert von, 103, 110
Chandler, Alfred, 565
Chapin, F. Stuart, 345, 628
Chaptal, J. A., 86
Charcot, J.-M., 212
Charity Organization Societies (COS), 612–13, 614
Chartists, and social surveys, 90–1
Chastenet, A.-M.-J., 140–1
Chateaubriand, François-René de, 55, 77, 107, 123
Chattopadhyay, Bankim Chandra, 485, 487
Chelpanov, Georgii Ivanovich, 433, 439–40, 441
Chen Da, 510
Chen Hanseng, 505
Chen Zhenhan, 510
Cheysson, Emile, 30
Chicago Juvenile Court, 657
Chicago School: of economics, 423–4; of political science, 323; of sociology, 219, 335–6, 344–5
Chicago Tribune, The (newspaper), 578–9
child development: and education, 134–7, 625–31; and gender, 680, 684, 686, 690; and psychology, 228, 258, 655–60
Child Study (journal), 656
Child Study Association of America (CSAA), 655–6
Chile, and economic development, 422–3
China: communism and reconstitution of social sciences, 509–13; and institutionalization of social sciences, 501–4; and modernization, 408, 410, 724–5; and native domains of learning, 499–501; and social sciences during 1930s, 504–7; and social sciences in Taiwan and Hong Kong, 507–509. *See also* indigenization; *specific disciplines*
Chodorow, Nancy, 690
Chorley, Richard J., 388
Christaller, Walter, 387